APPLIED MATHEMATICS

APPLIED MATHEMATICS

For

B.E. II Year (*III/IV Semester*)
Chhattisgarh Swami Vivekanand Technical University, Bhilai
(Strictly as per latest syllabus)

By

N.P. Bali
Former Principal
S.B. College, Gurgaon
Haryana

Dr. K.N. Mishra
Professor & Dean
Bhilai Institute of Technology, Durg
Chhattisgarh

UNIVERSITY SCIENCE PRESS
(An Imprint of Laxmi Publications (P) Ltd.)
BANGALORE • CHENNAI • COCHIN • GUWAHATI • HYDERABAD
JALANDHAR • KOLKATA • LUCKNOW • MUMBAI • PATNA
RANCHI • NEW DELHI

Published by :

UNIVERSITY SCIENCE PRESS

(An Imprint of Laxmi Publications (P) Ltd.)

113, Golden House, Daryaganj,

New Delhi-110002

Phone : 011-43 53 25 00

Fax : 011-43 53 25 28

www.laxmipublications.com

info@laxmipublications.com

First Edition : 2012

OFFICES

© **Bangalore**	080-26 75 69 30	© **Kolkata**	033-22 27 43 84
© **Chennai**	044-24 34 47 26	© **Lucknow**	0522-220 99 16
© **Cochin**	0484-237 70 04, 405 13 03	© **Mumbai**	022-24 91 54 15, 24 92 78 69
© **Guwahati**	0361-251 36 69, 251 38 81	© **Patna**	0612-230 00 97
© **Hyderabad**	040-24 65 23 33	© **Ranchi**	0651-221 47 64
© **Jalandhar**	0181-222 12 72		

UAM-9638-295-APPLIED MATHS-III (CG)-MIS **C—3748/011/06**

Typeset at : Goswami Associates, Delhi. Printed at Repro Knowledgecast Limited, Thane

CONTENTS

PREFACE

The time lag between development in science and their application to engineering practices has shrunk enormously in the past decades. The base of engineering and technology once largely empirical, is now scientific and mathematically approved.

This book has been designed unit wise to serve as a textbook B.E. 3rd and 4th semester of respective branches of engineering of Chhattisgarh Swami Vivekanand Technical University, Bhilai and other Technical Universities of India.

The student of B.E. find it difficulty to solve most of the problem in the exercise in the absence of an adequate number of solved examples. An outstanding and distinguishing feature of the book is the large number of typical solved examples and problems have been selected from paper of engineering examination of various University, so as to make a reader familiar with type of question usually set their in University examination.

As far as possible, it has been planned to make each unit an independent so that the various topic could be studied in any order. This book has been written with firm conviction that a good is one that can be read by a student with minimum guidance from his instructor.

We are greatly indebted to various authors whose medival text has been consulted in preparing the present edition of this book. The authors are indebted to the faculty and friends for advice and help in preparing this book.

We wish our reader good luck and bright future.

Suggestions for the improvement of text as well as intimation of misprint will be thankfully welcomed.

—AUTHORS

SYLLABUS

APPLIED MATHEMATICS–III

Chhattisgarh Swami Vivekanand Technical University, Bhilai (C.G.)

Semester : B.E. 3rd Sem.

Branch : Electronics and Telecom, Applied Electronics and Instrumentations

Subject : Mathematics–III

Code : 328311 (14)

Total Theory Periods : 40

Total Tutorial Periods : 10

Total Marks in End Semester Exam. : 80

Minimum Number of Class Test to be Conducted : 02

UNIT–1 : FOURIER SERIES **(Number of periods 8 + 2)**

Periodic functions, Definition of Fourier series, Euler's formulae, Dirichlet conditions, Change of interval, Even and Odd functions, Half range Fourier Sine and Cosine series, Parseval's identity, Practical harmonic analysis.

UNIT–2 : FOURIER TRANSFORMS **(Number of periods 8 + 2)**

Definition of Fourier integrals—Fourier Sine and Cosine integrals, Complex form of Fourier integral, Fourier Sine and Cosine transforms, Complex form of Fourier transform, Linearity, shifting and scaling properties, Modulation theorem, Inverse Fourier transform, Fourier transform of derivatives.

UNIT–3 : LAPLACE TRANSFORMS **(Number of periods 8 + 2)**

Definition, Linearity, shifting and scaling properties, Transform of elementary functions, Transform of derivatives and integrals, Multiplication by t and division by t. Inverse Laplace transform, Convolution theorem, Transform of periodic functions, Unit step function and Diract delta function, Initial value and final value theorems, Application to solution of ordinary differential equations.

UNIT–4 : COMPLEX VARIABLES **(Number of periods 8 + 2)**

Limit, Derivative, Analytic function, Cauchy-Riemann equations, Harmoic functions, Application to flow problems, Complex integration, Cauchy's integral theorem and integral formula, Taylor's and Laurent's series, Singular point, Poles and residues, Residue theorem and its application to contour integration.

UNIT–5 : CORRELATION AND REGRESSION **(Number of periods 8 + 2)**

Linear correlation, Measures of correlation, Karl Pearson's co-efficient of correlation, Spearman's rank correlation co-efficient, Bivariate frequency distribution, Regression, lines of regression and co-efficients of regression, Standard error estimate.

SYLLABUS

APPLIED MATHEMATICS–III

Chhattisgarh Swami Vivekanand Technical University, Bhilai (C.G.)

Semester : B.E. 3rd Sem. | Branch : Computer Science Engineering and IT

Subject : Mathematics–III | Code : 322311 (14)

Total Theory Periods : 40 | Total Tutorial Periods : 10

Total Marks in End Semester Exam. : 80

Minimum Number of Class Test to be Conducted : 02

UNIT–1 : FOURIER SERIES **(Number of periods 8 + 2)**

Expansion of function as Fourier series, Change of interval, Even and odd functions, Half-range Fourier series, Practical harmonic analysis.

UNIT–2 : LAPLACE TRANSFORMS **(Number of periods 8 + 2)**

Laplace transform of elementary functions, Properties of Laplace transform, Laplace transform of derivatives and integrals, multiplication by t^n and division by t, Laplace transform of periodic functions. Inverse Laplace transform, Convolution theorem, Application of Laplace transform to solution of ordinary differential equations.

UNIT–3 : THEORY OF COMPLEX VARIABLES **(Number of periods 8 + 2)**

Limit, Derivative and Analytic function; Cauchy-Riemann equations and its applications to flow problems; Complex Integration. Line and Contour integral, Cauchy integral theorem and Integral formula. Taylor's series, Laurent series; Singularities; Poles and their orders and residues; Evaluation of real definite integrals.

UNIT–4 : PARTIAL DIFFERENTIAL EQUATIONS **(Number of periods 8 + 2)**

Formation of partial differential equations, Equations solvable by direct integration, Lagrange's linear equations, Homogeneous linear differential equations with constant co-efficients, Non-homogeneous linear equations, Solution of partial differential equations by the method of separation of variables.

UNIT–5 : RANDOM VARIABLES **(Number of periods 8 + 2)**

Discrete and continuous probability distributions, Mathematical expectation, Mean and Variance, Moments, Moment generating function, probability distribution, Binomial, Poisson and Normal distribution.

SYLLABUS

APPLIED MATHEMATICS–III

Chhattisgarh Swami Vivekanand Technical University, Bhilai (C.G.)

Semester : B.E. 3rd Sem.

Branch : Electrical, Electrical and Electronics

Subject : Mathematics–III

Code : 324311 (14)

Total Theory Periods : 40

Total Tutorial Periods : 10

Total Marks in End Semester Exam. : 80

Minimum Number of Class Test to be Conducted : 02

UNIT–1 : FOURIER TRANSFORM **(Number of periods 8 + 2)**

Euler's Formula, Functions having points of discontinuity, Change of interval, Even and Odd functions, Half range series, Harmonic analysis.

UNIT–2 : LAPLACE TRANSFORMS **(Number of periods 8 + 2)**

Definition, Transform of elementary functions, Properties of Laplace transform, Transform of derivaties and integrals, Mulitplication by t^n, Division by t, Evaluation of integrals, Inverse Laplace Transform, Convolution theorem, Unit step founction, Unit impulse function, Periodic function, Application to solution of ordinary differential equations.

UNIT–3 : PARTIAL DIFFERENTIAL EQUATION **(Number of periods 8 + 2)**

Formation, Solution by direct integration method, Linear equation of first order, Homogeneous linear equation with constant co-efficients, Non-homogeneous linear equations, Method of separation of variables.

UNIT–4 : COMPLEX VARIABLES **(Number of periods 8 +2)**

Derivative, Cauchy-Riemann equations, Analytic functions, Harmonic functions, Flow problems, Complex integration, Cauchy theorem, Cauchy integral formula, Taylor and Laurent series, Singularity, Residue, Evaluation of real definite integrals.

UNIT–5 : Z-TRANSFORMS **(Number of periods 8 + 2)**

Definition, Properties, Initial value and final value theorems, Inverse Z-transform, Convolution theorem, Partial fraction, Residue method and Applications to solution of difference equations.

SYLLABUS

APPLIED MATHEMATICS–III

Chhattisgarh Swami Vivekanand Technical University, Bhilai (C.G.)

Semester : B.E. 3rd Sem. Branch : Mechanical, Mining, Civil and Chemical Engg.

Subject : Mathematics–III Code : 300311 (14)

Total Theory Periods : 40 Total Tutorial Periods : 10

Total Marks in End Semester Exam. : 80

Minimum Number of Class Test to be Conducted : 02

UNIT–1 : FOURIER SERIES **(Number of periods 8 + 2)**

Euler's Formula, Functions having points of discontinuity, Change of interval, Even and Odd functions, Half range series, Harmonic analysis.

UNIT–2 : LAPLACE TRANSFORMS **(Number of periods 8 + 2)**

Definition, Transform of elementary functions, Properties of Laplace transform, Transform of derivatives and integrals, Multplication by t^n, Division by t, Evaluation of integrals, Inverse Laplace Transform, Convolution theorem, Unit step function, Unit impulse function, Periodic function. Application to solution of ordinary differential equations.

UNIT–3 : PARTIAL DIFFERENTIAL EQUATIONS **(Number of periods 8 + 2)**

Formation, Solution by direct integration method, Linear equation of first order, Homogeneous linear equation with constant co-efficients, Non-homogeneous linear equations, Method of separation of variables.

UNIT–4 : COMPLEX VARIABLES **(Number of periods 8 + 2)**

Derivative, Cauchy-Riemann equations, Analytic functions, Hormonic functions, Flow problems, Complex integration, Cauchy theorem, Cauchy integral formula, Taylor and Laurent series, Singularity, Residue, Evaluation of real definite integrals.

UNIT–5 : STATISTICS **(Number of periods 8 + 2)**

Random variables, Discrete and continuous probability distribution, Expectation, Mean and Standard Deviation, Moments and moment generating function, Distributions–Binomial, Poisson and Normal distributions.

SYLLABUS

APPLIED MATHEMATICS–IV

Chhattisgarh Swami Vivekanand Technical University, Bhilai (C.G.)
B.E. Fourth Semester

UNIT–1 : SERIES SOLUTION OF DIFFERENTIAL EQUATIONS AND SPECIAL FUNCTIONS
(Number of periods 8 + 2)

Series solution of differential equations, The method of Frobenius, Bessel's differential equation, Bessel's function of the first kind—recurrence relations, generating function, orthogonality, Legendre's differential equation, Legendre's polynomial–Rodrigue's formula, generating function, recurrence relations, orthogonality.

UNIT–2 : PARTIAL DIFFERENTIAL EQUATIONS **(Number of periods 8 + 2)**

Formation, Solution of Lagrange's linear differential equation, Homogeneous linear differential equation with constant co-efficients, Non-homogeneous linear differential equations, Method of separation of variables.

UNIT–3 : APPLICATIONS OF PARTIAL DIFFERENTIAL EQUATIONS
(Number of periods 8 + 2)

Initial and boundary value problems, Vibrations of stretched string, D'Alembert's solution, One-dimensional heat flow, Transmission of signals along a Cable-telephone equation, Telegraph and radio equations, Vibrations of rectangular and circular membranes.

UNIT–4 : Z-TRANSFORMS **(Number of periods 8 + 2)**

Sequence, Basic operations on sequences. Definition of Z-transform, Linearity, Change of scale and shifting properties, Z-transform of standard sequences, Inverse Z-transform, Multiplication by n and division by n, Initial value and final value theorems, Convolution of sequences, Convolution theorem, Inverse Z-transform by partial fraction, power series and residue methods. Application to solution of difference equations.

UNIT–5 : RANDOM VARIABLE AND PROBABILITY DISTRIBUTIONS
(Number of periods 8 + 2)

Random variable, Discrete and continuous probability distribution, Mathematical Expectation, Mean and variance, Moments and moment generating function, Probability distributions–Binomial, Poisson and Normal distribution.

1

Fourier Series and Fourier Transforms

1.1. PERIODIC FUNCTIONS

A function $f(x)$ which satisfies the relation $f(x + \text{T}) = f(x)$ for all real x and some fixed T is called a periodic function. The *smallest positive number* T, for which this relation holds, is called the **period** of $f(x)$.

If T is the period of $f(x)$, then $f(x) = f(x + \text{T}) = f(x + 2\text{T}) = \ldots\ldots = f(x + n\text{T}) = \ldots\ldots$

Also, $f(x) = f(x - \text{T}) = f(x - 2\text{T}) = \ldots\ldots = f(x - n\text{T}) = \ldots\ldots$

$\therefore$ $f(x) = f(x \pm n\text{T})$, where n is a positive integer.

Thus, $f(x)$ repeats itself after periods of T.

For example, sin x, cos x, sec x and cosec x are periodic functions with period 2π while tan x and cot x are periodic functions with period π. The functions sin nx and cos nx are periodic with period $\dfrac{2\pi}{n}$.

The sum of a number of periodic functions is also periodic. If T_1 and T_2 are the periods of $f(x)$ and $g(x)$, then the period of $a\, f(x) + b\, g(x)$ is the least common multiple of T_1 and T_2.

For example, cos x, cos $2x$, cos $3x$ are periodic functions with periods 2π, π and $\dfrac{2\pi}{3}$ respectively.

$\therefore$ $f(x) = \cos x + \frac{1}{2}\cos 2x + \frac{1}{3}\cos 3x$ is also periodic with period 2π, the L.C.M. of 2π, π and $\dfrac{2\pi}{3}$.

1.2. FOURIER SERIES

Periodic functions are of common occurrence in many physical and engineering problems; for example, in conduction of heat and mechanical vibrations. It is useful to express these functions in a series of sines and cosines. Most of the single valued functions which occur in applied mathematics can be expressed in the form $\dfrac{a_0}{2} + a_1 \cos x + a_2 \cos 2x + a_3 \cos 3x + \ldots\ldots$
$+ b_1 \sin x + b_2 \sin 2x + b_3 \sin 3x + \ldots\ldots$
within a desired range of values of the variable. Such a series is known as *Fourier Series*. Thus, any function $f(x)$ defined in the interval $c_1 \le x \le c_2$ can be expressed in the Fourier Series

$$f(x) = \frac{a_0}{2} + \sum_{n=1}^{\infty} (a_n \cos nx + b_n \sin nx)$$

where, a_0, a_n, b_n (n = 1, 2, 3, ...) are constants, called the Fourier co-efficients of $f(x)$.

Note. To determine a_0, a_n and b_n, we shall need the following results : (m and n are integers)

(*i*) $\displaystyle\int_c^{c+2\pi} \sin nx\, dx = -\left[\frac{\cos nx}{n}\right]_c^{c+2\pi} = 0\,;\ \int_c^{c+2\pi} \cos nx\, dx = \left[\frac{\sin nx}{n}\right]_c^{c+2\pi} = 0,\ n \neq 0$

(*ii*) $\displaystyle\int_c^{c+2\pi} \sin mx \cos nx\, dx = \frac{1}{2}\int_c^{c+2\pi} [\sin (m+n)\, x + \sin (m-n)\, x]\, dx$

$$= -\frac{1}{2}\left[\frac{\cos (m+n)\, x}{m+n} + \frac{\cos (m-n)\, x}{m-n}\right]_c^{c+2\pi} = 0,\ m \neq n$$

(*iii*) $\displaystyle\int_c^{c+2\pi} \cos mx \cos nx\, dx = \frac{1}{2}\int_c^{c+2\pi} [\cos (m+n)\, x + \cos (m-n)\, x]\, dx$

$$= \frac{1}{2}\left[\frac{\sin (m+n)x}{m+n} + \frac{\sin (m-n)x}{m-n}\right]_c^{c+2\pi} = 0,\ m \neq n$$

(*iv*) $\displaystyle\int_c^{c+2\pi} \sin mx \sin nx\, dx = \frac{1}{2}\int_c^{c+2\pi} [\cos (m-n)\, x - \cos (m+n)\, x]\, dx$

$$= \frac{1}{2}\left[\frac{\sin (m-n)x}{m-n} - \frac{\sin (m+n)x}{m+n}\right]_c^{c+2\pi} = 0,\ m \neq n$$

(*v*) $\displaystyle\int_c^{c+2\pi} \cos^2 nx\, dx = \left[\frac{x}{2} + \frac{\sin 2nx}{4n}\right]_c^{c+2\pi} = \pi\,;\ \int_c^{c+2\pi} \sin^2 nx\, dx = \left[\frac{x}{2} - \frac{\sin 2nx}{4n}\right]_c^{c+2\pi} = \pi,\ n \neq 0$

(*vi*) $\displaystyle\int_c^{c+2\pi} \sin nx \cos nx\, dx = \frac{1}{2}\int_c^{c+2\pi} \sin 2nx\, dx = -\frac{1}{2}\left[\frac{\cos 2nx}{2n}\right]_c^{c+2\pi} = 0,\ n \neq 0$

(*vii*) To integrate the product of two functions, one of which is a power of x, we apply the *generalised rule of integration by parts.* If dashes denote differentiation and suffixes denote integration w.r.t. x, the rule can be stated as follows :

$\displaystyle\int uv\, dx = uv_1 - u'v_2 + u''v_3 - u'''v_4 + \ldots\ldots,$ where u and v are functions of x. *i.e.,* Integral of the product of two functions

= 1st function × integral of 2nd – go on differentiating 1st, integrating 2nd, signs alternately +ve and –ve.

[Simplification should be done only when the integration is over.]

For example, $\displaystyle\int x^3 e^{-2x}\, dx = x^3\left(\frac{e^{-2x}}{-2}\right) - 3x^2\left[\frac{e^{-2x}}{(-2)^2}\right] + 6x\left[\frac{e^{-2x}}{(-2)^3}\right] - 6\left[\frac{e^{-2x}}{(-2)^4}\right]$

$$= e^{-2x}\left[-\frac{1}{2}x^3 - \frac{3}{4}x^2 - \frac{3}{4}x - \frac{3}{8}\right] = -\frac{1}{8}e^{-2x}(4x^3 + 6x^2 + 6x + 3)$$

$$\int x^2 \cos nx\, dx = x^2\left(\frac{\sin nx}{n}\right) - 2x\left(\frac{-\cos nx}{n^2}\right) + 2\left(-\frac{\sin nx}{n^3}\right)$$

$$= \frac{x^2}{n}\sin nx + \frac{2x}{n^2}\cos nx - \frac{2}{n^3}\sin nx.$$

(*viii*) $\sin n\pi = 0$ and $\cos n\pi = (-1)^n$

$$\sin\left(n+\frac{1}{2}\right)\pi = (-1)^n \text{ and } \cos\left(n+\frac{1}{2}\right)\pi = \cos\left(n-\frac{1}{2}\right)\pi = 0, \text{ where } n \text{ is an integer.}$$

(*ix*) **Even and Odd Functions**

A function $f(x)$ is said to be *even* if $f(-x) = f(x)$ *e.g.*, x^2, $\cos x$, $\sin^2 x$ are even functions.

The graph of an even function is symmetrical about the y-axis.

A function $f(x)$ is said to be *odd* if $f(-x) = -f(x)$ *e.g.*, x^3, $\sin x$, $\tan^3 x$ are odd functions.

The graph of an odd function is symmetrical about the origin.

The product of two even functions or two odd functions is an even function while the product of an even function and an odd function is an odd function.

Also, $\int_{-c}^{c} f(x)\,dx = 0$, when $f(x)$ is an odd function

and $\int_{-c}^{c} f(x)\,dx = 2\int_{0}^{c} f(x)\,dx$, when $f(x)$ is an even function.

(*x*) $\int e^{ax}\sin bx\,dx = \frac{e^{ax}}{a^2+b^2}(a\sin bx - b\cos bx) + c$

$\int e^{ax}\cos bx\,dx = \frac{e^{ax}}{a^2+b^2}(a\cos bx + b\sin bx) + c$

1.3. EULER'S FORMULAE

The Fourier Series for the function $f(x)$ in the interval $c < x < c + 2\pi$ is given by

$$f(x) = \frac{a_0}{2} + \sum_{n=1}^{\infty} a_n \cos nx + \sum_{n=1}^{\infty} b_n \sin nx \qquad \text{...(1)}$$

In finding the co-efficients a_0, a_n and b_n, we assume that the series on the right hand side of (1) is uniformly convergent for $c < x < c + 2\pi$ and it can be integrated term by term in the given interval.

To find a_0, integrate both sides of (1) w.r.t. x, between the limits c to $c + 2\pi$.

$$\int_c^{c+2\pi} f(x)\,dx = \frac{a_0}{2}\int_c^{c+2\pi} dx + \int_c^{c+2\pi}\left(\sum_{n=1}^{\infty} a_n \cos nx\right)dx + \int_c^{c+2\pi}\left(\sum_{n=1}^{\infty} b_n \sin nx\right)dx$$

$$= \frac{a_0}{2}(c + 2\pi - c) + 0 + 0 \qquad \text{[By formulae } (i) \text{ above]}$$

$$= a_0\pi$$

$$\therefore \quad \boxed{a_0 = \frac{1}{\pi}\int_c^{c+2\pi} f(x)\,dx}$$

To find a_n, multiply both sides of (1) by $\cos nx$ and integrate w.r.t. x, between the limits c to $c + 2\pi$.

$$\int_c^{c+2\pi} f(x)\cos nx\,dx = \frac{a_0}{2}\int_c^{c+2\pi}\cos nx\,dx$$

$$+\int_c^{c+2\pi}\left(\sum_{n=1}^{\infty} a_n \cos nx\right)\cos nx\,dx + \int_c^{c+2\pi}\left(\sum_{n=1}^{\infty} b_n \sin nx\right)\cos nx\,dx$$

$$= 0 + a_n\pi + 0 \qquad \text{[By formulae (i), (v) and (vi)]}$$

$$= a_n\pi$$

$$\therefore \qquad \boxed{a_n = \frac{1}{\pi}\int_c^{c+2\pi} f(x)\cos nx\,dx}$$

To find b_n, multiply both sides of (1) by sin nx and integrate w.r.t. x, between the limits c to $c + 2\pi$.

$$\int_c^{c+2\pi} f(x)\sin nx\,dx = \frac{a_0}{2}\int_c^{c+2\pi}\sin nx\,dx + \int_c^{c+2\pi}\left(\sum_{n=1}^{\infty} a_n \cos nx\right)\sin nx\,dx$$

$$+\int_c^{c+2\pi}\left(\sum_{n=1}^{\infty} b_n \sin nx\right)\sin nx\,dx$$

$$= 0 + 0 + b_n\pi \qquad \text{[By formulae (i), (vi) and (v)]}$$

$$= b_n\pi$$

$$\therefore \qquad \boxed{b_n = \frac{1}{\pi}\int_c^{c+2\pi} f(x)\sin nx\,dx}$$

Hence, $a_0 = \frac{1}{\pi}\int_c^{c+2\pi} f(x)\,dx$; $a_n = \frac{1}{\pi}\int_c^{c+2\pi} f(x)\cos nx\,dx$; and $b_n = \frac{1}{\pi}\int_c^{c+2\pi} f(x)\sin nx\,dx$ ...(1)

These values of a_0, a_n and b_n are called Euler's formulae.

Cor. 1. *If c = 0, the interval becomes 0 < x < 2π, and the formulae (1) reduce to*

$$a_0 = \frac{1}{\pi}\int_0^{2\pi} f(x)\,dx;\quad a_n = \frac{1}{\pi}\int_0^{2\pi} f(x)\cos nx\,dx;\quad b_n = \frac{1}{\pi}\int_0^{2\pi} f(x)\sin nx\,dx$$

Cor. 2. *If c = – π, the interval becomes – π < x < π, and the formulae (1) reduce to*

$$a_0 = \frac{1}{\pi}\int_{-\pi}^{\pi} f(x)\,dx;\ a_n = \frac{1}{\pi}\int_{-\pi}^{\pi} f(x)\cos nx\,dx;\ b_n = \frac{1}{\pi}\int_{-\pi}^{\pi} f(x)\sin nx\,dx$$

Cor. 3. *When f(x) is an odd function* $a_0 = \frac{1}{\pi}\int_{-\pi}^{\pi} f(x)\,dx = 0$

Since cos nx is an even function, therefore, $f(x)$ cos nx is an odd function.

$$a_n = \frac{1}{\pi}\int_{-\pi}^{\pi} f(x)\cos nx\,dx = 0$$

Since sin nx is an odd function, therefore, $f(x)$ sin nx is an even function.

$$\therefore \qquad b_n = \frac{1}{\pi}\int_{-\pi}^{\pi} f(x)\sin nx\,dx = \frac{2}{\pi}\int_0^{\pi} f(x)\sin nx\,dx$$

Hence, if a periodic function $f(x)$ is odd, its Fourier expansion contains only sine terms,

i.e.,
$$f(x) = \sum_{n=1}^{\infty} b_n \sin nx, \quad \text{where} \quad b_n = \frac{2}{\pi}\int_0^{\pi} f(x) \sin nx \, dx$$

Cor. 4. *When f(x) is an even function* $a_0 = \frac{1}{\pi}\int_{-\pi}^{\pi} f(x)\, dx = \frac{2}{\pi}\int_0^{\pi} f(x)\, dx.$

Since $\cos nx$ is an even function, therefore, $f(x) \cos nx$ is an even function.

$$\therefore \qquad a_n = \frac{1}{\pi}\int_{-\pi}^{\pi} f(x) \cos nx \, dx = \frac{2}{\pi}\int_0^{\pi} f(x) \cos nx \, dx$$

Since $\sin nx$ is an odd function, therefore, $f(x) \sin nx$ is an odd function.

$$\therefore \qquad b_n = \frac{1}{\pi}\int_{-\pi}^{\pi} f(x) \sin nx \, dx = 0$$

Hence, if a periodic function $f(x)$ is even, its Fourier expansion contains only cosine terms,

i.e.,
$$f(x) = \frac{a_0}{2} + \sum_{n=1}^{\infty} a_n \cos nx, \text{where } a_0 = \frac{2}{\pi}\int_0^{\pi} f(x)\, dx \text{ and } a_0 = \frac{2}{\pi}\int_0^{\pi} f(x) \cos nx \, dx.$$

ILLUSTRATIVE EXAMPLES

Example 1. *Obtain the Fourier Series to represent* $f(x) = \frac{1}{4}(\pi - x)^2, 0 < x < 2\pi.$

Sol. Let $f(x) = \frac{1}{4}(\pi - x)^2 = \frac{a_0}{2} + \sum_{n=1}^{\infty} a_n \cos nx + \sum_{n=1}^{\infty} b_n \sin nx$

By Euler's formulae, we have

$$a_0 = \frac{1}{\pi}\int_0^{2\pi} f(x)\, dx = \frac{1}{\pi}\int_0^{2\pi} \frac{1}{4}(\pi - x)^2\, dx = \frac{1}{4\pi}\left[\frac{(\pi - x)^3}{-3}\right]_0^{2\pi} = -\frac{1}{12\pi}[-\pi^3 - \pi^3] = \frac{\pi^2}{6}$$

$$a_n = \frac{1}{\pi}\int_0^{2\pi} f(x) \cos nx \, dx = \frac{1}{\pi}\int_0^{2\pi} \frac{1}{4}(\pi - x)^2 \cos nx \, dx$$

$$= \frac{1}{4\pi}\left[(\pi - x)^2 \frac{\sin nx}{n} - \{-2(\pi - x)\}\left(-\frac{\cos nx}{n^2}\right) + 2\left(-\frac{\sin nx}{n^3}\right)\right]_0^{2\pi}$$

$$= \frac{1}{4\pi}\left[\left(0 + \frac{2\pi \cos 2n\pi}{n^2} + 0\right) - \left(0 - \frac{2\pi \cos 0}{n^2} + 0\right)\right] = \frac{1}{4\pi}\left[\frac{2\pi}{n^2} + \frac{2\pi}{n^2}\right] = \frac{1}{n^2}$$

$$b_n = \frac{1}{\pi}\int_0^{2\pi} f(x) \sin nx \, dx = \frac{1}{\pi}\int_0^{2\pi} \frac{1}{4}(\pi - x)^2 \sin nx \, dx$$

$$= \frac{1}{4\pi}\left[(\pi - x)^2\left(-\frac{\cos nx}{n}\right) - \{-2(\pi - x)\}\left(-\frac{\sin nx}{n^2}\right) + 2\left(\frac{\cos nx}{n^3}\right)\right]_0^{2\pi}$$

$$= \frac{1}{4\pi}\left[\left(-\frac{\pi^2 \cos 2n\pi}{n} - 0 + \frac{2 \cos 2n\pi}{n^3}\right) - \left(-\frac{\pi^2}{n} - 0 + \frac{2\cos 0}{n^3}\right)\right]$$

$$= \frac{1}{4\pi}\left[\left(-\frac{\pi^2}{n}+\frac{2}{n^3}\right)-\left(-\frac{\pi^2}{n}+\frac{2}{n^3}\right)\right] = 0$$

$$\therefore \quad f(x) = \frac{\pi^2}{12} + \sum_{n=1}^{\infty} \frac{\cos nx}{n^2} = \frac{\pi^2}{12} + \frac{\cos x}{1^2} + \frac{\cos 2x}{2^2} + \frac{\cos 3x}{3^2} + \ldots\ldots$$

Example 2. *Expand $f(x) = x \sin x$, $0 < x < 2\pi$ as a Fourier Series.*

Sol. Let $\quad f(x) = x \sin x = \dfrac{a_0}{2} + \displaystyle\sum_{n=1}^{\infty} a_n \cos nx + \sum_{n=1}^{\infty} b_n \sin nx$

By Euler's formulae, we have $a_0 = \dfrac{1}{\pi}\displaystyle\int_0^{2\pi} f(x)\, dx = \dfrac{1}{\pi}\int_0^{2\pi} x \sin x\, dx$

$$= \frac{1}{\pi}\Big[x(-\cos x) - 1.(-\sin x)\Big]_0^{2\pi} = \frac{1}{\pi}[-2\pi] = -2$$

$$a_n = \frac{1}{\pi}\int_0^{2\pi} f(x) \cos nx\, dx = \frac{1}{\pi}\int_0^{2\pi} x \sin x \cos nx\, dx = \frac{1}{2\pi}\int_0^{2\pi} x(2 \cos nx \sin x)\, dx$$

$$= \frac{1}{2\pi}\int_0^{2\pi} x[\sin (n+1)x - \sin (n-1)x]\, dx$$

$$= \frac{1}{2\pi}\left[x\left\{-\frac{\cos (n+1)x}{n+1} + \frac{\cos (n-1)x}{n-1}\right\} - 1.\left\{-\frac{\sin (n+1)x}{(n+1)^2} + \frac{\sin (n-1)x}{(n-1)^2}\right\}\right]_0^{2\pi}$$

$$= \frac{1}{2\pi}\left[2\pi\left\{-\frac{\cos 2(n+1)\pi}{n+1} + \frac{\cos 2(n-1)\pi}{n-1}\right\}\right]$$

$$= -\frac{1}{n+1} + \frac{1}{n-1} = \frac{2}{n^2-1},\ n \neq 1$$

When $n = 1$, we have

$$a_1 = \frac{1}{\pi}\int_0^{2\pi} x \sin x \cos x\, dx = \frac{1}{2\pi}\int_0^{2\pi} x \sin 2x\, dx$$

$$= \frac{1}{2\pi}\left[x\left(-\frac{\cos 2x}{2}\right) - 1.\left(-\frac{\sin 2x}{4}\right)\right]_0^{2\pi} = \frac{1}{2\pi}[-\pi] = -\frac{1}{2}$$

$$b_n = \frac{1}{\pi}\int_0^{2\pi} f(x) \sin nx\, dx = \frac{1}{\pi}\int_0^{2\pi} x \sin x \sin nx\, dx$$

$$= \frac{1}{2\pi}\int_0^{2\pi} x(2 \sin nx \sin x)\, dx = \frac{1}{2\pi}\int_0^{2\pi} x\,[\cos (n-1)x - \cos (n+1)x]\, dx$$

$$= \frac{1}{2\pi}\left[x\left\{\frac{\sin (n-1)x}{n-1} - \frac{\sin (n+1)x}{n+1}\right\} - 1.\left\{-\frac{\cos (n-1)x}{(n-1)^2} + \frac{\cos (n+1)x}{(n+1)^2}\right\}\right]_0^{2\pi}$$

$$= \frac{1}{2\pi}\left[\frac{\cos 2(n-1)\pi}{(n-1)^2} - \frac{\cos 2(n+1)\pi}{(n+1)^2} - \frac{1}{(n-1)^2} + \frac{1}{(n+1)^2}\right]$$

$$= \frac{1}{2\pi}\left[\frac{1}{(n-1)^2} - \frac{1}{(n+1)^2} - \frac{1}{(n-1)^2} + \frac{1}{(n+1)^2}\right] = 0, n \neq 1$$

When $n = 1$, we have

$$b_1 = \frac{1}{\pi}\int_0^{2\pi} x \sin x \sin x \, dx = \frac{1}{2\pi}\int_0^{2\pi} x(1 - \cos 2x)\, dx$$

$$= \frac{1}{2\pi}\left[x\left(x - \frac{\sin 2x}{2}\right) - 1.\left(\frac{x^2}{2} + \frac{\cos 2x}{4}\right)\right]_0^{2\pi}$$

$$= \frac{1}{2\pi}\left[2\pi(2\pi) - \frac{4\pi^2}{2} - \frac{1}{4} + \frac{1}{4}\right] = \frac{1}{2\pi}(2\pi^2) = \pi$$

$$\therefore \quad f(x) = \frac{a_0}{2} + a_1 \cos x + b_1 \sin x + \sum_{n=2}^{\infty} a_n \cos nx + \sum_{n=2}^{\infty} b_n \sin nx$$

$$= -1 - \frac{1}{2}\cos x + \pi \sin x + \sum_{n=2}^{\infty} \frac{2}{n^2 - 1}\cos nx + 0$$

$$= -1 + \pi \sin x - \frac{1}{2}\cos x + \frac{2}{2^2 - 1}\cos 2x + \frac{2}{3^2 - 1}\cos 3x + \ldots\ldots.$$

Example 3. *Find a Fourier Series to represent $x - x^2$ from $x = -\pi$ to $x = \pi$. Hence show that :* $\frac{1}{1^2} - \frac{1}{2^2} + \frac{1}{3^2} - \frac{1}{4^2} + \ldots\ldots = \frac{\pi^2}{12}$.

Sol. Let $x - x^2 = \frac{a_0}{2} + \sum_{n=1}^{\infty} a_n \cos nx + \sum_{n=1}^{\infty} b_n \sin nx$

By Euler's formulae, we have

$$a_0 = \frac{1}{\pi}\int_{-\pi}^{\pi} (x - x^2)\, dx = \frac{-2}{\pi}\int_0^{\pi} x^2\, dx = \frac{-2}{\pi}\left(\frac{x^3}{3}\right)_0^{\pi} = -\frac{2}{3}\pi^2$$

$$a_n = \frac{1}{\pi}\int_{-\pi}^{\pi} (x - x^2)\cos nx\, dx = -\frac{2}{\pi}\int_0^{\pi} x^2 \cos nx\, dx$$

$$= \frac{-2}{\pi}\left[\left(x^2\,\frac{\sin nx}{n}\right)_0^{\pi} - \int_0^{\pi} 2x\,.\,\frac{\sin nx}{n}\,dx\right]$$

$$= \frac{4}{n\pi}\left[\left\{x\,.\left(\frac{-\cos nx}{n}\right)\right\}_0^{\pi} - \int_0^{\pi} 1.\left(\frac{-\cos nx}{n}\right)dx\right] = \frac{-4}{n^2}(-1)^n$$

| $\because \cos n\pi = (-1)^n$ and second integral vanishes

$$b_n = \frac{1}{\pi}\int_{-\pi}^{\pi} (x - x^2)\sin nx\, dx$$

$$= \frac{2}{\pi}\int_0^{\pi} x \sin nx\, dx = \frac{2}{\pi}\left[\left\{x\,.\left(\frac{-\cos nx}{n}\right)\right\}_0^{\pi} - \int_0^{\pi} 1.\left(\frac{-\cos nx}{n}\right)dx\right]$$

$$= \frac{-2}{n}(-1)^n \quad | \because \cos n\pi = (-1)^n \text{ and second integral vanishes}$$

$$\therefore \; x - x^2 = -\frac{\pi^2}{3} - 4\sum_{n=1}^{\infty} \frac{(-1)^n}{n^2}\cos nx - 2\sum_{n=1}^{\infty} \frac{(-1)^n}{n}\sin nx$$

$$= -\frac{\pi^2}{3} - 4\left[-\frac{\cos x}{1^2} + \frac{\cos 2x}{2^2} - \frac{\cos 3x}{3^2} + ...\right]$$

$$-2\left[\frac{-\sin x}{1} + \frac{\sin 2x}{2} - \frac{\sin 3x}{3} + ...\right]$$

$$= -\frac{\pi^2}{3} + 4\left[\frac{\cos x}{1^2} - \frac{\cos 2x}{2^2} + \frac{\cos 3x}{3^2} - ......\right] + 2\left[\frac{\sin x}{1} - \frac{\sin 2x}{2} + \frac{\sin 3x}{3} - ......\right]$$

Putting $x = 0$, we get

$$0 = -\frac{\pi^2}{3} + 4\left(\frac{1}{1^2} - \frac{1}{2^2} + \frac{1}{3^2} - \frac{1}{4^2} + ...\right)$$

$$\Rightarrow \qquad \frac{1}{1^2} - \frac{1}{2^2} + \frac{1}{3^2} - \frac{1}{4^2} + ... = \frac{\pi^2}{12}.$$

Example 4. *Obtain the Fourier Series for $f(x) = e^{-x}$ in the interval $0 < x < 2\pi$.*

Sol. Let $f(x) = e^{-x} = \dfrac{a_0}{2} + \displaystyle\sum_{n=1}^{\infty} a_n \cos nx + \sum_{n=1}^{\infty} b_n \sin nx$

Then $$a_0 = \frac{1}{\pi}\int_0^{2\pi} f(x)\,dx = \frac{1}{\pi}\int_0^{2\pi} e^{-x}\,dx = \frac{1}{\pi}\Big[-e^{-x}\Big]_0^{2\pi} = \frac{1 - e^{-2\pi}}{\pi}$$

$$a_n = \frac{1}{\pi}\int_0^{2\pi} f(x)\cos nx\,dx = \frac{1}{\pi}\int_0^{2\pi} e^{-x}\cos nx\,dx$$

$$= \frac{1}{\pi}\left[\frac{e^{-x}}{1+n^2}(-\cos nx + n\sin nx)\right]_0^{2\pi} = \frac{1 - e^{-2\pi}}{\pi(1+n^2)}$$

$$b_n = \frac{1}{\pi}\int_0^{2\pi} f(x)\sin nx\,dx = \frac{1}{\pi}\int_0^{2\pi} e^{-x}\sin nx\,dx$$

$$= \frac{1}{\pi}\left[\frac{e^{-x}}{1+n^2}(-\sin nx - n\cos nx)\right]_0^{2\pi} = \frac{1 - e^{-2\pi}}{\pi}\cdot\frac{n}{1+n^2}$$

$$\therefore \quad e^{-x} = \frac{1 - e^{-2\pi}}{2\pi} + \frac{1 - e^{-2\pi}}{\pi}\sum_{n=1}^{\infty}\frac{\cos nx}{1+n^2} + \frac{1 - e^{-2\pi}}{\pi}\sum_{n=1}^{\infty}\frac{n\sin nx}{1+n^2}$$

$$= \frac{1 - e^{-2\pi}}{\pi}\left[\frac{1}{2} + \left(\frac{1}{2}\cos x + \frac{1}{5}\cos 2x + \frac{1}{10}\cos 3x + ...\right)\right.$$

$$\left. + \left(\frac{1}{2}\sin x + \frac{2}{5}\sin 2x + \frac{3}{10}\sin 3x + ...\right)\right]$$

Example 5. *Find the Fourier Series to represent e^{ax} in the interval $-\pi < x < \pi$.*

Sol. Let $f(x) = e^{ax} = \dfrac{a_0}{2} + \sum_{n=1}^{\infty} a_n \cos nx + \sum_{n=1}^{\infty} b_n \sin nx$

Then $a_0 = \dfrac{1}{\pi}\int_{-\pi}^{\pi} f(x)\,dx = \dfrac{1}{\pi}\int_{-\pi}^{\pi} e^{ax}\,dx = \dfrac{1}{\pi}\left[\dfrac{e^{ax}}{a}\right]_{-\pi}^{\pi} = \dfrac{1}{\pi a}(e^{a\pi} - e^{-a\pi}) = \dfrac{2\sinh a\pi}{\pi a}$

$$a_n = \frac{1}{\pi}\int_{-\pi}^{\pi} f(x)\cos nx\,dx = \frac{1}{\pi}\int_{-\pi}^{\pi} e^{ax}\cos nx\,dx$$

$$= \frac{1}{\pi}\left[\frac{e^{ax}}{a^2+n^2}(a\cos nx + n\sin nx)\right]_{-\pi}^{\pi} = \frac{1}{\pi(a^2+n^2)}[ae^{a\pi}\cos n\pi - ae^{-a\pi}\cos n\pi]$$

$$= \frac{a\cos n\pi\,(e^{a\pi} - e^{-a\pi})}{\pi(a^2+n^2)} = \frac{2a(-1)^n \sinh a\pi}{\pi(a^2+n^2)}$$

Similarly, $b_n = \dfrac{2n(-1)^n \sinh a\pi}{\pi(a^2+n^2)}$

$\therefore$
$$e^{ax} = \frac{\sinh a\pi}{a\pi} + \sum_{n=1}^{\infty} \frac{2a(-1)^n \sinh a\pi}{\pi(a^2+n^2)}\cos nx + \sum_{n=1}^{\infty} \frac{2n(-1)^n \sinh a\pi}{\pi(a^2+n^2)}\sin nx$$

$$= \frac{2\sinh a\pi}{\pi}\left[\frac{1}{2a} - a\left(\frac{\cos x}{a^2+1^2} - \frac{\cos 2x}{a^2+2^2} + \frac{\cos 3x}{a^2+3^2} - \ldots\ldots\right)\right.$$
$$\left. - \left(\frac{\sin x}{a^2+1^2} - \frac{2\sin 2x}{a^2+2^2} + \frac{3\sin 3x}{a^2+3^2} - \ldots\ldots\right)\right]$$

Example 6. *Express $f(x) = |\,x\,|$, $-\pi < x < \pi$, as Fourier Series.*

Sol. Since $f(-x) = |-x| = |x| = f(x)$

$\therefore$ $f(x)$ is an even function and hence $b_n = 0$

Let $f(x) = |\,x\,| = \dfrac{a_0}{2} + \sum_{n=1}^{\infty} a_n \cos nx$

Then, $a_0 = \dfrac{2}{\pi}\int_0^{\pi} f(x)\,dx = \dfrac{2}{\pi}\int_0^{\pi} |x|\,dx = \dfrac{2}{\pi}\int_0^{\pi} x\,dx = \dfrac{2}{\pi}\left[\dfrac{x^2}{2}\right]_0^{\pi} = \pi$

$$a_n = \frac{2}{\pi}\int_0^{\pi} f(x)\cos nx\,dx = \frac{2}{\pi}\int_0^{\pi} |x|\cos nx\,dx = \frac{2}{\pi}\int_0^{\pi} x\cos nx\,dx$$

$$= \frac{2}{\pi}\left[x\left(\frac{\sin nx}{n}\right) - 1.\left(-\frac{\cos nx}{n^2}\right)\right]_0^{\pi}$$

$$= \frac{2}{\pi}\left[\frac{\cos n\pi}{n^2} - \frac{1}{n^2}\right] = \frac{2}{\pi n^2}[(-1)^n - 1] = \begin{cases} 0, & \text{if } n \text{ is even} \\ \dfrac{-4}{\pi n^2}, & \text{if } n \text{ is odd} \end{cases}$$

$$\therefore \qquad |x| = \frac{\pi}{2} - \frac{4}{\pi}\left(\cos x + \frac{\cos 3x}{3^2} + \frac{\cos 5x}{5^2} + \ldots\ldots\right)$$

Note. Putting $x = 0$ in the above result, we get $\frac{1}{1^2} + \frac{1}{3^2} + \frac{1}{5^2} + \ldots\ldots = \frac{\pi^2}{8}$.

Example 7. *Expand the function $f(x) = x \sin x$ as a Fourier series in the interval $-\pi \le x \le \pi$.*

Deduce that $\frac{1}{1.3} - \frac{1}{3.5} + \frac{1}{5.7} - \frac{1}{7.9} + \ldots\ldots = \frac{\pi - 2}{4}$.

Sol. Since $x \sin x$ is an even function of x, $b_n = 0$

Let $\qquad f(x) = x \sin x = \frac{a_0}{2} + \sum_{n=1}^{\infty} a_n \cos nx$

Then, $\qquad a_0 = \frac{2}{\pi}\int_0^{\pi} x \sin x \, dx = \frac{2}{\pi}\Big[x(-\cos x) - 1.(-\sin x)\Big]_0^{\pi} = \frac{2}{\pi}(-\pi \cos \pi) = 2$

$$a_n = \frac{2}{\pi}\int_0^{\pi} x \sin x \cos nx \, dx = \frac{1}{\pi}\int_0^{\pi} x(2 \cos nx \sin x) \, dx$$

$$= \frac{1}{\pi}\int_0^{\pi} x[\sin (n+1)x - \sin (n-1)x] \, dx$$

$$= \frac{1}{\pi}\left[x\left\{-\frac{\cos (n+1)x}{n+1} + \frac{\cos (n-1)x}{n-1}\right\} - 1.\left\{-\frac{\sin (n+1)x}{(n+1)^2} + \frac{\sin (n-1)x}{(n-1)^2}\right\}\right]_0^{\pi}$$

$$= \frac{1}{\pi}\left[\pi\left\{-\frac{\cos (n+1)\pi}{n+1} + \frac{\cos (n-1)\pi}{n-1}\right\}\right], \; n \ne 1$$

$$= \frac{\cos (n-1)\pi}{n-1} - \frac{\cos (n+1)\pi}{n+1}$$

When n is odd, $n \ne 1$, $n - 1$ and $n + 1$ are even

$$\therefore \qquad a_n = \frac{1}{n-1} - \frac{1}{n+1} = \frac{2}{n^2 - 1}$$

When n is even, $n - 1$ and $n + 1$ are odd

$$\therefore \qquad a_n = \frac{-1}{n-1} + \frac{1}{n+1} = \frac{-2}{n^2 - 1}$$

When $n = 1$, we have $a_1 = \frac{2}{\pi}\int_0^{\pi} x \sin x \cos x \, dx = \frac{1}{\pi}\int_0^{\pi} x \sin 2x \, dx$

$$= \frac{1}{\pi}\left[x\left(-\frac{\cos 2x}{2}\right) - 1.\left(-\frac{\sin 2x}{4}\right)\right]_0^{\pi} = \frac{1}{\pi}\left[-\frac{\pi \cos 2\pi}{2}\right] = -\frac{1}{2}$$

$$\therefore \qquad x \sin x = 1 - \frac{1}{2}\cos x - 2\left(\frac{\cos 2x}{2^2 - 1} - \frac{\cos 3x}{3^2 - 1} + \frac{\cos 4x}{4^2 - 1} - \frac{\cos 5x}{5^2 - 1} + \ldots\ldots\right)$$

Putting $x = \frac{\pi}{2}$, we get $\frac{\pi}{2} = 1 - 2\left(\frac{-1}{2^2-1} + \frac{1}{4^2-1} - \frac{1}{6^2-1} + \ldots\ldots\right)$

$$\Rightarrow \qquad \frac{\pi}{2} - 1 = 2\left(\frac{1}{3} - \frac{1}{15} + \frac{1}{35} - \ldots\ldots\right)$$

$$\Rightarrow \qquad \frac{\pi - 2}{4} = \frac{1}{1.3} - \frac{1}{3.5} + \frac{1}{5.7} - \ldots\ldots$$

Example 8. *Obtain Fourier series for the function f(x) given by*

$$f(x) = \begin{cases} 1 + \frac{2x}{\pi}, & -\pi \le x \le 0 \\ 1 - \frac{2x}{\pi}, & 0 \le x \le \pi \end{cases}$$

Hence deduce that $\frac{1}{1^2} + \frac{1}{3^2} + \frac{1}{5^2} + \ldots\ldots = \frac{\pi^2}{8}$.

Sol. When $-\pi \le x \le 0,\ 0 \le -x \le \pi$

$$\therefore \qquad f(-x) = 1 - \frac{2(-x)}{\pi} = 1 + \frac{2x}{\pi} = f(x)$$

When $0 \le x \le \pi,\ -\pi \le -x \le 0$

$$\therefore \qquad f(-x) = 1 + \frac{2(-x)}{\pi} = 1 - \frac{2x}{\pi} = f(x)$$

$\Rightarrow$ $f(x)$ is an even function of x in $[-\pi, \pi]$. This is also clear from its graph which is symmetrical above the y-axis.

$$\therefore \qquad b_n = 0$$

Let $f(x) = \frac{a_0}{2} + \sum_{n=1}^{\infty} a_n \cos nx$

then,
$$a_0 = \frac{2}{\pi}\int_0^{\pi} f(x)\, dx = \frac{2}{\pi}\int_0^{\pi}\left(1 - \frac{2x}{\pi}\right) dx = \frac{2}{\pi}\left[x - \frac{x^2}{\pi}\right]_0^{\pi} = 0$$

$$a_n = \frac{2}{\pi}\int_0^{\pi} f(x)\cos nx\, dx = \frac{2}{\pi}\int_0^{\pi}\left(1 - \frac{2x}{\pi}\right)\cos nx\, dx$$

$$= \frac{2}{\pi}\left[\left(1 - \frac{2x}{\pi}\right)\frac{\sin nx}{n} - \left(-\frac{2}{\pi}\right)\left(-\frac{\cos nx}{n^2}\right)\right]_0^{\pi}$$

$$= \frac{2}{\pi}\left[-\frac{2\cos n\pi}{\pi n^2} + \frac{2}{\pi n^2}\right] = \frac{4}{\pi^2 n^2}[1 - (-1)^n]$$

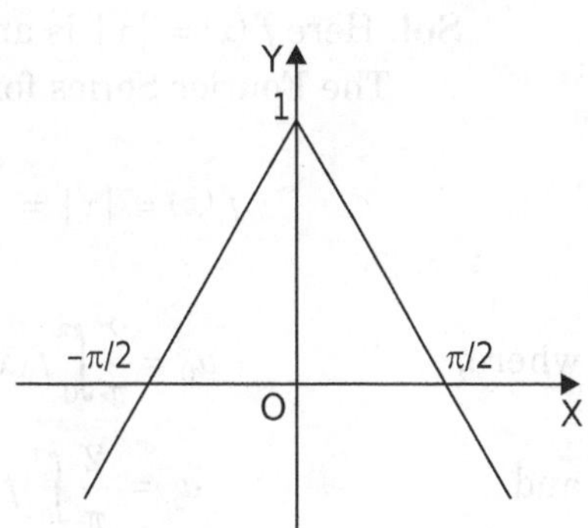

$$\therefore \qquad f(x) = \frac{4}{\pi^2}\sum_{n=1}^{\infty}[1 - (-1)^n]\frac{\cos nx}{n^2}$$

$$= \frac{4}{\pi^2}\left[\frac{2\cos x}{1^2} + \frac{2\cos 3x}{3^2} + \frac{2\cos 5x}{5^2} + \ldots\ldots\right]$$

$$= \frac{8}{\pi^2}\left(\frac{\cos x}{1^2} + \frac{\cos 3x}{3^2} + \frac{\cos 5x}{5^2} + \ldots\ldots\right)$$

Putting $x = 0$, we get $\frac{1}{1^2} + \frac{1}{3^2} + \frac{1}{5^2} + \ldots\ldots = \frac{\pi^2}{8}$, since $f(0) = 1$.

Example 9. *Find the Fourier series for the function $f(x) = x, -\pi < x < \pi$.*

Sol. Here $f(x) = x$ is an odd function.

$\therefore$ The Fourier Series for $f(x) = x$ is given by

$$f(x) = x = \sum_{n=1}^{\infty} b_n \sin nx$$

where
$$b_n = \frac{2}{\pi}\int_0^{\pi} f(x) \sin nx \, dx = \frac{2}{\pi}\int_0^{\pi} x \sin nx \, dx$$

$$= \frac{2}{\pi}\left\{\left[x\left(-\frac{\cos nx}{n}\right)\right]_0^{\pi} - \int_0^{\pi} 1\left(-\frac{\cos nx}{n}\right) dx\right\}$$

$$= \frac{2}{\pi}\left\{-\frac{\pi}{n}\cos n\pi + \frac{1}{n}\int_0^{\pi}\cos nx \, dx\right\} = \frac{2}{\pi}\left\{-\frac{\pi}{n}\cos n\pi + \frac{1}{n}\left[\frac{\sin nx}{n}\right]_0^{\pi}\right\}$$

$$= \frac{2}{\pi}\left\{-\frac{\pi}{n}\cos n\pi + 0\right\} = -\frac{2}{n}(-1)^n = \frac{2}{n}(-1)^{n+1}.$$

Hence the required Fourier Series is

$$x = \sum_{n=1}^{\infty} \frac{2}{n}(-1)^{n+1} \sin nx = 2\left[\sin x - \frac{1}{2}\sin 2x + \frac{1}{3}\sin 3x - \ldots\right].$$

Example 10. *Find the Fourier series for the periodic function $f(x)$ with period 2π, where $f(x) = |x|, -\pi < x < \pi$. Hence deduce that $\frac{1}{1^2} + \frac{1}{3^2} + \frac{1}{5^2} + \ldots = \frac{\pi^2}{8}$.*

Sol. Here $f(x) = |x|$ is an even function.

$\therefore$ The Fourier Series for $f(x) = |x|$ is given by

$$f(x) = |x| = \frac{a_0}{2} + \sum_{n=1}^{\infty} a_n \cos nx$$

where,
$$a_0 = \frac{2}{\pi}\int_0^{\pi} f(x)\, dx = \frac{2}{\pi}\int_0^{\pi} |x|\, dx = \frac{2}{\pi}\int_0^{\pi} x \, dx = \frac{2}{\pi}\frac{\pi^2}{2} = \pi$$

and
$$a_n = \frac{2}{\pi}\int_0^{\pi} f(x) \cos nx \, dx = \frac{2}{\pi}\int_0^{\pi} |x| \cos nx \, dx$$

$$= \frac{2}{\pi}\int_0^{\pi} x \cos nx \, dx = \frac{2}{\pi}\left\{\left[x\left(\frac{\sin nx}{n}\right)\right]_0^{\pi} - \int_0^{\pi} \cdot \left(\frac{\sin nx}{n}\right) dx\right\}$$

$$= \frac{2}{\pi}\left\{\frac{\pi}{n}(\sin n\pi - \sin 0) - \frac{1}{n}\int_0^{\pi} \sin nx\, dx\right\}$$

$$= \frac{2}{\pi}\left\{0 - \frac{1}{n}\left[-\frac{\cos nx}{n}\right]_0^{\pi}\right\} \qquad [\because \ \sin n\pi = 0]$$

$$= \frac{2}{n^2\pi}[\cos n\pi - 1] = \frac{2}{n^2\pi}[(-1)^n - 1] \qquad [\because \ \cos n\pi = (-1)^n]$$

$$= \begin{cases} -\dfrac{4}{n^2\pi}, \text{when } n \text{ is odd,} \\ 0 \ , \text{when } n \text{ is even.} \end{cases}$$

Putting above values in (1), the required Fourier Series is given by

$$|x| = \frac{\pi}{2} - \frac{4}{\pi}\left[\frac{\cos x}{1^2} - \frac{\cos 3x}{3^2} + \frac{\cos 5x}{5^2} - \ldots\right] \qquad \ldots(2)$$

Putting $x =$ in (2), we get

$$0 = |0| = \frac{\pi}{2} - \frac{4}{\pi}\left[\frac{1}{1^2} - \frac{(-1)}{3^2} + \frac{1}{5^2} - \frac{(-1)}{7^2} + \ldots\right]$$

$$\frac{1}{1^2} + \frac{1}{3^2} + \frac{1}{5^2} + \ldots = \frac{\pi^2}{8}.$$

Example 11. *Obtain Fourier's Series in the interval $(-\pi, \pi)$ for the function $f(z) = x \cos x$.*

Sol. Here $f(x) = x \cos x$. Clearly, $x \cos x$ is an odd function of x.

$\therefore$ The Fourier Series of the function $f(x) = x \cos x$ is given by

$$f(x) = \sum_{n=1}^{\infty} b_n \sin nx$$

where,

$$b_n = \frac{2\pi}{\pi}\int_0^{\pi} f(x)\sin nx\, dx = \frac{2}{\pi}\int_0^{\pi} x\cos x \sin nx\, dx$$

$$= \frac{1}{\pi}\int_0^{\pi} x\,\{\sin(n+1)x + \sin(n-1)x\}\,dx$$

$$= \frac{1}{\pi}\int_0^{\pi} [x\sin(n+1)x + x\sin(n-1)x]\,dx$$

$$= \frac{1}{\pi}\left[x\left\{-\frac{\cos(n+1)x}{n+1}\right\} - (1)\,x\left\{-\frac{\sin(n+1)x}{(n+1)^2}\right\}\right.$$

$$\left. + \left\{-\frac{\cos(n-1)x}{n-1}\right\} - (1)\,x\left\{-\frac{\sin(n-1)x}{(n-1)^2}\right\}\right]_0^{\pi}$$

$$= \frac{1}{\pi}\left[\pi\left\{\frac{-\cos(n+1)\pi}{(n+1)}\right\} + 0 + \pi\left\{\frac{-\cos(n-1)\pi}{n-1}\right\} + 0\right]$$

$$= \frac{\cos n\pi}{n+1} + \frac{\cos n\pi}{n-1} = \frac{2n\cos n\pi}{n^2-1} = \frac{2n(-1)^n}{n^2-1} \text{ if } n > 1.$$

If $n = 1$, then

$$b_1 = \frac{2}{\pi}\int_0^{\pi} x\cos x \sin x\, dx = \frac{1}{\pi}\int_0^{\pi} x \sin 2x\, dx$$

$$= \frac{1}{\pi}\left\{\left[x\left(-\frac{\cos 2x}{2}\right)\right]_0^{\pi} - \int_0^{\pi} 1\left(-\frac{\cos 2x}{2}\right)dx\right\}$$

$$= \frac{1}{\pi}\left\{-\frac{\pi}{2} + \frac{1}{2}\int_0^{\pi}\cos 2x\, dx\right\} = \frac{1}{\pi}\left\{-\frac{\pi}{2} + \frac{1}{2}\left[\frac{\sin 2x}{2}\right]_0^{\pi}\right\} = -\frac{1}{2}$$

Putting above values in (1), the required Fourier Series is

$$x\cos x = -\frac{1}{2}\sin x + \sum_{n=2}^{\infty}\frac{2n(-1)^n}{n^2-1}\sin nx.$$

TEST YOUR KNOWLEDGE

1. Expand in a Fourier Series the function $f(x) = x$ in the interval $0 < x < 2\pi$.
2. Express $f(x) = \frac{1}{2}(\pi - x)$ in a Fourier Series in the interval $0 < x < 2\pi$.
3. Find the Fourier Series for the function $f(x) = x + x^2$ in the interval $-\pi < x < \pi$. Hence show that
$$\frac{1}{1^2} - \frac{1}{2^2} + \frac{1}{3^2} - \frac{1}{4^2} + \ldots\ldots = \frac{\pi^2}{12}.$$
4. If $f(x) = \left(\frac{\pi - x}{2}\right)^2$ in the interval $0 < x < 2\pi$, show that $f(x) = \frac{\pi^2}{12} + \sum_{n=1}^{\infty}\frac{\cos nx}{n^2}$

   Hence obtain the following relations :

   (i) $\frac{1}{1^2} + \frac{1}{2^2} + \frac{1}{3^2} + \frac{1}{4^2} + \ldots\ldots = \frac{\pi^2}{6}$ (ii) $\frac{1}{1^2} - \frac{1}{2^2} + \frac{1}{3^2} - \frac{1}{4^2} + \ldots\ldots = \frac{\pi^2}{12}$

   (iii) $\frac{1}{1^2} + \frac{1}{3^2} + \frac{1}{5^2} + \ldots\ldots = \frac{\pi^2}{8}$.
5. Prove that for all values of x between $-\pi$ and π, $\frac{1}{2}x = \sin x - \frac{1}{2}\sin 2x + \frac{1}{3}\sin 3x - \frac{1}{4}\sin 4x + \ldots\ldots$
6. Obtain the Fourier Series to represent e^x in the interval $0 < x < 2\pi$.
7. Find the Fourier Series to represent e^x in the interval $-\pi < x < \pi$.
8. Find the Fourier Series to represent the function $f(x) = |\sin x|$, $-\pi < x < \pi$.
9. Expand $f(x) = |\cos x|$ as a Fourier Series in the interval $-\pi < x < \pi$.
10. Prove that in the interval $-\pi < x < \pi$, $x\cos x = -\frac{1}{2}\sin x + 2\sum_{n=2}^{\infty}\frac{n(-1)^n}{n^2-1}\sin nx$.

11. Prove that for $-\pi < x < \pi$, $\dfrac{x(\pi^2 - x^2)}{12} = \dfrac{\sin x}{1^3} - \dfrac{\sin 2x}{2^3} + \dfrac{\sin 3x}{3^3} - \dfrac{\sin 4x}{4^3} + \ldots\ldots$

12. Obtain a Fourier expansion for $\sqrt{1 - \cos x}$ in the interval $-\pi < x < \pi$.

13. Express $f(x) = \cos wx$, $-\pi < x < \pi$, where w is a fraction, as a Fourier Series. Hence prove that

$$\cot \theta = \frac{1}{\theta} + \frac{2\theta}{\theta^2 - \pi^2} + \frac{2\theta}{\theta^2 - 4\pi^2} + \ldots\ldots$$

14. Find the Fourier Series for $f(x)$ in the interval $(-\pi, \pi)$ when

$$f(x) = \begin{cases} \pi + x, & -\pi < x < 0 \\ \pi - x, & 0 < x < \pi . \end{cases}$$

15. Obtain a Fourier series to represent e^{-ax} from $x = -\pi$ to $x = \pi$. Hence derive series for $\dfrac{\pi}{\sinh \pi}$.

16. Prove that in the range $-\pi < x < \pi$, $\cosh ax = \dfrac{2a}{\pi} \sinh a\pi \left[\dfrac{1}{2a^2} + \displaystyle\sum_{n=1}^{\infty} \dfrac{(-1)^n}{n^2 + a^2} \cos nx\right]$.

Answers

1. $f(x) = \pi - 2 \displaystyle\sum_{n=1}^{\infty} \frac{\sin nx}{n}$ **2.** $f(x) = \displaystyle\sum_{n=1}^{\infty} \frac{\sin nx}{n}$

3. $f(x) = \dfrac{\pi^2}{3} - 4\left(\dfrac{\cos x}{1^2} - \dfrac{\cos 2x}{2^2} + \dfrac{\cos 3x}{3^2} - \ldots\ldots\right) + 2\left(\dfrac{\sin x}{1} - \dfrac{\sin 2x}{2} + \dfrac{\sin 3x}{3} - \ldots\ldots\right)$

6. $e^x = \dfrac{e^{2\pi} - 1}{2\pi} + \dfrac{e^{2\pi} - 1}{\pi} \displaystyle\sum_{n=1}^{\infty} \left(\frac{\cos nx}{1 + n^2} - \frac{n}{1 + n^2} \sin nx\right)$

7. $e^x = \dfrac{2 \sinh \pi}{\pi}$

$$\left[\frac{1}{2} - \left(\frac{1}{2} \cos x - \frac{1}{5} \cos 2x + \frac{1}{10} \cos 3x - \ldots\ldots\right) - \left(\frac{1}{2} \sin x - \frac{2}{5} \sin 2x + \frac{3}{10} \sin 3x - \ldots\ldots\right)\right]$$

8. $|\sin x| = \dfrac{2}{\pi} - \dfrac{4}{\pi}\left(\dfrac{\cos 2x}{3} + \dfrac{\cos 4x}{15} + \ldots\ldots + \dfrac{\cos 2nx}{4n^2 - 1} + \ldots\ldots\right)$

9. $|\cos x| = \dfrac{2}{\pi} + \dfrac{4}{\pi}\left(\dfrac{\cos 2x}{3} - \dfrac{\cos 4x}{15} + \ldots\ldots\right)$ **12.** $\sqrt{1 - \cos x} = \dfrac{2\sqrt{2}}{\pi} - \dfrac{4\sqrt{2}}{\pi} \displaystyle\sum_{n=1}^{\infty} \frac{\cos nx}{4n^2 - 1}$

13. $\cos wx = \dfrac{2w \sin w\pi}{\pi}\left(\dfrac{1}{2w^2} + \dfrac{\cos x}{1^2 - w^2} - \dfrac{\cos 2x}{2^2 - w^2} + \dfrac{\cos 3x}{3^2 - w^2} - \ldots\ldots\right)$

14. $f(x) = \dfrac{\pi}{2} + \dfrac{4}{\pi}\left(\dfrac{\cos x}{1^2} + \dfrac{\cos 3x}{3^2} + \dfrac{\cos 5x}{5^2} + \ldots\ldots\right)$

15. $e^{-ax} = \dfrac{2 \sinh a\pi}{\pi}\left[\left(\dfrac{1}{2a} - \dfrac{a \cos x}{1^2 + a^2} + \dfrac{a \cos 2x}{2^2 + a^2} - \ldots\ldots\right) - \left(\dfrac{\sin x}{1^2 + a^2} - \dfrac{2 \sin 2x}{2^2 + a^2} + \dfrac{3 \sin 3x}{3^2 + a^2} - \ldots\ldots\right)\right]$;

$$\frac{\pi}{\sinh \pi} = 2\left[\frac{1}{2^2 + 1} - \frac{1}{3^2 + 1} + \frac{1}{4^2 + 1} - \ldots\ldots\right].$$

1.4. DIRICHLET'S CONDITIONS

The sufficient conditions for the uniform convergence of a Fourier Series are called Dirichlet's conditions (after Dirichlet, a German mathematician). All the functions that normally arise in engineering problems satisfy these conditions and hence they can be expressed as a Fourier Series.

Any function f(x) can be expressed as a Fourier series $\frac{a_0}{2} + \sum_{n=1}^{\infty} a_n \cos nx + \sum_{n=1}^{\infty} b_n \sin nx$

where a_0, a_n, b_n *are constants, provided*

(i) $f(x)$ *is periodic, single valued and finite.*

(ii) $f(x)$ *has a finite number of finite discontinuities in any one period.*

(iii) $f(x)$ *has a finite number of maxima and minima.*

When these conditions are satisfied, the Fourier Series converges to $f(x)$ at every point of continuity. At a point of discontinuity, the sum of the series is equal to the mean of the limits on the right and left

i.e., $$\tfrac{1}{2}[f(x+0) + f(x-0)]$$

where $f(x+0)$ and $f(x-0)$ denote the limit on the right and the limit on the left respectively.

1.5. FOURIER SERIES FOR DISCONTINUOUS FUNCTIONS

In Art. 1.3, we derived Euler's formulae for a_0, a_n, b_n on the assumption that $f(x)$ is continuous in $(c, c + 2\pi)$. However, if $f(x)$ has finitely many points of finite discontinuity, even then it can be expressed as a Fourier Series. The integrals for a_0, a_n, b_n are to be evaluated by breaking up the range of integration.

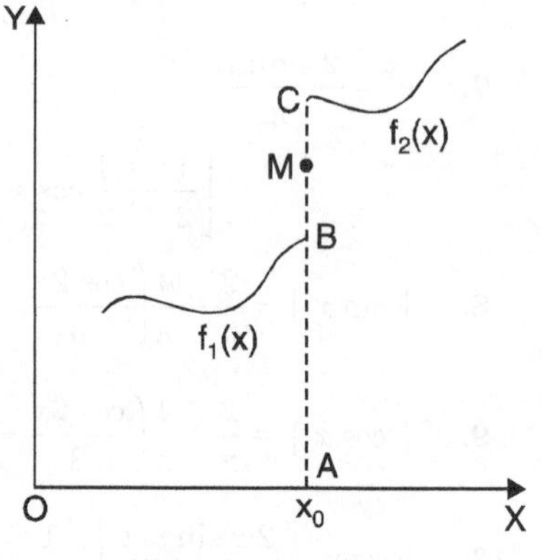

Let $f(x)$ be defined by $$f(x) = \begin{cases} f_1(x), & c < x < x_0 \\ f_2(x), & x_0 < x < c + 2\pi \end{cases}$$

where x_0 is the point of finite discontinuity in the interval $(c, c + 2\pi)$.

The values of a_0, a_n, b_n are given by

$$a_0 = \frac{1}{\pi}\left[\int_c^{x_0} f_1(x)\,dx + \int_{x_0}^{c+2\pi} f_2(x)\,dx\right]$$

$$a_n = \frac{1}{\pi}\left[\int_c^{x_0} f_1(x)\cos nx\,dx + \int_{x_0}^{c+2\pi} f_2(x)\cos nx\,dx\right]$$

$$b_n = \frac{1}{\pi}\left[\int_c^{x_0} f_1(x)\sin nx\,dx + \int_{x_0}^{c+2\pi} f_2(x)\sin nx\,dx\right]$$

At $x = x_0$, there is a finite jump in the graph of the function. Both the limits $f(x_0 - 0)$ and $f(x_0 + 0)$ exist but are unequal. The sum of the Fourier Series $= \frac{1}{2}[f(x_0 - 0) + f(x_0 + 0)] = \frac{1}{2}$ (AB + AC) = AM, where M is the mid-point of BC.

ILLUSTRATIVE EXAMPLES

Example 1. *Find the Fourier series to represent the function f(x) given by*

$$f(x) = \begin{cases} x & \text{for} \quad 0 \le x \le \pi \\ 2\pi - x & \text{for} \quad \pi \le x \le 2\pi. \end{cases}$$

and

Deduce that $\dfrac{1}{1^2} + \dfrac{1}{3^2} + \dfrac{1}{5^2} + \ldots\ldots = \dfrac{\pi^2}{8}$.

Sol. Let $\quad f(x) = \dfrac{a_0}{2} + \sum\limits_{n=1}^{\infty} a_n \cos nx + \sum\limits_{n=1}^{\infty} b_n \sin nx$

Then $\quad a_0 = \dfrac{1}{\pi}\displaystyle\int_0^{2\pi} f(x)\, dx = \dfrac{1}{\pi}\left[\int_0^{\pi} x\, dx + \int_{\pi}^{2\pi} (2\pi - x)\, dx\right]$

$$= \frac{1}{\pi}\left[\left|\frac{x^2}{2}\right|_0^{\pi} + \left|2\pi x - \frac{x^2}{2}\right|_{\pi}^{2\pi}\right] = \frac{1}{\pi}\left[\frac{\pi^2}{2} + (4\pi^2 - 2\pi^2) - \left(2\pi^2 - \frac{\pi^2}{2}\right)\right] = \pi$$

$$a_n = \frac{1}{\pi}\int_0^{2\pi} f(x) \cos nx\, dx = \frac{1}{\pi}\left[\int_0^{\pi} x \cos nx\, dx + \int_{\pi}^{2\pi} (2\pi - x) \cos nx\, dx\right]$$

$$= \frac{1}{\pi}\left[\left| x \cdot \frac{\sin nx}{n} - 1.\left(-\frac{\cos nx}{n^2}\right)\right|_0^{\pi} + \left|(2\pi - x)\left(\frac{\sin nx}{n}\right) - (-1)\left(-\frac{\cos nx}{n^2}\right)\right|_{\pi}^{2\pi}\right]$$

$$= \frac{1}{\pi}\left[\left(\frac{\cos n\pi}{n^2} - \frac{1}{n^2}\right) + \left(-\frac{\cos 2n\pi}{n^2} + \frac{\cos n\pi}{n^2}\right)\right]$$

$$= \frac{1}{\pi n^2}[(-1)^n - 1 - 1 + (-1)^n] = \frac{2}{\pi n^2}[(-1)^n - 1] = \begin{cases} -\dfrac{4}{\pi n^2}, \text{ if } n \text{ is odd} \\ 0, \text{ if } n \text{ is even} \end{cases}$$

$$b_n = \frac{1}{\pi}\int_0^{2\pi} f(x) \sin nx\, dx = \frac{1}{\pi}\left[\int_0^{\pi} x \sin nx\, dx + \int_{\pi}^{2\pi} (2\pi - x) \sin nx\, dx\right]$$

$$= \frac{1}{\pi}\left[\left| x\left(-\frac{\cos nx}{n}\right) - 1.\left(-\frac{\sin nx}{n^2}\right)\right|_0^{\pi} + \left|(2\pi - x) \times \left(-\frac{\cos nx}{n}\right) - 1.\left(-\frac{\sin nx}{n^2}\right)\right|_{\pi}^{2\pi}\right]$$

$$= \frac{1}{\pi}\left[-\frac{\pi \cos n\pi}{n} + \frac{\pi \cos n\pi}{n}\right] = 0$$

$$\therefore \quad f(x) = \frac{\pi}{2} - \frac{4}{\pi}\left(\frac{\cos x}{1^2} + \frac{\cos 3x}{3^2} + \frac{\cos 5x}{5^2} + \ldots\ldots\right)$$

Putting $x = 0$, we get $0 = \dfrac{\pi}{2} - \dfrac{4}{\pi}\left(\dfrac{1}{1^2} + \dfrac{1}{3^2} + \dfrac{1}{5^2} + \ldots\ldots\right)$

$$\Rightarrow \quad \frac{\pi^2}{8} = \frac{1}{1^2} + \frac{1}{3^2} + \frac{1}{5^2} + \ldots\ldots$$

Example 2. *If* $f(x)=\begin{cases}0, & -\pi \le x \le 0\\ \sin x, & 0 \le x \le \pi,\end{cases}$

prove that $f(x) = \dfrac{1}{\pi} + \dfrac{1}{2}\sin x - \dfrac{2}{\pi}\sum_{n=1}^{\infty} \dfrac{\cos 2nx}{4n^2-1}$.

Hence show that

(i) $\dfrac{1}{1.3} + \dfrac{1}{3.5} + \dfrac{1}{5.7} + \ldots\ldots = \dfrac{1}{2}$ *(ii)* $\dfrac{1}{1.3} - \dfrac{1}{3.5} + \dfrac{1}{5.7} - \ldots\ldots = \dfrac{\pi-2}{4}$.

Sol. Let $f(x) = \dfrac{a_0}{2} + \sum_{n=1}^{\infty} a_n \cos nx + \sum_{n=1}^{\infty} b_n \sin nx$

Then, $a_0 = \dfrac{1}{\pi}\int_{-\pi}^{\pi} f(x)\, dx = \dfrac{1}{\pi}\left[\int_{-\pi}^{0} 0\, dx + \int_0^{\pi} \sin x\, dx\right] = \dfrac{2}{\pi}$

$$a_n = \frac{1}{\pi}\int_{-\pi}^{\pi} f(x)\cos nx\, dx = \frac{1}{\pi}\left[\int_{-\pi}^{0} 0\, dx + \int_0^{\pi} \sin x \cos nx\, dx\right]$$

$$= \frac{1}{2\pi}\int_0^{\pi} 2\cos nx \sin x\, dx = \frac{1}{2\pi}\int_0^{\pi} [\sin(n+1)x - \sin(n-1)x]\, dx$$

$$= \frac{1}{2\pi}\left[-\frac{\cos(n+1)x}{n+1} + \frac{\cos(n-1)x}{n-1}\right]_0^{\pi}, \; n \ne 1$$

$$= \frac{1}{2\pi}\left[-\frac{\cos(n+1)\pi}{n+1} + \frac{\cos(n-1)\pi}{n-1} + \frac{1}{n+1} - \frac{1}{n-1}\right]$$

$$= \frac{1}{2\pi}\left[-\frac{(-1)^{n+1}}{n+1} + \frac{(-1)^{n-1}}{n-1} + \frac{1}{n+1} - \frac{1}{n-1}\right]$$

$$= \begin{cases}\dfrac{1}{2\pi}\left(-\dfrac{1}{n+1} + \dfrac{1}{n-1} + \dfrac{1}{n+1} - \dfrac{1}{n-1}\right), & \text{when } n \text{ is odd}\\[2ex] \dfrac{1}{2\pi}\left(\dfrac{1}{n+1} - \dfrac{1}{n-1} + \dfrac{1}{n+1} - \dfrac{1}{n-1}\right), & \text{when } n \text{ is even}\end{cases}$$

$$= \begin{cases}0, \text{ when } n \text{ is odd}, i.e., n = 3, 5, 7, \ldots\ldots\\ -\dfrac{2}{\pi(n^2-1)}, \text{ when } n \text{ is even}\end{cases}$$

When $n = 1$, we have

$$a_1 = \frac{1}{\pi}\int_0^{\pi} \sin x \cos x\, dx = \frac{1}{2\pi}\int_0^{\pi} \sin 2x\, dx = \frac{1}{2\pi}\left[-\frac{\cos 2x}{2}\right]_0^{\pi} = 0$$

$$b_n = \frac{1}{\pi}\int_{-\pi}^{\pi} f(x)\sin nx\, dx = \frac{1}{\pi}\left[\int_{-\pi}^{0} 0\, dx + \int_0^{\pi} \sin x \sin nx\, dx\right]$$

$$= \frac{1}{2\pi}\int_0^{\pi} 2\sin nx \sin x\, dx = \frac{1}{2\pi}\int_0^{\pi} [\cos(n-1)x - \cos(n+1)x]\, dx$$

$$= \frac{1}{2\pi}\left[\frac{\sin(n-1)x}{n-1} - \frac{\sin(n+1)x}{n+1}\right]_0^{\pi} = 0, n \neq 1$$

When $n = 1$, we have

$$b_1 = \frac{1}{\pi}\int_0^{\pi} \sin x \sin x \, dx = \frac{1}{2\pi}\int_0^{\pi}(1 - \cos 2x)\, dx = \frac{1}{2\pi}\left[x - \frac{\sin 2x}{2}\right]_0^{\pi} = \frac{1}{2}$$

$$\therefore\ f(x) = \frac{1}{\pi} - \frac{2}{\pi}\left[\frac{\cos 2x}{2^2-1} + \frac{\cos 4x}{4^2-1} + \frac{\cos 6x}{6^2-1} + \ldots\ldots\right] + \frac{1}{2}\sin x$$

$$= \frac{1}{\pi} + \frac{1}{2}\sin x - \frac{2}{\pi}\sum_{n=1}^{\infty}\frac{\cos 2nx}{(2n)^2 - 1} \quad \ldots(1)$$

Putting $x = 0$ in (1), we have $0 = \frac{1}{\pi} - \frac{2}{\pi}\sum_{n=1}^{\infty}\frac{1}{4n^2-1}$

$$\Rightarrow \frac{1}{2} = \sum_{n=1}^{\infty}\frac{1}{4n^2-1} = \sum_{n=1}^{\infty}\frac{1}{(2n-1)(2n+1)} = \frac{1}{1.3} + \frac{1}{3.5} + \frac{1}{5.7} + \ldots\ldots \quad \ldots(i)$$

Putting $x = \frac{\pi}{2}$ in (1), we have $1 = \frac{1}{\pi} + \frac{1}{2} - \frac{2}{\pi}\sum_{n=1}^{\infty}\frac{\cos n\pi}{4n^2-1}$

$$\Rightarrow \frac{1}{2} - \frac{1}{\pi} = -\frac{2}{\pi}\sum_{n=1}^{\infty}\frac{(-1)^n}{4n^2-1}$$

$$\Rightarrow \frac{\pi-2}{4} = -\sum_{n=1}^{\infty}\frac{(-1)^n}{(2n-1)(2n+1)} = -\left(-\frac{1}{1.3} + \frac{1}{3.5} - \frac{1}{5.7} + \ldots\ldots\right)$$

$$\Rightarrow \frac{1}{1.3} - \frac{1}{3.5} + \frac{1}{5.7} - \ldots\ldots = \frac{\pi-2}{4}. \quad \ldots(ii)$$

Example 3. *Find the Fourier Series of the periodic function f(x), where*

$$f(x) = \begin{cases} -\pi, & \text{when } -\pi < x < 0, \\ x, & \text{when } 0 < x < \pi. \end{cases}$$

Hence deduce that $\frac{\pi^2}{8} = \frac{1}{1^2} + \frac{1}{3^2} + \frac{1}{5^2} + \ldots$

Sol. Let the Fourier Series of the function $f(x)$ be given by

$$f(x) = \frac{a_0}{2} + \sum_{n=1}^{\infty}(a_n \cos nx + b_n \sin nx) \quad \ldots(1)$$

where

$$a_0 = \frac{1}{\pi}\int_{-\pi}^{\pi} f(x)\, dx = \frac{1}{\pi}\left[\int_{-\pi}^{0} f(x)\, dx + \int_0^{\pi} f(x)\, dx\right]$$

$$= \frac{1}{\pi}\int_{-\pi}^{0} -\pi\, dx + \frac{1}{\pi}\int_0^{\pi} x\, dx = -\left[x\right]_{-\pi}^{0} + \frac{1}{\pi}\left[\frac{x^2}{2}\right]_0^{\pi} = -\pi + \frac{1}{\pi}\cdot\frac{\pi^2}{2} = -\frac{\pi}{2}$$

$$a_n = \frac{1}{\pi}\int_{-\pi}^{\pi} f(x)\cos nx\, dx = \frac{1}{\pi}\left[\int_{-\pi}^{0} f(x)\cos nx\, dx + \int_0^{\pi} f(x)\cos nx\, dx\right]$$

$$= \frac{1}{\pi}\left[\int_{-\pi}^{0} -\pi \cos nx \, dx + \int_{0}^{\pi} x \cos nx \, dx\right]$$

$$= \frac{1}{\pi}\left[\left\{-\pi \frac{\sin nx}{n}\right\}_{-\pi}^{0} + \left\{x\left(\frac{\sin nx}{n}\right) - (1)\left(-\frac{\cos nx}{n^2}\right)\right\}_{0}^{\pi}\right]$$

$$= \frac{1}{\pi}\left[\frac{\cos n\pi}{n^2} - \frac{1}{n^2}\right] = \frac{\cos n\pi - 1}{\pi n^2} \qquad [\because \sin n\pi = 0]$$

$$= \frac{(-1)^n - 1}{\pi n^2} \qquad [\because \cos n\pi = (-1)^n]$$

$$= \begin{cases} 0, & \text{when } n \text{ is even} \\ -\dfrac{2}{\pi n^2}, & \text{when } n \text{ is odd} \end{cases}$$

and $$b_n = \frac{1}{\pi}\int_{-\pi}^{\pi} f(x) \sin nx \, dx = \frac{1}{\pi}\left[\int_{-\pi}^{0} -\pi \sin nx \, dx + \int_{0}^{\pi} x \sin nx \, dx\right]$$

$$= \frac{1}{\pi}\left[\left\{\frac{\pi \cos nx}{n}\right\}_{-\pi}^{0} + \left\{x\left(-\frac{\cos nx}{n}\right) - (1)\left(-\frac{\sin nx}{n^2}\right)\right\}_{0}^{\pi}\right]$$

$$= \frac{1}{\pi}\left[\frac{\pi}{n} - \frac{\pi \cos n\pi}{n} - \frac{\pi \cos n\pi}{n}\right] \qquad [\because \sin n\pi = 0]$$

$$= \frac{1 - 2\cos n\pi}{n} = \frac{1 - 2(-1)^n}{n} \qquad [\because \cos n\pi = (-1)^n]$$

$$= \begin{cases} -\dfrac{1}{n} & \text{when } n \text{ is even,} \\ \dfrac{3}{n}, & \text{when } n \text{ is odd.} \end{cases}$$

Thus putting values of a's and b's in (1), the required Fourier Series of $f(x)$ is obtained as follows :

$$f(x) = -\frac{\pi}{4} - \frac{2}{\pi}\left[\frac{1}{1^2}\cos x + \frac{1}{3^2}\cos 3x + \frac{1}{5^2}\cos 5x + \ldots\right] + \left[\frac{3}{1}\sin x - \frac{1}{2}\sin 2x + \frac{3}{3}\sin 3x - \ldots\right] \quad \ldots(2)$$

To find the sum : At $x = 0$, the series converges to

$$\frac{f(0+0) + f(0-0)}{2} = \frac{0 + (-\pi)}{2} = -\frac{\pi}{2}$$

and at $x = \pi$, the series converges to

$$\frac{f(-\pi+0) + f(\pi-0)}{2} = \frac{-\pi+\pi}{2} = 0$$

The L.H.S. of (2) is $-\dfrac{\pi}{2}$ at $x = 0$, and is 0 at $x = \pi$.

Therefore putting $x = 0$ in R.H.S. of (2), we get

$$-\frac{\pi}{2} = -\frac{\pi}{4} - \frac{2}{\pi}\left[\frac{1}{1^2} + \frac{1}{3^2} + \frac{1}{5^2} + \ldots\right] \quad \text{or} \quad \frac{\pi^2}{8} = \frac{1}{1^2} + \frac{1}{3^2} + \frac{1}{5^2} + \ldots$$

Similarly, if we put $x = \pi$ in (1), we obtain (3).

TEST YOUR KNOWLEDGE

1. Find the Fourier series to represent the function

$$f(x) = \begin{cases} -k & \text{when } -\pi < x < 0 \\ k & \text{when } \ 0 < x < \pi \end{cases}$$

Also deduce that $\frac{\pi}{4} = 1 - \frac{1}{3} + \frac{1}{5} - \frac{1}{7} + \ldots\ldots$

2. Develop $f(x)$ in a Fourier Series in the interval $(-\pi, \pi)$ if $f(x) = \begin{cases} 0 \text{ when } -\pi < x < 0 \\ 1 \text{ when } 0 < x < \pi. \end{cases}$

3. Find the Fourier Series expansion for $f(x)$, if $f(x) = \begin{cases} -\pi, & -\pi < x < 0 \\ x, & 0 < x < \pi \end{cases}$

Deduce that $\frac{1}{1^2} + \frac{1}{3^2} + \frac{1}{5^2} + \ldots\ldots = \frac{\pi^2}{8}$.

[**Hint.** For the deduction, put $x = 0$ in the expansion of $f(x)$.

$$f(0-0) = -\pi \text{ and } f(0+0) = 0 \quad \therefore \quad f(0) = \frac{1}{2}\{f(0-0) + f(0+0)\} = -\frac{\pi}{2}\Big]$$

4. Find the Fourier expansion of the function defined in one period by the relations

$$f(x) = \begin{cases} 1 \text{ for } 0 < x < \pi \\ 2 \text{ for } \pi < x < 2\pi \end{cases}$$

and deduce that $\frac{\pi}{4} = 1 - \frac{1}{3} + \frac{1}{5} - \frac{1}{7} + \ldots\ldots$

5. Find the Fourier Series of the following function :

$$f(x) = \begin{cases} x^2, & 0 \le x \le \pi \\ -x^2, & -\pi \le x \le 0. \end{cases}$$

6. An alternating current after passing through a rectifier has the form

$$i = \begin{cases} I_0 \sin x & \text{for} \quad 0 \le x \le \pi \\ 0 & \text{for} \quad \pi \le x \le 2\pi \end{cases}$$

where I_0 is the maximum current and the period is 2π. Express i as a Fourier series.

Answers

1. $f(x) = \frac{4k}{\pi}\left(\sin x + \frac{\sin 3x}{3} + \frac{\sin 5x}{5} + \ldots\ldots\right)$ **2.** $f(x) = \frac{1}{2} + \frac{2}{\pi}\left(\sin x + \frac{\sin 3x}{3} + \frac{\sin 5x}{5} + \ldots\ldots\right)$

3. $f(x) = -\frac{\pi}{4} - \frac{2}{\pi}\left(\cos x + \frac{\cos 3x}{3^2} + \frac{\cos 5x}{5^2} + \ldots\ldots\right) + \left(3\sin x - \frac{\sin 2x}{2} + \sin 3x - \frac{\sin 4x}{4} + \ldots\ldots\right)$

4. $f(x) = \frac{3}{2} - \frac{2}{\pi}\left(\sin x + \frac{\sin 3x}{3} + \frac{\sin 5x}{5} + \ldots\ldots\right)$

5. $f(x) = 2\left(\pi - \frac{4}{\pi}\right) \sin x - \pi \sin 2x + \frac{2}{3}\left(\pi - \frac{4}{9\pi}\right) \sin 3x - \frac{\pi}{2} \sin 4x + \ldots\ldots$

6. $i = \frac{I_0}{2} + \frac{I_0}{2} \sin x - \frac{2I_0}{\pi}\left(\frac{\cos 2x}{2^2 - 1} + \frac{\cos 4x}{4^2 - 1} + \frac{\cos 6x}{6^2 - 1} + \ldots\ldots\right).$

1.6. CHANGE OF INTERVAL

In many engineering problems, it is desired to expand a function in a Fourier series over an interval of length $2l$ and not 2π. In order to apply foregoing theory, this interval must be transformed into an interval of length 2π. This can be achieved by a transformation of the variable.

Consider a periodic function $f(x)$ defined in the interval $c < x < c + 2l$. To change the interval into one of length 2π, we put

$$\frac{x}{l} = \frac{z}{\pi} \quad \text{or} \quad z = \frac{\pi x}{l} \quad \text{so that when } x = c,\ z = \frac{\pi c}{l} = d \text{ (say)}$$

and when $x = c + 2l, \quad z = \frac{\pi(c + 2l)}{l} = \frac{\pi c}{l} + 2\pi = d + 2\pi.$

Thus the function $f(x)$ of period $2l$ in $(c, c + 2l)$ is transformed to the function $f\left(\frac{lz}{\pi}\right) = F(z)$, say, of period 2π in $(d, d + 2\pi)$ and the latter function can be expressed as the Fourier Series

$$F(z) = \frac{a_0}{2} + \sum_{n=1}^{\infty} a_n \cos nz + \sum_{n=1}^{\infty} b_n \sin nz \qquad \ldots(1)$$

where $a_0 = \frac{1}{\pi}\int_d^{d+2\pi} F(z)\, dz$; $a_n = \frac{1}{\pi}\int_d^{d+2\pi} F(z) \cos nz\, dz$; and $b_n = \frac{1}{\pi}\int_d^{d+2\pi} F(z) \sin nz\, dz$

Now making the inverse substitution $z = \frac{\pi x}{l}$, $dz = \frac{\pi}{l}\, dx$

When $z = d,\ x = c$ and when $z = d + 2\pi, x = c + 2l.$

The expression (1) becomes $F(z) = F\left(\frac{\pi x}{l}\right) = f(x) = \frac{a_0}{2} + \sum_{n=1}^{\infty} a_n \cos \frac{n\pi x}{l} + \sum_{n=1}^{\infty} b_n \sin \frac{n\pi x}{l}$ and the co-efficients a_0, a_n, b_n from (2) reduce to

$$a_0 = \frac{1}{l}\int_c^{c+2l} f(x)\, dx\ ; \quad a_n = \frac{1}{l}\int_c^{c+2l} f(x) \cos \frac{n\pi x}{l}\, dx\ ; \quad \text{and} \quad b_n = \frac{1}{l}\int_c^{c+2l} f(x) \sin \frac{n\pi x}{l} dx$$

Hence the Fourier Series f(x) in the interval c < x < c + 2l is given by

$$f(x) = \frac{a_0}{2} + \sum_{n=1}^{\infty} a_n \cos \frac{n\pi x}{l} + \sum_{n=1}^{\infty} b_n \sin \frac{n\pi x}{l}$$

where $a_0 = \int_c^{c+2l} f(x)\, dx, \quad a_n = \frac{1}{l}\int_c^{c+2l} f(x) \cos\frac{n\pi x}{l}\, dx$ and $b_n = \frac{1}{l}\int_c^{c+2l} f(x) \sin \frac{n\pi x}{l}\, dx.$

Cor. 1. If we put $c = 0$, the interval becomes $0 < x < 2l$ and the above results reduce to

$$a_0 = \frac{1}{l}\int_0^{2l} f(x)\, dx\ ; \quad a_n = \frac{1}{l}\int_0^{2l} f(x)\cos\frac{n\pi x}{l}\, dx; \quad \text{and} \quad a_n = \frac{1}{l}\int_0^{2l} f(x)\sin\frac{n\pi x}{l}\, dx.$$

Cor. 2. If we put $c = -l$, the interval becomes $-l < x < l$ and the above results reduce to

$$a_0 = \frac{1}{l}\int_{-l}^{l} f(x)\, dx, \quad a_n = \frac{1}{l}\int_{-l}^{l} f(x)\cos\frac{n\pi x}{l}\, dx \quad \text{and} \quad b_n = \frac{1}{l}\int_{-l}^{l} f(x)\sin\frac{n\pi x}{l}\, dx.$$

If f(x) is an even function, we have

$$a_0 = \frac{2}{l}\int_0^{l} f(x)\, dx, \quad a_n = \frac{2}{l}\int_0^{l} f(x)\cos\frac{n\pi x}{l}\, dx \quad \text{and} \quad b_n = 0$$

If f(x) is an odd function, we have $a_0 = 0,\ \ a_n = 0$

and
$$b_n = \frac{2}{l}\int_0^{l} f(x)\sin\frac{n\pi x}{l}\, dx.$$

ILLUSTRATIVE EXAMPLES

Example 1. *Find Fourier expansion for the function $f(x) = x - x^2,\ -1 < x < 1.$*

Sol. Let $f(x) = \dfrac{a_0}{2} + \sum\limits_{n=1}^{\infty} a_n \cos n\pi x + \sum\limits_{n=1}^{\infty} b_n \sin n\pi x$

Then, $$a_0 = \int_{-1}^{1}(x - x^2)\, dx = \int_{-1}^{1} x\, dx - \int_{-1}^{1} x^2\, dx = 0 - 2\int_0^1 x^2\, dx = -2\left[\frac{x^3}{3}\right]_0^1 = -\frac{2}{3}$$

$$a_n = \int_{-1}^{1}(x - x^2)\cos n\pi x\, dx = \int_{-1}^{1} x\cos n\pi x\, dx - \int_{-1}^{1} x^2\cos n\pi x\, dx$$

$$= 0 - 2\int_0^1 x^2\cos n\pi x\, dx = -2\left[x^2 \cdot \frac{\sin n\pi x}{n\pi} - 2x\left(-\frac{\cos n\pi x}{n^2\pi^2}\right) + 2\left(-\frac{\sin n\pi x}{n^3\pi^3}\right)\right]_0^1$$

$$= -2\left[\frac{2\cos n\pi}{n^2\pi^2}\right] = \frac{-4(-1)^n}{n^2\pi^2} = \frac{4(-1)^{n+1}}{n^2\pi^2}$$

$$b_n = \int_{-1}^{1}(x - x^2)\sin n\pi x\, dx = \int_{-1}^{1} x\sin n\pi x\, dx - \int_{-1}^{1} x^2\sin n\pi x\, dx$$

$$= 2\int_0^1 x\sin n\pi x\, dx - 0 = 2\left[x\left(-\frac{\cos n\pi x}{n\pi}\right) - 1.\left(-\frac{\sin n\pi x}{n^2\pi^2}\right)\right]_0^1$$

$$= 2\left[-\frac{\cos n\pi}{n\pi}\right] = \frac{-2(-1)^n}{n\pi} = \frac{2(-1)^{n+1}}{n\pi}$$

$$\therefore\ x - x^2 = -\frac{1}{3} + \frac{4}{\pi^2}\left(\frac{\cos \pi x}{1^2} - \frac{\cos 2\pi x}{2^2} + \frac{\cos 3\pi x}{3^2} - \ldots\ldots\right)$$

$$+ \frac{2}{\pi}\left(\frac{\sin \pi x}{1} - \frac{\sin 2\pi x}{2} + \frac{\sin 3\pi x}{3^2} - \ldots\ldots\right).$$

Example 2. *Find the Fourier Series to represent $f(x) = x^2 - 2$, when $-2 \le x \le 2$.*

Sol. Since $f(x)$ is an even function, $b_n = 0$.

Let $$f(x) = x^2 - 2 = \frac{a_0}{2} + \sum_{n=1}^{\infty} a_n \cos \frac{n\pi x}{2}$$

Then, $$a_0 = \frac{2}{2}\int_0^2 (x^2 - 2)\, dx = \left[\frac{x^3}{3} - 2x\right]_0^2 = \frac{8}{3} - 4 = -\frac{4}{3}$$

$$a_n = \frac{2}{2}\int_0^2 (x^2 - 2) \cos \frac{n\pi x}{2}\, dx$$

$$= \left[(x^2 - 2) . \frac{\sin \frac{n\pi x}{2}}{\frac{n\pi}{2}} - 2x\left(-\frac{\cos \frac{n\pi x}{2}}{\frac{n^2\pi^2}{4}}\right) + 2\left(-\frac{\sin \frac{n\pi x}{2}}{\frac{n^3\pi^3}{8}}\right)\right]_0^2$$

$$= \frac{16 \cos n\pi}{n^2\pi^2} = \frac{16(-1)^n}{n^2\pi^2}$$

$$\therefore \quad x^2 - 2 = -\frac{2}{3} - \frac{16}{\pi^2}\left(\cos \frac{\pi x}{2} - \frac{1}{4}\cos \pi x + \frac{1}{9}\cos \frac{3\pi x}{2} - \ldots\ldots\right).$$

Example 3. *Expand $f(x) = e^{-x}$ as a Fourier Series in the interval $(-l, l)$.*

Sol. Let $$(x) = e^{-x} = \frac{a_0}{2} + \sum_{n=1}^{\infty} a_n \cos \frac{n\pi x}{l} + \sum_{n=1}^{\infty} b_n \sin \frac{n\pi x}{l}$$

Then $$a_0 = \frac{1}{l}\int_{-l}^{l} e^{-x}\, dx = \frac{1}{l}\left[-e^{-x}\right]_{-l}^{l} = \frac{1}{l}(e^{l} - e^{-l}) = \frac{2 \sinh l}{l}$$

$$a_n = \frac{1}{l}\int_{-l}^{l} e^{-x} \cos \frac{n\pi x}{l}\, dx = \frac{1}{l}\left[\frac{e^{-x}}{1 + \left(\frac{n\pi}{l}\right)^2}\left(-\cos \frac{n\pi x}{l} + \frac{n\pi}{l}\sin \frac{n\pi x}{l}\right)\right]_{-l}^{l}$$

$$= \frac{l}{l^2 + (n\pi)^2}\left[-e^{-l}\cos n\pi + e^{l}\cos n\pi\right] = -\frac{2l \cos n\pi}{l^2 + (n\pi)^2}\left(\frac{e^{l} - e^{-l}}{2}\right) = \frac{2l\,(-1)^n \sinh l}{l^2 + (n\pi)^2}$$

$$b_n = \frac{1}{l}\int_{-l}^{l} e^{-x} \sin \frac{n\pi x}{l}\, dx$$

$$= \frac{1}{l}\left[\frac{e^{-x}}{1 + \left(\frac{n\pi}{l}\right)^2}\left(-\sin \frac{n\pi x}{l} - \frac{n\pi}{l}\cos \frac{n\pi x}{l}\right)\right]_{-l}^{l}$$

$$= -\frac{1}{l^2 + (n\pi)^2}\left[\frac{n\pi}{l}(e^{-l} - e^{l}) \cos n\pi\right] = \frac{2n\pi \cos n\pi}{l^2 + (n\pi)^2}\left(\frac{e^{l} - e^{-l}}{2}\right) = \frac{2n\pi\,(-1)^n \sinh l}{l^2 + (n\pi)^2}$$

$$\therefore \quad e^{-x} = \sinh l \left[\frac{1}{l} - 2l \left(\frac{1}{l^2+\pi^2} \cos \frac{\pi x}{l} - \frac{1}{l^2+2^2\pi^2} \cos \frac{2\pi x}{l} + \frac{1}{l^2+3^2\pi^2} \cos \frac{3\pi x}{l} - \ldots\ldots \right) \right.$$

$$\left. - 2\pi \left(\frac{1}{l^2+\pi^2} \sin \frac{\pi x}{l} - \frac{2}{l^2+2^2\pi^2} \sin \frac{2\pi x}{l} + \frac{3}{l^2+3^2\pi^2} \sin \frac{3\pi x}{l} - \ldots\ldots \right) \right].$$

Example 4. *Find the Fourier Series for the funtion f(x) defined by*

$$f(x) = \begin{cases} -1, & -3 < x < 0, \\ 0, & x = 0, \\ 1, & 0 < x < 3. \end{cases}$$

Sol. Let the Fourier Series for $f(x)$ be given by

$$f(x) = \frac{a_0}{2} + \sum_{n=1}^{\infty} \left(a_n \cos \frac{n\pi x}{3} + b_n \sin \frac{n\pi x}{3} \right)$$

[$\because$ the period, $2l = 6$, $\therefore l = 3$]

where
$$a_0 = \frac{1}{3} \int_{-3}^{3} f(x)\, dx = \frac{1}{3} \left[\int_{-3}^{3} f(x)\, dx + \int_0^3 f(x)\, dx \right]$$

$$= \frac{1}{3} \left[\int_{-3}^{0} -1\, dx + \int_0^3 1\, dx \right] = 0$$

$$a_n = \frac{1}{3} \int_{-3}^{3} f(x) \cos \frac{n\pi x}{3} dx = \frac{1}{3} \left[\int_{-3}^{0} (-1) \cos \frac{n\pi x}{3} dx + \int_0^3 1 . \cos \frac{n\pi x}{3} dx \right]$$

$$= \frac{1}{3} \left[\int_3^0 \cos \frac{n\pi t}{3} dt + \int_0^3 \cos \frac{n\pi x}{3} dx \right],$$

on putting $x = -t$, so that $dx = -dt$ in the Ist integral

$$= \frac{1}{3} \left[-\int_0^3 \cos \frac{n\pi t}{3} dt + \int_0^3 \cos \frac{n\pi x}{3} dx \right]$$

$$= \frac{1}{3} \left[-\int_0^3 \cos \frac{n\pi x}{3} dx + \int_0^3 \cos \frac{n\pi x}{3} dx \right] = 0$$

and
$$b_n = \frac{1}{3} \int_{-3}^{3} f(x) \sin \frac{n\pi x}{3} dx$$

$$= \frac{1}{3} \left[\int_{-3}^{0} (-1) . \sin \frac{n\pi x}{3} dx + \int_0^3 1 . \sin \frac{n\pi x}{3} dx \right]$$

$$= \frac{1}{3} \left[\left\{ \frac{3}{n\pi} \cos \frac{n\pi x}{3} \right\}_{-3}^{0} + \left\{ -\frac{3}{n\pi} \cos \frac{n\pi x}{3} \right\}_0^3 \right]$$

$$= \frac{1}{n\pi} [\{1 - \cos n\pi\} + \{-\cos n\pi + 1\}]$$

$$= \frac{2}{n\pi}[1 - \cos n\pi] = \frac{2}{n\pi}[1 - (-1)^n]$$

$$= \begin{cases} 0, & \text{when } n \text{ in even} \\ \dfrac{4}{n\pi}, & \text{when } n \text{ is odd.} \end{cases}$$

$\therefore$ The required Fourier Series for $f(x)$ be given by

$$f(x) = \frac{4}{\pi}\left\{\sin\frac{\pi x}{3} + \frac{1}{3}\sin \pi x + \frac{1}{5}\sin\frac{5\pi x}{3} + \ldots\right\}$$

TEST YOUR KNOWLEDGE

1. Find a Fourier series for $f(t) = 1 - t^2$ when $-1 \le t \le 1$.
2. Expand $f(x)$ in Fourier Series in the interval $(-2, 2)$ when $f(x) = \begin{cases} 0, & -2 < x < 0 \\ 1, & 0 < x < 2. \end{cases}$
3. Develop $f(x)$ in a Fourier Series in the interval $(0, 2)$ if $f(x) = \begin{cases} x, & 0 < x < 1 \\ 0, & 1 < x < 2. \end{cases}$
4. Find the Fourier expansion for $f(x) = \pi x$ from $x = -c$ to $x = c$.
5. Find the Fourier expansion for the function $f(x) = x - x^3$ in the interval $-1 < x < 1$.
6. Find the Fourier Series for the function given by $f(x) = \begin{cases} t, & 0 < t < 1 \\ 1 - t, & 1 < t < 2. \end{cases}$
7. Find the Fourier Series for the function $f(x) = \begin{cases} 0 & \text{when } -2 < x < -1 \\ k & \text{when } -1 < x < 1 \\ 0 & \text{when } 1 < x < 2. \end{cases}$
8. A sinusoidal voltage E sin ωt is passed through a half-wave rectifier which clips the negative portion of the wave.

   Expand the resulting periodic function $u(t) = \begin{cases} 0 & \text{when } -\dfrac{T}{2} < t < 0 \\ E \sin \omega t & \text{when } 0 < t < \dfrac{T}{2} \end{cases}$

   and $T = \dfrac{2\pi}{\omega}$, in a Fourier Series.

Answers

1. $1 - t^2 = \dfrac{2}{3} + \dfrac{4}{\pi^2}\left(\cos \pi t - \dfrac{\cos 2\pi t}{2^2} + \dfrac{\cos 3\pi t}{3^2} - \ldots\ldots\right)$

2. $f(x) = \dfrac{1}{2} + \dfrac{2}{\pi}\left(\sin\dfrac{\pi x}{2} + \dfrac{1}{3}\sin\dfrac{3\pi x}{2} + \dfrac{1}{5}\sin\dfrac{5\pi x}{2} + \ldots\ldots\right)$

3. $f(x) = \dfrac{1}{4} - \dfrac{2}{\pi^2}\left(\cos \pi x + \dfrac{\cos 3\pi x}{3^2} + \dfrac{\cos 5\pi x}{5^2} + \ldots\ldots\right) + \dfrac{1}{\pi}\left(\sin \pi x + \dfrac{\sin 2\pi x}{2} + \dfrac{\sin 3\pi x}{3} + \ldots\ldots\right)$

4. $f(x) = 2c\left[\sin\left(\frac{\pi x}{c}\right) - \frac{1}{2}\sin\left(\frac{2\pi x}{c}\right) + \frac{1}{3}\sin\left(\frac{3\pi x}{c}\right) - \ldots\ldots\right]$

5. $f(x) = \frac{12}{\pi^3}\left(\sin \pi x - \frac{\sin 2\pi x}{2^3} + \frac{\sin 3\pi x}{3^3} - \ldots\ldots\right)$

6. $f(t) = -\frac{4}{\pi^2}\left(\cos \pi t + \frac{\cos 3\pi t}{3^2} + \frac{\cos 5\pi t}{5^3} + \ldots\ldots\right) + \frac{2}{\pi}\left(\sin \pi t + \frac{\sin 3\pi t}{3} + \ldots\ldots\right)$

7. $f(x) = \frac{k}{2} + \frac{2k}{\pi}\left(\cos\frac{\pi t}{2} - \frac{1}{3}\cos\frac{3\pi t}{2} + \frac{1}{5}\cos\frac{5\pi t}{2} - \ldots\ldots\right)$

8. $u(t) = \frac{E}{\pi} + \frac{E}{2}\sin \omega t - \frac{2E}{\pi}\left(\frac{\cos 2\omega t}{1.3} + \frac{\cos 4\omega t}{3.5} + \frac{\cos 6\omega t}{5.7} + \ldots\ldots\right).$

1.7. HALF RANGE SERIES

Sometimes it is required to expand a function $f(x)$ in the range $(0, \pi)$ in a Fourier Series of period 2π or more generally in the range $(0, l)$ in a Fourier Series of period $2l$.

If it is required to expand $f(x)$ in the interval $(0, l)$, then it is immaterial what the function may be outside the range $0 < x < l$. We are free to choose it arbitrarily in the interval $(-l, 0)$.

If we extend the function $f(x)$ by reflecting it in the y-axis so that $f(-x) = f(x)$, then the extended function is even for which $b_n = 0$. The Fourier expansion of $f(x)$ will contain only cosine terms.

If we extend the function $f(x)$ by reflecting it in the origin so that $f(-x) = -f(x)$, then the extended function is odd for which $a_0 = a_n = 0$. The Fourier expansion of $f(x)$ will contain only sine terms.

For example, consider the function

$$f(x) = \begin{cases} x, & 0 < x < \frac{\pi}{2} \\ \pi - x, & \frac{\pi}{2} < x < \pi \end{cases}$$

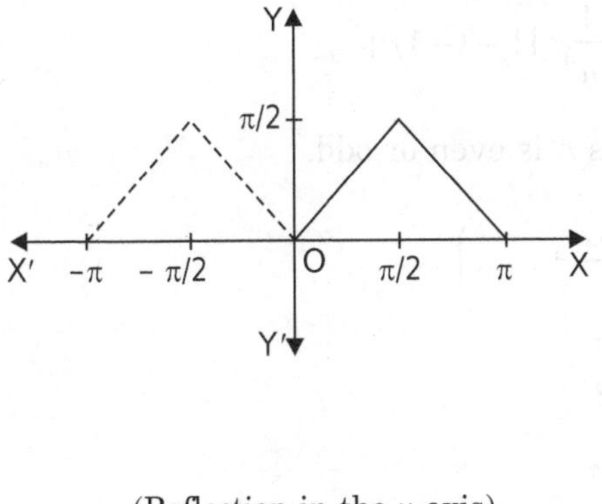

Y, π/2, X′, −π, −π/2, O, π/2, π, X, Y′

(Reflection in the y-axis)

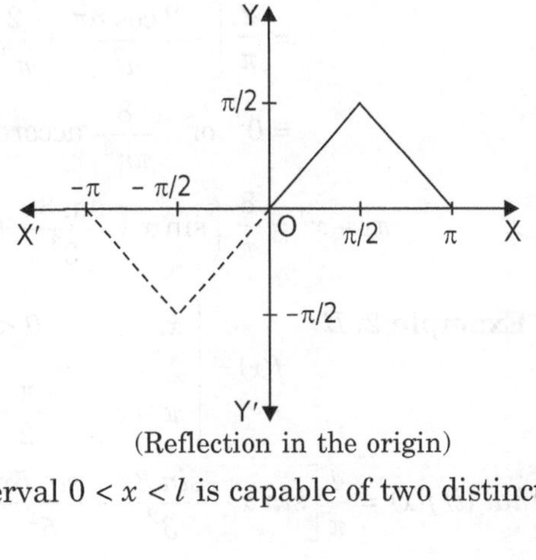

Y, π/2, −π, −π/2, X′, O, π/2, π, X, −π/2, Y′

(Reflection in the origin)

Hence a function $f(x)$ defined over the interval $0 < x < l$ is capable of two distinct half-range series.

The half-range cosine series is $f(x) = \frac{a_0}{2} + \sum_{n=1}^{\infty} a_n \cos \frac{n\pi x}{l}$

where $a_0 = \frac{2}{l}\int_0^l f(x)\,dx$; $a_n = \frac{2}{l}\int_0^l f(x) \cos \frac{n\pi x}{l}\,dx.$

The half-range sine series is $f(x) = \sum_{n=1}^{\infty} b_n \sin \frac{n\pi x}{l}$, where $b_n = \frac{2}{l}\int_0^l f(x) \sin \frac{n\pi x}{l}\,dx.$

Cor. If the range is $0 < x < \pi$, then

(*i*) The half-range cosine series is $f(x) = \frac{a_0}{2} + \sum_{n=1}^{\infty} a_n \cos nx$

where $a_0 = \frac{2}{\pi}\int_0^\pi f(x)\,dx$; $a_n = \frac{2}{\pi}\int_0^\pi f(x) \cos nx\,dx.$

(*ii*) The half-range sine series is $f(x) = \sum_{n=1}^{\infty} b_n \sin nx$; $b_n = \frac{2}{\pi}\int_0^\pi f(x) \sin nx\,dx.$

ILLUSTRATIVE EXAMPLES

Example 1. *Expand $\pi x - x^2$ in a half-range sine series in the interval $(0, \pi)$ upto the first three terms.*

Sol. Let $\pi x - x^2 = \sum_{n=1}^{\infty} b_n \sin nx$, then

$$b_n = \frac{2}{\pi}\int_0^\pi (\pi x - x^2) \sin nx\,dx$$

$$= \frac{2}{\pi}\left[(\pi x - x^2)\left(-\frac{\cos nx}{n}\right) - (\pi - 2x)\left(-\frac{\sin nx}{n^2}\right) + (-2)\left(\frac{\cos nx}{n^3}\right)\right]_0^\pi$$

$$= \frac{2}{\pi}\left[-\frac{2\cos n\pi}{n^3} + \frac{2}{n^3}\right] = \frac{4}{\pi n^3}[1 - (-1)^n]$$

$$= 0 \text{ or } \frac{8}{\pi n^3} \text{ according as } n \text{ is even or odd.}$$

$$\therefore \quad \pi x - x^2 = \frac{8}{\pi}\left(\sin x + \frac{\sin 3x}{3^3} + \frac{\sin 5x}{5^3} + ...\right).$$

Example 2. *If* $f(x) = \begin{cases} x, & 0 < x < \frac{\pi}{2} \\ \pi - x, & \frac{\pi}{2} < x < \pi \end{cases}$

show that (i) $f(x) = \frac{4}{\pi}\left[\sin x - \frac{\sin 3x}{3^2} + \frac{\sin 5x}{5^2} - .....\right]$

(*ii*) $f(x) = \frac{\pi}{4} - \frac{2}{\pi}\left[\frac{\cos 2x}{1^2} + \frac{\cos 6x}{3^2} + \frac{\cos 10x}{5^2} + ...\right].$

Sol. (*i*) **For the half-range sine series.**

Let $\quad f(x) = \sum_{n=1}^{\infty} b_n \sin nx$

Then $\quad b_n = \frac{2}{\pi}\int_0^{\pi} f(x)\sin nx\, dx = \frac{2}{\pi}\left[\int_0^{\pi/2} x \sin nx\, dx + \int_{\pi/2}^{\pi} (\pi - x)\sin nx\, dx\right]$

$$= \frac{2}{\pi}\left[x\left(-\frac{\cos nx}{n}\right) - 1.\left(-\frac{\sin nx}{n^2}\right)\right]_0^{\pi/2} + \frac{2}{\pi}\left[(\pi - x)\left(-\frac{\cos nx}{n}\right) - (-1)\left(-\frac{\sin nx}{n^2}\right)\right]_0^{\pi}$$

$$= \frac{2}{\pi}\left[-\frac{\pi}{2n}\cos\frac{n\pi}{2} + \frac{1}{n^2}\sin\frac{n\pi}{2}\right] + \frac{2}{\pi}\left[\frac{\pi}{2n}\cos\frac{n\pi}{2} + \frac{1}{n^2}\sin\frac{n\pi}{2}\right]$$

$$= \frac{2}{\pi}\left[\frac{2}{n^2}\sin\frac{n\pi}{2}\right] = \frac{4}{\pi n^2}\sin\frac{n\pi}{2}$$

When n is even, $b_n = 0$.

$\therefore \quad f(x) = \frac{4}{\pi}\left[\sin x - \frac{\sin 3x}{3^2} + \frac{\sin 5x}{5^2} - ......\right].$

(*ii*) **For the half-range cosine series.**

Let $\quad f(x) = \frac{a_0}{2} + \sum_{n=1}^{\infty} a_n \cos nx$

Then $\quad a_0 = \frac{2}{\pi}\int_0^{\pi} f(x)\, dx = \frac{2}{\pi}\left[\int_0^{\pi/2} x\, dx + \int_{\pi/2}^{\pi} (\pi - x)\, dx\right]$

$$= \frac{2}{\pi}\left[\left|\frac{x^2}{2}\right|_0^{\pi/2} + \left|\pi x - \frac{x^2}{2}\right|_{\pi/2}^{\pi}\right]$$

$$= \frac{2}{\pi}\left[\frac{\pi^2}{8} + \left(\pi^2 - \frac{\pi^2}{2}\right) - \left(\frac{\pi^2}{2} - \frac{\pi^2}{8}\right)\right] = \frac{2}{\pi}\left[\frac{\pi^2}{4}\right] = \frac{\pi}{2}$$

$$a_n = \frac{2}{\pi}\int_0^{\pi} f(x)\cos nx\, dx = \frac{2}{\pi}\left[\int_0^{\pi/2} x\cos nx\, dx + \int_{\pi/2}^{\pi} (\pi - x)\cos nx\, dx\right]$$

$$= \frac{2}{\pi}\left[x.\frac{\sin nx}{n} - 1.\left(-\frac{\cos nx}{n^2}\right)\right]_0^{\pi/2} + \frac{2}{\pi}\left[(\pi - x).\frac{\sin nx}{n} - (-1)\left(-\frac{\cos nx}{n^2}\right)\right]_{\pi/2}^{\pi}$$

$$= \frac{2}{\pi}\left[\frac{\pi}{2n}\sin\frac{n\pi}{2} + \frac{1}{n^2}\cos\frac{n\pi}{2} - \frac{1}{n^2}\right] + \frac{2}{\pi}\left[-\frac{\cos n\pi}{n^2} - \frac{\pi}{2n}\sin\frac{n\pi}{2} + \frac{1}{n^2}\cos\frac{n\pi}{2}\right]$$

$$= \frac{2}{\pi}\left[\frac{2}{n^2}\cos\frac{n\pi}{2} - \frac{\cos n\pi}{n^2} - \frac{1}{n^2}\right] = \frac{2}{\pi n^2}\left[2\cos\frac{n\pi}{2} - \cos n\pi - 1\right]$$

$$\therefore \quad a_1 = 0, a_2 = \frac{2}{\pi . 2^2}(2\cos\pi - \cos 2\pi - 1) = \frac{-2}{\pi . 1^2},$$

$$a_3 = 0, a_4 = 0, a_5 = 0, a_6 = \frac{2}{\pi . 6^2}(2\cos 3\pi - \cos 6\pi - 1) = \frac{-2}{\pi . 3^2},$$

$$a_7 = 0, a_8 = 0, a_9 = 0, a_{10} = \frac{2}{\pi . 10^2}(2\cos 5\pi - \cos 10\pi - 1) = \frac{-2}{\pi . 5^2}, \ldots\ldots$$

Hence $f(x) = \frac{\pi}{4} - \frac{2}{\pi}\left[\frac{\cos 2x}{1^2} + \frac{\cos 6x}{3^2} + \frac{\cos 10x}{5^2} + \ldots\ldots\right].$

Example 3. *Find a series of cosines of multiples of x which will represent x sin x in the interval (0, π) and show that* $\frac{1}{1.3} - \frac{1}{3.5} + \frac{1}{5.7} - \ldots\ldots = \frac{\pi - 2}{4}$.

Sol. Let $x \sin x = \frac{a_0}{2} + \sum_{n=1}^{\infty} a_n \cos nx$

Then $a_0 = \frac{2}{\pi}\int_0^{\pi} x \sin x \, dx = \frac{2}{\pi}\Big[x(-\cos x) - 1.(-\sin x)\Big]_0^{\pi} = \frac{2}{\pi}[-\pi\cos\pi] = 2$

$$a_n = \frac{2}{\pi}\int_0^{\pi} x \sin x \cos nx \, dx = \frac{1}{\pi}\int_0^{\pi} x(2\cos nx \sin x)\, dx$$

$$= \frac{1}{\pi}\int_0^{\pi} x[\sin(n+1)x - \sin(n-1)x]\, dx$$

$$= \frac{1}{\pi}\left[x\left\{-\frac{\cos(n+1)x}{n+1} + \frac{\cos(n-1)x}{n-1}\right\} - 1.\left\{-\frac{\sin(n+1)x}{(n+1)^2} + \frac{\sin(n-1)x}{(n-1)^2}\right\}\right]_0^{\pi}$$

$$= \frac{1}{\pi}\left[-\frac{\pi\cos(n+1)\pi}{n+1} + \frac{\pi\cos(n-1)\pi}{n-1}\right]$$

$$= -\frac{(-1)^{n+1}}{n+1} + \frac{(-1)^{n-1}}{n-1} = (-1)^{n-1}\left[\frac{1}{n-1} - \frac{1}{n+1}\right] = \frac{2(-1)^{n-1}}{(n-1)(n+1)}; \text{ when } n \neq 1$$

When $n = 1$, we have

$$a_1 = \frac{2}{\pi}\int_0^{\pi} x \sin x \cos x \, dx = \frac{1}{\pi}\int_0^{\pi} x \sin 2x \, dx = \frac{1}{\pi}\left[x\left(-\frac{\cos 2x}{2}\right) - 1.\left(-\frac{\sin 2x}{2^2}\right)\right]_0^{\pi}$$

$$= \frac{1}{\pi}\left[-\frac{\pi\cos 2\pi}{2}\right] = -\frac{1}{2}$$

$$\therefore \quad x \sin x = 1 - \frac{1}{2}\cos x - 2\left(\frac{\cos 2x}{1.3} - \frac{\cos 3x}{2.4} + \frac{\cos 4x}{3.5} - \ldots\ldots\right).$$

Putting $x = \frac{\pi}{2}$, we get $\frac{\pi}{2} = 1 - 2\left(\frac{-1}{1.3} + \frac{1}{3.5} + \frac{-1}{5.7} - \ldots\ldots\right)$

$$\Rightarrow \quad 1 + \frac{2}{1.3} - \frac{2}{3.5} + \frac{2}{5.7} - \ldots.. = \frac{\pi}{2}$$

$$\Rightarrow \qquad \frac{2}{1.3} - \frac{2}{3.5} + \frac{2}{5.7} - \ldots\ldots = \frac{\pi}{2} - 1$$

Hence
$$\frac{1}{1.3} - \frac{1}{3.5} + \frac{1}{5.7} - \ldots\ldots = \frac{\pi - 2}{4}.$$

Example 4. *Obtain a half-range cosine series for*

$$f(x) = \begin{cases} kx, & \text{for} \quad 0 \le x \le \dfrac{l}{2} \\ k(l - x), & \text{for} \quad \dfrac{l}{2} \le x \le l. \end{cases}$$

Deduce the sum of the series $\dfrac{1}{1^2} + \dfrac{1}{3^2} + \dfrac{1}{5^2} + \ldots\ldots$

Sol. Let $\quad f(x) = \dfrac{a_0}{2} + \displaystyle\sum_{n=1}^{\infty} a_n \cos \frac{n\pi x}{l}$

then
$$a_0 = \frac{2}{l}\int_0^l f(x)\, dx = \frac{2}{l}\left[\int_0^{l/2} kx\, dx + \int_{l/2}^{l} k(l - x)\, dx\right]$$

$$= \frac{2}{l}\left[\left|\frac{kx^2}{2}\right|_0^{l/2} + \left|k\left(lx - \frac{x^2}{2}\right)\right|_{l/2}^{l}\right]$$

$$= \frac{2}{l}\left[\frac{kl^2}{8} + k\left(l^2 - \frac{l^2}{2}\right) - k\left(\frac{l^2}{2} - \frac{l^2}{8}\right)\right] = \frac{2}{l}\left(\frac{kl^2}{4}\right) = \frac{kl}{2}$$

$$a_n = \frac{2}{l}\int_0^l f(x) \cos \frac{n\pi x}{l}\, dx = \frac{2}{l}\left[\int_0^{l/2} kx \,.\, \cos \frac{n\pi x}{l}\, dx + \int_{l/2}^{l} k(l - x) \,.\, \cos \frac{n\pi x}{l}\, dx\right]$$

$$= \frac{2}{l}\left[\left|kx \,.\, \frac{l}{n\pi} \sin \frac{n\pi x}{l} + k \,.\, \frac{l^2}{n^2\pi^2} \cos \frac{n\pi x}{l}\right|_0^{l/2}\right.$$

$$\left. + \left|k(l - x) \,.\, \frac{l}{n\pi} \sin \frac{n\pi x}{l} - k \,.\, \frac{l^2}{n^2\pi^2} \cos \frac{n\pi x}{l}\right|_{l/2}^{l}\right]$$

$$= \frac{2}{l}\left[\left|\frac{kl^2}{2n\pi} \sin \frac{n\pi}{2} + \frac{kl^2}{n^2\pi^2}\left(\cos \frac{n\pi}{2} - 1\right)\right|\right.$$

$$\left. + \left|\frac{-kl^2}{n^2\pi^2} \cos n\pi - \frac{kl^2}{2n\pi} \sin \frac{n\pi}{2} + \frac{kl^2}{n^2\pi^2} \cos \frac{n\pi}{2}\right|\right]$$

$$= \frac{2}{l}\left[\frac{2kl^2}{n^2\pi^2} \cos \frac{n\pi}{2} - \frac{kl^2}{n^2\pi^2} - \frac{kl^2}{n^2\pi^2} \cos n\pi\right] = \frac{2kl}{n^2\pi^2}\left[2 \cos \frac{n\pi}{2} - 1 - \cos n\pi\right]$$

When n is odd, $\cos \dfrac{n\pi}{2} = 0$ and $\cos n\pi = -1 \qquad \therefore \quad a_n = 0 \;\Rightarrow\; a_1 = a_3 = a_5 = \ldots\ldots = 0$

When n is even, $$a_2 = \frac{2kl}{2^2\pi^2}[2\cos\pi - 1 - \cos 2\pi] = -\frac{8kl}{2^2\pi^2};$$

$$a_4 = \frac{2kl}{4^2\pi^2}[2\cos 2\pi - 1 - \cos 4\pi] = 0$$

$$a_6 = \frac{2kl}{6^2\pi^2}[2\cos 3\pi - 1 - \cos 6\pi]$$

$$= \frac{2kl}{6^2\pi^2}(-2-1-1) = -\frac{8kl}{6^2\pi^2} \text{ and so on.}$$

$$\therefore \quad f(x) = \frac{kl}{4} - \frac{8kl}{\pi^2}\left(\frac{1}{2^2}\cos\frac{2\pi x}{l} + \frac{1}{6^2}\cos\frac{6\pi x}{l} + \ldots\ldots\right) \qquad \ldots(1)$$

Putting $x = l, f(l) = 0$

$\therefore$ From (1), we have $0 = \frac{kl}{4} - \frac{8kl}{\pi^2}\left(\frac{1}{2^2} + \frac{1}{6^2} + \ldots\ldots\right)$

$$\Rightarrow \quad \frac{1}{2^2} + \frac{1}{6^2} + \ldots\ldots = \frac{\pi^2}{32} \quad \Rightarrow \quad \frac{1}{2^2}\left[\frac{1}{1^2} + \frac{1}{3^2} + \ldots\ldots\right] = \frac{\pi^2}{32}$$

Hence $$\frac{1}{1^2} + \frac{1}{3^2} + \ldots\ldots = \frac{\pi^2}{8}.$$

Example 5. *Express $f(x) = x$ as a half range cosine series in the range $0 < x < 2$.*

Sol. Let the function $f(x) = x$ be reflected in y-axis such that in the range $(-2, 2)$ we get $f(-x) = f(x)$. Then the function $f(x)$ will be defined in the full range $(-2, 2)$ as given below :

$$f(x) = \begin{cases} x \text{ when} & 0 < x < 2 \\ -x \text{ when} & -2 < x < 0. \end{cases}$$

Hence the Fourier Series expansion for $f(x)$ over the full range $(-2, 2)$ will contain only cosine terms and is given by

$$f(x) = \frac{a_0}{2} + \sum_{n=1}^{\infty} a_n \cos\frac{n\pi x}{l}$$

$$f(x) = \frac{a_0}{2} + \sum_{n=1}^{\infty} a_n \cos\frac{n\pi x}{2} \qquad \ldots(1) \quad [\because \ 2l = 4, \ \therefore \ l = 2]$$

where $$a_0 = \frac{2}{l}\int_0^l f(x)\,dx = \frac{2}{2}\int_0^2 f(x)\,dx \qquad [\because \ l = 2]$$

$$= \int_0^2 x\,dx = 2$$

and $$a_n = \frac{2}{l}\int_0^l f(x)\cos\frac{n\pi x}{l}\,dx = \frac{2}{2}\int_0^2 x\cos\frac{n\pi x}{2}\,dx = \int_0^2 x\cos\frac{n\pi x}{2}\,dx$$

$$= \left[x \cdot \frac{2}{n\pi} \sin \frac{n\pi x}{2} \right]_0^2 - \int_0^2 1 \cdot \frac{2}{n\pi} \sin \frac{n\pi x}{2} dx = 0 - \frac{2}{n\pi} \cdot \frac{2}{n\pi} \left[- \cos \frac{n\pi x}{2} \right]_0^2$$

$$= \frac{4}{n^2\pi^2} [\cos n\pi - \cos 0] = \frac{4}{n^2\pi^2} [(-1)^n - 1].$$

Substituting these values in (1), we get

$$f(x) = \frac{2}{2} + \sum_{n=1}^{\infty} \frac{4}{n^2\pi^2} \{(-1)^n - 1\} \cos \frac{n\pi x}{2}$$

$$= 1 + \frac{4}{\pi^2} \left[-\frac{2}{1^2} \cos \frac{\pi x}{2} + 0 - \frac{2}{3^2} \cos \frac{4\pi x}{2} + 0 - \frac{2}{5^2} \cos \frac{5\pi x}{2} + \ldots \right]$$

$$= 1 - \frac{8}{\pi^2} \left[\frac{1}{1^2} \cos \frac{\pi x}{2} + \frac{1}{3^2} \cos \frac{3\pi x}{2} + \frac{1}{5^2} \cos \frac{5\pi x}{2} + \ldots \right]$$

which is the required Fourier Series expansion.

Example 6. *Represent the following function f(x) as a Fourier cosine half range series*

$$f(x) = \begin{cases} x, & 0 < x < \dfrac{\pi}{2}, \\ \pi - x, & \dfrac{\pi}{2} < x < \pi \end{cases}.$$

Sol. The graph of the function $f(x)$ in the interval $(0, \pi)$ is OAB as shown in the Fig. We now extend the function $f(x)$ in the the interval $(-\pi, 0)$, so that the new function is symmetrical about y-axis [graphically shown by $B'A'O$]. Thus $f(x)$ is an even function in the interval $(-\pi, \pi)$.

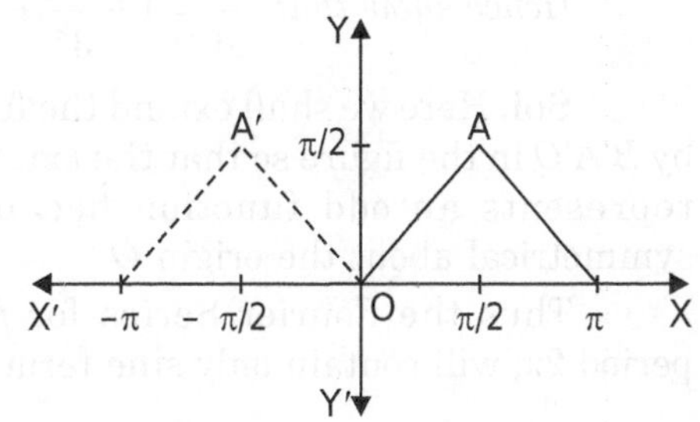

Hence the Fourier Series for $f(x)$ over the full period $(-\pi, \pi)$ will contain only cosine terms and is given by

$$f(x) = \frac{a_0}{2} + \sum_{n=1}^{\infty} a_n \cos \left(\frac{n\pi x}{l} \right)$$

i.e., $$f(x) = \frac{a_0}{2} + \sum_{n=1}^{\infty} a_n \cos nx \qquad [\because \; 2l = 2\pi, \therefore \; l = \pi]$$

where $$a_0 = \frac{2}{\pi} \int_0^{\pi} f(x)\, dx = \frac{2}{\pi} \left[\int_0^{\pi/2} x\, dx + \int_{\pi/2}^{\pi} (\pi - x)\, dx \right]$$

$$= \frac{2}{\pi} \left[\left\{ \frac{x^2}{2} \right\}_0^{\pi/2} + \left\{ \pi x - \frac{x^2}{2} \right\}_{\pi/2}^{\pi} \right] = \frac{2}{\pi} \left[\left\{ \frac{\pi^2}{8} - 0 \right\} + \left\{ \pi^2 - \frac{\pi^2}{2} - \frac{\pi^2}{2} + \frac{\pi^2}{8} \right\} \right] = \frac{\pi}{2}$$

and $$a_n = \frac{2}{\pi} \int_0^{\pi} f(x) \cos nx\, dx = \frac{2}{\pi} \int_0^{\pi/2} x \cos nx\, dx + \frac{2}{\pi} \int_{\pi/2}^{\pi} (\pi - x) \cos nx\, dx$$

$$= \frac{2}{\pi}\left[x.\left(\frac{\sin nx}{n}\right) - 1.\left(-\frac{\cos nx}{n^2}\right)\right]_0^{\pi/2} + \frac{2}{\pi}\left[(\pi - x).\left(\frac{\sin nx}{n}\right) - (-1).\left(-\frac{\cos nx}{n^2}\right)\right]_{\pi/2}^{\pi}$$

$$= \frac{2}{\pi}\left[\frac{\pi}{2n}\sin\frac{n\pi}{2} + \frac{1}{n^2}\cos\frac{n\pi}{2} - \frac{1}{n^2}\right] + \frac{2}{\pi}\left[-\frac{\cos n\pi}{n^2} - \frac{\pi}{2n}\sin\frac{n\pi}{2} + \frac{1}{n^2}\cos\frac{n\pi}{2}\right]$$

$$= \frac{2}{n^2\pi}\left[2\cos\frac{n\pi}{2} - \cos n\pi - 1\right].$$

$\therefore \quad a_1 = 0, a_2 = -\frac{8}{4\pi}, a_3 = 0, a_4 = 0, a_5 = 0, a_6 = \frac{8}{6^2\pi}, \ldots$ etc.

$\therefore \quad f(x) = \frac{\pi}{4} - \frac{8}{\pi}\left[\frac{1}{2^2}\cos 2x + \frac{1}{6^2}\cos 6x + \frac{1}{8^2}\cos 8x + \ldots\right].$

This is the Fourier series of the function $f(x)$ of full period 2π.

Hence it is the required Fourier cosine series of $f(x)$ over the half range $[0, \pi]$.

Example 7. *Represent the following function f(x) as a Fourier sine half range series*

$$f(x) = \begin{cases} x, & 0 \le x \le \frac{\pi}{2}, \\ \pi - x, & \frac{\pi}{2} \le x \le \pi. \end{cases}$$

Hence show that $\frac{\pi^2}{8} = 1 + \frac{1}{3^2} + \frac{1}{5^2} + \ldots.$

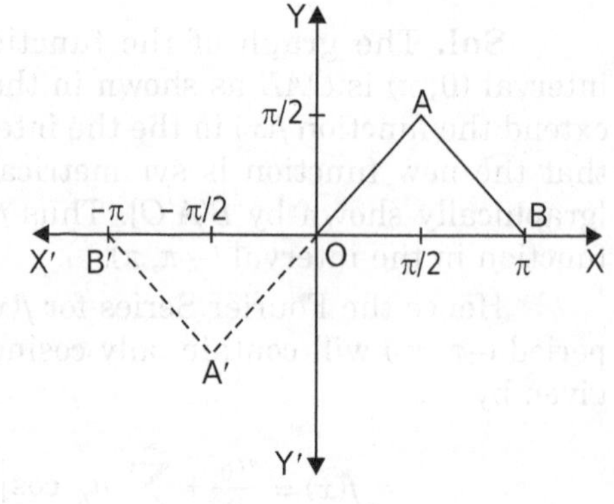

Sol. Here we shall extend the function $f(x)$ shown by $B'A'O$ in the figure so that the extended function $f(x)$ represents an odd function because the curve is symmetrical about the origin O.

Thus the Fourier Series for $f(x)$ over the full period 2π, will contain only sine terms, and is given by

$$f(x) = \sum_{n=1}^{\infty} b_n \sin nx$$

where $\quad b_n = \frac{2}{\pi}\int_0^{\pi} f(x)\sin nx\, dx = \frac{2}{\pi}\int_0^{\pi/2} x\sin nx\, dx + \frac{2}{\pi}\int_{\pi/2}^{\pi} (\pi - x)\sin nx\, dx$

$$= \frac{2}{\pi}\left[x.\left(-\frac{\cos nx}{n}\right) - (1).\left(-\frac{\sin nx}{n^2}\right)\right]_0^{\pi/2}$$

$$+ \frac{2}{\pi}\left[(\pi - x).\left(-\frac{\cos nx}{n}\right) - (-1).\left(-\frac{\sin nx}{n^2}\right)\right]_{\pi/2}^{\pi}$$

$$= \frac{2}{\pi}\left[-\frac{\pi}{2n}\cos\frac{n\pi}{2} + \frac{1}{n^2}\sin\frac{n\pi}{2}\right] + \frac{2}{\pi}\left[\frac{\pi}{2n}\cos\frac{n\pi}{2} + \frac{1}{n^2}\sin\frac{n\pi}{2}\right] = \frac{4}{\pi n^2}\sin\frac{n\pi}{2}.$$

$\therefore \quad b_2 = b_4 = b_6 = \ldots = 0$

and $\quad b_1 = \dfrac{4}{1^2\pi}, b_3 = -\dfrac{4}{3^2\pi}, b_5 = \dfrac{4}{5^2\pi}, \ldots$ etc.

Hence the required Fourier sine series of $f(x)$ over the half range $[0, \pi]$ is

$$f(x) = \frac{4}{\pi}\left[\frac{1}{1^2}\sin x - \frac{1}{3^2}\sin 3x + \frac{1}{5^2}\sin 5x - \ldots\right] \quad \ldots(1)$$

Putting $x = \dfrac{\pi}{2}$ on both sides of equation (1), we get

$$f\left(\frac{\pi}{2}\right) = \frac{\pi}{2} = \frac{4}{\pi}\left[\frac{1}{1^2}.1 - \frac{1}{3^2}.(-1) + \frac{1}{5^2}.1 - \frac{1}{7^2}.(-1) + \ldots\right]$$

or $\quad \dfrac{\pi^2}{8} = \dfrac{1}{1^2} + \dfrac{1}{3^2} + \dfrac{1}{5^2} + \ldots.$

TEST YOUR KNOWLEDGE

1. Obtain cosine and sine series for $f(x) = x$ in the interval $0 \le x \le \pi$. Hence show that $\dfrac{1}{1^2} + \dfrac{1}{3^2} + \dfrac{1}{5^2} + \ldots\ldots = \dfrac{\pi^2}{8}$.
2. Find the half-range cosine series for the function $f(x) = x^2$ in the range $0 \le x \le \pi$.
3. Find the half-range cosine series for the function $f(x) = (x-1)^2$ in the interval $0 < x < 1$. Hence show that

   (*i*) $\dfrac{1}{1^2} + \dfrac{1}{2^2} + \dfrac{1}{3^2} + \dfrac{1}{4^2} + \ldots\ldots = \dfrac{\pi^2}{6}$ $\quad$ (*ii*) $\dfrac{1}{1^2} - \dfrac{1}{2^2} + \dfrac{1}{3^2} - \dfrac{1}{4^2} + \ldots\ldots = \dfrac{\pi^2}{12}$

   (*iii*) $\dfrac{1}{1^2} + \dfrac{1}{3^2} + \dfrac{1}{5^2} + \dfrac{1}{7^2} + \ldots\ldots = \dfrac{\pi^2}{8}$.
4. Express $\sin x$ as a cosine series in $0 < x < \pi$.
5. If
$$f(x) = \begin{cases} \dfrac{\pi}{3}, & 0 \le x \le \dfrac{\pi}{3} \\ 0, & \dfrac{\pi}{3} \le x \le \dfrac{2\pi}{3} \\ -\dfrac{\pi}{3}, & \dfrac{2\pi}{3} \le x \le \pi \end{cases}$$
   then show that $\quad f(x) = \dfrac{2}{\sqrt{3}}\left[\cos x - \dfrac{\cos 5x}{5} + \dfrac{\cos 7x}{7} - \ldots\ldots\right]$.
6. If
$$f(x) = \begin{cases} mx, & 0 \le x \le \dfrac{\pi}{2} \\ m(\pi - x), & \dfrac{\pi}{2} \le x \le \pi \end{cases}$$
   then show that $f(x) = \dfrac{4m}{\pi}\left[\dfrac{\sin x}{1^2} - \dfrac{\sin 3x}{3^2} + \dfrac{\sin 5x}{5^2} - \ldots\ldots\right]$.

7. Express $f(x) = x$ as a half-range.

(*i*) sine series in $0 < x < 2$. (*ii*) cosine series in $0 < x < 2$.

8. Show that the series $\frac{1}{\pi}\sum_{n=1}^{\infty} \frac{1}{n} \sin - \frac{2n\pi x}{l}$ represents $\frac{l}{2} - x$ when $0 < x < l$.

9. Find the half-range sine series for

$$f(x) = \begin{cases} \frac{1}{4} - x, & 0 < x < \frac{1}{2} \\ x - \frac{3}{4}, & \frac{1}{2} < x < 1. \end{cases}$$

10. Represent the following function by Fourier sine series

$$f(x) = \begin{cases} 1 \text{ when } & 0 < x < \frac{l}{2} \\ 0 \text{ when } & \frac{l}{2} < x < l. \end{cases}$$

Answers

1. $\frac{\pi}{2} - \frac{4}{\pi}\left(\cos x + \frac{\cos 3x}{3^2} + \frac{\cos 5x}{5^2} + \ldots\ldots\right)$; $2\left(\sin x - \frac{\sin 2x}{2} + \frac{\sin 3x}{3} - \frac{\sin 4x}{4} + \ldots\ldots\right)$

2. $\frac{\pi^2}{3} - 4\left[\cos x - \frac{\cos 2x}{2^2} + \frac{\cos 3x}{3^2} - \frac{\cos 4x}{4^2} + \ldots\ldots\right]$

3. $\frac{1}{3} + \frac{1}{\pi^2}\left(\cos \pi x + \frac{\cos 2\pi x}{2^2} + \frac{\cos 3\pi x}{3^2} + \ldots\ldots\right)$ **4.** $\frac{2}{\pi} - \frac{4}{\pi}\left[\frac{\cos 2x}{1.3} + \frac{\cos 4x}{3.5} + \frac{\cos 6x}{5.7} + \ldots\ldots\right]$

7. (*i*) $\frac{4}{\pi}\left(\sin \frac{\pi x}{2} - \frac{1}{2}\sin \frac{2\pi x}{2} + \frac{1}{3}\sin \frac{3\pi x}{2} - \frac{1}{4}\sin \frac{4\pi x}{2} + \ldots\ldots\right)$

(*ii*) $1 - \frac{8}{\pi^2}\left[\cos \frac{\pi x}{2} + \frac{1}{3^2}\cos \frac{3\pi x}{2} + \frac{1}{5^2}\cos \frac{5\pi x}{2} + \ldots\ldots\right]$

9. $f(x) = \left(\frac{1}{\pi} - \frac{4}{\pi^2}\right)\sin \pi x + \left(\frac{1}{3\pi} - \frac{4}{3^2\pi^2}\right)\sin 3\pi x + \left(\frac{1}{5\pi} - \frac{4}{5^2\pi^2}\right)\sin 5\pi x + \ldots\ldots$

10. $f(x) = \frac{2}{\pi}\left[\sin \frac{\pi x}{l} + \sin \frac{2\pi x}{l} + \frac{1}{3}\sin \frac{3\pi x}{l} + \frac{1}{5}\sin \frac{5\pi x}{l} + \ldots\ldots\right]$.

1.8. ROOT MEAN SQUARE VALUE (R.M.S. VALUE)

The determination of root mean square value of a periodic function is a common problem in electric circuit theory of mechanical vibrations.

The R.M.S. value of a function $y = f(x)$ for $a < x < b$ is defined as

$$\bar{y} = \frac{\int_a^b y^2 dx}{b - a}$$

It is also called *effective value* of the function.

1.9. PARSEVAL'S THEOREM ON FOURIER CONSTANTS

If the Fourier Series of $f(x)$ over an interval $c < x < c + 2l$ is given as

$$f(x) = \frac{1}{2} a_0 + \sum_{n=1}^{\infty} \left\{ a_n \cos \frac{n\pi x}{l} + b_n \sin \frac{n\pi x}{l} \right\}$$

then $$\frac{1}{2l} \int_c^{c+2l} [f(x)]^2 \, dx = \frac{a_0^2}{4} + \frac{1}{2} \sum_{n=1}^{\infty} (a_n^2 + b_n^2).$$

Proof. The Fourier series of $f(x)$ in $c < x < c + 2l$ is given as

$$f(x) = \frac{a_0}{2} + \sum_{n=1}^{\infty} \left\{ a_n \cos \frac{n\pi x}{l} + b_n \sin \frac{n\pi x}{l} \right\} \quad ...(1)$$

where $a_0 = \frac{1}{l} \int_c^{c+2l} f(x) \, dx$; $a_n = \frac{1}{l} \int_c^{c+2l} f(x) \cos \frac{n\pi x}{l} dx$; $b_n = \frac{1}{l} \int_c^{c+2l} f(x) \sin \frac{n\pi x}{l} dx$

Multiplying both sides of (1) by $f(x)$, we have

$$[f(x)]^2 = \frac{a_0}{2} f(x) + \sum_{n=1}^{\infty} a_n f(x) \cos \frac{n\pi x}{l} + \sum_{n=1}^{\infty} b_n f(x) \sin \frac{n\pi x}{l} \quad ...(2)$$

Integrating both sides of (2) in between the limits c to $c + 2l$ w.r.t. x, we have

$$\int_c^{c+2l} [f(x)]^2 \, dx = \frac{a_0}{2} + \int_c^{c+2l} f(x) \, dx + \sum_{n=1}^{\infty} a_n \int_c^{c+2l} f(x) \cos \frac{n\pi x}{l} dx + \sum_{n=1}^{\infty} b_n \int_c^{c+2l} f(x) \sin \frac{n\pi x}{l} dx$$

$$= \frac{a_0}{2} . l a_0 + \sum_{n=1}^{\infty} a_n (l a_n) + \sum_{n=1}^{\infty} b_n (l b_n) \quad \text{[Using (1) and (2)]}$$

$$\Rightarrow \quad \int_c^{c+2l} [f(x)]^2 \, dx = \frac{l a_0^2}{2} + l \sum_{n=1}^{\infty} (a_n^2 + b_n^2)$$

$$\therefore \quad \frac{1}{2l} \int_c^{c+2l} [f(x)]^2 \, dx = \frac{1}{2l} \left\{ \frac{l a_0^2}{2} + l \sum_{n=1}^{\infty} (a_n^2 + b_n^2) \right\}$$

or $$\mathbf{\frac{1}{2l} \int_c^{c+2l} [f(x)]^2 \, dx = \frac{a_0^2}{4} + \frac{1}{2} \sum_{n=1}^{\infty} (a_n^2 + b_n^2)} \quad \text{(Parseval's identity)}$$

Hence the proof.

Note. *Parseval's identities in different cases :*

(*i*) If $f(x)$ is an even function in $(-l, l)$ then $\frac{2}{l} \int_0^l [f(x)]^2 \, dx = \frac{a_0^2}{2} + \sum_{n=1}^{\infty} a_n^2$

(*ii*) If $f(x)$ is an odd function in $(-l, l)$ then $\frac{2}{l}\int_0^l [f(x)]^2\, dx = \sum_{n=1}^{\infty} b_n^2$

(*iii*) If $f(x) = \frac{a_0}{2} + \sum_{n=1}^{\infty} a_n \cos\frac{n\pi x}{l}$ in $(0, l)$ then $\frac{2}{l}\int_0^l [f(x)]^2\, dx = \frac{a_0^2}{2} + \sum_{n=1}^{\infty} a_n^2$

(*iv*) If $f(x) = \sum_{n=1}^{\infty} b_n \sin\frac{n\pi x}{l}$ in $(0, l)$ then $\frac{2}{l}\int_0^l [f(x)]^2\, dx = \sum_{n=1}^{\infty} b_n^2$.

ILLUSTRATIVE EXAMPLES

Example 1. *Find the Fourier sine series for unity in $0 < x < \pi$ and hence show that*

$$1 + \frac{1}{3^2} + \frac{1}{5^2} + \frac{1}{7^2} + \ldots\ldots = \frac{\pi^2}{8}.$$

Sol. We require half-range Fourier sine series for 1 in $(0, \pi)$

Let $$1 = \sum_{n=1}^{\infty} b_n \sin nx \qquad \ldots(1)$$

Then $$b_n = \frac{2}{\pi}\int_0^{\pi} (1) \sin nx\, dx = \frac{2}{\pi}\left[-\frac{\cos nx}{n}\right]_0^{\pi} = -\frac{2}{n\pi}(\cos n\pi - 1)$$

$$= \frac{2}{n\pi}[1 - (-1)^n] \qquad [\because \cos n\pi = (-1)^n]$$

Now $b_n = 0$ when n is even ; and $b_n = \frac{4}{n\pi}$ when n is odd.

Substituting in (1), we get

$$\therefore \quad 1 = \sum_{m=1}^{\infty} \frac{4}{(2m-1)\pi} \sin(2m-1)x \quad \text{or} \quad 1 = \frac{4}{\pi}\left(\sin x + \frac{\sin 3x}{3} + \frac{\sin 5x}{5} + \ldots\ldots\right) \qquad \ldots(2)$$

Now from Parseval's theorem on Fourier constants

$$\int_c^{c+2l} [f(x)]^2\, dx = 2l\left[\frac{a_0^2}{4} + \frac{l}{2}\sum_{n=1}^{\infty}(a_n^2 + b_n^2)\right] \qquad \ldots(3)$$

Applying (3) to half-range sine series for 1 in $(0, \pi)$

$$c = 0,\ 2l = \pi,\ f(x) = 1,\ a_0 = 0,\ a_n = 0, \text{ and } b_n = \frac{4}{(2m-1)\pi},\ m = 1, 2, \ldots\ldots$$

We get, $$\int_0^{\pi} (1)^2\, dx = \pi \cdot \frac{1}{2}\sum_{m=1}^{\infty} \frac{16}{(2m-1)^2} \cdot \pi^2$$

$$\left[x\right]_0^{\pi} = \frac{8}{\pi}\left\{\frac{1}{1^2}+\frac{1}{3^2}+\frac{1}{5^2}+\ldots\ldots\right\} \quad \text{or} \quad \frac{\pi^2}{8} = 1+\frac{1}{3^2}+\frac{1}{5^2}+\ldots\ldots$$

Hence the result.

Example 2. *Find Fourier Series of x^2 in $(-\pi, \pi)$. Use Parseval's identity to prove that*

$$\frac{\pi^4}{90} = 1+\frac{1}{2^4}+\frac{1}{3^4}+\ldots\ldots$$

Sol. The Fourier Series of x^2 in $(-\pi, \pi)$ are

$$x^2 = \frac{\pi^2}{3}+\sum_{n=1}^{\infty}\frac{4(-1)^n}{n^2}\cos nx \qquad \ldots(1)$$

Here $\qquad a_0 = \dfrac{2\pi^2}{3},\ a_n = \dfrac{4(-1)^n}{n^2},\ b_n = 0,\ f(x) = x^2$

Now using Parseval's identity to (1)

$$\int_{-\pi}^{\pi}(x^2)^2\,dx = 2\pi\left[\frac{\pi^4}{9}+\frac{1}{2}\sum_{n=1}^{\infty}\frac{16}{n^4}\right]$$

$$\left[\frac{x^5}{5}\right]_{-\pi}^{\pi} = \frac{2\pi^5}{9}+\pi\sum_{n=1}^{\infty}\frac{16}{n^4} \quad \text{or} \quad \frac{2\pi^5}{5}-\frac{2\pi^5}{9} = \pi+\sum_{n=1}^{\infty}\frac{16}{n^4}$$

or $$\frac{\pi^4}{90} = \sum_{n=1}^{\infty}\frac{1}{n^4} \quad \text{or} \quad 1+\frac{1}{2^4}+\frac{1}{3^4}+\ldots\ldots = \frac{\pi^4}{90}.$$

TEST YOUR KNOWLEDGE

1. If $f(x)$ has the Fourier Series expansion

$$\frac{a_0}{2}+\sum_{n=1}^{\infty}\left(a_n\cos\frac{n\pi x}{l}+b_n\sin\frac{n\pi x}{l}\right) \text{ in } a \le x \le a+2l,$$

show that $\displaystyle\int_a^{a+2l}[f(x)]^2\,dx = 2l\left[\frac{a_0^2}{4}+\frac{1}{2}\sum_{n=1}^{\infty}(a_n^2+b_n^2)\right]$

2. If $f(x) = \dfrac{a_0}{2}+\displaystyle\sum_{n=1}^{\infty}a_n\cos\frac{n\pi x}{l}$ in $0 < x < l$, then show that

$$\int_0^l [f(x)]^2\,dx = \frac{l}{2}\left[\frac{a_0^2}{2}+\sum_{n=1}^{\infty}a_n^2\right].$$

3. If $f(x) = \displaystyle\sum_{n=1}^{\infty}b_n\sin\frac{n\pi x}{l}$ in $(0, l)$, then show that $\displaystyle\int_0^l [f(x)]^2\,dx = \frac{l}{2}\sum_{n=1}^{\infty}b_n^2$.

4. Prove that in the range $(0, l)$, $x = \frac{l}{2} - \frac{4l}{\pi^2} \sum_{n=1}^{\infty} \frac{1}{(2m-1)^2} \cos \frac{(2m-1)\pi x}{l}$ and deduce that

$$\frac{1}{1^4} + \frac{1}{3^4} + \frac{1}{5^4} + \ldots\ldots = \frac{\pi^4}{96}.$$

5. If an alternating current wave is represented by the series $i = \sum_{n=1}^{\infty} I_{2m-1} \sin [(2m-1)\,wt + \alpha_{2m-1}]$

show that the effective value of the current is $\sqrt{\frac{1}{2}(I_1^2 + I_3^2 + I_5^2 + \ldots\ldots)}$.

1.10. COMPLEX FORM OF FOURIER SERIES

We know that in the interval $c < x < c + 2l$, the Fourier Series expansion of $f(x)$ can be written as

$$f(x) = \frac{a_0}{2} + \sum_{n=1}^{\infty} \left(a_n \cos \frac{n\pi x}{l} + b_n \sin \frac{n\pi x}{l} \right) \qquad \ldots(1)$$

We also know that $\cos \theta = \frac{1}{2}(e^{i\theta} + e^{-i\theta})$ and $\sin \theta = \frac{1}{2i}(e^{i\theta} - e^{-i\theta})$

Substituting the exponential equivalents for the trigonometric terms in (1), we get

$$f(x) = \frac{a_0}{2} + \sum_{n=1}^{\infty} a_n \left(\frac{e^{\frac{in\pi x}{l}} + e^{\frac{in\pi x}{l}}}{2} \right) + \sum_{n=1}^{\infty} b_n \left(\frac{e^{\frac{in\pi x}{l}} - e^{-\frac{in\pi x}{l}}}{2i} \right)$$

$$= \frac{a_0}{2} + \sum_{n=1}^{\infty} \left(\frac{a_n - ib_n}{2} \right) e^{\frac{in\pi x}{l}} + \sum_{n=1}^{\infty} \left(\frac{a_n + ib_n}{2} \right) e^{-\frac{in\pi x}{l}} \qquad \left[\because \frac{1}{i} = -i \right]$$

$$= c_0 + \sum_{n=1}^{\infty} c_n e^{\frac{in\pi x}{l}} + \sum_{n=1}^{\infty} c_{-n}\, e^{-\frac{in\pi x}{l}} = \sum_{n=1}^{\infty} e^{\frac{in\pi x}{l}}$$

where $\qquad c_0 = \frac{a_0}{2}, c_n = \frac{a_n - ib_n}{2}, c_{-n} = \frac{a_n + ib_n}{2}$

Hence the *complex* or *exponential* form of the Fourier series is

$$\mathbf{f(x)} = \sum_{n=-\infty}^{\infty} \mathbf{c_n e^{\frac{in\pi x}{l}}} \qquad \ldots(2)$$

Now multiplying on both sides of (2) by $e^{-\frac{in\pi x}{l}}$ and integrating from c to $c + 2l$ w.r.t. x, we get

$$\int_c^{c+2l} f(x) \,.\, e^{-\frac{in\pi x}{l}}\, dx = c_n \int_c^{c+2l} e^{\frac{in\pi x}{l}}\, e^{-\frac{in\pi x}{l}}\, dx$$

$$= c_n \int_c^{c+2l} dx = c_n [x]_c^{c+2l} = c_n (2l)$$

$$\therefore \qquad \mathbf{c_n = \frac{1}{2l} \int_c^{c+2l} f(x)\, e^{-\frac{in\pi x}{l}}\, dx.}$$

ILLUSTRATIVE EXAMPLES

Example 1. *Find the complex form of Fourier series $f(x) = e^{ax}$ $(-\pi < x < \pi)$ in the form*

$$e^{ax} = \frac{\sinh a\pi}{\pi} \sum_{n=-\infty}^{\infty} (-1)^n \frac{a + in}{a^2 + n^2} e^{inx}$$

and hence prove that $\dfrac{\pi}{a \sinh a\pi} = \displaystyle\sum_{n=-\infty}^{\infty} \frac{(-1)^n}{n^2 + a^2}.$

Sol. The complex form of Fourier Series of $f(x)$ in $c < x < c + 2l$ is given by

$$f(x) = \sum_{n=-\infty}^{\infty} c_n e^{\frac{in\pi x}{l}} \quad \text{where } c_n = \frac{1}{2l}\int_c^{c+2l} f(x)\, e^{-\frac{in\pi x}{l}} dx$$

Here $2l = 2\pi \Rightarrow l = \pi \quad \therefore \quad f(x) = \displaystyle\sum_{n=-\infty}^{\infty} c_n e^{inx}$ ...(1)

Now
$$c_n = \frac{1}{2\pi} \int_{-\pi}^{\pi} f(x)\, e^{-\frac{in\pi x}{\pi}} dx$$

$$= \frac{1}{2\pi} \int_{-\pi}^{\pi} f(x)\, e^{-inx} dx = \frac{1}{2\pi} \int_{-\pi}^{\pi} e^{ax} . e^{-inx} dx \qquad [\because f(x) = e^{ax}]$$

$$= \frac{1}{2\pi} \int_{-\pi}^{\pi} e^{(a-in)x} dx = \frac{1}{2\pi} \left[\frac{e^{(a-in)x}}{(a-in)} \right]_{-\pi}^{\pi} = \frac{1}{2\pi} \left[\frac{e^{(a-in)\pi} - e^{-(a-in)\pi}}{a - in} \right]$$

$$= \frac{1}{2\pi} \left[\frac{e^{a\pi} (\cos n\pi - i \sin n\pi) - e^{-a\pi} (\cos n\pi + i \sin n\pi)}{a - in} \right]$$

$$(e^{\pm in\pi} = \cos n\pi \pm i \sin n\pi)$$

$$= \frac{1}{2\pi(a - in)} (e^{a\pi} - e^{-a\pi}) \cos n\pi = \frac{1}{2\pi(a - in)} (2 \sinh a\pi) \cos n\pi$$

$$\left(\because \frac{e^{a\pi} - e^{-a\pi}}{2} = \sinh a\pi \right)$$

$$= \frac{(-1)^n \sinh a\pi}{\pi(a - in)} \qquad [\because \cos n\pi = (-1)^n]$$

or
$$c_n = \frac{(-1)^n (a + in) \sinh a\pi}{\pi(a^2 + n^2)}$$

Substituting in (1), we get

$$f(x) = \sum_{n=-\infty}^{\infty} \frac{(-1)^n (a + in) \sinh a\pi}{\pi (a^2 + n^2)} e^{inx}$$

i.e.,
$$e^{ax} = \frac{\sinh a\pi}{\pi} \sum_{n=-\infty}^{\infty} \frac{(-1)^n (a + in)}{a^2 + n^2} e^{inx} \quad ...(2)$$

Putting $x = 0$ in (2), we get $1 = \dfrac{\sinh a\pi}{\pi} \displaystyle\sum_{n=-\infty}^{\infty} \frac{(-1)^n (a + in)}{a^2 + n^2}$

or
$$\frac{\pi}{\sinh a\pi} = \sum_{n=-\infty}^{\infty} (-1)^n \left\{\frac{a}{a^2 + n^2} + i \frac{n}{a^2 + n^2}\right\}$$

Equating the real part, we get

$$\frac{\pi}{\sinh a\pi} = \sum_{n=-\infty}^{\infty} (-1)^n \frac{a}{a^2 + n^2} \text{ or } \frac{\pi}{a \sinh a\pi} = \sum_{n=-\infty}^{\infty} \frac{(-1)^n}{a^2 + n^2}$$

which proves the result.

Example 2. *Find the complex form of the Fourier series of*

$f(x) = \cos ax$ in $-\pi < x < \pi$.

Sol. The complex form of Fourier series in $-\pi$ to π is given by

$$f(x) = \sum_{n=-\infty}^{\infty} C_n e^{inx} \quad ...(1) \quad \text{where } C_n = \frac{1}{2\pi} \int_{-\pi}^{\pi} f(x)\, e^{-inx}\, dx$$

$$C_n = \frac{1}{2\pi} \int_{-\pi}^{\pi} \cos ax \,.\, e^{-inx}\, dx = \frac{1}{2\pi} \left[\frac{e^{-inx}}{(-in)^2 + a^2} \{-in \cos ax + a \sin ax\}\right]_{-\pi}^{\pi}$$

$$\left[\because \int e^{ax} \cos (bx + c) = \frac{e^{ax}}{a^2 + b^2} \{a \cos (bx + c) + b \sin (bx + c)\}\right]$$

$$= \frac{1}{2\pi (a^2 - n^2)} [e^{-in\pi} (-in \cos a\pi + a \sin a\pi) - e^{in\pi} (-in \cos a\pi - a \sin a\pi)]$$

$$= \frac{1}{2\pi (a^2 - n^2)} [\cos n\pi \{-in \cos a\pi + a \sin a\pi + in \cos a\pi + a \sin a\pi\}]$$

$$[\because e^{\pm in\pi} = \cos n\pi \pm i \sin n\pi = \cos n\pi = (-1)^n]$$

$$\therefore \quad C_n = \frac{(-1)^n a \sin a\pi}{\pi (a^2 - n^2)}$$

Substituting in (1),

$$f(x) = \cos ax = \sum_{n=-\infty}^{\infty} \frac{(-1)^n a \sin a\pi}{\pi (a^2 - n^2)} e^{inx} = \frac{a \sin a\pi}{\pi} \sum_{n=-\infty}^{\infty} \frac{(-1)^n}{a^2 - n^2} e^{inx}.$$

TEST YOUR KNOWLEDGE

1. Derive the complex form (or exponential form) of Fourier Series for $f(x)$ in $-\pi < x < \pi$.
2. Find the complex form of Fourier Series for e^x in $-\pi < x < \pi$.
3. Find the complex form of Fourier Series for e^{-x} in $-1 < x < 1$.

Answers

1. $f(x) = \sum_{n=-\infty}^{\infty} c_n e^{-inx}$; $c_n = \frac{1}{2\pi}\int_{-\pi}^{\pi} f(x)\, e^{-inx}\, dx$ **2.** $e^x = \frac{\sinh \pi}{\pi} \sum_{n=-\infty}^{\infty} (-1)^n \left(\frac{1+in}{1+n^2}\right) e^{inx}$

3. $e^{-x} = \sum_{n=-\infty}^{\infty} (-1)^n (\sinh 1) \left(\frac{1-in\pi}{1+n^2\pi^2}\right) e^{in\pi x}$.

1.11. PRACTICAL HARMONIC ANALYSIS

When the function $f(x)$ is not given by an analytical expression, rather by its graph or a table of corresponding values at a number of equi-spaced points, we cannot evaluate the integrals for Fourier co-efficients. However, using the rules of approximate integration, we can find approximate values of the first few terms of the Fourier expansions.

Let the Fourier Series for $y = f(x)$ in $(0, 2\pi)$ be

$$y = \frac{a_0}{2} + a_1 \cos x + a_2 \cos 2x + \ldots\ldots + b_1 \sin x + b_2 \sin 2x + \ldots\ldots$$

where $$a_0 = \frac{1}{\pi}\int_0^{2\pi} y\, dx,\ a_n = \frac{1}{\pi}\int_0^{2\pi} y \cos nx\, dx\ ;\ b_n = \frac{1}{\pi}\int_0^{2\pi} y \sin nx\, dx.$$

Let the range $(0, 2\pi)$ be divided into m equal parts by the points $x = x_0, x_1, x_2 \ldots\ldots, x_m = 2\pi$ so that each sub-interval is of the length $\frac{2\pi}{m}$. Let the ordinates at these points be denoted by $y_0, y_1, y_2, \ldots\ldots, y_m$.

Using the trapezoidal rule of approximate integration, we get

$$a_0 = \frac{1}{\pi}\int_0^{2\pi} y dx \approx \frac{1}{\pi}\cdot\frac{2\pi}{m}(y_0 + y_1 + y_2 + \ldots\ldots + y_{m-1}) = 2 \times \frac{1}{m}\sum_{r=0}^{m-1} y_r$$

$$= 2\ [\text{mean value of } y \text{ in } (0, 2\pi)]$$

$$a_n = \frac{1}{\pi}\int_0^{2\pi} y \cos nx\, dx$$

$$\approx \frac{1}{\pi}\cdot\frac{2\pi}{m}(y_0 \cos nx_0 + y_1 \cos nx_1 + \ldots\ldots + y_{m-1} \cos nx_{m-1})$$

$$= 2 \times \frac{1}{m}\sum_{r=0}^{m-1} y_r \cos nx_r = 2\ [\text{mean value of } y \cos nx \text{ in } (0, 2\pi)]$$

$$b_n = \frac{1}{\pi}\int_0^{2\pi} y \sin nx \, dx \approx \frac{1}{\pi}\cdot\frac{2\pi}{m}(y_0 \sin nx_0 + y_1 \sin nx_1 + \ldots\ldots + y_{m-1} \sin nx_{m-1})$$

$$= 2 \times \frac{1}{m}\sum_{r=0}^{m-1} y_r \sin nx_r$$

$$= 2 \text{ [mean value of } y \sin nx \text{ in } (0, 2\pi)]$$

The process of finding the Fourier Series for a function given by numerical values is known as *harmonic analysis.*

The term $(a_1 \cos x + b_1 \sin x)$ is called the *fundamental* or *first harmonic,* the term $(a_2 \cos 2x + b_2 \sin 2x)$ is called the *second harmonic* and so on.

The number of ordinates used should not be less than twice the number of highest harmonic to be found.

ILLUSTRATIVE EXAMPLES

Example 1. *Analyse harmonically the data given below and express y in Fourier series upto the third harmonic:*

x :	0	$\frac{\pi}{3}$	$\frac{2\pi}{3}$	π	$\frac{4\pi}{3}$	$\frac{5\pi}{3}$	2π
y :	1.0	1.4	1.9	1.7	1.5	1.2	1.0

Sol. Since the last value of y is a repetition of the first, only the first six values will be used. The length of the interval is 2π.

Let $\quad y = \frac{a_0}{2} + (a_1 \cos x + b_1 \sin x) + (a_2 \cos 2x + b_2 \sin 2x) + (a_3 \cos 3x + b_3 \sin 3x)$

The values of x, y, $\sin x$, $\cos x$, $\sin 2x$, $\cos 2x$, $\sin 3x$, $\cos 3x$ are tabulated below :

x	y	$\cos x$	$\sin x$	$\cos 2x$	$\sin 2x$	$\cos 3x$	$\sin 3x$
0	1.0	1	0	1	0	1	0
$\frac{\pi}{3}$	1.4	0.5	0.866	– 0.5	0.866	– 1	0
$\frac{2\pi}{3}$	1.9	– 0.5	0.866	– 0.5	– 0.866	1	0
π	1.7	– 1	0	1	0	– 1	0
$\frac{4\pi}{3}$	1.5	– 0.5	0.866	– 0.5	0.866	1	0
$\frac{5\pi}{3}$	1.2	0.5	– 0.866	– 0.5	– 0.866	– 1	0

Using the values in the above table, we have

$$a_0 = \frac{2}{n}\ \Sigma y = \frac{2}{6}\ (8.7) = 2.9$$

$$a_1 = \frac{2}{n}\ \Sigma y \cos x = \frac{2}{6}\ [1(1.0 - 1.7) + 0.5\ (1.4 - 1.9 - 1.5 + 1.2)] = \frac{1}{3}\ (-1.1) \approx -0.37$$

$$b_1 = \frac{2}{n}\ \Sigma y \sin x = \frac{2}{6}\ [0.866\ (1.4 + 1.9 - 1.5 - 1.2)] = \frac{1}{3}\ [0.866\ (.6)] \approx\ 0.17 = 1.0392$$

$$a_2 = \frac{2}{n}\ \Sigma y \cos 2x = \frac{2}{6}\ [1(1.0 + 1.7) - 0.5(1.4 + 1.9 - 1.5 + 1.2)] = \frac{1}{3}\ [2.7 - 3] = -0.1$$

$$b_2 = \frac{2}{n}\ \Sigma y \sin 2x = \frac{2}{6}\ [0.866\ (1.4 - 1.9 + 1.5 - 1.2)] = \frac{1}{3}\ [0.866\ (-2)] = -0.06$$

$$a_3 = \frac{2}{n}\ \Sigma y \cos 3x = \frac{2}{6}\ [1.0 - 1.4 + 1.9 - 1.7 + 1.5 - 1.2] = \frac{1}{3}\ (0.1) = 0.03$$

$$b_3 = \frac{2}{n}\ \Sigma y \sin 3x = 0$$

$$\therefore \quad y = 1.45 + (-0.37 \cos x + 0.17 \sin x) - (0.1 \cos 2x + 0.06 \sin 2x) + 0.03 \cos 3x.$$

Example 2. *The following values of y give the displacement in inches of a certain machine part for the rotation x of the flywheel. Expand y in the form of a Fourier Series:*

x :	0	$\frac{\pi}{6}$	$\frac{2\pi}{6}$	$\frac{3\pi}{6}$	$\frac{4\pi}{6}$	$\frac{5\pi}{6}$
y :	0	9.2	14.4	17.8	17.3	11.7

Sol. Here length of interval is π, *i.e.*, $l = \frac{\pi}{2}$

$$\text{Let} \quad y = \frac{a_0}{2} + \left(a_1 \cos \frac{\pi x}{\pi/2} + b_1 \sin \frac{\pi x}{\pi/2}\right) + \left(a_2 \cos \frac{2\pi x}{\pi/2} + b_2 \sin \frac{\pi x}{\pi/2}\right) + \ldots\ldots$$

$$= \frac{a_0}{2} + (a_1 \cos 2x + b_1 \sin 2x) + (a_2 \cos 4x + b_2 \sin 4x) + \ldots\ldots$$

The values of x, y, $\sin 2x$, $\cos 2x$, $\sin 4x$, $\cos 4x$ are tabulated below:

x	y	$\cos 2x$	$\sin 2x$	$\cos 4x$	$\sin 4x$
0	0	1	0	1	0
$\frac{\pi}{6}$	9.2	.5	.87	– .5	.87
$\frac{2\pi}{6}$	14.4	– .5	.87	– .5	– .87
$\frac{3\pi}{6}$	17.8	– 1	0	1	0
$\frac{4\pi}{6}$	17.3	– .5	– .87	– .5	.87
$\frac{5\pi}{6}$	11.7	.5	– .87	– .5	– 8.7

Using the values in the above table, we have

$$a_0 = \frac{2}{n} \Sigma y = \frac{2}{6} (70.4) = 23.46$$

$$a_1 = \frac{2}{n} \Sigma y \cos 2x = \frac{2}{6} [.5(9.2 - 14.4 - 17.3 + 11.7) - 17.8]$$

$$= \frac{1}{3} (-23.2) = -7.73$$

$$b_1 = \frac{2}{n} \Sigma y \sin 2x = \frac{2}{6} [.87(9.2 + 14.4 - 17.3 - 11.7)]$$

$$= \frac{1}{3} [.87(-5.4)] = -1.566$$

$$a_2 = \frac{2}{n} \Sigma y \cos 4x = \frac{2}{6} [-.5 (9.2 + 14.4 + 17.3 + 11.7) + 17.8]$$

$$= \frac{1}{3} [-26.3 + 17.8] = \frac{1}{3} (-8.5) = -2.83$$

$$b_2 = \frac{2}{n} \Sigma y \sin 4x = \frac{2}{6} [87 (9.2 - 14.4 + 17.3 - 11.7)] = \frac{1}{3} [.87 (4)] = .116$$

$$\therefore \quad y = 11.73 - (7.73 \cos 2x + 1.566 \sin 2x) + (-2.83 \cos 4x + .116 \sin 4x).$$

Example 3. *Obtain the constant term and the co-efficients of the first sine and cosine terms in the Fourier expansion of y as given in the following table:*

x :	*0*	*1*	*2*	*3*	*4*	*5*
y :	*9*	*18*	*24*	*28*	*26*	*20*

Sol. Let the Fourier Series to represent y be

$$y = \frac{a_0}{2} + \left(a_1 \cos \frac{\pi x}{3} + b_1 \sin \frac{\pi x}{3} \right) + \left(a_2 \cos \frac{2\pi x}{3} + b_2 \sin \frac{2\pi x}{3} \right) + \ldots\ldots$$

The values of x, y, $\sin \frac{\pi x}{3}$, $\cos \frac{\pi x}{3}$ are tabulated below:

x	y	$\cos \frac{\pi x}{3}$	$\sin \frac{\pi x}{3}$
0	9	1	0
1	18	$\frac{1}{2}$	$\frac{\sqrt{3}}{2}$
2	24	$-\frac{1}{2}$	$\frac{\sqrt{3}}{2}$
3	28	-1	0
4	26	$-\frac{1}{2}$	$-\frac{\sqrt{3}}{2}$
5	20	$\frac{1}{2}$	$-\frac{\sqrt{3}}{2}$

Using the values in the above table, we have

$$a_0 = \frac{2}{n}\Sigma y = \frac{2}{6}(125) = 41.67$$

$$a_1 = \frac{2}{n}\Sigma y \cos\frac{\pi x}{3} = \frac{2}{6}\left[9 - 28 + \tfrac{1}{2}(18 - 24 - 26 + 20)\right] = \tfrac{1}{3}[-19 - 6] = -8.33$$

$$b_1 = \frac{2}{n}\Sigma y \sin\frac{\pi x}{3} = \frac{2}{6}\left[\frac{\sqrt{3}}{2}(18 + 24 - 26 - 20)\right] = \frac{\sqrt{3}}{6}(-4) = -\frac{2\sqrt{3}}{3} = -1.15.$$

Example 4. *The following table gives the variations of periodic current over a period:*

t (secs.):	*0*	*T/6*	*T/3*	*T/2*	*2T/3*	*5T/6*	*T*
A (amp.):	*1.98*	*1.30*	*1.05*	*1.30*	*– 0.88*	*– 0.25*	*1.98*

Show, by numerical analysis, that there is a direct current part of 0.75 amp. in the variable current and obtain the amplitude of the first harmonic.

Sol. Hence length of interval is T *i.e.*, $l = \frac{T}{2}$

Let
$$A = \frac{a_0}{2} + a_1 \cos\frac{2\pi t}{T} + b_1 \sin\frac{2\pi t}{T} + \ldots\ldots$$

where $\frac{a_0}{2}$ represents the direct current part and $\sqrt{a_1^2 + b_1^2}$ gives the amplitude of the first harmonic.

The values of t, A, $\sin\frac{2\pi t}{T}$, $\cos\frac{2\pi t}{T}$ are tabulated below:

t	A	$\cos\frac{2\pi t}{T}$	$\sin\frac{2\pi t}{T}$
0	1.98	1	0
T/6	1.30	0.5	0.866
T/3	1.05	– 0.5	0.866
T/2	1.30	– 1	0
2T/3	– 0.88	– 0.5	– 0.866
5T/6	– 0.25	0.5	– 0.866

Using the values in the above table, we have

$$a_0 = \frac{2}{n}\Sigma A = \frac{2}{6}(4.5) = 1.50$$

$$a_1 = \frac{2}{n}\Sigma A \cos\frac{2\pi t}{T} = \frac{2}{6}[1.98 - 1.30 + 0.5(1.30 - 1.05 + 0.88 - 0.25)]$$

$$= \frac{1}{3}[0.68 + 0.5(.88)] = \frac{1}{3}(1.12) = 0.373$$

$$b_1 = \frac{2}{n}\,\Sigma A \sin\frac{2\pi t}{T} = \frac{2}{6}\,[0.866\,(1.30 + 1.05 + 0.88 + 0.25)]$$

$$= \frac{1}{3}\,[0.866 \times 3.48] = 1.005$$

$$\therefore \qquad A = 0.75 + 0.373\cos\frac{2\pi t}{T} + 1.005\sin\frac{2\pi t}{T} + \ldots\ldots$$

Direct current part $= \frac{a_0}{2} = 0.75$ amp.

The amplitude of the first harmonic $= \sqrt{a_1^2 + b_1^2} = \sqrt{(0.373)^2 + (1.005)^2} = 1.072$.

Example 5. *In a machine the displacement of a given points is given for a certain angle θ° as follows:*

θ°	0°	30°	60°	90°	120°	150°	180°	210°	240°	270°	300°	330°
y	7.9	8.0	7.2	5.6	3.6	1.7	0.5	0.2	0.9	2.5	4.7	6.8

Find the coefficient of sin 2θ in the Fourier Series representing the above variation.

Sol. Let the Fourier Series of y be

$$y = \frac{a_0}{2} + a_1\cos\theta + b_1\sin\theta + a_2\cos 2\theta + b_2\sin 2\theta + \ldots .$$

θ°	θ *in radians*	*sin 2θ*	*y*	*y sin 2θ*
0°	0	0	7.9	0
30°	π/6	0.87	8.0	6.96
60°	2π/6	0.87	7.2	6.264
90°	3π/6	0	5.6	0
120°	4π/6	– 0.87	3.6	– 3.132
150°	5π/6	– 0.87	1.7	– 1.479
180°	6π/6	0	0.5	0
210°	7π/6	0.87	0.2	0.174
240°	8π/6	0.87	0.9	0.783
270°	9π/6	0	2.5	0
300°	10π/6	– 0.87	4.7	– 4.089
330°	11π/6	– 0.87	6.8	– 5.916
		Σ =		– 0.435

Now b_2 = coefficient of sin 2θ in the Fourier Series

= 2 Mean of y sin 2θ

$$= 2 \times \frac{-0.435}{12} = -0.0725.$$

Example 6. *The following table gives the variations of periodic current over a period*

t sec	*0*	*T/6*	*T/3*	*T/2*	*2T/3*	*5T/6*	*T*
A amp.	*1.98*	*1.30*	*1.05*	*1.30*	*– 0.88*	*– 0.25*	*1.98*

Show that there is a direct current part of 0.75 amp in the variable current and obtain the amplitude of the first harmonic. (CSVTU Bhilai (Third Sem.) 2006)

Sol. Let the Fourier series to represent A in $(0, T)$ be

$$A = \frac{a_0}{2} + a_1 \cos \frac{2\pi t}{T} + b_1 \sin \frac{2\pi t}{T} + a_1 \cos \frac{4\pi t}{T} + b_2 \sin \frac{4\pi t}{T} + \ldots .$$

To evaluate the coefficients, we form the following table:

t	$\frac{2\pi t}{T}$	$\cos \frac{2\pi t}{T}$	$\sin \frac{2\pi t}{T}$	A	$A \cos \frac{2\pi t}{T}$	$A \sin \frac{2\pi t}{T}$
0	0	1.0	0.00	1.98	1.980	0.000
$T/6$	$\pi/3$	0.5	0.866	1.30	0.650	1.126
$T/3$	$2\pi/3$	– 0.5	0.866	1.05	– 0.525	0.909
$T/2$	π	– 1.0	0.000	1.30	– 1.300	0.000
$2T/3$	$4\pi/3$	– 0.5	– 0.866	– 0.88	0.440	0.762
$5T/6$	$3\pi/3$	0.5	– 0.866	– 0.25	– 0.125	0.217
			$\Sigma =$	4.5	1.12	3.014

$$f\left(\frac{\pi}{2}\right) = \frac{\pi}{2} = \frac{4}{\pi}\left[\frac{1}{1^2}.1 - \frac{1}{3^2}.(-1) + \frac{1}{5^2}.1 - \frac{1}{7^2}.(-1) + \ldots\right]$$

or

$$\frac{\pi^2}{8} = \frac{1}{1^2} + \frac{1}{3^2} + \frac{1}{5^2} + \ldots .$$

$$a_0 = 2.\frac{1}{6}\sum A = \frac{1}{3}(4.5) = 1.5$$

$$a_1 = 2.\frac{1}{6}\sum A \cos\frac{2\pi t}{T} = \frac{1}{3}(1.12) = 0.373$$

$$b_1 = 2.\frac{1}{6}\sum A \sin\frac{2\pi t}{T} = \frac{1}{3}(3.014) = 1.005.$$

Thus the direct current part in the variable current $= \frac{a_0}{2} = \frac{1.5}{2} = 0.75$

and amplitude of the first harmonic $= \sqrt{{a_1}^2 + {b_1}^2} = \sqrt{(0.373)^2 + (1.005)^2} = 1.072.$

Example 7. *Find the Fourier series as far as the second harmonic to represent the function given by table below:*

x	*0°*	*30°*	*60°*	*90°*	*120°*	*150°*	*180°*	*210°*	*240°*	*270°*	*300°*	*330°*
f(x)	*2.34*	*3.01*	*3.69*	*4.15*	*3.69*	*2.20*	*0.83*	*0.51*	*0.88*	*1.09*	*1.19*	*1.64*

Sol.

$x°$	*sin x*	*sin 2x*	*cos x*	*cos 2x*	*f(x)*	*f(x). sin x*	*f(x). sin 2x*	*f(x). cos x*	*f(x). cos 2x*
0°	0	0	1	1	2.34	0	0	2.340	2.340
30°	0.50	0.87	0.87	0.50	3.01	1.505	2.619	2.619	1.505
60°	0.87	0.87	0.50	– 0.50	3.69	3.210	3.210	1.845	– 1.845
90°	1.00	0	0	– 1.00	4.15	4.150	0	0	– 4.150
120°	0.87	– 0.87	– 0.50	– 0.50	3.69	3.120	– 3.210	– 1.845	– 1.845
150°	0.50	– 0.87	– 0.87	0.50	2.20	1.100	– 1.914	– 1.914	1.100
180°	0	0	– 1	1.00	0.83	0	0	– 0.830	0.830
210°	– 0.50	0.87	– 0.87	0.50	0.51	– 0.255	0.444	– 0.444	0.255
240°	– 0.87	0.87	– 0.50	– 0.50	0.88	– 0.766	0.766	– 0.440	– 0.440
270°	– 1.00	0	0	– 1.00	1.09	– 1.090	0	0	– 1.090
300°	– 0.87	–0.87	0.50	–0.50	1.19	– 1.035	– 1.035	0.595	– 0.595
330°	– 0.50	– 0.87	0.87	0.50	1.64	– 0.820	– 1.427	1.427	0.820
				Σ =	25.22	9.209	– 0.547	3.353	– 3.115

Now $$a_0 = 2 \times \text{Mean of } f(x) = 2 \times \frac{25.22}{12} = 4.203$$

$$a_1 = 2 \times \text{Mean of } f(x) \cos x = 2 \times \frac{3.353}{12} = 0.559$$

$$a_2 = 2 \times \text{Mean of } f(x) \cos 2x = 2 \times \frac{-3.115}{12} = 0.519$$

$$b_1 = 2 \times \text{Mean of } f(x) \sin x = 2 \times \frac{9.209}{12} = 1.535$$

$$b_2 = 2 \times \text{Mean of } f(x) \sin 2x = 2 \times \frac{-0.547}{12} = -0.091.$$

∴ The required Fourier series is

$$f(x) = \frac{a_0}{2} + a_1 \cos x + a_2 \cos 2x + \ldots + b_1 \sin x + b_2 \sin 2x + \ldots$$

$$= 2.1015 + 0.559 \cos x - 0.519 \cos 2x + \ldots + 1.535 \sin x - 0.091 \sin 2x + \ldots .$$

Example 8. *The following moment T on the crank-shaft of a steam engine for the crank angle θ° is given as follows:*

θ	*0*	*30*	*60*	*90*	*120*	*150*	*180*
T	*0*	*5224*	*8097*	*7850*	*5499*	*2626*	*0*

Obtain the first four terms in a series of sines to represent T and calculate T for θ = 75°.

Sol. Let the Fourier sine series to represent T in (0, 180) be

$$T = b_1 = \sin \theta + b_2 \sin 2\theta + b_3 \sin 3\theta + b_4 \sin 4\theta + \ldots .$$

To evaluate the coefficients, we form the following table:

θ	T	sin θ	sin 2θ	sin 3θ	sin 4θ
0	0	0	0	0	0
30	5224	0.500	0.866	1	0.866
60	8097	0.866	0.866	0	– 0.866
90	7850	1.000	0	– 1	0
120	5499	0.866	– 0.866	0	0.866
150	2626	0.500	– 0.866	1	– 0.866

$$\therefore \quad b_1 = \frac{\Sigma T \sin\theta}{6} = \frac{1}{3}\{(5224 + 2626)(0.5) + (8097 + 5499)(0.8666) + 7850(1)$$

$$= 785$$

$$b_2 = 2.\frac{\Sigma T \sin 2\theta}{6} = \frac{1}{3}\{(5224 + (8097)(0.866) + (5499 + 2626)(-0.866)\} = 150$$

$$b_3 = 2.\frac{\Sigma T \sin 3\theta}{6} = \frac{1}{3}\{5224 - 7850 + 2626)\} = 0$$

$$b_4 = 2.\frac{\Sigma T \sin 4\theta}{6} = \frac{1}{3}\{(5224 + 5499)(0.866) + (8097 + 2626)(-0.866)\} = 0.$$

Hence $\quad T = 785 \sin\theta + 150 \sin 2\theta.$

For $\theta = 75°$, we have

$$T = 785 \sin 75° + 150 \sin 150°$$

$$= 785\,(0.9659) + 150\,(0.5) = 8332.$$

Example 9. *The turning moment T units of the crank shaft of a steam engine is given for a series of values of the crank-angle θ in degrees:*

θ :	*0*	*30*	*60*	*90*	*120*	*150*	*180*
T :	*0*	*5224*	*8097*	*7850*	*5499*	*2626*	*0*

Find the first four terms in a series of sines to represent T. Also calculate T when θ = 75°.

Sol. Let the half-range sine series to represent T be

$$T = b_1 \sin\theta + b_2 \sin 2\theta + b_3 \sin 3\theta + b_4 \sin 4\theta$$

The values of θ, T, sinθ, sin 2θ, sin 3θ, sin 4θ are tabulated below:

θ	T	sin θ	sin 2θ	sin 3θ	sin 4θ
0	0	0	0	0	0
30	5224	0.5	0.866	1	0.866
60	8097	0.866	0.866	0	– 0.866
90	7850	1	0	– 1	0
120	5499	0.866	– 0.866	0	0.866
150	2626	0.5	– 0.866	1	– 0.866

Using the values in the above table, we have

$$b_1 = \frac{2}{n}\Sigma T \sin\theta$$

$$= \frac{2}{6} [0.5 (5224 + 2626) + 0.866 (8097 + 5499) + 7850]$$

$$= \frac{1}{3} (3925 + 11774 + 7850) = \frac{1}{3} (23549) = 7850$$

$$b_2 = \frac{2}{n} \Sigma T \sin 2\theta = \frac{2}{6} [0.866 (5224 + 8097 - 5499 - 2626)]$$

$$= \frac{1}{3} [0.866 \times 5196] = 1500$$

$$b_3 = \frac{2}{n} \Sigma T \sin 3\theta = \frac{2}{6} [5224 - 7850 + 2626] = 0$$

$$b_4 = \frac{2}{n} \Sigma T \sin 4\theta = \frac{2}{6} [0.866 (5224 - 8097 + 5499 - 2626)] = 0$$

$$\therefore \quad T = 7850 \sin \theta + 1500 \sin 2\theta$$

When $\theta = 75°$, $T = 7850 \sin 75° + 1500 \sin 150°$

$$= 7850 \times 0.9659 + 1500 \times 0.5 = 8332.$$

TEST YOUR KNOWLEDGE

1. The following values of y give the displacement of a certain machine part for the rotation x of the flywheel

x :	0°	60°	120°	180°	240°	300°	360°
y :	1.98	2.15	2.77	– 0.22	– 0.31	1.43	1.98

Express y in Fourier series upto the third harmonic.

2. The displacement y of a part of a mechanism is tabulated with corresponding angular movement $x°$ of the crank. Express y as a Fourier series neglecting the harmonics above the third.

$x°$:	0	30	60	90	120	150	180	210	240	270	300	330
y :	1.80	1.10	0.30	0.16	0.50	1.30	2.16	1.25	1.30	1.52	1.76	2.00

3. Determine the first two harmonics of the Fourier series for the following values:

$x°$:	30	60	90	120	150	180	210	240	270	300	330	360
y :	2.34	3.01	3.68	4.15	3.69	2.20	0.83	0.51	0.88	1.09	1.19	1.64

4. In a machine the displacement y of a given point is given for a certain angle θ as follows:

$\theta°$ =	0	30	60	90	120	150	180	210	240	270	300	330
y =	7.9	8.0	7.2	5.6	3.6	1.7	0.5	0.2	0.9	2.5	4.7	6.8

Find the co-efficient of $\sin 2\theta$ in the Fourier series representing the above variation.

5. Obtain the first three co-efficients in the Fourier cosine series for y, where y is given in the following table:

x :	0	1	2	3	4	5
y :	4	8	15	7	6	2

6. The turning moment T on the crank-shaft of a steam engine for the crank angle θ degrees is given as follows:

θ :	0	15	30	45	60	75	90	105	120	135	150	165	180
T :	0	2.7	5.2	7.0	8.1	8.3	7.9	6.8	5.5	4.1	2.6	1.2	0

Expand T in a series of sines upto third harmonics.

Answers

1. $y = 1.3 + (0.92 \cos x + 1.097 \sin x) - (0.42 \cos 2x + 0.681 \sin 2x) + 0.36 \cos 3x$
2. $y = 1.26 + (0.04 \cos x - 0.63 \sin x) + (0.53 \cos 2x - 0.23 \sin 2x) + (-0.1 \cos 3x + 0.085 \sin 3x)$
3. $y = 2.102 - 0.283 \cos x + 1.60 \sin x - 0.18 \cos 2x - 0.49 \sin 2x$
4. -0.072
5. $y = 7 - 2.8 \cos \theta - 1.5 \cos 2\theta + 2.7 \cos 3\theta$
6. $T = 7.8 \sin \theta + 1.5 \sin 2\theta - 0.03 \sin 3\theta$.

FOURIER TRANSFORMS

1.12. INTEGRAL TRANSFORMS

The integral transform of a function $f(x)$ is defined by the equation

$$I\{f(x)\} = \bar{f}(s) = \int_a^b f(x)\, K(s, x)\, dx,$$

where $K(s, x)$ is a known function of s and x, called the **kernel** of the transform; s is called the **parameter** of the transform and $f(x)$ is called the **inverse transform** of $\bar{f}(s)$.

Some of the well-known transforms are given below:

(*i*) **Laplace Transform.** When $K(s, x) = e^{-sx}$, we have the Laplace transform of $f(x)$.

Thus
$$L\{f(x)\} = \bar{f}(s) = \int_0^\infty f(x)\, e^{-sx}\, dx$$

(*ii*) **Fourier Transform.** When $K(s, x) = e^{isx}$, we have the Fourier transform of $f(x)$.

Thus
$$F(s) = \int_{-\infty}^\infty f(x)\, e^{isx}\, dx$$

(*iii*) **Hankel Transform.** When $K(s, x) = x\, J_n(sx)$, we have the Hankel transform of $f(x)$.

Thus
$$H_n(s) = \int_0^\infty f(x)\, x\, J_n(sx)\, dx$$

where $J_n(sx)$ is the Bessel function of the first kind and order n.

(*iv*) **Mellin Transform.** When $K(s, x) = x^{s-1}$, we have the Mellin transform of $f(x)$. Thus

$$M(s) = \int_0^\infty f(x)\, x^{s-1}\, dx$$

(*v*) **Fourier Sine Transform.** When $K(s, x) = \sin sx$, we have the Fourier sine transform of $f(x)$. Thus

$$F_s(s) = \int_0^\infty f(x) \sin sx\, dx$$

(*vi*) **Fourier Cosine Transform.** When $K(s, x) = \cos sx$, we have the Fourier cosine transform of $f(x)$. Thus

$$F_c(s) = \int_0^\infty f(x) \cos sx\, dx$$

We have already discussed Laplace transform and its applications to the solution of ordinary differential equations. In the present chapter, we shall discuss Fourier transforms which are useful in solving boundary value problems arising in engineering *e.g.*, conduction of heat, theory of communication, wave propagation etc.

The effect of applying an integral transform to a partial differential equation is to reduce the number of independent variables by one. The choice of a particular transform is decided by the nature of the boundary conditions and the facility with which the transform can be inverted to give $f(x)$.

1.13. FOURIER TRANSFORMS AND INVERSION FORMULAE

(1) *Fourier sine transform and its inversion formula*

Fourier sine integral is $f(x) = \frac{2}{\pi}\int_0^\infty \sin \lambda x \int_0^\infty f(t) \sin \lambda t \, dt \, d\lambda$

Replacing λ by s, we get $f(x) = \frac{2}{\pi}\int_0^\infty \sin sx \int_0^\infty f(t) \sin st \, dt \, ds$

Denoting the value of the inner integral by $F_s(s)$, we have

$$f(x) = \frac{2}{\pi}\int_0^\infty F_s(s) \sin sx \, ds \quad \text{...(1)}$$

where
$$F_s(s) = \int_0^\infty f(x) \sin sx \, dx \quad \text{...(2)}$$

The function $F_s(s)$ as defined by equation (2) is known as the **Fourier sine transform** of $f(x)$ in $0 < x < \infty$.

The function $f(x)$ as defined by equation (1) is called the **inverse Fourier sine transform** of $F_s(s)$. Thus equation (1) gives the inversion formula for the Fourier sine transform.

(2) *Fourier cosine transform and its inversion formula*

Fourier cosine integral is $f(x) = \frac{2}{\pi}\int_0^\infty \cos \lambda x \int_0^\infty f(t) \cos \lambda t \, dt \, d\lambda$

Replacing λ by s, we get $f(x) = \frac{2}{\pi}\int_0^\infty \cos sx \int_0^\infty f(t) \cos st \, dt \, ds$

Denoting the value of the inner integral by $F_c(s)$, we have

$$f(x) = \int_0^\infty F_c(s) \cos sx \, dx \quad \text{...(3)}$$

where
$$F_c(s) = \int_0^\infty f(x) \cos sx \, dx \quad \text{...(4)}$$

The function $F_c(s)$ as defined by equation (4) is known as the **Fourier cosine transform** of $f(x)$ in $0 < x < \infty$.

The function $f(x)$ as defined by equation (3) is called the **inverse Fourier cosine transform** of $F_c(s)$. Thus equation (3) gives the inversion formula for the Fourier cosine transform.

(3) *Complex Fourier transform and its inversion formula*

Complex form of Fourier integral is $f(x) = \frac{1}{2\pi}\int_{-\infty}^\infty e^{-i\lambda x} \int_{-\infty}^\infty f(t) e^{i\lambda t} \, dt \, d\lambda$

Replacing λ by s, we get $f(x) = \frac{1}{2\pi}\int_{-\infty}^\infty e^{-isx} \int_{-\infty}^\infty f(t) e^{ist} \, dt \, ds$

Denoting the value of the inner integral by $F(s)$, we have

$$f(x) = \frac{1}{2\pi}\int_{-\infty}^\infty F(s) \, e^{-isx} \, ds \quad \text{...(5)}$$

where
$$F(s) = \int_{-\infty}^\infty f(x) \, e^{isx} \, dx \quad \text{...(6)}$$

The function F(s) as defined by equation (6) is known as the **Fourier transform** of $f(x)$.

The function $f(x)$ as defined by equation (5) is called the **inverse Fourier transform** of F(s). Thus equation (5) gives the inversion formula for the Fourier transform.

(4) *Finite Fourier sine transform and its inversion formula*

The **finite Fourier sine transform** of $f(x)$ in $0 < x < c$ is defined as

$$F_s(n) = \int_0^c f(x) \sin \frac{n\pi x}{c} dx \text{ where } n \text{ is an integer.}$$

The function $f(x)$ is then called the **inverse finite Fourier sine transform** of $F_s(n)$ and is given by

$$f(x) = \frac{2}{c} \sum_{n=1}^{\infty} F_s(n) \sin \frac{n\pi x}{c}.$$

(5) *Finite Fourier cosine transform and its inversion formula*

The **finite Fourier cosine transform** of $f(x)$ in $0 < x < c$ is defined as

$$F_c(n) = \int_0^c f(x) \cos \frac{n\pi x}{c} dx$$

where n is an integer.

The function $f(x)$ is then called the **inverse finite Fourier cosine transform** of $F_c(n)$ and is given by

$$f(x) = \frac{1}{c} F_c(0) + \frac{2}{c} \sum_{n=1}^{\infty} F_c(n) \cos \frac{n\pi x}{c}.$$

1.14. FOURIER INTEGRAL THEOREM

Consider a function $f(x)$ satisfying Dirichlet's conditions in every interval $(-c, c)$, however large. Then

$$f(x) = \frac{a_0}{2} + \sum_{n=1}^{\infty} \left(a_n \cos \frac{n\pi x}{c} + b_n \sin \frac{n\pi x}{x} \right) \quad \ldots(1)$$

where, $$a_0 = \frac{1}{c} \int_{-c}^{c} f(t)\, dt, a_n = \frac{1}{c} \int_{-c}^{c} f(t) \cos \frac{n\pi t}{c} dt \quad \text{and} \quad b_n = \frac{1}{c} \int_{-c}^{c} f(t) \sin \frac{n\pi t}{c} dt$$

Substituting the values of a_0, a_n and b_n in (1), we get

$$f(x) = \frac{1}{2c} \int_{-c}^{c} f(t)\, dt + \frac{1}{c} \sum_{n=1}^{\infty} \int_{-c}^{c} \left[\cos \frac{n\pi x}{c} \cos \frac{n\pi t}{c} + \sin \frac{n\pi x}{c} \sin \frac{n\pi t}{c} \right] f(t)\, dt$$

$$= \frac{1}{2c} \int_{-c}^{c} f(t)\, dt + \frac{1}{c} \sum_{n=1}^{\infty} \int_{-c}^{c} \cos \frac{n\pi(t-x)}{c} . f(t)\, dt \quad \ldots(2)$$

If we assume that the integral $\int_{-\infty}^{\infty} | f(x) |\, dx$ converges, then

$$\underset{c \to \infty}{\text{Lt}} \left[\frac{1}{2c} \int_{-c}^{c} f(t)\, dt \right] = 0$$

Putting $\frac{\pi}{c} = \Delta\lambda$, the second term in (2) becomes

$$\frac{1}{\pi} \sum_{n=1}^{\infty} \Delta\lambda \int_{-c}^{c} \cos \{n\Delta\lambda(t - x)\} f(t)\, dt$$

This is of the form $\sum_{n=1}^{\infty} F(n\Delta\lambda)\, \Delta\lambda$ whose limit as $\Delta\lambda \to 0$, is $\int_0^{\infty} F(\lambda)\, d\lambda$.

Hence as $c \to \infty$, (2) reduces to

$$f(x) = \frac{1}{\pi} \int_0^{\infty} \int_{-\infty}^{\infty} f(t) \cos \lambda(t - x)\, dt\, d\lambda \qquad ...(3)$$

which is known as **Fourier integral** of $f(x)$.

Equation (3) is true at a point of continuity. At a point of discontinuity, the value of the integral on the right is

$$\frac{1}{2} [f(x + 0) + f(x - 0)].$$

1.15. FOURIER SINE AND COSINE INTEGRALS

We know that $\cos \lambda(t - x) = \cos \lambda t \cos \lambda x + \sin \lambda t \sin \lambda x$

$\therefore$ Fourier integral of $f(x)$ can be written as

$$f(x) = \frac{1}{\pi} \int_0^{\infty} \int_{-\infty}^{\infty} f(t)\{\cos \lambda t \cos \lambda x + \sin \lambda t \sin \lambda x\}\, dt\, d\lambda$$

$$= \frac{1}{\pi} \int_0^{\infty} \cos \lambda x \int_{-\infty}^{\infty} f(t) \cos \lambda t\, dt\, d\lambda + \frac{1}{\pi} \int_0^{\infty} \sin \lambda x \int_{-\infty}^{\infty} f(t) \sin \lambda t\, dt\, d\lambda \qquad ...(4)$$

When f(x) is an odd function, $f(t) \cos \lambda t$ is odd while $f(t) \sin \lambda t$ is even. Thus the first integral in (4) vanishes and we get

$$f(x) = \frac{2}{\pi} \int_0^{\infty} \sin \lambda x \int_0^{\infty} f(t) \sin \lambda t\, dt\, d\lambda \qquad ...(5)$$

This is called **Fourier sine integral.**

When f(x) is an even function, $f(t) \cos \lambda t$ is even while $f(t) \sin \lambda t$ is odd. Thus the second integral in (4) vanishes and we get

$$f(x) = \frac{2}{\pi} \int_0^{\infty} \cos \lambda x \int_0^{\infty} f(t) \cos \lambda t\, dt\, d\lambda \qquad ...(6)$$

This is called **Fourier cosine integral.**

1.16. COMPLEX FORM OF FOURIER INTEGRAL

Since $\cos \lambda(t - x)$ is an even function of λ, we have from (3),

$$f(x) = \frac{1}{2\pi} \int_{-\infty}^{\infty} \int_{-\infty}^{\infty} f(t) \cos \lambda(t - x)\, dt\, d\lambda \qquad ...(7)$$

$$\left[\because\ 2\int_0^{a} f(x)\, dx = \int_{-a}^{a} f(x)\, dx, \text{ if } f(x) \text{ is even}\right]$$

Also $\sin \lambda(t-x)$ is an odd function of λ, so that

$$0 = \frac{1}{2\pi}\int_{-\infty}^{\infty}\int_{-\infty}^{\infty} f(t)\sin\lambda(t-x)\,dt\,d\lambda \qquad ...(8)$$

$$\left[\because \int_{-a}^{a} f(x)\,dx = 0, \text{ if } f(x) \text{ is odd}\right]$$

Multiplying (8) by i and adding to (7), we get

$$f(x) = \frac{1}{2\pi}\int_{-\infty}^{\infty}\int_{-\infty}^{\infty} f(t)\,\{\cos\lambda(t-x) + i\sin\lambda(t-x)\}\,dt\,d\lambda$$

$$= \frac{1}{2\pi}\int_{-\infty}^{\infty}\int_{-\infty}^{\infty} f(t)\,e^{i\lambda(t-x)}\,dt\,d\lambda$$

$$= \frac{1}{2\pi}\int_{-\infty}^{\infty} e^{-i\lambda x}\int_{-\infty}^{\infty} f(t)\,e^{i\lambda t}\,dt\,d\lambda$$

which is known as the **complex form of Fourier integral.**

Example. *Express the function* $f(x) = \begin{cases} 1 & \text{for } |x| \le 1, \\ 0 & \text{for } |x| > 1, \end{cases}$

as a Fourier integral. Hence evaluate $\int_0^{\infty} \frac{\sin\lambda\cos\lambda x}{\lambda}\,d\lambda$.

Sol. The Fourier integral for $f(x)$ is

$$\frac{1}{\pi}\int_0^{\infty}\int_{-\infty}^{\infty} f(t)\cos\lambda(t-x)\,dt\,d\lambda = \frac{1}{\pi}\int_0^{\infty}\int_{-1}^{1}\cos\lambda(t-x)\,dt\,d\lambda$$

[$\because$ $f(t) = 0$ when $-\infty < t < -1$, $f(t) = 1$ when $-1 < t < 1$ and $f(t) = 0$ when $1 < t < \infty$]

$$= \frac{1}{\pi}\int_0^{\infty}\left[\frac{\sin\lambda(t-x)}{\lambda}\right]_{-1}^{1} d\lambda = \frac{1}{\pi}\int_0^{\infty}\frac{\sin\lambda(1-x) - \sin\lambda(-1-x)}{\lambda}\,d\lambda$$

$$= \frac{1}{\pi}\int_0^{\infty}\frac{\sin\lambda(1+x) + \sin\lambda(1-x)}{\lambda}\,d\lambda = \frac{2}{\pi}\int_0^{\infty}\frac{\sin\lambda\cos\lambda x}{\lambda}\,d\lambda$$

$$\therefore \quad \int_0^{\infty}\frac{\sin\lambda\cos\lambda x}{\lambda}\,d\lambda = \frac{1}{2}\pi\,f(x) = \begin{cases} \frac{1}{2}\pi & \text{for } |x| < 1, \\ 0 & \text{for } |x| > 1 \end{cases}$$

At $|x| = 1$, *i.e.*, $x = \pm 1$, $f(x)$ is discontinuous and the integral has the value $\frac{1}{2}\left(\frac{\pi}{2} + 0\right) = \frac{\pi}{4}$.

Note. Putting $x = 0$, we get $\int_0^{\infty}\frac{\sin\lambda}{\lambda}\,d\lambda = \frac{\pi}{2}$ or $\int_0^{\infty}\frac{\sin x}{x}\,dx = \frac{\pi}{2}$.

1.17. INVERSION FORMULAE FOR FOURIER TRANSFORMS

(1) *Fourier sine transform and its inversion formula*

Fourier sine integral is $\quad f(x) = \frac{2}{\pi}\int_0^{\infty}\sin\lambda x\int_0^{\infty} f(t)\sin\lambda t\,dt\,d\lambda$

Replacing λ by s, we get $\quad f(x) = \frac{2}{\pi}\int_0^{\infty}\sin sx\int_0^{\infty} f(t)\sin st\,dt\,ds$

Denoting the value of the inner integral by $\sqrt{\frac{\pi}{2}}\ F_s\ \{f(x)\}$, we have

$$f(x) = \sqrt{\frac{2}{\pi}} \int_0^{\infty} F_s\{f(x)\} \sin sx\, ds \qquad ...(1)$$

where $$F_s\{f(x)\} = \bar{f}(s) = \sqrt{\frac{2}{\pi}} \int_0^{\infty} f(t) \sin st\, dt \qquad ...(2)$$

Equation (2) is defined as Fourier sine transform of the function $f(x)$.Equation (1) which determines the function $f(x)$ is known as the inversion formula for the Fourier sine transform.

(2) *Fourier cosine transform and its inversion formula*

Fourier cosine integral is $f(x) = \frac{2}{\pi} \int_0^{\infty} \cos \lambda x \int_0^{\infty} f(t) \cos \lambda t\, dt\, d\lambda$

Replacing λ by s, we get $f(x) = \frac{2}{\pi} \int_0^{\infty} \cos sx \int_0^{\infty} f(t) \cos st\, dt\, ds$

Denoting the value of the inner integral by $\sqrt{\frac{\pi}{2}}\ F_c\ \{f(x)\}$, we have

$$f(x) = \sqrt{\frac{2}{\pi}} \int_0^{\infty} F_c\ \{f(x)\} \cos sx\, ds \qquad ...(3)$$

where $$F_c\ \{f(x)\} = \bar{f}(s) = \sqrt{\frac{2}{\pi}} \int_0^{\infty} f(t) \cos st\, dt \qquad ...(4)$$

Equation (4) is defined as Fourier cosine transform of the function $f(x)$. Equation (3) which determines the function $f(x)$ is known as the inversion formula for the Fourier cosine transform.

(3) *Complex Fourier transform and its inversion formula*

Complex form of Fourier integral is $f(x) = \frac{1}{2\pi} \int_{-\infty}^{\infty} e^{-i\lambda x} \int_{-\infty}^{\infty} f(t)\, e^{i\lambda t}\, dt\, d\lambda$

Replacing λ by s, we get $f(x) = \frac{1}{2\pi} \int_{-\infty}^{\infty} e^{-isx} \int_{-\infty}^{\infty} f(t)\, e^{ist}\, dt\, ds$

Denoting the value of the inner integral by $\sqrt{2\pi}\ F\{f(x)\}$, we have

$$f(x) = \frac{1}{\sqrt{2\pi}} \int_{-\infty}^{\infty} F\{f(x)\}\, e^{-isx}\, ds \qquad ...(5)$$

where $$F\{f(x)\} = \bar{f}(s) = \frac{1}{\sqrt{2\pi}} \int_{-\infty}^{\infty} f(t)\, e^{ist}\, dt \qquad ...(6)$$

Equation (6) is defined as complex Fourier transform of the function $f(x)$. Equation (5) which determines the function $f(x)$ is known as the inversion formula for the complex Fourier transform.

ILLUSTRATIVE EXAMPLES

Example 1. *Using Fourier sine integral, show that*

$$\int_0^\infty \frac{1-\cos \pi\lambda}{\lambda} \sin x\lambda \, d\lambda = \begin{cases} \frac{1}{2}\pi, & \text{when } 0 < x < \pi \\ 0, & \text{when } x > \pi \end{cases}$$

Sol. Let
$$f(x) = \begin{cases} \frac{1}{2}\pi, & \text{when } 0 < x < \pi \\ 0, & \text{when } x > \pi \end{cases}$$

Using Fourier sine integral, we have

$$f(x) = \frac{2}{\pi}\int_0^\infty \sin \lambda x \int_0^\infty f(t) \sin \lambda t \, dt \, d\lambda$$

$$= \frac{2}{\pi}\int_0^\infty \sin \lambda x \left[\int_0^\pi f(t)\sin \lambda t \, dt + \int_\pi^\infty f(t) \sin \lambda t \, dt\right] d\lambda$$

$$= \frac{2}{\pi}\int_0^\infty \sin \lambda x \int_0^\pi \frac{1}{2}\pi \sin \lambda t \, dt \, d\lambda \qquad \text{[on substituting for } f(t)]$$

$$= \int_0^\infty \sin \lambda x \left[-\frac{\cos \lambda t}{\lambda}\right]_0^\pi d\lambda = \int_0^\pi \frac{1-\cos \pi\lambda}{\lambda} \sin x\lambda \, d\lambda$$

$$\therefore \quad \int_0^\infty \frac{1-\cos \pi\lambda}{\lambda} . \sin x\lambda \, d\lambda = f(x) = \begin{cases} \frac{1}{2}\pi, & \text{when } 0 < x < \pi \\ 0, & \text{when } x > \pi \end{cases}.$$

Example 2. *Find the Fourier sine transform of* $e^{-|x|}$. *Hence evaluate* $\int_0^\infty \frac{x \sin mx}{1+x^2} dx$.

Sol. In the interval $(0, \infty)$, x is positive so that $e^{-|x|} = e^{-x}$. Fourier sine transform of $f(x) = e^{-x}$ is given by

$$F_s\{f(x)\} = \sqrt{\frac{2}{\pi}}\int_0^\infty f(x) \sin sx \, dx = \sqrt{\frac{2}{\pi}}\int_0^\infty e^{-x} \sin sx \, dx$$

$$= \sqrt{\frac{2}{\pi}}\left[\frac{e^{-x}}{1+s^2}(-\sin sx - s\cos sx)\right]_0^\infty = \sqrt{\frac{2}{\pi}}\left(\frac{s}{1+s^2}\right)$$

Using inversion formula for Fourier sine transform, we get

$$f(x) = \sqrt{\frac{2}{\pi}}\int_0^\infty F_s\{f(x)\} \sin sx \, ds \quad \text{or} \quad e^{-x} = \frac{2}{\pi}\int_0^\infty \frac{s}{1+s^2} \sin sx \, ds$$

Replacing x by m, we have
$$e^{-m} = \frac{2}{\pi}\int_0^\infty \frac{s}{1+s^2} \sin ms \, ds = \frac{2}{\pi}\int_0^\infty \frac{x \sin mx}{1+m^2} dx$$

Hence
$$\int_0^\infty \frac{x \sin mx}{1+m^2} dx = \frac{\pi}{2} e^{-m}.$$

Example 3. *Find the Fourier cosine transform of* e^{-x^2}.

Sol. Fourier cosine transform of e^{-x^2} is given by

$$F_c\{e^{-x^2}\} = \sqrt{\frac{2}{\pi}}\int_0^\infty e^{-x^2} \cos sx \, dx = \text{I (say)} \qquad \text{...(1)}$$

Differentiating w.r.t. s, we have

$$\frac{dI}{ds} = -\sqrt{\frac{2}{\pi}}\int_0^\infty xe^{-x^2}\sin sx\,dx = \frac{1}{2}\sqrt{\frac{2}{\pi}}\int_0^\infty (\sin sx)(-2x\,e^{-x^2})\,dx$$

$$= \frac{1}{2}\sqrt{\frac{2}{\pi}}\left[\left\{\sin sx\,e^{-x^2}\right\}_0^\infty - s\int_0^\infty \cos sx\,e^{-x^2}\,dx\right] \quad \text{(Integrating by parts)}$$

$$= -\frac{s}{2}\sqrt{\frac{2}{\pi}}\int_0^\infty e^{-x^2}\cos sx\,dx = -\frac{s}{2}I$$

or $$\frac{dI}{I} = -\frac{s}{2}\,ds$$

Integrating, we have

$$\log I = -\frac{s^2}{4} + \log A \quad \text{or} \quad I = A\,e^{-\frac{s^2}{4}} \qquad \ldots(2)$$

Now when $s = 0$, from (1), $I = \sqrt{\frac{2}{\pi}}\int_0^\infty e^{-x^2}\,dx = \sqrt{\frac{2}{\pi}}\cdot\frac{\sqrt{\pi}}{2} = \frac{1}{\sqrt{2}}$

$\therefore$ From (2), $$\frac{1}{\sqrt{2}} = A$$

Hence $$I = F_c\{e^{-x^2}\} = \frac{1}{\sqrt{2}}\,e^{-\frac{s^2}{4}}.$$

Example 4. *Find the Fourier transform of* $f(x) = \begin{cases} 1 - x^2, & \text{if } |x| < 1 \\ 0, & \text{if } |x| > 1 \end{cases}$

and use it to evaluate $\int_0^\infty \left(\frac{x\cos x - \sin x}{x^3}\right)\cos\frac{x}{2}\,dx.$

Sol. Fourier transform of $f(x)$ is given by

$$F\{f(x)\} = \frac{1}{\sqrt{2\pi}}\int_{-\infty}^\infty f(x)\,e^{isx}\,dx = \frac{1}{\sqrt{2\pi}}\int_{-1}^1 (1 - x^2)\,e^{isx}dx$$

$$= \frac{1}{\sqrt{2\pi}}\left[(1 - x^2)\cdot\frac{e^{isx}}{is} - (-2x)\frac{e^{isx}}{(is)^2} + (-2)\frac{e^{isx}}{(is)^3}\right]_{-1}^1$$

$$= \frac{1}{\sqrt{2\pi}}\left[-\frac{2}{s^2}(e^{is} + e^{-is}) + \frac{2}{is^3}(e^{is} - e^{-is})\right]$$

$$= \frac{1}{\sqrt{2\pi}}\left[-\frac{4}{s^2}\left(\frac{e^{is} + e^{-is}}{2}\right) + \frac{4}{s^3}\left(\frac{e^{is} - e^{-is}}{2i}\right)\right]$$

$$= \frac{4}{\sqrt{2\pi}}\left[-\frac{\cos s}{s^2} + \frac{\sin s}{s^3}\right] = -2\sqrt{\frac{2}{\pi}}\left(\frac{s\cos s - \sin s}{s^3}\right)$$

By inversion formula for Fourier transform, we have

$$f(x) = \frac{1}{\sqrt{2\pi}}\int_{-\infty}^\infty F\{f(x)\}\,e^{-isx}\,ds = -\frac{2}{\pi}\int_{-\infty}^\infty \left(\frac{s\cos s - \sin s}{s^3}\right)(\cos sx - i\sin sx)\,ds$$

$$= -\frac{2}{\pi}\int_{-\infty}^{\infty}\left(\frac{s\cos s - \sin s}{s^3}\right)\cos sx\, ds + \frac{2i}{\pi}\int_{-\infty}^{\infty}\left(\frac{s\cos s - \sin s}{s^3}\right)\sin sx\, ds$$

$$= -\frac{2}{\pi}\int_{-\infty}^{\infty}\left(\frac{s\cos s - \sin s}{s^3}\right)\cos sx\, ds$$

(since the integrand in the second integral on R.H.S is odd)

$$= -\frac{4}{\pi}\int_{0}^{\infty}\left(\frac{s\cos s - \sin s}{s^3}\right)\cos sx\, ds$$

or
$$\int_{0}^{\infty}\left(\frac{s\cos s - \sin s}{s^3}\right)\cos sx\, ds = -\frac{\pi}{4}f(x) = \begin{cases} -\frac{\pi}{4}(1-x^2), & \text{if } |x|<1 \\ 0, & \text{if } |x|>1 \end{cases}$$

Taking $x = \frac{1}{2}$, we have $\int_{0}^{\infty}\left(\frac{s\cos s - \sin s}{s^3}\right)\cos\frac{s}{2}\, ds = -\frac{3\pi}{16}$

Hence
$$\int_{0}^{\infty}\left(\frac{x\cos x - \sin x}{x^3}\right)\cos\frac{x}{2}\, dx = -\frac{3\pi}{16}.$$

Example 5. *Solve the integral equation* $\int_0^\infty F(x)\cos px\, dx = \begin{cases} 1-p, & 0 \le p \le 1 \\ 0, & p>1 \end{cases}$

Hence deduce that $\int_0^\infty \frac{\sin^2 t}{t^2}\, dt = \frac{\pi}{2}$.

Sol. Let
$$\int_0^\infty F(x)\cos px\, dx = f(p), \quad \text{then } f(p) = \begin{cases} 1-p, & 0 \le p \le 1 \\ 0, & p>1 \end{cases}$$

$\therefore$
$$\sqrt{\frac{2}{\pi}}\int_0^\infty F(x)\cos px\, dx = \sqrt{\frac{2}{\pi}}\, f(p) = \bar{f}(p)$$

By inversion formula for Fourier cosine transform, we have

$$F(x) = \sqrt{\frac{2}{\pi}}\int_0^\infty \bar{f}(p)\cos px\, dp = \frac{2}{\pi}\int_0^\infty f(p)\cos px\, dp$$

$$= \frac{2}{\pi}\left[\int_0^1 (1-p)\cos px\, dp + \int_1^\infty 0\,.\cos px\, dp\right]$$

$$= \frac{2}{\pi}\left[(1-p)\,.\frac{\sin px}{x} - (-1)\,.\frac{-\cos px}{x^2}\right]_0^1 = \frac{2}{\pi}\left[-\frac{\cos x}{x^2} + \frac{1}{x^2}\right] = \frac{2(1-\cos x)}{\pi x^2}.$$

Deduction. Since $\int_0^\infty F(x)\cos px\, dx = \begin{cases} 1-p, & 0 \le p \le 1 \\ 0, & p>1 \end{cases}$ where $F(x) = \frac{2(1-\cos x)}{\pi x^2}$

$\therefore$
$$\frac{2}{\pi}\int_0^\infty \frac{1-\cos x}{x^2}\,.\cos px\, dx = \begin{cases} 1-p, & 0 \le p \le 1 \\ 0, & p>1 \end{cases}$$

When $p = 0$, we have $\frac{2}{\pi}\int_0^\infty \frac{1-\cos x}{x^2}\, dx = 1$ or $\int_0^\infty \frac{2\sin^2\frac{x}{2}}{x^2}\, dx = \frac{\pi}{2}$

Putting $x = 2t$ so that $dx = 2dt$, we get $\int_0^\infty \frac{\sin^2 t}{t^2}\, dt = \frac{\pi}{2}$.

Example 6. *Find the Fourier sine transform of*

$$f(x) = \begin{cases} x, & \text{for } 0 < x < 1 \\ 2 - x, & \text{for } 1 < x < 2 \\ 0, & \text{for } x > 2 \end{cases}.$$

Sol. Fourier transform of $f(x)$ is

$$F_s\{f(x)\} = \int_0^\infty f(x) \sin sx \, dx$$

$$= \int_0^1 f(x) \sin sx \, dx + \int_1^2 f(x) \sin sx \, dx + \int_2^\infty f(x) \sin sx \, dx$$

$$= \int_0^1 x \sin sx \, dx + \int_1^2 (2 - x) \sin sx \, dx + \int_2^\infty 0 \, dx$$

$$= \left\{ x . \frac{-\cos sx}{s} - 1 . \frac{-\sin sx}{s^2} \right\}_0^1 + \left\{ (2 - x) . \frac{-\cos sx}{s} - (-1) . \frac{-\sin sx}{s^2} \right\}_1^2$$

$$= \left(-\frac{\cos s}{s} + \frac{\sin s}{s^2} \right) + \left(-\frac{\sin 2s}{s^2} + \frac{\cos s}{s} + \frac{\sin s}{s^2} \right)$$

$$= \frac{2 \sin s - \sin 2s}{s^2} = \frac{2 \sin s - 2 \sin s \cos s}{s^2} = \frac{2 \sin s (1 - \cos s)}{s^2}.$$

Example 7. *Find the Fourier sine transform of* $\dfrac{e^{-ax}}{x}$.

Sol. Fourier sine transform of $f(x) = \dfrac{e^{-ax}}{x}$ is

$$F_s\{f(x)\} = \int_0^\infty f(x) \sin sx \, dx$$

$$= \int_0^\infty \frac{e^{-ax}}{x} \sin sx \, dx = I \text{ (say)} \qquad \ldots(1)$$

Differentiating w.r.t. s, we have

$$\frac{dI}{ds} = \int_0^\infty \frac{\partial}{\partial s} \left(\frac{e^{-ax}}{x} \sin sx \right) dx$$

$$= \int_0^\infty \frac{e^{-ax}}{x} . x \cos sx \, dx = \int_0^\infty e^{-ax} \cos sx \, dx$$

$$= \left[\frac{e^{-ax}}{a^2 + s^2} (-a \cos sx + s \sin sx) \right]_0^\infty$$

$$= 0 - \frac{1}{a^2 + s^2} (-a) = \frac{a}{s^2 + a^2}$$

Integrating w.r.t. s, we get

$$I = \tan^{-1} \frac{s}{a} + c \qquad \ldots(2)$$

Now, when $s = 0$, from (1), $I = 0$

$\therefore$ From (2), $\quad 0 = 0 + c \;\Rightarrow\; c = 0$

$$\therefore \qquad I = \tan^{-1}\frac{s}{a} \quad \text{or} \quad F_s\left\{\frac{e^{-ax}}{x}\right\} = \tan^{-1}\frac{s}{a}.$$

Example 8. *Find the Fourier cosine transform of* e^{-x^2}.

Sol. Fourier cosine transform of e^{-x^2} is given by

$$F_c\{e^{-x^2}\} = \int_0^\infty e^{-x^2} \cos sx \, dx = I \text{ (say)} \qquad \text{...(1)}$$

Differentiating w.r.t. s, we have

$$\frac{dI}{ds} = -\int_0^\infty x e^{-x^2} \sin sx \, dx = \frac{1}{2}\int_0^\infty (\sin sx)(-2x\, e^{-x^2})\, dx$$

$$= \frac{1}{2}\left[\left\{\sin sx\, e^{-x^2}\right\}_0^\infty - s\int_0^\infty \cos sx\, e^{-x^2}\, dx\right]$$

(Integrating by parts)

$$= -\frac{s}{2}\int_0^\infty e^{-x^2} \cos sx\, dx = -\frac{s}{2} I$$

or

$$\frac{dI}{I} = -\frac{s}{2}\, ds$$

Integrating, we have $\log I = -\dfrac{s^2}{4} + \log A$ or $I = A\, e^{-\frac{s^2}{4}}$ ...(2)

Now when $s = 0$, from (1), $I = \displaystyle\int_0^\infty e^{-x^2}\, dx = \frac{\sqrt{\pi}}{2}$

$\therefore$ From (2), $\quad \dfrac{\sqrt{\pi}}{2} = A$

Hence $\quad I = F_c\{e^{-x^2}\} = \dfrac{\sqrt{\pi}}{2} e^{-\frac{s^2}{4}}$.

Note. $\quad I = \displaystyle\int_0^\infty e^{-x^2} dx \quad$ Put $x^2 = t$ *i.e.,* $x = \sqrt{t}$

$$= \int_0^\infty \frac{e^{-t}}{2\sqrt{t}}\, dt = \frac{1}{2}\int_0^\infty t^{-1/2} e^{-t} dt = \frac{1}{2}\Gamma\left(\frac{1}{2}\right) = \frac{\sqrt{\pi}}{2}.$$

Example 9. *Find the Fourier cosine transform of* $f(x) = \dfrac{1}{1+x^2}$.

Hence derive Fourier sine transform of $\phi(x) = \dfrac{1}{x(1+x^2)}$.

Sol. Fourier cosine transform of $\quad f(x) = \dfrac{1}{1+x^2}$ is

$$F_c\{f(x)\} = \int_0^\infty \frac{\cos sx}{1+x^2}\, dx = I \quad \text{(say)} \qquad \text{...(1)}$$

Differentiating w.r.t. s, we have

$$\frac{dI}{ds} = \int_0^{\infty} \frac{-x \sin sx}{1+x^2}\,dx = \int_0^{\infty} \frac{-x^2 \sin sx}{x(1+x^2)}\,dx$$

$$= \int_0^{\infty} \frac{[1-(1+x^2)] \sin sx}{x(1+x^2)}\,dx = \int_0^{\infty} \frac{\sin sx}{x(1+x^2)}\,dx - \int_0^{\infty} \frac{\sin sx}{x}\,dx$$

$$= \int_0^{\infty} \frac{\sin sx}{x(1+x^2)}\,dx - \frac{\pi}{2} \qquad \text{...(2)}$$

Differentiating again w.r.t. s, we have

$$\frac{d^2I}{ds^2} = \int_0^{\infty} \frac{x \cos sx}{x(1+x^2)}\,dx = \int_0^{\infty} \frac{\cos sx}{1+x^2}\,dx = I$$

$\Rightarrow \qquad (D^2 - 1)\, I = 0, \qquad \text{where} \qquad D = \dfrac{d}{ds}$

Its A.E. is $\qquad D^2 - 1 = 0 \qquad \text{whence} \qquad D = \pm 1$

$$\therefore \qquad I = c_1 e^{s} + c_2 e^{-s} \qquad \text{...(3)}$$

$$\frac{dI}{ds} = c_1 e^{s} - c_2 e^{-s} \qquad \text{...(4)}$$

When $\quad s = 0$, from (1), $I = \displaystyle\int_0^{\infty} \frac{dx}{1+x^2} = \left[\tan^{-1} x\right]_0^{\infty} = \frac{\pi}{2}$

From (3), $\qquad I = c_1 + c_2$

$$\therefore \qquad c_1 + c_2 = \frac{\pi}{2} \qquad \text{...(5)}$$

When $\quad s = 0, \quad$ from (2), $\quad \dfrac{dI}{ds} = -\dfrac{\pi}{2}$

From (4), $\qquad \dfrac{dI}{ds} = c_1 - c_2$

$$\therefore \qquad c_1 - c_2 = -\frac{\pi}{2} \qquad \text{...(6)}$$

Solving (5) and (6), $\quad c_1 = 0, \quad c_2 = \dfrac{\pi}{2}$

$$\therefore \qquad I = \frac{\pi}{2} e^{-s} \Rightarrow F_c\{f(x)\} = \frac{\pi}{2} e^{-s}$$

Putting the values of c_1 and c_2 in (4)

$$\frac{dI}{ds} = -\frac{\pi}{2} e^{-s}$$

$$\therefore \quad \text{From (2),} \qquad -\frac{\pi}{2} e^{-s} = \int_0^{\infty} \frac{\sin sx}{x(1+x^2)}\,dx - \frac{\pi}{2}$$

$$\Rightarrow \qquad \int_0^{\infty} \frac{\sin sx}{x(1+x^2)}\,dx = \frac{\pi}{2}\,(1 - e^{-s})$$

$$\Rightarrow \qquad F_s\{\phi(x)\} = \frac{\pi}{2}\,(1 - e^{-s}).$$

Example 10. *Find the Fourier transform of* $f(x) = \begin{cases} 1-x^2, & \text{if } |x|<1 \\ 0, & \text{if } |x|>1 \end{cases}$

and use it to evaluate $\int_0^\infty \left(\frac{x\cos x - \sin x}{x^3}\right)\cos\frac{x}{2}\,dx.$

Sol. Fourier transform of $f(x)$ is given by

$$F\{f(x)\} = \int_{-\infty}^{\infty} f(x)e^{isx}\,dx$$

$$= \int_{-\infty}^{-1} f(x)\,e^{isx}dx + \int_{-1}^{1} f(x)\,e^{isx}\,dx + \int_{1}^{\infty} f(x)\,e^{isx}\,dx$$

$$= \int_{-\infty}^{-1} 0\,dx + \int_{-1}^{1} (1-x^2)\,e^{isx}\,dx + \int_{1}^{\infty} 0\,dx$$

$$= \left[(1-x^2).\frac{e^{isx}}{is} - (-2x)\frac{e^{isx}}{(is)^2} + (-2)\frac{e^{isx}}{(is)^3}\right]_{-1}^{1}$$

$$= \left[-\frac{2}{s^2}(e^{is}+e^{-is}) + \frac{2}{is^3}(e^{is}-e^{-is})\right]$$

$$= \left[-\frac{4}{s^2}\left(\frac{e^{is}+e^{-is}}{2}\right) + \frac{4}{s^3}\left(\frac{e^{is}-e^{-is}}{2i}\right)\right]$$

$$= 4\left[-\frac{\cos s}{s^2} + \frac{\sin s}{s^3}\right] = -\frac{4}{s^3}\,(s\cos s - \sin s)$$

By inversion formula for Fourier transform, we have

$$f(x) = \frac{1}{2\pi}\int_{-\infty}^{\infty} F\{f(x)\}\,.\,e^{-isx}\,ds$$

$$= -\frac{2}{\pi}\int_{-\infty}^{\infty}\left(\frac{s\cos s - \sin s}{s^3}\right)(\cos sx - i\sin sx)\,ds$$

$$= -\frac{2}{\pi}\int_{-\infty}^{\infty}\left(\frac{s\cos s - \sin s}{s^3}\right)\cos sx\,ds + \frac{2i}{\pi}\int_{-\infty}^{\infty}\left(\frac{s\cos s - \sin s}{s^3}\right)\sin sx\,ds$$

$$= -\frac{4}{\pi}\int_{0}^{\infty}\left(\frac{s\cos s - \sin s}{s^3}\right)\cos sx\,ds$$

(Since the integrand in the first integral is even and that in the second integral is odd)

$$\Rightarrow \int_0^\infty \left(\frac{s\cos s - \sin s}{s^3}\right)\cos sx\,ds = -\frac{\pi}{4}f(x) = \begin{cases} -\frac{\pi}{4}(1-x^2), & \text{if } |x|<1 \\ 0, & \text{if } |x|>1 \end{cases}$$

Putting $x = \frac{1}{2}$, we have $\int_0^\infty \left(\frac{s\cos s - \sin s}{s^3}\right)\cos\frac{s}{2}\,ds = -\frac{\pi}{4}\left(1-\frac{1}{4}\right) = -\frac{3\pi}{16}$

Hence $\int_0^\infty \left(\frac{x\cos x - \sin x}{x^3}\right)\cos\frac{x}{2}\,dx = -\frac{3\pi}{16}.$

Example 11. *Solve the integral equation* $\int_0^\infty f(x) \cos px \, dx = \begin{cases} 1-p, & 0 \le p \le 1 \\ 0, & p > 1 \end{cases}$

Hence deduce that $\displaystyle\int_0^\infty \frac{\sin^2 t}{t^2} dt = \frac{\pi}{2}$.

Sol. Here $\displaystyle\int_0^\infty f(x) \cos px \, dx = F_c(p)$

$$\therefore \quad F_c(p) = \begin{cases} 1-p, & 0 \le p \le 1 \\ 0, & p > 1 \end{cases}$$

By inversion formula for Fourier cosine transform, we have

$$f(x) = \frac{2}{\pi} \int_0^\infty F_c(p) \cos px \, dp$$

$$= \frac{2}{\pi} \left[\int_0^1 F_c(p) \cos px \, dp + \int_1^\infty F_c(p) \cos px \, dp \right]$$

$$= \frac{2}{\pi} \left[\int_0^1 (1-p) \cos px \, dp + \int_1^\infty 0 \, dp \right] \qquad \text{(Integrating by parts)}$$

$$= \frac{2}{\pi} \left[(1-p) . \frac{\sin px}{x} - (-1) . \frac{-\cos px}{x^2} \right]_0^1 = \frac{2}{\pi} \left[-\frac{\cos x}{x^2} + \frac{1}{x^2} \right] = \frac{2(1-\cos x)}{\pi x^2}.$$

Deduction. Since $\displaystyle\int_0^\infty f(x) \cos px \, dx = \begin{cases} 1-p, & 0 \le p \le 1 \\ 0, & p > 1 \end{cases}$, where $f(x) = \dfrac{2(1-\cos x)}{\pi x^2}$

$$\therefore \quad \frac{2}{\pi} \int_0^\infty \left(\frac{1-\cos x}{x^2} \right) \cos px \, dx = \begin{cases} 1-p, & 0 \le p \le 1 \\ 0, & p > 1 \end{cases}$$

When $p = 0$, we have $\displaystyle\frac{2}{\pi} \int_0^\infty \frac{1-\cos x}{x^2} dx = 1$ or $\displaystyle\int_0^\infty \frac{2 \sin^2 \frac{x}{2}}{x^2} dx = \frac{\pi}{2}$

Putting $\dfrac{x}{2} = t$ so that $dx = 2dt$, we have $\displaystyle\int_0^\infty \frac{\sin^2 t}{t^2} = \frac{\pi}{2}$.

Example 12. *Find the finite Fourier sine transform of*

$$f(x) = \begin{cases} \dfrac{2k}{l} x, & 0 \le x \le \dfrac{l}{2} \\ \dfrac{2k}{l}(l-x), & \dfrac{l}{2} \le x \le l \end{cases}.$$

Sol. Finite Fourier sine transform of $f(x)$ in $0 \le x \le l$ is

$$F_s(n) = \int_0^l f(x) \sin \frac{n\pi x}{l} dx$$

$$= \int_0^{l/2} f(x) \sin \frac{n\pi x}{l} dx + \int_{l/2}^l f(x) \sin \frac{n\pi x}{l} dx$$

$$= \int_0^{l/2} \frac{2k}{l} x \sin \frac{n\pi x}{l} dx + \int_{l/2}^l \frac{2k}{l} (l-x) \sin \frac{n\pi x}{l} dx$$

$$= \frac{2k}{l} \int_0^{l/2} x \sin \frac{n\pi x}{l} dx + \frac{2k}{l} \int_{l/2}^l (l-x) \sin \frac{n\pi x}{l} dx \qquad \text{(Integrating by parts)}$$

$$= \frac{2k}{l}\left[x \cdot \frac{-\cos\frac{n\pi x}{l}}{\frac{n\pi}{l}} - 1 \cdot \frac{-\sin\frac{n\pi x}{l}}{\left(\frac{n\pi}{l}\right)^2}\right]_0^{l/2} + \frac{2k}{l}\left[(l-x) \cdot \frac{-\cos\frac{n\pi x}{l}}{\frac{n\pi}{l}} - (-1) \cdot \frac{-\sin\frac{n\pi x}{l}}{\left(\frac{n\pi}{l}\right)^2}\right]_{l/2}^{l}$$

$$= \frac{2k}{l}\left[-\frac{l^2}{2n\pi}\cos\frac{n\pi}{2} + \frac{l^2}{n^2\pi^2}\sin\frac{n\pi}{2}\right] + \frac{2k}{l}\left[\frac{l^2}{2n\pi}\cos\frac{n\pi}{2} + \frac{l^2}{n^2\pi^2}\sin\frac{n\pi}{2}\right]$$

$$= \frac{2k}{l} \cdot \frac{2l^2}{n^2\pi^2}\sin\frac{n\pi}{2} = \frac{4kl}{n^2\pi^2}\sin\frac{n\pi}{2}.$$

Example 13. *Find the Fourier transform of* $f(x) = \begin{cases} 1 & \text{for } |x| < 1 \\ 0 & \text{for } |x| > 1 \end{cases}$.

Hence evaluate $\int_0^\infty \frac{\sin x}{x}\,dx$.

Sol. By the definition of complex Fourier transform of $f(x)$, we have for $s \neq 0$,

$$F[f(x)] = \int_{-\infty}^{\infty} e^{isx} f(x)\,dx = \int_{-1}^{1} e^{isx} \cdot 1\,dx = \left[\frac{e^{isx}}{is}\right]_{-1}^{1}$$

$$= \frac{e^{is} - e^{is}}{is} = 2\,\frac{\sin s}{s}.$$

Thus $\quad F[f(x)] = F(s) = 2\,\frac{\sin s}{s},\ s \neq 0.$

For $s = 0$, we have

$$F[f(x)] = F(s) = \int_{-1}^{1} dx = 2$$

Now by the inversion formula, we get

$$f(x) = \frac{1}{2\pi}\int_{-\infty}^{\infty} F(s)\,e^{-isx}\,ds$$

or $$\frac{1}{2\pi}\int_{-\infty}^{\infty} \frac{2\sin s}{s} e^{-isx}\,ds = \begin{cases} 1 & \text{for } |x| < 1 \\ 0 & \text{for } |x| > 1 \end{cases}$$

Putting $x = 0$, we get

$$\int_{-\infty}^{\infty} \frac{\sin s}{s}\,ds = \pi$$

or $$2\int_0^{\infty} \frac{\sin s}{s}\,ds = \pi, \text{ since the integrand is even}$$

$$\therefore \quad \int_0^{\infty} \frac{\sin s}{s}\,ds = \frac{\pi}{2}.$$

Since variables are dummy in any definite integral, hence on changing s to x, we get

$$\int_0^{\infty} \frac{\sin x}{x}\,dx = \frac{\pi}{2}.$$

Example 14. *Find the Fourier transform of f(x) =* $\begin{cases} 1, |x| < a \\ 0, |x| > a \end{cases}$.

Hence evaluate $\int_{-\infty}^{\infty} \frac{\sin \lambda a \cos \lambda x}{\lambda} d\lambda$.

Sol. By the definition of complex Fourier transform of $f(x)$, we have for $s \neq 0$,

$$F[f(x)] = \int_{-\infty}^{\infty} e^{isx} f(x)\, dx = \int_{-a}^{a} e^{isx} . 1\, dx = \left[\frac{e^{isx}}{is}\right]_{-a}^{a} = \frac{e^{isa} - e^{-isa}}{is}.$$

Thus $\quad F[f(x) = F(s) = 2 \frac{\sin sa}{s}, s \neq 0.$

For $s = 0$, we have

$$F[f(x)] = F(s) = \int_{-a}^{a} dx = 2a .$$

Now by the inversion formula, we get

$$f(x) = \frac{1}{2\pi} \int_{-\infty}^{\infty} F(s) e^{-isx} ds$$

or $$\frac{1}{2\pi} \int_{-\infty}^{\infty} 2 \frac{\sin sa}{s} e^{-isx} dx = \begin{cases} 1, |x| < a \\ 0, |x| > a \end{cases}.$$

But $$\text{L.H.S.} = \frac{1}{\pi} \int_{-\infty}^{\infty} \frac{\sin sa \cos sx}{s} ds - \frac{i}{\pi} \int_{-\infty}^{\infty} \frac{\sin sa \sin sx}{s} ds$$

$$= \frac{1}{\pi} \int_{-\infty}^{\infty} \frac{\sin sa \cos sx}{s} ds,$$

since the integrand in the second integral is an odd function

$$\therefore \quad \int_{-\infty}^{\infty} \frac{\sin sa \cos sx}{s} ds = \begin{cases} \pi, |x| < a \\ 0, |x| > a \end{cases}.$$

Since variables are dummy in any definite integral, hence on changing s to λ, we get

$$= \int_{-\infty}^{\infty} \frac{\sin \lambda a \cos \lambda x}{\lambda} d\lambda = \begin{cases} \pi, |x| < a \\ 0, |x| > a \end{cases}.$$

Example 15. *Find the Fourier sine transform of* $e^{-|x|}$. *Hence show that*

$$\int_0^{\infty} \frac{x \sin mx}{1 + x^2} dx = \frac{\pi e^{-m}}{2}, m > 0.$$

Sol. Since x is positive in the interval $(0, \infty)$, hence we find that

$$e^{-|x|} = e^{-x}, x \in (0, \infty).$$

$\therefore$ Fourier sine transform of $f(x) = e^{-|x|}$ is given by

$$F_s [f(x)] = \int_0^{\infty} f(x) \sin sx\, dx = \int_0^{\infty} e^{-x} \sin sx\, dx$$

$$= \left[\frac{e^{-x}}{1 + s^2} (-\sin sx - s \cos sx)\right]_0^{\infty} = \frac{s}{1 + s^2}$$

Using inversion formula for Fourier sine transforms, we get

$$f(x) = \frac{2}{\pi}\int_0^\infty F_s\,[f(x)\sin sx\,ds$$

or $$e^{-x} = \frac{2}{\pi}\int_0^\infty \frac{s}{1+s^2}\sin sx\,ds.$$

Now changing x to m, we get

$$e^{-m} = \frac{2}{\pi}\int_0^\infty \frac{s\sin ms}{1+s^2}\,ds$$

$$= \frac{2}{\pi}\int_0^\infty \frac{x\sin mx}{1+x^2}\,dx, \text{ since variable are dummy in a definite integral.}$$

Hence $$\int_0^\infty \frac{x\sin mx}{1+x^2}\,dx = \frac{\pi e^{-m}}{2}$$

Example 26. *Find the Fourier cosine transform of* $f(x) = \dfrac{1}{1+x^2}$.

Hence derive Fourier sine transform of $\phi(x) = \dfrac{x}{1+x^2}$

Sol. By the definition of Fourier cosine transform of $f(x)$, we have

$$F_c = \{f(x)\} = \int_0^\infty f(x)\cos sx\,dx = \int_0^\infty \frac{\cos sx}{1+x^2}\,dx = I, \text{ say} \qquad ...(i)$$

$$\therefore \quad \frac{dI}{ds} = \int_0^\infty \frac{-x\sin sx}{1+x^2}\,dx = -\int_0^\infty \frac{x^2\sin sx}{x(1+x^2)} \qquad ...(ii)$$

$$= -\int_0^\infty \frac{[(1+x^2)-1]\sin sx}{x(1+x^2)}\,dx = -\int_0^\infty \frac{\sin sx}{x}\,dx + \int_0^\infty \frac{\sin sx}{x(1+x^2)}\,dx$$

or $$\frac{dI}{ds} = -\frac{\pi}{2} + \int_0^\infty \frac{\sin sx}{x(1+x^2)}\,dx \qquad ...(iii)$$

$$\therefore \quad \frac{d^2I}{ds^2} = \int_0^\infty \frac{x\cos sx}{x(1+x^2)}\,dx \quad \text{or} \quad \frac{d^2I}{ds^2} = \int_0^\infty \frac{\cos sx}{1+x^2}\,dx = I$$

or $$\frac{d^2I}{ds^2} - I = 0 \quad \text{or} \quad (D^2-1)\,I = 0, \text{ where } D = \frac{dI}{ds} \qquad ...(iv)$$

which is a linear differential equation with constant coefficients.

$\therefore$ The auxiliary equation of (*iv*) is

$D^2 - 1 = 0$; *i.e.*, $D = 1, -1$.

Hence, the general solution of (*iv*) is given by

$$I = c_1e^s + c_2e^{-s}. \qquad ...(v)$$

$$\therefore \quad \frac{dI}{ds} = c_1e^s - c_2e^{-s}. \qquad ...(vi)$$

When $s = 0$, (i) and (v) give

$$I(0) = \int_0^\infty \frac{dx}{1+x^2} = \left[\tan^{-1} x\right]_0^\infty = \frac{\pi}{2} \quad \text{and} \quad I(0) = c_1 + c_2$$

or $$c_1 + c_2 = \frac{\pi}{2}. \qquad ...(vii)$$

Also when $s = 0$, (iii) and (iv) give

$$\left(\frac{dI}{ds}\right)_{s=0} = -\frac{\pi}{2} + \int_0^\infty 0\, dx = -\frac{\pi}{2} \quad \text{and} \quad \left(\frac{dI}{ds}\right)_0 = c_1 - c_2$$

or $$c_1 - c_2 = -\frac{\pi}{2}.$$

Solving (vii) and $(viii)$, we obtain

$$c_1 = 0,\ c_2 = \frac{\pi}{2}.$$

Thus from (i) and (v), we have

$$F_c\{f(x)\} = I = \frac{\pi}{2} e^{-s}.$$

Now $$F_s[\phi(x)] = \int_0^\infty \frac{x \sin sx}{1+x^2} dx = -\frac{dI}{ds}, \text{ from } (ii)$$

$$= \frac{\pi}{2} e^{-s}, \text{ from } (vi), \text{ with } c_1 = 0,\ c_2 = \frac{\pi}{2}.$$

TEST YOUR KNOWLEDGE

1. Find the Fourier sine and cosine transform of the function $f(x) = e^{-x}$ and hence show that

$$\int_0^\infty \frac{x \sin mx}{1+x^2} dx = \frac{\pi}{2} e^{-m} \quad \text{and} \quad \int_0^\infty \frac{\cos mx}{1+x^2} dx = \frac{\pi}{2} e^{-m}.$$

2. Find the Fourier sine transform of

$(i)\ f(x) = \begin{cases} \sin x, & 0 < x < a \\ 0, & x > a \end{cases}$ $\qquad (ii)\ f(x) = \begin{cases} 0, & 0 \le x < a \\ x, & a \le x \le b \\ 0, & x > b \end{cases}$

3. Find the Fourier cosine transform of

$(i)\ f(x) = e^{-x} + e^{-2x}, x > 0$ $\qquad (ii)\ f(x) = e^{-\frac{x^2}{2}}$

$(iii)\ f(x) = \begin{cases} x, & \text{for } 0 < x < 1 \\ 2-x, & \text{for } 1 < x < 2 \\ 0, & \text{for } x > 2 \end{cases}$ $\qquad (iv)\ f(x) = 2e^{-5x}, + 5e^{-2x}$

4. Find the Fourier sine and cosine transforms of

$(i)\ f(x) = \begin{cases} 1, & \text{for } 0 \le x < a \\ 0, & \text{for } x \ge a \end{cases}$

$(ii)\ f(x) = e^{-ax}, a > 0$

[**Note.** Remember the results of Q. 8.]

5. Find the Fourier transforms of the following functions:

(*i*) $f(x) = \begin{cases} 1, & |x| < a \\ 0, & |x| > a \end{cases}$ (*ii*) $f(x) = \begin{cases} x, & |x| \le 4 \\ 0, & |x| > 4 \end{cases}$ [*A.M.I.E. (S) 2005*]

(*iii*) $f(x) = \begin{cases} x^2, & |x| < a \\ 0, & |x| > a \end{cases}$ (*iv*) $f(x) = e^{-|x|}$

(*v*) $f(x) = e^{-\frac{x^2}{2}}, -\infty < x < \infty.$

6. Find the Fourier transform of $f(x) = \begin{cases} 1 & \text{for } |x| < 1 \\ 0 & \text{for } |x| > 1 \end{cases}$

Hence evaluate $\int_0^\infty \frac{\sin x}{x} dx$.

7. Using inverse Fourier sine transform, find *f*(*x*) if

(*i*) $F_s(\lambda) = \frac{1}{\lambda} e^{-a\lambda}$ (*ii*) $F_s(\lambda) = \frac{\lambda}{1 + \lambda^2}$.

8. Find the finite Fourier sine and cosine transforms of the following functions:

(*i*) $f(x) = 2x, 0 \le x \le 4$ (*ii*) $f(x) = x(l - x), 0 \le x \le l$

(*iii*) $f(x) = x^2, 0 \le x \le 2$ (*iv*) $f(x) = a\left(1 - \frac{x}{l}\right), 0 \le x \le l$

(*v*) $f(x) = \begin{cases} kx, & 0 \le x \le \frac{\pi}{2} \\ k(\pi - x), & \frac{\pi}{2} \le x \le \pi \end{cases}$.

9. Find the finite Fourier sine transform of the following functions:

(*i*) $f(x) = \cos x, 0 \le x \le \pi$ (*ii*) $f(x) = \begin{cases} \frac{2x}{3}, & 0 \le x \le \frac{\pi}{3} \\ \frac{\pi - x}{3}, & \frac{\pi}{3} \le x \le \pi \end{cases}$.

10. Find the finite Fourier cosine transform of the following functions:

(*i*) $f(x) = \sin x, 0 \le x \le \pi$ (*ii*) $f(x) = \begin{cases} kx, & 0 \le x \le \frac{l}{2} \\ k(l - x), & \frac{l}{2} \le x \le l \end{cases}$.

11. Solve the following integral equations:

(*i*) $\int_0^\infty f(x) \sin tx \, dx = \begin{cases} 1, & 0 \le t < 1 \\ 2, & 1 \le t < 2 \\ 0, & t \ge 2 \end{cases}$ (*ii*) $\int_0^\infty f(x) \cos \lambda x \, dx = e^{-\lambda}, \lambda > 0$

(*iii*) $\int_0^\infty f(x) \sin \lambda x \, dx = \begin{cases} 1 - \lambda, & 0 \le \lambda \le 1 \\ 0, & \lambda \ge 1 \end{cases}$ (*iv*) $\int_0^\infty f(x) \sin \lambda x \, dx = \frac{\lambda}{\lambda^2 + k^2}$.

Answers

2. (*i*) $\frac{1}{2}\left[\frac{\sin(1 - s)a}{1 - s} - \frac{\sin(1 + s)a}{1 + s}\right]$ (*ii*) $\frac{a \cos sa - b \cos sb}{s} + \frac{\sin sb - \sin sa}{s^2}$

3. (*i*) $\frac{6 + 3s^2}{4 + 5s^2 + s^4}$ (*ii*) $\sqrt{\frac{\pi}{2}} e^{-\frac{s^2}{2}}$

(*iii*) $\frac{2\cos s(1-\cos s)}{s^2}$ (*iv*) $10\left(\frac{1}{s^2+25}+\frac{1}{s^2+4}\right)$

4. (*i*) $\frac{1-\cos as}{s}, \frac{\sin as}{s}$ (*ii*) $\frac{s}{s^2+a^2}, \frac{a}{s^2+a^2}$

5. (*i*) $\frac{2\sin as}{s}$ (*ii*) $2i\left(\frac{as\cos as-\sin as}{s^2}\right)$

(*iii*) $\frac{2}{s^3}\,[(a^2s^2-2)\sin as+2as\cos as]$ (*iv*) $\frac{2}{1+s^2}$

(*v*) $\sqrt{2\pi}\, e^{-\frac{s^2}{2}}$ **6.** $\frac{2\sin s}{s}$

7. (*i*) $f(x)=\frac{2}{\pi}\tan^{-1}\frac{x}{a}$ (*ii*) $f(x)=e^{-x}$

8. (*i*) $-\frac{32}{n\pi}\cos n\pi=-\frac{32(-1)^n}{n\pi}\,;\,\frac{32[(-1)^n-1]}{n^2\pi^2},\ F_c(0)=16$

(*ii*) $\frac{2l^3}{n^3\pi^3}[1-(-1)^n]\,;\,-\frac{l^3}{n^2\pi^2}[(-1)^n+1],\ F_c(0)=\frac{l^3}{6}$

(*iii*) $\frac{8}{n^3\pi^3}[(-1)^n-2]-\frac{8}{n\pi}(-1)^n\,;\,\frac{16}{n^2\pi^2}(-1)^n,\ F_c(0)=\frac{8}{3}$

(*iv*) $\frac{al}{n\pi}\,;\,\frac{al}{n^2\pi^2}[1-(-1)^n],\ F_c(0)=\frac{al}{2}$

(*v*) $\frac{2k}{n^2}\sin\frac{n\pi}{2}\,;\,\frac{k}{n^2}\left[2\cos\frac{n\pi}{2}-(-1)^n-1\right],\ F_c(0)=\frac{k\pi^2}{4}$

9. (*i*) $\frac{n}{n^2-1}[1+(-1)^n]$ (*ii*) $\frac{1}{n^2}\sin\frac{n\pi}{3}$

10. (*i*) $-\frac{1}{n^2-1}\,[1+(-1)^n],\ F_c(0)=2$ (*ii*) $\frac{kl^2}{n^2\pi^2}\left(\cos\frac{n\pi}{2}-1\right),\ F_c(0)=\frac{kl^2}{4}$

11. (*i*) $f(x)=\frac{2}{\pi x}\,(1+\cos x-2\cos 2x)$ (*ii*) $f(x)=\frac{2}{\pi(1+x^2)}$

(*iii*) $f(x)=\frac{2}{\pi}\left(\frac{x-\sin x}{x^2}\right)$ (*iv*) $f(x)=e^{-kx},\ x>0.$

1.18. PROPERTIES OF FOURIER TRANSFORMS

1. Linearity Property. *If F(s) and G(s) are Fourier transforms of f(x) and g(x) respectively, then F[af(x) + bG(x)] = aF(s) + bG(s) where a and b are constants.*

Proof. By definition of Fourier transform, we have

$$F(s)=F\{f(x)\}=\int_{-\infty}^{\infty} f(x)\,.\,e^{isx}\,dx$$

and

$$G(s)=F\{g(x)\}=\int_{-\infty}^{\infty} g(x)\,.\,e^{isx}\,dx$$

$\therefore \quad F[af(x) + bg(x)] = \int_{-\infty}^{\infty} [af(x) + bg(x)]\, e^{isx} dx$

$$= a\int_{-\infty}^{\infty} f(x)\,.\,e^{isx} dx + b\int_{-\infty}^{\infty} g(x)\,.\,e^{isx}\,dx$$

$$= aF(s) + bG(s).$$

Cor. (*i*) If $F_s(s)$ and $G_s(s)$ are the Fourier sine transforms of $f(x)$ and $g(x)$ respectively, then $F_s[af(x) + bg(x)] = aF_s(s) + bG_s(s)$ where a and b are constants.

(*ii*) If $F_c(s)$ and $G_c(s)$ are the Fourier cosine transforms of $f(x)$ and $g(x)$ respectively, then $F_c[af(x) + bg(x)] = aF_c(s) + bG_c(s)$ where a and b are constants.

2. Change of scale property. *If F(s) is the complex Fourier transform of f(x), then*

$$\mathbf{F[f(ax)] = \frac{1}{a}F\left(\frac{s}{a}\right), a \neq 0.}$$

Proof. By definition of complex Fourier transform, we have

$$F(s) = F\{f(x)\} = \int_{-\infty}^{\infty} f(x)\,.\,e^{isx}\,dx \qquad ...(1)$$

$$\therefore \quad F[f(ax)] = \int_{-\infty}^{\infty} f(ax)\,.\,e^{isx}\,dx \qquad ...(2)$$

Putting $ax = t$ *i.e.*, $x = \frac{t}{a}$, we have $dx = \frac{dt}{a}$.

When $x \to -\infty$, $t \to -\infty$ and when $x \to \infty$, $t \to \infty$

$\therefore$ From (2) we have

$$F[f(ax)] = \int_{-\infty}^{\infty} f(t)\,e^{ist/a}\,.\,\frac{dt}{a}$$

$$= \frac{1}{a}\int_{-\infty}^{\infty} f(t)\,e^{i(s/a)t}\,dt = \frac{1}{a}F\left(\frac{s}{a}\right) \qquad \text{[by (2)]}$$

Cor. (*i*) If $F_s(s)$ is the Fourier sine transform of $f(x)$, then

$$F_s[f(ax)] = \frac{1}{a}F_s\left(\frac{s}{a}\right).$$

(*ii*) If $F_c(s)$ is the Fourier cosine transform of $f(x)$, then

$$F_c[f(ax)] = \frac{1}{a}F_c\left(\frac{s}{a}\right).$$

3. Shifting Property. *If F(s) is the complex Fourier transform of f(x), then*

$$\mathbf{F[f(x - a)] = e^{isa}F(s).}$$

Proof. By definition of complex Fourier transform, we have

$$F(s) = F\{f(x)\} = \int_{-\infty}^{\infty} f(x)\,.\,e^{isx}\,dx$$

$$\therefore \quad F[f(x - a)] = \int_{-\infty}^{\infty} f(x - a)\,e^{isx}\,dx \qquad ...(1)$$

Putting $x - a = t$ *i.e.*, $x = a + t$, we have $dx = dt$

When $x \to -\infty$, $t \to -\infty$ and when $x \to \infty$, $t \to \infty$

$\therefore$ From (1) we have

$$F[f(x-a)] = \int_{-\infty}^{\infty} f(t)\, e^{is(a+t)}\, dt = \int_{-\infty}^{\infty} f(t) . e^{isa} . e^{ist}\, dt$$

$$= e^{isa} \int_{-\infty}^{\infty} f(t) . e^{ist}\, dt = e^{isa}\, F(s) .$$

4. Modulation Theorem. *If F(s) is the complex Fourier transform of f(x), then*

$$\mathbf{F[f(x) \cos ax] = \frac{1}{2} [F(s+a) + F(s-a)].}$$

Proof. By definition of complex Fourier transform, we have

$$F(s) = F\{f(x)\} = \int_{-\infty}^{\infty} f(x)\, e^{isx}\, dx$$

$$\therefore \qquad F[f(x) \cos ax] = \int_{-\infty}^{\infty} f(x) \cos ax . e^{isx}\, dx$$

$$= \int_{-\infty}^{\infty} f(x) \left(\frac{e^{iax} + e^{-iax}}{2} \right) e^{isx}\, dx$$

$$= \frac{1}{2} \left[\int_{-\infty}^{\infty} f(x)\, e^{i(s+a)x}\, dx + \int_{-\infty}^{\infty} f(x) . e^{i(s-a)x}\, dx \right]$$

$$= \frac{1}{2} \; [F(s+a) + F(s-a)].$$

Cor. If $F_s(s)$ and $F_c(s)$ are Fourier sine and cosine transforms of $f(x)$ respectively, then

(*i*) $F_s[f(x) \cos ax] = \frac{1}{2} [F_s(s+a) + F_s(s-a)]$

(*ii*) $F_s[f(x) \sin ax] = \frac{1}{2} [F_c(s-a) - F_c(s+a)]$

(*iii*) $F_c[f(x) \cos ax] = \frac{1}{2} [F_c(s+a) + F_c(s-a)]$

(*iv*) $F_c[f(x) \sin ax] = \frac{1}{2} [F_s(s+a) - F_s(s-a)]$.

1.19. FOURIER TRANSFORMS OF THE DERIVATIVES OF A FUNCTION

Let $\bar{u}(s, t)$ be the Fourier transform of the function $u(x, t)$.

Then $$\bar{u}(s, t) = \frac{1}{\sqrt{2\pi}} \int_{-\infty}^{\infty} u(x, t)\, e^{isx}\, dx$$

Suppose u and $\frac{\partial u}{\partial x}$ both vanish as $x \to \pm\infty$. Then the Fourier transform of $\frac{\partial^2 u}{\partial x^2}$ is given by

$$F\left\{\frac{\partial^2 u}{\partial x^2}\right\} = \frac{1}{\sqrt{2\pi}}\int_{-\infty}^{\infty} \frac{\partial^2 u}{\partial x^2} e^{isx}\, dx$$

$$= \frac{1}{\sqrt{2\pi}}\left[\left\{e^{isx}\frac{\partial u}{\partial x} - ise^{isx}\,.\,u\right\}_{-\infty}^{\infty} + \int_{-\infty}^{\infty} (is)^2 e^{isx}\,.\,u\,dx\right]$$

$$= -\frac{s^2}{\sqrt{2\pi}}\int_{-\infty}^{\infty} ue^{isx}\,dx = -s^2\,\overline{u}$$

If $\overline{u}_s\,(s, t)$ and $\overline{u}_c\,(s, t)$ be the Fourier sine and cosine transforms of $u(x, t)$, then

$$\overline{u}_s(s, t) = \sqrt{\frac{2}{\pi}}\int_0^{\infty} u \sin sx\, dx \quad \text{and} \quad \overline{u}(s, t) = \sqrt{\frac{2}{\pi}}\int_0^{\infty} u \cos sx\, dx$$

The Fourier sine transform of $\frac{\partial^2 u}{\partial x^2}$ is given by

$$F_s\left\{\frac{\partial^2 u}{\partial x^2}\right\} = \sqrt{\frac{2}{\pi}}\int_0^{\infty} \frac{\partial^2 u}{\partial x^2}\sin sx\, dx$$

$$= \sqrt{\frac{2}{\pi}}\left[\left\{\sin sx\frac{\partial u}{\partial x} - s\cos sx\,.\,u\right\}_0^{\infty} - \int_0^{\infty} s^2 \sin sx\,.\,u\,dx\right]$$

$$= \sqrt{\frac{2}{\pi}}\left[s\,\{u\}_{x=0} - s^2\int_0^{\infty} u\sin sx\,dx\right] = \sqrt{\frac{2}{\pi}}\,[s\,\{u\}_{x=0} - s^2\,\overline{u}_s]$$

Similarly, we have

$$F_c\left\{\frac{\partial^2 u}{\partial x^2}\right\} = \sqrt{\frac{2}{\pi}}\left[-\left\{\frac{\partial u}{\partial x}\right\}_{x=0} - s^2\,\overline{u}_c\right].$$

1.20. FOURIER TRANSFORM OF DERIVATIVES

We have already seen that,

$$F\{f^n(x)\} = (-is)^n\, F(s)$$

(*i*) $\therefore\quad F\left(\frac{\partial^2 u}{\partial x^2}\right) = (-is)^2\, F\{u(x)\} = -s^2\,\overline{u}$, where $\overline{u}$ is Fourier transform of u *w.r.t.x.*

(*ii*) $F_c\{f'(x)\} = -\sqrt{\frac{2}{\pi}}\, f(o) + sF_s(s)$

$$\text{L.H.S.} = \sqrt{\frac{2}{\pi}}\int_0^{\infty} f'(x)\,.\,\cos sx\,dx = \sqrt{\frac{2}{\pi}}\int_0^{\infty}\cos sx\, d\,\{f(x)\}$$

$$= \sqrt{\frac{2}{\pi}}\left[\{f(x)\cos sx\}_0^{\infty} + s\int_0^{\infty} f(x)\sin sx\,dx\right]$$

$$= s\,F_s(s) - \sqrt{\frac{2}{\pi}}\,f(0) \text{ assuming } f(x) \to 0 \text{ as } x \to \infty$$

(*iii*) $$F_s\{f'(x)\} = \sqrt{\frac{2}{\pi}}\int_0^\infty \sin sx \, d\,[f(x)]$$

$$= \sqrt{\frac{2}{\pi}}\left[\{f(x)\sin sx\}_0^\infty - s\int_0^\infty f(x)\cos sx\,dx\right] = -\,s\,f_c(s)$$

(*iv*) $$F_c\{f''(x)\} = \sqrt{\frac{2}{\pi}}\int_0^\infty \cos sxd\,f'(x)]$$

$$= \sqrt{\frac{2}{\pi}}\left[\{f'(x)\cos sx\}_0^\infty + s\int_0^\infty f'(x)\sin sx\,dx\right]$$

$$= -\sqrt{\frac{2}{\pi}}\,f'(0) + sF_s\{f'(x)\}$$

$$= -\,s^2\,F_c(s) - \sqrt{\frac{2}{\pi}}\,f'(0) \text{ assuming } f(x), f'(x) \to 0 \text{ as } x \to \infty$$

(*v*) $$F_c\{f'\}(x) = \sqrt{\frac{2}{\pi}}\left[\int_0^\infty \sin sxd\,[f'(x)]\right]$$

$$= \sqrt{\frac{2}{\pi}}\left[[(f'(x)\sin sx]_0^\infty - s\int_0^\infty f'(x)\cos sx\,dx\right]$$

$$= -\,s\,F_c\,\{f'(x)\} = -s\left[s\,F_s(s) - \sqrt{\frac{2}{\pi}}\,f(0)\right]$$

$$= -\,s^2\,F_s(s) + \sqrt{\frac{2}{\pi}}\,s\,f(0) \text{ assuming } f(x), f'(x) \to 0 \text{ as } x \to \infty.$$

1.21. APPLICATION OF FOURIER TRANSFORMS TO BOUNDARY VALUE PROBLEMS

Application of Fourier transforms to boundary value problems can be illustrated through the following examples.

ILLUSTRATIVE EXAMPLES

Example 1. *Solve the equation* $\dfrac{\partial u}{\partial t} = \dfrac{\partial^2 u}{\partial x^2}$, $x > 0, t > 0$

subject to the conditions

(i) $u = 0$, *when* $x = 0, t > 0$ *(ii)* $u = \begin{cases} 1, & 0 < x < 1 \\ 0, & x \geq 1 \end{cases}$ *when* $t = 0$

and (iii) $u(x, t)$ *is bounded.*

Sol. Since $(u)_{x=0}$ is given, taking Fourier sine transform of both sides of the given equation, we have

$$\sqrt{\frac{2}{\pi}}\int_0^\infty \frac{\partial u}{\partial t}\sin sx\,dx = \sqrt{\frac{2}{\pi}}\int_0^\infty \frac{\partial^2 u}{\partial x^2}\sin sx\,dx$$

or $$\sqrt{\frac{2}{\pi}}\frac{\partial}{\partial t}\int_0^\infty u\sin sx\,dx = \sqrt{\frac{2}{\pi}}\,[s(u)_{x=0} - s^2\,\overline{u}_s]$$

or $$\frac{d\overline{u}_s}{dt} + s^2\,\overline{u}_s = \sqrt{\frac{2}{\pi}}\,s(u)_{x=0}$$

or $$\frac{d\overline{u}_s}{dt} + s^2\,\overline{u}_s = 0 \quad \text{since } u = 0, t > 0 \text{ when } x = 0, t > 0$$

Its solution is $\overline{u}_s = Ae^{-s^2 t}$, where A is a constant ...(1)

Now, when $t = 0$, Fourier sine transform of $u(x, t)$

i.e., $$\overline{u}_s(s,0) = \sqrt{\frac{2}{\pi}}\int_0^\infty u(x,0)\sin sx\,dx = \sqrt{\frac{2}{\pi}}\int_0^1 1.\sin sx\,dx + \sqrt{\frac{2}{\pi}}\int_1^\infty 0.\sin sx\,dx$$

$$= \sqrt{\frac{2}{\pi}}\left[-\frac{\cos sx}{s}\right]_0^1 = \sqrt{\frac{2}{\pi}}\left(\frac{1-\cos s}{s}\right)$$

$\therefore$ From (1), $A = \overline{u}_s(s,0) = \sqrt{\frac{2}{\pi}}\left(\frac{1-\cos s}{s}\right)$ so that $\overline{u}_s = \sqrt{\frac{2}{\pi}}\left(\frac{1-\cos s}{s}\right)e^{-s^2 t}$

Applying the inverse Fourier sine transform, we have

$$u(x,t) = \frac{2}{\pi}\int_0^\infty \frac{1-\cos s}{s}\,e^{-s^2 t}\sin sx\,ds \quad \text{which is the required solution.}$$

Example 2. *If the initial temperature of an infinite bar is given by*

$$\theta(x) = \begin{cases} \theta_0 & \text{for } |x| < a \\ 0 & \text{for } |x| > a, \end{cases}$$

determine the temperature at any instant x and at any instant t.

Sol. To determine the temperature $\theta(x, t)$, we have to solve the heat-flow equation

$$\frac{\partial \theta}{\partial t} = c^2\frac{\partial^2 \theta}{\partial t^2}, \quad t > 0 \qquad ...(1)$$

subject to the initial condition $\theta(x,0) = \begin{cases} \theta_0 & \text{for } |x| < a \\ 0 & \text{for } |x| > a \end{cases}$ ...(2)

Taking Fourier transform of (1), we get

$$\frac{1}{\sqrt{2\pi}}\int_{-\infty}^\infty \frac{\partial \theta}{\partial t}e^{isx}\,dx = \frac{c^2}{\sqrt{2\pi}}\int_{-\infty}^\infty \frac{\partial^2 \theta}{\partial t^2}e^{isx}\,dx$$

or $$\frac{1}{\sqrt{2\pi}}\frac{d}{dt}\int_{-\infty}^\infty \theta\,e^{isx}\,dx = -c^2 s^2\overline{\theta}$$

or $$\frac{d\overline{\theta}}{dt} = -c^2 s^2\overline{\theta} \qquad ...(3)$$

Now taking the Fourier transform of (2), we get

$$\overline{\theta}(s,0)=\frac{1}{\sqrt{2\pi}}\int_{-\infty}^{\infty}\theta(x,0)\,e^{isx}\,dx=\frac{1}{\sqrt{2\pi}}\int_{-a}^{a}\theta_0\,e^{isx}\,dx=\frac{\theta_0}{\sqrt{2\pi}}\left[\frac{e^{isx}}{is}\right]_{-a}^{a}$$

$$=\frac{\theta_0}{\sqrt{2\pi}}\left[\frac{e^{isa}-e^{-isa}}{is}\right]=\frac{2\theta_0}{s\sqrt{2\pi}}\left[\frac{e^{isa}-e^{-isa}}{2i}\right]$$

$$=\frac{2\theta_0\sin sa}{s\sqrt{2\pi}}$$

From (3), $\frac{d\overline{\theta}}{\overline{\theta}}=-\,c^2s^2\,dt$ ...(4)

Integrating $\log\overline{\theta}=-\,c^2s^2t+\log A$ or $\overline{\theta}=Ae^{-c^2s^2t}$

Since $\overline{\theta}=\frac{2\theta_0\sin sa}{s\sqrt{2\pi}}$ when $t=0$, from (4), we get $A=\frac{2\theta_0\sin sa}{s\sqrt{2\pi}}$

$$\therefore\qquad \overline{\theta}=\frac{2\theta_0\sin sa}{s\sqrt{2\pi}}\,e^{-c^2s^2t}$$

Taking its inverse Fourier transform, we get

$$\theta(x,t)=\frac{1}{2\pi}\int_{-\infty}^{\infty}\frac{2\theta_0\sin as}{s}\,.\,e^{-c^2s^2t}\,.\,e^{-sx}\,ds$$

$$=\frac{\theta_0}{\pi}\int_{-\infty}^{\infty}\frac{\sin as}{s}\,.\,e^{-c^2s^2t}\,(\cos xs-i\sin xs)\,ds$$

$$=\frac{\theta_0}{\pi}\left[\int_{-\infty}^{\infty}\frac{\sin as}{s}\,.\,e^{-c^2s^2t}\cos xs\,ds-i\int_{-\infty}^{\infty}\frac{\sin as}{s}\,.\,e^{-c^2s^2t}\sin xs\,ds\right]$$

$$=\frac{\theta_0}{\pi}\int_{-\infty}^{\infty}\frac{\sin as}{s}\,.\,e^{-c^2s^2t}\cos xs\,ds$$

(The second integral vanishes since its integrand is an odd function)

$$=\frac{2\theta_0}{\pi}\int_{0}^{\infty}\frac{\sin as}{s}\,.\,e^{-c^2s^2t}\cos xs\,ds=\frac{\theta_0}{\pi}\int_{0}^{\infty}\frac{e^{-c^2s^2t}}{s}\,.\,2\sin as\cos xs\,ds$$

$$=\frac{\theta_0}{\pi}\int_{0}^{\infty}e^{-c^2s^2t}\left(\frac{\sin(a+x)s+\sin(a-x)s}{s}\right)ds$$

$$=\frac{\theta_0}{\pi}\int_{0}^{\infty}e^{-v^2}\left\{\sin\frac{(a+x)v}{c\sqrt{t}}+\sin\frac{(1-x)v}{c\sqrt{t}}\right\}\frac{dv}{v}\quad\text{where } v^2=c^2s^2t$$

$$=\frac{\theta_0}{2}\left\{erf\left(\frac{a+x}{2c\sqrt{t}}\right)+erf\left(\frac{a-x}{2c\sqrt{t}}\right)\right\}.$$

1.22. FOURIER TRANSFORMS OF THE DERIVATIVES OF A FUNCTION

The Fourier transform of the function $u(x, t)$ is given by

$$F[u(x, t) = \int_{-\infty}^{\infty} u e^{isx}\, dx$$

Then the Fourier transform of $\partial^2 u/\partial x^2$, *i.e.*

$$F\left[\frac{\partial^2 u}{\partial x^2}\right] = \int_{-\infty}^{\infty} \frac{\partial^2 u}{\partial x^2} e^{isx}\, dx = \left| e^{isx} \frac{\partial u}{\partial x} - is\, e^{isx} . u \right|_{-\infty}^{\infty} + (is)^2 \int_{-\infty}^{\infty} u e^{isx}\, dx.$$

on applying the general rule of integration by parts. If u and $\frac{\partial u}{\partial x}$ tend to zero as x tends to $\pm\infty$, then

$$F\left[\frac{\partial^2 u}{\partial x^2}\right] = -s^2 F[u] \qquad ...(1)$$

Similarly, in the case of Fourier sine and cosine transforms, we have

$$F_s\left[\frac{\partial^2 u}{\partial x^2}\right] = s(u)_{x=0} - s^2 F_s[u] \qquad ...(2)$$

and

$$F_c\left[\frac{\partial^2 u}{\partial x^2}\right] = -\left(\frac{\partial u}{\partial x}\right)_{x=0} - s^2 F_c[u] \qquad ...(3)$$

In general, *the Fourier transform of the nth derivative of f(x) is given by*

$$F\left[\frac{d^n f}{dx^n}\right] = (-is)^n F[f(x)] \qquad ...(4)$$

Provided the first $n - 1$ derivative vanish as $x \to \pm\infty$.

For

$$F[f^n(x)] = \int_{-\infty}^{\infty} f^n(x) e^{isx}\, dx$$

$$= \left| e^{isx} f^{n-1} - ise^{isx} f^{n-2} + (is)^2 e^{isx} f^{n-3} - ... \right|_{-\infty}^{\infty}$$

$$+ (-is)^n \int_{-\infty}^{\infty} f . e^{isx}\, dx$$

by the general rule of integration by parts, whence follows (4).

ILLUSTRATIVE EXAMPLES

Example 1. *Solve* $\partial u\partial t = 2\partial^2 u/\partial x^2$, *if* $u(0, t) = 0$, $u(x, 0) = e^{-x}$ $(x > 0)$, $u(x, t)$ *is bounded where* $x > 0, t > 0$.

Sol. Given $\partial u/\partial t = 2\partial^2 u/\partial x^2$, $x > 0$, $t > 0$ ...(*i*)

with boundary conditions: $u(0, t) = 0$, $u(x, t)$ is bounded ...(*ii*)

and initial condition $u(x, 0) = e^{-x}$, $x > 0$ ...(*iii*)

Since $u(0, t)$ is given, we take Fourier sine transform of both sides of (*i*) so that

$$\int_0^\infty \frac{\partial u}{\partial t} \sin px\, dx = 2\int_0^\infty \frac{\partial^2 u}{\partial x^2} \sin px\, dx$$

or $$\frac{d}{dt}\int_0^\infty u(x,t) \sin px\, dx = 2\left[\left|\frac{\partial u}{\partial x} \sin px\right|_0^\infty - \int_0^\infty \frac{\partial u}{\partial x}\, .\, p\cos px\, dx\right] \text{ (Integrating by parts)}$$

or $$\frac{d\bar{u}_s}{dt} = -2p\int_0^\infty \frac{\partial u}{\partial x} \cos px\, dx, \text{ if } \frac{\partial u}{\partial x} \to 0 \text{ as } x \to \infty$$

where $\bar{u}_s(p,t) = \int_0^\infty u(x,t) \sin px\, dx$

$$= -2p\left[\,|\,u(x,t)\cos px\,|_0^\infty - \int_0^\infty u(x,t) - (-p\sin px)\, dx\right]$$

[Again integrating by parts]

$$= -2p\left[0 - u(0,t)\, p\int_0^\infty u(x,t)\sin px\, dx\right]$$

[$\because$ $u(x, t) \to 0$ as $x \to \infty$ by (*ii*)]

$$= 2pu\,(0,t) - 2p^2\bar{u}_s$$

or $$\frac{d\bar{u}_s}{dt} = -2p^2\bar{u}_s \qquad \text{[by (ii)]}$$

Integrating $\int \frac{d\bar{u}_s}{\bar{u}_s} - \log c = -2p^2\int dt$ or $\log \bar{u}_s - \log c = -2p^2 t$

$$\therefore \qquad \bar{u}_s(p,t) = ce^{-2p^2 t} \qquad ...(iv)$$

Taking Fourier sine transform of both sides of (*iii*), we get

$$\int_0^\infty u(x,0)\sin px\, dx = \int_0^\infty e^{-x}\sin px\, dx$$

or $$\bar{u}_s(p,0) = \left|\frac{e^{-x}}{1+p^2}(-\sin px - p\cos px)\right|_0^\infty = \frac{p}{1+p^2} \qquad ...(v)$$

Putting $t = 0$ in (*iv*) and using (*v*), we obtain $p/(1 + p^2) = c$.

Thus (*iv*) becomes $\bar{u}_s(p,t) = \frac{p}{1+p^2}e^{-2p^2 t}$

Now taking inverse fourier sine transform, we get

$$u(x,t) = \frac{2}{\pi}\int_0^\infty \frac{pe^{-2p^2 t}}{1+p^2}\sin px\, dp.$$

Example 2. *A bar of length a is at zero temperature. At t = 0, the end x = a is suddenly raised to temperature u_0 and the end x = 0 is insulated. Find the temperature at any point x of the bar at any time t > 0, assuming that the surface of the bar is insulated.*

Sol. Here we have to solve the differential equation

$$\frac{\partial u}{\partial t} = c^2\frac{\partial^2 u}{\partial x^2} \quad (0 < x < a,\ t > 0) \qquad ...(i)$$

subject to the conditions

$$u(x, 0) = 0 \quad ...(ii); \quad u_x(0, t) = 0 \quad ...(iii) \quad \text{and } u(a, t) = u_0 \qquad ...(iv)$$

The Laplace transform of (*i*), if L $[u(x, t)] = \bar{u}(x, s)$, is

$$s\bar{u} - u(x, 0) = c^2 \frac{d^2\bar{u}}{dx^2}$$

Using (*ii*), we get $\dfrac{d^2\bar{u}}{dx^2} - \dfrac{s}{c^2}\bar{u} = 0$ ...(*v*)

Similarly, the Laplace transforms of (*iii*) and (*iv*) are

$$\bar{u}_x(0, s) = 0 \quad ...(vi); \quad \bar{u}(a, s) = \frac{u_0}{s} \qquad ...(vii)$$

Solving (*v*), we have $\bar{u} = C_1 e^{x\sqrt{s}/c} + C_2 e^{-x\sqrt{s}/c}$

Using (*vi*), we find $C_1 = C_2$ so that

$$\bar{u} = C_1(e^{\sqrt{s}x/c} + e^{-\sqrt{s}x/c}) = 2C_1 \cosh(\sqrt{s}x/c)$$

Now using (*vii*), we have

$$\bar{u} = \frac{u_0 \cosh(\sqrt{s}x/c)}{s \cosh(\sqrt{s}a/c)}$$

By the inversion formula, we get

$u(x, t)$ = sum of the residues of $\left(\dfrac{e^{st} . u_0 \cosh(\sqrt{s}x/c}{s \cosh(\sqrt{s}a/c)}\right)$ at all the poles which occur at $s = 0$

and $\cosh(\sqrt{s}a/c) = 0$ *i.e.*, at $s = 0$, $\sqrt{s}a/c = \left(n - \dfrac{1}{2}\right)\pi i$, $n = 0, \pm 1, \pm 2, ...$

or at $s = 0$, $\quad s (= s_n) = -\dfrac{(2n-1)^2 c^2\pi^2}{-4a^2} = 0, 1, 2, ...$

$$\text{Now} \quad (\text{Res})_{s=0} = \underset{s \to 0}{\text{Lt}} \left\{ s . \frac{u_0 e^{st} \cosh(\sqrt{s}x/c)}{s \cosh(\sqrt{s}a/c)} \right\} = u_0$$

$$(\text{Res})_{s=s_n} = u_0 \underset{s \to s_n}{\text{Lt}} \left\{ (s - s_n) . \frac{u_0 e^{st} \cosh(\sqrt{s}x/c)}{s \cosh(\sqrt{s}a/c)} \right\}$$

$$= u_0 \underset{s \to s_n}{\text{Lt}} \left\{ \frac{s - s_n}{\cosh(\sqrt{s}a/c)} \right\} . \underset{s \to s_n}{\text{Lt}} \left\{ \frac{e^{st} \cosh(\sqrt{s}x/c)}{s} \right\} \qquad \left[\frac{0}{0}\text{ form}\right]$$

$$= u_0 \underset{s \to s_n}{\text{Lt}} \frac{1}{\sinh(\sqrt{s}a/c) . (a/2\sqrt{s}/c)} . \underset{s \to s_n}{\text{Lt}} \left\{ \frac{e^{st} \cosh(\sqrt{s}x/c)}{s} \right\}$$

$$= \frac{4u_0(-1)^n}{(2n-1)\pi} e^{-(2n-1)^2 \pi^2 c^2 t/4a^2} \cos\frac{(2n-1)\pi x}{2a}$$

Thus we get $u(x, t) = u_0 + \frac{4u_0}{\pi} \sum_{n=1}^{\infty} \frac{(-1)^n}{2n-1} e^{-(2n-1)^2 \pi^2 c^2 t/4a^2} \cos \frac{(2n-1)\pi x}{2a}$.

Vibrations of a string

Example 3. *An infinite string is initially at rest and that the initial displacement is f(x), $(-\infty < x < \infty)$. Determine the displacement y(x, t) of the string.*

Sol. The equation for the vibration of the string is

$$\partial^2 y/\partial t^2 = c^2 \partial^2 y/\partial x^2 \quad ...(i)$$

and the initial conditions are

$$(\partial y/\partial t)_{t=0} = 0; \quad y(x, 0) = f(x) \quad ...(ii)$$

Multiplying (*i*) by e^{isx} and integrating w.r.t. x from $-\infty$ to ∞, we get

$$\partial^2 Y/\partial t^2 = c^2(-s^2 Y) \text{ provided } y \text{ and } \partial y/\partial x \to 0 \text{ as } x \to \infty$$

$\therefore$ A solution of $d^2Y/dt^2 + c^2s^2Y = 0$ is $Y = A_1 \cos cst + A_2 \sin cst$ ...(*iii*)

Also Fourier transforms of (*ii*) are

$$\partial y/\partial t = 0 \quad \text{and} \quad Y = F(s) \quad \text{when } t = 0$$

Applying these to (*iii*), we get

$$A_2 = 0 \quad \text{and} \quad A_1 = F(s)$$

Thus $Y = F(s) \cos cst$

Now taking inverse Fourier transform, we get

$$y(x, t) = \frac{1}{2\pi} \int_{-\infty}^{\infty} F(s) \cos cst \, . \, e^{-isx} \, dx$$

$$= \frac{1}{2\pi} \int_{-\infty}^{\infty} F(s) \frac{e^{icst} + e^{-icst}}{2} . e^{-isx} \, dx$$

$$= \frac{1}{4\pi} \int_{-\infty}^{\infty} [F(s) e^{is(x-ct)} + F(s) e^{-is(x+ct)}] \, ds$$

$$= \frac{1}{2} [f(x-ct) + f(x+ct)] \qquad \left[\because f(x) = \frac{1}{2\pi} \int_{-\infty}^{\infty} F(s) e^{-isx} \, ds\right]$$

Example 4. *An infinitely long string having one end at x = 0, is initially at rest along the x-axis. The end x = 0 is given a transverse displacement f (t), t > 0. Find the displacement of any point of the string at any time.*

Sol. Let $y(x, t)$ be the transverse displacement of any point x of the string at any time t. Then we have to solve the wave equation

$$\frac{\partial^2 y}{\partial t^2} = c^2 \frac{\partial^2 y}{\partial x^2} \quad (x > 0, t > 0) \quad ...(i)$$

subject to the conditions $y(x, 0) = 0$, $y_t(x, 0) = 0$, $y(0, t) = f(t)$ and the displacement $y(x, t)$ is bounded.

The Laplace transform of (*i*), writing $L[y(x, t)] = \bar{y}(x, s)$ is

$$s^2 \bar{y} - sy(x, 0) - \frac{\partial y(x, 0)}{\partial t} = c^2 \frac{d^2 \bar{y}}{dx^2}$$

Using the first two conditions, we have

$$\frac{d^2 \bar{y}}{dx^2} = \left[\frac{s}{c}\right]^2 \bar{y} \quad ...(ii)$$

Similarly the Laplace transforms of the third and fourth conditions are

$$\bar{y}(0, s) = \bar{f}(s) \quad \text{at} \quad x = 0 \qquad ...(iii)$$

and $\bar{y}$ (x, s) is bounded. ...(iv)

Solving (ii), we get

$$\bar{y}(x, s) = C_1 e^{sx/c} + C_2 e^{-sx/c}$$

To satisfy condition (iv), we must have $C_1 = 0$.

Using the condition (iii), we get $C_2 = \bar{f}(s)$.

$$\therefore \qquad \bar{y}(x, s) = \bar{f}(s)\, e^{-sx/c}$$

Using the complex inversion formula, we obtain

$$y = \frac{1}{2\pi i}\int_{a-i\infty}^{a+i\infty} e^{(t-x/c)s}\,\bar{f}(s)\,ds = f(t - x/c).$$

Example 5. *A tightly stretched flexible string has its ends fixed at x = 0 and x = l. At time t = 0, the string is given a shape defined by F(x) = μx (l – x), where μ is a constant and then released. Find the displacement of any point x of the string at any time t > 0.*

Sol. We have to solve the wave equation

$$\frac{\partial^2 y}{\partial t^2} = c^2 \frac{\partial^2 y}{\partial x^2} \qquad (x > 0, t > 0)$$

subject to the conditions $y(0, t) = 0, y(l, t) = 0$

and $\qquad y(x, 0) = \mu x(l - x), y_t(x, 0) = 0$

Now taking Laplace transform, writing $L[y(x, t)] = \bar{y}(x, s)$, we get

$$s^2\bar{y} - s\bar{y}(x, 0) - \frac{\partial y(x, 0)}{\partial t} = c^2 \frac{d^2\bar{y}}{dx^2} \qquad ...(i)$$

where $\qquad \bar{y}(0, s) = 0, \; \bar{y}(l, s) = 0$ ...(ii)

$$\therefore \quad (i) \text{ reduces to } \frac{d^2\bar{y}}{dx^2} - \left(\frac{s}{c}\right)^2 \bar{y} = -\frac{\mu s x(l - x)}{c^2}$$

Its solution is $\qquad \bar{y}(x, s) = c_1 \cosh(sx/c) + c_2 \sinh(sx/c) + \dfrac{\mu x(l - x)}{s} - \dfrac{2c^2\mu}{s^3}$

Applying the conditions (ii), we get

$$c_1 = 2c^2\mu/s^2 \quad \text{and} \quad c_2 = \frac{2c^2\mu}{s^3}\left\{\frac{1 - \cosh(sl/c)}{\sinh(sl/c)}\right\} - \frac{2c^2\mu}{s^3}\tanh(s/2c)$$

Thus, $\qquad \bar{y}(x, s) = \dfrac{2c^2\mu}{s^3}\left[\dfrac{\cosh\{s(2x - l)/2c}{\cosh(sl/2c)}\right] + \dfrac{\mu x(l - x)}{s} - \dfrac{2c^2\mu}{s^3}$

Now using the inversion formula (3), we get

$y(x, t)$ = sum of the residues of

$$2c^2\mu\left[e^{st}\frac{\cosh\{s(2x - l)/2c}{s^3\cosh(sl/2c)}\right] \text{ at all the poles } + \mu x(l - x) - c^2\mu t^2 \text{, we have}$$

Sum of the residues of $2c^2\mu\left[\dfrac{e^{st}\cosh\{s(2x-l)/2c}{s^3\cosh sl/2c}\right]$ at all the poles

$$= c^2\mu\left[t^2+\left(\frac{2x-l}{2c}\right)^2-\left(\frac{l}{2c}\right)^2\right]$$

$$-\frac{32c^2\mu}{\pi^3}\left(\frac{l}{2c}\right)^2\sum_{n=1}^{\infty}\left[\frac{(-1)^n}{(2n-1)^3}\cos\left\{\frac{(2n-1)\pi(2x-l)}{2l}\right\}\cos\left\{\frac{(2n-1)\pi ct}{l}\right\}\right]$$

$$= c^2\mu t^2-\mu x(l-x)+\frac{8\mu l^2}{\pi^3}\sum_{n=1}^{\infty}\left[\frac{1}{(2n-1)^3}\sin\frac{(2n-1)\pi x}{l}\cos\frac{(2n-1)\pi ct}{l}\right]$$

Hence $y(x,t)=\dfrac{8\mu l^2}{\pi^2}\displaystyle\sum_{n=1}^{\infty}\left[\frac{1}{(2n-1)^3}\sin\frac{(2n-1)\pi x}{l}\cos\frac{(2n-1)\pi ct}{l}\right]$

Transmission lines

Example 6. *A transmission line of length l has negligible inductance and leakance. A constant voltage v_0 is applied at the sending end ($x = 0$) and is open circuited at the far end. Assuming the initial voltage and current to be zero, determine the voltage and current.*

Sol. For a transmission line with L = G = 0, the voltage v and current i are given by the equations

$$\frac{\partial^2 v}{\partial x^2}=\text{RC}\frac{\partial v}{\partial t}\quad\text{and}\quad\frac{\partial v}{\partial x}+\text{R}i=0 \qquad ...(i)$$

The boundary conditions are (for $t > 0$)

$$v = v_0 \text{ at } x = 0 \quad\text{and}\quad i=\frac{\partial v}{\partial x}=0 \text{ at } x = l \qquad ...(ii)$$

The initial condition is $v = 0$ at $t = 0$ $(x > 0)$

Laplace transforms of (*i*) and (*ii*) are

$$\frac{\partial^2\bar{v}}{\partial x^2}=\text{RC}\,(s\bar{v}-0) \qquad ...(iii)$$

and
$$\bar{v}=v_0/s \text{ at } x=0,\ \frac{\partial\bar{v}}{\partial x}=0 \text{ at } x=l \qquad ...(iv)$$

∴ The solution of (*iii*) is

$$\bar{v}=c_1\cosh\sqrt{(\text{RC}s)}\,x+c_2\sinh\sqrt{(\text{RC}s)}\,x$$

Applying conditions (*iv*), it gives

$$v_0/s=c_1,\ 0=c_1\sinh\sqrt{(\text{RC}s)}\,l+c_2\cosh\sqrt{(\text{RC}s)}\,l$$

$$\bar{v}=\frac{v_0}{s}\left[\cosh\sqrt{(\text{RC}s)}\,x-\frac{\sinh\sqrt{(\text{RC}s)}\,l}{\cosh\sqrt{(\text{RC}s)}\,l}\sinh\sqrt{(\text{RC}s)}\,x\right]$$

$$= \frac{v_0}{s} \frac{\cosh pq\sqrt{s}}{\cosh p\sqrt{s}} \text{ where } p = \sqrt{(\text{RC})}\, l \quad \text{and} \quad q = (l - x)/l$$

By the inversion formula (3), we get

$$v\,(x, t) = \text{sum of the residues of } (e^{st}\bar{v}) \text{ at all poles of } e^{st}\bar{v}. \quad \text{...}(iv)$$

These poles are at $\quad s = 0 \quad$ and $\quad p\sqrt{s} = \pm\, i\,(2n - 1)\,\pi/2 = \pm\, ipk$ (say)

Now $\quad \text{Res}\,(e^{st}\bar{v})_{s=0} = \underset{s \to 0}{\text{Lt}} \dfrac{se^{st}\, v_0 \cosh pq\sqrt{s}}{s \cosh p\sqrt{s}} = v_0$

and $\quad \text{Res}\,(e^{st}\bar{v})_{s=-k^2} = \underset{s \to -k^2}{\text{Lt}} \dfrac{(s + k^2)\, e^{st}\, v_0 \cosh pq\sqrt{s}}{s \cosh p\sqrt{s}} \quad \left(\dfrac{0}{0} \text{ form}\right)$

$$= \underset{s \to -k^2}{\text{Lt}} \frac{v_0 . e^{st} \cosh pq\sqrt{s} + (s + k^2)\,(\ldots\ldots)}{\cosh p\sqrt{s} + s \sinh p\sqrt{s} . \frac{1}{2} ps^{-1/2}}$$

$$= \frac{v_0 e^{-k^2 t} \cosh\,(ipqk) + 0}{0 + 1/2(ipk) \sinh\,(ipk)} = \frac{2v_0 e^{-k^2 t} \cos\,(pqk)}{-\, pk \sin pk}$$

Adding up all the residues, (iv) gives

$$u\,(x, t) = v_0 + \frac{4v_0}{\pi} \sum_{n=1}^{\infty} \frac{(-1)^n}{2n - 1} e^{-[(2n-1)^2 \pi^2 t / 4\text{RC}l^2]} \cos\,[(2n - 1)\,\pi\,(l - x)/2l]$$

$[\because\ pk = (2n - 1)\,\pi/2,\ -\sin pk = (-1)^n,\ pqk$

$= \dfrac{1}{2}\,(2n - 1)\,\pi\,(l - x)/l,\ k^2 = (2n - 1)^2\pi^2/4\text{RC}l^2]$

Also $\quad i = -\dfrac{1}{\text{R}} \dfrac{\partial v}{\partial x} \quad$ [by (i).]

Example 7. *Determine the distribution of temperature in the semi-infinite medium $x \geq 0$, when the end $x = 0$ is maintained at zero temperature and the initial distribution of temperature is $f(x)$.*

Sol. Let $u(x, t)$ be the temperature at any point x and at any time t. Then we have to solve the heat-flow equation

$$\frac{\partial u}{\partial t} = c^2 \frac{\partial^2 u}{\partial x^2} \;(x > 0, t > 0). \quad \text{...}(i)$$

Subject to initial condition,

I.C.: $\quad u(x, 0) = f(x) \quad$ ...(ii)

and the boundary condition,

B.C.: $\quad u(0, t) = 0. \quad$ ...(iii)

Taking Fourier sine transform of (*i*) and denoting $F_s[u(x, t)]$ by $\overline{u}_s$, we have

$$\frac{d\overline{u}_s}{dt} = c^2 F_s\left[\frac{\partial^2 u}{\partial x^2}\right] = c^2 [s(\overline{u})_{x=0} - s^2 F_s(\overline{u})]$$

$$= c^2 [su(0, t) - s^2 u_s] = -c^2 s^2 u_s, \qquad \text{[by } (iii)\text{]}$$

or $$\frac{d\overline{u}_s}{dt} + c^2 s^2 \overline{u}_s = 0. \qquad \ldots(iv)$$

Also the Fourier sine transform of (*ii*) is

$$\overline{u}_s = \bar{f}_s(s) \quad \text{at} \quad t = 0. \qquad \ldots(v)$$

Solving (*iv*), we have

$$\frac{d\overline{u}_s}{\overline{u}_s} = -c^2 s^2\, dt$$

where variables are separated. Hence, on integration, we get

$$\int \frac{d\overline{u}_s}{\overline{u}_s} - \log C = -c^2 s^2 \int dt$$

or $$\log \overline{u}_s - \log C = -c^2 s^2 t$$

or $$\overline{u}_s = Ce^{-c^2 s^2 t}$$

At $t = 0$, we have by (*v*) : $\overline{u}_s = \bar{f}_s(s)$.

$\therefore$ $$\bar{f}_s(s) = C.$$

Thus, we have $$\overline{u}_s = \bar{f}_s(s)\, e^{-c^2 s^2 t}.$$

Hence taking its inverse Fourier sine transform, we obtain

$$u(x, t) = \frac{2}{\pi} \int_0^{\infty} \bar{f}_s(s)\, e^{-c^2 s^2 t} \sin xs\, ds.$$

Example 8. *An infinite string is initially at rest and that the initial displacement is f (x) (– ∞ < x < ∞). Determine the displacement y(x, t) of the string.*

Sol. The equation for the vibration of string is

$$\frac{\partial^2 y}{\partial t^2} = c^2 \frac{\partial^2 y}{\partial x^2} \qquad \ldots(i)$$

and the given initial conditions are

I.C : $$\left(\frac{\partial y}{\partial t}\right)_{t=0} = 0;\ y(x, 0) = f(x). \qquad \ldots(ii)$$

Multiplying (*i*) by e^{isx} and integrating w.r.t. x from $-\infty$ to ∞, we get

$$\int_{-\infty}^{\infty} \frac{\partial^2 y}{\partial t^2} e^{isx}\, dx = c^2 \int_{-\infty}^{\infty} \frac{\partial^2 y}{\partial x^2} e^{isx}\, dx$$

or $$\frac{d^2}{dt^2}\int_{-\infty}^{\infty} y(x,t)e^{isx}\,dx = c^2 \text{F}\left[\frac{\partial^2 y}{\partial x^2}\right]$$

or $$\frac{d^2\text{Y}}{dt^2} = c^2\,(-\,s^2\text{Y}), \text{ where}$$

$$\text{Y} = \int_{-\infty}^{\infty} y(x,t)\,e^{isx}\,dx \text{ provided } y \text{ and } \frac{\partial y}{\partial x} \to 0 \text{ as } x \to \infty$$

or $$\frac{d^2\text{Y}}{dt^2} + c^2 s^2 \text{Y} = 0 \quad \text{or} \quad (\text{D}^2 + c^2 s^2)\,\text{Y} = 0 \qquad ...(iii)$$

which is a linear differential equation with constant coefficient.

$\therefore$ its A.E. is $\Delta^2 + c^2s^2 = 0$; *i.e.*, $\text{D} = \pm\, ics$.

Hence the general solution of (*iii*) is

$$\text{Y} = \text{A}_1 \cos cst + \text{A}_2 \sin cst. \qquad ...(iv)$$

Also Fourier transforms of (*ii*) are

$$\int_{-\infty}^{\infty} \frac{\partial y}{\partial t}\, e^{isx}\, dx = \int_{-\infty}^{\infty} 0\,.\,e^{isx}\, dx$$

and $$\int_{-\infty}^{\infty} ye^{isx}\,dx = \int_{-\infty}^{\infty} f(x)e^{isx}\,dx \text{ when } t = 0$$

or $$\frac{d}{dt}\int_{-\infty}^{\infty} ye^{isx}\,dx = 0 \quad \text{and} \quad \int_{-\infty}^{\infty} ye^{isx}\,dx = \text{F}\,[f(x)] \text{ when } t = 0$$

or $$\frac{d\text{Y}}{dt} = 0 \text{ and } \text{Y} = \text{F}(s) \text{ when } t = 0,$$

where F(s) denotes the Fourier transform of $f(x)$.

Applying these to (*iv*), we get

$$0 = \frac{d\text{Y}}{dt} = -\,cs\,\text{A}_1 \sin cst + cs\,\text{A}_2 \cos cst$$

and $$\text{F}(s) = \text{Y} = \text{A}_1 \cos st + \text{A}_2 \sin st \text{ when } t = 0$$

i.e., $$0 = cs\text{A}_2 \quad \text{and} \quad \text{F}(s) = \text{A}_1.$$

$\therefore$ $$\text{A}_2 = 0 \quad \text{and} \quad \text{A}_1 = \text{F}(s).$$

Thus, (*iv*) becomes $\text{Y} = \text{F}(s) \cos cst$.

Now taking inverse Fourier transforms, we get

$$y(x,t) = \frac{1}{2\pi}\int_{-\infty}^{\infty} \text{F}(s) \cos cst\,.\,e^{-isx}\,dx = \frac{1}{2\pi}\int_{-\infty}^{\infty} \text{F}(s)\frac{e^{icst} + e^{-icst}}{2}\,.\,e^{-isx}\,dx$$

$$= \frac{1}{4\pi}\int_{-\infty}^{\infty} [\text{F}(s)\,e^{-is(x-ct)} + \text{F}(s)\,e^{-is(x+ct)}]\,ds$$

$$= \frac{1}{2}\,[f(x-ct) + \text{F}(x+ct)], \text{ since } f(x) = \frac{1}{2\pi}\int_{-\infty}^{\infty} \text{F}(s)e^{-ist}\,ds$$

TEST YOUR KNOWLEDGE

Use the method of residues to find the inverse of the following Laplace transforms:

1. $\dfrac{1}{(s-2)(s^2+1)}$
2. $\dfrac{1}{s(s-2)^2}$
3. $\dfrac{1}{s^2(s^2-a^2)}$
4. $\dfrac{s^2+3}{(s+1)(s^2-2s+5)}$
5. $\dfrac{s^2-1}{(s^2+1)^2}$
6. $\dfrac{1+2s}{(s+2)^2(s-1)^2}$
7. $\dfrac{1}{(s^2+2s+5)^2}$.

Answers

1. $\dfrac{1}{5}(e^{2x}-2\sin x-\cos x)$
2. $\dfrac{1}{4}(1-e^{2x}+2xe^{2x})$
3. $(\sinh ax - ax)/a^2$
4. $\dfrac{1}{2}e^{-x}+\dfrac{1}{2}e^{x}(\cos x+\sin x)$
5. $x\cos x$
6. $\dfrac{1}{3}x(e^{x}-e^{-2x})$
7. $\dfrac{1}{16}e^{-x}(\sin 2x-2x\cos 2x)$.

2

Laplace Transforms and Z-Transforms

2.1. INTRODUCTION

A transformation is a mathematical device which converts one function into another. For example, when the differential operator D $\left(\equiv \dfrac{d}{dx}\right)$ operates on $f(x) = \sin x$, it gives a new function

$$g(x) = \text{D}\, f(x) = \cos x.$$

Laplace transform or Laplace transformation is widely used by scientists and engineers. It is particularly effective in solving linear differential equations—ordinary as well as partial. It reduces an ordinary differential equation into an algebraic equation.

Laplace transform directly gives the solution of differential equations with given initial conditions without the necessity of first finding the general solution and then evaluating the arbitrary constants.

2.2. DEFINITION

Let $f(t)$ be a function of t defined for all $t \geq 0$. Then the *Laplace transform* of $f(t)$, denoted by $\text{L}\{f(t)\}$, is defined by

$$\text{L}\{f(t)\} = \int_0^{\infty} e^{-st} f(t)\, dt$$

provided that the integral exists, 's' is a parameter which may be real or complex.

$\text{L}\{f(t)\}$ is clearly a function of s and is briefly written as $\bar{f}(s)$ *i.e.*, $\text{L}\{f(t)\} = \bar{f}(s)$.

2.3. LINEARITY PROPERTY

If c_1, c_2 are constants and f, g are functions of t, then

$$\text{L}\{c_1 f(t) + c_2 g(t)\} = c_1 \text{L}\{f(t)\} + c_2 \text{L}\{g(t)\}$$

By definition, $\text{L}\{c_1 f(t) + c_2 g(t)\} = \int_0^{\infty} e^{-st} \{c_1 f(t) + c_2 g(t)\}\, dt$

$$= c_1 \int_0^{\infty} e^{-st} f(t)\, dt + c_2 \int_0^{\infty} e^{-st} g(t)\, dt = c_1 \text{L}\{f(t)\} + c_2 \text{L}\{g(t)\}$$

The result can easily by generalised.

2.4. LAPLACE TRANSFORMS OF SOME ELEMENTARY FUNCTIONS

(1) $\mathbf{L\{1\} = \frac{1}{s}}$, $\mathbf{s > 0}$

$$L\{1\} = \int_0^\infty e^{-st} \,.\, 1\, dt = \left[-\frac{e^{-st}}{s} \right]_0^\infty = \frac{1}{s}, \text{ if } s > 0.$$

(2) $\mathbf{L\{t^n\} = \frac{n!}{s^{n+1}}}$, where n is a positive integer.

$$L\{t^n\} = \int_0^\infty e^{-st} \,.\, t^n \, dt = \int_0^\infty e^{-x} \left(\frac{x}{s}\right)^n \frac{dx}{s}, \text{ on putting } st = x$$

$$= \frac{1}{s^{n+1}} \int_0^\infty x^n e^{-x} \, dx = \frac{\Gamma(n+1)}{s^{n+1}}$$

provided that $s > 0$ and $n + 1 > 0$ *i.e.*, $n > -1$.

If n is a positive integer, $\Gamma(n + 1) = n\,!$

$$\therefore \qquad L\{t^n\} = \frac{n!}{s^{n+1}}.$$

(3) $\mathbf{L\{e^{at}\} = \frac{1}{s-a}}$, $s > a$

$$L\{e^{at}\} = \int_0^\infty e^{-st} \,.\, e^{at}\, dt = \int_0^\infty e^{-(s-a)t}\, dt = \left[-\frac{e^{-(s-a)t}}{s-a} \right]_0^\infty = \frac{a}{s-a}, \text{ if } s > a.$$

(4) $\mathbf{L\{\sin at\} = \frac{a}{s^2 + a^2}}$, $s > 0$

$$L\{\sin at\} = \int_0^\infty e^{-st} \sin at \, dt = \left[\frac{e^{-st}}{s^2 + a^2} (-s \sin at - a \cos at) \right]_0^\infty = \frac{a}{s^2 + a^2}.$$

(5) $\mathbf{L\{\cos at\} = \frac{s}{s^2 + a^2}}$, $s > 0$.

$$L\{\cos at\} = \int_0^\infty e^{-st} \,.\, \cos at \, dt = \left[\frac{e^{-st}}{s^2 + a^2} (-s \cos at + a \sin at) \right]_0^\infty = \frac{s}{s^2 + a^2}.$$

(6) $\mathbf{L\{\sinh at\} = \frac{a}{s^2 - a^2}}$, $s > |\, a \,|$.

$$L[\sinh at] = \int_0^\infty e^{-st} \sinh at \, dt = \int_0^\infty e^{-st} . \left[\frac{e^{at} - e^{-at}}{2} \right] dt$$

$$= \frac{1}{2} \left[\int_0^\infty e^{-(s-a)t} \, dt - \int_0^\infty e^{-(s+a)t} \, dt \right]$$

$$= \frac{1}{2} \left[\frac{1}{s-a} - \frac{1}{s+a} \right] = \frac{a}{s^2 - a^2}, \text{ for } s > |\, a \,|.$$

Note. We can also prove it by using linearity property.

Thus $L\{\sinh at\} = L\{\frac{1}{2}(e^{at} - e^{-at})\} = \frac{1}{2} L(e^{at}) - \frac{1}{2} L\{e^{-at}\}$

$$= \frac{1}{2}\left(\frac{1}{s-a}\right) - \frac{1}{2}\left(\frac{1}{s+a}\right) = \frac{a}{s^2 - a^2}$$

(7) $\mathbf{L\{\cosh at\} = \dfrac{s}{s^2 - a^2}}$, $s > |a|$.

$$L\{\cosh at\} = L\{\tfrac{1}{2}(e^{at} + e^{-at})\} = \tfrac{1}{2} L\{e^{at}\} + \frac{1}{2} L\{e^{-at}\} = \frac{1}{2}\left(\frac{1}{s-a}\right) + \frac{1}{2}\left(\frac{1}{s+a}\right) = \frac{s}{s^2 - a^2}.$$

2.5. (a) FIRST SHIFTING THEOREM

If $\mathbf{L\{f(t)\} = \bar{f}(s)}$, then $\mathbf{L\{e^{at} f(t)\} = \bar{f}(s - a)}$.

By definition, $L\{e^{at} f(t)\} = \int_0^\infty e^{-st} \cdot e^{at} f(t)\, dt = \int_0^\infty e^{-(s-a)t} f(t)\, dt$

$$= \int_0^\infty e^{-pt} f(t)\, dt, \text{ where } p = s - a$$

$$= \bar{f}(p) = \bar{f}(s - a)$$

Similarly, $L\{e^{-at} f(t)\} = \bar{f}(s + a)$

Applying this theorem to the elementary functions of Art. 2.4, we get the following useful results :

(1) $\mathbf{L\{e^{at} t^n\} = \dfrac{n!}{(s-a)^{n+1}}}$ $\qquad \because L(t^n) = \dfrac{n!}{s^{n+1}}$

(2) $\mathbf{L\{e^{at} \sin bt\} = \dfrac{b}{(s-a)^2 + b^2}}$ (3) $\mathbf{L\{e^{at} \cos bt\} = \dfrac{s-a}{(s-a)^2 + b^2}}$

(4) $\mathbf{L\{e^{at} \sinh bt\} = \dfrac{b}{(s-a)^2 - b^2}}$ (5) $\mathbf{L\{e^{at} \cosh bt\} = \dfrac{s-a}{(s-a)^2 - b^2}}$.

2.5. (b) CHANGE OF SCALE PROPERTY

If $L\{f(t)\} = \bar{f}(s)$, *then* $L\{f(at)\} = \dfrac{1}{a} \bar{f}\left(\dfrac{s}{a}\right)$.

Proof. $L\{f(at)\} = \int_0^\infty e^{-st} f(at)dt$ $\qquad$ put $at = u$ so that $dt = \dfrac{du}{a}$

$$= \int_0^\infty e^{-\frac{su}{a}} f(u) . \frac{du}{a}$$

$$= \frac{1}{a}\int_0^\infty e^{-\lambda u} f(u)\, du \quad \text{where } \lambda = \frac{s}{a}$$

$$= \frac{1}{a} \text{ L } \{f(u)\} = \frac{1}{a}\bar{f}(\lambda) = \frac{1}{a}\bar{f}\left(\frac{s}{a}\right).$$

ILLUSTRATIVE EXAMPLES

Example 1. *Find the Laplace transforms of*

(i) $\sin 2t \cos 3t$ *(ii)* $\sin^3 2t$

(iii) $\cosh^3 2t$.

Sol. (*i*) Since $\sin 2t \cos 3t = \frac{1}{2}(2 \cos 3t \sin 2t) = \frac{1}{2}(\sin 5t - \sin t)$

$$\therefore \quad \text{L } \{\sin 2t \cos 3t\} = \text{L } \{\tfrac{1}{2}(\sin 5t - \sin t)\} = \tfrac{1}{2}\,[\text{L } \{\sin 5t\} - \text{L } \{\sin t\}]$$

$$= \frac{1}{2}\left[\frac{5}{s^2+5^2} - \frac{1}{s^2+1^2}\right] = \frac{2(s^2-5)}{(s^2+25)(s^2+1)}$$

(*ii*) Since $\sin 6t = 3 \sin 2t - 4 \sin^3 2t$

$$\Rightarrow \quad \sin^3 2t = \frac{3}{4}\sin 2t - \frac{1}{4}\sin 6t$$

$$\therefore \quad \text{L } \{\sin^3 2t\} = \text{L } \{\frac{3}{4}\sin 2t - \frac{1}{4}\sin 6t\} = \frac{3}{4}\text{L } \{\sin 2t\} - \frac{1}{4}\text{L } \{\sin 6t\}$$

$$= \frac{3}{4}\cdot\frac{2}{s^2+2^2} - \frac{1}{4}\cdot\frac{6}{s^2+6^2} = \frac{48}{(s^2+4)(s^2+36)}.$$

(*iii*) Since $\cosh 6t = 4 \cosh^3 2t - 3 \cosh 2t$

$$\Rightarrow \quad \cosh^3 2t = \frac{3}{4}\cosh 2t + \frac{1}{4}\cosh 6t$$

$$\therefore \quad \text{L } \{\cosh^3 2t\} = \text{L } \{\frac{3}{4}\cosh 2t + \frac{1}{4}\cosh 6t\} = \frac{3}{4}\text{ L } \{\cosh 2t\} + \frac{1}{4}\text{ L } \{\cosh 6t\}$$

$$= \frac{3}{4}\cdot\frac{s}{s^2-2^2} + \frac{1}{4}\cdot\frac{s}{s^2-6^2} = \frac{s\,(s^2-28)}{(s^2-4)\,(s^2-36)}.$$

Example 2. *Find the Laplace transform of* $e^{-3t}(\cos 4t + 3 \sin 4t)$.

Sol. $\text{L } \{e^{-3t}(\cos 4t + 3\sin 4t)\} = \text{L } \{e^{-3t}\cos 4t\} + 3\text{L } \{e^{-3t}\sin 4t\}$

$$= \frac{s+3}{(s+3)^2+4^2} + 3.\frac{4}{(s+3)^2+4^2} = \frac{s+15}{s^2+6s+25}.$$

Example 3. *Find the Laplace transform of* $t^2 \sin at$.

Sol. Since $\text{L}\{t^2\} = \frac{2\,!}{s^3} = \frac{2}{s^3}$

$\therefore$ By first shifting theorem, we have

$$L\,\{t^2 e^{iat}\} = \frac{2}{(s-ia)^3} = \frac{2(s+ia)^3}{[(s-ia)(s+ia)]^3}$$

or $$L\,\{t^2(\cos at + i\sin at)\} = \frac{2\,[(s^3-3a^2s)+i(3as^2-a^3)]}{(s^2+a^2)^3}$$

Equating the imaginary parts on both sides, we get

$$L\{t^2\sin at\} = \frac{2a\,(3s^2-a^2)}{(s^2+a^2)^3}.$$

Example 4. *Find the Laplace transform of* $te^{-4t}\sin 3t$.

Sol. Since $$L\,\{t\} = \frac{1}{t^2}$$

$\therefore$ By first shifting theorem, we have

$$L\,\{\,te^{3it}\,\} = \frac{1}{(s-3i)^2} = \frac{(s+3i)^2}{[(s-3i)(s+3i)]^2}$$

or $$L\,\{t(\cos 3t + i\sin 3t\} = \frac{(s^2-9)+6is}{(s^2+9)^2}$$

Equating the imaginary parts on both sides, we get

$$L\,[t\sin 3t] = \frac{6s}{(s^2+9)^2}$$

Again applying the first shifting theorem, we have

$$L\{e^{-4t}\,.\,t\sin 3t\} = \frac{6(s+4)}{[(s+4)^2+9]^2} = \frac{6(s+4)}{(s^2+8s+25)^2}.$$

Example 5. *Find the Laplace transform of f(t) defined as*

$$f(t) = \begin{cases} \dfrac{t}{T}, & \text{when } 0<t<T \\ 1, & \text{when } t>T. \end{cases}$$

Sol. $$L\,\{f(t)\} = \int_0^\infty e^{-st} f(t)\,dt = \int_0^T e^{-st}\,.\,\frac{t}{T}\,dt + \int_T^\infty e^{-st}\,.\,1\,dt$$

$$= \frac{1}{T}\left[\left|t\,.\,\frac{e^{-st}}{-s}\right|_0^T - \int_0^T 1\,.\,\frac{e^{-st}}{-s}\,dt\right] + \left[\frac{e^{-st}}{-s}\right]_T^\infty$$

$$= \frac{1}{T}\left[-\frac{T}{s}e^{-sT} + \frac{1}{s}\left\{\frac{e^{-st}}{-s}\right\}_0^T\right] + \frac{e^{-sT}}{s}$$

$$= -\frac{e^{-sT}}{s} - \frac{e^{-sT}-1}{Ts^2} + \frac{e^{-sT}}{s} = \frac{1-e^{-sT}}{Ts^2}.$$

Example 6. *Find the Laplace transform of the function F(t), where*

$$F(t) = \begin{cases} \sin t, & 0 < t < \pi \\ 0\ , & t > \pi \end{cases}.$$

Sol. By definition,

$$L\{F(t)\} = \int_0^\infty e^{-st}\, F(t)\, dt = \int_0^\pi e^{-st} \sin t\, dt + \int_\pi^\infty e^{-st} . 0\, dt$$

$$= \left[\frac{e^{-st}}{s^2+1}(-s\sin t - \cos t)\right]_0^\pi + 0 = \frac{e^{-s\pi}+1}{s^2+1},\ s > 0.$$

Example 7. *Evaluate L{F(t)} if* $F(t) = \begin{cases} (t-1)^2, & t > 1 \\ 0\ , & 0 < t < 1. \end{cases}$

Sol. By definition,

$$L\{F(t)\} = \int_0^\infty e^{-st}\, F(t)\, dt = \int_0^1 e^{-st} . 0\, dt + \int_1^\infty e^{-st} . (t-1)^2\, dt$$

$$= 0 + \left[-\frac{(t-1)^2}{s} e^{-st}\right]_1^\infty + \frac{2}{s}\int_1^\infty (t-1)\, e^{-st}\, dt$$

$$= \frac{2}{s}\left[-\frac{(t-1)}{s} e^{-st} - \frac{e^{-st}}{s^2}\right]_1^\infty = \frac{2e^{-s}}{s^3},\ s > 0.$$

Example 8. *Find (i)* $L\{\sin\sqrt{t}\}$ *(ii)* $L\{J_0(t)\}$.

Sol. (*i*) Since

$$\sin\sqrt{t} = \sqrt{t} - \frac{(\sqrt{t})^3}{3!} + \frac{(\sqrt{t})^5}{5!} - \frac{(\sqrt{t})^7}{7!} + \ldots\ldots \infty = t^{1/2} - \frac{t^{3/2}}{3!} + \frac{t^{5/2}}{5!} - \frac{t^{7/2}}{7!} + \ldots\ldots \infty$$

$$\therefore \quad L\{\sin\sqrt{t}\} = L\{t^{1/2}\} - \frac{1}{3!} L\{t^{3/2}\} + \frac{1}{5!} L\{t^{5/2}\} - \frac{1}{7!} L\{t^{7/2}\} + \ldots\ldots$$

$$= \frac{\Gamma(3/2)}{s^{3/2}} - \frac{\Gamma(5/2)}{3!\, s^{5/2}} + \frac{\Gamma(7/2)}{5!\, s^{7/2}} - \frac{\Gamma(9/2)}{7!\, s^{9/2}} + \ldots\ldots$$

$$= \frac{\frac{1}{2}\sqrt{\pi}}{s^{3/2}} - \frac{1}{6} . \frac{\frac{3}{2} . \frac{1}{2} . \sqrt{\pi}}{s^{5/2}} + \frac{1}{120} . \frac{\frac{5}{2} . \frac{3}{2} . \frac{1}{2} . \sqrt{\pi}}{s^{7/2}} - \ldots\ldots$$

$$= \frac{\sqrt{\pi}}{2s^{3/2}}\left\{1 - \left(\frac{1}{4s}\right) + \frac{1}{2!}\left(\frac{1}{4s}\right)^2 - \frac{1}{3!}\left(\frac{1}{4s}\right)^3 + \ldots\ldots\right\} = \frac{\sqrt{\pi}}{2s^{3/2}} . e^{-1/4s}.$$

(*ii*) By definition,

$$J_0(t) = 1 - \frac{t^2}{2^2} + \frac{t^4}{2^2 . 4^2} - \frac{t^6}{2^2 . 4^4 . 6^2} + \ldots\ldots$$

$$\therefore \quad L\{J_0(t)\} = L\{1\} - \frac{1}{2^2} L\{t^2\} + \frac{1}{2^2 . 4^2} L\{t^4\} - \frac{1}{2^2 . 4^2 . 6^2} L\{t^6\} + \ldots\ldots$$

$$= \frac{1}{s} - \frac{1}{2^2} . \frac{2!}{s^3} + \frac{1}{2^2 . 4^4} . \frac{4!}{s^5} - \frac{1}{2^2 . 4^2 . 6^2} . \frac{6!}{s^7} + \ldots\ldots$$

$$= \frac{1}{s}\left\{1 - \frac{1}{2}\left(\frac{1}{s^2}\right) + \frac{1.3}{2.4}\left(\frac{1}{s^2}\right)^2 - \frac{1.3.5}{2.4.6}\left(\frac{1}{s^2}\right)^3 + \ldots\ldots\right\}$$

$$= \frac{1}{s}\left(1 + \frac{1}{s^2}\right)^{-1/2} = \frac{1}{\sqrt{1+s^2}}, \ s > 0.$$

Example 9. *Find $L\{t^3e^{-3t}\}$.*

Sol. We know that $L\{t^3\} = \frac{3!}{s^4} = \frac{6}{s^4}$.

$\therefore$ From first shifting theorem

$$L\{t^3e^{-3t}\} = \frac{6}{(s+3)^4}.$$

Example 10. *Find the value of $L\{e^{-t} (3 \sin 2t - 5 \cos 2t)\}$.*

Sol. By linearity of L, we have

$$L\{3 \sin 2t - 5 \cos 2t\} = 3L \{\sin 2t\} - 5L \{\cos 2t\}$$

$$= 3 \frac{2}{s^2 + 2^2} - 5 \frac{s}{s^2 + 2^2} = \frac{6 - 5s}{s^2 + 4} = f(x), \text{ say.}$$

Now from first shifting theorem, we have

$$L \{e^{-t} (3 \sin 2t - 5 \cos 2t)\} = f(s+1) = \frac{6 - 5(s+1)}{(s+1)^2 + 4} = \frac{-5s + 1}{s^2 + 2s + 5}.$$

Example 11. *Find the Laplace transform of*

(i) $e^{2t} (2t^2 - 3t + 4)$ *(ii)* $e^{-3t} (2 \cos 5t - 3 \sin 5t)$.

Sol. By applying linearity property, we have

(i) $L\{e^{2t}(2t^2 - 3t + 4\} = 2L\{e^{2t} t^2\} - 3L\{e^{2t} t\} + 4L\{e^{2t}\}$

$$= 2 . \frac{2}{(s-2)^3} - 3 . \frac{1}{(s-2)^2} + 4 . \frac{1}{s-2},$$

[By applying the first shifting property]

$$= \frac{4s^2 - 11s + 26}{(s-2)^3}$$

(ii) $L \{e^{-3t} (2 \cos 5t - 3 \sin 5t)\} = 2L\{e^{-3t} \cos 5t\} - 3L \{e^{-3t} \sin 5t\}$

$$= 2 . \frac{s+3}{(s+3)^2 + 25} - 3 . \frac{5}{(s+3)^2 + 25},$$

[By applying the first shifting property]

$$= \frac{25 - 9}{(s+3)^2 + 25}.$$

Example 12. *Find $L\{e^{-2t}(3\cos 6t - 5\sin 6t)\}$.*

Sol. By definition

$$L\{3\cos 6t - 5\sin 6t\} = 3L\{\cos 6t\} - 5L\{\sin 6t\}$$

$$= 3\frac{s}{s^2+6^2} - 5\frac{6}{s^2+6^2} = \frac{3s-30}{s^2+36} = f(s), \text{ say}$$

Therefore, from first shifting theorem, we have

$$L\{e^{-2t}(3\cos 6t - 5\sin 6t)\} = f(s+2) = \frac{3(s+2)-30}{(s+2)^2+36} = \frac{3s-24}{s^2+4s+40}.$$

Example 13. *Find the transform of the function $F(t) = (t+3)^2 e^t$.*

Sol. By linearity property, we have

$$L\{(t+3)^2\} = L\{t^2+6t+9\} = L(t^2) + 6L\{t\} + 9L\{1\}$$

$$= \frac{2!}{s^3} + \frac{6}{s^2} + \frac{9}{s} = \frac{2+6s+9s^2}{s^3} = f(s), \text{ say}$$

Therefore, from first shifiting theorem, we have

$$L\{F(t)\} = L\{(t+3)^2 e^t\} = f(s-1) = \frac{2+6(s-1)+9(s-1)^2}{(s-1)^3} = \frac{9s^2-12s+5}{(s-1)^3}.$$

Example 14. *Find $L\{e^t \sin^2 t\}$.*

Sol. We have

$$L\{\sin^2 t\} = L\left\{\frac{1}{2}(1-\cos 2t)\right\} = \frac{1}{2}L\{1\} - \frac{1}{2}\{\cos 2t\}$$

$$= \frac{1}{2}\left[\frac{1}{s} - \frac{s}{s^2+2^2}\right] = \frac{2}{s(s^2+4)} = f(s), \text{ say.}$$

Therefore, from shifting theorem, we have

$$L\{e^t \sin^2 t\} = f(s-1) = \frac{2}{(s-1)\{(s-1)^2+4\}} = \frac{2}{(s-1)(s^2-2s+5)}.$$

Example 15. *Find $L\{e^{-t}(3\sinh 2t - 5\cosh 2t)\}$.*

Sol. By linearity property, we have

$$L\{3\sinh 2t - 5\cosh 2t\} = 3L\{\sinh 2t\} - 5L\{\cosh 2t\}$$

$$= 3 \cdot \frac{2}{s^2-2^2} - 5 \cdot \frac{s}{s^2-2^2} = \frac{6-5s}{5^2-4} = f(s), \text{ say.}$$

Therefore, from first shifting property, we have

$$L\{e^{-t}(3\sinh 2t - 5\cosh 2t\} = f(s+1) = \frac{6-5(s+1)}{(s+1)^2-4} = \frac{1-5s}{s^2+2s-3}.$$

Example 16. *Show that*

(i) $L\{t\sin at\} = \dfrac{2as}{(s^2+a^2)^2}$. *(ii)* $L\{t\cos at\} = \dfrac{s^2-a^2}{(s^2+a^2)^2}$.

Sol. We have $\quad \text{L}(t) = \dfrac{1}{s^2}$

$\therefore \quad \text{L}\{te^{-iat}\} = \dfrac{1}{(s-ia)^2}$, using first shifting property

$$= \frac{(s+ia)^2}{(s-ia)^2\,(s+ia)^2} = \frac{(s+ia)^2}{(s^2+a^2)^2} = \frac{(s^2-a^2)+i2as}{(s^2+a^2)^2}$$

Thus $\text{L}\{t \cos at + it \sin at\} = \dfrac{s^2-a^2}{(s^2+a^2)^2} + i\dfrac{2as}{(s^2+a^2)^2}$.

Applying linearity property in the left hand side, we get

$$\text{L}\{t \cos at\} + i\text{L}\,\{t \sin at\} = \frac{s^2-a^2}{(s^2+a^2)^2} + i\frac{2as}{(s^2+a^2)^2}.$$

Equating real and imaginary parts on both sides, we get

$$\text{L}\{t \cos at\} = \frac{s^2-a^2}{(s^2+a^2)^2} \quad \text{and} \quad \text{L}\{t \sin at\} = \frac{2as}{(s^2+a^2)^2}.$$

TEST YOUR KNOWLEDGE

Find the Laplace transforms of (1 to 14):

1. $\cosh at - \cos at$ **2.** $\cos(at+b)$ **3.** $\sin at \sin bt$

4. $\sin^2 3t$ **5.** $\cos^3 2t$ **6.** $\sinh^3 2t$

7. $e^{-2t} \sin 4t$ **8.** $t^3 e^{-3t}$ **9.** $\cosh at \cos at$

10. $e^{-3t}(2\cos 5t - 3\sin 5t)$ **11.** $\sin at - at\cos at$

12. $e^{-t}(\sin 2t - 2t\cos 2t)$ **13.** $\sinh\dfrac{t}{2}\sin\dfrac{\sqrt{3}\,t}{2}$ **14.** $t\,e^{at}\sin at$

15. If $\text{L}\,\{f(t)\} = \bar{f}(s)$, show that $\text{L}\left\{f\left(\dfrac{t}{a}\right)\right\} = a\bar{f}(as)$.

16. Show that $\text{L}\,\{\sin kt \sinh kt\} = \dfrac{2k^2 s}{s^4+4k^4}$.

17. Show that

(*i*) $\text{L}\{t \sin at\} = \dfrac{2as}{(s^2+a^2)^2}$ (*ii*) $\text{L}\{t\cos at\} = \dfrac{s^2-a^2}{(s^2+a^2)^2}$.

18. Find the Laplace transform of $f(t)$ defined as

(*i*) $f(t) = \begin{cases} t, & 0<t<4 \\ 5, & t>4 \end{cases}$ (*ii*) $f(t) = \begin{cases} \sin t, & 0<t<\pi \\ 0, & t>\pi \end{cases}$

(*iii*) $f(t) = \begin{cases} \cos t, & 0<t<2\pi \\ 0, & t>2\pi \end{cases}$ (*iv*) $f(t) = \begin{cases} t^2 & 0<t<2 \\ t-1, & 2<t<3 \\ 7, & t>3. \end{cases}$

19. Find the Laplace transform of $f(t)$ defined as $f(t) = |t-1| + |t+1|, t \geq 0$.

$$\left[\textbf{Hint. } f(t) = \begin{cases} 2, & 0 \leq t \leq 1 \\ 2t, & t > 1. \end{cases}\right]$$

Answers

1. $\dfrac{2a^2 s}{s^4 - a^4}$ **2.** $\dfrac{s \cos b - a \sin b}{s^2 + a^2}$

3. $\dfrac{2abs}{[s^2 + (a+b)^2][s^2 + (a-b)^2]}$ **4.** $\dfrac{18}{s(s^2 + 36)}$ **5.** $\dfrac{s(s^2 + 28)}{(s^2 + 4)(s^2 + 36)}$

6. $\dfrac{48}{(s^2 - 4)(s^2 - 36)}$ **7.** $\dfrac{4}{s^2 + 4s + 20}$ **8.** $\dfrac{6}{(s+3)^4}$

9. $\dfrac{1}{2}\left[\dfrac{s-a}{(s-a)^2 + a^2} + \dfrac{s+a}{(s+a)^2 + a^2}\right]$ **10.** $\dfrac{2s - 9}{s^2 + 6s + 34}$ **11.** $\dfrac{2a^3}{(s^2 + a^2)^2}$

12. $\dfrac{16}{(s^2 + 2s + 5)^2}$ **13.** $\dfrac{\sqrt{3}\, s}{2(s^4 + s^2 + 1)}$ **14.** $\dfrac{2a(s-a)}{(s^2 - 2as + 2a^2)^2}$

18. (*i*) $\dfrac{1}{s^2} + e^{-4s}\left(\dfrac{1}{s} - \dfrac{1}{s^2}\right)$ (*ii*) $\dfrac{1 + e^{-\pi s}}{s^2 + 1}$ (*iii*) $\dfrac{s(1 - e^{-2\pi s})}{s^2 + 1}$

(*iv*) $\dfrac{2}{s^3} - \dfrac{e^{-2s}}{s^3}(2 + 3s + 3s^2) + \dfrac{e^{-3s}}{s^2}(5s - 1)$

19. $\dfrac{2}{s}\left(1 + \dfrac{e^{-s}}{s}\right)$.

2.6. INVERSE LAPLACE TRANSFORM

Definition. If $L\{f(t)\} = \bar{f}(s)$, then $f(t)$ is called the inverse Laplace transform of $\bar{f}(s)$ and is denoted by

$$L^{-1}\{\bar{f}(s)\} = f(t)$$

Here L^{-1} denotes the inverse Laplace transform.

For example, since $L\{e^{5t}\} = \dfrac{1}{s-5}$

$$\therefore \quad L^{-1}\left\{\frac{1}{s-5}\right\} = e^{5t}$$

The inverse Laplace transforms given below follow at once from the results of Laplace transforms given earlier :

(1) $\mathbf{L^{-1}\left\{\dfrac{1}{s}\right\} = 1}$ (2) $\mathbf{L^{-1}\left\{\dfrac{1}{s-a}\right\} = e^{at}}$

(3) $L^{-1}\left\{\frac{1}{s^n}\right\} = \frac{t^{n-1}}{(n-1)!}$, **if n is a positive integer.** $\left[\text{otherwise} = \frac{t^{n-1}}{\Gamma(n)}\right]$

(4) $L^{-1}\left\{\frac{1}{(s-a)^n}\right\} = e^{at}\frac{t^{n-1}}{(n-1)!}$ (5) $L^{-1}\left\{\frac{1}{s^2+a^2}\right\} = \frac{1}{a}\sin at$

(6) $L^{-1}\left\{\frac{s}{s^2+a^2}\right\} = \cos at$ (7) $L^{-1}\left\{\frac{1}{s^2-a^2}\right\} = \frac{1}{a}\sinh at$

(8) $L^{-1}\left\{\frac{s}{s^2-a^2}\right\} = \cosh at$ (9) $L^{-1}\left\{\frac{1}{(s-a)^2+b^2}\right\} = \frac{1}{b}e^{at}\sin bt$

(10) $L^{-1}\left\{\frac{s-a}{(s-a)^2+b^2}\right\} = e^{at}\cos bt$ (11) $L^{-1}\left[\frac{s}{(s^2+a^2)^2}\right] = \frac{1}{2a}t\sin at$

(12) $L^{-1}\left[\frac{1}{(s^2+a^2)^2}\right] = \frac{1}{2a^3}(\sin at - at\cos at)$.

All the above results must be remembered.

ILLUSTRATIVE EXAMPLES

Example 1. *Find the inverse Laplace transforms of*

(*i*) $\frac{3(s^2-1)^2}{2s^5}$ (*ii*) $\frac{3s+5\sqrt{2}}{s^2+8}$

(*iii*) $\frac{4s+15}{16s^2-25}$ (*iv*) $\frac{s+1}{s^2+s+1}$.

Sol. (*i*) Since $\frac{3(s^2-1)^2}{2s^5} = \frac{3s^4-6s^2+3}{2s^5} = \frac{3}{2}\cdot\frac{1}{s} - 3\cdot\frac{1}{s^3} + \frac{3}{2}\cdot\frac{1}{s^5}$

$$\therefore \quad L^{-1}\left\{\frac{3(s^2-1)^2}{2s^5}\right\} = \frac{3}{2}L^{-1}\left\{\frac{1}{s}\right\} - 3L^{-1}\left\{\frac{1}{s^3}\right\} + \frac{3}{2}L^{-1}\left\{\frac{1}{s^5}\right\}$$

$$= \frac{3}{2}(1) - 3\left\{\frac{t^2}{2!}\right\} + \frac{3}{2}\left\{\frac{t^4}{4!}\right\} = \frac{3}{2} - \frac{3}{2}t^2 + \frac{1}{16}t^4.$$

(*ii*) Since $\frac{3s+5\sqrt{2}}{s^2+8} = 3\frac{s}{s^2+(2\sqrt{2})^2} + 5\sqrt{2}\frac{1}{s^2+(2\sqrt{2})^2}$

$$\therefore \quad L^{-1}\left\{\frac{3s+5\sqrt{2}}{s^2+8}\right\} = 3L^{-1}\left\{\frac{s}{s^2+(2\sqrt{2})^2}\right\} + 5\sqrt{2}\,L^{-1}\left\{\frac{1}{s^2+(2\sqrt{2})^2}\right\}$$

$$= 3\cos(2\sqrt{2}t) + 5\sqrt{2}\,.\,\frac{1}{2\sqrt{2}}\sin(2\sqrt{2}\,t)$$

$$= 3\cos(2\sqrt{2}t) + \frac{5}{2}\sin(2\sqrt{2}t).$$

(*iii*) Since $\dfrac{4s+15}{16s^2-25} = \dfrac{4s+15}{16\left(s^2-\dfrac{25}{16}\right)} = \dfrac{1}{4}\cdot\dfrac{s}{s^2-(\frac{5}{4})^2} + \dfrac{15}{16}\cdot\dfrac{1}{s^2-(\frac{5}{4})^2}$

$$\therefore \quad L^{-1}\left\{\frac{4s+15}{16s^2-25}\right\} = \frac{1}{4}L^{-1}\left[\frac{s}{s^2-(\frac{5}{4})^2}\right] + \frac{15}{16}L^{-1}\left[\frac{1}{s^2-(\frac{5}{4})^2}\right]$$

$$= \frac{1}{4}\cosh\left(\frac{5}{4}t\right) + \frac{15}{16}\cdot\frac{1}{\frac{5}{4}}\sinh\left(\frac{5}{4}t\right)$$

$$= \frac{1}{4}\cosh\left(\frac{5}{4}t\right) + \frac{3}{4}\sinh\left(\frac{5}{4}t\right)$$

(*iv*) Since $\dfrac{s+1}{s^2+s+1} = \dfrac{\left(s+\frac{1}{2}\right)+\frac{1}{2}}{\left(s+\frac{1}{2}\right)^2+\frac{3}{4}} = \dfrac{s+\frac{1}{2}}{\left(s+\frac{1}{2}\right)^2+\left(\frac{\sqrt{3}}{2}\right)^2} + \dfrac{1}{2}\cdot\dfrac{1}{\left(s+\frac{1}{2}\right)^2+\left(\frac{\sqrt{3}}{2}\right)^2}$

$$\therefore \quad L^{-1}\left\{\frac{s+1}{s^2+s+1}\right\} = L^{-1}\left[\frac{s+\frac{1}{2}}{\left(s+\frac{1}{2}\right)^2+\left(\frac{\sqrt{3}}{2}\right)^2}\right] + \frac{1}{2}L^{-1}\left(\frac{1}{\left(s+\frac{1}{2}\right)^2+\left(\frac{\sqrt{3}}{2}\right)^2}\right)$$

$$= e^{-\frac{1}{2}t}\cos\left(\frac{\sqrt{3}}{2}t\right) + \frac{1}{2}\cdot\frac{\sqrt{3}}{2}\,e^{-\frac{1}{2}t}\sin\left(\frac{\sqrt{3}}{2}t\right)$$

$$= \frac{1}{\sqrt{3}}e^{-\frac{1}{2}t}\left(\sqrt{3}\cos\frac{\sqrt{3}}{2}t + \sin\frac{\sqrt{3}}{2}t\right).$$

Example 2. *Find the inverse Laplace transforms of*

(*i*) $\dfrac{2s^2-4}{(s+1)(s-2)(s-3)}$ (*ii*) $\dfrac{2s^2-1}{(s^2+1)(s^2+4)}$ (*iii*) $\dfrac{21s-33}{(s+1)(s-2)^3}$.

Sol. (*i*) Here the denominator has non-repeated linear factors.

Let $\dfrac{2s^2-4}{(s+1)(s-2)(s-3)} = \dfrac{A}{s+1} + \dfrac{B}{s-2} + \dfrac{C}{s-3}$

Multiplying both sides by $(s+1)(s-2)(s-3)$, we get

$$2s^2 - 4 = A(s-2)(s-3) + B(s+1)(s-3) + C(s+1)(s-2)$$

Putting $s = -1$, we get $\quad -2 = A(-3)(-4) \quad \Rightarrow \quad A = -\frac{1}{6}$

Putting $s = 2$, we get $\quad 4 = B(3)(-1) \quad \Rightarrow \quad B = -\frac{4}{3}$

Putting $s = 3$, we get $\quad 14 = C(4)(1) \quad \Rightarrow \quad C = \frac{7}{2}$

Thus
$$\frac{2s^2-4}{(s+1)(s-2)(s-3)} = -\frac{1}{6(s+1)} - \frac{4}{3(s-2)} + \frac{7}{2(s-3)}$$

$$\therefore \quad L^{-1}\left\{\frac{2s^2-4}{(s+1)(s-2)(s-3)}\right\} = -\frac{1}{6}L^{-1}\left\{\frac{1}{s+1}\right\} - \frac{4}{3}L^{-1}\left\{\frac{1}{s-2}\right\} + \frac{7}{2}L^{-1}\left\{\frac{1}{s-3}\right\}$$

$$= -\frac{1}{6}e^{-t} - \frac{4}{3}e^{2t} + \frac{7}{2}e^{3t}.$$

(*ii*) Since the fraction involves only even powers of s, we put $s^2 = p$ so that the given fraction becomes $\frac{2p-1}{(p+1)(p+4)}$ which has only non-repeated linear factors in the denominator.

Let
$$\frac{2p-1}{(p+1)(p+4)} = \frac{A}{p+1} + \frac{B}{p+4} \qquad ...(1)$$

Multiplying both sides by $(p + 1)(p + 4)$, we get

$$2p - 1 = A(p + 4) + B(p + 1)$$

Putting $p = -1$, we get $\quad -3 = A(3) \quad \Rightarrow \quad A = -1$

Putting $p = -4$, we get $\quad -9 = B(-3) \quad \Rightarrow \quad B = 3$

Putting the values of A and B in (1) and replacing p by s^2, we have

$$\frac{2s^2-1}{(s^2+1)(s^2+4)} = \frac{-1}{s^2+1} + \frac{3}{s^2+4}$$

$$\therefore \quad L^{-1}\left\{\frac{2s^2-1}{(s^2+1)(s^2+4)}\right\} = -L^{-1}\left\{\frac{1}{s^2+1^2}\right\} + 3L^{-1}\left\{\frac{1}{s^2+2^2}\right\} = -\sin t + \frac{3}{2}\sin 2t.$$

(*iii*) Here the denominator has a linear factor repeated thrice.

Let
$$\frac{21s-33}{(s+1)(s-2)^3} = \frac{A}{s+1} + \frac{B}{s-2} + \frac{C}{(s-2)^2} + \frac{D}{(s-2)^3}$$

Multiplying both sides $(s + 1)(s - 2)^3$, we get

$$21s - 33 = A(s-2)^3 + B(s+1)(s-2)^2 + C(s+1)(s-2) + D(s+1)$$

Putting $s = -1$, we get $\quad -54 = A(-27) \quad \Rightarrow \quad A = 2$

Putting $s = 2$, we get $\quad 9 = D(3) \quad \Rightarrow \quad D = 3$

Equating co-efficients of s^3, $\quad 0 = A + B \quad \Rightarrow \quad B = -A = -2$

Putting $s = 0$, we get $\quad -33 = -8A + 4B - 2C + D$

or $\quad -33 = 16 - 8 - 2C + 3 \quad$ or $\quad C = 6$

Thus
$$\frac{21s-33}{(s+1)(s-2)^3} = \frac{2}{s+1} - \frac{2}{s-2} + \frac{6}{(s-2)^2} + \frac{3}{(s-2)^3}$$

$$\therefore \quad L^{-1}\left\{\frac{21s-33}{(s+1)(s-2)^3}\right\} = 2L^{-1}\left\{\frac{1}{s+1}\right\} - 2L^{-1}\left\{\frac{1}{s-2}\right\} + 6L^{-1}\left\{\frac{1}{(s-2)^2}\right\} + 3L^{-1}\left\{\frac{1}{(s-2)^3}\right\}$$

$$= 2e^{-t} - 2e^{2t} + 6.\, e^{2t} \,.\, \frac{t}{1!} + 3.\, e^{2t} \,.\, \frac{t^2}{2!}$$

$$= 2e^{-t} - 2e^{2t} + 6te^{2t} + \frac{3}{2}\, t^2 e^{2t}.$$

Example 3. *Find the inverse Laplace transforms of*

(*i*) $\dfrac{5s+3}{(s-1)(s^2+2s+5)}$ (*ii*) $\dfrac{s}{s^4+s^2+1}$

(*iii*) $\dfrac{s}{s^4+4a^4}$ (*iv*) $\dfrac{s^3}{s^4-a^4}$.

Sol. (*i*) Here the denominator has a non-factorisable factor $s^2 + 2s + 5$.

Let $$\frac{5s+3}{(s-1)(s^2+2s+5)} = \frac{A}{s-1} + \frac{Bs+C}{s^2+2s+5}$$

Multiplying both sides by $(s - 1)(s^2 + 2s + 5)$, we get

$$5s + 3 = A(s^2 + 2s + 5) + (Bs + C)(s - 1)$$

Putting $s = 1$, we get $8 = 8A \Rightarrow A = 1$

Equating the co-efficients of s^2,

$0 = A + B \Rightarrow B = -A = -1$

Putting $s = 0$, we get $3 = 5A - C \Rightarrow C = 5A - 3 = 5 \times 1 - 3 = 2$

Thus $$\frac{5s+3}{(s-1)(s^2+2s+5)} = \frac{1}{s-1} + \frac{-s+2}{s^2+2s+5} = \frac{1}{s-1} - \frac{(s+1)-3}{(s+1)^2+4}$$

$$= \frac{1}{s-1} - \frac{s+1}{(s+1)^2+2^2} + 3.\frac{1}{(s+1)^2+2^2}$$

$$\therefore \quad L^{-1}\left\{\frac{5s+3}{(s-1)(s^2+2s+5)}\right\} = L^{-1}\left\{\frac{1}{s-1}\right\} - L^{-1}\left\{\frac{s+1}{(s+1)^2+2^2}\right\} + 3L^{-1}\left\{\frac{1}{(s+1)^2+2^2}\right\}$$

$$= e^t - e^{-t}\cos 2t + \tfrac{3}{2}\, e^{-t}\sin 2t.$$

(*ii*) Since $s^4 + s^2 + 1 = (s^4 + 2s^2 + 1) - s^2 = (s^2 + 1)^2 - s^2$

$= (s^2 + 1 + s)(s^2 + 1 - s)$

$$\therefore \quad \frac{s}{s^4+s^2+1} = \frac{s}{(s^2+s+1)(s^2-s+1)}$$

Let $$\frac{s}{(s^2+s+1)(s^2-s+1)} = \frac{As+B}{s^2+s+1} + \frac{Cs+D}{s^2-s+1}$$

Multiplying both sides by $(s^2 + s + 1)(s^2 - s + 1)$, we get

$$s = (As + B)(s^2 - s + 1) + (Cs + D)(s^2 + s + 1)$$

Equating co-efficients of s^3, $0 = A + C$ or $C = -A$ ...(1)

Equating co-efficients of s^2, $0 = -A + B + C + D$ ...(2)

Equating co-efficients of s, $1 = A - B + C + D$ ...(3)

Putting $s = 0$, $0 = B + D$ or $D = -B$ ...(4)

From (2), $0 = -A + B - A - B$

$\Rightarrow$ $\quad$ A = 0 $\quad \therefore$ From (1), C = 0

From (3), $\quad 1 = A - B - A - B$

$\Rightarrow$ $\quad B = -\frac{1}{2} \quad \therefore$ From (4), $D = \frac{1}{2}$

Thus
$$\frac{s}{s^4+s^2+1} = \frac{-\frac{1}{2}}{s^2+s+1} + \frac{\frac{1}{2}}{s^2-s+1} = \frac{1}{2}\left[\frac{1}{(s-\frac{1}{2})^2+\frac{3}{4}} - \frac{1}{(s+\frac{1}{2})^2+\frac{3}{4}}\right]$$

$$\therefore \quad L^{-1}\left[\frac{s}{s^4+s^2+1}\right] = \frac{1}{2}\left[L^{-1}\left\{\frac{1}{\left(s-\frac{1}{2}\right)^2+\left(\frac{\sqrt{3}}{2}\right)^2}\right\} - L^{-1}\left\{\frac{1}{\left(s+\frac{1}{2}\right)^2+\left\{\frac{\sqrt{3}}{2}\right\}^2}\right\}\right]$$

$$= \frac{1}{2}\left[\frac{1}{\sqrt{3}/2}e^{t/2}\sin\frac{\sqrt{3}}{2}t - \frac{1}{\sqrt{3}/2}e^{-t/2}\sin\frac{\sqrt{3}}{2}t\right]$$

$$= \frac{2}{\sqrt{3}}\sin\frac{\sqrt{3}}{2}t\left(\frac{e^{t/2}-e^{-t/2}}{2}\right) = \frac{2}{\sqrt{3}}\sin\frac{\sqrt{3}}{2}t\sinh\frac{t}{2}.$$

Note. In certain cases, with a little insight, the partial fractions can be written down directly.

Thus
$$\frac{s}{(s^2+s+1)(s^2-s+1)} = \frac{1}{2}\left[\frac{1}{s^2-s+1} - \frac{1}{s^2+s+1}\right].$$

(*iii*) Since $\quad s^4 + 4a^4 = (s^2 + 2a^2)^2 - (2as)^2 = (s^2 + 2as + 2a^2)(s^2 - 2as + 2a^2)$

$$\therefore \quad \frac{s}{s^4+4a^4} = \frac{s}{(s^2+2as+2a^2)(s^2-2as+2a^2)}$$

$$= \frac{1}{4a}\left(\frac{1}{s^2-2as+2a^2} - \frac{1}{s^2+2as+2a^2}\right)$$

$$= \frac{1}{4a}\left[\frac{1}{(s-a)^2+a^2} - \frac{1}{(s+a)^2+a^2}\right]$$

$$L^{-1}\left[\frac{s}{s^4+4a^4}\right] = \frac{1}{4a}\left[L^{-1}\left\{\frac{1}{(s-a)^2+a^2}\right\} - L^{-1}\left\{\frac{1}{(s+a)^2+a^2}\right\}\right]$$

$$= \frac{1}{4a}\left\{\frac{1}{a}e^{at}\sin at - \frac{1}{a}e^{-at}\sin at\right\}$$

$$= \frac{1}{2a^2}\sin at\left\{\frac{e^{at}-e^{-at}}{2}\right\} = \frac{1}{2a^2}\sin at\sinh at.$$

(*iv*)
$$\frac{s^3}{s^4-a^4} = s\left[\frac{s^2}{(s^2-a^2)(s^2+a^2)}\right] = \frac{1}{2}s\left[\frac{1}{s^2-a^2} + \frac{1}{s^2+a^2}\right]$$

$$= \frac{1}{2}\left[\frac{s}{s^2-a^2} + \frac{s}{s^2+a^2}\right]$$

$$\therefore \quad L^{-1}\left\{\frac{s^3}{s^4-a^4}\right\} = \frac{1}{2}\left[L^{-1}\left\{\frac{s}{s^2-a^2}\right\} + L^{-1}\left\{\frac{s}{s^2+a^2}\right\}\right] = \frac{1}{2}\,(\cosh at + \cos at).$$

Example 4. *Prove that* $L^{-1}\left\{\dfrac{e^{-1/s}}{s}\right\} = J_0(2\sqrt{t})$.

Sol.
$$L^{-1}\left\{\frac{e^{-1/s}}{s}\right\} = L^{-1}\left\{\frac{1}{s}e^{-1/s}\right\} = L^{-1}\left\{\frac{1}{s}\left(1-\frac{1}{s}+\frac{1}{2!\,s^2}-\frac{1}{3!\,s^3}+\ldots\ldots\right)\right\}$$

$$= L^{-1}\left\{\frac{1}{s}-\frac{1}{s^2}+\frac{1}{2!\,s^3}-\frac{1}{3!\,s^4}+\ldots\ldots\right\}$$

$$= 1 - t + \frac{t^2}{(2!)^2} - \frac{t^3}{(3!)^2} + \ldots\ldots = J_0\,(2\sqrt{t}).$$

Example 5. *Find the inverse Laplace transforms of the following:*

(i) $\dfrac{s+2}{s^2-4s+13}$. *(ii)* $\dfrac{s+1}{s^2+2s-8}$.

Sol. *(i)* $L^{-1}\left\{\dfrac{s+2}{s^2-4s+13}\right\}$

Consider $s^2 - 4s + 13 = s^2 - 4s + 4 + 9 = (s-2)^2 + 3^2$ (Completing the square)

$$\therefore \quad L^{-1}\left\{\frac{s+2}{s^2-4s+13}\right\} = L^{-1}\left\{\frac{s+2}{(s-2)^2+3^2}\right\} = L^{-1}\left\{\frac{s-2+4}{(s-2)^2+3^2}\right\}$$

$$= L^{-1}\left\{\frac{s-2}{(s-2)^2+3^2}\right\} + L^{-1}\left\{\frac{4}{(s-2)^2+3^2}\right\}$$

$$= L^{-1}\left\{\frac{s-2}{(s-2)^2+3^2}\right\} + \frac{4}{3}L^{-1}\left\{\frac{3}{(s-2)^2+3^2}\right\}$$

$$= e^{2t}\cos 3t + \frac{4}{3}\,e^{2t}\sin 3t.$$

(ii) $L^{-1}\left\{\dfrac{s+1}{s^2+2s-8}\right\}$.

Consider $s^2 + 2s - 8 = s^2 + 2s + 1 - 9 = (s+1)^2 - 3^2$. (Completing the square)

$$\therefore \quad L^{-1}\left\{\frac{s+1}{s^2+2s-8}\right\} = L^{-1}\left\{\frac{s+1}{(s+1)^2-3^2}\right\} = e^{-t}\cosh 3t.$$

Example 6. *Find*

(i) $L^{-1}\left\{\dfrac{s}{s^2+2}+\dfrac{6s}{s^2-16}+\dfrac{3}{s-3}\right\}$ *(ii)* $L^{-1}\left\{\dfrac{6}{2s-3}-\dfrac{3+4s}{9s^2-16}+\dfrac{8-6s}{16s^2+9}\right\}$.

Sol. (*i*) $L^{-1}\left\{\frac{s}{s^2+2}+\frac{6s}{s^2-16}+\frac{3}{s-3}\right\} = L^{-1}\left\{\frac{s}{s^2+(\sqrt{2})^2}\right\}+6L^{-1}\left\{\frac{s}{s^2-4^2}\right\}+3L^{-1}\left\{\frac{1}{s-3}\right\}$

$$= \cos\sqrt{2}\,t + 6\cosh 4t + 3e^{3t}.$$

(*ii*) $L^{-1}\left\{\frac{6}{2s-3}-\frac{3+4s}{9s^2-16}+\frac{8-6s}{16s^2+9}\right\} = 3L^{-1}\left\{\frac{1}{s-(3/2)}\right\}-3L^{-1}\left\{\frac{1}{9s^2-16}\right\}$

$$-4L^{-1}\left\{\frac{s}{9s^2-16}\right\}+8L^{-1}\left\{\frac{1}{16s^2+9}\right\}-6L^{-1}\left\{\frac{s}{16s^2+9}\right\}$$

$$= 3L^{-1}\left\{\frac{1}{s-3/2}\right\}-\frac{1}{3}L^{-1}\left\{\frac{1}{s^2-(4/3)^2}\right\}-\frac{4}{9}L^{-1}\left\{\frac{s}{s^2-(4/3)^2}\right\}$$

$$+\frac{1}{2}L^{-1}\left\{\frac{1}{s^2+(3/4)^2}\right\}-\frac{3}{8}L^{-1}\left\{\frac{s}{s^2+(3/4)^2}\right\}$$

$$= 3e^{(3/2)t}-\frac{1}{3}\cdot\frac{1}{(4/3)}\sinh\frac{4t}{3}-\frac{4}{9}\cosh\frac{4t}{3}+\frac{1}{2}\frac{1}{(3/4)}\sin\frac{3t}{4}-\frac{3}{8}\cos\frac{3t}{4}$$

$$= 3e^{3t/2}-\frac{1}{4}\sinh\frac{4t}{3}-\frac{4}{9}\cosh\frac{4t}{3}+\frac{2}{3}\sin\frac{3t}{4}-\frac{3}{8}\cos\frac{3t}{4}.$$

TEST YOUR KNOWLEDGE

Find the inverse Laplace transforms of:

1. $\left(\frac{1-\sqrt{s}}{s^2}\right)^2$
2. $\frac{2s+6}{s^2+4}$
3. $\frac{s+2}{s^2-4s+13}$
4. $\frac{4s+12}{s^2+8s+16}$
5. $\frac{s-a}{(s-a)^2+b^2}$
6. $\frac{s+8}{s^2+4s+5}$
7. $\frac{s}{(s+a)^2}$
8. $\frac{s+1}{s^2+2s}$
9. $\frac{s^2+1}{s^3+3s^2+2s}$
10. $\frac{2s^2-6s+5}{s^3-6s^2+11s-6}$
11. $\frac{s^2+2s-3}{s(s-3)(s+2)}$
12. $\frac{4s+5}{(s-1)^2(s+2)}$
13. $\frac{s}{(s^2-1)^2}$
14. $\frac{1+2s}{(s+2)^2\,(s-1)^2}$
15. $\frac{1}{s(s^2+4)}$
16. $\frac{1}{(s+1)(s^2+2s+2)}$
17. $\frac{s}{(s+1)^2(s^2+1)}$
18. $\frac{1}{s^2(s^2+a^2)}$
19. (*i*) $\frac{s+2}{(s^2+4s+5)^2}$ (*ii*) $\frac{s+3}{(s^2+6s+13)^2}$ (*iii*) $\frac{1}{(s^2+2s+5)^2}$
20. $\frac{1}{s^3-a^3}$
21. $\frac{s}{(s^2+1)(s^2+4)}$
22. $\frac{3s+1}{(s+1)^4}$
23. $\frac{a(s^2-2a^2)}{s^4+4a^4}$.

Answers

1. $\frac{t^3}{6} + \frac{t^2}{2} - \frac{16}{15\sqrt{\pi}} t^{5/2}$
2. $2 \cos 2t + 3 \sin 2t$
3. $e^{2t} \cos 3t + \frac{4}{3} e^{2t} \sin 3t$
4. $4e^{-4t} (1 - t)$
5. $e^{at} \cos bt$
6. $e^{-2t} (\cos t + 6 \sin t)$
7. $(1 - at)e^{-at}$
8. $\frac{1}{2} (1 + e^{-2t})$
9. $\frac{1}{2} - 2e^{-t} + \frac{5}{2} e^{-2t}$
10. $\frac{1}{2} e^t - e^{2t} + \frac{5}{2} e^{3t}$
11. $\frac{1}{2} + \frac{4}{5} e^{3t} - \frac{3}{10} e^{-2t}$
12. $\frac{1}{3} e^t + 3te^t - \frac{1}{3} e^{-2t}$
13. $\frac{1}{2} t \sinh t$
14. $\frac{1}{3} t(e^t - e^{-2t})$
15. $\frac{1}{4} (1 - \cos 2t)$
16. $e^{-t} (1 - \cos t)$
17. $\frac{1}{2} (\sin t - t\, e^{-t})$
18. $\frac{1}{a^3} (at - \sin at)$
19. (*i*) $\frac{1}{2} te^{-2t} \sin t$ (*ii*) $\frac{1}{4} te^{-3t} \sin 2t$ (*iii*) $\frac{1}{16} e^{-t} (\sin 2t - 2t \cos 2t)$
20. $\frac{1}{3a^2}\left[e^{at} - e^{-\frac{1}{2}at}\left\{\cos \frac{\sqrt{3}}{2} at + \sqrt{3} \sin \frac{\sqrt{3}}{2} at\right\}\right]$
21. $\frac{1}{3} (\cos t - \cos 2t)$
22. $e^{-t}\left(\frac{3}{2}t^2 - \frac{1}{3}t^3\right)$
23. $\cos at \sinh at$.

2.7. LAPLACE TRANSFORMS OF DERIVATIVES

Theorem. *If $f'(t)$ be continuous and $L\{f(t)\} = \bar{f}(s)$, then*

$$L\{f'(t)\} = s\bar{f}(s) - f(0), \text{ provided } \underset{t\to\infty}{\text{Lt}} [e^{-st} f(t)] = 0.$$

Proof. $L\{f'(t)\} = \int_0^\infty e^{-st} f'(t)\, dt$

$$= \left[e^{-st} f(t)\right]_0^\infty - \int_0^\infty - se^{-st} . f(t)\, dt = \underset{t\to\infty}{\text{Lt}}\, e^{-st} f(t) - f(0) + s \int_0^\infty e^{-st} f(t)\, dt$$

$$= 0 - f(0) + s\text{L}\{f(t)\} = s\bar{f}(s) - f(0).$$

Note 1. If $f(t)$ be such that $\underset{t\to\infty}{\text{Lt}} [e^{-st} f(t)] = 0$, then $f(t)$ is said to be of exponential order s.

Note 2. The above theorem can be generalised.

Generalisation. *If $f(t)$ and its first $(n - 1)$ derivatives be continuous, then*

$$L\{f^n(t)\} = s^n\bar{f}(s) - s^{n-1} f(0) - s^{n-2} f'(0) - s^{n-3} f''(0) - \ldots\ldots - f^{n-1}(0).$$

Using the general rule of integration by parts,

$$L\{f^n(t)\} = \int_0^\infty e^{-st} f^n(t)\, dt$$

$$= \Big[e^{-st} f^{n-1}(t) - (-s)e^{-st} f^{n-2}(t) + (-s)^2 e^{-st} f^{n-3}(t) - \ldots$$

$$\ldots + (-1)^{n-1} (-s)^{n-1} e^{-st} f(t) \Big]_0^\infty + (-1)^n (-s)^n \int_0^\infty e^{-st} f(t)\, dt$$

$$= - sf^{n-1}(0) - f^{n-2}(0) - s^2 f^{n-3}(0) - \dots - s^{n-1}f(0) + s^n L\{f(t)\}$$

$$= s^n \bar{f}(s) - s^{n-1}f(0) - s^{n-2}f'(0) - s^{n-3}f''(0) - \dots\dots - f^{n-1}(0)$$

assuming that $\underset{t\to\infty}{\text{Lt}} [e^{-st} f^m(t)] = 0$, where $m = 0, 1, 2, \dots\dots, n - 1$.

Note. In particular, when $n = 2, 3, 4, \dots\dots$, we have

$$L\{f''(t)\} = s^2 \bar{f}(s) - sf(0) - f'(0)$$

$$L\{f'''(t)\} = s^3 \bar{f}(s) - s^2 f(0) - sf'(0) - f''(0)$$

$$L\{f^{iv}(t)\} = s^4 \bar{f}(s) - s^3 f(0) - s^2 f'(0) - sf''(0) - f'''(0) \text{ and so on.}$$

These results will be used in the solution of differential equations.

2.8. LAPLACE TRANSFORMS OF INTEGRALS

Theorem. *If* $L\{f(t)\} = \bar{f}(s)$, *then*

$$L\left\{\int_0^t f(u)\,du\right\} = \frac{1}{s}\bar{f}(s) \quad or \quad L^{-1}\left\{\frac{1}{s}\bar{f}(s)\right\} = \int_0^t f(u)\,du.$$

Proof. Let $\phi(t) = \int_0^t f(u)\,du$, then $\phi'(t) = f(t)$ and $\phi(0) = 0$

$$\therefore \quad L\{\phi'(t)\} = s\bar{\phi}(s) - \phi(0)$$

$$\Rightarrow \quad L\{f(t)\} = s\bar{\phi}(s) \Rightarrow \bar{f}(s) = sL\{\phi(t)\} \Rightarrow L\{\phi(t)\} = \frac{1}{s}\bar{f}(s)$$

$$\therefore \quad L\left[\int_0^t f(u)\,du\right] = \frac{1}{s}\bar{f}(s) \quad \text{or} \quad L^{-1}\left\{\frac{1}{s}\bar{f}(s)\right\} = \int_0^t f(u)\,du$$

Hence, **if** $\mathbf{L^{-1}\{\bar{f}(s)\} = f(t)}$ **then** $\mathbf{L^{-1}\left[\frac{1}{s}\bar{f}(s)\right] = \int_0^t f(u)\,du.}$

ILLUSTRATIVE EXAMPLES

Example 1. *If* $L\{t \sin \omega t\} = \dfrac{2\omega s}{(s^2+\omega^2)^2}$, *evaluate*

(*i*) $L\{\omega t \cos \omega t + \sin \omega t\}$ (*ii*) $L\{2 \cos \omega t - \omega t \sin \omega t\}$.

Sol. Let $f(t) = t \sin \omega t$, then $f'(t) = \omega t \cos \omega t + \sin \omega t$

and $f''(t) = 2\omega \cos \omega t - \omega^2 t \sin \omega t$

Also $f(0) = 0, f'(0) = 0, f''(0) = 2\omega$

$$\bar{f}(s) = \frac{2\omega s}{(s^2+\omega^2)^2}$$

Since $L\{f^n(t)\} = s^n\bar{f}(s) - s^{n-1}f(0) - s^{n-2}f'(0) - s^{n-3}f''(0) - \dots\dots$

(*i*) $L\{f'(t)\} = s\bar{f}(s) - f(0)$

$$\Rightarrow \quad L\{\omega t \cos \omega t + \sin \omega t\} = s \cdot \frac{2\omega s}{(s^2+\omega^2)^2} - 0 = \frac{2\omega s^2}{(s^2+\omega^2)^2}.$$

(*ii*) $$L\{f''(t)\} = s^2 \bar{f}(s) - sf(0) - f'(0)$$

$$\Rightarrow \quad L[2\omega \cos \omega t - \omega^2 t \sin \omega t] = s^2 \cdot \frac{2\omega s}{(s^2+\omega^2)^2} - s \cdot 0 - 0 = \frac{2\omega s^3}{(s^2+\omega^2)^2}$$

$$\therefore \quad L\{2 \cos \omega t - \omega t \sin \omega t\} = \frac{2s^3}{(s^2+\omega^2)^2}.$$

Example 2. *Find L{t^2 sin at}.*

Sol. Since $$L\{\sin at\} = \frac{a}{s^2+a^2},\ s > 0,$$

$$\therefore \quad L\{t^2 \sin at\} = (-1)^2 \frac{d^2}{ds^2} L\{\sin at\}$$

$$= \frac{d^2}{ds^2}\left\{\frac{a}{s^2+a^2}\right\} = \frac{d}{ds}\left\{\frac{-2as}{(s^2+a^2)^2}\right\} = \frac{2a(3s^2-s^2)}{(s^2+a^2)^3},\ s > 0.$$

Example 3. *Find L{t^3 cos t}.*

Sol. Since $$L\{\cos t\} = \frac{s}{s^2+1},\ s > 0,$$

$$\therefore \quad L\{t^3 \cos t\} = (-1)^3 \frac{d^3}{ds^3}\left(\frac{s}{s^2+1}\right) = -\frac{d^2}{ds^2}\left\{\frac{-s^2+1}{(s^2+1)^2}\right\}$$

$$= \frac{d^2}{ds^2}\left\{\frac{s^2-1}{(s^2+1)^2}\right\} = \frac{d^2}{ds^2}\left\{\frac{s^2+1-2}{(s^2+1)^2}\right\} = \frac{d^2}{ds^2}\left\{\frac{1}{s^2+1} - \frac{2}{(s^2+1)^2}\right\}$$

$$= \frac{d}{ds}\left\{-\frac{2s}{(s^2+1)^2} + \frac{8s}{(s^2+1)^3}\right\} = \frac{6s^4-36s^2+6}{(s^2+1)^4}.$$

Example 4. *Show that* $L\{t^n e^{at}\} = \dfrac{n!}{(s-a)^{n+1}},\ s > a.$

Sol. Since $L\{e^{at}\} = \dfrac{1}{s-a},\ s > a,$

$$\therefore \quad L\{t^n e^{at}\} = (-1)^n \frac{d^n}{ds^n}\left(\frac{1}{s-a}\right) = (-1)^n \cdot \frac{(-1)^n n!}{(s-a)^{n+1}} = \frac{n!}{(s-a)^{n+1}},\ s > a.$$

Example 5. *Find L{sin at – at cos at}.*

Sol. Let F(t) = sin at – at cos at, then

$$F'(t)\} = a^2 t \sin at \quad \text{and} \quad F(0) = 0.$$

$$\therefore \quad L\{F'(t)\} = sL\{F(t)\} - F(0) \text{ gives}$$

$$L\{a^2 t \sin at\} = sL\{\sin at - at \cos at\} - 0$$

or $$\text{L}\{\sin at - at\cos at\} = \frac{1}{s}\,\text{L}\{a^2 t\sin at\} = \frac{a^2}{s}\,\text{L}\{t\sin at\}$$

$$= -\frac{a^2}{s}\frac{d}{ds}\,\text{L}\{\sin at\} = -\frac{a^2}{s}\cdot\frac{d}{ds}\left(\frac{a}{s^2+a^2}\right)$$

$$= -\frac{a^2}{s}\cdot\left[\frac{-2as}{(s^2+a^2)^2}\right] = \frac{2a^3}{(s^2+a^2)^2}.$$

Example 6. *Find the Laplace transformation of the following functions*
(i) $f(t) = t\sin^2 t$. *(ii)* $f(t) = t\cos^2 t$.

Sol. (*i*) Since $\sin^2 t = \frac{1}{2}(1-\cos 2t)$, so

$$\text{L}\{f(t)\} = \text{L}\{t\sin^2 t\} = \text{L}\left\{t\,\frac{(1-\cos 2t)}{2}\right\} = \frac{1}{2}\,\text{L}\{t\} - \frac{1}{2}\,\text{L}\{t\cos 2t\} \quad ...(1)$$

Since $\text{L}\{t^n\} = \frac{n!}{s^{n+1}}$.

$\therefore$ $\text{L}\{t\} = \frac{1}{s^2}$. ...(2)

Again, since $\text{L}\{\cos at\} = \frac{s}{s^2+a^2}$.

$$\therefore \quad \text{L}\{t\cos 2t\} = (-1)\frac{d}{ds}\left[\frac{s}{s^2+2^2}\right] = -\left[\frac{(s^2+2^2).1-s.2s}{(s^2+2^2)^2}\right] = \frac{s^2-4}{(s^2+2^2)^2}. \quad ...(3)$$

From equation (1), (2) and (3), we get

$$\text{L}\{t\sin^2 t\} = \frac{1}{2s^2} - \frac{(s^2-4)}{2(s^2+2^2)^2} = \frac{6s^2+8}{s^2(s^2+4)^2}.$$

(*ii*) Since $\cos^2 t = \frac{1}{2}(1+\cos 2t)$, so

$$\text{L}\{f(t)\} = \text{L}\{t\cos^2 t\} = \text{L}\left\{t\,\frac{(1+\cos 2t)}{2}\right\} = \frac{1}{2}\,\text{L}\{t\} + \frac{1}{2}\,\text{L}\{t\cos 2t\} \quad ...(1)$$

Since $\text{L}\{t^n\} = \frac{n!}{s^{n+1}}$.

$\therefore$ $\text{L}\{t\} = \frac{1}{s^2}$. ...(2)

Again, since $\text{L}\{\cos at\} = \frac{s}{s^2+a^2}$,

$$\therefore \quad \text{L}\{t\cos 2t\} = (-1)\frac{d}{dt}\left[\frac{s}{s^2+2^2}\right] = -\left[\frac{(s^2+2^2)-2s^2}{(s^2+2^2)^2}\right] = \frac{s^2-4}{(s^2+2^2)^2}.$$

From equations (1), (2) and (3), we get

$$L\{t \cos^2 t\} = \frac{1}{2s^2} + \frac{s^2-4}{2(s^2+4)^2} = \frac{s^4+8s^2+16+s^2-4s^2}{2s^2(s^2+4)^2} = \frac{s^4+2s^2+8}{s^2(s^2+4)^2}.$$

Example 7. *Find the value of $L\{t^2 e^{-at}\}$.*

Sol. Since $\quad L\{e^{-at}\} = \frac{1}{s+a}$, so

$$L\{t^2e^{-at}\} = (-1)^2 \frac{d^2}{ds^2}\left\{\frac{1}{s+a}\right\} = \frac{2}{(s+a)^3}.$$

Example 8. *Find the value of $L\{te^{-t} \sin^2 t\}$.*

Sol. Since $\sin^2 t = \frac{1}{2}\{1-\cos 2t)$, so

$$L\{te^{t} \sin^2 t\} = L\left\{te^{-t}\frac{(1-\cos 2t)}{2}\right\} = \frac{1}{2}L\{te^{-t}\} - \frac{1}{2}L\{te^{-t}\cos 2t\}. \quad ...(1)$$

Now, $$L\{te^{-t}\} = (-1)\frac{d}{ds}L\{e^{-t}\} = (-1)\frac{d}{ds}\left(\frac{1}{s+1}\right) = \frac{1}{(s+1)^2} \quad ...(2)$$

and $$L\{\{te^{-t}\cos 2t\} = (-1)\frac{d}{ds}L\{e^{-t}\cos 2t\} = (-1)\frac{d}{ds}\left[\frac{(s+1)}{(s+1)^2+2^2}\right]$$

$$= (-1)\frac{(s^2+2s+5).1-(s+1).2(s+1)}{(s^2+2s+5)^2} = \frac{s^2+2s+3}{(s^2+2s+5)^2} \quad ...(3)$$

From equations (1), (2) and (3), we get

$$L\{te^{-1}\sin^2 t\} = \frac{1}{2(s+1)^2} - \frac{s^2+2s+3}{2(s^2+2s+5)^2} = \frac{s^4+4s^3+13s^2+18s+22}{2(s+1)^2(s^2+2s+5)^2}.$$

Example 9. *Find the Laplace transformation of the following functions:*

(*i*) $\frac{1-\cos at}{t}$ (*ii*) $\frac{1-e^{-t}}{t}$.

(*iii*) $\frac{\sinh t}{t}$ (*iv*) $\frac{e^{at}-\cos bt}{t}$.

Sol. (*i*) Let $F(t) = 1 - \cos at$. Then

$$\underset{t\to 0}{\text{Lt}}\frac{F(t)}{t} = \underset{t\to 0}{\text{Lt}}\frac{1-\cos at}{t} = 0$$

and $$f(s) = L\{F(t)\} = L\{1-\cos at\} = L\{1\} - L\{\cos at\}$$

$$= \frac{1}{s} - \frac{s}{s^2+a^2} = \frac{a^2}{s(s^2+a^2)}, \; s > 0.$$

We know that if $L\{F(t)\} = f(s)$, then $L\left\{\frac{1}{t}F(t)\right\} = \int_s^\infty f(x)\,dx$ provided $\underset{t\to 0}{\text{Lt}}\frac{F(t)}{t}$ exists.

Therefore, $$L\left\{\frac{1-\cos at}{t}\right\} = \int_s^\infty f(x)\,dx = \int_s^\infty \frac{a^2}{x(x^2+a^2)}\,dx$$

$$= \int_s^\infty \left[\frac{1}{x} - \frac{x}{x^2+a^2}\right] dx = \left[\log x - \frac{1}{2}\log(x^2+a^2)\right]_s^\infty$$

$$= \left[\frac{1}{2}\log\frac{x^2}{x^2+a^2}\right]_s^\infty = \frac{1}{2}\operatorname*{Lt}_{x\to\infty}\log\frac{x^2}{x^2+a^2} - \frac{1}{2}\log\frac{s^2}{s^2+a^2}$$

$$= \frac{1}{2}\log 1 + \frac{1}{2}\log\frac{s^2+a^2}{s^2} = \frac{1}{2}\log\left(1+\frac{a^2}{s^2}\right).$$

(*ii*) Let $F(t) = 1 - e^{-t}$. Then

$$\operatorname*{Lt}_{t\to 0}\frac{F(t)}{t} - \operatorname*{Lt}_{t\to 0}\frac{1-e^{-t}}{t} = 1$$

and $$f(s) = L\{F(t)\} = L\{1-e^{-t}\} = L\{1\} - L\{e^{-t}\} = \frac{1}{s} - \frac{1}{s+1} = \frac{1}{s(s+1)},\ s > 0.$$

We know that if $L\{F(t)\} = f(s)$, then $L\left\{\frac{1}{t}F(t)\right\} = \int_s^\infty f(x)\,dx$ provided $\operatorname*{Lt}_{t\to 0}\frac{F(t)}{t}$ exists.

Therefore,

$$L\left\{\frac{1-e^{-t}}{t}\right\} = \int_s^\infty f(x)\,dx = \int_s^\infty \frac{1}{x(x+1)}\,dx = \int_s^\infty \left[\frac{1}{x} - \frac{1}{x+1}\right]dx$$

$$= \left[\log x - \log(x+1)\right]_s^\infty = \left[\log\frac{x}{x+1}\right]_s^\infty$$

$$= \operatorname*{Lt}_{x\to\infty}\log\frac{x}{x+1} - \log\frac{s}{s+1} = \log 1 - \log\frac{s}{s+1}$$

$$= \log\frac{s+1}{s} = \log\left(1+\frac{1}{s}\right).$$

(*iii*) Let $F(t) = \sinh t$. Then

$$\operatorname*{Lt}_{t\to 0}\frac{F(t)}{t} = \operatorname*{Lt}_{t\to 0}\frac{\sinh t}{t} = 1$$

$$f(s) = L\{F(t)\} = L\{\sinh t\} = \frac{1}{s^2-1},\ s > 1.$$

$$\therefore \quad L\left\{\frac{\sinh t}{t}\right\} = \int_s^\infty F(x)\,dx = \int_s^\infty \frac{1}{x^2-1}\,dx = \left[\frac{1}{2}\log\frac{x-1}{x+1}\right]_s^\infty$$

$$= \frac{1}{2}\operatorname*{Lt}_{x\to\infty}\log\frac{x-1}{x+1} - \frac{1}{2}\log\frac{s-1}{s+1}$$

$$= \frac{1}{2}\log 1 - \frac{1}{2}\log\frac{s-1}{s+1} = \frac{1}{2}\log\frac{s+1}{s-1}.$$

(*iv*) Let $F(t) = e^{at} - \cos bt$. Then

$$\operatorname*{Lt}_{t \to 0} \frac{F(t)}{t} = \operatorname*{Lt}_{t \to 0} \frac{e^{at} - \cos bt}{t} = a$$

and
$$f(s) = L\{F(t)\} = L\{e^{at} - \cos bt\} - L\{e^{at}\} - L\{\cos bt\}$$

$$= \frac{1}{s-a} - \frac{s}{s^2+b^2} = \frac{as+b^2}{(s-a)(s^2+b^2)}, \; s > 0.$$

$$\therefore \quad L\left\{\frac{e^{at} - \cos bt}{t}\right\} = \int_s^\infty f(x)\,dx = \int_s^\infty \frac{ax+b^2}{(x-a)(x^2+b^2)}\,dx$$

$$= \int_s^\infty \left[\frac{1}{x-a} - \frac{x}{x^2+b^2}\right] dx = \left[\log(x-a) - \frac{1}{2}\log(x^2+b^2)\right]_s^\infty$$

$$= \left[\frac{1}{2}\log\frac{(x-a)^2}{x^2+b^2}\right]_s^\infty = \frac{1}{2}\operatorname*{Lt}_{x\to\infty}\log\frac{(x-a)^2}{x^2+b^2} - \frac{1}{2}\log\frac{(s-a)^2}{s^2+b^2}$$

$$= \frac{1}{2}\log 1 - \frac{1}{2}\log\frac{(s-a)^2}{s^2+b^2} = \frac{1}{2}\log\frac{s^2+b^2}{(s-a)^2}.$$

Example 10. *Evaluate* $\int_0^\infty \frac{e^{-at} - e^{-bt}}{t}\,dt.$

Sol. Let $F(t) = e^{-at} - e^{-bt}$. Then

$$f(s) = L\{F(t)\} = L\{e^{-at}\} - L\{e^{-bt}\} = \frac{1}{s+a} - \frac{1}{s+b}.$$

$$\therefore \quad L\left\{\frac{F(t)}{t}\right\} = \int_s^\infty f(x)\,dx$$

or
$$\int_0^\infty e^{-st}\,.\,\frac{e^{-at} - e^{-bt}}{t}\,dt = \int_s^\infty \left(\frac{1}{x+a} - \frac{1}{x+b}\right)dx$$

$$= [-\log(x+a) - \log(x+b)]_s^\infty = -\log\frac{s+a}{s+b} = \log\frac{s+b}{s+a}.$$

Now taking limit as $s \to 0$, we have

$$\int_0^\infty \frac{e^{-at} - e^{-bt}}{t}\,dt = \log\frac{b}{a}.$$

Example 11. *Find* (*i*) $L^{-1}\left\{\frac{1}{s^2+8s+16}\right\}$ (*ii*) $L^{-1}\left\{\frac{3s+2}{4s^2+12s+9}\right\}$

(*iii*) $L^{-1}\left\{\frac{3s+7}{s^2-2s-3}\right\}$ (*iv*) $L^{-1}\left\{\frac{3s-2}{s^2-4s+20}\right\}$.

Sol. (*i*) Given expression

$$= L^{-1}\left\{\frac{1}{s^2+8s+16}\right\} = L^{-1}\left\{\frac{1}{(s+4)^2}\right\} = e^{-4t} L^{-1}\left\{\frac{1}{s^2}\right\},$$

| Using first shifting theorem

$$= e^{-4t}\frac{t}{1!} = te^{-4t}.$$

(*ii*) Given expression $= L^{-1}\left\{\frac{3s+2}{4s^2+12s+9}\right\} = L^{-1}\left\{\frac{3(s+3/2)-5/2}{4(s+3/2)^2}\right\}$

$$= e^{-3t/2} L^{-1}\left\{\frac{3s-5/2}{4s^2}\right\}, \text{ using first shifting theorem}$$

$$= e^{-3t/2}\left[\frac{3}{4}L^{-1}\left\{\frac{1}{s}\right\} - \frac{5}{8}L^{-1}\left\{\frac{1}{s^2}\right\}\right] = e^{-3t/2}\left[\frac{3}{4}.1 - \frac{5}{8}.\frac{t}{1!}\right] = \frac{1}{8}e^{-3t/2}(2-5t).$$

(*iii*) Given expression $= L^{-1}\left\{\frac{3s+7}{s^2-2s-3}\right\} = L^{-1}\left\{\frac{3(s-1)+10}{(s-1)^2-4}\right\}$

$$= L^{-1}\left\{\frac{3(s-1)}{(s-1)^2-4} + \frac{10}{(s-1)^2-4}\right\}$$

$$= 3L^{-1}\left\{\frac{s-1}{(s-1)^2-2^2}\right\} + 10L^{-1}\left\{\frac{1}{(s-1)^2-2^2}\right\}$$

$$= 3e^t L^{-1}\left\{\frac{s}{s^2-2^2}\right\} + 10e^t L^{-1}\left\{\frac{1}{s^2-2^2}\right\}, \text{ using first shifting theorem}$$

$$= 3e^t \cosh 2t + 5e^t \sinh 2t = 4e^{3t} - e^{-t}.$$

(*iv*) Given expression $= L^{-1}\left\{\frac{3s-2}{s^2-4s+20}\right\} = L^{-1}\left\{\frac{3s-2}{(s-2)^2+16}\right\}$

$$= L^{-1}\left\{\frac{3(s-2)}{(s-2)^2+16} + \frac{4}{(s-2)^2+16}\right\}$$

$$= 3L^{-1}\left\{\frac{s-2}{(s-2)^2+4^2}\right\} + 4L^{-1}\left\{\frac{1}{(s-2)^2+4^2}\right\}$$

$$= 3e^{2t}L^{-1}\left\{\frac{s}{s^2+4^2}\right\} + 4e^{2t}L^{-1}\left\{\frac{1}{s^2+4^2}\right\}, \text{ by first shifting theorem}$$

$$= 3e^{2t}\cos 4t + e^{2t}\sin 4t.$$

Example 12. *Evaluate*

(*i*) $L^{-1}\left\{\dfrac{1}{s(s+1)(s+2)}\right\}$ (*ii*) $L^{-1}\left\{\dfrac{2s+3}{(s-1)(s+2)^2}\right\}$

(*iii*) $L^{-1}\left\{\dfrac{2s^2-6s+5}{s^3-6s^2+11s-6}\right\}$

Sol. (*i*) To find the inverse Laplace transform, first we shall resolve given rational function into partial fractions.

Let $$\frac{1}{s(s+1)(s+2)} = \frac{A}{s} + \frac{B}{s+1} + \frac{C}{s+2}.$$

Multiplying both sides by $s(s+1)(s+2)$, we get

$$1 = A(s+1)(s+2) + B(s+2)s + C(s+1)s.$$

Put $s = 0;\ 1 = A(0+1)(0+2) + 0 + 0 \quad \therefore\ A = \frac{1}{2}$

$s = -1,\ 1 = B(-1)(-1+2) \quad \therefore\ B = -1$

$s = -2;\ 1 = C(-2+1)(-2) \quad \therefore\ C = \frac{1}{2}$

$$\therefore\ L^{-1}\left\{\frac{1}{s(s+1)(s+2)}\right\} = L^{-1}\left\{\frac{1}{2}\cdot\frac{1}{s} - \frac{1}{s+1} + \frac{1}{2}\cdot\frac{1}{s+2}\right\}$$

$$= \frac{1}{2}L^{-1}\left\{\frac{1}{s}\right\} - L^{-1}\left\{\frac{1}{s+1}\right\} + \frac{1}{2}L^{-1}\left\{\frac{1}{s+2}\right\}$$

$$= \frac{1}{2}\cdot 1 - e^{-t} + \frac{1}{2}e^{-2t} = \frac{1}{2} - e^{-t} + \frac{1}{2}e^{-2t}.$$

(*ii*) Let $$\frac{2s+3}{(s-1)(s+2)^2} = \frac{A}{s-1} + \frac{B}{s+2} + \frac{C}{(s+2)^2}.$$

Multiplying both side by $(s-1)(s+2)^2$, we get

$$2s + 3 = A(s+2)^2 + B(s-1)(s+2) + C(s-1)$$

Put $s = 1;\ 5 = 9A \quad \therefore\ A = 5/9$

$s = -2;\ -1 = -3C \quad \therefore\ C = 1/3.$

Since there is no common factor in the first and last term of right-hand side we cannot make these two terms simultaneously zero to find the value of B. Therefore, we shall give some value of s, other than 1 and -2, say $s = 0$.

Put $s = 0;\ 3 = 4A - 2B - C$ or $2B = \frac{20}{9} - \frac{1}{3} - 3 \quad \therefore\ B = -10/9$

Hence $$L^{-1}\left\{\frac{2s+3}{(s-1)(s+2)^2}\right\} = \frac{5}{9}L^{-1}\left\{\frac{1}{s+1}\right\} - \frac{10}{9}L^{-1}\left\{\frac{1}{s+2}\right\} + \frac{1}{3}L^{-1}\left\{\frac{1}{(s+2)^2}\right\}$$

$$= \frac{5}{9}e^{t} - \frac{10}{9}e^{-2t} + \frac{1}{3}te^{-2t}.$$

(*iii*) Consider $s^3 - 6s^2 + 11s - 6 = (s - 1)(s - 2)(s - 3)$.

Let $$\frac{2s^2 - 6s + 5}{(s-1)(s-2)(s-3)} = \frac{A}{s-1} + \frac{B}{s-2} + \frac{C}{s-3}$$

or $$2s^2 - 6s + 5 = A(s-2)(s-3) + B(s-1)(s-3) + C(s-2)(s-2)$$

Put $s = 1;\quad 1 = 2A \qquad \therefore\ A = \frac{1}{2}$

$s = 2;\quad 1 = -B \qquad \therefore\ B = -1$

$s = 3;\quad 5 = 2C \qquad \therefore\ C = \frac{5}{2}$

$$\therefore\quad L^{-1}\left\{\frac{2s^2 - 6s + 5}{s^3 - 6s^2 + 11s - 6}\right\} = \frac{1}{2}L^{-1}\left\{\frac{1}{s-1}\right\} - L^{-1}\left\{\frac{1}{s-2}\right\} + \frac{5}{2}L^{-1}\left\{\frac{1}{s-3}\right\}$$

$$= \frac{1}{2}e^t - e^{2t} + \frac{5}{2}e^{3t}.$$

Example 13. *Find* $L^{-1}\left\{\frac{5s^2 - 15s - 11}{(s+1)(s-2)^3}\right\}$.

Sol. We have

$$L^{-1}\left\{\frac{5s^2 - 15s - 11}{(s+1)^2(s-2)^3}\right\} = L^{-1}\left\{\frac{-1}{3(s+1)} + \frac{1}{3(s-2)} + \frac{4}{(s-2)^2} - \frac{7}{(s-2)^3}\right\}$$

$$= -\frac{1}{3}L^{-1}\left\{\frac{1}{s+1}\right\} + \frac{1}{3}L^{-1}\left\{\frac{1}{s-2}\right\} + 4L^{-1}\left\{\frac{1}{(s-2)^2}\right\} - 7L^{-1}\left\{\frac{1}{(s-2)^3}\right\}$$

$$= -\frac{1}{3}e^{-t} + \frac{1}{3}e^{2t} + 4e^{2t}L^{-1}\left\{\frac{1}{s^2}\right\} - 7e^{2t}L^{-1}\left\{\frac{1}{s^3}\right\}$$

$$= -\frac{1}{3}e^{-t} + \frac{1}{3}e^{2t} + 4e^{2t}\frac{t}{1!} - 7e^{2t}\frac{t^2}{2!}$$

$$= -\frac{1}{3}e^{-t} + \frac{1}{3}e^{2t} + 4te^{2t} - \frac{7}{2}t^2e^{2t}.$$

Example 14. *Find* (*i*) $L^{-1}\left\{\log\left(1 + \frac{1}{s^2}\right)\right\}$ (*ii*) $L^{-1}\left\{\log\left(1 - \frac{1}{s^2}\right)\right\}$.

Sol. (*i*) Let $$f(s) = \log\left(1 + \frac{1}{s^2}\right) = -\log\frac{s^2}{s^2+1} = -2\log s + \log(s^2 + 1)$$

$$\therefore\quad f'(s) = -2\left(\frac{1}{s} - \frac{s}{s^2+1}\right)$$

$$\therefore\quad L^{-1}\{f'(s)\} = -2\left[L^{-1}\left\{\frac{1}{s}\right\} - L^{-1}\left\{\frac{s}{s^2+1^2}\right\}\right] = -2[1 - \cos t]$$

or $$(-1)tL^{-1}\{f(s)\} = -2(1 - \cos t)$$

or $$L^{-1}\left\{\log\left(1 + \frac{1}{s^2}\right)\right\} = \frac{2}{t}(1 - \cos t).$$

(*ii*) Let $$f(s) = \log\left(1 - \frac{1}{s^2}\right) = \log\frac{s^2 - 1}{s^2} = -2\log s + \log(s^2 - 1)$$

$$\therefore \quad f'(s) = 2\left(\frac{1}{s} - \frac{s}{s^2 - 1}\right)$$

$$\therefore \quad L^{-1}\{f'(s)\} = -2\left[L^{-1}\left\{\frac{1}{s}\right\} - L^{-1}\left\{\frac{s}{s^2 - 1^2}\right\}\right] = -2\,[1 - \cosh t]$$

or $$(-1)\,tL^{-1}\{f(s)\} = -2\,(1 - \cosh t)$$

or $$L^{-1}\left\{\log\left(1 - \frac{1}{s^2}\right)\right\} = \frac{2}{t}(1 - \cosh t).$$

Example 15. *Find the inverse Laplace transform of* $\frac{s}{s^4 + 4a^2}$.

Sol. Since $s^4 + 4a^4 = (s^2 + 2a^2)^2 - (2as)^2 = (s^2 + 2as + 2a^2)\,(s^2 - 2as + 2a^2)$,

so let $$\frac{s}{s^4 + 4a^2} = \frac{As + B}{s^2 + 2as + 2a^2} + \frac{Cs + D}{s^2 - 2as + 2a^2}.$$

Multiplying both sides by $s^4 + 4a^2$,

$$s = (As + B)\,(s^2 - 2as + 2a^2) + (Cs + D)\,(s^2 + 2as + 2a^2).$$

Equating coefficient of s^3, s^2, s, we get

$$0 = A + C \qquad ...(1)$$

$$0 = -2aA + B + 2aC + D \qquad ...(2)$$

$$1 = 2a^2A - 2aB + 2a^2C + 2aD \qquad ...(3)$$

Also putting $s = 0$, we get

$$0 = 2a^2B + 2a^2D; \quad i.e., \quad 0 = B + D. \qquad ...(4)$$

From (2) and (4), we get

$$0 = -A + C. \qquad ...(5)$$

Solving (1) and (5), we get $A = C = 0$.

Then (3) reduces to $D - B = -\frac{1}{2a}$ and by (4), we get

$$B = -\frac{1}{4a},\ D = \frac{1}{4a}.$$

$$\therefore \quad L^{-1}\left\{\frac{s}{s^4 + 4a^4}\right\} = -\frac{1}{4a}L^{-1}\left\{\frac{1}{s^2 + 2as + 2a^2}\right\} + \frac{1}{4a}L^{-1}\left\{\frac{1}{s^2 - 2as + 2a^2}\right\}$$

$$-\frac{1}{4a}L^{-1}\left\{\frac{1}{(s + a)^2 + a^2}\right\} + \frac{1}{4a}L^{-1}\left\{\frac{1}{(s - a)^2 + a^2}\right\}$$

$$= -\frac{1}{4a}\cdot\frac{1}{a}e^{-at}\sin at + \frac{1}{4a}\cdot\frac{1}{a}e^{at}\sin at$$

$$= \frac{1}{2a^2}\sin at\left(\frac{e^{at}-e^{-at}}{2}\right)$$

$$= \frac{1}{2a^2}\sin at \sinh at.$$

TEST YOUR KNOWLEDGE

1. Given $L\left\{2\sqrt{\frac{t}{\pi}}\right\} = \frac{1}{s^{3/2}}$, show that $L\left\{\frac{1}{\sqrt{\pi t}}\right\} = \frac{1}{\sqrt{s}}$.
2. If $L\{f(t)\} = \bar{f}(s)$, show that $L\{f'(t)\} = s\bar{f}(s) - f(0)$.
   Use it to evaluate $L\{f'(t)\}$, when $f(t) = e^{-5t}\sin t$.
3. If $L\{f(t)\} = \bar{f}(s)$, show that $L\{f''(t)\} = s^2\bar{f}(s) - sf(0) - f'(0)$.
   If $f(t) = e^{-2t}\sin 4t$, find $L\{f''(t)\}$.
4. If $L^{-1}\{\bar{f}(s)\} = f(t)$, show that $L^{-1}\left\{\frac{1}{s}\bar{f}(s)\right\} = \int_0^t f(x)\,dx$.

   Use this result to obtain $L^{-1}\left\{\frac{1}{s(s+a)}\right\}$ and $L^{-1}\left\{\frac{1}{s^2(s+a)}\right\}$.
5. Find the inverse Laplace transform of $\frac{1}{s(s^2+a^2)}$.

Answers

2. $\frac{s}{s^2+10s+26}$ 3. $\frac{-16(s+5)}{s^2+4s+20}$

4. $\frac{1}{a}(1-e^{-at})$; $\frac{1}{a^2}(at+e^{-at}-1)$ 5. $\frac{1}{a^2}(1-\cos at)$.

2.9. MULTIPLICATION BY t^n

If $L\{f(t)\} = \bar{f}(s)$, *then* $\mathbf{L\{t^n f(t)\} = (-1)^n \frac{d^n}{ds^n}[\bar{f}(s)]}$, where $n = 1, 2, 3, \ldots\ldots$

Proof. We prove the theorem by induction

$$L\{f(t)\} = \bar{f}(s) \Rightarrow \int_0^\infty e^{-st} f(t)\,dt = \bar{f}(s)$$

Differentiating both sides w.r.t. s (using Leibnitz's rule for differentiation under the integral sign), we have

$$\frac{d}{ds}\int_0^\infty e^{-st} f(t)\,dt = \frac{d}{ds}[\bar{f}(s)]$$

or $$\int_0^\infty \frac{\partial}{\partial s} [e^{-st} f(t)]\, dt = \frac{d}{ds} [\bar{f}(s)]$$

or $$\int_0^\infty - t\, e^{-st} f(t)\, dt = \frac{d}{ds} [\bar{f}(s)]$$

or $$\int_0^\infty e^{-st} [t\, f(t)]\, dt = -\frac{d}{ds} [\bar{f}(s)]$$

or $$L\{t\, f(t)\} = -\frac{d}{ds} [\bar{f}(s)]$$

which confirms the truth of the theorem for $n = 1$.

Now assume the theorem to be true for $n = m$, so that

$$L\{t^m f(t)\} = (-1)^m \frac{d^m}{ds^m} [\bar{f}(s)] \quad \text{or} \quad \int_0^\infty e^{-st} t^m f(t)\, dt = (-1)^m \frac{d^m}{ds^m} [\bar{f}(s)]$$

Differentiating both sides w.r.t. s, we have

$$\frac{d}{ds} \int_0^\infty e^{-st} t^m f(t)\, dt = (-1)^m \frac{d^{m+1}}{ds^{m+1}} [\bar{f}(s)]$$

or $$\int_0^\infty \frac{\partial}{\partial s} [e^{-st} t^m f(t)]\, dt = (-1)^m \frac{d^{m+1}}{ds^{m+1}} [\bar{f}(s)]$$

or $$\int_0^\infty - te^{-st} t^m f(t)\, dt = (-1)^m \frac{d^{m+1}}{ds^{m+1}} [\bar{f}(s)]$$

or $$\int_0^\infty e^{-st} [t^{m+1} f(t)]\, dt = (-1)^{m+1} \frac{d^{m+1}}{ds^{m+1}} [\bar{f}(s)]$$

or $$L\{t^{m+1} f(t)\} = (-1)^{m+1} \frac{d^{m+1}}{ds^{m+1}} [\bar{f}(s)]$$

which shows that the theorem is true for $n = m + 1$.

Hence by mathematical induction, the theorem is true for all positive integers n.

ILLUSTRATIVE EXAMPLES

Example 1. *Find the Laplace transforms of*

(*i*) *t* $\sin^2 3t$ (*ii*) $t^3 e^{-3t}$.

Sol. (*i*) Since $$L\{\sin^2 3t\} = L\left\{\frac{1-\cos 6t}{2}\right\} = \frac{1}{2}\left(\frac{1}{s} - \frac{s}{s^2+36}\right) = \frac{18}{s(s^2+36)}$$

$\therefore$ $$L\{t \sin^2 3t\} = (-1) \frac{d}{ds}\left[\frac{18}{s(s^2+36)}\right] = -18 \frac{d}{ds} [(s^3 + 36s)^{-1}]$$

$$= -18\,(-1)\,(s^3 + 36s)^{-2}\,(3s^2 + 36) = \frac{54(s^2+12)}{s^2(s^2+36)^2}.$$

(*ii*) Since $$L\{e^{-3t}\} = \frac{1}{s+3}$$

$\therefore$ $$L\{t^3 e^{-3t}\} = (-1)^3 \frac{d^3}{ds^3}\left(\frac{1}{s+3}\right) = -\frac{(-1)^3 . (3\,!)}{(s+3)^4} = \frac{6}{(s+3)^4}.$$

Example 2. *Find the inverse Laplace transform of* $\log \dfrac{s+1}{s-1}$.

Sol. Let $L^{-1}\left\{\log \dfrac{s+1}{s-1}\right\} = f(t)$ then $L\{f(t)\} = \bar{f}(s) = \log \dfrac{s+1}{s-1}$.

$$\therefore \quad L\{tf(t)\} = (-1)\frac{d}{ds}\left\{\log \frac{s+1}{s-1}\right\} = -\frac{d}{ds}[\log(s+1) - \log(s-1)]$$

$$= -\left[\frac{1}{s+1} - \frac{1}{s-1}\right] = \frac{1}{s-1} - \frac{1}{s+1} = L\{e^t - e^{-t}\} = L\{2\sinh t\}$$

$$\Rightarrow \quad tf(t) = 2\sinh t \quad \text{or} \quad f(t) = \frac{2\sinh t}{t}$$

Hence $$L^{-1}\left\{\log \frac{s+1}{s-1}\right\} = \frac{2\sinh t}{t}.$$

2.10. DIVISION BY t

If $L\{f(t)\} = \bar{f}(s)$, *then* $\mathbf{L}\left\{\dfrac{1}{t}\mathbf{f(t)}\right\} = \displaystyle\int_s^\infty \bar{\mathbf{f}}(\mathbf{s})\,\mathbf{ds}$ *provided the integral exists.*

Proof. We have $\bar{f}(s) = \displaystyle\int_0^\infty e^{-st} f(t)\,dt$

Integrating both sides w.r.t. s from s to ∞, we have

$$\int_0^\infty \bar{f}(s)\,ds = \int_0^\infty\left[\int_0^\infty e^{-st} f(t)\,dt\right] ds$$

Since s and t are independent, changing the order of integration on the right-hand side, we have

$$\int_0^\infty \bar{f}(s)\,ds = \int_0^\infty\left[\int_s^\infty e^{-st}\,ds\right] f(t)\,dt$$

$$= \int_0^\infty\left[\frac{e^{-st}}{-t}\right]_s^\infty f(t)\,dt = \int_0^\infty e^{-st}\frac{f(t)}{t}\,dt = L\left\{\frac{1}{t}f(t)\right\}.$$

Example 3. *Find the Laplace transforms of*

(*i*) $\dfrac{e^{-at} - e^{-bt}}{t}$ (*ii*) $\dfrac{\cos at - \cos bt}{t}$.

Sol. (*i*) Since $L\{e^{-at} - e^{-bt}\} = \dfrac{1}{s+a} - \dfrac{1}{s+b}$

$$\therefore \quad L\left\{\frac{e^{-at} - e^{-bt}}{t}\right\} = \int_s^\infty\left(\frac{1}{s+a} - \frac{1}{s+b}\right) ds$$

$$= \Big[\log(s+a) - \log(s+b)\Big]_s^\infty = \left[\log\frac{s+a}{s+b}\right]_s^\infty = \left[\log\frac{1+\frac{a}{s}}{1+\frac{b}{s}}\right]_s^\infty$$

$$= \log 1 - \log \frac{1+\frac{a}{s}}{1+\frac{b}{s}} = 0 - \log \frac{s+a}{s+b} = \log \frac{s+b}{s+a}.$$

(ii) Since $L\{\cos at - \cos bt\} = \frac{s}{s^2+a^2} - \frac{s}{s^2+b^2}$

$$\therefore \quad L\left\{\frac{\cos at - \cos bt}{t}\right\} = \int_s^\infty \left(\frac{s}{s^2+a^2} - \frac{s}{s^2+b^2}\right) ds$$

$$= \left[\frac{1}{2}\log(s^2+a^2) - \frac{1}{2}\log(s^2+b^2)\right]_s^\infty$$

$$= \frac{1}{2}\left[\log \frac{s^2+a^2}{s^2+b^2}\right]_s^\infty = \frac{1}{2}\left[\log \frac{1+\frac{a^2}{s^2}}{1+\frac{b^2}{s^2}}\right]_s^\infty$$

$$= \frac{1}{2}\left[\log 1 - \log \frac{s^2+a^2}{s^2+b^2}\right] = \frac{1}{2}\log \frac{s^2+b^2}{s^2+a^2}.$$

Example 4. *Find the Laplace transforms of the following functions:*

(i) $t \sin 3t \cos 2t$ *(ii)* $t\, e^{-t} \cosh t$

(iii) $t^2\, e^{-2t} \cos t$.

Sol. *(i)* $L(\sin 3t \cos 2t) = \frac{1}{2} L(\sin 5t + \sin t)$

$$= \frac{1}{2}\left(\frac{5}{s^2+5^2} + \frac{1}{s^2+1}\right)$$

$$\Rightarrow \quad L(t \sin 3t \cos 2t) = -\frac{d}{ds}\left[\frac{1}{2}\left(\frac{5}{s^2+25} + \frac{1}{s^2+1}\right)\right]$$

$$= -\frac{1}{2}\left[\frac{-5}{(s^2+25)^2}(2s) - \frac{1}{(s^2+1)^2}(2s)\right] = \frac{5s}{(s^2+25)^2} + \frac{s}{(s^2+1)^2}.$$

(ii)
$$L(e^{-t}\cosh t) = L\left[e^{-t}\left(\frac{e^t+e^{-t}}{2}\right)\right] = \frac{1}{2} L(1+e^{-2t}) = \frac{1}{2}\left(\frac{1}{s} + \frac{1}{s+2}\right)$$

$$\Rightarrow \quad L(te^{-t}\cosh t) = -\frac{d}{ds}\left[\frac{1}{2}\left(\frac{1}{s} + \frac{1}{s+2}\right)\right]$$

$$= -\frac{1}{2}\left[-\frac{1}{s^2} - \frac{1}{(s+2)^2}\right] = \frac{1}{2}\,\frac{(s+2)^2+s^2}{s^2(s+2)^2} = \frac{s^2+2s+2}{s^2(s+2)^2}.$$

(iii)
$$L(e^{-2t}\cos t) = \frac{s+2}{(s+2)^2+1} = \frac{s+2}{s^2+4s+5}$$

$$\Rightarrow L\,(t^2 e^{-2t} \cos t) = (-1)^2 \frac{d^2}{ds^2}\left[\frac{s+2}{s^2+4s+5}\right]$$

$$= \frac{d}{ds}\left[\frac{(s^2+4s+5)\times 1-(s+2)(2s+4)}{(s^2+4s+5)^2}\right] = \frac{d}{ds}\left[-\frac{s^2+4s+3}{(s^2+4s+5)^2}\right]$$

$$= -\frac{(s^2+4s+5)^2(2s+4)-(s^2+4s+3)\,.\,2(s^2+4s+5)(2s+4)}{(s^2+4s+5)^4}$$

$$= -\frac{(s^2+4s+5)(2s+4)-2(2s+4)(s^2+4s+3)}{(s^2+4s+5)^3}$$

$$= \frac{2(s^3+6s^2+9s+2)}{(s^2+4s+5)^3}.$$

Example 5. *Find the Laplace transforms of the following functions:*

(*i*) $\dfrac{e^{-t}\sin t}{t}$ (*ii*) $\dfrac{1-\cos 2t}{t}$.

Sol. (*i*) $L\,(e^{-t}\sin t) = \dfrac{1}{(s+1)^2+1}$

$$\Rightarrow \quad L\left(\frac{e^{-t}\sin t}{t}\right) = \int_s^\infty \frac{1}{(s+1)^2+1}\,ds = \Big[\tan^{-1}(s+1)\Big]_s^\infty$$

$$= \frac{\pi}{2} - \tan^{-1}(s+1) = \cot^{-1}(s+1).$$

(*ii*) $L\,(1-\cos 2t) = \dfrac{1}{s} - \dfrac{s}{s^2+2^2}$

$$\Rightarrow \quad L\left(\frac{1-\cos 2t}{t}\right) = \int_s^\infty \left(\frac{1}{s}-\frac{s}{s^2+4}\right) ds$$

$$= \left[\log s - \frac{1}{2}\log(s^2+4)\right]_s^\infty = \frac{1}{2}\left[\log\frac{s^2}{s^2+4}\right]_s^\infty$$

$$= \frac{1}{2}\left[\lim_{s\to\infty}\log\frac{1}{1+\frac{4}{s^2}} - \log\frac{s^2}{s^2+4}\right]$$

$$= \frac{1}{2}\left[\log 1 + \log\frac{s^2+4}{s^2}\right] = \frac{1}{2}\log\left(\frac{s^2+4}{s^2}\right).$$

Example 6. *Evaluate*

(*i*) $\displaystyle\int_0^\infty t^3 e^{-t} \sin t\, dt$ (*ii*) $\displaystyle\int_0^\infty \frac{e^{-t}\sin^2 t}{t}\,dt.$

Sol. (*i*) $\int_0^{\infty} t^3 e^{-t} \sin t\, dt = \int_0^{\infty} e^{-st}(t^3 \sin t)\, dt$ where $s = 1$

$= \text{L}(t^3 \sin t)$ by definition

$$= (-1)^3 \frac{d^3}{ds^3}\left(\frac{1}{s^2+1}\right) = \frac{d^2}{ds^2}\left[\frac{2s}{(s^2+1)^2}\right]$$

$$= \frac{d}{ds}\left[\frac{(s^2+1)^2 \times 2 - 2s \times 2(s^2+1) \times 2s}{(s^2+1)^4}\right]$$

$$= \frac{d}{ds}\left[\frac{2(s^2+1) - 8s^2}{(s^2+1)^3}\right] = \frac{d}{ds}\left[\frac{2(1-3s^2)}{(s^2+1)^3}\right]$$

$$= 2 \times \frac{(s^2+1)^3(-6s) - (1-3s^2) \times 3(s^2+1)^2 \times 2s}{(s^2+1)^6}$$

$$= 2 \times \frac{-6s(s^2+1) - 6s(1-3s^2)}{(s^2+1)^4} = \frac{24s(s^2-1)}{(s^2+1)^4} = 0 \text{ as } s \to 1.$$

(*ii*) $\int_0^{\infty} \frac{e^{-2t} \sin^2 t}{t}\, dt = \int_0^{\infty} e^{-st}\left(\frac{\sin^2 t}{t}\right) dt$ where $s = 2$

$= \text{L}\left(\frac{\sin^2 t}{t}\right)$ by definition

$= \text{L}\left(\frac{1-\cos 2t}{t}\right) = \frac{1}{2}\log\left(\frac{s^2+4}{s^2}\right)$ as in example 3 (*ii*)

$= \frac{1}{2}\log 2$ as $s \to 2$.

2.11. CONVOLUTION THEOREM

If $L^{-1}\{\bar{f}(s)\} = f(t)$ *and* $L^{-1}\{\bar{g}(s)\} = g(t)$, *then* $L^{-1}\{\bar{f}(s)\, \bar{g}(s)\} = \int_0^t f(u)\, g(t-u)\, du.$

Proof. Let $\phi(t) = \int_0^t f(u)\, g(t-u)\, du$

then $$\text{L}\{\phi(t)\} = \int_0^{\infty} e^{-st}\left[\int_0^t f(u)\, g(t-u)\, du\right] dt = \int_0^{\infty}\int_0^t e^{-st} f(u)\, g(t-u)\, du\, dt$$

On changing the order of integration, we get

$$\text{L}\{\phi(t)\} = \int_0^{\infty}\int_u^{\infty} e^{-st} f(u)\, g(t-u)\, dt\, du$$

$$= \int_0^\infty e^{-su} f(u) \left[\int_u^\infty e^{-s(t-u)} g(t-u)\, dt \right] du$$

$$= \int_0^\infty e^{-su} f(u) \left[\int_0^\infty e^{-sv} g(v)\, dv \right] du$$

on putting $t - u = v$

$$= \int_0^\infty e^{-su} f(u)\, \bar{g}(s)\, du = \bar{g}(s) \int_0^\infty e^{-su} f(u)\, du$$

$$= \bar{g}(s) \,.\, \bar{f}(t) = \bar{f}(s)\, \bar{g}(s)$$

$$\Rightarrow \quad L^{-1}\{\bar{f}(s)\, \bar{g}(s)\} = \phi(t) = \int_0^t f(u)\, g(t-u)\, du.$$

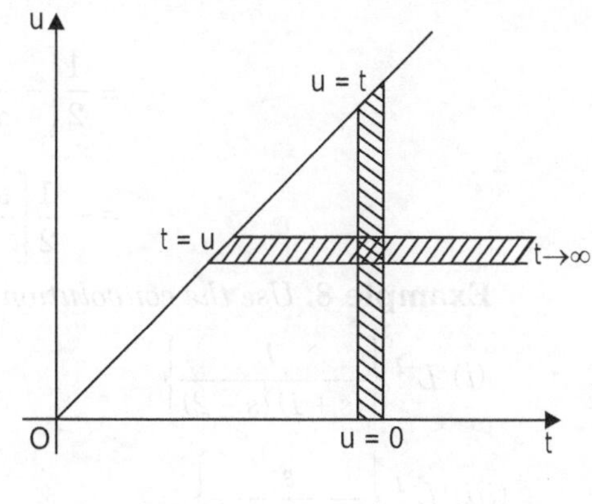

Example 7. *Apply convolution theorem to evaluate:*

(*i*) $L^{-1}\left\{\dfrac{s}{(s^2+a^2)^2}\right\}$ (*ii*) $L^{-1}\left\{\dfrac{s^2}{(s^2+a^2)(s^2+b^2)}\right\}$.

Sol. (*i*) Since $L^{-1}\left\{\dfrac{s}{s^2+a^2}\right\} = \cos at$ and $L^{-1}\left\{\dfrac{1}{s^2+a^2}\right\} = \dfrac{1}{a}\sin at$

∴ By convolution theorem, we have

$$L^{-1}\left[\frac{s}{(s^2+a^2)^2}\right] = L^{-1}\left[\frac{s}{s^2+a^2} \cdot \frac{1}{s^2+a^2}\right] = \int_0^t \cos a(t-u)\, \frac{\sin au}{a}\, du$$

$$= \frac{1}{2a}\int_0^t 2 \sin au \cos a(t-u)\, du$$

$$= \frac{1}{2a}\int_0^t [\sin at + \sin(2au - at)]\, du$$

$$= \frac{1}{2a}\left[u \sin at - \frac{1}{2a}\cos(2au - at)\right]_0^t = \frac{1}{2a}\, t \sin at.$$

(*ii*) Since $L^{-1}\left[\dfrac{s}{s^2+a^2}\right] = \cos at$ and $L^{-1}\left[\dfrac{s}{s^2+b^2}\right] = \cos bt$

∴ By convolution theorem, we have

$$L^{-1}\left[\frac{s^2}{(s^2+a^2)(s^2+b^2)}\right] = L^{-1}\left[\frac{s}{s^2+a^2} \cdot \frac{s}{s^2+b^2}\right]$$

$$= \int_0^t \cos a(t-u) \cos bu\, du$$

$$= \frac{1}{2}\int_0^t [\cos\{at - (a-b)u\} + \cos\{at - (a+b)u\}]\, du$$

$$= \frac{1}{2}\left[-\frac{1}{a-b} \sin \{at-(a+b)u\} - \frac{1}{a+b} \sin \{at-(a+b)u\} \right]_0^t$$

$$= -\frac{1}{2}\left[\frac{\sin bt - \sin at}{a-b} - \frac{\sin at + \sin bt}{a+b} \right] = \frac{a \sin at - b \sin bt}{a^2 - b^2}.$$

Example 8. *Use the convolution theorem to find*

(*i*) $L^{-1}\left\{\frac{1}{(s+1)(s-2)}\right\}$ (*ii*) $L^{-1}\left\{\frac{1}{(s-1)(s+2)}\right\}$.

(*iii*) $L^{-1}\left\{\frac{s}{(s^2+a^2)^2}\right\}$

(*iv*) $L^{-1}\left\{\frac{1}{(s-2)(s^2+1)}\right\}$ (*v*) $L^{-1}\left\{\frac{1}{s(s^2+4)^2}\right\}$.

Sol. (*i*) Since $L^{-1}\left\{\frac{1}{s+1}\right\} = e^{-t}$ and $L^{-1}\left\{\frac{1}{s-2}\right\} = e^{2t}$

∴ By the convolution theorem, we have

$$L^{-1}\left\{\frac{1}{s+1} \cdot \frac{1}{s-2}\right\} = \int_0^t e^{-x} e^{2(t-x)}\, dx = \int_0^t e^{2t} . e^{-3x}\, dx$$

$$= e^{2t}\left[\frac{-e^{-3x}}{3}\right]_0^t = \frac{1}{3}[e^{2t} - e^{-t}].$$

(*ii*) Since $L^{-1}\left\{\frac{1}{s-1}\right\} = e^{t}$ and $L^{-1}\left\{\frac{1}{s+2}\right\} = e^{-2t}$

∴ By the convolution theorem, we have

$$L^{-1}\left\{\frac{1}{s-1} \cdot \frac{1}{s+2}\right\} = \int_0^t e^{x} . e^{-2(t-x)}\, dx = \int_0^t e^{-2t} e^{3x} dx$$

$$= e^{-2t}\left[\frac{e^{3x}}{3}\right]_0^t = \frac{1}{3}[e^{t} - e^{-2t}].$$

(*iii*) Since $L^{-1}\left\{\frac{s}{s^2+a^2}\right\} = \cos at$ and $L^{-1}\left\{\frac{1}{s^2+a^2}\right\} = \frac{1}{a}\sin at$.

∴ By the convolution theorem, we have

$$L^{-1}\left\{\frac{s}{(s^2+a^2)^2}\right\} = L^{-1}\left\{\frac{s}{s^2+a^2} \cdot \frac{1}{s^2+a^2}\right\} = \int_0^t \cos ax . \frac{1}{a} \sin a(t-x)\, dx$$

$$= \frac{1}{a}\int_0^t \cos ax\, (\sin at \cos ax - \cos at \sin ax)\, dx$$

$$= \frac{1}{a}\sin at \int_0^t \cos^2 ax\, dx - \frac{1}{a}\cos at \int_0^t \sin ax \cos ax\, dx$$

$$= \frac{1}{2a}\sin at \int_0^t (1+\cos 2ax)\, dx - \frac{1}{2a}\cos at \int_0^t \sin 2ax\, dx$$

$$= \frac{1}{2a}\sin at \left[x + \frac{\sin 2ax}{2a}\right]_0^t - \frac{1}{2a}\cos at \left[\frac{-\cos 2ax}{2a}\right]_0^t$$

$$= \frac{1}{2a}\sin at \left[t + \frac{\sin 2at}{2a}\right] - \frac{1}{2a}\cos at \left[\frac{1-\cos 2at}{2a}\right]$$

$$= \frac{t \sin at}{2a} + \frac{\sin at.\, 2\sin at \cos at}{4a^2} - \frac{\cos at\, .\, 2\sin^2 at}{4a^2}$$

$$= \frac{1}{2a} t \sin at.$$

(*iv*) Since $L^{-1}\left\{\frac{1}{s-2}\right\} = e^{2t}$ and $L^{-1}\left\{\frac{1}{s^2+1}\right\} = \sin t.$

∴ By the convolution theorem, we have

$$L^{-1}\left\{\frac{1}{(s-2)(s^2+1)}\right\} = L^{-1}\left\{\frac{1}{s^2+1}\cdot\frac{1}{s-2}\right\}$$

$$= \int_0^t \sin x\, e^{2(t-x)}\, dx = e^{2t}\int_0^t e^{-2x} \sin x\, dx$$

$$= e^{2t}\left[\frac{e^{-2x}}{2^2+1}\{-2\sin x - \cos x\}\right]_0^t = \frac{1}{5}(e^{2t} - 2\sin t - \cos t).$$

(*v*) We have $L^{-1}\left\{\frac{1}{s(s^2+4)^2}\right\} = L^{-1}\left\{\frac{1}{s^2}\cdot\frac{s}{(s^2+4)^2}\right\}.$

Let $f(s) = \frac{1}{s^2}$ and $g(s) = \frac{s}{(s^2+4)^2}$. Then

$$L^{-1}\{f(s)\} = L^{-1}\left\{\frac{1}{s^2}\right\} = t = F(t), \text{ say}$$

and

$$L^{-1}\{g(s)\} = L^{-1}\left\{\frac{s}{(s^2+4)^2}\right\} = \frac{t \sin 2t}{4} = G(t) \text{ say}$$

∴ By the convolution theorem, we have

$$L^{-1}\left\{\frac{1}{s^2}\cdot\frac{s}{(s^2+4)^2}\right\} = L^{-1}\{f(s)\, g(s)\} = L^{-1}\{g(s)\, f(s)\}$$

$$= \int_0^t G(x)F(t-x)\, dx = \int_0^t \frac{x}{4}\sin 2x\, .\, (t-x)\, dx$$

$$= \frac{t}{4}\int_0^t x \sin 2x\, dx - \frac{1}{4}\int_0^t x^2 \sin 2x\, dx$$

$$= \frac{t}{4}\left[-\frac{x}{2}\cos x + \frac{1}{4}\sin 2x\right]_0^t - \frac{1}{4}\left[-\frac{x^2}{2}\cos 2x + \frac{x}{2}\sin 2x + \frac{1}{4}\cos 2x\right]_0^t$$

$$= \frac{1}{16}(1 - t\sin 2t - \cos 2t).$$

Example 9. *Find* $L^{-1}\left\{\frac{1}{(s-1)^5\,(s+2)}\right\}$.

Sol. Since $L^{-1}\left\{\frac{1}{(s-1)^5}\right\} = e^t L^{-1}\left\{\frac{1}{s^5}\right\} = e^t \cdot \frac{t^4}{4!} = \frac{t^4 e^t}{24}$ and $L^{-1}\left\{\frac{1}{s+2}\right\} = e^{-2t}$.

$\therefore$ By the convolution theorem, we have

$$L^{-1}\left\{\frac{1}{(s-1)^5\,(s+2)}\right\} = L^{-1}\left\{\frac{1}{(s-1)^5}\cdot\frac{1}{s+2}\right\} = \int_0^t \frac{x^4 e^x}{24}\,.\,e^{-2(t-x)}\,dx = \frac{e^{-2t}}{24}\int_0^t x^4 e^{3x}\,dx$$

$$= \frac{e^{-2t}}{24}\left[\frac{t^4 e^{3t}}{3} - \frac{4}{9}t^3e^{3t} + \frac{4}{9}t^2e^{3t} - \frac{8}{27}te^{3t} + \left(\frac{8}{81}e^{3t} - \frac{8}{81}\right)\right]$$

$$= \frac{e^t}{72}\left(t^4 - \frac{4}{3}t^3 + \frac{4}{3}t^2 - \frac{8}{9}t + \frac{8}{27}\right) - \frac{e^{-2t}}{243}.$$

Example 10. *Use the convolution theorem to find*

(*i*) $L^{-1}\left\{\frac{1}{s^2\,(s+1)^2}\right\}$ (*ii*) $L^{-1}\left\{\frac{s^2}{(s^2+4)^2}\right\}$ (*iii*) $L^{-1}\left\{\frac{s}{(s^2+4)^3}\right\}$.

Sol. (*i*) Since $L^{-1}\left\{\frac{1}{s^2}\right\} = t$ and $L^{-1}\left\{\frac{1}{(s+1)^2}\right\} = e^{-t}L^{-1}\left\{\frac{1}{s^2}\right\} = te^{-1}$.

$\therefore$ By the convolution theorem, we have

$$L^{-1}\left\{\frac{1}{s^2\,(s+1)^2}\right\} = \int_0^t xe^{-x}\,(t-x)\,dx = \int_0^t (xt - x^2)e^{-x}\,dx$$

$$= \left[(xt - x^2)\frac{e^{-x}}{-1} - (t-2x)\,e^{-x} + (-2)\,(-e^{-x})\right]_0^t$$

$$= te^{-1} + 2e^{-t} + t - 2.$$

(*ii*) Since $L^{-1}\left\{\frac{s}{s^2+4}\right\} = \cos 2t$, therefore by the convolution theorem, we have

$$L^{-1}\left\{\frac{s^2}{(s^2+4)^2}\right\} = L^{-1}\left\{\frac{s}{s^2+4}\cdot\frac{s}{s^2+4}\right\} = \int_0^t \cos 2x\cos 2(t-x)\,dx$$

$$= \int_0^t \cos 2x\,(\cos 2t\cos 2x + \sin 2t\sin 2x)\,dx$$

$$= \cos 2t \int_0^t \cos^2 2x\, dx + \sin 2t \int_0^t \cos 2x \sin 2x\, dx$$

$$= \frac{1}{2} \cos 2t \int_0^t (1 + \cos 4x)\, dx + \frac{\sin 2t}{2} \int_0^t \sin 4x\, dx$$

$$= \frac{\cos 2t}{2}\left(t + \frac{\sin 4t}{4}\right) + \frac{\sin 2t}{2}\left(\frac{1 - \cos 4t}{4}\right)$$

$$= \frac{t \cos 2t}{2} + \frac{\sin 2t}{8} + \frac{1}{8}(\sin 4t \cos 2t - \cos 4t \sin 2t)$$

$$= \frac{t \cos 2t}{2} + \frac{\sin 2t}{8} + \frac{1}{8}\sin (4t - 2t)$$

$$= \frac{t \cos 2t}{2} + \frac{\sin 2t}{8} + \frac{\sin 2t}{8} = \frac{t \cos 2t}{2} + \frac{\sin 2t}{4}$$

(*iii*) We have $L^{-1}\left\{\frac{s}{(s^2+4)^2}\right\} = \frac{t \sin 2t}{4}$ and $L^{-1}\left\{\frac{1}{s^2+4}\right\} = \frac{1}{2}\sin 2t$

By the convolution theorem, we have

$$L^{-1}\left\{\frac{s}{(s^2+4)^3}\right\} = L^{-1}\left\{\frac{s}{(s^2+4)^2} \cdot \frac{1}{s^2+4}\right\} = \int_0^t \frac{x \sin 2x}{4} \cdot \frac{1}{2} \sin 2(t - x)\, dx$$

$$= \frac{1}{8}\int_0^t x \sin 2x\, (\sin 2t \cos 2x - \cos 2t \sin 2x)\, dx$$

$$= \frac{\sin 2t}{8} \int_0^t x \sin 2x \cos 2x\, dx - \frac{\cos 2t}{8} \int_0^t x \sin^2 2x\, dx$$

$$= \frac{\sin 2t}{16} \int_0^t x \sin 4x\, dx - \frac{\cos 2t}{16} \int_0^t x(1 - \cos 4x)\, dx$$

$$= \frac{\sin 2t}{16}\left[-\frac{x \cos 4x}{4} + \frac{\sin 4x}{16}\right]_0^t - \frac{\cos 2t}{16}\left[\frac{x^2}{2} - \frac{x \sin 4x}{4} - \frac{\cos 4x}{16}\right]_0^t$$

$$= -\frac{t \sin 2t \cos 4t}{64} + \frac{\sin 2t \sin 4t}{256} - \frac{t^2}{32}\cos 2t + \frac{t \cos 2t \sin 4t}{64} + \frac{\cos 2t}{256}(\cos 4t - 1)$$

$$= \frac{t}{64}(\sin 4t \cos 2t - \sin 2t \cos 4t) + \frac{1}{256}(\cos 2t \cos 4t + \sin 2t \sin 4t) - \frac{t^2}{32}\cos 2t - \frac{\cos 2t}{256}$$

$$= \frac{t}{64}\sin(4t - 2t) + \frac{1}{256}\cos(4t - 2t) - \frac{t^2}{32}\cos 2t - \frac{\cos 2t}{256}$$

$$= \frac{t}{64}(\sin 2t - 2t \cos 2t).$$

TEST YOUR KNOWLEDGE

1. If $L\{f(t)\} = \bar{f}(s)$, show that $L\{tf(t)\} = -\frac{d}{ds}[\bar{f}(s)]$.

2. If $L\{f(t)\} = \bar{f}(s)$, show that $L\{t^2f(t)\} = \frac{d^2}{ds^2}[\bar{f}(s)]$. Use this result to obtain $L\{t^2 \cos at\}$.

3. Find the Laplace transforms of

(*i*) $t \sinh at$ (*ii*) $t^2 \sin 2t$

(*iii*) t^2e^{-2t} (*iv*) $t\, e^{at} \sin at$.

4. If $L\left\{\frac{1-\cos at}{a^2}\right\} = \frac{1}{s(s^2+a^2)}$, show that $L\left\{\frac{t(1-\cos at)}{a^2}\right\} = \frac{3s^2+a^2}{s^2(s^2+a^2)^2}$.

5. Find the inverse Laplace transforms of

(*i*) $\log\left(1+\frac{1}{s^2}\right)$ (*ii*) $\log\left(\frac{1+s}{s}\right)$.

6. Find the Laplace transforms of

(*i*) $\frac{\sin 2t}{t}$ (*ii*) $\frac{1-e^t}{t}$

(*iii*) $\frac{\cos 2t - \cos 3t}{t}$ (*iv*) $\frac{\sin^2 t}{t}$

(*v*) $\frac{\sinh t}{t}$ (*vi*) $\frac{1-\cos t}{t^2}$.

7. Evaluate:

(*i*) $\int_0^\infty t\, e^{-3t} \sin t\, dt$ (*ii*) $\int_0^\infty \frac{\sin t}{t}\, dt$

(*iii*) $\int_0^\infty \frac{e^{-t}-e^{-3t}}{t}\, dt$.

8. Evaluate:

(*i*) $L\left[\int_0^t e^{-t} \cos t\, dt\right]$ (*ii*) $L\left[\int_0^t \frac{\sin t}{t}\, dt\right]$.

9. Apply convolution theorem to evaluate:

(*i*) $L^{-1}\left[\frac{s^2}{(s^2+4)^2}\right]$ (*ii*) $L^{-1}\left[\frac{1}{s(s^2+4)}\right]$

(*iii*) $L^{-1}\left[\frac{1}{s(s^2-a^2)}\right]$ (*iv*) $L^{-1}\left[\frac{s^2}{s^4-a^4}\right]$.

10. Find the inverse transforms of:

(*i*) $\cot^{-1}\left(\frac{s}{a}\right)$ (*ii*) $\tan^{-1}\left(\frac{2}{s}\right)$

(*iii*) $\cot^{-1}(s+1)$ (*iv*) $\log \frac{s(s+1)}{s^2+4}$.

11. Show that:

(*i*) $L^{-1}\left(\frac{1}{s}\sin\frac{1}{s}\right) = t - \frac{t^3}{(3!)^2} + \frac{t^5}{(5!)^2} - \frac{t^7}{(7!)^2} + \ldots\ldots$

(*ii*) $L^{-1}\left(\frac{1}{s}\cos\frac{1}{s}\right) = 1 - \frac{t^2}{(2!)^2} + \frac{t^4}{(4!)^2} - \frac{t^6}{(6!)^2} + \ldots\ldots$

Answers

2. $\dfrac{2s(s^2 - 3a^2)}{(s^2 + a^2)^3}$ 3. (i) $\dfrac{2as}{(s^2 - a^2)^2}$ (ii) $\dfrac{4(3s^2 - 4)}{(s^2 + 4)^3}$

(iii) $\dfrac{2}{(s + 2)^3}$ (iv) $\dfrac{2a(s - a)}{(s^2 - 2as + 2a^2)^2}$ 5. (i) $\dfrac{2(1 - \cos t)}{t}$

(ii) $\dfrac{1}{t}(1 - e^{-t})$ 6. (i) $\cot^{-1}\left(\dfrac{s}{2}\right)$ (ii) $\log \dfrac{s - 1}{s}$

(iii) $\dfrac{1}{2} \log \dfrac{s^2 + 9}{s^2 + 4}$ (iv) $\dfrac{1}{4} \log \dfrac{s^2 + 4}{s^2}$ (v) $\dfrac{1}{2} \log \left(\dfrac{s + 1}{s - 1}\right)$

(vi) $\cot^{-1} s - \dfrac{1}{2} s \log \left(1 + \dfrac{1}{s^2}\right)$

7. (i) $\dfrac{3}{50}$ (ii) $\dfrac{\pi}{2}$ (iii) $\log 3$

8. (i) $\dfrac{s + 1}{s(s^2 + 2s + 2)}$ (ii) $\dfrac{1}{s} \cot^{-1} s$ 9. (i) $\dfrac{1}{4}(\sin 2t + 2t \cos 2t)$

(ii) $\dfrac{1}{4}(1 - \cos 2t)$ (iii) $\dfrac{1}{a^2}(\cosh at - 1)$ (iv) $\dfrac{1}{2a}(\sinh at + \sin at)$

10. (i) $\dfrac{\sin at}{t}$ (ii) $\dfrac{\sin 2t}{t}$ (iii) $e^{-t} \dfrac{\sin t}{t}$

(iv) $\dfrac{2 \cos 2t - e^{-t} - 1}{t}$

2.12. APPLICATION TO DIFFERENTIAL EQUATIONS

Laplace transforms can be used to solve ordinary as well as partial differential equations. We shall apply this method to solve only ordinary linear differential equations with constant co-efficients. The advantage of this method is that it yields the particular solution directly without the necessity of first finding the general solution and then evaluating the arbitrary constants.

Method. 1. Take Laplace transforms of both sides of the given differential equation, using initial conditions. This gives an algebraic equation.

2. Solve the algebraic equation to get $\bar{y}$ in terms of s.

3. Take inverse Laplace transform of both sides. This gives y as a function of t which is the desired solution.

Remember. $L\{f^n(t)\} = s^n \bar{f}(s) - s^{n-1}f(0) - s^{n-2}f'(0) \ldots\ldots - f^{n-1}(0)$.

ILLUSTRATIVE EXAMPLES

Example 1. *Solve the equation:*

$$\frac{d^3y}{dt^3} + 2\frac{d^2y}{dt^2} - \frac{dy}{dt} - 2y = 0, \text{ where } y = 1, \ \frac{dy}{dt} = 2, \ \frac{d^2y}{dt^2} = 2 \text{ at } t = 0.$$

Sol. The given equation is $y''' + 2y'' - y' - 2y = 0$

Taking the Laplace transform of both sides, we get

$$[s^3\bar{y} - s^2y(0) - sy'(0) - y''(0)] + 2[s^2\bar{y} - sy(0) - y'(0)] - [s\bar{y} - y(0)] - 2\bar{y} = 0 \quad ...(1)$$

Using the given conditions $y(0) = 1, y'(0) = 2, y''(0) = 2$, equation (1) reduces to

$$(s^3 + 2s^2 - s - 2)\bar{y} - s^2 - 2s - 2 - 2s - 4 + 1 = 0$$

or
$$(s^3 + 2s^2 - s - 2)\bar{y} = s^2 + 4s + 5$$

$$\therefore \quad \bar{y} = \frac{s^2 + 4s + 5}{s^3 + 2s^2 - s - 2} = \frac{s^2 + 4s + 5}{(s-1)(s+1)(s+2)}$$

$$= \frac{5}{3(s-1)} - \frac{1}{s+1} + \frac{1}{3(s+2)} \quad \text{(Partial Fractions)}$$

Taking the inverse Laplace transforms of both sides, we get

$$y = \frac{5}{3}L^{-1}\left\{\frac{1}{s-1}\right\} - L^{-1}\left\{\frac{1}{s+1}\right\} + \frac{1}{3}L^{-1}\left\{\frac{1}{s+2}\right\} = \frac{5}{3}e^t - e^{-t} + \frac{1}{3}e^{-2t}$$

or
$$y = \frac{1}{3}(5e^t + e^{-2t}) - e^{-t}$$ which is the required solution.

Example 2. *Solve the equation*

$y'' - 3y' + 2y = 4t + e^{3t}$, when $y(0) = 1$ and $y'(0) = -1$.

Sol. Taking the Laplace transform of both sides, we get

$$[s^2\bar{y} - sy(0) - y'(0)] - 3[s\bar{y} - y(0)] + 2\bar{y} = \frac{4}{s^2} + \frac{1}{s-3}$$

Using the given conditions, it reduces to

$$(s^2 - 3s + 2)\bar{y} - s + 1 + 3 = \frac{4}{s^2} + \frac{1}{s-3}$$

or
$$(s^2 - 3s + 2)\bar{y} = \frac{4}{s^2} + \frac{1}{s-3} + s - 4$$

$$\therefore \quad \bar{y} = \frac{s^4 - 7s^3 + 13s^2 + 4s - 12}{s^2(s-1)(s-2)(s-3)}$$

$$= \frac{3}{s} + \frac{2}{s^2} - \frac{1}{2(s-1)} - \frac{2}{s-2} + \frac{1}{2(s-3)} \quad \text{(Partial Fractions)}$$

Taking the inverse Laplace transforms of both sides, we get

$$y = 3L^{-1}\left\{\frac{1}{s}\right\} + 2L^{-1}\left\{\frac{1}{s^2}\right\} - \frac{1}{2}L^{-1}\left\{\frac{1}{s-1}\right\} - 2L^{-1}\left\{\frac{1}{s-2}\right\} + \frac{1}{2}L^{-1}\left\{\frac{1}{s-3}\right\}$$

$$= 3 + 2t - \frac{1}{2}e^t - 2e^{2t} + \frac{1}{2}e^{3t}$$

or $$y = 3 + 2t + \frac{1}{2}(e^{3t} - e^t) - 2e^{2t}$$ which is the required solution.

Example 3. *Solve the equation* $\frac{d^2x}{dt^2} + 2\frac{dx}{dt} + 5x = e^{-t}\sin t,\ x(0) = 0,\ x'(0) = 1.$

Sol. The given equation is $x'' + 2x' + 5x = e^{-t}\sin t$

Taking the Laplace transform of both sides, we get

$$[s^2\bar{x} - sx(0) - x'(0)] + 2[s\bar{x} - x(0)] + 5\bar{x} = \frac{1}{(s+1)^2+1}$$

Using the given conditions, it reduces to

$$(s^2 + 2s + 5)\,\bar{x} - 1 = \frac{1}{s^2+2s+2}$$

or $$\bar{x} = \frac{1}{(s^2+2s+2)(s^2+2s+5)} + \frac{1}{s^2+2s+5}$$

$$= \frac{1}{3}\left[\frac{1}{s^2+2s+2} - \frac{1}{s^2+2s+5}\right] + \frac{1}{s^2+2s+5}$$

$$= \frac{1}{3}\left[\frac{1}{s^2+2s+2} + \frac{2}{s^2+2s+5}\right] = \frac{1}{3}\left[\frac{1}{(s+1)^2+1} + \frac{2}{(s+1)^2+2^2}\right]$$

Taking the inverse Laplace transform of both sides, we get

$$x = \frac{1}{3}L^{-1}\left\{\frac{1}{(s+1)^2+1} + 2\cdot\frac{1}{(s+1)^2+2^2}\right\}$$

$$= \frac{1}{3}\left[e^{-t}\sin t + 2\cdot\frac{1}{2}e^{-t}\sin 2t\right]$$

or $$x = \frac{1}{3}e^{-t}(\sin t + \sin 2t)$$ which is the required solution.

Example 4. *Solve:* $(D+1)^2y = t$ *given that* $y = -3$ *when* $t = 0$ *and* $y = -1$, *when* $t = 1$.

Sol. The given differential equation can be written as

$$(D^2 + 2D + 1)\,y = t \quad \text{or} \quad y'' + 2y' + y = t \qquad ...(1)$$

with boundary conditions.

BC: $y(0) = -3,\ y(1) = -1.$

Taking the Laplace transform of both sides of (1), we have

$$L\{y''\} + 2L\{y'\} + L\{y\} = L\{t\}$$

$$s^2 L\{y\} - sy(0) - y'(0) + 2[sL\{y\} - y(0)] + L\{y\} = \frac{1}{s^2}$$

or $$(s^2 + 2s + 1)\, L\{y\} - s(-3) - A - 2(-3) = \frac{1}{s^2} \text{ where } y'(0) = A$$

or $$(s+1)^2\, L\{y\} = \frac{1}{s^2} - 3s - 6 + A$$

or $$L\{y\} = \frac{1}{s^2(s+1)^2} - \frac{3s+6}{(s+1)^2} + \frac{A}{(s+1)^2}$$

$$= -\frac{2}{s} + \frac{1}{s^2} + \frac{2}{s+1} + \frac{1}{(s+1)^2} - \frac{3}{s+1} - \frac{3}{(s+1)^2} + \frac{A}{(s+1)^2}$$

$$= -\frac{2}{s} + \frac{1}{s^2} - \frac{1}{s+1} + \frac{A-2}{(s+1)^2}.$$

$\therefore$ $$y = -2L^{-1}\left\{\frac{1}{s}\right\} + L^{-1}\left\{\frac{1}{s^2}\right\} - L^{-1}\left\{\frac{1}{s+1}\right\} + (A-2)L^{-1}\left\{\frac{1}{(s+1)^2}\right\}$$

$$= -2 \,.\, 1 + \frac{t}{1!} - e^{-t} + (A-2)\, e^{-t} L^{-1}\left\{\frac{1}{s^2}\right\}$$

$$= -2 + t - e^{-t} + (A-2)\, te^{-t}$$

Now, since $y = -1$, when $t = 1$, we have

$$-1 = -2 + 1 - e^{-1} + (A-2)\, e^{-4} \text{ or } A = 3.$$

Hence the required complete solution is

$$y = -2 + t - e^{-t} + te^{-t}.$$

Example 5. *Solve* $(D^2 + 1)\, y = t \cos 2t,\ y = 0,\ \dfrac{dy}{dt} = 0$ *when* $t = 0$.

Sol. The given differential equation can be written as

$$y'' + y = t \cos 2t$$

with the initial conditions

$$\text{IC: } y(0) = 0,\ y'(0) = 0.$$

Taking the Laplace transform of both sides of (1), we have

$$L\{y''\} + L\{y\} = L\{t \cos 2t\}$$

or $$s^2 L\{y\} - sy(0) - y'(0) + L\{y\} = -\frac{d}{ds}\,(L\{\cos 2t\})$$

or $$(s^2 + 1)\, L\{y\} = -\frac{d}{ds}\left(\frac{s}{s^2+4}\right) = -\frac{1}{s^2+4} + \frac{2s^2}{(s^2+4)^2}$$

or $$L\{y\} = \frac{s^2 - 4}{(s^2+1)(s^2+4)^2} = -\frac{5}{9(s^2+1)} + \frac{5}{9(s^2+4)} + \frac{8}{3(s^2+4)^2}.$$

Taking the inverse Laplace transform, we have

$$y = -\frac{5}{9}L^{-1}\left\{\frac{1}{s^2+1}\right\} + \frac{5}{9}L^{-1}\left\{\frac{1}{s^2+4}\right\} + \frac{8}{3}L^{-1}\left\{\frac{1}{(s^2+4)^2}\right\}$$

$$= -\frac{5}{9}\sin t + \frac{5}{18}\sin 2t + \frac{8}{3}\int_0^t \frac{1}{2}\sin 2x \,.\, \frac{1}{2}\sin 2(t-x)\,dx$$

by the convolution theorem since

$$L^{-1}\left\{\frac{1}{s^2+4}\right\} = \frac{1}{2}\sin 2t = -\frac{5}{9}\sin t + \frac{5}{18}\sin 2t + \int_0^t \{\cos 2(t-2x) - \cos 2t\}\,dx$$

$$= -\frac{5}{9}\sin t + \frac{5}{18}\sin 2t + \frac{1}{3}\left[-\frac{1}{4}\sin 2(t-2x) - x\sin 2t\right]_0^t$$

$$= -\frac{5}{9}\sin t + \frac{5}{18}\sin 2t + \frac{1}{12}\sin 2t - \frac{t}{3}\cos 2t + \frac{1}{12}\sin 2t$$

or $$y = -\frac{5}{9}\sin t + \frac{4}{9}\sin 2t - \frac{t}{3}\cos 2t$$

which is the required solution.

Example 6. *Solve the simultaneous equations:*

$$\frac{dx}{dt} - y = e^t, \quad \frac{dy}{dt} + x = \sin t,$$

given x(0) = 1, y(0) = 0.

Sol. Taking the Laplace transforms of the given equations, we get

$$[s\bar{x} - x(0)] - \bar{y} = \frac{1}{s-1}$$

i.e., $$s\bar{x} - 1 - \bar{y} = \frac{1}{s-1} \qquad [\because\ x(0) = 1]$$

i.e., $$s\bar{x} - \bar{y} = \frac{s}{s-1} \qquad \text{...(1)}$$

and $$[s\bar{y} - y(0)] + \bar{x} = \frac{1}{s^2+1}$$

i.e., $$\bar{x} + s\bar{y} = \frac{1}{s^2+1} \qquad \text{...(2)} \quad [\because\ y(0) = 0]$$

Solving (1) and (2) for $\bar{x}$ and $\bar{y}$, we have

$$\bar{x} = \frac{s^2}{(s-1)(s^2+1)} + \frac{1}{(s^2+1)^2} = \frac{1}{2}\left[\frac{1}{s-1} + \frac{s}{s^2+1} + \frac{1}{s^2+1}\right] + \frac{1}{(s^2+1)^2}$$

and $$\bar{y} = \frac{s}{(s^2+1)^2} - \frac{s}{(s-1)(s^2+1)} = \frac{s}{(s^2+1)^2} - \frac{1}{2}\left[\frac{1}{s-1} - 1\frac{s}{s^2+1} + \frac{1}{s^2+1}\right]$$

Taking the inverse Laplace transforms of both sides, we get

$$x = \frac{1}{2} L^{-1}\left[\frac{1}{s-1} + \frac{s}{s^2+1} + \frac{1}{s^2+1}\right] + L^{-1}\left[\frac{1}{(s^2+1)^2}\right]$$

$$= \frac{1}{2}[e^t + \cos t + \sin t] + \frac{1}{2}(\sin t - t\cos t)$$

$$\left[\because L^{-1}\left[\frac{1}{(s^2+a^2)^2}\right] = \frac{1}{2a^2}(\sin at - at\cos at)\right]$$

$$= \frac{1}{2}[e^t + \cos t + 2\sin t - t\cos t]$$

$$y = L^{-1}\left[\frac{s}{(s^2+1)^2}\right] - \frac{1}{2}L^{-1}\left[\frac{1}{s-1} - \frac{s}{s^2+1} + \frac{1}{s^2+1}\right]$$

$$= \frac{1}{2}t\sin t - \frac{1}{2}[e^t - \cos t + \sin t] \qquad \left[\because L^{-1}\left[\frac{s}{(s^2+a^2)^2}\right] = \frac{1}{2a}t\sin at\right]$$

$$= \frac{1}{2}[t\sin t - e^t + \cos t - \sin t]$$

Hence $\quad x = \frac{1}{2}(e^t + \cos t + 2\sin t - t\cos t)$

$$y = \frac{1}{2}(t\sin t - e^t + \cos t - \sin t).$$

Example 7. *Solve the simultaneous equations*

$$(D^2 - 3)x - 4y = 0$$

$$x + (D^2 + 1)y = 0$$

for $t > 0$, *given that* $x = y = \frac{dy}{dt} = 0$ *and* $\frac{dx}{dt} = 2$ *at* $t = 0$.

Sol. Taking the Laplace transforms of the given equations, we get

$$s^2\bar{x} - sx(0) - x'(0) - 3\bar{x} - 4\bar{y} = 0$$

i.e., $\quad (s^2 - 3)\bar{x} - 4\bar{y} = 2 \quad$ ...(1) $\quad [\because x(0) = 0, x'(0) = 2]$

and $\quad \bar{x} + s^2\bar{y} - sy(0) - y'(0) + \bar{y} = 0$

i.e., $\quad \bar{x} + (s^2 + 1)\bar{y} = 0 \quad$ ...(2) $\quad [\because y(0) = y'(0) = 0]$

Solving (1) and (2) for $\bar{x}$ and $\bar{y}$, we get

$$\bar{x} = \frac{2(s^2+1)}{(s^2-1)^2} = \frac{1}{(s-1)^2} + \frac{1}{(s+1)^2}$$

and $\quad \bar{y} = -\frac{2}{(s^2-1)^2} = -\frac{1}{2}\left[\frac{1}{s+1} - \frac{1}{s-1} + \frac{1}{(s+1)^2} + \frac{1}{(s-1)^2}\right]$

Taking the inverse Laplace transforms of both sides, we get

$$x = L^{-1}\left[\frac{1}{(s-1)^2}+\frac{1}{(s+1)^2}\right] = te^t + te^{-t} = 2t\left(\frac{e^t+e^{-t}}{2}\right) = 2t\cosh t$$

and $$y = -\frac{1}{2}L^{-1}\left(\frac{1}{s+1}-\frac{1}{s-1}-\frac{1}{(s+1)^2}+\frac{1}{(s+1)^2}\right)$$

$$= -\frac{1}{2}(e^{-t}-e^t-te^{-t}+te^t) = \frac{e^t-e^{-t}}{2} - t\left(\frac{e^t-e^{-t}}{2}\right) = (1-t)\sinh t$$

Hence $x = 2t\cosh t, y = (1-t)\sinh t.$

Example 8. *Solve* $(D-2)x-(D+1)y = 6e^{3t}$

$$(2D-3)x+(D-3)y = 6e^{3t}$$

if $x = 3, y = 0$ *when* $t = 0$.

Sol. Taking the Laplace transform of both sides of the two equations, we have

$$L\{x'\} - 2L\{x\} - L\{y'\} - L\{y\} = 6L\{e^{3t}\}$$

and $$2L\{x'\} - 3L\{x\} + L\{y'\} - 3L\{y\} = 6L\{e^{3t}\}$$

or $$s\bar{x} - x(0) - 2\bar{x} - [s\bar{y} - y(0)] - \bar{y} = \frac{6}{s-3}$$

and $$2[s\bar{x} - x(0)] - 3\bar{x} + s\bar{y} - y(0) - 3\bar{y} = \frac{6}{s-3}$$

or $$(s-2)\bar{x} - (s+1)\bar{y} = 3 + \frac{6}{s-3} = \frac{3s-3}{s-3} \quad ...(1)$$

and $$(2x-3)\bar{x} + (s-3)\bar{y} = 6 + \frac{6}{s-3} = \frac{6s-12}{s-3}. \quad ...(2)$$

Solving (1) and (2) for $\bar{x}$ and $\bar{y}$, we have

$$\bar{x} = \frac{3s^2-6s-1}{(s-3)(s-1)^2} = \frac{1}{s-1}+\frac{2}{(s-1)^2}+\frac{2}{s-3}$$

and $$\bar{y} = \frac{-3s+5}{(s-3)(s-1)^2} = \frac{1}{s-1}-\frac{1}{(s-1)^2}-\frac{1}{s-3}.$$

Now applying inverse Laplace transform, we get

$$x = L^{-1}\{\bar{x}\} = L^{-1}\left\{\frac{1}{s-1}\right\} + 2L^{-1}\left\{\frac{1}{(s-1)^2}\right\} + 2L^{-1}\left\{\frac{1}{s-3}\right\} = e^t + 2t\,e^t + 2e^{3t}$$

and $$y = L^{-1}\{\bar{y}\} = L^{-1}\left\{\frac{1}{s-1}\right\} - L^{-1}\left\{\frac{1}{(s-1)^2}\right\} - L^{-1}\left\{\frac{1}{s-3}\right\} = e^t - te^t - e^{3t}.$$

TEST YOUR KNOWLEDGE

Solve the following equations by the transform method (1 – 17):

1. $(D^2 - 2D + 2)x = 0$, $x = Dx = 1$ at $t = 0$.

2. $\frac{d^2x}{dt^2} - 2\frac{dx}{dt} + x = e^t$ with $x = 2$, $\frac{dx}{dt} = -1$ at $t = 0$.

3. $y'' + 4y' + 3y = e^{-t}$, $y(0) = y'(0) = 1$.

4. $x'' - 3x' + 2x = 1 - e^{2t}$, $x(0) = 1$, $x'(0) = 0$.

5. $y'' + y' - 2y = t$, $y(0) = 1$, $y'(0) = 0$.

6. $\frac{dx}{dt} + x = \sin \omega t$, $x(0) = 2$.

7. $(D^2 + 5D + 6)x = 5e^t$, $x(0) = 2$, $x'(0) = 1$.

8. $(D^2 - 1)\, x = a \cosh t$, $x(0) = x'(0) = 0$.

9. $\frac{d^2x}{dt^2} + 9x = \cos 2t$, $x(0) = 1$, $x\left(\frac{\pi}{2}\right) = -1$.

10. $(D^2 - D - 2)\, x = 20 \sin 2t$, $x(0) = -1$, $x'(0) = 2$.

11. $\frac{d^2y}{dt^2} + 2\frac{dy}{dt} + 2y = \frac{17}{2} \sin 5t$, $y = 2$ and $\frac{dy}{dt} = -4$ when $t = 0$.

12. $\frac{d^2y}{dt^2} + \frac{dy}{dt} - 2y = 3 \cos 3t - 11 \sin 3t$ given that $y(0) = 0$ and $y'(0) = 6$.

13. $\frac{d^2x}{dt^2} + x = t \cos 2t$ given that $x(0) = x'(0) = 0$.

14. $(D^3 + D)x = 2$, $x = 3$, $Dx = 1$, $D^2x = -2$ at $t = 0$.

15. $(D^3 + D^2)\, x = 6t^2 + 4$, $x = D^2 x = 0$, $Dx = 2$ at $t = 0$.

16. $(D^3 - D)x = 2 \cos t$, $x = 3$, $Dx = 2$, $D^2x = 1$ at $t = 0$.

17. $(D^3 - 3D^2 + 3D - 1)\, y = t^2e^t$, $y = 1$, $Dy = 0$, $D^2y = -2$ at $t = 0$.

18. Constant voltage E is applied at $t = 0$ to a circuit with an inductance L, capacitance C and resistance R. Find the current I at time t, if the initial current and charge are zero.

Solve the following simultaneous equations by using Laplace transforms (19 – 22):

19. $\frac{dx}{dt} - 2x + 3y = 0$, $\frac{dy}{dt} + 2x - y = 0$ given that $x(0) = 8$ and $y(0) = 3$.

20. $(D^2 - D)y + z = 0$, $(D - 1)y + Dz = 0$ given that for $t = 0$, $y = 0$, $z = 1$ and $Dy = 0$.

21. $D^2x + y = -5 \cos 2t$, $D^2y + x = 5 \cos 2t$ where $x(0) = x'(0) = y'(0) = 1$ and $y(0) = -1$.

22. $\frac{dx}{dt} - \frac{dy}{dt} - 2x + 2y = 1 - 2t$, $\frac{d^2x}{dt^2} + 2\frac{dy}{dt} + x = 0$,

    subject to the conditions $x = 0$, $y = 0$, $\frac{dx}{dt} = 0$ when $t = 0$.

23. The coordinates (x, y) of a particle moving along a plane curve at any time t are given by

    $$\frac{dy}{dt} + 2x = \sin 2t, \quad \frac{dx}{dt} - 2y = \cos 2t, \ (t > 0).$$

    If at $t = 0$, $x = 1$ and $y = 0$, show by using transforms, that the particle moves along the curve $4x^2 + 4xy + 5y^2 = 4$.

24. The currents i_1 and i_2 in mesh are given by the differential equations:

$$\frac{di_1}{dt} - \omega i_2 = a \cos pt, \quad \frac{di_2}{dt} + \omega i_1 = a \sin pt.$$

Find the currents i_1 and i_2 by Laplace transform, if $i_1 = i_2 = 0$ at $t = 0$.

25. A mechanical system, with two degrees of freedom, satisfies the equations:

$$2\frac{d^2x}{dt^2} + 3\frac{dy}{dt} = 4, \quad 2\frac{d^2y}{dt^2} - 3\frac{dx}{dt} = 0.$$

Use Laplace transforms to determine x and y at any instant, given that $x, y, \frac{dx}{dt}, \frac{dy}{dt}$ all vanish at $t = 0$.

Answers

1. $x = e^t \cos t$ **2.** $x = \left(2 - 3t + \frac{1}{2}t^2\right)e^t$ **3.** $y = \frac{7}{4}e^{-t} - \frac{3}{4}e^{-3t} + \frac{1}{2}te^{-t}$

4. $x = \frac{1}{2} + \left(\frac{1}{2} - t\right)e^{2t}$ **5.** $y = e^t + \frac{1}{4}(e^{-2t} - 1) - \frac{1}{2}t$

6. $x = \left(2 + \frac{w}{w^2+1}\right)e^{-t} + \frac{\sin wt - w\cos wt}{w^2+1}$ **7.** $x = \frac{5}{12}e^t + \frac{16}{3}e^{-2t} - \frac{15}{4}e^{-3t}$

8. $x = \frac{1}{2}at \sinh t$ **9.** $x = \frac{1}{5}(4\cos 3t + 4\sin 3t + \cos 2t)$

10. $x = 2e^{2t} - 4e^{-t} + \cos 2t - 3\sin 2t$ **11.** $y = e^{-t}(4\cos 2t - \frac{1}{2}\sin 2t) - 2\cos 2t + \frac{1}{2}\sin 2t$

12. $y = e^t - e^{-2t} + \sin 3t$ **13.** $x = \frac{1}{9}(4\sin 2t - 5\sin t - 3t\cos 2t)$

14. $x = 1 + 2t + 2\cos t - \sin t$ **15.** $x = \frac{1}{2}t^4 - 2t^3 + 8t^2 - 14t + 16(1 - e^{-t})$

16. $x = 3\sinh t + \cosh t - \sin t + 2$ **17.** $y = e^t\left(1 - t - \frac{1}{2}t^2 + \frac{1}{60}t^5\right)$

18. $I = \frac{E}{R}\left(1 - e^{-Rt/L}\right)$ **19.** $x = 5e^{-t} + 3e^{4t}$, $y = 5e^{-t} - 2e^{4t}$

20. $y = \frac{1}{2}(\sinh t - t\,e^t)$, $z = \cosh t$ **21.** $x = \sin t + \cos 2t$, $y = \sin t - \cos 2t$

22. $x = 2(1 - e^{-t} - te^{-t})$, $y = -t(1 + 2e^{-t}) + 2(1 - e^{-t})$

24. $i_1 = \frac{a}{p+\omega}(\sin \omega t + \sin pt)$, $i_2 = \frac{a}{p+\omega}(\cos \omega t - \cos pt)$

25. $x = \frac{8}{9}\left(1 - \cos\frac{3}{2}t\right)$, $y = \frac{8}{9}\left(\frac{3}{2}t - \sin\frac{3}{2}t\right)$.

2.13. LAPLACE TRANSFORMS OF SOME OTHER USEFUL FUNCTIONS

(1) Unit Step Function (or Heaviside's Unit Function)

Def. The unit step function $u(t-a)$ is defined as

$$u(t-a) = \begin{cases} 0, & \text{for } t < a \\ 1, & \text{for } t \geq a \end{cases} \quad \text{where } a \geq 0.$$

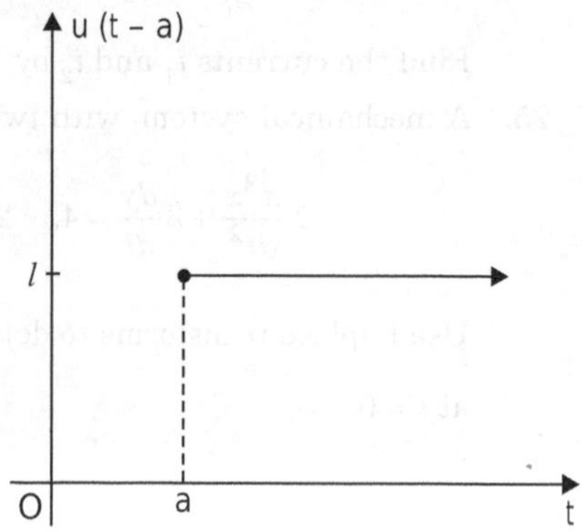

As a particular case,

$$u(t) = \begin{cases} 0, & \text{for } t < 0 \\ 1, & \text{for } t \geq 0 \end{cases}$$

The product

$$f(t)\,.\,u(t-a) = \begin{cases} 0, & \text{for } t < a \\ f(t), & \text{for } t \geq a \end{cases}$$

The function $f(t-a)\,.\,u(t-a)$ represents the graph of $f(t)$ shifted through a distance 'a' to the right.

Laplace Transform of Unit Step Function

$$L\{u(t-a)\} = \int_0^\infty e^{-st}\, u(t-a)\, dt$$

$$= \int_0^a e^{-st}\,.\,0\,dt + \int_0^\infty e^{-st}\,.\,1\,dt = 0 + \left[\frac{e^{-st}}{-s}\right]_a^\infty = \frac{1}{s}e^{-as}$$

In particular, $L\{u(t)\} = \frac{1}{s}$.

Second Shifting Theorem

If $L\{f(t)\} = \bar{f}(s)$, *then* $\mathbf{L\{f(t-a)\,.\,u(t-a)\} = e^{-as}\,\bar{f}(s)}$.

$$L\{f(t-a)\,.\,u(t-a)\} = \int_0^\infty e^{-st}\, f(t-a)\,.\,u(t-a)\,dt$$

$$= \int_0^\infty e^{-st}\, f(t-a)\,dt = \int_0^\infty e^{-s(u+a)} f(u)\,du, \text{ where } u = t-a$$

$$= e^{-as}\int_0^\infty e^{-su} f(u)\,du = e^{-as}\,\bar{f}(s)$$

Cor. 1. $L^{-1}\{e^{-as}\,\bar{f}(s)\} = f(t-a)\,.\,u(t-a)$

Cor. 2. If $a = 0$, $L\{f(t)\,u(t)\} = \bar{f}(s) = L\{f(t)\}$.

ILLUSTRATIVE EXAMPLES

Example 1. *Find the Laplace transforms of*

(i) $(t-1)^2\,u(t-1)$ *(ii)* $\sin t\; u(t-\pi)$ *(iii)* $e^{-3t}\,u(t-2)$.

Sol. (*i*) Comparing $(t-1)^2\,u(t-1)$ with $f(t-a)\,u(t-a)$, we have

$$a = 1 \quad \text{and} \quad f(t) = t^2$$

$$\therefore \qquad \bar{f}(s) = L\{f(t)\} = \frac{2}{s^3}$$

$\therefore \quad L\{(t-1)^2 u(t-1)\} = e^{-s}\bar{f}(s)$ [By second shifting property]

$$= \frac{2e^{-s}}{s^3}.$$

(*ii*) Expressing sin t as a function of $(t-\pi)$, we have

$$\sin t = \sin[(t-\pi)+\pi] = -\sin(t-\pi)$$

Comparing $-\sin(t-\pi)\, u(t-\pi)$ with $f(t-a)\, u(t-a)$, we get

$$a = \pi \quad \text{and} \quad f(t) = -\sin t$$

$$\therefore \quad \bar{f}(s) = L\{f(t)\} = -\frac{1}{s^2+1}$$

$$\therefore \quad L\{\sin t\, u(t-\pi)\} = L\{-\sin(t-\pi)\, u(t-\pi)\}$$

$= e^{-\pi s} f(s)$ [By second shifting property]

$$= -\frac{e^{-\pi s}}{s^2+1}.$$

(*iii*) Expressing e^{-3t} as a function of $(t-2)$, we have

$$e^{-3t} = e^{-3[(t-2)+2]} = e^{-6}.\, e^{-3(t-2)}$$

$$\therefore \quad e^{-3t} u(t-2) = e^{-6}.\, e^{-3(t-2)}\, u(t-2)$$

Comparing $e^{-3(t-2)}\, u(t-2)$ with $f(t-a)\, u(t-a)$, we get

$$a = 2 \quad \text{and} \quad f(t) = e^{-3t}$$

$$\therefore \quad \bar{f}(s) = L\{f(t)\} = \frac{1}{s+3}$$

$$\therefore \quad L\{e^{-3t} u(t-2)\} = e^{-6} L\{e^{-(3t-2)} u(t-2)\} = e^{-6}.\, e^{-2s}\, \bar{f}(s) = \frac{e^{-2(s+3)}}{s+3}.$$

(2) Unit Impulse Function (or Dirac-delta Function)

In mechanics, we come across problems where a very large force acts for a very small time. In the study of bending of beams, we have point loads which is equivalent to large pressure acting over a very small area. To deal with such problems, we introduce the unit impulse function or Dirac-delta function.

Unit impulse function is considered as the limiting form of the function

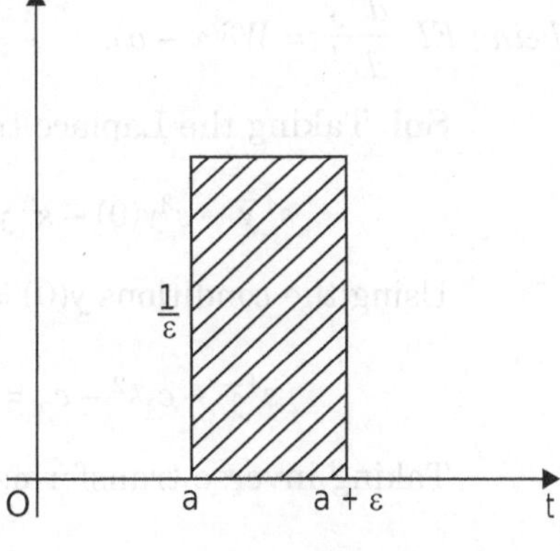

$$\delta\varepsilon(t-a) = \begin{cases} 0, & \text{for } t < a \\ \dfrac{1}{\varepsilon}, & \text{for } a \le t \le a+\varepsilon \\ 0, & \text{for } t > a \end{cases}$$

This function is represented in the figure. Integrating this function, we get

$$\int_0^\infty \delta\varepsilon(t-a)\, dt = \int_a^{a+\varepsilon} \frac{1}{\varepsilon} dt = \frac{1}{\varepsilon}(a+\varepsilon-a) = 1$$

As $\varepsilon \to 0$, the function $\delta\varepsilon(t-a)$ tends to be infinite at $x = a$ and zero elsewhere, with the characteristic property that its integral across $t = a$ is unity. If $\delta\varepsilon(t-a)$ represents a force acting for short duration ε at time $t = a$, then the integral

$$\underset{\varepsilon \to 0}{\text{Lt}} \int_0^{a+\varepsilon} \delta_\varepsilon (t-a)\, dt = 1$$

represents unit impulse at $t = a$. Hence the limiting form of $\delta_\varepsilon(t - a)$ as $\varepsilon \to 0$ is expressed as unit impulse function denoted by $\delta(t - a)$.

Thus the unit impulse function $\delta(t - a)$ is defined as follows:

$$\delta(t-a) = \begin{cases} \infty, & \text{for } t = a \\ 0, & \text{for } t \neq a \end{cases}$$

such that $\displaystyle\int_0^\infty \delta(t-a)\, dt = 1.$

Laplace Transform of Unit Impulse Function

If $f(t)$ be a function of t continuous at $t = a$, then

$$\int_0^\infty f(t)\, \delta_\varepsilon (t-a)\, dt = \int_\varepsilon^{a+\varepsilon} f(t) \,.\, \frac{1}{\varepsilon} dt$$

$$= (a + \varepsilon - a)\, f(c) \,.\, \frac{1}{\varepsilon} = f(c) \quad \text{where } a < c < a + \varepsilon$$

(By mean value theorem for integrals)

As $\varepsilon \to 0$, we get $\displaystyle\int_0^\infty f(t)\, \delta(t-a)\, dt = f(a).$

Cor. 1. $\quad \displaystyle \text{L}\, \{\delta(t-a)\} = \int_0^\infty e^{-st}\, \delta(t-a)\, dt = e^{-sa}.$

Cor. 2. $\quad \text{L}\, \{\delta(t)\} = e^{-s} = e^0 = 1.$

Example 2. *Obtain the deflection of a weightless beam of length l and freely supported at ends, when a concentrated load W acts at x = a. The differential equation for deflection being* $\text{EI}\, \dfrac{d^4 y}{dx^4} = \text{W}\delta(x - a).$

Sol. Taking the Laplace transform of the differential equation, we get

$$s^4 \bar{y} - s^3 y(0) - s^2\, y'(0) - sy''(0) - y'''(0) = \frac{\text{W}}{\text{EI}}\, e^{-as}$$

Using the conditions $y(0) = y''(0) = 0$ and taking $y'(0) = c_1$, $y'''(0) = c_2$, we get

$$s^4 \bar{y} - c_1 s^2 - c_2 = \frac{\text{W}}{\text{EI}} \,.\, e^{-as} \quad \text{or} \quad \bar{y} = \frac{c_1}{s^2} + \frac{c_2}{s^4} + \frac{\text{W}}{\text{EI}} \,.\, \frac{e^{-as}}{s^4}$$

Taking inverse transforms, we have

$$y = c_1 x + c_2 \,.\, \frac{x^3}{3!} + \frac{\text{W}}{\text{EI}} \,.\, \frac{(x-a)^3}{3!}\, u(x-a) \qquad \text{...(1)}$$

$$\therefore \qquad y' = c_1 + \frac{1}{2}\, c_2 x^2 + \frac{\text{W}}{2\text{EI}}\, (x-a)^2\, u(x-a)$$

$$y'' = c_2 x + \frac{\text{W}}{\text{EI}}\, (x-a)\, u(x-a)$$

Using the conditions $y(l) = 0, y''(l) = 0$, we get

$$0 = c_1 l + c_2 \frac{l^3}{6} + \frac{W}{EI} \cdot \frac{b^3}{6} \quad \text{or} \quad 0 = c_2 l + \frac{W}{EI} b$$

$[\because \; u(x-a) = 1, \text{ for } x \geq a \text{ and } l - a = b]$

a b = l - a

W

$$\therefore \qquad c_2 = -\frac{W}{EI} \cdot \frac{b}{l}$$

and
$$c_1 = -\frac{1}{l}\left[c_2 \frac{l^3}{6} + \frac{W}{EI} \cdot \frac{b^3}{6}\right]$$

$$= -\frac{1}{l}\left[-\frac{W}{EI} \cdot \frac{bl^2}{6} + \frac{W}{EI} \cdot \frac{b^3}{6}\right]$$

$$= \frac{Wb}{6EIl}(l^2 - b^2) = \frac{Wb}{6EIl}(l+b)(l-b) = \frac{Wab}{6EIl}(l+b)$$

$\therefore$ From (1) the solution is

$$y = \frac{Wab}{6EIl}(l+b)x - \frac{W}{6EI} \cdot \frac{b}{l} x^3 + \frac{W}{EI} \cdot \frac{(x-a)^3}{3!} u(x-a)$$

$$= \frac{W}{6EI}\left[\frac{ab(l+b)}{l} x - \frac{b}{l} x^3 + (x-a)^3 u(x-a)\right]$$

When $0 < x < a$, $u(x-a) = 0$

$$\therefore \qquad y = \frac{W}{6EI}\left[\frac{ab(l+b)}{l} - \frac{b}{l} x^3\right] \qquad \text{...(2)}$$

When $a < x < l$, $u(x-a) = 1$

$$\therefore \qquad y = \frac{W}{6EI}\left[\frac{ab(l+b)}{l} x - \frac{b}{l} x^3 + (x-a)^3\right] \qquad \text{...(3)}$$

From (2) and (3), we have

$$y(a) = \frac{W}{6EIl}\,[a^2 b(l+b) - ba^3]$$

$$= \frac{W}{6EIl} a^2 b\,[(l+b) - a] = \frac{W}{6EIl} a^2 b\,.\,(2b) = \frac{1}{3}\frac{W}{EI} \cdot \frac{a^2 b^2}{l}.$$

(3) Periodic Functions

If f(t) is a periodic function with period T, i.e., f(t + T) = f(t), then

$$L\{f(t)\} = \frac{1}{1 - e^{-sT}} \int_0^T e^{-st} f(t)\, dt.$$

$$L\{f(t)\} = \int_0^\infty e^{-st} f(t)\, dt = \int_0^T e^{-st} f(t)\, dt + \int_T^{2T} e^{-st} f(t)\, dt + \int_{2T}^{3T} e^{-st} f(t)\, dt + \ldots\ldots$$

Putting $t = u, t = u + T, t = u + 2T$, ... in the successive integrals

$$L\{f(t)\} = \int_0^T e^{-su} f(u)\, du + \int_0^T e^{-s(u+T)} f(u+T)\, du + \int_0^T e^{-s(u+2T)} f(u+2T)\, du + ...$$

Since $f(u) = f(u + T) = f(u + 2T) = ...$, we have

$$L\{f(t)\} = \int_0^T e^{-su} f(u)\, du + e^{-sT} \int_0^T e^{-su} f(u)\, du + e^{-2sT} \int_0^T e^{-su} f(u)\, du + ...$$

$$= (1 + e^{-sT} + e^{-2sT} + ...) \int_0^T e^{-su} f(u)\, du = \frac{1}{1-e^{-sT}} \int_0^T e^{-st} f(t)\, dt.$$

Example 3. *Find the Laplace transform of the triangular wave function of period 2c given by*

$$f(t) = t, \qquad 0 < t < c$$
$$= 2c - t, \qquad c < t < 2c.$$

Sol.

$$L\{f(t)\} = \frac{1}{1-e^{-2cs}} \int_0^{2c} e^{-st} f(t)\, dt = \frac{1}{1-e^{-2cs}} \left[\int_0^c e^{-st} .\, t\, dt + \int_c^{2c} e^{-st} (2c - t) dt \right]$$

$$= \frac{1}{1-e^{-2cs}} \left[\left\{ t . \frac{e^{-st}}{-s} - 1 . \frac{e^{-st}}{s^2} \right\}_0^c + \left\{ (2c - t) . \frac{e^{-st}}{-s} - (-1) . \frac{e^{-st}}{s^2} \right\}_c^{2c} \right]$$

$$= \frac{1}{1-e^{-2cs}} \left[\left\{ -\frac{ce^{-cs}}{s} - \frac{e^{-cs}}{s^2} + \frac{1}{s^2} \right\} + \left\{ \frac{e^{-2cs}}{s^2} + \frac{ce^{-cs}}{s} - \frac{e^{-cs}}{s^2} \right\} \right]$$

$$= \frac{1}{1-e^{-2cs}} \left(\frac{1 - 2e^{-cs} + e^{-2cs}}{s^2} \right) = \frac{1}{s^2} . \frac{(1-e^{-cs})^2}{(1+e^{-cs})(1-e^{-cs})} = \frac{1}{s^2} . \frac{1-e^{-cs}}{1+e^{-cs}}$$

$$= \frac{1}{s^2} \left(\frac{e^{cs/2} - e^{-cs/2}}{e^{cs/2} + e^{-cs/2}} \right) = \frac{1}{s^2} \tanh \frac{cs}{2}.$$

Example 4. *Find the Laplace transform of the rectified semi-wave function defined by*

$$f(t) = \sin \omega t, \qquad 0 < t < \frac{\pi}{\omega}$$

$$= 0, \qquad \frac{\pi}{\omega} < t < \frac{2\pi}{\omega}.$$

Sol. Here $f(t)$ is a periodic function with period $\frac{2\pi}{\omega}$

$$\therefore \quad L\{f(t)\} = \frac{1}{1-e^{-\frac{2\pi s}{\omega}}} \int_0^{2\pi/\omega} e^{-st} f(t)\, dt = \frac{1}{1-e^{-\frac{2\pi s}{\omega}}} \left[\int_0^{\pi/\omega} e^{-st} \sin \omega t\, dt + \int_{\pi/\omega}^{2\pi/\omega} e^{-st} .\, 0\, dt \right]$$

$$= \frac{1}{1-e^{-\frac{2\pi s}{\omega}}} \left[\frac{e^{-st}(-s \sin \omega t - \omega \cos \omega t)}{s^2 + \omega^2} \right]_0^{\pi/\omega}$$

$$= \frac{\omega e^{\frac{\pi s}{\omega}} + \omega}{\left(1 - e^{-\frac{\pi s}{\omega}}\right)\left(1 + e^{-\frac{\pi s}{\omega}}\right)\left(s^2 + \omega^2\right)} = \frac{\omega}{\left(1 - e^{-\frac{\pi s}{\omega}}\right)\left(s^2 + \omega^2\right)}.$$

Example 5. *Find the Laplace transform of the square-wave function of period a defined as*

$$f(t) = 1, \quad \text{when} \quad 0 < t < a/2$$
$$= -1, \text{ when } a/2 < t < a.$$

(CSVTU Bhilai (Third Sem.))

Sol. Since $f(t)$ is a periodic function with period a.

$$\therefore \quad L[f(t)] = \frac{1}{1 - e^{-as}} \int_0^a e^{-st} f(t)\, dt = \frac{1}{1 - e^{-as}} \left\{ \int_0^{a/2} e^{-st} \,.\, 1 dt + \int_{a/2}^a e^{-st} \,.(-1) dt \right\}$$

$$= \frac{1}{1 - e^{-as}} \left\{ \left[\frac{e^{-st}}{-s} \right]_0^{a/2} - \left[\frac{e^{-st}}{-s} \right]_{a/2}^a \right\}$$

$$= \frac{1}{1 - e^{-as}} \left[\left(-\frac{e^{as/2}}{s} + \frac{1}{s} \right) + \left(\frac{e^{-as}}{s} - \frac{e^{-as/2}}{s} \right) \right]$$

$$= \frac{1}{s} \cdot \frac{1 - 2e^{-as/2} + e^{-as}}{1 - e^{-as}} = \frac{1}{s} \cdot \frac{(1 - e^{-as/2})^2}{(1 - e^{-as/2})(1 + e^{-as/2})}$$

$$= \frac{1}{s} \cdot \frac{1 - e^{-as/2}}{1 + e^{-as/2}} = \frac{1}{s} \cdot \frac{e^{as/4} - e^{-as/4}}{e^{as/4} + e^{-as/4}} = \frac{1}{s} \,.\, \tanh\left(\frac{as}{4}\right).$$

TEST YOUR KNOWLEDGE

1. Find the Laplace transforms of
   (*i*) $e^{t-2}\, u(t-2)$ (*ii*) $t^2\, u(t-3)$ (*iii*) $\sin 2t\; u(t-\pi)$ (*iv*) $e^{2t}\, u(t-3)$.
2. Find the inverse Laplace transforms of
   (*i*) $\dfrac{e^{-\pi s}}{s^2+1}$ (*ii*) $\dfrac{e^{-\pi s}}{s^2}$ (*iii*) $\dfrac{se^{-2s}}{s^2-1}$ (*iv*) $\dfrac{e^{-cs}}{s^2(s+a)}, c > 0.$
3. A light beam of length l has its ends $x = 0$ and $x = l$ hinged. If concentrated load W acts at the point $x = \dfrac{l}{3}$, show that the deflection is given by
$$\text{EI}y = \frac{\text{W}}{81} x\,(3l^2 - 7x^2) + \frac{\text{W}}{6}\left(x - \frac{l}{3}\right)^3 u\left(x - \frac{l}{3}\right).$$
4. The deflection of a beam of length l, clamped horizontally at both ends and loaded at $x = \dfrac{l}{4}$ by a weight W is given by $\text{EI}\,\dfrac{d^4y}{dx^4} = \text{W}\,\delta\left(x - \dfrac{l}{4}\right)$.
   Find the deflection curve, given that $y = \dfrac{dy}{dx} = 0$, when $x = 0$ and $x = l$.

5. Find the Laplace transform of the square-wave function of period a defined as

$$f(t) = 1, \quad \text{when } 0 < t < \frac{a}{2}$$
$$= -1, \text{ when } \frac{a}{2} < t < a.$$

6. Find the Laplace transform of the saw-toothed wave function of period T, defined by

$$f(t) = \frac{t}{T}, \text{ for } 0 < t < T.$$

Answers

1. (*i*) $\dfrac{e^{-2s}}{s-1}$ (*ii*) $\dfrac{e^{-3s}(2+6s+9s^2)}{s^3}$

(*iii*) $\dfrac{2e^{-\pi s}}{s^2+4}$ (*iv*) $\dfrac{e^{-3(s-2)}}{s-2}$.

2. (*i*) $-\sin t \,.\, u(t-\pi)$ (*ii*) $(t-\pi)\,.\,u(t-\pi)$

(*iii*) $\cosh(t-2)\,.\,u(t-2)$ (*iv*) $\dfrac{1}{a^2}[a(t-c)-1+e^{-a(t-c)}]\,.\,u(t-c)$.

4. $EIy = \dfrac{W}{6}\left(x-\dfrac{l}{4}\right)^3 u\left(x-\dfrac{l}{4}\right)+\dfrac{9}{128}WLx^2-\dfrac{9}{64}W^2x^3$.

5. $\dfrac{1}{s}\tanh\left(\dfrac{as}{4}\right)$ 6. $\dfrac{1}{s^2T}-\dfrac{e^{-sT}}{s(1-e^{-sT})}$.

Z-TRANSFORMS

2.14. INTRODUCTION

The Z-transform plays an important role in the field of Communication Engineering and Control Engineering at the stage of analysis and representation of discrete-time linear shift invariance system. When continuous signals are sampled, discrete-time functions arise. The application of Z-transform in discrete time systems is similar to that of the Laplace transform in continuous time systems.

2.15. DEFINITION OF THE Z-TRANSFORM

Let $\{f(n)\}$ be a sequence defined for all positive integers 'n'. Then the Z-transform of $f(n)$ is defined as

$$Z\{f(n)\} = \sum_{n=0}^{\infty} f(n)\, z^{-n} = F(z) \qquad \text{...(1)}$$

where z is an arbitrary complex number.

This is one sided Z-transform.

Note 1. If $\{f(n)\}$ is defined for $n = 0, \pm 1, \pm 2, \ldots\ldots$ then $Z\{f(n)\} = \sum_{n=-\infty}^{\infty} f(n)\, z^{-n}$ and is called **two sided Z-transforms.**

2. If $f(n) = 0$ for $n < 0$ then $\{f(n)\}$ is called a **casual sequence.**

3. The bracket { } represents sequence.

The infinite series on R.H.S. of (1) will be convergent only for certain values of z depending on the sequence $f(n)$. The inverse Z-transform of $Z\{f(n)\} = F(z)$ is defined as $Z^{-1}\{F(z)\} = f(n)$.

2.16. UNIT SAMPLE SEQUENCE

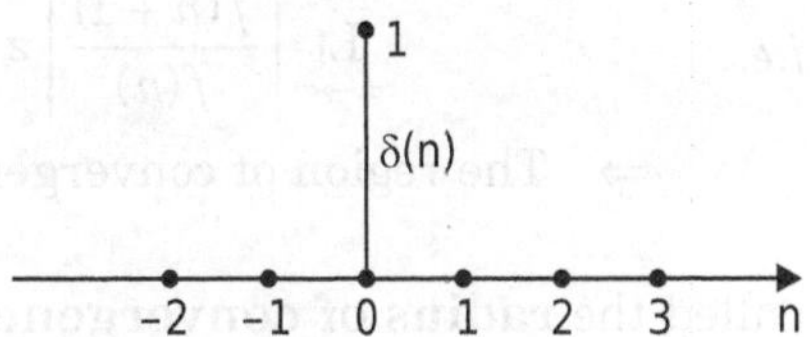

It is defined as $\delta(n) = \begin{cases} 1 & \text{for } n = 0 \\ 0 & \text{for } n \neq 0 \end{cases}$

2.16.1. Unit Step Sequence

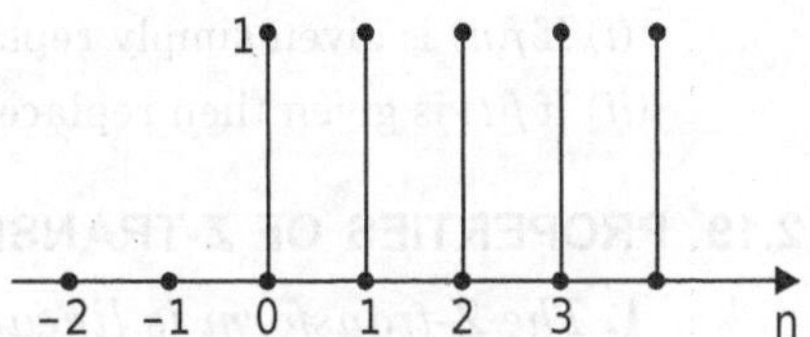

It is defined as $u(n) = \begin{cases} 1 & \text{for } n \geq 0 \\ 0 & \text{for } n < 0 \end{cases}$

2.16.2. Relation between Unit Sample Sequence and Unit Step Sequence

$$u(n) = \sum_{k=-\infty}^{\infty} \delta(k) \text{ and } \delta(n) = u(n) - u(n-1)$$

We have $\delta(n-k) = \begin{cases} 1 & \text{if } k = n \\ 0 & \text{if } k \neq n \end{cases}$ $\quad\Big|\quad u(n-k) = \begin{cases} 1 & \text{for } (n-k) \geq 0 \\ 0 & \text{for } (n-k) < 0 \end{cases}$

Also $f(n) = \sum_{k=-\infty}^{\infty} f(k)\, \delta(n-k).$

2.17. Z-TRANSFORM FOR DISCRETE VALUES OF t

If $f(t)$ is a function defined for discrete values of t where $t = n$T, $n = 0, 1, 2, \ldots\ldots$, T being the sampling period, then Z-transform of $f(t)$ is defined as

$$Z\{f(t)\} = \sum_{n=0}^{\infty} f(nT) z^{-n} = F(z).$$

Note. The important element of discrete-time systems is the samples in which a switch close to admit an input signal in every T seconds. A **samples** is a conversion device which converts a continuous signal into a sequence of pulses occurring at sampling instants 0, T, 2T, ... where T is the sampling period.

2.18. CONVERGENCE OF SERIES

Consider, $$Z\{f(n)\} = \sum_{n=0}^{\infty} f(n)z^{-n}$$

Now for convergence by Cauchy's ratio test

$$\underset{n\to\infty}{\text{Lt}} \left|\frac{u_{n+1}}{u_n}\right| < 1 \quad i.e., \quad \underset{n\to\infty}{\text{Lt}} \left|\frac{f(n+1)\,z^{-(n+1)}}{f(n)\,z^{-n}}\right| < 1$$

i.e., $$\underset{n\to\infty}{\text{Lt}} \left|\frac{f(n+1)}{f(n)}\right| z^{-1} < 1 \quad \Rightarrow \quad \underset{n\to\infty}{\text{Lt}} \left|\frac{f(n+1)}{f(n)}\right| < |z|$$

$\Rightarrow$ The region of convergence is the exterior of the circle of radius R (say). Hence R is called the **radius of convergence** of the series $\sum_{n=0}^{\infty} f(n)\, z^{-n}$.

Important Note. While finding Z-transforms

(*i*) If $f(n)$ is given simply replace $f(n)$.

(*ii*) If $f(t)$ is given then replace t by nT.

2.19. PROPERTIES OF Z-TRANSFORMS

1. *The Z-transform is linear*

i.e., (*i*) $Z\{af(n) \pm bg(n)\} = aZ\{f(n)\} \pm bZ\{g(n)\}$

(*ii*) $Z\{af(t) \pm bg(t)\} = aZ\{f(t)\} \pm bZ\{g(t)\}$.

Proof. (*i*) $$Z\{af(n) \pm bg(n)\} = \sum_{n=0}^{\infty} \{af(n) \pm bg(n)\}\, z^{-n}$$

$$= a\sum_{n=0}^{\infty} f(n)z^{-n} \pm b\sum_{n=0}^{\infty} g(n)z^{-n} = af(z) \pm bg(z)$$

$$= aZ\{f(n)\} \pm bZ\{g(n)\}$$

(*ii*) Proof is similar to (*i*).

Note. Proof is similar in both the cases of $Z\{f(n)\}$ and $Z\{f(t)\}$.

2. $Z\{\delta(n)\} = 1$.

Proof. $$Z\{\delta(n)\} = \sum_{n=0}^{\infty} \delta(n)z^{-n} = 1 \qquad \text{(Using the definition)}$$

3. $Z\{u(n)\} = \dfrac{z}{z-1}$ *for* $|\, z\,| > 1$.

Proof. $$Z\{u(n)\} = \sum_{n=0}^{\infty} u(n)z^{-n} = \sum_{n=0}^{\infty} z^{-n} \qquad \text{(Using the definition)}$$

$$= \sum_{n=0}^{\infty} \frac{1}{z^n} = \frac{1}{1-z^{-1}} \; if \left| z^{-1} \right| < 1$$

$\therefore \quad Z\{u(n)\} = \dfrac{z}{z-1}$ if $|z| > 1$.

4. (*i*) $Z\{a^n f(t)\} = F\left(\dfrac{z}{a}\right)$ (*ii*) $Z\{a^n f(n)\} = F\left(\dfrac{z}{a}\right)$.

Proof. (*i*) $Z\{a^n f(t)\} = \sum_{n=0}^{\infty} a^n f(nT)\, z^{-n} = \sum_{n=0}^{\infty} f(nT)\left(\dfrac{z}{a}\right)^{-n} = F\left(\dfrac{z}{a}\right)$

(*ii*) The proof is similar to (*i*).

5. $Z\{a^n u(n)\} = \dfrac{z}{z-a}, \; |z| > a.$

Proof. $Z\{a^n u(n)\} = \sum_{n=0}^{\infty} a^n (1) z^{-n} \qquad \because u(n) = \begin{cases} 1 & \text{for } n \ge 0 \\ 0 & \text{for } n < 0 \end{cases}$

Now $\text{R.H.S.} = \sum_{n=0}^{\infty} (az^{-1})^n = \dfrac{1}{1-az^{-1}}$ if $|az^{-1}| < 1$

$\therefore \quad Z\{a^n u(n)\} = \dfrac{z}{z-a}$ if $|z| > |a|$.

6. $Z\{nf(t)\} = -z\dfrac{dF(z)}{dz}$.

Proof. $F(z) = Z\{f(t)\} = \sum_{n=0}^{\infty} f(nT) z^{-n}$

$$\frac{dF(z)}{dz} = \sum_{n=0}^{\infty} [-nf(z)\, z^{-n-1}] = -\sum_{n=0}^{\infty} nf(nT) z^{-n} = -Z\{nf(t)\}$$

$\therefore \quad Z\{nf(t)\} = -z\dfrac{dF(z)}{dz}$.

7. $Z\{f(t + pT)\} = Z\{f((n + p)T)\}$

$$= z^k\left[F(z) - f(0.\,T) - \frac{f(T)}{z} - \frac{f(2T)}{z^2} - \ldots\ldots - \frac{f((k-1)T)}{z^{p-1}}\right].$$

Proof. $Z\{f(n + p)T\} = \sum_{n=0}^{\infty} f((n+p)T) z^{-n} = z^p \sum_{n=0}^{\infty} f((n+p)T) z^{-(n+p)}$

$$= z^k \sum_{m=i}^{\infty} f(mT) z^{-m} \quad \text{where } m = n + p$$

$$= z^k\left[\sum_{m=0}^{\infty} f(mT) z^{-m} - \sum_{m=0}^{p-1} f(mT) z^{-m}\right]$$

$$= z^k\left[F(z) - f(0) - \frac{f(T)}{z} - \frac{f(2T)}{z^2} - \ldots\ldots - \frac{f((k-1)T)}{z^{k-1}}\right]$$

Note. If $f\{(n+p)T\}$ is denoted by f_{n+k}, then

$$Z[f(t+pT)] = Z[f_{n+p}] = z^k\left[F(z) - f_0 - \frac{f_1}{z} - \frac{f_2}{z^2} - \ldots\ldots - \frac{f_{k-1}}{z^{k-1}}\right].$$

2.20. Z-TRANSFORM OF BASIC SEQUENCES

1. $$Z(k) = \frac{kz}{z-1} \text{ if } |z| > 1$$

$$Z(k) = \sum_{n=0}^{\infty} kz^{-n} = k\{1 + z^{-1} + z^{-2} + \ldots\ldots + \infty\} = k\,\frac{1}{1-z^{-1}} = \frac{kz}{z-1} \text{ if } |z| > 1.$$

Note. $Z(1) = \dfrac{z}{z-1}$ if $|z| > 1$.

2. $$Z\{(-1)^n\} = \frac{z}{z+1} \text{ if } |z| > 1$$

$$Z\{(-1)^n\} = \sum_{n=0}^{\infty} (-1)^n z^{-n} = \sum_{n=0}^{\infty} (-z)^n = \frac{1}{1+1/z} = \frac{z}{z+1} \text{ if } |z| > 1.$$

3. $$Z\{a^n\} = \frac{z}{z-a} \text{ if } |z| > |a|$$

$$Z\{a^n\} = \sum_{n=0}^{\infty} a^n z^{-n} = \sum_{n=0}^{\infty}\left(\frac{a}{z}\right)^n = \frac{1}{1-a/z} = \frac{z}{z-a} \text{ if } |z| > |a|.$$

4. $$Z\{n\} = \frac{z}{(z-1)^2}$$

$$Z\{n\} = \sum_{n=0}^{\infty} (n)z^{-n} = -z\,\frac{d}{dz}\{Z(1)\} = -z\,\frac{d}{dz}\left(\frac{z}{z-1}\right) = \frac{z}{(z-1)^2}.$$

5. $$Z\{na^n\} = \frac{az}{(z-a)^2}$$

$$Z\{na^n\} = -z\,\frac{d}{dz}\left\{\frac{z}{z-a}\right\} = \frac{az}{(z-a)^2}.$$

Note. $Z\{n^{m+1}\} = (-z)^m \dfrac{d^m}{dz^m} z(n)$

6. $$Z\{\cos n\theta\} = \frac{z(z-\cos\theta)}{z^2 - 2z\cos\theta + 1}\ ;\ Z\{\sin n\theta\} = \frac{z\sin\theta}{z^2 - 2z\cos\theta + 1} \text{ if } |z| > 1$$

We know that $\quad Z\{a^n\} = \dfrac{z}{z-a}$ if $|z| > |a|$

Let $\quad a = e^{i\theta}\ \therefore\ Z\{e^{in\theta}\} = \dfrac{z}{z-e^{i\theta}} = \dfrac{z}{z-\{\cos\theta + i\sin\theta\}}$

$$Z\{\cos n\theta + i \sin n\theta\} = \frac{z\{z - \cos\theta + i \sin\theta\}}{\{z - \cos\theta + i\sin\theta\}\{z - \cos\theta - i \sin\theta\}}$$

$$= \frac{z(z - \cos\theta) + iz\sin\theta}{z^2 - 2z\cos\theta + 1}$$

Equating real and imaginary parts, we get

$$Z\{\cos n\theta\} = \frac{z(z - \cos\theta)}{z^2 - 2z\cos\theta + 1}\ ;\ Z\{\sin n\theta\} = \frac{z\sin\theta}{z^2 - 2z\cos\theta + 1}\ ;\ |\ z\ | > 1.$$

7. $$Z\{r^n \cos n\theta\} = \frac{z(z - r\cos\theta)}{z^2 - 2zr\cos\theta + r^2}\ ;\ Z\{r^n \sin n\theta\} = \frac{zr\sin\theta}{z^2 - 2zr\cos\theta + r^2} \text{ if } |\ z\ | > |\ r\ |$$

$$Z\{a^n\} = \frac{z}{z - a} \text{ if } |\ z\ | > |\ a\ |$$

Let $a = re^{i\theta}$ in $Z\{a^n\}$ then $Z\{r^n\ e^{in\theta}\} = \dfrac{z}{z - re^{i\theta}}$

$$\therefore \quad Z\{r^n (\cos n\theta + i\sin n\theta)\} = \frac{(z/r)}{(z/r) - e^{i\theta}} = \frac{(z/r)[(z/r) - \cos\theta + i\sin\theta]}{\{(z/r) - \cos\theta\}^2 + \sin^2\theta}$$

Equating real and imaginary parts, we get

$$Z\{r^n \cos n\theta\} = \frac{z(z - r\cos\theta)}{z^2 - 2zr\cos\theta + r^2}\ ;\ Z\{r^n \sin n\theta\} = \frac{zr\sin\theta}{z^2 - 2zr\cos\theta + r^2} \text{ if } |\ z\ | > |\ r\ |.$$

2.21. Z-TRANSFORMS OF STANDARD DISCRETE FUNCTIONS

1. $$Z(t) = \frac{Tz}{(z-1)^2}$$

$$Z(t) = \sum_{n=0}^{\infty} (nT)z^{-n} = T \sum_{n=0}^{\infty} nz^{-n} = T(-z)\frac{d}{dz}Z(1)$$

$$= -Tz\frac{d}{dz}\left(\frac{z}{z-1}\right) = \frac{Tz}{(z-1)^2}.$$

2. $$Z(e^{-at}) = \frac{z}{z - e^{-aT}}\ ;\ |\ z\ | > |\ e^{-aT}\ |\ ;\ Z(e^{at}) = \frac{z}{z - e^{aT}}\ ;\ |\ z\ | > |\ e^{aT}\ |$$

$$Z(e^{-at}) = \sum_{n=0}^{\infty} e^{-anT} z^{-n} = \sum_{n=0}^{\infty} (e^{-aT})^n z^{-n} = \frac{z}{z - e^{-atT}}.$$

Similarly, we can prove for $Z(e^{at})$.

3. $$Z(\cos\omega t) = \frac{z(z - \cos\omega T)}{z^2 - 2z\cos\omega T + 1}$$

$$Z(\cos\omega t) = \sum_{n=0}^{\infty} \cos n\omega T\ z^{-n} = Z(\cos n\theta) \quad \text{where } \theta = \omega T$$

$$= \frac{z(z - \cos\omega T)}{z^2 - 2z\cos\omega T + 1}.$$

Aliter: $Z(\cos \omega t) = Z\left[\dfrac{e^{i\omega t} + e^{-i\omega t}}{2}\right]$

$$= \frac{1}{2}\left[\frac{z}{z - e^{i\omega T}} + \frac{z}{z - e^{-i\omega T}}\right] = \frac{z}{2}\left[\frac{2z - (e^{i\omega T} + e^{-i\omega T})}{(z - e^{i\omega T})(z - e^{-i\omega T})}\right]$$

$$= \frac{z}{2}\left[\frac{2z - 2\cos \omega T}{z^2 - (e^{i\omega T} + e^{-i\omega T})z + 1}\right] = \frac{z(z - \cos \omega T)}{z^2 - 2z\cos \omega T + 1}.$$

4. $$Z\{\sin \omega t\} = \frac{z \sin \omega T}{z^2 - 2z\cos \omega T + 1}$$

$$Z\{\sin \omega t\} = Z\left\{\frac{e^{i\omega t} - e^{-i\omega t}}{2}\right\} = \frac{1}{2i}\left\{\frac{z}{z - e^{i\omega T}} - \frac{z}{z - e^{-i\omega T}}\right\}$$

$$= \frac{z}{2i}\left\{\frac{e^{i\omega T} - e^{-i\omega T}}{z^2 - z(e^{i\omega T} + e^{-i\omega T}) + 1}\right\} = \frac{z \sin \omega T}{z^2 - 2z\cos \omega T + 1}.$$

5. $$Z(t^k) = -\,Tz\,\frac{d}{dz}\,\{Z(t^{k-1})\}$$

$$Z(t^k) = \sum_{n=0}^{\infty} (nT)^k z^{-n} = Tz \sum_{n=0}^{\infty} n^k\, T^{k-1}\, z^{-(n+1)}$$

$$= Tz \sum_{n=0}^{\infty} (nT)^{k-1}\, n z^{-(n+1)} \qquad \text{...(1)}$$

Similarly, $Z(t^{k-1}) = \sum_{n=0}^{\infty} (nT)^{k-1}\, z^{-n}$

$$\frac{d}{dz}\,[Z(t^{k-1})] = \sum_{n=0}^{\infty} (nT)^{k-1}\,(-\,n)\, z^{-(n+1)}$$

$$= -\sum_{n=0}^{\infty} (nT)^{k-1}\, n z^{(n+1)} \qquad \text{...(2)}$$

Using (2) in (1), we get $Z(t^k) = -\,Tz\,\dfrac{d}{dz}\,[Z(t^{k-1})]$ ...(3)

Setting $k = 1, 2, 3, \ldots\ldots$ we get $Z(t), Z(t^2), \ldots\ldots$

$$Z(t) = -\,Tz\,\frac{d}{dz}\,[Z(1)] = -\,Tz\,\frac{d}{dz}\left(\frac{z}{z-1}\right) = \frac{Tz}{(z-1)^2} \qquad \text{...(4)}$$

$$Z(t^2) = -\,Tz\,\frac{d}{dz}\,z(t) = -\,Tz\,\frac{d}{dz}\left[\frac{Tz}{(z-1)^2}\right] = \frac{T^2 z(z+1)}{(z-1)^3} \qquad \text{...(5)}$$

and so on.

2.22. FIRST SHIFTING THEOREM

If $Z\{f(t)\} = F(z)$ *then* $\quad Z\{e^{-at} f(t)\} = F\{ze^{aT}\}.$

Proof.
$$Z\{e^{-at} f(t)\} = \sum_{n=0}^{\infty} e^{-anT} f(nT) z^{-n} = \sum_{n=0}^{\infty} f(nT)(ze^{aT})^{-n}$$

$$= F\{ze^{aT}\} \quad \because \quad F(z) = \sum_{n=0}^{\infty} f(nT) z^{-n}$$

i.e.,
$$Z\{e^{-at} f(t)\} = Z\{f(t)\}_{z \to ze^{aT}}$$
$$= F\{z\} \text{ where } z \to ze^{aT}.$$

2.22.1. Deductions

1.
$$Z\{e^{-at}\} = Z\{e^{-at} \,.\, 1\} = z\{1\}_{z \to ze^{aT}}$$

$$= \left(\frac{z}{z-1}\right)_{z \to ze^{aT}} = \frac{ze^{aT}}{ze^{aT}-1} \qquad \therefore \quad Z\{e^{-at}\} = \frac{z}{z - e^{-aT}}.$$

2.
$$Z\{te^{-at}\} = Z\{e^{-at} \,.\, t\} = Z\{t\}_{z \to ze^{aT}}$$

$$= \left\{\frac{Tz}{(z-1)^2}\right\}_{z \to ze^{aT}} = \frac{Tze^{aT}}{(ze^{aT}-1)^2} \qquad \therefore \quad Z\{te^{-at}\} = \frac{Tze^{-aT}}{(z - e^{-aT})^2}.$$

3.
$$Z\{e^{-iat}\} = Z\{e^{-iat} \,.\, 1\} = Z\{1\}_{z \to ze^{iaT}}$$

$$= \left[\frac{z}{z-1}\right]_{z \to ze^{iaT}} = \frac{z}{z - e^{-iaT}}$$

$$= \frac{z(z - e^{iaT})}{(z - e^{-iaT})(z - e^{iaT})} = \frac{z[z - (\cos aT + i \sin aT)]}{z^2 - z(e^{iaT} + e^{-iaT}) + 1}$$

$$\therefore \quad Z\{e^{-iat}\} = \frac{z[(z - \cos aT) - i \sin aT]}{z^2 - 2z \cos aT + 1}$$

$$\Rightarrow \quad Z\{\cos at - i \sin at\} = \frac{z[(z - \cos aT) - i \sin aT]}{z^2 - 2z \cos aT + 1}$$

$$\Leftrightarrow \quad Z[\cos at] = \frac{z(z - \cos aT)}{z^2 - 2z \cos aT + 1} \; ; \; Z[\sin at] = \frac{-z \sin aT}{z^2 - 2z \cos aT + 1}.$$

4.
$$Z\{e^{-at} \cos bt\} = Z\{\cos bt\}_{z \to ze^{aT}}$$

$$= \left[\frac{z(z - \cos bT)}{z^2 - 2z \cos bT + 1}\right]_{z \to ze^{bT}} = \frac{ze^{aT}\{ze^{aT} - \cos bT\}}{z^2 e^{2aT} - 2ze^{aT} \cos bT + 1}.$$

5.
$$Z\{e^{-at} \sin bt\} = Z\{\sin bt\}_{z \to ze^{aT}}$$

$$= \left\{\frac{z \sin bt}{z^2 - 2z \cos \omega T + 1}\right\}_{z \to ze^{aT}} = \frac{ze^{aT} \sin bt}{z^2 e^{2aT} - 2ze^{aT} \cos bT + 1}.$$

2.23. SECOND SHIFTING THEOREM

If $Z\{f(t)\} = F(z)$ *then* $Z\{f(t + T)\} = z[F(z) - f(0)]$.

Proof.
$$Z\{f(t + T)\} = \sum_{n=0}^{\infty} f(nT + T)z^{-n}$$

$$= z \sum_{n=0}^{\infty} f[(n + 1)T]\, z^{-(n+1)} = z \sum_{k=1}^{\infty} f(kT)\, z^{-k}$$

$$= z \left[\sum_{k=0}^{\infty} f(kT)\, z^{-k} - f(0) \right] = z[F(z) - f(0)].$$

2.24. INITIAL VALUE THEOREM

If $z\{f(t)\} = F(z)$ *then* $f(0) = \underset{z\to\infty}{Lt} F(z)$.

Proof.
$$F(z) = Z[f(t)] = \sum_{n=0}^{\infty} f(nT)\, z^{-n}$$

$$= f(0 \,.\, T) + \frac{f(1\,.\,T)}{z} + \frac{f(2\,.\,T)}{z^2} + \ldots\ldots \infty = f(0) + \frac{1}{z} f(T) + \frac{1}{z^2} f(2T) + \ldots\ldots\infty$$

$$\therefore \quad \underset{z\to\infty}{\text{Lt}} F(z) = f(0).$$

2.25. FINAL VALUE THEOREM

If $Z[f(t)] = F(z)$ *then* $\underset{z\to\infty}{Lt} f(t) = \underset{z\to 1}{Lt} (z - 1) F(z)$.

Proof.
$$Z[f(t + T) - f(t)] = \sum_{n=0}^{\infty} [f(nT + T) - f(nT)]\, z^{-n}$$

$$Z[f(t + T)] - Z[f(t)] = \sum_{n=0}^{\infty} [f(nT + T) - f(nT)]\, z^{-n}$$

$$zF(z) - zf(0) - F(z) = \sum_{n=0}^{\infty} [f(nT + T) - f(nT)]\, z^{-n}$$

$$\underset{z\to 1}{\text{Lt}} [(z - 1) F(z) - zf(0)] = \underset{z\to 1}{\text{Lt}} \left[\sum_{n=0}^{\infty} \{f(nT + T) - f(nT)\} z^{-n} \right]$$

$$\underset{z\to 1}{\text{Lt}} [(z - 1) F(z)] - f(0) = \sum_{n=0}^{\infty} [f(nT + T) - f(nT)]$$

$$= \underset{n\to\infty}{\text{Lt}} [f(T) - f(0) + f(2T) - f(T) + \ldots\ldots + f((n + 1)T) - f(nT)]$$

$$= \underset{n\to\infty}{\text{Lt}} f((n + 1)T) - f(0) = f(\infty) - f(0) = \underset{t\to\infty}{\text{Lt}} f(t) - f(0)$$

$$\therefore \quad \underset{z\to 1}{\text{Lt}} (z - 1)F(z) = \underset{t\to\infty}{\text{Lt}} f(t).$$

2.26. SHIFT PROPERTY

If f(n) is a casual sequence and m is any positive integer then

$$Z\{f(n-m)\} = z^{-m}F(z) \quad \text{where } F(z) = Z\{f(n)\}.$$

Proof.
$$Z\{f(n-m)\} = \sum_{n=0}^{\infty} f(n-m)z^{-n}$$

$$= \sum_{k=-m}^{\infty} f(k)z^{-(m+k)} = \sum_{k=0}^{\infty} f(k)\, z^{-k} \,.\, z^{-m} = z^{-m}\, \mathrm{F}(x)$$

$\therefore$ $\quad Z\{f(n-m)\} = z^{-m}\mathrm{F}(z).$

Cor. $\quad Z^{-1}[z^{-m}\mathrm{F}(z)] = f(n-m) = Z^{-1}\{\mathrm{F}(n)\}_{n \to n-m}$

ILLUSTRATIVE EXAMPLES

Example 1. *Find the Z-transform of*

(*i*) n^2 (*ii*) $n(n-1)$ (*iii*) $\dfrac{1}{n}$

(*iv*) $\dfrac{1}{n(n+1)}$ (*v*) $\cos \dfrac{n\pi}{2}$.

Sol. (*i*) $\quad Z(n^2) = Z(n\,.\,n) = -\dfrac{d}{dz} z(n) = -\dfrac{d}{dz}\left\{\dfrac{z}{(z-1)^2}\right\} = \dfrac{z(z+1)}{(z-1)^3}$.

(*ii*) $\quad Z\{n(n-1)\} = Z\{n^2 - n\} = Z(n^2) - Z(n)$

$$= \frac{z(z+1)}{(z-1)^3} - \frac{z}{(z-1)^2} = \frac{z^2 + z - z(z-1)}{(z-1)^3} = \frac{2z}{(z-1)^3}.$$

(*iii*)
$$Z\left\{\frac{1}{n}\right\} = \sum_{n=1}^{\infty} \frac{1}{n} z^{-n} = \frac{1}{z} + \frac{1}{2z^2} + \frac{1}{3z^3} + \ldots\ldots \infty = -\log\left(1 - \frac{1}{z}\right)$$

if $\left(\dfrac{1}{z}\right) < 1 = \log\left(\dfrac{z}{z-1}\right)$ if $|\, z \,| > 1$.

(*iv*)
$$Z\left\{\frac{1}{n(n+1)}\right\} = Z\left\{\frac{1}{n} - \frac{1}{n+1}\right\} = Z\left\{\frac{1}{n}\right\} - Z\left\{\frac{1}{n+1}\right\}$$

$$= \sum_{n=1}^{\infty} \frac{1}{n} z^{-n} - \sum_{n=0}^{\infty} \frac{1}{n+1} z^{-n}$$

$$= \log\left(\frac{z}{z-1}\right) - \left\{1 + \frac{1}{2z} + \frac{1}{3z^2} + \ldots\ldots \infty\right\}$$

$$= \log\left(\frac{z}{z-1}\right) - z\left\{\frac{1}{z} + \frac{1}{2}\left(\frac{1}{z}\right)^2 + \frac{1}{3}\left(\frac{1}{z}\right)^3 + \ldots\ldots \infty\right\}$$

$$= \log\left(\frac{z}{z-1}\right) - z\left\{-\log\left(1-\frac{1}{z}\right)\right\} = (z-1)\log\left\{\frac{z-1}{z}\right\}.$$

(v) $$Z\left\{\cos\frac{n\pi}{2}\right\} = \sum_{n=0}^{\infty} \cos\frac{n\pi}{2} z^{-n} = 1 - \frac{1}{z^2} + \frac{1}{z^4} - \ldots\ldots \infty$$

$$= \left[1 + \frac{1}{z^2}\right]^{-1} = \frac{z^2}{z^2+1} \text{ if } |z| > 1.$$

Example 2. *Find the Z-transform of*

(i) $\frac{1}{2}(n+1)(n+2)$ (ii) $ab^n (a \neq 0, b \neq 0)$ (iii) $f(n) = \begin{cases} n, & n \geq 0 \\ 0, & n < 0 \end{cases}$

(iv) $f(n) = \begin{cases} 1; & n = k \\ 0; & n \neq k \end{cases}$ (v) $f(n) = \begin{cases} 0 & \text{if } n > 0 \\ 1 & \text{if } n \leq 0 \end{cases}$ (vi) $f(n) = \begin{cases} \frac{a^n}{n!}, & n \geq 0 \\ 0, & \text{otherwise} \end{cases}$.

Sol. (i) $$Z\left\{\frac{1}{2}(n+1)(n+2)\right\} = \frac{1}{2}\{Z(n^2) + 3Z(n) + 2Z(1)\}$$

$$= \frac{1}{2}\left[\frac{z(z+1)}{(z-1)^3} + \frac{3z}{(z-1)^2} + \frac{2z}{(z-1)}\right] \text{ if } |z| > 1.$$

(ii) $$Z\{ab^n\} = aZ\{b^n\} = a\sum_{n=0}^{\infty} b^n z^{-n} = a\sum_{n=0}^{\infty}\left(\frac{b}{z}\right)^n$$

$$= a\left\{\frac{z}{z-b}\right\} = \frac{az}{z-b} \text{ if } |z| > |b|.$$

(iii) $$Z\{f(n)\} = \sum_{n=0}^{\infty} f(n)^{-n} = \sum_{n=0}^{\infty} nz^{z-n} = \frac{1}{z} + \frac{2}{z^2} + \frac{3}{z^3} + \ldots\ldots \infty$$

$$= \frac{1}{z}\left(1 - \frac{1}{z}\right)^{-2} = \frac{z}{(z-1)^2} \text{ if } |z| > 1.$$

(iv) $$Z\{f(n)\} = \sum_{n=0}^{\infty} f(n) z^{-n} = \sum_{\substack{n=0 \\ n\neq k}}^{\infty} (0) \, . \, z^{-n} + 1 \, . \, z^{-k} = \frac{1}{z^k}.$$

(v) $$Z\{f(n)\} = \sum_{n=-\infty}^{\infty} f(n) z^{-n} = \sum_{n=-\infty}^{\infty} (1) z^{-n} = \sum_{n=0}^{\infty} z^n = \frac{1}{1-z} \text{ if } |z| < 1.$$

(vi) $$Z\{f(n)\} = \sum_{n=0}^{\infty} f(n) z^{-n} = \sum_{n=0}^{\infty} \frac{a^n}{n!} z^{-n} = \sum_{n=0}^{\infty} \frac{(az^{-1})^n}{n!} = e^{az^{-1}} = e^{a/z}.$$

Example 3. *Find the Z-transform of*

(i) $u(n-1)$ (ii) $4^n \delta(n-1)$ (iii) $\delta(n-k)$.

Sol. (*i*) $Z\{u(n-1)\} = \sum_{n=1}^{\infty} 1 \,.\, z^{-n}$

$$= \frac{1}{z} + \frac{1}{z^2} + \frac{1}{z^3} + \ldots\ldots \infty = \frac{1}{z}\left\{1 + \frac{1}{z} + \frac{1}{z^2} + \ldots\ldots\right\} = \frac{1}{z}\left\{\frac{1}{1-\frac{1}{z}}\right\} \text{ if } \left|\frac{1}{z}\right| < 1$$

$$= \frac{1}{z-1} \text{ if } |\, z \,| > 1.$$

(*ii*) $Z\{4^n \delta(n-1)\} = \sum_{n=0}^{\infty} 4^n \delta(n-1) z^{-n} = \frac{4}{z}.$

(*iii*) $Z\{\delta(n-k)\} = \sum_{n=0}^{\infty} \delta(n-k)\, z^{-n} = \frac{1}{z^k}$, k is a (+) ve integer.

Example 4. *Find the Z-transform of* $\left\{\frac{a^k}{k!}\right\}$, *(k ≥ 0).*

Sol. By definition,

$$z\left[\left\{\frac{a^k}{k!}\right\}\right] = \sum_{k=0}^{\infty} \frac{a^k}{k!} z^{-k} = \sum_{k=0}^{\infty} \frac{(az^{-1})^k}{k!} = 1 + \frac{az^{-1}}{1!} + \frac{(az^{-1})^2}{2!} + \frac{(az^{-1})^3}{3!} + \ldots.$$

This is an exponential series,

$$\therefore \quad Z\left[\left(\left\{\frac{a^k}{k!}\right\}\right)\right] = e^{az^{-1}} = e^{a/z}.$$

Example 5. *Find the Z-transform of*

(*i*) $e^{4t} \cos t$.

(*ii*) $2^k \sin \alpha k$, $k \geq 0$.

Sol. (*i*) We know that

$$Z\,[\{\cos t\}] = \frac{z(z-\cos 1)}{z^2 - 2z\cos 1 + 1}.$$

Applying damping rule, we have

$$Z\,[\{e^{4t}\cos t\}] = Z\,[\{(e^{-4})^{-t}\cos t\}]$$

$$= F(e^{-4}z), \text{ where } F(z) = Z[\{f(t)\}] = Z\,[\{\cos t\}]$$

$$= \frac{e^{-4}z\,(e^{-4}z - \cos 1)}{(e^{-4}z)^2 - 2(e^{-4}\,z)\cos 1 + 1} = \frac{z\,(z - e^4\cos 1)}{z^2 - 2e^4\,z\cos 1 + e^8}.$$

(*ii*) We know that $Z\,[\{\sin \alpha k\}] = \dfrac{z\sin\alpha}{z^2 - 2z\cos\alpha + 1}.$

Applying damping rule, we have

$$Z\,[\{2^k \sin \alpha k\}] = Z\,[\{(2^{-1})^{-k} \sin \alpha k\}]$$

$$= F(2^{-1}z), \text{ where } F(z) = Z\,[\{f(k)\}] = Z[\{\sin \alpha k\}]$$

$$= \frac{(2^{-1}z)\sin\alpha}{(2^{-1}z)^2 - 2(2^{-1}z)\cos\alpha + 1} = \frac{2z\sin\alpha}{z^2 - 4z\cos\alpha + 4}.$$

Example 6. *Find the Z-transform of*

(*i*) $e^{-t}\, t^2$ (*ii*) $e^{-2t}\, t^3$ (*iii*) e^{2t+5}

(*iv*) $e^{3t} \sin 2t$ (*v*) $e^{4t} \cos t$ (*vi*) $a^n \cos n\pi$.

Sol. (*i*) $Z\{e^{-t}\, t^2\} = Z\{t^2\}_{z \to ze^{T}}$

$$= \left\{\frac{T^2 z(z+1)}{(z-1)^3}\right\}_{z \to ze^{T}} = \frac{T^2 ze^{T}\,(ze^{T}+1)}{(ze^{T}-1)^3}$$

(*ii*) $Z\{e^{-2t}\, t^3\} = Z\{t^3\}_{z \to ze^{2T}}$

$$= \left[\frac{T^3 z(1+4z+z^2)}{(z-1)^4}\right]_{z \to ze^{2T}} = \frac{T^3 ze^{2T}[1+4ze^{2T}+z^2e^{4T}]}{(ze^{2T}-1)^4}$$

(*iii*) $Z\{e^{2t+5}\} = e^5 Z\{e^{2t}\} = e^5 \cdot \dfrac{z}{z - e^{2T}}$

(*iv*) $Z\{e^{3t} \sin 2t\} = Z\{\sin 2t\}_{z \to ze^{-3T}}$

$$= \left[\frac{z\sin 2T}{z^2 - 2z\cos 2T + 1}\right]_{z \to ze^{-3T}} = \frac{ze^{-3T}\sin 2T}{z^2e^{-6T} - 2ze^{-3T}\cos 2T + 1}$$

(*v*) $Z\{e^{4t} \cos t\} = Z\{\cos t\}_{z \to ze^{-4T}}$

$$= \left[\frac{z(z-\cos T)}{z^2 - 2z\cos T + 1}\right]_{z \to ze^{-4T}} = \frac{ze^{-4T}(ze^{-4T}-\cos T)}{z^2e^{-8T} - 2ze^{-4T}\cos T + 1}$$

(*vi*) $Z\{a^n \cos n\pi\} = \sum_{n=0}^{\infty} a^n \cos n\pi \,.\, z^{-n}$

$$= 1 - \frac{a}{z} + \left(\frac{a}{z}\right)^2 - \left(\frac{a}{z}\right)^3 + \ldots\ldots \infty = \left(1+\frac{a}{z}\right)^{-1}$$

if $\left|\dfrac{a}{z}\right| < 1 = \dfrac{z}{z+a}$ if $|z| > |a|$.

Example 7. *Find the Z-transform of*

(*i*) ${}^nC_p\ (0 \le p \le n)$ (*ii*) ${}^{n+p}C_p$.

Sol. (*i*) By definition of Z-transform, we have

$$Z\,[\{{}^nC_p\}] = \sum_{p=0}^{\infty} {}^nC_p\, z^{-p} = \sum_{p=0}^{\infty} {}^nC_p\, z^{p} \qquad [\because\ {}^nC_p = 0 \text{ if } p > n]$$

$$= 1 + {}^nC_1\, z^{-1} + {}^nC_2 z^{-2} + \ldots + {}^nC_n\, z^{-n}$$

$$= (1 + z^{-1})^n.$$

(*ii*) We have

$$Z\,[\{{}^{n+p}C_p\}] = \sum_{p=0}^{\infty} {}^{n+p}C_p\, z^{-p} = 1 + {}^{n+1}C_1 z^{-1} + {}^{n+2}C_2 z^{-2} + {}^{n+3}C_3 z^{-3} + \ldots \infty$$

$$= 1 + (n+1)\,z^{-1} + \frac{(n+2)(n+1)}{2!} z^{-2} + \frac{(n+3)(n+2)(n+1)}{3!} z^{-3} + \ldots \infty$$

$$= 1 + (-n-1)(-z^{-1}) + \frac{(-n-1)(-n-2)}{2!} (-z^{-1})^2 + \frac{(-n-1)(-n-2)(-n-3)}{3!} (-2)^{-1})^3 + \ldots \infty$$

$$= (1 - z^{-1})^{-n-1}.$$

Example 8. *Show that* $Z\,({}^{n+p}C_p) = (1 - 1/z)^{-(p+1)}$. *Using the damping rule, show that* $Z({}^{n+p}C_p a^n) = (1 - a/z)^{-(p+1)}$.

Sol. We have

$$Z({}^{n+p}C_p) = \sum_{n=0}^{\infty} {}^{n+p}C_p z^{-n} = 1 + {}^{p+1}C_p z^{-1} + {}^{p+2}C_p z^{-2} + {}^{p+3}C_p z^{-3} + \ldots \infty$$

$$= 1 + {}^{p+1}C_1 z^{-1} + {}^{p+2}C_2 z^{-2} + {}^{p+3}C_3\, z^{-3}$$

$$= 1 + (p+1)z^{-1} + \frac{(p+2)(p+1)}{2!} z^{-2} + \frac{(p+3)(p+2)(p+1)}{3!} z^{-3} + \ldots \infty$$

$$= 1 + (-p-1)(-z^{-1}) + \frac{(-p-1)(-p-2)}{2!} (-z^{-1})^2 + \frac{(-p-1)(-p-2)(-p-3)}{3!} (-z^{-1})^3 + \ldots \infty$$

$$= (1 - z^{-1})^{-p-1}.$$

Thus, $Z\,({}^{n+p}C_p) = (1 - z^{-1})^{-(p+1)}$.

By damping rule, we have

$$Z({}^{n+p}C_p\, a^n) = Z\,((a^{-1})^{-n}\, {}^{n+p}C_p)$$

$$= U\,(a^{-1} z) \text{ where } U(z) = Z({}^{n+p}C_p) = (1 - z)^{-(p+1)}$$

$$= (1 - (a^{-1}z)^{-1})^{-(p+1)} = \left(1 - \frac{a}{z}\right)^{-(p+1)}$$

2.27. DIFFERENTIATION

Let $Z[\{f(n)\}] = F(z)$. An infinite series can be differentiated term by term with in its region of convergence. $F(z)$ may be treated as a function of z^{-1}.

$$F(z) = \sum_{n=0}^{\infty} f(n)\, z^{-n} = \sum_{n=0}^{\infty} f(n)(z^{-1})^n$$

Differentiating on both sides w.r.t. z^{-1}

$$\frac{d}{dz^{-1}} F(z) = \sum_{n=0}^{\infty} n f(n)\,(z^{-1})^{n-1} \qquad \ldots(1)$$

$$z^{-1} \frac{dF(z)}{dz^{-1}} = \sum_{n=0}^{\infty} nf(n)\, z^{-n} = Z\{nf(n)\}$$

$$\therefore \qquad Z\{nf(n)\} = z^{-1} \frac{dF(z)}{dz^{-1}} \qquad ...(2)$$

Differentiating (1) w.r.t. z^{-1} again, we get

$$\frac{d^2F(z)}{d(z^{-1})^2} = \sum_{n=0}^{\infty} n(n-1)\, f(n)\, (z^{-1})^{n-2}$$

$$z^{-2} \frac{d^2F(z)}{d(z^{-1})^2} = \sum_{n=0}^{\infty} n(n-1)\, f(n)\, z^{-n} = Z\{n(n-1)\, f(n)\}$$

$$\therefore \qquad Z[n(n-1)\, f(n)] = z^{-2} \frac{d^2F(z)}{d(z^{-1})^2}. \qquad ...(3)$$

Example 9. *Find the Z-transform of*

(*i*) $na^n\, u(n)$ (*ii*) $n(n-1)\, a^n\, u(n)$.

Sol. Let $f(n) = a^n u(n)$

(*i*) $$Z\{na^n u(n)\} = z^{-1} \frac{d}{dz^{-1}} \left(\frac{z}{z-a} \right) = z^{-1} \frac{d}{dz^{-1}} [1 - az^{-1}]^{-1}$$

$$= z^{-1} \frac{a}{(1-az^{-1})^2} = \frac{az^{-1}}{(1-az^{-1})^2}.$$

(*ii*) $$Z\{n(n-1)\, a^n\, u(n)\} = z^{-2} \frac{d^2}{d(z^{-1})^2} \{1 - az^{-1}\}^{-1} = \frac{2a^2z^2}{(1-az^{-1})^3}.$$

Note. If $a = 1$ then $Z[nu(n)] = \frac{z^{-1}}{(1-z^{-1})^2}$; $Z[n(n-1)\, u(n)] = \frac{2z^{-2}}{(1-z^{-1})^3}$.

Example 10. *Show that* $Z\left[\frac{1}{n!}\right] = e^{1/z}$.

Hence evaluate $Z\left[\frac{1}{(n+1)!}\right]$ *and* $Z\left[\frac{1}{(n+2)!}\right]$.

Sol. By definition,

$$Z\left[\frac{1}{n!}\right] = \sum_{n=0}^{\infty} \frac{1}{n!} z^{-n} = 1 + \frac{z^{-1}}{1!} + \frac{z^{-2}}{2!} + \frac{z^{-3}}{3!} + \ldots = e^{1/z}.$$

Shifting $\frac{1}{n!}$ are unit to the left gives

$$Z\left[\frac{1}{(n+1)!}\right] = z\left[Z\left(\frac{1}{n!}\right) - 1\right] = z\,(e^{1/z} - 1).$$

Similarly shifting $\frac{1}{n!}$ two units to the left gives

$$Z\left[\frac{1}{(n+2)!}\right] = z^2 (e^{1/z} - 1 - z^{-1}).$$

Example 11. *Find the Z-transform of*

(i) $n \sin n\theta$ *(ii)* $n^2 e^{n\theta}$.

Sol. *(i)* Let $u_n = \sin n\theta$. Then $Z[u_n] = Z[\sin n\theta]$

$$= \frac{z \sin \theta}{z^2 - 2z \cos \theta + 1} = U(z), \text{ say.}$$

We know that $Z[nu]_n = -z \frac{d}{dz} U(z)$.

$$\therefore \quad Z[n \sin n\theta] = -z \frac{d}{dz} Z[\sin n\theta] = -z \frac{d}{dz}\left(\frac{z \sin \theta}{z^2 - 2z \cos \theta + 1}\right)$$

$$= -z \frac{\sin \theta - z^2 \sin \theta}{(z^2 - 2z \cos \theta + 1)^2} = \frac{z(z^2 - 1) \sin \theta}{(z^2 - 2z \cos \theta + 1)^2}.$$

(ii) Let $u_n = e^{n\theta}$. Then applying damping rule, we get

$$Z[u_n] = Z[e^{n\theta}]$$

$$= Z[(e^{-\theta})^{-n} . 1] = \frac{ze^{-\theta}}{ze^{-\theta} - 1} \qquad \left[\because Z[1] = \frac{1}{z-1}\right]$$

$$= \frac{z}{z - e^{\theta}} = U(z), \text{ say.}$$

We know that $Z[n^2 u_n] = (-z)^2 \frac{d^2 U(z)}{dz^2}$.

$$\therefore \quad Z[n^2 e^{n\theta}] = z^2 \frac{d^2}{dz^2} Z(e^{n\theta}) = z^2 \frac{d^2}{dz^2}\left(\frac{z}{z - e^{\theta}}\right) = z^2 \frac{d}{dz}\left\{\frac{-e^{\theta}}{(z - e^{\theta})^2}\right\}$$

$$= z^2 . \frac{2e^{\theta}}{(z - e^{\theta})^3} = \frac{2e^{\theta} z^2}{(z - e^{\theta})^3}.$$

Example 12. *Find the Z-transforms of the following:*

(i) e^{-an} *(ii)* e^{-an} *(iii)* $e^{-an} n^2$.

Sol. *(i)* Let $u_n = e^{-an}$. Then applying damping rule, we get

$$Z(u_n) = Z[e^{-an}] = Z[(e^a)^{-n} . 1]$$

$$= \frac{ze^a}{ze^a - 1} \qquad \left[\because Z[1] = \frac{z}{z-1}\right]$$

$$= \frac{z}{z - e^{-a}}.$$

(*ii*) Let $u_n = e^{-an}$. Then applying the damping rule, we get

$$Z[u_n] = Z[e^{-an}] = Z[(e^a)^{-n} \,.\, 1]$$

$$= \frac{ze^a}{ze^a - 1} \qquad \left[\because\ Z(1) = \frac{z}{z-1}\right]$$

$$= \frac{z}{z - e^{-a}} = U(z), \text{ say.}$$

We know that $Z[nu_n] = -z\frac{d}{dz}U(z)$.

$$\therefore \qquad Z[e^{-an}\, n] = -z\frac{d}{dz}\left(\frac{z}{z-e^{-a}}\right) = -z\frac{-e^{-a}}{(z-e^{-a})^2} = \frac{ze^{-a}}{(z-e^{-a})^2}$$

(*iii*) Let $u_n = e^{-an}$. Then applying damping rule, we get

$$Z[u_n] = Z[e^{-an}] = \frac{z}{z-e^{-a}} = U(z), \text{ say.}$$

We know that $Z[n^2 u_n] = (-z)^2 \frac{d^2 U(z)}{dz^2}$

$$\therefore \qquad Z[e^{-an} n^2] = z^2 \frac{d^2}{dz^2}\left(\frac{z}{z-e^{-a}}\right) = z^2 \frac{d}{dz}\left[\frac{-e^{-a}}{(z-e^{-a})^2}\right]$$

$$= z^2 \frac{2e^{-a}}{(z-e^{-a})^3} = \frac{2z^2 e^{-a}}{(z-e^{-a})^3}.$$

Example 13. *If* $Z[u_n] = \frac{z}{z-1} + \frac{z}{z^2+1}$, *find the Z-transform of* u_{n+2}.

Sol. We have

$$Z[u_n] = \frac{z}{z-1} + \frac{z}{z^2+1} = \left(1 - \frac{1}{z}\right)^{-1} + \frac{1}{z}\left(1 + \frac{1}{z^2}\right)^{-1}$$

$$= (1 + z^{-1} + z^{-2} + \dots \infty) + z^{-1}(1 - z^{-2} + z^{-4} - \dots \infty)$$

$$= 1 + 2z^{-1} + z^{-2} + z^{-4} + 2z^{-5} + \dots \infty. \qquad \dots(i)$$

But we know that

$$U(z) = Z[u_n] = u_0 + u_1 z^{-1} + u_2 z^{-2} + u_3 z^{-3} + \dots \infty. \qquad \dots(ii)$$

Thus comparing (*i*) and (*ii*), we get

$$u_0 = 1,\ u_1 = 2,\ u_2 = 1 \text{ etc.}$$

Now shifting u_n two units to the left, we get

$$Z[u_{n+2}] = z^2 [U(z) - u_0 - u_1 z^{-1}]$$

$$= z^2\left[\frac{z}{z-1} + \frac{z}{z^2+1} - 1 - 2z^{-1}\right] = \frac{z^3}{z-1} + \frac{z^3}{z^2+1} - z^2 - 2z$$

$$= \frac{z^3(z^2+1) + z^3(z-1) - z^2(z^3 - z^2 + z - 1) - 2z(z^3 - z^2 + z - 1)}{(z-1)(z^2+1)}$$

$$= \frac{z^3 - z^2 + 2z}{(z-1)(z^2+1)}.$$

Example 14. *If* $U(z) = \dfrac{2z^2 + 5z + 14}{(z^2-1)^4}$, *evaluate* u_2 *and* u_3.

Sol. Writing U(z) as follows:

$$U(z) = \frac{1}{z^2} \cdot \frac{2 + 5z^{-1} + 14z^{-2}}{(1 - z^{-1})^4}$$

By initial value theorem, we have

$$u_0 = \underset{z \to \infty}{\text{Lt}} \ U(z) = 0 \text{ as } \underset{z \to \infty}{\text{Lt}} \frac{1}{z} = 0 \text{ and } \underset{z \to \infty}{\text{Lt}} \frac{1}{z^2} = 0.$$

Similarly, $u_1 = \underset{z \to \infty}{\text{Lt}} \{z [U(z) - u_0]\} = 0.$

Now $u_2 = \underset{z \to \infty}{\text{Lt}} \{z^2 [U(z) - u_0 - u_1 z^{-1}]\}$

$$= \underset{z \to \infty}{\text{Lt}} \left\{ \frac{2 + 5z^{-1} + 14z^{-2}}{(1 - z^{-1})^4} \right\} = 2 - 0 - 0 = 2$$

and $u_3 = \underset{z \to \infty}{\text{Lt}} \{z^3 [U(z) - u_0 - u_1 z^{-1} - u_2 z^{-2}]\} = \underset{z \to \infty}{\text{Lt}} z^3 [U(z) - 0 - 0 - 2z^{-2}]$

$$= \underset{z \to \infty}{\text{Lt}} z^3 \left[\frac{2z^2 + 5z + 14}{(z-1)^4} - \frac{2}{z^2} \right] = \underset{z \to \infty}{\text{Lt}} z^3 \left\{ \frac{13z^3 + 2z^2 + 8z - 2}{z^2 (z-1)^4} \right\} = 13.$$

Example 15. *Given that* $Z[u_n] = \dfrac{2z^2 + 3z + 4}{(z-3)^3}$, $|z| > 3$, *show that* $u_1 = 2$, $u_2 = 21$, $u_3 = 139$.

Sol. Writing Z $[u_n]$ as follows:

$$Z[u_n] = \frac{1}{z} \cdot \frac{2 + 3z^{-1} + 4z^{-2}}{(1 - 3z^{-1})^3} = U(z), \text{ say.}$$

By initial value theorem, we have

$$u_0 = \underset{z \to \infty}{\text{Lt}} \ U(z) = 0$$

$$u_1 = \underset{z \to \infty}{\text{Lt}} \{z [U(z) - u_0\} = \underset{z \to \infty}{\text{Lt}} \frac{2 + 3z^{-1} + 4z^{-2}}{(1 - 3z^{-1})^3} = 2, \text{ as } |z| > 3$$

$$u_2 = \underset{z \to \infty}{\text{Lt}} \{z^2 [U(z) - u_0 - u_1 z^{-1}]\}$$

$$= \underset{z \to \infty}{\text{Lt}} \{z^2 [U(z) - 0 - 2z^{-1}]\} = \underset{z \to \infty}{\text{Lt}} z^2 \left[\frac{2z^2 \ 3z + 4}{(z-3)^3} - \frac{2}{z} \right]$$

$$= \underset{z\to\infty}{\text{Lt}}\ z^2\left[\frac{21z^2-50z+54}{z(z-3)^3}\right\} = \underset{z\to\infty}{\text{Lt}}\ \frac{21-50z^{-1}+54z^{-2}}{(1-3z^{-1})^3} \qquad [\because\ |z|>3]$$

$$= 21$$

$$u_3 = \underset{z\to\infty}{\text{Lt}}\ \{z^3\,[\text{U}(z) - u_0 - u_1z^{-1} - u_2z^{-2}]\}$$

$$= \underset{z\to\infty}{\text{Lt}}\ \{z^3\,[\text{U}(z) - 0 - 2z^{-1} - 21z^{-2}]\}$$

$$= \underset{z\to\infty}{\text{Lt}}\ z^3\left[\frac{2z^2+3z+4}{(z-3)^3} - \frac{2}{z} - \frac{21}{z^2}\right] = \underset{z\to\infty}{\text{Lt}}\ z^3\left\{\frac{139z^2-513z+567}{z^2\,(z-3)^3}\right\}$$

$$= \underset{z\to\infty}{\text{Lt}}\ \frac{139-513z^{-1}+567z^{-2}}{(1-3z^{-1})^3} \qquad [\because\ |z|>3]$$

$$= 139.$$

2.28. CONVOLUTION OF SEQUENCES

The convolution of two sequences $\{f(n)\}$ and $\{g(n)\}$ is defined as

$$\omega(n) = \sum_{k=-\infty}^{\infty} f(k)\,g(n-k) = f * g$$

Note. If it is one sided (right) sequence, take $f(k) = 0$, $g(k) = 0$ for $k < 0$, then

$$\omega(n) = \sum_{k=0}^{\infty} f(k)\,g(n-k) = f * g.$$

2.29. CONVOLUTION THEOREM

If $w(n)$ is the convolution of two sequences $f(n)$ and $g(n)$, then

$$Z\{w(n)\} = W(z) = Z\{f(n)\}\,.\,Z\{g(n)\} = F(z)\,G(z)$$

Proof. $$\text{W}(z) = \text{Z}\{w(n)\} = \text{Z}\left[\sum_{k=0}^{\infty} f(k)\,g(n-k)\right] = \sum_{n=0}^{\infty}\left[\sum_{k=0}^{\infty} f(k)\,g(n-k)\right]z^{-n}$$

$$= \sum_{k=0}^{\infty} f(k)\left[\sum_{n=0}^{\infty} g(n-k)z^{-n}\right] \qquad \text{(By changing the order of summation)}$$

$$= \sum_{k=0}^{\infty} f(k)\left[\sum_{p=0}^{\infty} g(p)z^{-(p+k)}\right], \qquad \text{(Putting } n-k=p\text{)}$$

$$= \left[\sum_{k=0}^{\infty} f(k)z^{-k}\right]\left[\sum_{p=0}^{\infty} g(p)z^{-p}\right] = \text{F}(z)\,\text{G}(z).$$

Note. This result will be true only for those values of z inside the region of convergence.

2.29.1. Another Form of Convolution Theorem

If Z{f(t)} = F(z), Z{g(t)} = G(z), then the convolution product is

$$w(t) = \sum_{k=0}^{n} f(kT)\, g(nT - kT) = f * g$$

and
$$Z\{w(t)\} = W(z) = Z\{f(t)\}\, Z\{g(t)\} = F(z)\, G(z).$$

Proof. (Here we are dealing with one sided Z-transforms only)

$$F(z) = \sum_{m=0}^{\infty} f(mT)\, z^{-m} \; ; \; G(z) = \sum_{n=0}^{\infty} g(nT)\, z^{-n}$$

$$F(z)\, G(z) = \left\{\sum_{m=0}^{\infty} f(mT) z^{-m}\right\}\left\{\sum_{n=0}^{\infty} g(nT) z^{-n}\right\}$$

$$= \sum_{n=0}^{\infty} \sum_{m=0}^{\infty} f(mT)\, g(nT)\, z^{-m-n} = \sum_{n=0}^{\infty} \left(\sum_{p=0}^{n} f(pT) g\{(n-p)T\}\right) z^{-n}$$

$$= Z\left\{\sum_{p=0}^{n} f(pT) g\{(n-p)T\}\right\} z^{-n} = Z\{f * g\}.$$

Example 16. *Find the Z-transform of f * g where*
(i) $f(n) = u(n)$, $g(n) = 2^n u(n)$
(ii) $f(n) = 3^n u(n)$, $g(n) = 4^n u(n)$ using convolution theorem.

Sol. *(i)*
$$F(z) = Z\{u(n)\} = \sum_{n=0}^{\infty} 1 \,.\, z^{-n} = \frac{z}{z-1} \text{ if } |\, z \,| > 1$$

$$G(z) = Z\{2^n u(n)\} = \sum_{n=0}^{\infty} 2^n z^{-n} = \frac{z}{z-2} \text{ if } |\, z \,| > |\, 2 \,|$$

By convolution theorem

$$Z\{f*g\} = Z\{w(n)\} = W(z) = F(z)\,.\,G(z)$$

$$= \frac{z}{z-1} \cdot \frac{z}{z-2} = \frac{z^2}{(z-1)(z-2)} \text{ if } |\, z \,| > |\, 2 \,|.$$

(ii)
$$F(z) = Z\{3^n u(n)\} = \frac{z}{z-3} \text{ if } |\, z \,| > |\, 3 \,|$$

$$G(z) = Z\{4^n u(n)\} = \frac{z}{z-4} \text{ if } |\, z \,| > |\, 4 \,|$$

By convolution theorem

$$Z\{f*g\} = Z\{w(n)\} = W(z) = F(z)\, G(z)$$

$$= \frac{z}{z-3} \cdot \frac{z}{z-4} = \frac{z^2}{(z-3)(z-4)} \text{ if } |\, z \,| > |\, 4 \,|.$$

Example 17. *Use convolution to evaluate* $Z^{-1}\left\{\dfrac{z^2}{(z-a)(z-b)}\right\}$.

Sol. We know that

$$Z^{-1}\left\{\frac{z}{z-a}\right\} = a^n \quad \text{and} \quad Z^{-1}\left\{\frac{z}{z-b}\right\} = b^n.$$

$$\therefore \quad Z^{-1}\left\{\frac{z^2}{(z-a)(z-b)}\right\} = Z^{-1}\left\{\frac{z}{z-a}\cdot\frac{z}{z-b}\right\} = a^n * b^n$$

$$= \sum_{m=0}^{n} a^m \cdot b^{n-m} = b^n \sum_{m=0}^{n} \left(\frac{a}{b}\right)^m \text{ which is a G.P.}$$

$$= b^n \cdot \frac{(a/b)^{n+1}-1}{(a/b)-1} = \frac{a^{n+1}-b^{n+1}}{a-b}.$$

Example 18. *Using convolution theorem, evaluate the inverse Z-transform of* $\left(\dfrac{z}{z-a}\right)^2$.

Sol. By convolution theorem, we have

$$Z^{-1}\left[\left(\frac{z}{z-a}\right)^2\right] = Z^{-1}\left[\frac{z}{z-a}\cdot\frac{z}{z-a}\right] = u_n * v_n$$

where $$u_n = Z^{-1}\left(\frac{z}{z-a}\right) = a^n \text{ and } v_n = Z^{-1}\left(\frac{z}{z-a}\right) = a^n$$

$$= \sum_{m=0}^{n} u_n v_{n-m} = \sum_{m=0}^{n} a^n a^{n-m} = a^n \sum_{m=0}^{n} 1 = (n+1)a^n.$$

Example 19. *Show that* $\dfrac{1}{n!} * \dfrac{1}{n!} = \dfrac{2^n}{n!}$.

Sol. Let $u_n = v_n = \dfrac{1}{n!}$. Then by convolution theorem, we have

$$\frac{1}{n!} * \frac{1}{n!} = u_n * v_n = \sum_{m=0}^{n} u_m \cdot v_{n-m}$$

$$= \sum_{m=0}^{n} \frac{1}{m!}\cdot\frac{1}{(n-m)!} = \frac{1}{0!\,n!} + \frac{1}{1!(n-1)!} + \frac{1}{2!(n-2)!} + \ldots + \frac{1}{n!\,0!}$$

$$= \frac{1}{n!}\left[1 + \frac{n}{1!} + \frac{n(n-1)}{2!} + \ldots + 1\right] = \frac{1}{n!}\left[{}^nC_0 + {}^nC_1 + {}^nC_2 + \ldots + {}^nC_n\right]$$

$$= \frac{1}{n!}(1+1)^n = \frac{2^n}{n!}.$$

Example 20. *Using convolution theorem, find inverse Z-transform of* $\dfrac{18z^2}{(2z-1)(4z+1)}$.

Sol. By convolution theorem, we have

$$Z^{-1}\left[\frac{18z^2}{(2z-1)(4z+1)}\right] = Z^{-1}\left[\frac{9}{4}\cdot\frac{z}{z-\frac{1}{2}}\cdot\frac{z}{z+\frac{1}{4}}\right] = \frac{9}{4}Z^{-1}\left[\frac{z}{z-\frac{1}{2}}\cdot\frac{z}{z+\frac{1}{4}}\right]$$

$$= \frac{9}{4}u_n * v_n \text{ where } u_n = Z^{-1}\left(\frac{z}{z-\frac{1}{2}}\right) = \left(\frac{1}{2}\right)^n \text{ and } v_n = Z^{-1}\left(\frac{z}{z+\frac{1}{4}}\right) = \left(-\frac{1}{4}\right)^n$$

$$= \frac{9}{4}\sum_{m=0}^{n} u_n v_{n-m} = \frac{9}{4}\sum_{m=0}^{n}\left(\frac{1}{2}\right)^n\left(-\frac{1}{4}\right)^{n-m}$$

$$= \frac{9}{4}\left(\frac{1}{2}\right)^n\sum_{m=0}^{n}\left(-\frac{1}{4}\right)^{n-m} = \frac{9}{4}\left(\frac{1}{2}\right)^n\left[1+\left(-\frac{1}{4}\right)+\left(-\frac{1}{4}\right)^2+\ldots+\left(-\frac{1}{4}\right)^n\right]$$

$$= \frac{9}{4}\left(\frac{1}{2}\right)^n\cdot\frac{1-\left(-\frac{1}{4}\right)^{n+1}}{1-\left(-\frac{1}{4}\right)} = \frac{9}{5}\left(\frac{1}{2}\right)^n\left[1-\left(-\frac{1}{4}\right)^{n+1}\right].$$

Table of Z-Transforms

$f(n)$	$Z\{f(n)\}$	$f(n)$	$Z\{f(n)\}$
1	$\dfrac{z}{z-1}$	$a^n u(n)$ or a^n	$\dfrac{z}{z-a}$
n	$\dfrac{z}{(z-1)^2}$	$na^n u(n)$ or na^n	$\dfrac{az}{(z-a)^2}$
n^2	$\dfrac{z^2+z}{(z-1)^3}$	$(n+1)\,a^n u(n)$ or $(n+1)a^n$	$\dfrac{z^2}{(z-a)^2}$
n^3	$\dfrac{z^3+4z^2+z}{(z-1)^4}$	$n(n-1)a^n u(n)$ or $n(n-1)a^n$	$\dfrac{2a^2z}{(z-a)^3}$
$n(n-1)$	$\dfrac{2z}{(z-1)^2}$	$\cos\dfrac{n\pi}{2}$ or $\cos\dfrac{n\pi}{2}\,.\,u(n)$	$\dfrac{z^2}{z^2+1}$
$n^{(k)}$	$\dfrac{k!\,z}{(z-1)^{k+1}}$	$\sin\dfrac{n\pi}{2}$ or $\sin\dfrac{n\pi}{2}\,.\,u(n)$	$\dfrac{z}{z^2+1}$
$(-1)^n$	$\dfrac{z}{z+1}$	$a^n\cos\dfrac{n\pi}{2}\,.\,u(n)$ or $a^n\cos\dfrac{n\pi}{2}$	$\dfrac{z^2}{z^2+a^2}$

$u(n)$	$\frac{z}{z-1}, \lvert z \rvert > 1$	$a^n \sin \frac{n\pi}{2} u(n)$ or $a^n \sin \frac{n\pi}{2}$	$\frac{az}{z^2 + a^2}$
$u(n-1)$	$\frac{1}{z-1}$	$\cos n\theta \,.\, u(n)$ or $\cos n\theta$	$\frac{z(z - \cos\theta)}{z^2 - 2z\cos\theta + 1}$
$u(n-k)$	$z^{-k} . \frac{z}{z-1}$	$\sin n\theta \,.\, u(n)$ or $\sin n\theta$	$\frac{z \sin\theta}{z^2 - 2z\cos\theta + 1}$
$\frac{1}{n}$	$\log\left(\frac{z}{z-1}\right), \lvert z \rvert > 1$	$r^n \cos n\theta \,.\, u(n)$ or $r^n \cos n\theta$	$\frac{z(z - r\cos\theta)}{z^2 - 2zr\cos\theta + r^2}$
$\frac{1}{n+1}$	$z \log\left(\frac{z}{z-1}\right)$	$r^n \sin\theta\, u(n)$ or $r^n \sin n\theta$	$\frac{rz\sin\theta}{z^2 - 2zr\cos\theta + r^2}$
$\frac{1}{n!}$	$e^{1/z}$	$a^n f(n)$	$F(z/a)$
$\delta(n)$	1	$nf(n)$	$-z \frac{dF(z)}{dz}$
$\delta(n-k)$	z^{-k}	n^{m+1}	$(-z)^m \frac{d^m}{dz^m} z(n)$

$f(t)$	$Z\{f(t)\}$	$f(t)$	$Z\{f(t)\}$
t^k	$-Tz \frac{d}{dz} [Z(t^{k-1})]$	$nf(nT) = nf(t)$	$-z \frac{d}{dz} F(z)$
t	$\frac{Tz}{(z-1)^2}$	e^{-at}	$\frac{z}{z - e^{-aT}}$ if $\lvert z \rvert > \lvert e^{-aT} \rvert$
t^2	$\frac{T^2 z(z+1)}{(z-1)^3}$	e^{at}	$\frac{z}{z - e^{aT}}$ if $\lvert z \rvert > \lvert e^{-aT} \rvert$
t^3	$\frac{T^3 z(1 + 4z - z^2)}{(z-1)^4}$	$\sin \omega t$	$\frac{z \sin \omega t}{z^2 - 2z\cos\omega T + 1}$ if $\lvert z \rvert > 1$
$a^n f(t)$	$F(z/a)$	$\cos \omega t$	$\frac{z(z - \cos\omega T)}{z^2 - 2z\cos\omega T + 1}$ if $\lvert z \rvert > 1$

TEST YOUR KNOWLEDGE

1. Find the Z-transform (one sided) of the following sequences $\{f(n)\}$ where $f(n)$ is

   (*i*) $\left(\frac{1}{4}\right)^n u(n)$ (*ii*) $\delta(n-3)$ (*iii*) $(-1)^n u(n)$

   (*iv*) $3^n \sin \frac{n\pi}{2}$ (*v*) $2^n \cos \frac{n\pi}{2}$ (*vi*) $4^n + \left(\frac{1}{2}\right)^n + u(n-3)$.

2. Find the Z-transforms of $f(t)$

(*i*) e^{-3t+4} (*ii*) $\sin 3T$ (*iii*) $\cos 2T$

(*iv*) $\sin^2 3t$ (*v*) $\cos^3 t$ (*vi*) $e^{-at} \sin bt$

(*vii*) $e^{-t} . t^2$

3. Find the Z-transform of

(*i*) $n2^n$ (*ii*) $n \cos n\theta$.

4. Show that

(*i*) $Z(\cosh n\theta) = \dfrac{z(z - \cosh \theta)}{z^2 - 2z \cosh \theta + 1}$ (*ii*) $Z(\sinh \theta) = \dfrac{z \sinh \theta}{z^2 - 2z \cosh \theta + 1}$.

5. Show that

(*i*) $Z(e^{-an} \cos n\theta) = \dfrac{Ze^a(ze^a - \cos \theta)}{Z^2e^{2a} - 2ze^a \cos \theta + 1}$ (*ii*) $Z(e^{-an} \sin n\theta) = \dfrac{Ze^a \sin \theta}{Z^2e^{2a} - 2Ze^a \cos \theta + 1}$.

6. Using $Z(n^2) = \dfrac{z(z + 1)}{(z - 1)^3}$, show that $Z(n + 1)^2 = \dfrac{z^2(z + 1)}{(z - 1)^3}$.

7. Given $Z(u_n) = \dfrac{2z^2 + 3z + 4}{(z - 3)^3}$, $|z| > 3$, show that $u_1 = 2$, $u_2 = 21$, $u_3 = 139$.

8. Using $Z(n) = \dfrac{z}{(z - 1)^2}$, show that $Z(n \cos n\theta) = \dfrac{(z^3 + z) \cos \theta - 2z^2}{(z^2 - 2z \cos \theta + 1)^2}$.

9. Find the convolutions of

(*i*) $n(n - 1) * 3^n$ (*ii*) $3^n * \cos n\theta$ (*iii*) $\cos \dfrac{n\pi}{2} * \sin \dfrac{n\pi}{2}$.

Answers

1. (*i*) $\dfrac{4z}{4z - 1}$ (*ii*) $\dfrac{1}{z^3}$ (*iii*) $\dfrac{z}{z + 1}$

(*iv*) $\dfrac{3z}{z^2 + 9}$ (*v*) $\dfrac{z^2}{z^2 + 4}$ (*vi*) $\dfrac{z}{z - 4} + \dfrac{2z}{2z - 1} + \dfrac{1}{z^2(z - 1)}$.

2. (*i*) $e^4 . \dfrac{z}{z - e^{-3T}}$ (*ii*) $\dfrac{z \sin 3T}{z^2 - 2z \cos 3T + 1}$ (*iii*) $\dfrac{z(z - \cos 2T)}{z^2 - 2z \cos 2T + 1}$

(*iv*) $\dfrac{1}{2}\dfrac{z}{z - 1} - \dfrac{1}{2}\left[\dfrac{z(z - \cos 6T)}{z^2 - 2z \cos 6T + 1}\right]$

(*v*) $\dfrac{3}{4}\left[\dfrac{z(z - \cos T)}{z^2 - 2z \cos T + 1}\right] + \dfrac{1}{4}\left[\dfrac{z(z - \cos 3T)}{z^2 - 2z \cos 3T + 1}\right]$

(*vi*) $\dfrac{ze^{eT} \sin bt}{z^2e^{2aT} - 2ze^{aT} \cos bT + 1}$ (*vii*) $\dfrac{T^2 \, ze^T \, (ze^T + 1)}{(2e^T - 1)^3}$.

3. (*i*) $\dfrac{2z}{(z - 2)^2}$ (*ii*) $z^{-1} \dfrac{d}{dz^{-1}}\left[\dfrac{1 - \cos \theta \, z^{-1}}{1 - 2 \cos \theta . z^{-1} + z^{-2}}\right]$.

9. (*i*) $\dfrac{2z^2}{(z - 1)^2(z - 3)}$ (*ii*) $\dfrac{z^2(z - \cos \theta)}{(z - 3)(z^2 - 2z \cos \theta + 1)}$ (*iii*) $\dfrac{z^3}{(z^2 + 1)^2}$.

2.30. INVERSE Z-TRANSFORMS

The inverse Z-transform is a process of determining the sequence which generates a given Z-transform. We have the following methods to find inverse Z-transforms:

(i) Convolution method *(ii) Long division method*

(iii) Partial fraction method *(iv) Inverse integral method*

(v) Power series method.

2.30.1. Convolution Method

We know that $Z\{f * g\} = F(z)\, G(z)$

$$\therefore \quad Z^{-1}\{F(z)\, G(z)\} = f * g.$$

Example. *Find* $Z^{-1}\left\{\dfrac{z^2}{(z-a)(z-b)}\right\}$ *using convolution theorem.*

Sol. We know that $Z^{-1}\{F(z)\, G(z)\} = f * g$

Let $F(z) = \dfrac{z}{(z-a)} \quad \therefore \quad f(n) = Z^{-1}\{F(z)\} = Z^{-1}\left\{\dfrac{z}{z-a}\right\} = a^n$

$$G(z) = \frac{z}{z-b} \quad \therefore \quad g(n) = Z^{-1}\{G(z)\} = Z^{-1}\left\{\frac{z}{z-b}\right\} = b^n$$

$$Z^{-1}\{F(z)\, G(z)\} = f * g = a^n * b^n$$

$$= \sum_{m=0}^{\infty} a^m b^{n-m} = b^n \sum_{m=0}^{n} \left(\frac{a}{b}\right)^m \qquad \text{(a G.P.)}$$

$$= b^n \left\{\frac{\left(\frac{a}{b}\right)^{n+1} - 1}{\frac{a}{b} - 1}\right\} = \frac{a^{n+1} - b^{n+1}}{a - b}.$$

2.30.2. Long Division Method

Since Z-transform is defined by the series $F(z) = \sum_{n=0}^{\infty} f(n)\, z^{-n}$ (one sided), to find the inverse Z-transform *i.e.*, $Z^{-1}[F(z)]$, expand F(z) in the proper power series and collect the coefficient of z^{-n} to get $f(n)$.

Example. *Find inverse Z-transform of*

(i) $\dfrac{10z}{z^2 - 3z + 2}$ *(ii)* $\dfrac{2(z^3 - z)}{(z^2 + 1)^2}$.

Sol. *(i)* $$F(z) = \frac{10z}{z^2 - 3z + 2} = \frac{10z^{-1}}{1 - 3z^{-1} + 2z^{-2}}$$

By actual division,

$$\begin{array}{r|l} & 10z^{-1} + 30z^{-2} + 70z^{-3} + \ldots\ldots \\ \hline 1 - 3z^{-1} + 2z^{-2}) & 10z^{-1} \\ & 10z^{-1} - 30z^{-2} + 20z^{-3} \\ \hline & 30z^{-2} - 20z^{-3} \\ & 30z^{-2} - 90z^{-3} + 60z^{-4} \\ \hline & 70z^{-3} - 60z^{-4} \\ & 70z^{-3} - 210z^{-4} + 140z^{-5} \\ \hline & + 150z^{-4} - 140z^{-5} \end{array}$$

$$\therefore \qquad F(z) = 10z^{-1} + 30z^{-2} + 70z^{-3} + \ldots\ldots$$

Now comparing the quotient with

$$\sum_{n=0}^{\infty} f(n)\, z^{-n} = f(0) + f(1)z^{-1} + f(2)\, z^{-2} + \ldots\ldots$$

We get the sequence $f(n)$ as $\quad f(0) = 0, f(1) = 10, f(2) = 30, f(3) = 70, \ldots\ldots$

i.e., we can get $\quad f(n) = 10(2^n - 1), n = 0, 1, 2, \ldots\ldots$

(*ii*) $$F(z) = \frac{2z(z^2 - 1)}{(z^2 + 1)^2} = \frac{2z^{-1} - 2z^{-3}}{1 + 2z^{-2} + z^{-4}}$$

By actual division, we get $\quad F(z) = 2z^{-1} - 6z^{-3} + 10z^{-5} - 14z^{-7} + \ldots\ldots$

Comparing the quotient with

$$\sum_{n=0}^{\infty} f(n)\, z^{-n} = f(0) + f(1)z^{-1} + f(2)z^{-2} + \ldots\ldots$$

We get $\quad f(0) = 0, f(1) = 2, f(2) = 0, f(3) = -6,$

$f(4) = 0, f(5) = 10, f(6) = 0, \ldots\ldots$

In general $\quad f(n) = 2n \sin \frac{n\pi}{2}, n = 0, 1, 2, \ldots\ldots$

Note. Long division method is very useful to get $f(n)$.

2.30.3. Partial Fractional Method

Here we split the given F(z) into partial fractions whose inverse transforms can be written directly.

Example 1. *Find the inverse Z-transform of*

(*i*) $\dfrac{z}{z^2 + 7z + 10}$ $\qquad$ (*ii*) $\dfrac{z^3 - 20z}{(z-2)^3(z-4)}$ $\qquad$ (*iii*) $\dfrac{8z^2}{(2z-1)(4z-1)}$.

Sol. (*i*) $$\frac{F(z)}{z} = \frac{1}{z^2 + 7z + 10} = \frac{1}{(z+2)(z+5)} = \frac{A}{z+2} + \frac{B}{z+5}$$

$$= \frac{1}{3}\,\frac{1}{z+2} - \frac{1}{3}\cdot\frac{1}{z+5} \qquad \left(\because\ A = \frac{1}{3}, B = -\frac{1}{3}\right)$$

$$\therefore \quad F(z) = \frac{1}{3}\frac{z}{z+2} - \frac{1}{3}\frac{z}{z+5}$$

$$\therefore \quad f(n) = Z^{-1}\{F(z)\} = \frac{1}{3}\, Z^{-1}\left\{\frac{z}{z+2}\right\} - \frac{1}{3}\, Z^{-1}\left\{\frac{z}{z+5}\right\}$$

$$= \frac{1}{3}(-2)^n - \frac{1}{3}(-5)^n \qquad \left[\because\ z(a^n) = \frac{z}{z-a}\right]$$

(*ii*)
$$F(z) = \frac{z^3 - 20z}{(z-2)^3(z-4)} \quad \text{or} \quad \frac{F(z)}{z} = \frac{z^2-20}{(z-2)^3(z-4)}$$

Now
$$\frac{F(z)}{z} = \frac{z^2-20}{(z-2)^3(z-4)} = \frac{A+Bz+Cz^2}{(z-2)^3} + \frac{D}{z-4}$$

$$\Rightarrow \quad D = -\frac{1}{2},\ A = 6,\ B = 0,\ C = \frac{1}{2}$$

$$\therefore \quad \frac{F(z)}{z} = \frac{6+\frac{1}{2}z^2}{(z-2)^3} + \frac{\left(-\frac{1}{2}\right)}{z-4}$$

or
$$F(z) = \frac{1}{2}\left\{\frac{12z+z^3}{(z-2)^3} - \frac{z}{z-4}\right\}$$

$$= \frac{1}{2}\left\{\frac{z(z-2)^2 + 4z^2 + 8z}{(z-2)^3} - \frac{z}{z-4}\right\}$$

$$= \frac{1}{2}\left[\frac{z}{z-2} + 2\,.\,\frac{2z^2+4z}{(z-2)^3} - \frac{z}{z-4}\right]$$

Now
$$f(n) = Z^{-1}\{F(z)\} = \frac{1}{2}\{2^n + 2n^2 2^n - 4^n\}, \qquad \left[\because\ Z^{-1}\left\{\frac{az^2+a^2z}{(z-a)^3}\right\} = n^2a^n\right]$$

$$= 2^{n-1} + 2^n\,.\,n^2 - 2^{2n-1}.$$

(*iii*)
$$F(z) = \frac{8z^2}{(2z-1)(4z-1)} = \frac{z^2}{\left(z-\frac{1}{2}\right)\left(z-\frac{1}{4}\right)}$$

$$\frac{F(z)}{z} = \frac{z}{\left(z-\frac{1}{2}\right)\left(z-\frac{1}{4}\right)} = \frac{A}{z-\frac{1}{2}} + \frac{B}{z-\frac{1}{4}} = \frac{2}{z-\frac{1}{2}} - \frac{1}{z-\frac{1}{4}}$$

$$\therefore \quad F(z) = \frac{2z}{z-1/2} - \frac{z}{z-1/4}$$

$$f(n) = Z^{-1}\{F(z)\} = 2\left(\frac{1}{2}\right)^n - \left(\frac{1}{4}\right)^n,\ n = 0, 1, 2, \ldots\ldots$$

Example 2. *Find the inverse Z-transform of*

(*i*) $\dfrac{2z^2+3z}{(z+2)(z-4)}$ (*ii*) $\dfrac{5z}{(2-z)(3z-1)}$.

(*iii*) $\dfrac{3z^2+2}{(5z-1)(5z+2)}$.

Sol. (*i*) We write $\quad F(z) = \dfrac{2z^2+3z}{(z+2)(z-4)}$ as

$$\frac{F(z)}{z} = \frac{2z+3}{(z+2)(z-4)} = \frac{A}{z+2} + \frac{B}{z-4}.$$

Multiplying throughout by $(z+2)(z-4)$, we get

$$2z + 3 = A(z-4) + B(z+2).$$

Putting $z = -2, 4$ successively and solving the resulting equations, we get

$$A = \frac{1}{6} \text{ and } B = \frac{11}{6}$$

$$\therefore \qquad F(z) = \frac{1}{6}\frac{z}{z+2} + \frac{11}{6}\frac{z}{z-4}.$$

On inversion, we have

$$f(k) = \frac{1}{6}(-2)^k + \frac{\pi}{6}(4)^k \qquad \left[\text{Using } Z^{-1}\left(\frac{z}{z-a}\right) = a^k\right]$$

(*ii*) We write $F(z) = \dfrac{5z}{(2-z)(3z-1)}$ as

$$\frac{F(z)}{z} = \frac{5}{(2-z)(3z-1)} = \frac{A}{2-z} + \frac{B}{3z-1}$$

Multiplying throughout by $(2-z)(3z-1)$, we get

$$5 = A(3z-1) + B(2-z).$$

Putting $z = 2, \frac{1}{3}$ successively and solving the resulting equations, we get

$A = 1, B = 3.$

$$\therefore \qquad F(z) = \frac{z}{2-z} + \frac{3z}{3z-1} \quad \text{or} \quad F(z) = -\frac{2}{z-2} + \frac{z}{z-1/3}.$$

On inversion, we have

$$f(k) = -(2^k) + \left(\frac{1}{3}\right)^k.$$

(*iii*) We write $F(z) = \dfrac{3z^2+2}{(5z-1)(5z+2)}$ as

$$\frac{F(z)}{z} = \frac{3z^2+2}{z(5z-1)(5z+2)} = \frac{A}{z} + \frac{B}{5z-1} + \frac{C}{5z+2}$$

Multiplying throughout by $z(5z-1)(5z+2)$, we get

$$3z^2 + 2 = A(25z^2 + 5z - 2) + B(5z^2 + 2z) + C(5z^2 - z).$$

Putting $z = 0, \frac{1}{5}, -\frac{2}{5}$ successively, we get

$$2 = -2A, \frac{53}{25} = \frac{3}{5}B, \frac{62}{25} = \frac{6}{5}C.$$

$$\therefore \quad A = -1, B = \frac{53}{15}, C = \frac{31}{15}.$$

Thus,
$$F(z) = -1 + \frac{53}{15}\frac{z}{5z-1} + \frac{31}{15}\frac{z}{5z+2}$$

or
$$F(z) = -1 + \frac{53}{75}\frac{z}{z-0.2} + \frac{31}{75}\frac{z}{z-(-0.4)}.$$

On inversion, we have

$$f(k) = -\delta(k) + \frac{53}{75}(0.2)^k + \frac{31}{75}(-0.4)^k$$

where
$$\delta(k) = \begin{cases} 1, & k = 0 \\ 0, & k \neq 0. \end{cases}$$

Example 3. *Find the inverse Z-transform of* $\frac{z}{z^2 + 7z + 10}$.

Sol. We write $F(z) = \frac{z}{z^2 + 7z + 10}$ as

$$\frac{F(z)}{z} = \frac{1}{z^2 + 75 + 10} = \frac{1}{(z+2)(z+5)} = \frac{A}{z+2} + \frac{B}{z+5}.$$

Multiplying throughout by $(z + 2)(z + 5)$, we get

$$1 = A(z + 5) + B(z + 2).$$

Putting $z = -2, -5$ successively and solving the resulting equations, we get

$$A = \frac{1}{3}, B = -\frac{1}{3}.$$

$$\therefore \quad F(z) = \frac{1}{3}\frac{z}{z+2} - \frac{1}{3}\frac{z}{z+5}.$$

On inversion, we have

$$f(k) = \frac{1}{3}(-2)^k - \frac{1}{3}(-5)^k.$$

Example 4. *Find the inverse Z-transform of* $\frac{z^3 - 20z}{(z-2)^3 (z-4)}$.

Sol. We write $F(z) = \frac{z^3 - 20z}{(z-2)^3 (z-4)}$ as

$$\frac{F(z)}{z} = \frac{z^2 - 20}{(z-2)^3 (z-4)} = \frac{A + Bz + Cz^2}{(z-2)^3} + \frac{D}{z-4}.$$

Multiplying throughout by $(z - 2)^3 (z - 4)$, we get

$$z^2 - 20 = (A + Bz + Cz^2)(z - 4) D(z - 2)^3.$$

Putting $z = 4$, we get

$$-4 = D \cdot 8.$$

$$\therefore \quad D = -\frac{1}{2}.$$

Putting $z = 0, 1, -1$ successively, we get

$$-20 = -4A - 8D, \; -19 = -3(A + B + C) - D, \; -19 = -5(A - B + C) - 27D.$$

Solving these equations, we get

$$A = 6, B = 0, C = \frac{1}{2}.$$

Thus
$$F(z) = \frac{1}{2}\frac{12z + z^2}{(z-2)^3} - \frac{z}{z-4}$$

$$= \frac{z(z-2)^2 + 4z^2 + 8z}{(z-2)^3} - \frac{z}{z-4} = \frac{1}{2}\left\{\frac{z}{z-2} + 2 \cdot \frac{2z^2 + 4z}{(z-2)^3}\right\} - \frac{z}{z-4}.$$

On inversion, we get

$$f(k) = \frac{1}{2}(e^k + 2 \cdot k^2 2^k) - 4^k \quad \text{or} \quad f(k) = 2^{k-1} + k^2 2^k - 4^k.$$

Example 5. *Obtain* $Z^{-1}\left\{\frac{1}{(z-2)(z-3)}\right\}$ *for*

(i) $|z| < 2$; *(ii)* $2 < |z| < 3$; *(iii)* $|z| > 3$.

Sol. (*i*) Splitting into partial fractions, we obtain

$$F(z) = \frac{1}{(z-2)(z-3)} = \frac{A}{z-2} + \frac{B}{z+3} \text{ where } A = -1, B = 1.$$

$$\therefore \quad F(z) = \frac{1}{z-3} - \frac{1}{z-2} = -\frac{1}{3}\left(1 - \frac{z}{3}\right)^{-1} + \frac{1}{2}\left(1 - \frac{z}{2}\right)^{-1} \text{ so that } \frac{z}{3} < 1 \text{ and } \frac{z}{2} < 1$$

$$= -\frac{1}{3}\left(1 - \frac{z}{3} + \frac{z^2}{3^2} + \frac{z^3}{3^3} + \ldots\right) + \frac{1}{2}\left(1 + \frac{z}{2} + \frac{z^2}{2^2} + \frac{z^3}{2^3} + \ldots\right)$$

$$= \left(-\frac{1}{3} - \frac{z}{3^2} - \frac{z^2}{3^3} - \frac{z^3}{3^4} - \ldots\right) + \left(\frac{1}{2} + \frac{z}{2^2} + \frac{z^2}{2^3} + \frac{z^3}{2^4} + \ldots\right).$$

(*ii*) For $2 < |z| < 3$: Splitting into partial fractions, we obtain

$$F(z) = \frac{1}{z-3} - \frac{1}{z-2} = -\frac{1}{3}\left(1 - \frac{z}{3}\right)^{-1} - \frac{1}{z}\left(1 - \frac{2}{z}\right)^{-1} \text{ so that } \frac{z}{3} < 1 \text{ and } \frac{2}{z} < 1$$

$$= -\frac{1}{3}\left(1 + \frac{z}{3} + \frac{z^2}{3^2} + \frac{z^3}{3^3} + \ldots\right) - \frac{1}{z}\left(1 + \frac{2}{z} + \frac{2^2}{z^2} + \frac{2^3}{z^3} + \ldots\right)$$

$$= \sum_{k=0}^{\infty} \{-3^{k-1}\} z^k + \sum_{k=1}^{\infty} \{-2^{k-1}\} z^{-k}$$

$$= \sum_{k=-\infty}^{0} -3^{-(k+1)} z^{-k} + \sum_{k=1}^{\infty} \{-2^{k-1}\} z^{-k},$$

Replacing k with $-k$ in the first sum.

On inversion, we get

$$f(k) = -2^{k-1},\ k > 0 \text{ and } u_k = -3^{-(k+1)},\ k \leq 0.$$

(*iii*) For $|z| > 3$: Splitting into partial fractions, we obtain

$$F(z) = \frac{1}{z-3} - \frac{1}{z-2} = \frac{1}{z}\left(1-\frac{3}{z}\right)^{-1} - \frac{1}{z}\left(1-\frac{2}{z}\right)^{-1} \text{ so that } \frac{3}{z} < 1 \text{ and } \frac{2}{z} < 1$$

$$= \frac{1}{z}\left(1+\frac{3}{z}+\frac{3^2}{z^2}+\frac{3^3}{z^3}+\ldots\right) - \frac{1}{z}\left(1+\frac{2}{z}+\frac{2^2}{z^2}+\frac{2^3}{z^3}+\ldots\right)$$

$$= \sum_{k=1}^{\infty} 3^{k-1} z^{-k} - \sum_{k=1}^{\infty} 2^{k-1} z^{-k} = \sum_{k=1}^{\infty} (3^{k-1} - 2^{k-1})\, z^{-k}.$$

On inversion, we get

$$f(k) = 3^{k-1} - 2^{k-1},\ k \geq 1 \text{ and } u_k = 0,\ k \leq 0.$$

2.30.4. Inverse Integral Method (or Residue Method)

By using the theory of complex variables, it can be shown that the inverse Z-transform is given by $f(n) = \frac{1}{2\pi i}\oint_C F(z)\, z^{n-1}\, dz$, where C is the circle (may be even closed contour) which contains all the isolated singularities of F(z) and containing the origin of the z-plane in the region of convergence. Hence by Cauchy's Residue theorem,

$f(n)$ = sum of the residues of the singularities of F(z).

Example 6. *Find the inverse z-transform of*

(*i*) $\dfrac{z}{z^2+7z+10}$ (*ii*) $\dfrac{z}{z^2-2z+2}$ (*iii*) $\dfrac{z^2+z}{(z-1)(z^2+1)}$

(*iv*) $\dfrac{2z}{z^3-z^2+z-1}$ (*v*) $\dfrac{z(z+1)}{(z-1)^3}$ (*vi*) $\dfrac{z(z^2-1)}{(z^2+1)^2}$

Sol. (*i*)

$$F(z) = \frac{z}{z^2+7z+10}$$

$$f(n) = \frac{1}{2\pi i}\oint_C z^{n-1} F(z)dz = \text{sum of residues}$$

$$= \frac{1}{2\pi i}\oint_C z^{n-1} \frac{z}{z^2+7z+10}\, dz = \frac{1}{2\pi i}\oint_C \frac{z^n}{(z+2)(z+5)}\, dz$$

Poles are $z = -2, -5$. These are simple poles.

$$\text{Residue at } z = -2 = \lim_{z\to -2} (z+2)\frac{z^n}{(z+2)(z+5)} = \frac{(-2)^n}{3}$$

Residue at $z = -5 = \underset{z \to -5}{\text{Lt}} (z+5) \dfrac{z^n}{(z+2)(z+5)} = \dfrac{(-5)^n}{-3}$

$$\therefore \quad f(n) = \text{sum of residues} = \frac{(-2)^n}{3} + \left\{\frac{(-5)^n}{-3}\right\} = \frac{1}{3}\{(-2)^n - (-5)^n\}$$

(*ii*) $\quad F(z) = \dfrac{z}{z^2 - 2z + 2}$

$$f(n) = \frac{1}{2\pi i}\oint_C z^{n-1} F(z)\, dz = \text{sum of the residues}$$

$$= \frac{1}{2\pi i}\oint_C z^{n-1} \cdot \frac{z}{z^2 - 2z + 2}\, dz = \frac{1}{2\pi i}\oint_C \frac{z^n}{z^2 - 2z + 2}\, dz$$

The poles are given by $z^2 - 2z + 2 = 0 \therefore z = 1 \pm i$ both are simple poles.
Residue at $z = 1 + i$ is

$$\underset{z \to 1+i}{\text{Lt}} [z - (1+i)] \frac{z^n}{[z-(1+i)][z-(1-i)]} = \frac{(1+i)^n}{2i}$$

Similarly, residue at $z = 1 - i$ is

$$\underset{z \to 1-i}{\text{Lt}} [z - (1-i)] \frac{z^n}{[z-(1+i)][z-(1-i)]} = \frac{(1-i)^n}{-2i}$$

$$\therefore \quad f(n) = \text{sum of the residues} = \frac{(1+i)^n}{2i} - \frac{(1-i)^n}{-2i}$$

$$= \frac{1}{2i}[(1+i)^n - (1-i)^n] \qquad \text{...(1)}$$

We know that $\quad (1+i)^n = (\sqrt{2})^n \left\{\cos\dfrac{n\pi}{4} + i \sin\dfrac{n\pi}{4}\right\}$

$$(1-i)^n = (\sqrt{2})^n \left\{\cos\frac{n\pi}{4} - i \sin\frac{n\pi}{4}\right\}$$

$$\therefore \quad (1+i)^n - (1-i)^n = (\sqrt{2})^n \left\{2i \sin\frac{n\pi}{4}\right\}$$

Substituting in (1), we get $f(n) = \left\{ (\sqrt{2})^n \sin\dfrac{n\pi}{4} \right\}$.

(*iii*) $\quad F(z) = \dfrac{z^2 + z}{(z-1)(z^2+1)}$

$$f(n) = \frac{1}{2\pi i}\int_C z^{n-1} F(z)\, dz = \text{sum of the residues}$$

$$= \frac{1}{2\pi i}\int_C z^{n-1} \frac{z^2 + z}{(z-1)(z^2+1)}\, dz = \frac{1}{2\pi i}\int_C z^n \frac{(z+1)}{(z-1)(z^2+1)}\, dz$$

Poles are given by $z = 1, \pm i$

Residue at $z = 1 = \underset{z \to 1}{\text{Lt}} (z-1) z^n \dfrac{(z+1)}{(z-1)(z^2+1)} = 1$

Residue at $z = i = \underset{z \to i}{\text{Lt}} (z-i) \dfrac{z^n (z+1)}{(z-1)(z-i)(z+i)} = \dfrac{i^n (i+1)}{(i-1)(2i)} = \dfrac{i^n (1+i)}{-2(1+i)} = -\dfrac{1}{2} i^n$

Similarly, residue at $z = -i$ is $= -\dfrac{1}{2} (-i)^n$

$\therefore$ $f(n)$ = sum of the residues

$$= 1 - \frac{1}{2} i^n - \frac{1}{2} (-i)^n = 1 - \frac{1}{2} \{i^n + (-i)^n\} \quad \text{...(1)}$$

$$i^n = \cos \frac{n\pi}{2} + i \sin \frac{n\pi}{2} \;;\; (-i)^n = \cos \frac{n\pi}{2} - i \sin \frac{n\pi}{2}$$

$i^n + (-i)^n = 2 \cos \dfrac{n\pi}{2}$. Substituting in (1), we get

$$f(n) = 1 - \frac{1}{2} \left\{2 \cos \frac{n\pi}{2}\right\} = 1 - \cos \frac{n\pi}{2}.$$

(*iv*)
$$F(z) = \frac{2z}{z^3 - z^2 + z - 1} = \frac{2z}{(z-1)(z^2+1)}$$

$$f(n) = \frac{1}{2\pi i} \int_C z^{n-1} F(z)\, dz = \text{sum of the residues}$$

$$= \frac{1}{2\pi i} \int_C z^{n-1} \frac{2z}{(z-1)(z^2+1)}\, dz = \frac{1}{2\pi i} \int_C z^n \cdot \frac{2}{(z-1)(z^2+1)}\, dz$$

Poles are $z = 1, z = \pm i$

Residue at $(z = 1) = \underset{z \to 1}{\text{Lt}} (z-1) \dfrac{2z^n}{(z-1)(z^2+1)} = 1$

Residue at $(z = i) = \underset{z \to i}{\text{Lt}} (z-i) \dfrac{2z^n}{(z-1)(z-i)(z+i)} = \dfrac{2i^n}{(i-1)(2i)} = -\dfrac{i^n}{1+i}$

Residue at $(z = -i) = -\dfrac{(-i)^n}{1-i}$

$\therefore$ $f(n) = \text{sum of the residues} = 1 - \left\{\dfrac{i^n}{1+i} + \dfrac{(-i)^n}{1-i}\right\}$

$$= 1 - 2\,\text{R.P} \left(\frac{i^n}{1+i}\right) = 1 - 2\text{R.P} \frac{e^{in\pi/2}}{\sqrt{2}\, e^{i\pi/4}}$$

$$= 1 - \sqrt{2}\ \text{R.P}\ e^{in\pi/2 - i\pi/4}$$

$$= 1 - \sqrt{2} \cos\left(\frac{n\pi}{2} - \frac{\pi}{4}\right) = 1 - \left(\cos \frac{n\pi}{2} + \sin \frac{n\pi}{2}\right).$$

(*v*) $$F(z) = \frac{z(z+1)}{(z-1)^3}$$

$$f(n) = \frac{1}{2\pi i}\int_C z^{n-1}\, F(z)\, dz = \frac{1}{2\pi i}\int_C z^n \frac{z(z+1)}{(z-1)^3}\, dz$$

= sum of the residues

$z = 1$ is a pole of order 3

$$\therefore \quad \text{Residue at } (z = 1) = \frac{1}{2!} \underset{z\to 1}{\text{Lt}} \frac{d^2}{dz^2}\left\{\frac{(z-1)^3 \,.\, z^n (z+1)}{(z-1)^3}\right\}$$

$$= \frac{1}{2!} \underset{z\to 1}{\text{Lt}} \{n(n+1)z^{n-1} + n(n-1)z^{n-2}\}$$

$$= \frac{1}{2}\{n^2 + n + n^2 - n\} = n^2$$

$$\therefore \quad f(n) = n^2,\ n = 0, 1, 2, \ldots\ldots$$

(*vi*) $$F(z) = \frac{z(z^2-1)}{(z^2+1)^2}$$

$$f(n) = \frac{1}{2\pi i}\int_C z^{n-1}\, F(z)\, dz = \frac{1}{2\pi i}\int_C \frac{z^n (z^2-1)}{(z^2+1)^2}\, dz = \text{sum of the residues}$$

$z = \pm i$ are poles and each is a pole of order 2.

$$\text{Residue at } (z = i) = \underset{z\to i}{\text{Lt}} \frac{d}{dz}\left\{(z-i)^2 \frac{z^n (z^2-1)}{(z-i)^2 (z+i)^2}\right\}$$

$$= \underset{z\to i}{\text{Lt}} \frac{d}{dz}\left\{\frac{z^n (z^2-1)}{(z+i)^2}\right\}$$

$$= \underset{z\to i}{\text{Lt}} \frac{(z+i)^2 [z^n \,.\, 2z + nz^{n-1}(z^2-1)] - z^n (z^2-1)\,.\,2(z+i)}{(z+i)^4} = \frac{n}{2} i^{n-1}$$

Similarly, $R_2 = \text{Residue of } f(z) \text{ at } (z = -i) = \frac{n}{2}(-i)^{n-1}$

$\therefore$ $f(n)$ = sum of the residues

$$= \frac{n}{2}\{i^{n-1} + (-i)^{n-1}\},\ n = 0, 1, 2, \ldots\ldots$$

2.30.5. Power Series Method

In this method we find the inverse Z-transform by expanding F(*z*) in power series as illustrated in the following example.

Example. *Find* $Z^{-1}\left\{\log\left(\frac{z}{z+1}\right)\right\}$ *by power series method*

Sol. $$F(z) = \log\left\{\frac{z}{z+1}\right\}, \quad \text{Let } z = \frac{1}{y}$$

then $$F(z) = \log\left\{\frac{1/y}{1/y+1}\right\} = -\log(1+y)$$

$$= -y + \frac{1}{2}y^2 - \frac{1}{3}y^3 + \ldots\ldots = -\frac{1}{z} + \frac{1}{2z^2} - \frac{1}{3z^3} + \ldots.. + \frac{(-1)^n}{n} z^{-n}$$

$$\therefore \quad f(n) = z^{-1}\{F(z)\} = \begin{cases} 0 & \text{for } n = 0 \\ \dfrac{(-1)^n}{n} & \text{otherwise} \end{cases}$$

2.31. APPLICATIONS OF Z-TRANSFORMS TO DIFFERENCE EQUATIONS

We know that Laplace transforms method is quite effective in solving linear differential equations. Z-transform is useful in solving difference equations.

The given difference equation can be converted to the form $\overline{u} = \phi(z)$ by taking Z-transform on it, provided the initial values of y are known. Using inversion, we can get the value of u_k, the solution of the given difference equation.

Theorem. *To prove that* $Z\{u_{n+k}\} = z^k\left\{\overline{u} - u_0 - \frac{u_1}{z} - \ldots.. - \frac{u_{k-1}}{z^{k-1}}\right\}$, *where* $Z\{u_k\} = \overline{u}$.

Proof. $$Z\{u_{n+k}\} = \sum_{n=0}^{\infty} u_{n+k}\, z^{-n} = \sum_{n=0}^{\infty} u_{n+k}\, z^{-(n+k)} \cdot z^k = z^k \sum_{n=0}^{\infty} u_{n+k}\, z^{-(n+k)}$$

Setting $m = n + k$, we get

$$Z\{u_{n+k}\} = z^k \sum_{m=k}^{\infty} u_m\, z^{-m} = z^k \left[\sum_{m=0}^{\infty} u_m z^{-m} - \sum_{m=0}^{k-1} u_m z^{-m}\right]$$

$$= z^k\left\{\overline{u} - u_0 - \frac{u_1}{z} - \frac{u_2}{z^2} - \ldots.. - \frac{u_{k-1}}{z^{k-1}}\right\}.$$

Note. For $n = 1, 2, 3$, we have the following relationships :

$$Z\{u_{k+1}\} = z\{\overline{u} - u_0\}$$

$$Z\{u_{k+2}\} = z^2\left\{\overline{u} - u_0 - \frac{u_1}{z}\right\}$$

$$Z\{u_{k+3}\} = z^3\left\{\overline{u} - u_0 - \frac{u_1}{z} - \frac{u_2}{z^2}\right\} \text{ and so on.}$$

We can also write the above as

$$Z\{u(k+1)\} = z\{\overline{u} - u(0)\}$$

$$Z\{u(k+2)\} = z^2\left\{\overline{u} - u(0) - \frac{u(1)}{z}\right\}$$

$$Z\{u(k+3)\} = z^3\left\{\overline{u} - u(0) - \frac{u(1)}{z} - \frac{u(2)}{z^2}\right\} \text{ and so on.}$$

Example 7. *Find the inverse Z-transforms of* $\dfrac{2z}{(z-1)(z^2+1)}$

Sol. Let $\quad F(z) = \dfrac{2z}{(z-1)(z^2+1)} = \dfrac{2z}{(z-1)(z+i)(z-i)}$

It has three simple poles at $z = 1, z = \pm i$.

Using F(z) in the inversion integral, we have

$$f(k) = \frac{1}{2\pi i}\int_C F(z)\, z^{k-1}\, dz,$$

where C is a circle large enough to enclose the poles of F(z)

$= \text{sum of residues of } F(z)\, z^{k-1} \text{ at } z = 1, z = \pm i.$

Now $\text{Res}\,[F(z)\, z^{k-1}]_{z=1} = \underset{z\to 1}{\text{Lt}}\left\{(z-1)\,.\,\dfrac{2z^k}{(z-1)(z^2+1)}\right\} = \underset{z\to 1}{\text{Lt}}\left\{\dfrac{2z^k}{z^2+1}\right\} = 1$

$\text{Res}\,[F(z)\, z^{k-1}]_{z=i} = \underset{z\to i}{\text{Lt}}\left\{(z-i)\,.\,\dfrac{2z^k}{(z-1)(z+i)(z-i)}\right\} = \underset{z\to i}{\text{Lt}}\left\{\dfrac{2z^k}{(z-1)(z+i)}\right\} = -\dfrac{(i)^k}{1+i}$

$\text{Res}\,[F(z)\, z^{k-1}]_{z=-i} = \underset{z\to 1}{\text{Lt}}\left\{(z+i)\,.\,\dfrac{2z^k}{(z-1)(z+i)(z-i)}\right\} = \underset{z\to -i}{\text{Lt}}\left\{\dfrac{2z^k}{(z-1)(z-i)}\right\} = \dfrac{(-i)^k}{i-1}$

Thus the required inverse Z-transform

$$f(k) = 1 - \frac{(i)^k}{1+i} - \frac{(-i)^k}{1-i}.$$

Example 8. *Find* $Z^{-1}\left[\dfrac{z(3z^2-6z+4)}{(z-1)^2(z-2)}\right]$, *by residue method.*

Sol. Let $\quad F(z) = \dfrac{z(3z^2-6z+4)}{(z-1)^2(z-2)}$. The poles are determined by

$(z-1)^2(z-2) = 0 \;\Rightarrow\; z = 1, 1, 2.$

There are two poles, simple pole at $z = 2$ and pole of order 2 at $z = 1$. Let us consider the contour $|z| > 2$.

Using F(z) in the inverse integral, we have

$$f(k) = \frac{1}{2\pi i}\int_C F(z)\, z^{k-1}\, dz,$$

where C is a circle large enough to enclose the poles of F(z)

$= \text{sum of residues of } F(z)\, x^{k-1} \text{ at } z = 1, 2$

Now $\text{Res}\,[F(z)\, z^{k-1}]_{z=2} = \underset{z\to 2}{\text{Lt}}\left\{\dfrac{(z-2)\,.\,z^{k-1}\,.\,z(3z^2-6z+4)}{(z-1)^2(z-2)}\right\}$

$$= \underset{z\to 2}{\text{Lt}}\left\{\frac{3z^{k+2} - 6z^{k+1} + 4z^k}{(z-1)^2}\right\}$$

$$= 3 \,.\, 2^{k+2} - 6 \,.\, 2^{k+1} + 4 \,.\, 2^k$$
$$= 12 \,.\, 2^k - 12 \,.\, 2^k + 4 \,.\, 2^k = 4 \,.\, 2^k = 2^{k+2}$$

$$\text{Res } \{F(z)\, z^{k-1}\}_{z=1} = \left[\frac{d}{dz}\left\{(z-1)^2 \,.\, \frac{z^{k-1} \,.\, z\,(3z^2 - 6z + 4)}{(z-1)^2\,(z-2)}\right\}\right]_{z=1}$$

$$= \left[\frac{d}{dz}\left\{\frac{3z^{k+2} - 6z^{k+1} + 4z^k}{z-2}\right\}\right]_{z=1}$$

$$= \left[\frac{(z-2)\,\{3(k+2)\,z^{k+1} - 6(k+1)\,z^k + 4kz^{k-1}\} - (3z^{k+2} - 6z^{k+1} + 4z^k)\,.\,1}{(z-2)^2}\right]_{z=1}$$

$$= -\,\{(3k+6) - 6k - 6 + 4k\} - \{3 - 6 + 4\}$$
$$= -\,3k - 6 + 6k + 6 - 4k - 3 + 6 - 4 = -\,k - 1.$$

$\therefore \quad f(k)$ = sum of residues

$$= [2^{k+2} - k - 1]\ U(k).$$

Example 9. *Find* $Z^{-1}\left[\dfrac{9z^3}{(3z-1)^2\,(z-2)}\right]$*, by residue method.*

Sol. Let $\quad F(z) = \dfrac{9z^3}{(3z-1)^2\,(z-2)}$. Its poles are determined by

$$(3z-1)^2\,(z-2) = 0 \;\Rightarrow\; z = \frac{1}{3}, \frac{1}{3}, 2.$$

Therefore are two poles; *i.e.*, one simple pole at $z = 2$ and second pole of order 2 at $z = \dfrac{1}{3}$.

Let us consider the contour $|z| > 2$.

$$\text{Now} \quad \text{Res } [F(z)\, z^{k-1}]_{z=2} = \underset{z\to 2}{\text{Lt}}\left\{(z-2)\,.\,\frac{z^{k-1}\,.\,9z^3}{(3z-1)^2\,(z-2)}\right\}$$

$$= \underset{z\to 2}{\text{Lt}}\left\{\frac{9z^{k+2}}{(3z-1)^2}\right\} = \frac{9\,.\,2^{k+2}}{25}$$

$$\text{Res } [F(z)\, z^{k-1}]_{z=\frac{1}{3}} = \left[\frac{d}{dz}\left\{\frac{\left(z-\frac{1}{3}\right)^2 .\, z^{k-1}\,.\,9z^3}{(3z-1)^2\,(z-2)}\right\}\right]_{z=\frac{1}{3}}$$

$$= \left[\frac{(z-2)\,(k+2)\,z^{k+1} - z^{k+2}}{(z-2)^2}\right]_{z=\frac{1}{3}}$$

$$= \left[\frac{d}{dz}\left\{\frac{9z^{k+2}}{9\,(z-2)}\right\}\right]_{z=\frac{1}{3}} = \frac{d}{dz}\left(\frac{z^{k+2}}{z-2}\right)_{z=\frac{1}{3}}$$

$$= \left[\frac{(k+2)\,z^{k+2} - 2\,(k+2)\,z^{k+1} - z^{k+2}}{(z-2)^2}\right]_{z=\frac{1}{3}}$$

$$= \left[\frac{(k+1)\,z^{k+2} - 2\,(k+2)\,z^{k+1}}{(z-2)^2}\right]_{z=\frac{1}{3}}$$

$$= \left[\frac{z^{k+1}\,\{(k+1)\,z - 2k - 4\}}{(z-2)^2}\right]_{z=\frac{1}{3}} = \frac{\left(\frac{1}{3}\right)^{k+1}\left\{(k+1)\frac{1}{3} - 2k - 4\right\}}{\frac{25}{9}}$$

$$= \frac{\left(\frac{1}{3}\right)^{k+1} .\, 9 \,.\, \frac{1}{3}\,[k+1-6k-12]}{25} = \frac{\left(\frac{1}{3}\right)^{k} (-5k-11)}{25}.$$

Hence, $\quad f(k) = \text{sum of residues}$

$$= \frac{9}{25}\,.\,2^{k+2} - \left(\frac{1}{3}\right)^k \frac{(5k+11)}{25}.$$

Example 10. *Solve:* $u_{k+1} + u_k = 1$ *if* $u_0 = 0$.

Sol. Taking Z-transforms on both sides, we get

$$Z\{u_{k+1}\} + Z\{u_k\} = Z\{1\}$$

$$z(\overline{u} - u_0) + \overline{u} = \frac{z}{z-1} \quad \text{where } \overline{u} = Z\{u_k\}$$

$$\overline{u}\,(z+1) = \frac{z}{(z-1)} \qquad (\because \quad u_0 = 0)$$

or
$$\overline{u} = \frac{z}{(z-1)(z+1)}$$

or
$$\overline{u} = \frac{1}{2}\left[\frac{z}{z-1} - \frac{z}{z+1}\right]$$

Inverting, we get
$$u_k = \frac{1}{2}\,Z^{-1}\left\{\frac{z}{z-1}\right\} - \frac{1}{2}\,Z^{-1}\left\{\frac{z}{z+1}\right\} = \frac{1}{2} - \frac{1}{2}\,(-1)^k.$$

Example 11. *Solve:* $f(n+2) - 3f(n+1) + 2f(n) = 0$; $f(0) = 0$ *and* $f(1) = 1$.

Sol. Taking Z-transform on both sides of the given equation, we get

$$Z\{f(n+2)\} - 3Z\{f(n+1)\} + 2Z\{f(n)\} = 0$$

$$z^2\left\{\overline{f} - f(0) - \frac{f(1)}{z}\right\} - 3z\{\overline{f} - f(0)\} + 2\overline{f} = 0 \; ; \; \overline{f} = Z(f(n))\}$$

$$(z^2 - 3z + 2)\overline{f} = z \quad \text{or} \quad \overline{f} = \frac{z}{z^2 - 3z + 2}$$

or $$\frac{\bar{f}}{z} = \frac{1}{(z-1)(z-2)} = \frac{1}{z-2} - \frac{1}{z-1}$$

or $$\bar{f} = \frac{z}{z-2} - \frac{z}{z-1} \quad \therefore f(n) = Z^{-1}\left\{\frac{z}{z-2} - \frac{z}{z-1}\right\}$$

Taking inverse Z-transform, we get $f(n) = 2^n - 1$, where $n = 0, 1, 2, \ldots\ldots$

Example 12. *Solve:* $y_{n+2} + 6y_{n+1} + 9y_n = 2^n$ *given* $y_0 = y_1 = 0$.

Sol. Taking Z-transforms on both sides

$$Z\{y_{n+2}\} + 6Z\{y_{n+1}\} + 9Z\{y_n\} = Z(2^n\}$$

$$z^2\left\{\bar{y} - y_0 - \frac{y_1}{z}\right\} + 6z\{\bar{y} - y_0\} + 9\bar{y} = \frac{z}{z-2}$$

$$\bar{y}(z^2 + 6z + 9) = \frac{z}{z-2} \quad \text{or} \quad \bar{y} = \frac{z}{(z+3)^2(z-2)}$$

or $$\frac{\bar{y}}{z} = \frac{1}{(z-2)(z+3)^2} = \frac{A}{z-2} + \frac{B}{z+3} + \frac{C}{(z+3)^2} \qquad \ldots(1)$$

Equation (1) gives $\quad A = \frac{1}{25}, B = -\frac{1}{25}, C = -\frac{1}{5}$

$$\therefore \quad \bar{y} = \frac{1}{25}\frac{z}{z-2} - \frac{1}{25}\frac{z}{z+3} - \frac{1}{5}\frac{z}{(z+3)^2}$$

Taking inverse Z-transforms on both sides, we get

$$y_n = Z^{-1}\{\bar{y}\} = \frac{1}{25}Z^{-1}\left\{\frac{z}{z-2}\right\} - \frac{1}{25}Z^{-1}\left\{\frac{z}{z+3}\right\} - \frac{1}{5}Z^{-1}\left\{\frac{z}{(z+3)^2}\right\}$$

$$= \frac{1}{25}(2)^n - \frac{1}{25}(-3)^n - \frac{1}{5}n(-3)^{n-1}.$$

Example 13. *Solve the equation* $u_{n+2} - 2u_{n+1} + u_n = 3n + 5$.

Sol. Taking Z-transforms on both sides, we get

$$Z\{u_{n+2}\} - 2Z\{u_{n+1}\} + z\{u_n\} = Z\{3n + 5\}$$

$$z^2\left\{\bar{u} - u_0 - \frac{u_1}{z}\right\} - 2z\{\bar{u} - u_0\} + \bar{u} = \frac{3z}{(z-1)^2} + \frac{5z}{z-1}$$

or $$\bar{u} = \frac{5z^2 - 2z}{(z-1)^4} + u_0\frac{z^2 - 2z}{(z-1)^2} + u_1\frac{z}{(z-1)^2}$$

Taking inverse transform on both sides, we get

$$u_n = Z^{-1}\{\bar{u}\} = Z^{-1}\left\{\frac{5z^2 - 2z}{(z-1)^4}\right\} + u_0 Z^{-1}\left\{\frac{z^2 - 2z}{(z-1)^2}\right\} + u_1 Z^{-1}\left\{\frac{z}{(z-1)^2}\right\} \qquad \ldots(1)$$

Writing $$\frac{5z^2 - 2z}{(z-1)^4} = \frac{1}{2}\frac{z^3 + 4z^2 + z}{(z-1)^4} + \frac{z^2 + z}{(z-1)^3} - \frac{3}{2}\frac{z}{(z-1)^2}$$

$$\frac{z^2 - 2z}{(z-1)^2} = \frac{z}{z-1} - \frac{z}{(z-1)^2}$$

and noting that $\quad Z(1) = \dfrac{z}{z-1},\ Z(n) = \dfrac{z}{(z-1)^2},\ Z(n^2) = \dfrac{z^2+z}{(z-1)^3}$

and $\quad Z(n^3) = \dfrac{z^3+4z^2+z}{(z-1)^4}$ from (1), we get

$$u_n = \left(\frac{1}{2}n^3 + n^2 - \frac{3}{2}n\right) + u_0(1-n) + u_1(n)$$

$$= \frac{1}{2}n(n-1)(n+3) + C_0 + C_1 n, \text{ where } C_0 = u_0,\ C_1 = u_1 - u_0.$$

Note. In the above problem u_0 and u_1 are not given.

Example 14. *Solve* $x_{n+1} = 7x_n + 10y_n\ ;\ y_{n+1} = x_n + 4y_n$, *where* $x_0 = 3,\ y_0 = 2.$

Sol. $\quad x_{n+1} + 7x_n + 10y_n\ ;\ x_0 = 3$

Taking Z-transforms, we get

$$Z(x_{n+1}) = 7Z(x_n) + 10Z(y_n)$$

$$z(\bar{x} - x_0) = 7\bar{x} + 10\bar{y} \text{ or } (7-z)\bar{x} + 10\bar{y} = -3z \qquad \ldots(1)$$

Again taking Z-transforms, to $y_{n+1} = x_n + 4y_n,\ y_0 = 2$, we get

$$Z\{y_{n+1}\} = Z\{x_n\} + 4Z\{(y_n\}$$

$$z(\bar{y} - y_0) = \bar{x} + 4\bar{y} \quad \text{or} \quad \bar{x} - (z-4)\bar{y} = -2z \qquad \ldots(2)$$

By eliminating $\bar{x}$ from (1) and (2), we get

$$(z^2 - 11z + 18)\bar{y} = 2z^2 - 11z$$

$$\bar{y} = \frac{2z^2 - 11z}{(z-9)(z-2)} = \frac{z}{z-9} + \frac{z}{z-2}$$

$$y_n = Z^{-1}\{\bar{y}\} = Z^{-1}\left\{\frac{z}{z-9}\right\} + Z^{-1}\left\{\frac{z}{z-2}\right\} = 9^n + 2^n$$

From the equation (2), we get

$$x_n = y_{n+1} - 4y_n = 9^{n+1} + 2^{n+1} - 4(9^n + 2^n)$$

$$= 9 \,.\, 9^n + 2 \,.\, 2^n - 4(9^n + 2^n) = 5 \,.\, 9^n - 2 \,.\, 2^n$$

$$\therefore \quad x_n = 5 \,.\, 9^n - 2 \,.\, 2^n \quad \text{and} \quad y_n = 9^n + 2^n.$$

Example 15. *Solve* $y_{x+2} - 2\cos\alpha\, y_{x+1} + y_x = 0,\ y_0 = 0,\ y_1 = 1.$

Sol. Taking Z-transform on both sides, we get

$$z^2\left\{\bar{y} - y_0 - \frac{y_1}{z}\right\} - 2\cos\alpha \,.\, z\,\{\bar{y} - y_0\} + \bar{y} = 0\ ;\ \bar{y} - z\{y_x\}$$

or

$$z^2\left(\bar{y} - \frac{1}{z}\right) - 2\cos\alpha z\ \bar{y} + \bar{y} = 0$$

$$\bar{y} = \frac{z}{z^2 - 2\cos\alpha z + 1}$$

$$\bar{y} = \frac{z}{(z - e^{i\alpha})(z - e^{-i\alpha})} = \frac{Az}{z - e^{i\alpha}} + \frac{Bz}{z - e^{-i\alpha}}$$

$$= \frac{-i}{2\sin\alpha}\frac{z}{z - e^{i\alpha}} + \frac{i}{2\sin\alpha}\frac{z}{z - e^{-i\alpha}}$$

$$y_x = \frac{-i}{2\sin\alpha} Z^{-1}\left[\frac{z}{z - e^{i\alpha}}\right] + \frac{i}{2\sin\alpha} Z^{-1}\left[\frac{z}{z - e^{-i\alpha}}\right]$$

$$= \frac{-i}{2\sin\alpha}[(e^{i\alpha})^x - (e^{-i\alpha})^x]$$

$$= \frac{-i}{2\sin\alpha}[(\cos\alpha x + i\sin\alpha x) - (\cos\alpha x - i\sin\alpha x)] = \frac{\sin\alpha x}{\sin\alpha}.$$

Example 16. *Using Z-transform, solve*

$u_{k+1} + u_k = 1$ *if* $u_0 = 0$.

Sol. The given difference equation is

$$u_{k+1} + u_k = 1. \qquad ...(1)$$

If $Z(u_k) = U(z)$, then $Z(u_k + 1) = z\,[U(z) - u_0]$. Also $Z(1) = \dfrac{z}{z-1}$

$\therefore$ Taking the z-transform of both sides of (1), we get

$$Z[u_{k+1} + u_k] = Z(1) \quad \text{or} \quad Z(u_{k+1}) + Z(u_k) = Z(1)$$

or $$z\,[U(z) - u_0] + U(z) = \frac{z}{z-1}. \qquad ...(2)$$

On putting the value of u_0 in (2), we get

$$U(z) = \frac{z}{(z-1)(z+1)}$$

$\therefore$ $$\frac{U(z)}{z} = \frac{1}{(z-1)(z+1)} = \frac{1}{2}\cdot\frac{1}{z-1} - \frac{1}{2}\cdot\frac{1}{z+1}$$ are breaking into partial fractions

or $$U(z) = \frac{1}{2}\cdot\frac{z}{z-1} - \frac{1}{2}\cdot\frac{z}{z+1}.$$

On inversion, we obtain

$$u_k = \frac{1}{2}\cdot z^{-1}\left(\frac{z}{z-1}\right) - \frac{1}{2}\cdot z^{-1}\left(\frac{z}{z+1}\right) = \frac{1}{2}\cdot 1^k - \frac{1}{2}(-1)^k.$$

Hence the requred solution of (1) is

$$u_k = \frac{1}{2} - \frac{1}{2}\cdot(-1)^k.$$

Example 17. *Using Z-transform, solve* $6y_{k+2} - y_{k+1} - y_k = 0$, *given that* $y(0) = 0, y(1) = 1$.

Sol. The given difference equation is

$$6y_{k+2} - y_{k+1} - y_k = 0. \quad ...(1)$$

Taking the Z-transform of both sides of (1), we get

$$Z[6y_{k+2} - y_{k+1} - y_k] = 0 \quad \text{or} \quad 6Z(y_{k+2}) - Z(y_{k+1}) - Z(y_k) = 0$$

or
$$6z^2 [Y(z) - y(0) - y(1) z^{-1}] - z [Y(z) - y(0)] - Y(z) = 0.$$

On putting the values of $y(0)$ and $y(1)$, we get

$$6z^2 [Y(z) - z^{-1}] - z Y(z) - Y(z) = 0$$

or
$$(6z^2 - z - 1) Y(z) = 6z$$

or
$$Y(z) = \frac{6z}{6z^2 - z - 1} = \frac{6z}{(3z+1)(2z-1)} = \frac{z^{-1}}{\left(1 - \frac{z^{-1}}{3}\right)\left(1 - \frac{z^{-1}}{2}\right)}$$

$$= \frac{6/5}{1 - z^{-1}/2} - \frac{6/5}{1 + z^{-1/3}}.$$

On inversion, we get

$$y_k = \frac{6}{5} Z^{-1}\left[\frac{z}{z - 1/2}\right] - \frac{6}{5} Z^{-1}\left[\frac{z}{z + 1/3}\right] = \frac{6}{5}\left[\left(\frac{1}{2}\right)^k - \left(-\frac{1}{3}\right)^k\right].$$

Example 18. *using Z-transform, solve*

$$u_{k+2} + 4u_{k+1} + 3u_k = 3^k$$

given that $u_0 = 0,\ u_1 = 1.$

Sol. The given difference equation is

$$u_{k+2} + 4u_{k+1} + 3u_k = 3^k. \quad ...(1)$$

If $Z[u_k] = U(z)$, then $Z[u_{k+1}] = z[U(z) - u_0]$, $Z[u_{k+2}] = z^2 [U(z) - u_0 - u_1 z^{-1}]$.

Also
$$Z[3^k] = \frac{z}{z-3}.$$

Taking the Z-transform of both sides of (1), we get

$$Z[u_{k+2}] + 4Z[u_{k+1}] + 3Z[u_k] = Z[3^k]$$

or
$$Z^2 [U(z) - u_0 - u_1 z^{-1}] + 4z [U(z) - u_0] + 3U(z) = \frac{z}{z-3}. \quad ...(2)$$

Putting the values of u_0 and u_1 in (2), we get

$$z^2 [U(z) - 0 - z^{-1}] + 4z [U(z) - 0] + 3U(z) = \frac{z}{z-3}$$

or
$$(z^2 + 4z + 3) U(z) = z + \frac{z}{z-3}.$$

$$\therefore \quad \frac{U(z)}{z} = \frac{1}{(z+1)(z+3)} + \frac{1}{(z-3)(z+1)(z+3)}$$

$$= \frac{3}{8}\frac{1}{z+1} + \frac{1}{24}\frac{1}{z-3} - \frac{5}{12}\frac{1}{z+3}, \text{ on breaking into partial fractions}$$

$$U(z) = \frac{3}{8}\frac{z}{z+1} + \frac{1}{24}\frac{z}{z-3} - \frac{5}{12}\frac{z}{z+3}.$$

On inversion, we obtain

$$u_k = \frac{3}{8} Z^{-1}\left(\frac{z}{z+1}\right) + \frac{1}{24} Z^{-1}\left(\frac{z}{z-3}\right) - \frac{5}{12} Z^{-1}\left(\frac{z}{z+3}\right)$$

$$= \frac{3}{8}(-1)^k + \frac{1}{24}3^k - \frac{5}{12}(-3)^k.$$

Example 19. *Using Z-transform, solve* $y_{k+2} + 6y_{k+1} + 9y_k = 2^k$ *given that* $y_0 = y_1 = 0$.

Sol. The given difference equation is

$$y_{k+2} + 6y_{k+1} + 9y_k = 2^k. \quad \text{...(1)}$$

Taking the Z-transforms of both sides, we get

$$Z[y_{k+2}] + 6Z[y_{k+1}] + 9Z[y_k] = Z[2^k]$$

or
$$z^2[Y(z) - y_0 - y_1 z^{-1}] + 6z[Y(z) - y_0] + 9Y(z) = \frac{z}{z-2} \quad \text{...(2)}$$

Putting the values of y_0 and y_1 in (2), we get

$$(z^2 + 6z + 9)\,Y(z) = \frac{z}{z-2}$$

or
$$Y(z) = \frac{z}{(z-2)(z^2+6z+9)}$$

or
$$\frac{Y(z)}{z} = \frac{1}{(z-2)(z+3)^2} = \frac{1}{25}\left[\frac{1}{z-2} - \frac{1}{z+3} - \frac{5}{(z+3)^2}\right],$$

on splitting into partial fractions

or
$$Y(z) = \frac{1}{25}\left[\frac{z}{z-2} - \frac{z}{z+3} - 5\frac{z}{(z+3)^2}\right].$$

On taking inverse Z-transforms of both sides, we obtain

$$y_k = \frac{1}{25}\left[Z^{-1}\left(\frac{z}{z-2}\right) - Z^{-1}\left(\frac{z}{z+3}\right) + \frac{5}{3}Z^{-1}\left(-\frac{3z}{(z+3)^2}\right)\right]$$

$$= \frac{1}{2}\left[2^k - (-3)^k + \frac{5}{3}k(-3)^k\right] \qquad \left[\because\ Z^{-1}\left\{\frac{az}{(z-a)^2}\right\} = ka^k\right]$$

Example 20. *Using Z-transform, solve $6y_{k+2} + 5y_{k+1} - y_k = 6U(k)$ given that $y(0) = y(1) = 0$.*

Sol. The given difference equation is

$$6y_{k+2} + 5y_{k+1} = 6\text{U}(k) \qquad \text{...(1)}$$

where U(k) is unit step sequence.

Taking the Z-transform of both sides of (1), we get

$$\text{Z}\,[6y_{k+2} + 5y_{k+1} - y_k] = \text{Z}\,[6\text{U}(k)]$$

or $$6z^2\,[\text{Y}(z) - y(0) - y(1)\,z^{-1}] + 5z\,[\text{Y}(z) - y(0)] - \text{Y}(z) = 6\,.\,\frac{z}{z-1},\ k \geq 0 \text{ and } 0,\ k < 0.$$

On putting the values of $y(0)$ and $y(1)$, we get

$$6z^2\,\text{Y}(z) - 5z\,\text{Y}(z) - \text{Y}(z) = \frac{6z}{z-1}$$

or $$(6z^2 - 5z - 1)\,\text{Y}(z) = \frac{6z}{z-1}$$

or $$\text{Y}(z) = \frac{6z}{(6z+1)\,(z-1)^2},\ k \geq 0 \text{ and } 0,\ k < 0.$$

Ita poles are determined by

$$(6z + 1)\,(z - 1)^2 = 0 \ \Rightarrow\ z = -\frac{1}{6},\ 1,\ 1.$$

There are two poles; *i.e.*, one simple pole at $z = -\dfrac{1}{6}$ and second pole or order 2 at $z = 1$.

Let us consider the contour $|z| > 1$

$$\text{Now Res}\,[\text{Y}(z)\,z^{k-1}]_{z=-\frac{1}{6}} = \underset{z \to -\frac{1}{6}}{\text{Lt}} \left\{\left(z + \frac{1}{6}\right).\frac{z^{k-1}\,.6z}{(6z+1)\,(z-1)^2}\right\}$$

$$= \underset{z \to -\frac{1}{6}}{\text{Lt}} \left\{\frac{z^k}{(z-1)^2}\right\} = \frac{36}{49}\left(-\frac{1}{6}\right)^k,\ k \geq 0 \text{ and } 0,\ k < 0$$

$$= \frac{36}{49}\left(-\frac{1}{6}\right)^k \text{U}(k)$$

$$\text{Res}\,[\text{Y}(z)\,z^{k-1}]_{z=1} = \frac{d}{dz}\left[\frac{(z-1)^2\,.\,6z^k}{(6z+1)\,(z-1)^2}\right]_{z=1} = \left[\frac{d}{dz}\,\frac{6z^k}{6z+1}\right]_{z=1}$$

$$= \left[\frac{(6z+1)\,.\,6kz^{k-1} - 6z^k\,.\,6}{(6z+1)^2}\right]_{z=1} = \left[\frac{6kz^{k-1}}{6z+1} - \frac{36z^k}{(6z+1)^2}\right]_{z=1}$$

$$= \left[\frac{6}{7}k - \frac{36}{49}\right] = \frac{6}{7}(k+1) - \frac{78}{49},\ k \geq 0 \text{ and } 0,\ k < 0$$

$$= \left[\frac{6}{7}(k+1) - \frac{78}{49}\right] U(k).$$

$\therefore$ $$y_k = \text{sum of residues}$$

$$= \left[\frac{6}{7}(k+1) - \frac{78}{49} + \frac{36}{49}\left(-\frac{1}{6}\right)^k\right] U(k).$$

TEST YOUR KNOWLEDGE

1. Find the inverse Z-transform of $f(n)$ by any method:

(i) $\dfrac{z^2+z}{(z-1)^2}$ (ii) $\dfrac{z^2+2z}{(z-1)(z-2)(z-3)}$ (iii) $\dfrac{z}{(z-1)(z-2)(z-3)}$

(iv) $\dfrac{z^2}{(z-a)(z-b)}$ (v) $\dfrac{z}{z^2+11z+30}$ (vi) $\dfrac{5z}{(2z-1)(z-3)}$

(vii) $\dfrac{z^2}{(z+2)(z^2+4)}$ (viii) $\dfrac{4z^{-1}}{(1-z^{-1})^2}$ (ix) $\dfrac{4-8z^{-1}+6z^{-2}}{(1-2z^{-1})^2(1+z^{-1})}$

(x) $\dfrac{4z^2-2z}{z^3-5z^2+8z-4}$.

2. Using convolution theorem, find the Z^{-1} of

(i) $\dfrac{z^2}{(z-4)(z-3)}$ (ii) $\dfrac{8z^2}{(2z-1)(4z-1)}$ (iii) $\dfrac{z^2}{(z-a)^2}$.

3. Solve the following difference equations:

(i) $u(k+2) - 5u(k+1) + 6u(k) = 6k$ if $u(0) = u(1) = 0$

(ii) $y(n) + y(n-1) - 4y(n-2) = 0,\ n \ge 2,\ y(0) = 3,\ y(1) = -2$

(iii) $u_{x+2} + u_x = 5 \,.\, 2^x,\ u_0 = 1,\ u_1 = 0$

(iv) $y_{n+3} - 3y_{n+1} + 2y_n = 0,\ y_1 = 0,\ y_2 = 8,\ y_3 = -8$

(v) $u_{n+2} - 2\cos \alpha u_{n+1} + \alpha u_n = 0,\ y_0 = 1,\ y_1 = \cos\alpha$

(vi) $y_{n+2} - 4y_n = 0$

(vii) $y(n) - y(n-1) = u(n) + u(n-1)$

(viii) $u_{n+2} - 4u_{n+1} + 4u_n = 0$ given $u_0 = 1,\ u = 0$

(ix) $y_{n+3} + y_{n+2} - 8y_{n+1} - 12y_n = 0,\ y_0 = 1,\ y_1 = y_2 = 0$

(x) $x_{n+1} - y_n = 1;\ y_{n+1} - x_n = 0;\ x_0 = 0,\ y_0 = -1.$

Answers

1. (i) $2n+1$ (ii) $\dfrac{3}{2} - 4 \,.\, 2^n + \dfrac{5}{2} \,.\, 3^n$ (iii) $\dfrac{1}{2} - \dfrac{1}{2} \,.\, 2^n + \dfrac{1}{2} \,.\, 3^n$

(iv) $\dfrac{1}{ab}\,[a^{n+1} - b^{n+1}]$ (v) $(-5)^n - (-6)^n$ (vi) $3^n - \left(\dfrac{1}{2}\right)^n$

(*vii*) $\frac{(-2)^{n+1}}{8} + \frac{(2i)^{n+1}}{8i(1+i)} - \frac{(2i)^{n+1}}{2i(1-i)}$ (*viii*) $2n(n-1)u(n)$

(*ix*) $2(-1)^n + 2^n(n+2)$ (*x*) $2 + 2^n(3n-2)$

2. (*i*) $4^{n+1} - 3^{n+1}$ (*ii*) $\left(\frac{1}{2}\right)^{n-1} - \left(\frac{1}{2}\right)^{2n}$ (*iii*) $(n+1)a^n u(n)$

3. (*i*) $u(k) = \frac{1}{12} 6^k - \frac{1}{3} . 3^k + \frac{1}{4} 2^k$ (*ii*) $y(n) = \frac{1}{5} 4^n + \frac{14}{5} (-1)^n$

(*iii*) $u_x = 2^x - 2 \sin \frac{\pi x}{2}$ (*iv*) $y_n = \frac{8}{3} + \frac{4}{3} (-2)^n$ (*v*) $y_n = \cos n\alpha$

(*vi*) $y_n = A2^n + B(-2)^n$ (*vii*) $y(n) = 1 + 2n$ (*viii*) $u(n) = 2^n(1-n)$

(*ix*) $y_n = \frac{4}{25} . 3 + \frac{1}{25} (21 - 15n)(-2)^n$

(*x*) $x_n = \frac{1}{2}\left[1 - \cos \frac{n\pi}{2} - \sin \frac{n\pi}{2}\right]; y_n = \frac{1}{2}\left[\sin \frac{n\pi}{2} - \cos \frac{n\pi}{2} - 1\right].$

3

Partial Differential Equations and Their Applications

3.1. INTRODUCTION

A differential equation which involves partial derivatives is called a *partial differential equation*.

For example,
$$x\frac{\partial z}{\partial x} + y\frac{\partial z}{\partial y} = z \qquad ...(1)$$

$$\frac{\partial^2 z}{\partial x^2} + \frac{\partial^2 z}{\partial y^2} = 0 \qquad ...(2)$$

$$\frac{\partial^2 u}{\partial x \partial y} = \left(\frac{\partial u}{\partial z}\right)^3 \qquad ...(3)$$

are partial differential equations.

The *order* of a partial differential equation is the order of the highest partial derivative in the equation. The *degree* of a partial differential equation is the degree of the highest order partial derivative occurring in the equation.

Thus, equation (1) is of first order, equations (2) and (3) are of second order. The degree of all the above equations is one.

If z is a function of two independent variables x and y, then we shall use the following notation for the partial derivatives of z :

$$\frac{\partial z}{\partial x} = p, \frac{\partial z}{\partial y} = q, \frac{\partial^2 z}{\partial x^2} = r, \frac{\partial^2 z}{\partial x \partial y} = s, \frac{\partial^2 z}{\partial y^2} = t.$$

3.2. FORMATION OF PARTIAL DIFFERENTIAL EQUATIONS

Partial differential equations can be formed either by the elimination of arbitrary constants or by the elimination of arbitrary functions. If the number of arbitrary constants to be eliminated is equal to the number of independent variables, the partial differential equations that arise are of the first order. If the number of arbitrary constants to be eliminated is more than the number of independent variables, the partial differential equations obtained are of second or higher order. If the partial differential equation is obtained by elimination of arbitrary functions, then the order of the partial differential equation is, in general, equal to the number of arbitrary functions eliminated.

The following examples illustrate the method.

ILLUSTRATIVE EXAMPLES

Example 1. *Form partial differential equations from the following equations by eliminating the arbitrary constants:*

(*i*) $z = ax + by + ab$ (*ii*) $z = ax + a^2y^2 + b$.

Sol. (*i*) Differentiating z partially w.r.t. x and y

$$p = \frac{\partial z}{\partial x} = a, \; q = \frac{\partial z}{\partial y} = b$$

Substituting for a and b in the given equation, we get

$$z = px + qy + pq$$

which is a partial differential equation of the first order.

(*ii*) Differentiating z partially w.r.t. x and y

$$p = \frac{\partial z}{\partial x} = a, \; q = \frac{\partial z}{\partial y} = 2a^2y$$

Eliminating a between these results, we get $q = 2p^2y$

which is a partial differential equation of the first order.

Example 2. *Form partial differential equations from the following equations by eliminating the arbitrary constants:*

(*i*) $z = (x^2 + a)(y^2 + b)$ (*ii*) $2z = \frac{x^2}{a^2} + \frac{y^2}{b^2}$ (*iii*) $\frac{x^2}{a^2} + \frac{y^2}{b^2} + \frac{z^2}{c^2} = 1$.

Sol. (*i*) Differentiating z partially w.r.t. x and y

$$p = \frac{\partial z}{\partial x} = 2x(y^2 + b) \quad ...(1)$$

$$q = \frac{\partial z}{\partial y} = 2y(x^2 + a) \quad ...(2)$$

Multiplying (1) and (2), we get

$$pq = 4xy(x^2 + a)(y^2 + b) \quad \text{or} \quad pq = 4xyz$$

which is a partial differential equation of the first order.

(*ii*) Here $$2z = \frac{x^2}{a^2} + \frac{y^2}{b^2}$$

Differentiating partially w.r.t. x and y

$$2\frac{\partial z}{\partial x} = \frac{2x}{a^2} \text{ or } \frac{1}{a^2} = \frac{p}{x} \text{ and } 2\frac{\partial z}{\partial y} = \frac{2y}{b^2} \text{ or } \frac{1}{b^2} = \frac{q}{y}$$

Substituting for $\frac{1}{a^2}$ and $\frac{1}{b^2}$ in the given equation, we get

$$2z = px + qy$$

which is a partial differential equation of the first order.

(*iii*) Here the number of arbitrary constants (a, b, c) is greater than the number of independent variables (x, y).

Differentiating partially w.r.t. x and y, we have

$$\frac{2x}{a^2} + \frac{2z}{c^2} \cdot \frac{\partial z}{\partial x} = 0 \quad \text{or} \quad c^2x + a^2z\frac{\partial z}{\partial x} = 0 \quad ...(1)$$

and $$\frac{2y}{b^2} + \frac{2z}{c^2} \cdot \frac{\partial z}{\partial y} = 0 \quad \text{or} \quad c^2 y + b^2 z \frac{\partial z}{\partial y} = 0 \qquad ...(2)$$

Again differentiating (1) partially w.r.t. x, we have

$$c^2 + a^2 \left(\frac{\partial z}{\partial x}\right)^2 + a^2 z \frac{\partial^2 z}{\partial x^2} = 0 \quad \text{or} \quad \frac{c^2}{a^2} + \left(\frac{\partial z}{\partial x}\right)^2 + z \frac{\partial^2 z}{\partial x^2} = 0$$

Substituting $\frac{c^2}{a^2} = -\frac{z}{x}\frac{\partial z}{\partial x}$ from (1), we have

$$-\frac{z}{x}\frac{\partial z}{\partial x} + \left(\frac{\partial z}{\partial x}\right)^2 + z \frac{\partial^2 z}{\partial x^2} = 0 \quad \text{or} \quad xz \frac{\partial^2 z}{\partial x^2} + x\left(\frac{\partial z}{\partial x}\right)^2 - z \frac{\partial z}{\partial x} = 0$$

which is a partial differential equation of the second order.

Note. Instead of differentiating (1) partially w.r.t. x, if we differentiate (2) partially w.r.t. y and substitute for $\frac{c^2}{b^2}$ from (2), we shall get $yz \frac{\partial^2 z}{\partial y^2} + y\left(\frac{\partial z}{\partial y}\right)^2 - z\frac{\partial z}{\partial y} = 0.$

Example 3. *Form the partial differential equation by eliminating the function f from the relation*

$$z = y^2 + 2f\left(\frac{1}{x} + \log y\right).$$

Sol. Differentiating z partially w.r.t. x and y

$$p = \frac{\partial z}{\partial x} = 2f'\left(\frac{1}{x} + \log y\right).\left(-\frac{1}{x^2}\right)$$

or $$-px^2 = 2f'\left(\frac{1}{x} + \log y\right) \qquad ...(1)$$

and $$q = \frac{\partial z}{\partial y} = 2y + 2f'\left(\frac{1}{x} + \log y\right)\left(\frac{1}{y}\right)$$

or $$qy - 2y^2 = 2f'\left(\frac{1}{x} + \log y\right) \qquad ...(2)$$

From (1) and (2), we have $-px^2 = qy - 2y^2$ or $x^2 p + yq = 2y^2$

which is a partial differential equation of the first order.

Example 4. *Form the partial differential equation by eliminating the arbitrary functions from*

(*i*) $z = f(x + it) + g(x - it)$, *where* $i = \sqrt{-1}$

(*ii*) $z = xf_1(x + t) + f_2(x + t)$

(*iii*) $f(x + y + z, x^2 + y^2 + z^2) = 0.$

Sol. (*i*) Given $z = f(x + it) + g(x - it)$

Differentiating z twice partially w.r.t. x and t, we have

$$\frac{\partial z}{\partial x} = f'(x + it) + g'(x - it)$$

$$\frac{\partial^2 z}{\partial x^2} = f''(x + it) + g''(x - it) \qquad ...(1)$$

$$\frac{\partial z}{\partial t} = if'(x+it) - ig'(x-it)$$

$$\frac{\partial^2 z}{\partial t^2} = i^2 f''(x+it) + i^2 g''(x-it)$$

or $$\frac{\partial^2 z}{\partial t^2} = -f''(x+it) - g''(x-it) \quad ...(2)$$

Adding (1) and (2), we have $\frac{\partial^2 z}{\partial x^2} + \frac{\partial^2 z}{\partial t^2} = 0$

which is a partial differential equation of the second order.

(*ii*) Given $z = xf_1(x+t) + f_2(x+t)$

$$\therefore \quad \frac{\partial z}{\partial x} = f_1(x+t) + xf_1'(x+t) + f_2'(x+t)$$

$$\frac{\partial^2 z}{\partial x^2} = f_1'(x+t) + f_1'(x+t) + xf_1''(x+t) + f_2''(x+t)$$

or $$\frac{\partial^2 z}{\partial x^2} = 2f_1'(x+t) + xf_1''(x+t) + f_2''(x+t) \quad ...(1)$$

$$\frac{\partial z}{\partial t} = xf_1'(x+t) + f_2'(x+t)$$

$$\frac{\partial^2 z}{\partial t^2} = xf_1''(x+t) + f_2''(x+t) \quad ...(2)$$

Subtracting (2) from (1) $\frac{\partial^2 z}{\partial x^2} - \frac{\partial^2 z}{\partial t^2} = 2f_1'(x+t) \quad ...(3)$

Also $$\frac{\partial^2 z}{\partial x \partial t} = f_1'(x+t) + xf_1''(x+t) + f_2''(x+t)$$

$$= \frac{1}{2}\left[\frac{\partial^2 z}{\partial x^2} - \frac{\partial^2 z}{\partial t^2}\right] + \frac{\partial^2 z}{\partial t^2} \quad \text{[Using (2) and (3)]}$$

or $$\frac{\partial^2 z}{\partial x^2} - 2\frac{\partial^2 z}{\partial x \partial t} + \frac{\partial^2 z}{\partial t^2} = 0$$

which is a partial differential equation of the second order.

(*iii*) Let $x + y + z = u$, $x^2 + y^2 + z^2 = v$, then $f(u, v) = 0$

Differentiating partially w.r.t. x and y, we have

$$\frac{\partial f}{\partial u}\left(\frac{\partial u}{\partial x} + \frac{\partial u}{\partial z}p\right) + \frac{\partial f}{\partial v}\left(\frac{\partial v}{\partial x} + \frac{\partial v}{\partial z}p\right) = 0$$

or $$\frac{\partial f}{\partial u}(1+p) + \frac{\partial f}{\partial v}(2x + 2zp) = 0 \quad ...(1)$$

and $$\frac{\partial f}{\partial u}\left(\frac{\partial u}{\partial y} + \frac{\partial u}{\partial z}q\right) + \frac{\partial f}{\partial v}\left(\frac{\partial v}{\partial y} + \frac{\partial v}{\partial z}q\right) = 0$$

or $$\frac{\partial f}{\partial u}(1+q) + \frac{\partial f}{\partial v}(2y+2zq) = 0 \quad ...(2)$$

Eliminating $\frac{\partial f}{\partial u}$ and $\frac{\partial f}{\partial v}$ from (1) and (2), we get

$$(1+p)(2y+2zq) = (1+q)(2x+2zp)$$

or $$(y-z)p + (z-x)q = x - y$$

which is a partial differential equation of the first order.

Example 5. *Eliminate the arbitrary function f from the relation*

$$z = e^{ny} f(x-y). \quad \text{(P.T.U. 2002)}$$

Sol. The given relation is $z = e^{ny} f(x-y)$. ...(1)

Differentiating equation (1) partially w.r.t x and y, respectively we get

$$p = \frac{\partial z}{\partial x} = e^{ny} f'(x-y)\,.\,1 \quad ...(2)$$

and $$q = \frac{\partial z}{\partial y} = ne^{ny} f(x-y) + e^{ny} f'(x-y)\,.\,(-1)$$

$$= ne^{ny} f(x-y) - e^{ny} f'(x-y)$$

$$= nz - p \quad \text{[By equations (1) and (2)]}$$

$$\Rightarrow \quad p + q = nz.$$

This is the required partial differential equation.

Example 6. *Find the partial differential equation by eliminating the arbitrary functions from the following relations:*

(*i*) $z = f(x+ay)$. (*ii*) $z = f(x-at) + \phi(x+at)$.

(*iii*) $z = f(x+at) + x\,F(x+at)$.

Sol. (*i*) Here $z = f(x+ay)$.

Differentiating partially w.r.t x and y, we get

$$\frac{\partial z}{\partial x} = f'(x+ay)\,.\,1 \quad \text{or} \quad p = f'(x+ay) \quad ...(1)$$

and $$\frac{\partial z}{\partial y} = f'(x+ay)\,.\,a \quad \text{or} \quad q/a = f'(x+ay). \quad ...(2)$$

Eliminating arbitrary function from equations (1) and (2), we get

$$p = q/a \quad \text{or} \quad ap = q.$$

This is the required partial differential equation.

(*ii*) Here $z = f(x-at) + \phi(x+at)$.

Differentiating partially w.r.t. x and t, we get

$$\frac{\partial z}{\partial x} = f'(x-at) + \phi'(x+at) \quad ...(1)$$

and $$\frac{\partial z}{\partial t} = -af'(x-at) + a\phi'(x+at) \quad ...(2)$$

Differentiating partially again to equations (1) and (2), we get

$$\frac{\partial^2 z}{\partial x^2} = f''(x-at) + \phi''(x+at) \quad ...(3)$$

and $$\frac{\partial^2 z}{\partial t^2} = a^2 f''(x - at) + a^2 \phi''(x + at). \quad ....(4)$$

Now, from equations (3) and (4), we obtain

$$\frac{\partial^2 z}{\partial t^2} = a^2 \frac{\partial^2 z}{\partial x^2}.$$

This is the required partial differential equation.

(*iii*) Here $z = f(x + at) + x\text{F}(x + at)$.

Differentiating partially w.r.t x and t, we get

$$\frac{\partial z}{\partial x} = f'(x + at) + \text{F}(x + at) + x\text{F}'(x + at)$$

and $$\frac{\partial z}{\partial t} = af'(x + at) + ax\text{F}'(x + at).$$

Again differentiating partially, we get

$$\frac{\partial^2 z}{\partial x^2} = f''(x + at) + 2\text{F}'(x + at) + x\text{F}''(x + at) \quad ...(1)$$

$$\frac{\partial^2 z}{\partial t^2} = a^2 f''(x + at) + a^2 x\text{F}''(x + at) \quad ...(2)$$

and $$\frac{\partial^2 z}{\partial x \partial t} = af''(x + at) + a\text{F}'(x + at) + ax\text{F}''(x + at). \quad ...(3)$$

Now, from equations (1), (2) and (3), we have

$$a^2 \frac{\partial^2 z}{\partial x^2} - 2a \frac{\partial^2 z}{\partial x\, \partial t} + \frac{\partial^2 z}{\partial t^2} = 0$$

or $$\left(a^2 \frac{\partial^2}{\partial x^2} - 2a \frac{\partial^2}{\partial x\, \partial t} + \frac{\partial^2}{\partial t^2}\right) z = 0 \quad \text{or} \quad \left(a \frac{\partial}{\partial x} - \frac{\partial}{\partial t}\right)^2 z = 0.$$

Example 7. *From the partial differential equation by eliminating the arbitrary functions* ϕ *from*

$$f(x + y + z, x^2 + y^2 - z^2) = 0.$$

What is the order of this partial differential equation?

Sol. Given that,

$$f(x + y + z, x^2 + y^2 - z^2) = 0 \quad ...(1)$$

Let $u = x + y + z$ and $v = x^2 + y^2 - z^2$. ...(2)

Then equation (1) reduces to the following form:

$$f(u, v) = 0. \quad ...(3)$$

Differentiating equation (3) partially w.r.t x, we get

$$\frac{\partial f}{\partial u}\left(\frac{\partial u}{\partial x} + p \frac{\partial u}{\partial z}\right) + \frac{\partial f}{\partial v}\left(\frac{\partial v}{\partial x} + p \frac{\partial v}{\partial z}\right) = 0. \quad ...(4)$$

From equation (2),

$$\left.\begin{array}{l}\dfrac{\partial u}{\partial x}=1, \dfrac{\partial u}{\partial z}=1, \dfrac{\partial v}{\partial x}=2x, \\ \dfrac{\partial v}{\partial z}=-2z, \dfrac{\partial u}{\partial y}=1, \dfrac{\partial v}{\partial y}=2y\end{array}\right\}. \qquad \ldots(5)$$

From equations (4) and (5), we have

$$\frac{\partial f}{\partial u}(1+p)+2\frac{\partial f}{\partial v}(x-pz)=0$$

$$\Rightarrow \qquad \frac{\partial f}{\partial u}\Big/\frac{\partial f}{\partial v}=-2(x-pz)/(1+p). \qquad \ldots(6)$$

Again, differentiating (3) partially w.r.t y, we get

$$\frac{\partial f}{\partial u}\left(\frac{\partial u}{\partial y}+q\frac{\partial u}{\partial z}\right)+\frac{\partial f}{\partial v}\left(\frac{\partial v}{\partial y}+q\frac{\partial v}{\partial z}\right)=0$$

or $$\frac{\partial f}{\partial u}(1+q)+2\frac{\partial f}{\partial v}(y-zq)=0 \qquad \text{[by equation (5)]}$$

$$\Rightarrow \qquad \frac{\partial f}{\partial u}\Big/\frac{\partial f}{\partial v}=-2(y-qz)/(1+q) \qquad \ldots(7)$$

Hence eliminating ϕ from equations (6) and (7), we obtain

$$(x-pz)/(1+p)=(y-qz)/(1+q)$$

or $$(1+q)(x-pz)=(1+p)(y-qz) \quad \text{or} \quad (y+z)p-(x+z)q=x-y$$

which is the required partial differential equation of first order.

TEST YOUR KNOWLEDGE

Form the partial differential equations by eliminating the arbitrary constants from the following:

1. $z = ax + by + a^2 + b^2$
2. $(x-a)^2 + (y-b)^2 + z^2 = 1$
3. $z = xy + y\sqrt{x^2 - a^2} + b$
4. $z = ax^2 + bxy + cy^2$
5. $z = axe^y + \frac{1}{2}a^2e^{2y} + b$.
6. Find the differential equation of all planes which are at a constant distance 'a' from the origin.
7. (*a*) Find the differential equation of all spheres whose centres lie on the z-axis.
   (*b*) Find the differential equation of all spheres of radius 3 units having their centres in the xy-plane.

Form partial differential equations by eliminating the arbitrary functions from the following:

8. $z = f(x^2 - y^2)$
9. $z = f(y/x)$
10. $xyz = \phi(x + y + z)$
11. $z = f_1(x)\, f_2(y)$
12. $z = yf(x) + xg(y)$
13. $z = f(x + ay) + g(x - ay)$
14. $z = f(x + 4t) + g(x - 4t)$
15. $z = f_1(y + 2x) + f_2(y - 3x)$
16. $f(xy + z^2, x + y + z) = 0$
17. $f(x^2 + y^2, z - xy) = 0$

Answers

1. $z = px + qy + p^2 + q^2$
2. $z^2(p^2 + q^2 + 1) = 1$
3. $px + qy = pq$
4. $x^2r + 2xys + y^2t = 2z$
5. $q = px + p^2$
6. $z = px + qy \pm a\sqrt{1 + p^2 + q^2}$
7. (*a*) $py - qx = 0$ (*b*) $z^2(p^2 + q^2 + 1) = 9$

8. $py + qx = 0$ **9.** $px + qy = 0$ **10.** $x(y - z)p + y(z - x)q = z(x - y)$

11. $z \frac{\partial^2 z}{\partial x \partial y} - \frac{\partial z}{\partial x} \cdot \frac{\partial z}{\partial y} = 0$ **12.** $xy \frac{\partial^2 z}{\partial x \partial y} = x\frac{\partial z}{\partial x} + y\frac{\partial z}{\partial y} - z$ **13.** $\frac{\partial^2 z}{\partial y^2} = a^2 \frac{\partial^2 z}{\partial x^2}$

14. $16 \frac{\partial^2 z}{\partial x^2} - \frac{\partial^2 z}{\partial t^2} = 0$ **15.** $\frac{\partial^2 z}{\partial x^2} + \frac{\partial^2 z}{\partial x \partial y} - 6\frac{\partial^2 z}{\partial y^2} = 0$ **16.** $p(x - 2z) + q(2z - y) = y - x$

17. $py - qx = y^2 - x^2$.

3.3. EQUATIONS SOLVABLE BY DIRECT INTEGRATION

Those equations which contain **only one** partial derivative can be solved by direct integration. However, in place of the constants of integration, we must use arbitrary functions of the variable kept constant.

ILLUSTRATIVE EXAMPLES

Example 1. *Solve* $\frac{\partial^2 z}{\partial x \partial y} = \sin x \sin y$, *given that* $\frac{\partial z}{\partial y} = -2 \sin y$ *when* $x = 0$ *and* $z = 0$ *when* y *is an odd multiple of* $\frac{\pi}{2}$.

Sol. Given $$\frac{\partial^2 z}{\partial x \partial y} = \sin x \sin y$$

Integrating w.r.t. x, keeping y as constant, we have

$$\frac{\partial z}{\partial y} = -\cos x \sin y + f(y) \quad ...(1)$$

When $x = 0$, $$\frac{\partial z}{\partial y} = -2 \sin y \quad | \text{ given}$$

$\therefore$ $-2 \sin y = -\sin y + f(y)$ or $f(y) = -\sin y$

$\therefore$ From (1), $$\frac{\partial z}{\partial y} = -\cos x \sin y - \sin y$$

Integrating w.r.t. y, keeping x as constant, we have

$$z = \cos x \cos y + \cos y + \phi(x) \quad ...(2)$$

When y is an odd multiple of $\frac{\pi}{2}$, $z = 0$ | given

$\therefore$ $0 = 0 + 0 + \phi(x)$, since $\cos(2n + 1)\frac{\pi}{2} = 0$ or $\phi(x) = 0$

$\therefore$ From (2), $z = (1 + \cos x) \cos y$ which is the required particular solution.

Example 2. *Solve* $\frac{\partial^2 z}{\partial x^2} + z = 0$, *given that when* $x = 0$, $z = e^y$ *and* $\frac{\partial z}{\partial x} = 1$.

Sol. If z were a function of **x alone**, the solution would have been $z = c_1 \cos x + c_2 \sin x$, where c_1 and c_2 are arbitrary constants.

But here z is a function of x and y, therefore, c_1 and c_2 are arbitrary functions of y, the independent variable kept constant.

$\therefore$ The solution of the given equation is

$$z = f(y) \cos x + \phi(y) \sin x \quad \text{and} \quad \frac{\partial z}{\partial x} = -f(y) \sin x + \phi(y) \cos x$$

When $x = 0$, $z = e^y$ $\quad\therefore\quad e^y = f(y)$

When $x = 0$, $\frac{\partial z}{\partial x} = 1$ $\quad\therefore\quad 1 = \phi(y)$

$\therefore$ The required particular solution is $z = e^y \cos x + \sin x$.

TEST YOUR KNOWLEDGE

Solve the following equations:

1. $\frac{\partial^2 z}{\partial x^2} = xy$
2. $\frac{\partial^2 z}{\partial x \partial y} = \frac{1}{xy}$
3. $\frac{\partial^2 z}{\partial x \partial y} = 2x + 2y$
4. $\frac{\partial^2 z}{\partial y^2} = \sin(xy)$
5. $\frac{\partial^2 u}{\partial y \partial x} = 4x \sin(3xy)$
6. $\log \left[\frac{\partial^2 z}{\partial x \partial y}\right] = x + y$
7. $\frac{\partial^3 z}{\partial x^2 \partial y} = \cos(2x + 3y)$
8. $\frac{\partial^2 u}{\partial x \partial t} = e^{-t} \cos x$
9. $\frac{\partial^2 z}{\partial x^2} = a^2 z$ given that when $x = 0$, $\frac{\partial z}{\partial x} = a \sin y$ and $\frac{\partial z}{\partial y} = 0$.

Answers

1. $z = \frac{1}{6} x^3 y + x f(y) + \phi(y)$
2. $z = \log x \log y + g(y) + \phi(x)$
3. $z = xy(x + y) + g(y) + \phi(x)$
4. $z = -\frac{1}{x^2} \sin xy + y f(x) + \phi(x)$
5. $u = -\frac{4}{9y} \sin(3xy) + f(x) + \phi(y)$
6. $z = e^{x+y} + g(y) + \phi(x)$
7. $z = f(x) + x\phi(y) + g(y) - \frac{1}{12} \sin(2x + 3y)$
8. $u = -e^{-t} \sin x + f(x) + g(t)$
9. A $\cosh ax + \sin y \sinh ax$.

3.4. LINEAR PARTIAL DIFFERENTIAL EQUATIONS OF THE FIRST ORDER

A differential equation involving first order partial derivatives p and q only is called a *partial differential equation of the first order*. If p and q both occur in the first degree only and are not multiplied together, then it is called a *linear partial differential equation of the first order*.

3.5. LAGRANGE'S LINEAR EQUATION

The partial differential equation of the form $Pp + Qq = R$, where P, Q and R are functions of x, y, z is the standard form of a linear partial differential equation of the first order and is called *Lagrange's Linear Equation*.

Now Lagrange's linear equation $\quad Pp + Qq = R \quad$ ...(1)

is obtained by eliminating an arbitrary function ϕ from $\quad \phi(u, v) = 0 \quad$ ...(2)

where u, v are some definite functions of x, y, z.

Differentiating (2) partially w.r.t. x and y

$$\frac{\partial\phi}{\partial u}\left(\frac{\partial u}{\partial x}+\frac{\partial u}{\partial z}p\right)+\frac{\partial\phi}{\partial v}\left(\frac{\partial v}{\partial x}+\frac{\partial v}{\partial z}p\right)=0$$

and
$$\frac{\partial\phi}{\partial u}\left(\frac{\partial u}{\partial y}+\frac{\partial u}{\partial z}q\right)+\frac{\partial\phi}{\partial v}\left(\frac{\partial v}{\partial y}+\frac{\partial v}{\partial z}q\right)=0$$

Eliminating $\frac{\partial\phi}{\partial u}$ and $\frac{\partial\phi}{\partial v}$, we get

$$\left(\frac{\partial u}{\partial x}+\frac{\partial u}{\partial z}p\right)\left(\frac{\partial v}{\partial y}+\frac{\partial v}{\partial z}q\right)-\left(\frac{\partial v}{\partial x}+\frac{\partial v}{\partial z}p\right)\left(\frac{\partial u}{\partial y}+\frac{\partial u}{\partial z}q\right)=0$$

or
$$\left(\frac{\partial u}{\partial y}.\frac{\partial v}{\partial z}-\frac{\partial u}{\partial z}.\frac{\partial v}{\partial y}\right)p+\left(\frac{\partial u}{\partial z}.\frac{\partial v}{\partial x}-\frac{\partial u}{\partial x}.\frac{\partial v}{\partial z}\right)q=\frac{\partial u}{\partial x}.\frac{\partial v}{\partial y}-\frac{\partial u}{\partial y}.\frac{\partial v}{\partial x}$$

which is the same as (1) with

$$\mathrm{P}=\frac{\partial u}{\partial y}.\frac{\partial v}{\partial z}-\frac{\partial u}{\partial z}.\frac{\partial v}{\partial y},\quad \mathrm{Q}=\frac{\partial u}{\partial z}.\frac{\partial v}{\partial x}-\frac{\partial u}{\partial x}.\frac{\partial v}{\partial z},\quad \mathrm{R}=\frac{\partial u}{\partial x}.\frac{\partial v}{\partial y}-\frac{\partial u}{\partial y}.\frac{\partial v}{\partial x}.$$

To determine u, v from P, Q, R, suppose $u = a$ and $v = b$, where a, b are constants, so that

$$\frac{\partial u}{\partial x}dx+\frac{\partial u}{\partial y}dy+\frac{\partial u}{\partial z}dz=du=0$$

$$\frac{\partial v}{\partial x}dx+\frac{\partial v}{\partial y}dy+\frac{\partial v}{\partial z}dz=dv=0$$

By cross-multiplication, we have

$$\frac{dx}{\frac{\partial u}{\partial y}.\frac{\partial v}{\partial z}-\frac{\partial u}{\partial z}.\frac{\partial v}{\partial y}}=\frac{dy}{\frac{\partial u}{\partial z}.\frac{\partial v}{\partial x}-\frac{\partial u}{\partial x}.\frac{\partial v}{\partial z}}=\frac{dz}{\frac{\partial u}{\partial x}.\frac{\partial v}{\partial y}-\frac{\partial u}{\partial y}.\frac{\partial v}{\partial x}}$$

or
$$\frac{dx}{\mathrm{P}}=\frac{dy}{\mathrm{Q}}=\frac{dz}{\mathrm{R}} \qquad ...(3)$$

The solutions of these equations are $u = a$ and $v = b$. Thus determining u, v from the simultaneous equations (3), we have the solution of the partial differential equation.

$$\mathrm{P}p+\mathrm{Q}q=\mathrm{R} \quad \text{as} \quad \phi(u, v)=0 \quad \text{or} \quad u=f(v).$$

Note. Equations (3) are called **Lagrange's auxiliary equations** or **subsidiary equations.**

3.6. WORKING METHOD

To solve the equation $\mathrm{P}p + \mathrm{Q}q = \mathrm{R}$

(*i*) form the auxiliary equations $\frac{dx}{\mathrm{P}}=\frac{dy}{\mathrm{Q}}=\frac{dz}{\mathrm{R}}$

(*ii*) solve the auxiliary equations by the method of grouping or the method of multipliers or both to get two independent solutions $u = a$ and $v = b$, where a, b are arbitrary constants.

(*iii*) then $\phi(u, v) = 0$ or $u = f(v)$ is the general solution of the equation $\mathrm{P}p + \mathrm{Q}q = \mathrm{R}$.

ILLUSTRATIVE EXAMPLES

Example 1. *Solve the following differential equations:*

(*i*) $\dfrac{y^2 z}{x} p + xzq = y^2$ (*ii*) $pz - qz = z^2 + (x + y)^2$

(*iii*) $(x^2 - yz)p + (y^2 - zx)q = z^2 - xy$.

Sol. (*i*) The given equation can be written as $y^2zp + x^2zq = xy^2$

Comparing with $Pp + Qq = R$, we have $P = y^2z$, $Q = x^2z$, $R = xy^2$

$\therefore$ The auxiliary equations are $\dfrac{dx}{y^2 z} = \dfrac{dy}{x^2 z} = \dfrac{dz}{xy^2}$

Taking the first two members, we have $x^2dx = y^2dy$

which on integration gives $x^3 - y^3 = a$ ...(1)

Again taking the first and third members, we have $xdx = zdz$

which on integration gives $x^2 - z^2 = b$ ...(2)

From (1) and (2), the general solution is $\phi(x^3 - y^3, x^2 - z^2) = 0$.

(*ii*) Here the auxiliary equations are $\dfrac{dx}{z} = \dfrac{dy}{-z} = \dfrac{dz}{z^2 + (x + y)^2}$

Taking the first two members, we have $dx + dy = 0$

which on integration gives $x + y = a$ ...(1)

Again taking the first and third members, we have

$$dx = \frac{zdz}{z^2 + a^2}, \text{ since } x + y = a$$

or
$$\frac{2z\,dz}{z^2 + a^2} = 2dx$$

which on integration gives $\log (z^2 + a^2) = 2x + b$ or $\log [z^2 + (x + y)^2] - 2x = b$ ...(2)

From (1) and (2), the general solution is

$$\phi[x + y, \log (x^2 + y^2 + z^2 + 2xy) - 2x] = 0.$$

(*iii*) Here the auxiliary equations are

$$\frac{dx}{x^2 - yz} = \frac{dy}{y^2 - zx} = \frac{dz}{z^2 - xy} \quad ...(1)$$

$$\therefore \quad \frac{dx - dy}{(x - y)(x + y + z)} = \frac{dy - dz}{(y - z)(x + y + z)}$$

or
$$\frac{dx - dy}{x - y} = \frac{dy - dz}{y - z}$$

which on integration gives $\log (x - y) = \log (y - z) + \log a$

or
$$\log \left(\frac{x - y}{y - z}\right) = \log a \quad \text{or} \quad \frac{x - y}{y - z} = a \quad ...(2)$$

Using x, y, z as multipliers,

each fraction of (1) $= \dfrac{x\,dx + y\,dy + z\,dz}{x^3 + y^3 + z^3 - 3xyz}$

$$= \frac{x\,dx + y\,dy + z\,dz}{(x + y + z)(x^2 + y^2 + z^2 - xy - yz - zx)} \qquad ...(3)$$

Also, each fraction of (1)

$$= \frac{dx + dy + dz}{x^2 + y^2 + z^2 - xy - yz - zx} \qquad ...(4)$$

From (3) and (4), $\dfrac{x\,dx + y\,dy + z\,dz}{x + y + z} = dx + dy + dz$

or $xdx + ydy + zdz = (x + y + z)\, d(x + y + z)$

Integrating $\dfrac{x^2}{2} + \dfrac{y^2}{2} + \dfrac{z^2}{2} = \dfrac{(x + y + z)^2}{2} + c$

or $x^2 + y^2 + z^2 = (x + y + z)^2 + 2c$

or $2xy + 2yz + 2zx + 2c = 0$

or $xy + yz + zx = b,$ where $b = -c$ ...(5)

From (2) and (5), the general solution is

$$\phi\left(\frac{x - y}{y - z},\, xy + yz + zx\right) = 0.$$

Example 2. *Solve the following differential equations:*

(i) $(mz - ny)p + (nx - lz)q = ly - mx$

(ii) $x^2(y - z)p + y^2(z - x)q = z^2(x - y).$

Sol. (*i*) Here the auxiliary equations are $\dfrac{dx}{mz - ny} = \dfrac{dy}{nx - lz} = \dfrac{dz}{ly - mx}$

Using x, y, z as multipliers, we get

each fraction $= \dfrac{x\,dx + y\,dy + z\,dz}{0}$

$\therefore$ $xdx + ydy + zdz = 0$

which on integration gives $x^2 + y^2 + z^2 = a$ ...(1)

Again using l, m, n as multipliers, we get

each fraction $= \dfrac{ldx + mdy + ndz}{0}$

$\therefore$ $ldx + mdy + ndz = 0$

which on integration gives $lx + my + nz = b$ ...(2)

From (1) and (2), the general solution is

$$x^2 + y^2 + z^2 = f(lx + my + nz).$$

(*ii*) Here the auxiliary equations are

$$\frac{dx}{x^2(y - z)} = \frac{dy}{y^2(z - x)} = \frac{dz}{z^2(x - y)}$$

Using $\frac{1}{x^2}, \frac{1}{y^2}, \frac{1}{z^2}$ as multipliers, we get

$$\text{each fraction} = \frac{\frac{1}{x^2}dx + \frac{1}{y^2}dy + \frac{1}{z^2}dz}{0}$$

$$\therefore \quad \frac{dx}{x^2} + \frac{dy}{y^2} + \frac{dz}{z^2} = 0$$

which on integration gives $\frac{1}{x} + \frac{1}{y} + \frac{1}{z} = a$ ...(1)

Again using $\frac{1}{x}, \frac{1}{y}, \frac{1}{z}$ as multipliers, we get

$$\text{each fraction} = \frac{\frac{1}{x}dx + \frac{1}{y}dy + \frac{1}{z}dz}{0}$$

$$\therefore \quad \frac{dx}{x} + \frac{dy}{y} + \frac{dz}{z} = 0$$

which on integration gives $\log x + \log y + \log z = \log b$

or $\log (xyz) = \log b$ or $xyz = b$ ...(2)

From (1) and (2), the general solution is $\phi\left(\frac{1}{x} + \frac{1}{y} + \frac{1}{z}, xyz\right) = 0.$

Example 3. *Solve the following differential equations:*

(*i*) $(x^2 - y^2 - z^2)p + 2xyq = 2xz$

(*ii*) $(z^2 - 2yz - y^2)p + (xy + zx)q = xy - zx.$

Sol. (*i*) Here the auxiliary equations are

$$\frac{dx}{x^2 - y^2 - z^2} = \frac{dy}{2xy} = \frac{dz}{2xz}$$

Taking the last two members, we have

$$\frac{dy}{y} = \frac{dz}{z}$$

which on integration gives $\log y = \log z + \log a$

or $\log \frac{y}{z} = \log a$ or $\frac{y}{z} = a$ ...(1)

Using x, y, z as multipliers, we get

$$\text{each fraction} = \frac{x\,dx + y\,dy + z\,dz}{x(x^2 + y^2 + z^2)}$$

$$\therefore \quad \frac{x\,dx + y\,dy + z\,dz}{x(x^2 + y^2 + z^2)} = \frac{dz}{2xz} \quad \text{or} \quad \frac{2x\,dx + 2y\,dy + 2z\,dz}{x^2 + y^2 + z^2} = \frac{dz}{z}$$

which on integration gives $\log (x^2 + y^2 + z^2) = \log z + \log b$

or $$\log\left(\frac{x^2+y^2+z^2}{z}\right)=\log b \quad \text{or} \quad \frac{x^2+y^2+z^2}{z}=b \qquad \text{...(2)}$$

From (1) and (2), the general solution is

$$\frac{x^2+y^2+z^2}{z}=f\left(\frac{y}{z}\right) \quad \text{or} \quad x^2+y^2+z^2=z\,f\left(\frac{y}{z}\right).$$

(*ii*) Here the auxiliary equations are

$$\frac{dx}{z^2-2yz-y^2}=\frac{dy}{xy+zx}=\frac{dz}{xy-zx}$$

Taking x, y, z as multipliers, we have

$$\text{each fraction}=\frac{x\,dx+y\,dy+z\,dz}{0}$$

$\therefore$ $x\,dx+y\,dy+z\,dz=0$

which on integration gives $x^2+y^2+z^2=a$ ...(1)

Again, taking the last two members, we have

$$\frac{dy}{y+z}=\frac{dz}{y-z} \quad \text{or} \quad (y-z)dy=(y+z)dz$$

or $$y\,dy-(z\,dy+y\,dz)-z\,dz=0 \quad \text{or} \quad y\,dy-d(yz)-z\,dz=0$$

which on integration gives $y^2-2yz-z^2=b$ ...(2)

From (1) and (2), the general solution is

$$\phi(x^2+y^2+z^2,\ y^2-2yz-z^2)=0.$$

Example 4. *Solve:* $x(y^2-z^2)\frac{\partial z}{\partial x}+y(z^2-x^2)\frac{\partial z}{\partial y}=z(x^2-y^2).$

Sol. The given equation is

$$x(y^2-z^2)\,p+y(z^2-x^2)\,q=z\,(x^2-y^2)$$

The subsidiary equations are

$$\frac{dx}{x(y^2-z^2)}=\frac{dy}{y(z^2-x^2)}=\frac{dz}{z(x^2-y^2)}$$

Using multipliers x, y, z ; each member

$$=\frac{x\,dx+y\,dy+z\,dz}{x^2(y^2-z^2)+y^2(z^2-x^2)+z^2(x^2-y^2)}$$

$$=\frac{x\,dx+y\,dy+z\,dz}{0}$$

$\therefore$ $x\,dx+y\,dy+z\,dz=0$

Integrating $$\frac{x^2}{2}+\frac{y^2}{2}+\frac{z^2}{2}=\frac{a}{2}$$

$$x^2+y^2+z^2=a \qquad \text{...(1)}$$

Using multipliers $\frac{1}{x}, \frac{1}{y}, \frac{1}{z}$; each member

$$= \frac{\frac{1}{x}\,dx + \frac{1}{y}\,dy + \frac{1}{z}\,dz}{(y^2 - z^2) + (z^2 - x^2) + (x^2 - y^2)} = \frac{\frac{1}{x}\,dx + \frac{1}{y}\,dy + \frac{1}{z}\,dz}{0}$$

$\therefore \quad \frac{1}{x}\,dx + \frac{1}{y}\,dy + \frac{1}{z}\,dz = 0.$

Integrating $\log x + \log y + \log z = \log b$

or $\qquad \log (xyz) = \log b \quad \text{or} \quad xyz = b \qquad$ ...(2)

From (1) and (2), the general solution is

$$f(x^2 + y^2 + z^2, xyz) = 0.$$

Example 5. *Solve:* $xp - yq + x^2 - y^2 = 0.$

Sol. The given equation

$$xp - yq = y^2 - x^2$$

The subsidiary equations are

$$\frac{dx}{x} = \frac{dy}{-y} + \frac{dz}{y^2 - x^2}$$

From the first two members, we have

$$\frac{dx}{x} + \frac{dy}{y} = 0$$

Integrating $\log x + \log y = \log a$

or $\qquad \log (xy) = \log a \quad \text{or} \quad xy = a \qquad$ ...(1)

Using multipliers $x, y, 1$; each member

$$= \frac{x\,dx + y\,dy + dz}{x^2 - y^2 + y^2 - x^2} = \frac{x\,dx + y\,dy + dz}{0}$$

$\therefore \qquad x\,dx + y\,dy + dz = 0$

Integrating $\qquad \frac{x^2}{2} + \frac{y^2}{2} + z = b \qquad$ ...(2)

From (1) and (2), the general solution is

$$f\left(\frac{x^2}{2} + \frac{y^2}{2} + z, xy\right) = 0.$$

Example 6. *Solve:* $(z^2 - 2yz - y^2)p + (xy + zx)q = xy - zx.$

Sol. The subsidiary equations are

$$\frac{dx}{z^2 - 2yz - y^2} = \frac{dy}{xy + zx} = \frac{dz}{xy - zx}.$$

Using multipliers x, y, z ; each member

$$= \frac{x\,dx + y\,dy + z\,dz}{0}$$

∴ $$x\,dx + y\,dy + z\,dz = 0$$

Integrating $$\frac{x^2}{2} + \frac{y^2}{2} + \frac{z^2}{2} = \frac{a}{2} \quad \text{or} \quad x^2 + y^2 + z^2 = a \qquad ...(1)$$

From the last two members, we have

$$\frac{dy}{x(y+z)} = \frac{dz}{x(y-z)}$$

or $$(y - z)\,dy = (y + z)\,dz \quad \text{or} \quad y\,dy - (z\,dy + y\,dz) - z\,dz = 0$$

Integrating $$\frac{y^2}{2} - yz - \frac{y^2}{2} = \frac{b}{2}$$

or $$y^2 - 2yz - z^2 = b \qquad ...(2)$$

From (1) and (2), the general solution is

$$f(x^2 + y^2 + z^2, y^2 - 2yz - z^2) = 0.$$

Example 7. *Solve:* $px\,(z - 2y^2) = (z - qy)(z - y^2 - 2x^3)$.

Sol. The given equation can be written as

$$px\,(z - 2y^2) = z(z - y^2 - 2x^3) - qy(z - y^2 - 2x^3)$$

or $$x(z - 2y^2)\,p + y(z - y^2 - 2x^3)q = z(z - y^2 - 2x^3)$$

The subsidiary equations are

$$\frac{dx}{x(z-2y^2)} = \frac{dy}{y(z-y^2-2x^3)} = \frac{dz}{z(z-y^2-2x^3)}$$

From the last two members, we have

$$\frac{dy}{y} = \frac{dz}{z}$$

Integrating $$\log y = \log z + \log a$$

or $$\log \frac{y}{z} = \log a \quad \text{or} \quad \frac{y}{z} = a \qquad ...(1)$$

Taking the first and third members, we have

$$z(z - y^2 - 2x^3)\,dx = x(z - 2y^2)dz$$

Putting $y = az$ [From (1)]

$$z(z - a^2z^2 - 2x^3)dx = x(z - 2a^2z^2)\,dz$$

or $$(z - a^2z^2 - 2x^3)dx = x(1 - 2a^2z)\,dz$$

or $$(xdz - zdx) - a^2(2x\,zdz - z^2\,dx) + 2x^3\,dx = 0$$

Dividing by x^2, we get

$$\left(\frac{xdz - zdx}{x^2}\right) - a^2\left(\frac{2x\,zdz - z^2\,dx}{x^2}\right) + 2x\,dx = 0$$

or $$d\left(\frac{z}{x}\right) - a^2 d\left(\frac{z^2}{x}\right) + d(x^2) = 0$$

Integrating $$\frac{z}{x} - a^2\left(\frac{z^2}{x}\right) + x^2 = b$$

or $$\frac{z}{x} - \frac{y^2}{z^2} \cdot \frac{z^2}{x} + x^2 = b \qquad \left(\because\ a = \frac{y}{z}\right)$$

or $$\frac{z}{x} - \frac{y^2}{x} + x^2 = b \qquad ...(2)$$

From (1) and (2), the general solution is

$$\frac{z}{x} - \frac{y^2}{x} + x^2 = f\left(\frac{y}{z}\right).$$

Example 8. *Solve* $x\frac{\partial u}{\partial x} + y\frac{\partial u}{\partial y} + z\frac{\partial u}{\partial z} = xyz$.

Sol.[In the differential equation Pp + Qq = R, there are **two independent variables** x and y. To solve it, we have to find **two independent solutions** satisfying the equation $\frac{dx}{P} = \frac{dy}{Q} = \frac{dz}{R}$. This procedure can be extended to linear partial differential equations of the first order involving more than two independent variables.

If $u_1 = c_1$, $u_2 = c_2$, ..., $u_n = c_n$ are n independent solutions of the equations

$$\frac{dx_1}{P_1} = \frac{dx_2}{P_2} = \text{......} = \frac{dx_n}{P_n} = \frac{dz}{R}$$

then the general solution of the partial differential equation

$$P_1\frac{\partial z}{\partial x_1} + P_2\frac{\partial z}{\partial x_2} + \text{......} + P_n\frac{\partial z}{\partial x_n} = R$$

is $\phi(u_1, u_2, \text{......}, u_n) = 0$, where ϕ is an arbitrary function.]

Here the auxiliary equations are $$\frac{dx}{x} = \frac{dy}{y} = \frac{dz}{z} = \frac{du}{xyz}$$

Taking first and second members, we have

$$\frac{dx}{x} = \frac{dy}{y}$$

which on integration gives $$\log x = \log y + \log c_1$$

or $$\log\frac{x}{y} = \log c_1 \text{ or } \frac{x}{y} = c_1 \qquad ...(1)$$

Similarly, taking second and third members, we have

$$\frac{y}{z} = c_2 \qquad ...(2)$$

Also, we have $$\frac{yz\,dx + zx\,dy + xy\,dz}{3xyz} = \frac{du}{xyz}$$

or $$d(xyz) = 3du$$

which on integration gives $$xyz - 3u = c_3 \quad ...(3)$$

From (1), (2) and (3), the general solution is $\phi\left(\frac{x}{y}, \frac{y}{z}, xyz - 3u\right) = 0.$

Example 9. *Solve: $xzp + yzq = xy$.*

Sol. Here the Lagrange's subsidiary equations are

$$\frac{dx}{xz} = \frac{dy}{zy} = \frac{dz}{xy}.$$

Taking first two members, we get

$$\frac{dx}{x} = \frac{dy}{y}; \quad \therefore \quad \frac{x}{y} = c_1 \quad ...(1)$$

From the last two members, we get

$$\frac{dy}{z} = \frac{dz}{x} \quad \text{or} \quad \frac{dy}{z} = \frac{dz}{c_1 y} \quad \text{[by equation (1)]}$$

or $$c_1 y dy - z dz = 0; \quad \therefore \quad c_1 y^2 - z^2 = c_2$$

or $$\frac{x}{y} \cdot y^2 - z^2 = c_2 \quad \text{or} \quad xy - z^2 = c_2. \quad ...(2)$$

Hence the required general solution is

$$\phi\left(\frac{x}{y}, xy - z^2\right) = 0$$

where ϕ is an arbitrary function.

Example 10. *Solve: $yzp + zxq = xy$.*

Sol. Given $yzp + zxq = xy$. ...(1)

Comparing (1) with the standard from $Pp + Qq = R$, we have

$$P = yz, Q = zx, R = xy.$$

$\therefore$ The Lagrange's auxiliary equations for (1) are

$$\frac{dx}{P} = \frac{dy}{Q} = \frac{dz}{R} \quad i.e., \quad \frac{dx}{yz} = \frac{dy}{zx} = \frac{dz}{xy} \quad ...(2)$$

Taking first two members of (2), we have

$x\,dx - y\,dy = 0$. Integrating $x^2 - y^2 = a$.

Again taking first and last members of (2), we have

$x\,dx - z\,dz = 0$. Integrating $x^2 - z^2 = b$.

Thus two independent solutions of (2) are

$$x^2 - y^2 = a \quad \text{and} \quad x^2 - z^2 = b$$

Hence the required general integral (or solution) of (1) is

$$\phi(x^2 - y^2, x^2 - z^2) = 0$$

where ϕ is an arbitrary function.

Example 11. *Solve:* $x(y^2 + z)\,p - y\,(x^2 + z)q = z(x^2 - y^2)$.

Sol. The lagrange's subsidiary equations of the given partial differential equation are

$$\frac{dx}{x(y^2+z)} = \frac{dy}{-y\,(x^2+z)} = \frac{dz}{z(x^2-y^2)} \quad ...(1)$$

Each ratio of equation (1) = $\dfrac{x\,dx + y\,dy - dz}{0} = \dfrac{\frac{dx}{x}+\frac{dy}{y}+\frac{dz}{z}}{0}$.

Hence $2xdx + 2ydy - 2dz = 0$,

$$\frac{dx}{x}+\frac{dy}{y}+\frac{dz}{z} = 0 \quad \text{or} \quad d\,(x^2 + y^2 - 2z) = 0, \text{ and } d\,(\log\,(xyz)) = 0$$

Integrating, $x^2 + y^2 - 2z = a$ and $\log\,(xyz) = \log b$

or $x^2 + y^2 - 2z = a$ and $xyz = b$.

Hence the required general solution is

$$\phi\,(x^2 + y^2 - 2z,\ xyz) = 0.$$

Example 12. *Solve:* $(z^2 - 2yz - y^2)\,p + (xy + xz)\,q = xy - xz$.

Sol. Here Lagrange's auxiliary equations are:

$$\frac{dx}{z^2-2yz-y^2} = \frac{dy}{xy+zx} = \frac{dz}{xy-zx}.$$

Using x, y, z as multipliers, each fraction of (1) is equal to

$$\frac{x\,dx + y\,dy + z\,dz}{0}.$$

Thus $x\,dx + y\,dy + z\,dz = 0$.

Integrating, $x^2 + y^2 + z^2 = c_1$. ...(2)

Again, taking last two members of (1), we have

$$\frac{dy}{x(y+z)} = \frac{dz}{x(y-z)} \quad \text{or} \quad \frac{dy}{y+z} = \frac{dz}{y-z}$$

or $(y - z)\,dy = (y + z)\,dz$ or $y\,dy - (z\,dy + y\,dz) - z\,dz = 0$.

Integrating, $\frac{1}{2}y^2 - yz - \frac{1}{2}z^2 = c_2/2$ or $y^2 - 2yz - z^2 = c_2$. ...(3)

Hence the general solution of the given equation is

$$\phi(x^2 + y^2 + z^2,\ y^2 - 2yz - z^2) = 0.$$

Example 13. *Solve:* $p \tan x + q \tan y = \tan z$.

Sol. Here the Lagrange's subsidiary equations are

$$\frac{dx}{\tan x} = \frac{dy}{\tan y} = \frac{dz}{\tan z} \quad ...(1)$$

From the first two fractions of equation (1),

$$\frac{dx}{\tan x} = \frac{dy}{\tan y}, \quad i.e., \quad \cot x\,dx = \cot y\,dy.$$

Integrating, $\log \sin x = \log \sin y + \log c_1$; $\therefore$ $\dfrac{\sin x}{\sin y} = c_1$ ...(2)

Again, from last two fractions, we have

$$\frac{dy}{\tan y} = \frac{dz}{\tan z}; \quad i.e., \quad \cot y \, dy = \cot z \, dz.$$

Integrating, $\log \sin y = \log \sin z + \log c_2$; $\therefore$ $\dfrac{\sin y}{\sin z} = c_2$. ..(3)

Hence from equations (2) and (3), the required solution of the given equation is

$$\frac{\sin x}{\sin y} = \phi\left(\frac{\sin y}{\sin z}\right).$$

Example 14. *Solve:* $y^2p - xyq = x(z - 2y)$. [CSVTU Bhilai B.E. (Third Sem.) 2006]

Sol. Here the Lagrange's subsidiary equations are

$$\frac{dx}{y^2} = \frac{dy}{-xy} = \frac{dz}{x(x-2y)} \quad ...(1)$$

Taking first two fractions of equation (1) and then solving, we get

$$2x\,dx + 2y\,dy = 0.$$

$\therefore$ $x^2 + y^2 = c_1$. ...(2)

Now, taking last two fractions and then solving we get

$$\frac{dz}{dy} = -\frac{z-2y}{y} \quad \text{or} \quad \frac{dz}{dy} + \frac{1}{y}z = 2 \quad ...(3)$$

which is a linear equation is z.

Hence its

$$\text{I.F.} = e^{\int (1/y)dy} = e^{\log y} = y.$$

Hence the solution of equation (3) is

$$zy = \int 2y\,dy + c_2 \quad \text{or} \quad zy - y^2 = c_2. \quad ...(4)$$

Hence the required general solution is

$$\phi(x^2 + y^2, zy - y^2) = 0$$

where ϕ is an arbitrary function.

Example 15. *Solve:* $(mz - ny)\dfrac{\partial z}{\partial x} + (nx - lz)\dfrac{\partial z}{\partial y} = ly - mx$.

Sol. Here the subsidiary equations are

$$\frac{dx}{mz - ny} = \frac{dy}{nx - lz} = \frac{dz}{ly - mx}.$$

Using x, y, z as multipliers, each fraction $= \dfrac{xdx + ydy + zdz}{0}$

$\therefore$ $xdx + ydy + zdz = 0$.

Integrating, $x^2 + y^2 + z^2 = c_1$.

Again using l, m, n as multipliers, each fraction $= \dfrac{ldx + mdy + ndz}{0}$.

$\therefore \quad ldx + mdy + ndz = 0$

Integrating, $\quad lx + my + nz = c_2$.

Hence the required general solution is

$$\phi(lx + my + nz, x^2 + y^2 + z^2) = 0.$$

Example 16. *Solve: $x(y - z)p + y(z - x)q = z(x - y)$.*

Sol. The given equation can be written as

$$(xy - zx)p + (yz - xy)q = zx - yz. \qquad ...(1)$$

Hence the subsidiary equations of equation (1) are

$$\frac{dx}{xy - zx} = \frac{dy}{yz - xy} = \frac{dz}{xz - yz}.$$

Now using 1, 1, 1 as multipliers, each fraction $= \dfrac{dx \,.\, dy + dz}{0}$.

$\therefore \quad dx + dy + dz = 0.$

Integrating, $\quad x + y + z = c_1$.

Again using $1/x$, $1/y$, $1/z$ as multiplies, each fraction

$$= \frac{\frac{1}{x}dx + \frac{1}{y}dy + \frac{1}{z}dz}{0}.$$

$\therefore \quad \dfrac{1}{x}dx + \dfrac{1}{y}dy + \dfrac{1}{z}dz = 0$ or $d(\log x + \log y + \log z) = 0$.

Integrating, $\log x + \log y + \log z = \log c_2$.

$\therefore \quad xyz = c_2$.

Hence the required general solution is

$$\phi(x + y + z, xyz) = 0.$$

Example 17. *Solve: $p \cos(x + y) + q \sin(x + y) = z$.*

Sol. Here Lagrange's subsidiary equations are

$$\frac{dx}{\cos(x + y)} = \frac{dy}{\sin(x + y)} = \frac{dz}{z} \qquad ...(1)$$

$$\text{Each fraction of (1)} = \frac{dx - dy}{\cos(x + y) - \sin(x + y)} = \frac{dx - dy}{\cos(x + y) - \sin(x + y)} \qquad ...(2)$$

Taking $\quad \dfrac{dx + dy}{\cos(x + y) + \sin(x + y)} = \dfrac{dx - dy}{\cos(x + y) - \sin(x + y)}$

or $\quad \dfrac{\cos(x + y) - \sin(x + y)}{cs(x + y) + \sin(x + y)}(dx + dy) = dx - dy.$

Integrating, we get

$$\log[\cos(x + y) + \sin(x + y)] = (x - y) + \log c_1$$

or $$\cos(x + y) + \sin(x + y) = c_1 e^{x - y}. \qquad ...(3)$$

Also taking $\dfrac{dx + dy}{\cos(x + y) + \sin(x + y)} = \dfrac{dz}{z}$.

Let $x + y = u \Rightarrow dx + dy = du$, so that $\dfrac{du}{\cos u + \sin u} = \dfrac{dz}{z}$

or $$\frac{du}{\sqrt{2}\left(\frac{1}{\sqrt{2}}\cos u + \frac{1}{\sqrt{2}}\sin u\right)} = \frac{dz}{z} \quad \text{or} \quad \frac{du}{\sin\left(u + \frac{1}{4}\pi\right)} = \sqrt{2}\,\frac{dz}{z}$$

or $$\operatorname{cosec}\left(u + \frac{1}{4}\pi\right) du = \sqrt{2}\left(\frac{dz}{z}\right).$$

Integrating $$\log \tan\left(\frac{1}{2}u + \frac{1}{8}\pi\right) = \sqrt{2}\log z - \log c_2$$

or $$\log c_2 = \log\left[z^{\sqrt{2}} \cot\left(\frac{1}{2}u + \frac{1}{8}\pi\right)\right]$$

or $$c_2 = z^{\sqrt{2}} \cot\left\{\frac{1}{2}(x + y) + \frac{1}{8}\pi\right\} \quad \text{...(4)}$$

Hence general solution [in view of (3) and (4)] is

$$\cos(x + y) + \sin(x + y) = e^{x-y}\,\phi\left[z^{\sqrt{2}} \cot\left\{\frac{1}{2}(x + y) + \frac{1}{8}\pi\right\}\right]$$

$$\phi\left[z^{\sqrt{2}} \cot\left\{\frac{1}{2}(x + y) + \frac{1}{8}\pi\right\}\right] = e^{x-y}\,\phi\left[z^{\sqrt{2}} \tan\left\{\frac{3}{8}\pi - \frac{1}{2}(x + y)\right\}\right].$$

TEST YOUR KNOWLEDGE

Solve the following equations:

1. $2p + 3q = 1$
2. $xp + yq = 3z$
3. $(y - z)p + (x - y)q = z - x$
4. $p \tan x + q \tan y = \tan z$
5. $p\sqrt{x} + q\sqrt{y} = \sqrt{z}$
6. $(z - y)p + (x - z)q = y - x$
7. $yzp + zxq = xy$
8. $(y + z)p + (z + x)q = x + y$
9. $y^2p - xyq = x(z - 2y)$
10. $x(y - z)p + y(z - x)q = z(x - y)$
11. $\left(\dfrac{y - z}{yz}\right)p + \left(\dfrac{z - x}{zx}\right)q = \dfrac{x - y}{xy}$
12. $x(z^2 - y^2)p + y(x^2 - z^2)q = z(y^2 - x^2)$
13. $x^2p + y^2q = (x + y)z$
14. $z(xp - yq) = y^2 - x^2$
15. $(y^3x - 2x^4)\,p - (2y^4 - x^3y)q = 9z(x^3 - y^3)$
16. $x_1p_1 + 2x_2p_2 + 3x_3p_3 + 4x_4p_4 = 0.$

[**Hint.** The auxiliary equations are $\dfrac{dx_1}{x_1} = \dfrac{dx_2}{2x_2} = \dfrac{dx_3}{3x_3} = \dfrac{dx_4}{4x_4} = \dfrac{dz}{0}$.]

Answers

1. $\phi(3x - 2y, y - 3z) = 0$
2. $\phi\left(\dfrac{x}{y}, \dfrac{x^3}{z}\right) = 0$
3. $\phi\left(x + y + z, \dfrac{x^2}{2} + yz\right) = 0$
4. $\phi\left(\dfrac{\sin x}{\sin y}, \dfrac{\sin y}{\sin z}\right) = 0$
5. $\sqrt{x} - \sqrt{y} = f(\sqrt{x} - \sqrt{z})$
6. $x^2 + y^2 + z^2 = f(x + y + z)$

7. $\phi(x^2 - y^2, x^2 - z^2) = 0$ 8. $\phi\left[\frac{x-y}{y-z}, (x-y)^2(x+y+z)\right] = 0$

9. $\phi(x^2 + y^2, yz - y^2) = 0$ 10. $\phi(x + y + z, xyz) = 0$ 11. $\phi(x + y + z, xyz) = 0$

12. $\phi(x^2 + y^2 + z^2, xyz) = 0$ 13. $\phi\left(\frac{x-y}{z}, \frac{1}{y} - \frac{1}{x}\right) = 0$ 14. $\phi(xy, x^2 + y^2 + z^2) = 0$

15. $\phi\left(\frac{y}{x^2} + \frac{x}{y^2}, xyz^{1/3}\right) = 0$ 16. $\phi\left(\frac{x_1^2}{x_2}, \frac{x_1^3}{x_3}, \frac{x_1^4}{x_4}\right) = 0$

3.7. NON-LINEAR EQUATIONS OF THE FIRST ORDER

A partial differential equation which involves first order partial derivatives p and q with degree higher than one and the products of p and q is called a non-linear partial differential equation of the first order. The *complete solution* of such an equation involves only two arbitrary constants (*i.e.*, equal to the number of independent variables).

There is a general method for solving such equations. Before giving this method, we shall deal with some special types of such equations which can be solved easily by methods other than the general method.

3.8. (a) EQUATIONS OF THE FORM f(p, q) = 0

i.e., equations involving only **p** *and* **q** *and no* **x, y, z**

The complete solution is $z = ax + by + c$ ...(1)

where a and b are connected by the relation

$$f(a, b) = 0 \quad ...(2) \qquad \left[\text{Since } p = \frac{\partial z}{\partial x} = a \text{ and } q = \frac{\partial z}{\partial y} = b\right]$$

From (2), we can find b in terms of a. Let $b = \phi(a)$.

Putting this value of b in (1), the complete solution is

$$z = ax + \phi(a)y + c, \quad \text{where } a \text{ and } c \text{ are arbitrary constants.}$$

ILLUSTRATIVE EXAMPLES

Example 1. *Solve:*

(*i*) $\sqrt{p} + \sqrt{q} = 1$ (*ii*) $pq = p + q$.

Sol. (*i*) The equation is of the form $f(p, q) = 0$

The complete solution is $z = ax + by + c$ ...(1)

where $\sqrt{a} + \sqrt{b} = 1$ or $b = (1 - \sqrt{a})^2$

$\therefore$ From (1), the complete solution is

$$z = ax + (1 - \sqrt{a})^2 y + c$$

(*ii*) The equation is of the form $f(p, q) = 0$

The complete solution is $z = ax + by + c$ ...(1)

where $ab = a + b$ or $b = \frac{a}{a-1}$

∴ From (1), the complete solution is $z = ax + \frac{a}{a-1} y + c.$

Example 2. *Solve* $x^2p^2 + y^2q^2 = z^2$.

Sol. The given equation can be written as

$$\left(\frac{x}{z}\cdot\frac{\partial z}{\partial x}\right)^2 + \left(\frac{y}{z}\cdot\frac{\partial z}{\partial y}\right)^2 = 1 \qquad ...(1)$$

Let $\frac{dx}{x} = dX, \frac{dy}{y} = dY, \frac{dz}{z} = dZ$

then $X = \log x, Y = \log y, Z = \log z$

$$\frac{\partial Z}{\partial X} = \frac{x}{z}\cdot\frac{\partial z}{\partial x} \text{ and } \frac{\partial Z}{\partial Y} = \frac{y}{z}\cdot\frac{\partial z}{\partial y}$$

∴ (1) can be written as $\left(\frac{\partial Z}{\partial X}\right)^2 + \left(\frac{\partial Z}{\partial Y}\right)^2 = 1$

or $P^2 + Q^2 = 1$, where $P = \frac{\partial Z}{\partial X}$ and $Q = \frac{\partial Z}{\partial Y}$

It is of the form $f(P, Q) = 0$

Its complete solution is $Z = aX + bY + c$ ...(2)

where $a^2 + b^2 = 1$ or $b = \sqrt{1-a^2}$

∴ From (2), the complete solution is $\log z = a \log x + \sqrt{1-a^2} \log y + c.$

Example 3. *Solve* $(y - x)(qy - px) = (p - q)^2$.

Sol. Let $x + y = X$ and $xy = Y$

so that $p = \frac{\partial z}{\partial x} = \frac{\partial z}{\partial X}\cdot\frac{\partial X}{\partial x} + \frac{\partial z}{\partial Y}\cdot\frac{\partial Y}{\partial x} = \frac{\partial z}{\partial X} + y\frac{\partial z}{\partial Y}$

and $q = \frac{\partial z}{\partial y} = \frac{\partial z}{\partial X}\cdot\frac{\partial X}{\partial y} + \frac{\partial z}{\partial Y}\cdot\frac{\partial Y}{\partial y} = \frac{\partial z}{\partial X} + x\frac{\partial z}{\partial Y}$

∴ $qy - px = (y - x)\frac{\partial z}{\partial X}$ and $p - q = (y - x)\frac{\partial z}{\partial Y}$

∴ The given equation can be written as $\frac{\partial z}{\partial X} = \left(\frac{\partial z}{\partial Y}\right)^2$

or $P = Q^2$, where $P = \frac{\partial Z}{\partial X}$ and $Q = \frac{\partial Z}{\partial Y}$

It is of the form $f(P, Q) = 0$

Its complete solution is $z = aX + bY + c$ ...(1)

where $a = b^2$ or $b = \sqrt{a}$

∴ From (1), the complete solution is $z = a(x + y) + \sqrt{a}\, xy + c.$

3.8. (b) EQUATIONS OF THE FORM z = px + qy + f(p, q)

The complete solution is $z = ax + by + f(a, b)$ obtained by writing a for p and b for q.

Example 4. *Solve* $z = px + qy + \sqrt{1 + p^2 + q^2}$.

Sol. The given equation is of the form $z = px + qy + f(p, q)$

Its complete solution is $z = ax + by + \sqrt{1 + a^2 + b^2}$.

Example 5. *Solve* $4xyz = pq + 2px^2y + 2qxy^2$.

Sol. Let $x^2 = X$ and $y^2 = Y$

so that $p = \frac{\partial z}{\partial x} = \frac{\partial z}{\partial X} \cdot \frac{\partial X}{\partial x} = 2x \frac{\partial z}{\partial X}$ and $q = \frac{\partial z}{\partial y} = \frac{\partial z}{\partial Y} \cdot \frac{\partial Y}{\partial y} = 2y \frac{\partial z}{\partial Y}$

$\therefore$ The given equation becomes

$$4xyz = 4xy \frac{\partial z}{\partial X} \cdot \frac{\partial z}{\partial Y} + 4x^3y \frac{\partial z}{\partial X} + 4xy^3 \frac{\partial z}{\partial Y}$$

or $$z = x^2 \frac{\partial z}{\partial X} + y^2 \frac{\partial z}{\partial Y} + \frac{\partial z}{\partial X} \cdot \frac{\partial z}{\partial Y} = X \frac{\partial z}{\partial X} + Y \frac{\partial z}{\partial Y} + \frac{\partial z}{\partial X} \cdot \frac{\partial z}{\partial Y}$$

or $$z = PX + QY + PQ, \quad \text{where } P = \frac{\partial z}{\partial X} \text{ and } Q = \frac{\partial z}{\partial Y}$$

It is of the form $z = PX + QY + f(P, Q)$

Its complete solution is $z = aX + bY + ab$ or $z = ax^2 + by^2 + ab$.

3.8. (c) EQUATIONS OF THE FORM f (z, p, q) = 0

i.e., equations not containing **x** *and* **y**.

Let us assume $z = \phi(x + ay) = \phi(u)$, where $u = x + ay$ as a trial solution of the given equation.

$$\therefore \quad p = \frac{\partial z}{\partial x} = \phi'(x + ay) = \phi'(u) = \frac{dz}{du}$$

$$q = \frac{\partial z}{\partial y} = \phi'(x + ay)a = a\phi'(u) = a\frac{dz}{du}$$

Substituting these values of p and q in the given equation, we get $f\left(z, \frac{dz}{du}, a\frac{dz}{du}\right) = 0$ which is an ordinary differential equation of the first order. Integrating it, we get the complete solution.

Method. (*i*) Assume $u = x + ay$ so that $p = \frac{dz}{du}$ and $q = a\frac{dz}{du}$.

(*ii*) Substitute these values of p and q in the given equation.

(*iii*) Solve the resulting ordinary differential equation in z and u.

(*iv*) Replace u by $x + ay$.

Example 6. *Solve* $z^2(p^2 + q^2 + 1) = a^2$.

Sol. The given equation is of the form $f(z, p, q) = 0$

Let $u = x + by$ | note the use of b instead of a, since a is a given constant

so that $p = \frac{dz}{du}$ and $q = b\frac{dz}{du}$.

Substituting these values of p and q in the given equation, we get

$$z^2\left[\left(\frac{dz}{du}\right)^2 + b^2\left(\frac{dz}{du}\right)^2 + 1\right] = a^2 \quad \text{or} \quad z^2(1+b^2)\left(\frac{dz}{du}\right)^2 = a^2 - z^2$$

or $$z\sqrt{1+b^2}\,\frac{dz}{du} = \pm\sqrt{a^2-z^2} \quad \text{or} \quad \pm\sqrt{1+b^2}\,.\frac{z}{\sqrt{a^2-z^2}}dz = du$$

Integrating, we have $$\pm\sqrt{1+b^2}\sqrt{a^2-z^2} = u + c$$

or $$(1+b^2)(a^2-z^2) = (x+by+c)^2$$

which is the required complete solution.

Example 7. *Solve $z^2(p^2x^2+q^2) = 1$.*

Sol. The given equation can be written as

$$z^2\left[\left(x\frac{\partial z}{\partial x}\right)^2 + \left(\frac{\partial z}{\partial y}\right)^2\right] = 1 \qquad \text{...(1)}$$

Let $$X = \log x$$

so that $$\frac{\partial z}{\partial x} = \frac{\partial z}{\partial X}.\frac{\partial X}{\partial x} = \frac{\partial z}{\partial X}.\frac{1}{x} \quad i.e., \quad x\frac{\partial z}{\partial x} = \frac{\partial z}{\partial X}$$

∴ Equation (1) reduces to $z^2\left[\left(\frac{\partial z}{\partial X}\right)^2 + \left(\frac{\partial z}{\partial y}\right)^2\right] = 1$

or $$z^2(P^2+q^2) = 1, \quad \text{where } P = \frac{\partial z}{\partial X} \qquad \text{...(2)}$$

Let $$u = X + ay \text{ so that } P = \frac{dz}{du} \text{ and } q = a\frac{dz}{du}$$

∴ From (2), we have $z^2\left[\left(\frac{dz}{du}\right)^2 + a^2\left(\frac{dz}{du}\right)^2\right] = 1$

or $$(1+a^2)z^2\left(\frac{dz}{du}\right)^2 = 1 \quad \text{or} \quad \sqrt{1+a^2}\,.\,z\,dz = \pm\,du$$

Integrating, $$\sqrt{1+a^2}\,.\frac{z^2}{2} = \pm\,u + b$$

or $$\sqrt{1+a^2}\,.\,z^2 = \pm\,2(X+ay) + 2b$$

or $$\sqrt{1+a^2}\,.\,z^2 = \pm\,2(\log x + ay) + c$$

which is the required complete solution.

3.8. (d) EQUATIONS OF THE FORM $f_1(x, p) = f_2(y, q)$

i.e., equations in which **z** *is absent and the terms involving* **x** *and* **p** *can be separated from those involving* **y** *and* **q.**

As a trial solution, let us put each side equal to an arbitrary constant a. Then

$$f_1(x, p) = f_2(y, q) = a$$

Solving these equations for p and q, let $p = F_1(x)$ and $q = F_2(y)$

Since $dz = \dfrac{\partial z}{\partial x}\,dx + \dfrac{\partial z}{\partial y}\,dy = p\,dx + q\,dy$

$\therefore$ $dz = F_1(x)\,dx + F_2(y)\,dy$

Integrating, $z = \int F_1(x)\,dx + \int F_2(y)\,dy + b$

which is the required complete solution.

Example 8. *Solve* $p^2 - q^2 = x - y$.

Sol. The given equation is $p^2 - x = q^2 - y$ which is of the form $f_1(x, p) = f_2(y, q)$

Let $p^2 - x = q^2 - y = a$

then $p^2 = x + a$ and $q^2 = y + a$ *i.e.*, $p = \sqrt{x+a}$ and $q = \sqrt{y+a}$

Substituting these values of p and q in $dz = pdx + qdy$, we get

$$dz = \sqrt{x+a}\;dx + \sqrt{y+a}\;dy$$

Integrating, $z = \dfrac{2}{3}(x+a)^{3/2} + \dfrac{2}{3}(y+a)^{3/2} + b$

which is the required complete solution.

Example 9. *Solve* $yp = 2yx + \log q$.

Sol. The given equation can be written as

$$p = 2x + \frac{1}{y}\log q \quad \text{or} \quad p - 2x = \frac{1}{y}\log q$$

which is of the form $f_1(x, p) = f_2(y, q)$.

Let $p - 2x = \dfrac{1}{y}\log q = a,$ then $p = 2x + a$ and $\log q = ay$, *i.e.*, $q = e^{ay}$

Substituting these values of p and q in $dz = pdx + qdy$, we get

$$dz = (2x + a)\,dx + e^{ay}\,dy$$

Integrating, $z = x^2 + ax + \dfrac{1}{a}e^{ay} + b$ which is the required complete solution.

Example 10. *Solve* $z^2(p^2 + q^2) = x^2 + y^2$.

Sol. The given equation can be written as

$$\left(z\frac{\partial z}{\partial x}\right)^2 + \left(z\frac{\partial z}{\partial y}\right)^2 = x^2 + y^2 \qquad \text{...(1)}$$

Let $zdz = dZ$ so that $Z = \frac{1}{2}z^2$

Now $\dfrac{\partial Z}{\partial x} = \dfrac{\partial Z}{\partial z}\cdot\dfrac{\partial z}{\partial x} = z\dfrac{\partial z}{\partial x}$ and $\dfrac{\partial Z}{\partial y} = \dfrac{\partial Z}{\partial z}\cdot\dfrac{\partial z}{\partial y} = z\dfrac{\partial z}{\partial y}$

$\therefore$ Equation (1) becomes $\left(\dfrac{\partial Z}{\partial x}\right)^2 + \left(\dfrac{\partial Z}{\partial y}\right)^2 = x^2 + y^2$

or $P^2 + Q^2 = x^2 + y^2$, where $P = \dfrac{\partial Z}{\partial x}$ and $Q = \dfrac{\partial Z}{\partial y}$

or $P^2 - x^2 = y^2 - Q^2$ which is of the form $f_1(x, P) = f_2(y, Q)$.

Let $P^2 - x^2 = y^2 - Q^2 = a,$ then $P = \sqrt{x^2 + a}$ and $Q = \sqrt{y^2 - a}$

Substituting these values of P and Q in

$$dZ = Pdx + Qdy, \quad \text{we get} \quad dZ = \sqrt{x^2 + a}\, dx + \sqrt{y^2 - a}\, dy$$

Integrating,

$$Z = \frac{1}{2} x\sqrt{x^2 + a} + \frac{a}{2} \log (x + \sqrt{x^2 + a}) + \frac{1}{2} y\sqrt{y^2 - a} - \frac{a}{2} \log (y + \sqrt{y^2 - a}) + b$$

or
$$z^2 = x\sqrt{x^2 + a} + y\sqrt{y^2 - a} + a \log \frac{x + \sqrt{x^2 + a}}{y + \sqrt{y^2 - a}} + c, \quad (\text{where } c = 2b)$$

is the required complete solution.

TEST YOUR KNOWLEDGE

Solve the following equations:

1. $p^2 + q^2 = 2$
2. $p^3 - q^3 = 0$
3. $p^2 + p = q^2$
4. $p = e^q$
5. $(x + y)(p + q)^2 + (x - y)(p - q)^2 = 1$ **[Hint.** Put $x + y = X^2$ and $x - y = Y^2$]
6. $z = px + qy - 2\sqrt{pq}$
7. $z = px + qy + \sin (p + q)$
8. $(pq - p - q)(z - px - qy) = pq$
9. $p^2 = zq$
10. $p(1 + q^2) = q(z - a)$
11. $p(1 + q) = qz$
12. $z^2 (p^2 + q^2 + 1) = 1$
13. $z = p^2 + q^2$
14. $p + q = z$
15. $z^2 (p^2z^2 + q^2) = 1$
16. $9(p^2z + q^2) = 4$
17. $p^2 + q^2 = x + y$
18. $p - x^2 = q + y^2$
19. $\sqrt{p} + \sqrt{q} = x + y$
20. $\sqrt{p} + \sqrt{q} = 2x$
21. $pe^y = qe^x$
22. $q = xyp^2$
23. $q(p - \cos x) = \cos y$
24. $yp + xq + pq = 0$
25. $z(p^2 - q^2) = x - y.$
26. $p + q = \sin x + \sin y.$

Answers

1. $z = ax \pm \sqrt{2 - a^2}\; y + c$
2. $z = a(x + y) + c$
3. $z = ax \pm \sqrt{a^2 + a}\; y + c$
4. $z = ax + y \log a + c$
5. $z = a\sqrt{x + y} + \sqrt{1 - a^2}\sqrt{x - y} + c$
6. $z = ax + by - 2\sqrt{ab}$
7. $z = ax + by + \sin (a + b)$
8. $z = ax + by + \dfrac{ab}{ab - a - b}$
9. $z = be^{ax + a^2 y}$
10. $4(bz - ab - 1) = (x + by + c)^2$
11. $\log (az - 1) = x + ay + b$
12. $(1 + a^2)(1 - z^2) = (x + ay + b)^2$
13. $4z(1 + a^2) = (x + ay + b)^2$
14. $(1 + a) \log z = x + ay + b$
15. $(z^2 + a^2)^3 = 9(x + ay + b)^2$
16. $(z + a^2)^3 = (x + ay + b)^2$
17. $z = \frac{2}{3}(x + a)^{3/2} + \frac{2}{3}(y - a)^{3/2} + b$
18. $z = \frac{1}{3}(x^3 - y^3) + a(x + y) + b$

19. $z = \frac{1}{3}(x+a)^3 + \frac{1}{3}(y-a)^3 + b$

20. $z = \frac{1}{6}(2x+a)^3 + a^2y + b$

21. $z = ae^x + ae^y + b$

22. $z = 2\sqrt{ax} + \frac{1}{2}ay^2 + b$

23. $z = ax + \sin x + \frac{1}{a}\sin y + b$

24. $2z = ax^2 - \frac{a}{a+1}y^2 + b$

25. $z^{3/2} = (x+a)^{3/2} + (y+a)^{3/2} + b$ $\left[\textbf{Hint.} \text{ Put } \sqrt{z}dz = dZ \text{ so that } Z = \frac{2}{3}z^{3/2}\right]$

26. $z = a(x-y) - (\cos x + \cos y) + c.$

3.9. CHARPIT'S METHOD

This is a general method for finding the complete solution of non-linear partial differential equations of the first order.

Let the given equation be $\quad f(x, y, z, p, q) = 0 \quad$ ...(1)

If we can find another relation $\quad F(x, y, z, p, q) = 0 \quad$ ...(2)

involving x, y, z, p and q, then we can solve (1) and (2) for p and q and substitute in

$$dz = pdx + qdy \quad ...(3)$$

Solution of (3), if it exists, is the complete solution of (1).

To determine F, we differentiate (1) and (2) partially with respect to x and y. Thus

$$\frac{\partial f}{\partial x} + \frac{\partial f}{\partial z}.p + \frac{\partial f}{\partial p}.\frac{\partial p}{\partial x} + \frac{\partial f}{\partial q}.\frac{\partial q}{\partial x} = 0 \quad ...(4)$$

$$\frac{\partial F}{\partial x} + \frac{\partial F}{\partial z}.p + \frac{\partial F}{\partial p}.\frac{\partial p}{\partial x} + \frac{\partial F}{\partial q}.\frac{\partial q}{\partial x} = 0 \quad ...(5)$$

$$\frac{\partial f}{\partial y} + \frac{\partial f}{\partial z}.q + \frac{\partial f}{\partial p}.\frac{\partial p}{\partial y} + \frac{\partial f}{\partial q}.\frac{\partial q}{\partial y} = 0 \quad ...(6)$$

$$\frac{\partial F}{\partial y} + \frac{\partial F}{\partial z}.q + \frac{\partial F}{\partial p}.\frac{\partial p}{\partial y} + \frac{\partial F}{\partial q}.\frac{\partial q}{\partial y} = 0 \quad ...(7)$$

Eliminating $\frac{\partial p}{\partial x}$ between (4) and (5), we get

$$\left(\frac{\partial f}{\partial x}.\frac{\partial F}{\partial p} - \frac{\partial F}{\partial x}.\frac{\partial f}{\partial p}\right) + \left(\frac{\partial f}{\partial z}.\frac{\partial F}{\partial p} - \frac{\partial F}{\partial z}.\frac{\partial f}{\partial p}\right)p + \left(\frac{\partial f}{\partial q}.\frac{\partial F}{\partial p} - \frac{\partial F}{\partial q}.\frac{\partial f}{\partial p}\right)\frac{\partial q}{\partial x} = 0 \quad ...(8)$$

Eliminating $\frac{\partial q}{\partial y}$ between (6) and (7), we get

$$\left(\frac{\partial f}{\partial y}.\frac{\partial F}{\partial q} - \frac{\partial F}{\partial y}.\frac{\partial f}{\partial q}\right) + \left(\frac{\partial f}{\partial z}.\frac{\partial F}{\partial q} - \frac{\partial F}{\partial z}.\frac{\partial f}{\partial q}\right)q + \left(\frac{\partial f}{\partial p}.\frac{\partial F}{\partial q} - \frac{\partial F}{\partial p}.\frac{\partial f}{\partial q}\right)\frac{\partial p}{\partial y} = 0 \quad ...(9)$$

Since $\frac{\partial q}{\partial x} = \frac{\partial^2 z}{\partial x \partial y} = \frac{\partial^2 z}{\partial y \partial x} = \frac{\partial p}{\partial y}$, the last terms in (8) and (9) differ in sign only. Adding (8) and (9), we get

$$\left(\frac{\partial f}{\partial x} + p\frac{\partial f}{\partial z}\right)\frac{\partial F}{\partial p} + \left(\frac{\partial f}{\partial y} + q\frac{\partial f}{\partial z}\right)\frac{\partial F}{\partial q} + \left(-p\frac{\partial f}{\partial p} - q\frac{\partial f}{\partial q}\right)\frac{\partial F}{\partial z} + \left(-\frac{\partial f}{\partial p}\right)\frac{\partial F}{\partial x} + \left(-\frac{\partial f}{\partial q}\right)\frac{\partial F}{\partial y} = 0$$

...(10)

which is a linear partial differential equation of the first order with x, y, z, p, q as independent variables and F as the dependent variable.

$\therefore$ The auxiliary equations are

$$\frac{dp}{\frac{\partial f}{\partial x}+p\frac{\partial f}{\partial z}}=\frac{dq}{\frac{\partial f}{\partial y}+q\frac{\partial f}{\partial z}}=\frac{dz}{-p\frac{\partial f}{\partial p}-q\frac{\partial f}{\partial q}}=\frac{dx}{-\frac{\partial f}{\partial p}}=\frac{dy}{-\frac{\partial f}{\partial q}}=\frac{\partial F}{0} \quad ...(11)$$

Any integral of (11) which involves p or q or both can be taken as the assumed relation (2). In practice, we choose the simplest of the integrals of (11).

ILLUSTRATIVE EXAMPLES

Example 1. *Solve:* $2zx - px^2 - 2qxy + pq = 0.$

Sol. Here $f = 2zx - px^2 - 2qxy + pq = 0$ ...(1)

$$\therefore \quad \frac{\partial f}{\partial x} = 2z - 2px - 2qy, \frac{\partial f}{\partial y} = -2qx, \frac{\partial f}{\partial z} = 2x, \frac{\partial f}{\partial p} = -x^2 + q, \frac{\partial f}{\partial q} = -2xy + p$$

Charpit's auxiliary equations are

$$\frac{dp}{\frac{\partial f}{\partial x}+p\frac{\partial f}{\partial z}}=\frac{dq}{\frac{\partial f}{\partial y}+q\frac{\partial f}{\partial z}}=\frac{dz}{-p\frac{\partial f}{\partial p}-q\frac{\partial f}{\partial q}}=\frac{dx}{-\frac{\partial f}{\partial p}}=\frac{dy}{-\frac{\partial f}{\partial q}}=\frac{dF}{0}$$

or
$$\frac{dp}{2z-2qy}=\frac{dq}{0}=\frac{dz}{px^2-2pq+2qxy}=\frac{dx}{x^2-q}=\frac{dy}{2xy-p}=\frac{dF}{0}$$

$\therefore$ $dq = 0$ or $q = a$

Putting $q = a$ in (1), we get $p = \dfrac{2x(z-ay)}{x^2-a}$

$$\therefore \quad dz = pdx + qdy = \frac{2x(z-ay)}{x^2-a}dx + ady$$

or
$$\frac{dz - ady}{z-ay} = \frac{2x}{x^2-a}dx$$

Integrating, $\log (z - ay) = \log (x^2 - a) + \log b$ or $z - ay = b (x^2 - a)$

$\therefore$ $z = ay + b(x^2 - a)$ which is the required complete integral.

Example 2. *Solve by Charpit's method:*

$p_x + qy = pq.$

Sol. The given differential equation is

$f \equiv px + qy - pq = 0$ ...(1)

$$\therefore \quad \frac{\partial f}{\partial x} = p, \frac{\partial f}{\partial y} = q; \frac{\partial f}{\partial z} = 0; \frac{\partial f}{\partial p} = x - q; \frac{\partial f}{\partial q} = y - p.$$

Hence the auxiliary equations of Charpit's are

$$\frac{dp}{\frac{\partial f}{dx}+\frac{\partial f}{\partial z}p}=\frac{dq}{\frac{\partial f}{\partial y}+\frac{\partial f}{\partial z}q}=\frac{dz}{-p\frac{\partial f}{\partial p}-q\frac{\partial f}{\partial q}}=\frac{dx}{-\frac{\partial f}{\partial q}}=\frac{dy}{-\frac{\partial f}{\partial q}}=\frac{dF}{0}$$

$$\Rightarrow \quad \frac{dp}{p}=\frac{dq}{q}=\frac{dz}{-p(x-q)-q(y-p)}=\frac{dx}{q-x}=\frac{dy}{p-y}=\frac{dF}{0}.$$

Taking first two ratios, we have

$$\frac{dp}{p} = \frac{dq}{q}.$$

On integration, we get

$$\log p = \log q + \log a$$

or $\quad p = aq$, were a is any arbitrary constant.

Substituting the above values of p in (1), we obtain

$$q\,(ax + y) = aq^2 \;\Rightarrow\; q = \frac{ax + y}{a}.$$

$$\therefore \quad p = ax + y.$$

Hence $dz = pdx + qdy$ gives

$$dz = (ax + y)\,dx + \frac{ax + y}{a}\,dy \;\Rightarrow\; adz = a\,(ax + y) + dz + (ax + y)\,dy$$

$$\Rightarrow \quad adz = (ax + y)\,(adx + dy) \;\Rightarrow\; adz = (ax + y)\,d\,(ax + y).$$

On integration, we get

$$az = \frac{1}{2}(ax + b)^2 + b,$$

where b is any arbitrary constant.

Hence the required complete integral of (1) is

$$az = \frac{1}{2}(ax + y)^2 + b.$$

Example 3. *Solve the Charpit's method $(p^2 + q^2)\,y = qz$.*

Sol. The given differential equation is

$$f \equiv (p^2 + q^2)\,y - qz = 0 \qquad ...(1)$$

$$\therefore \quad \frac{\partial f}{\partial x} = 0,\; \frac{\partial f}{\partial y} = p^2 + q^2;\; \frac{\partial f}{\partial z} = -\,q;\; \frac{\partial f}{\partial p} = 2py;\; \frac{\partial f}{\partial q} = 2qy - z.$$

Hence the Charpit's auxiliary equations are

$$\frac{dp}{-\,pq} = \frac{dq}{(p^2 + q^2) - q^2} = \frac{dz}{-\,p\,.\,(2qy) - q\,(2qy - z)} = \frac{dx}{-\,2py} = \frac{dy}{-\,2qy + z} = \frac{dF}{0}. \qquad ...(2)$$

Taking first two ratios of (2), we have

$$\frac{dp}{-\,pq} = \frac{dq}{p^2} \;\Rightarrow\; p\,dp + q\,dq = 0 \;\Rightarrow\; d\,(p^2 + q^2) = 0.$$

On integration, we get

$$p^2 + q^2 = a^2 \qquad ...(3)$$

where a is any arbitrary constant.

Substituting the above value in (1), we obtain

$$a^2y = qz \;\Rightarrow\; q = \frac{a^2 y}{z}.$$

So, by (1) $\quad p^2 + \dfrac{a^4 y^2}{z^2} = a^2 \;\Rightarrow\; p = \sqrt{a^2 - \dfrac{a^4 y^2}{z^2}} = \dfrac{a}{z}\sqrt{z^2 - a^2 y^2}.$

Now substituting the value of p and q in

$$dz = pdx + qdy,$$

we get
$$dz = \frac{a}{z}\sqrt{z^2 - a^2y^2}\, dx + \frac{a^2y}{z}\, dy$$

$$\Rightarrow \quad \frac{zdz - a^2y\,dy}{\sqrt{z^2 - a^2y^2}} = adx \quad \Rightarrow \quad d[\sqrt{z^2 - a^2y^2}\,] = d\,(ax)$$

On integration, we get

$$\sqrt{z^2 - a^2y^2} = ax + b,$$

where b is any arbitrary constant.

i.e., $\quad z^2 - a^2y^2 = (ax + b)^2 \quad$ *i.e.*, $\quad z^2 = a^2y^2 + (ax + b)^2 \qquad$ ...(4)

which is the required complete integral of (1).

General Solution. Putting $b = \phi(a)$ in equation (4), we get

$$z^2 = a^2y^2 + [ax + \phi(a)]^2. \qquad ...(5)$$

Differentiating equation (5) partially w.r.t. a, we get

$$0 = 2ay^2 + 2\,[ax + \phi(a)]\,[x + \phi'(a)]. \qquad ...(6)$$

But eliminating a from equations (5) and (6) we get the required general solution.

Singular Solution. Differentiating (4) partially w.r.t. a and b, we get

$$0 = 2ay^2 + 2\,(ax + b)\,x \quad \text{and} \quad 0 = 2\,(ax + b) \;\Rightarrow\; ax + b = 0 \quad \text{and} \quad ay^2 = 0$$

Substituting above values in (4), we get

$$z^2 = 0, \quad \textit{i.e.}, \quad z = 0$$

$$\therefore \quad p = \frac{\partial z}{\partial x} = 0 \quad \text{and} \quad q = \frac{\partial z}{\partial y} = 0.$$

The above values of p and y satisfied (1).

Hence the required singular solution of (1) is

$$z = 0.$$

Example 4. *Solve by Charpit's method:* $z^2 = pqxy$.

Sol. The given differential equation is

$$f = z^2 - pqxy = 0 \qquad ...(1)$$

$$\therefore \quad \frac{\partial f}{\partial x} = -\,pqy; \; \frac{\partial f}{\partial y} = -\,pqx; \; \frac{\partial f}{\partial z} = 2z; \; \frac{\partial f}{\partial p} = -\,qxy\,; \; \frac{\partial f}{\partial q} = -\,pxy.$$

Hence Charpit's auxiliary equations are

$$\frac{dp}{-\,pqy + p\,.\,2z} = \frac{dq}{-\,pqx + q\,.\,2z} = \frac{dz}{-\,p\,(-\,qxy) - q\,(-\,pxy)} = \frac{dx}{qxy} = \frac{dy}{pxy}. \qquad ...(2)$$

Each ratio of the above is

$$\frac{xdp + pdx}{2pzx} = \frac{ydq + qdy}{2qyz} \quad \Rightarrow \quad \frac{xdp + pdx}{px} = \frac{ydq + qdy}{qy}.$$

On integration, we get

$$\int \frac{d\,(px)}{px} = \int \frac{d\,(qy)}{qy} + \text{C} \quad \text{or} \quad \log px = \log qy + \log c, \text{ when } \text{C} = \log c$$

or
$$px = cqy \qquad ...(3)$$

where c is any arbitrary constant.

From (1) and (3), we have

$$z^2 = cqy \,.\, qy = cq^2y^2.$$

$$\therefore \qquad q^2 = \frac{z^2}{cy^2} \quad \text{or} \quad q = \frac{z}{y\sqrt{c}}. \qquad \text{...(4)}$$

From (3) and (4),

$$p = \frac{cz}{y\sqrt{c}} \cdot \frac{y}{x} = \frac{\sqrt{c}\,.\,z}{x}. \qquad \text{...(5)}$$

Substituting the above values of p and q in $dz = p\ dx = q\ dy$, we obtain

$$dz = \sqrt{c}\,\frac{z}{x}\,dx + \frac{z}{\sqrt{cy}}\,dy \quad \Rightarrow \quad \frac{dz}{z} = \frac{\sqrt{c}}{x}\,dx + \frac{1}{\sqrt{cy}}\,dy.$$

On integration, we get

$$\log z = a \log x + \frac{1}{a} \log y + \log b, \quad \text{where } b \text{ is any arbitrary constant}$$

or $\qquad \log z = \log (bx^a y^{1/a}).$

Hence the required compcete integral of (1) is

$$z = bx^a\, y^{1/a}.$$

TEST YOUR KNOWLEDGE

Solve the following equations by Charpit's method:

1. $px + qy = pq$
2. $z^2 = pqxy$
3. $pxy + pq + qy = yz$
4. $z = p^2x + q^2y$
5. $2(z + px + qy) = yp^2$
6. $q + xp = p^2$.

Answers

1. $az = \frac{1}{2}(y + ax)^2 + b$
2. $z = ax^b y^{1/b}$
3. $\log (z - ax) = y - a \log (a + y) + b$
4. $\sqrt{(1+a)z} = \sqrt{ax} + \sqrt{y} + b$
5. $z = \dfrac{ax}{y^2} - \dfrac{a^2}{4y^3} + \dfrac{b}{y}$
6. $z = axe^{-y} - \frac{1}{2}a^2e^{-2y} + b$.

3.10. HOMOGENEOUS LINEAR EQUATIONS WITH CONSTANT CO-EFFICIENTS

An equation of the form

$$a_0 \frac{\partial^n z}{\partial x^n} + a_1 \frac{\partial^n z}{\partial x^{n-1} \partial y} + a_2 \frac{\partial^n z}{\partial x^{n-2} \partial y^2} + \ldots\ldots + a_n \frac{\partial^n z}{\partial y^n} = F(x, y) \qquad \text{...(1)}$$

where $a_0, a_1, a_2, \ldots\ldots, a_n$ are constants, is called a *homogeneous linear partial differential equation of the nth order with constant co-efficients.*

Here, all the partial derivatives are of the nth order

Writing D for $\dfrac{\partial}{\partial x}$ and D′ for $\dfrac{\partial}{\partial y}$, (1) can be written as

$$(D^n + a_1 D^{n-1} D' + a_2 D^{n-2} D'^2 + \ldots + a_n D'^n) z = F(x, y)$$

or $$\phi(D, D')z = F(x, y) \qquad \text{...(2)}$$

As in the case of ordinary linear differential equations with constant co-efficients the complete solution of (2) consists of two parts:

(*i*) the **complementary function** (C.F.) which is the complete solution of the equation ϕ (D, D′)z = 0. It must contain n arbitrary functions, where n is the order of the differential equation.

(*ii*) the **particular integral** (P.I.) which is a particular solution (free from arbitrary constants) of

$$\phi(D, D')z = F(x, y)$$

The complete solution of (2) is z = C.F. + P.I.

3.11. RULES FOR FINDING THE C.F.

We explain the procedure for finding the complementary function of a differential equation of the second order. It can be easily extended to differential equations of higher order.

Consider the equation $\dfrac{\partial^2 z}{\partial x^2} + a_1 \dfrac{\partial^2 z}{\partial x \partial y} + a_2 \dfrac{\partial^2 z}{\partial y^2} = 0$

which in symbolic form is $\quad (D^2 + a_1 DD' + a_2 D'^2)z = 0$ ...(1)

Its symbolic operator equated to zero, *i.e.*, $D^2 + a_1 DD' + a_2 D'^2 = 0$ ...(2)

is called the **auxiliary equation** (A.E.)

Considered as a quadratic in D/D′, let its roots be m_1, m_2.

Case I. *When the A.E. has distinct roots i.e.,* $m_1 \neq m_2$, then (2) can be written as

$$(D - m_1 D')(D - m_2 D')z = 0 \qquad ...(3)$$

Now the solution of $(D - m_2 D')z = 0$ will also be a solution of (3).

But $\quad (D - m_2 D')z = 0 \Rightarrow p - m_2 q = 0$

which is of Lagrange's form and the auxiliary equations are $\dfrac{dx}{1} = \dfrac{dy}{-m_2} = \dfrac{dz}{0}$

The first two members give $dy + m_2 dx = 0$ or $y + m_2 x = a$

Also $\quad dz = 0$ or $z = b$

$\therefore \quad z = f_2(y + m_2 x)$ is a solution of $(D - m_2 D')z = 0$

Similarly, (3) will also be satisfied by the solution of

$$(D - m_1 D')z = 0 \quad \textit{i.e.,} \quad \text{by } z = f_1(y + m_1 x).$$

Hence the complete solution of (2) is

$$z = f_1(y + m_1 x) + f_2(y + m_2 x).$$

Case II. *When the A.E. has equal roots,* each = m, then (2) can be written as

$$(D - mD')(D - mD')z = 0 \qquad ...(4)$$

Let $(D - mD')z = u$, then (4) becomes $(D - mD')u = 0$

Its solution is $u = f(y + mx)$, as proved in Case I.

$\therefore \quad (D - mD')z = u$ takes the form $(D - mD')z = f(y + mx)$

or $\quad p - mq = f(y + mx)$ [Lagrange's form]

The Auxiliary equations are $\dfrac{dx}{1} = \dfrac{dy}{-m} = \dfrac{dz}{f(y + mx)}$

Now $\quad \dfrac{dx}{1} = \dfrac{dy}{-m}$ gives $dy + mdx = 0$ or $y + mx = a$

Also $\frac{dx}{1} = \frac{dz}{f(a)}$ gives $dz = f(a)\,dx$ or $z = f(a)\,x + b$ *i.e.,* $z - xf\,(y + mx) = b$

$\therefore$ The complete solution of (2) is

$$z - x\phi\,(y + mx) = \phi(y + mx) \quad \text{or} \quad z = f(y + mx) + xf(y + mx).$$

Note 1. We know that for the differential equation

$$(D^2 + a_1DD' + a_2D'^2)\,z = 0,$$

the A.E. is

$$D^2 + a_1DD' + a_2D'^2 = 0 \qquad ...(1)$$

Now the roots of (1), considered as a quadratic in D/D′, are the same as those of

$$m^2 + a_1m + a_2 = 0 \qquad ...(2)$$

$\therefore$ This quadratic in m can also be called the A.E. Henceforth we shall write the A.E. in terms of m.

Thus the A.E. is obtained by putting $D = m$ and $D' = 1$ in $\phi(D, D') = 0$.

Hence the A.E. of the differential equation $\phi(D, D')z = F(x, y)$ is $\phi(m, 1) = 0$.

Note 2. Generalising the results of Case I and Case II, we have (*i*) if the roots of A.E. are m_1, m_2, m_3, ... (all distinct roots), then

$$\text{C.F.} = f_1\,(y + m_1x) + f_2\,(y + m_2x) + f_3\,(y + m_3x) + ...$$

(*ii*) if the roots of A.E. are m_1, m_1, m_2, ... (two equal roots), then

$$\text{C.F.} = f_1\,(y + m_1x) + xf_2\,(y + m_1x) + f_3\,(y + m_2x) + ...$$

(*iii*) if the roots of A.E. are m_1, m_1, m_1, ... (three equal roots), then

$$\text{C.F.} = f_1\,(y + m_1x) + xf_2\,(y + m_1x) + x^2f_3\,(y + m_1x) + ...$$

Example. *Solve:*

(*i*) $\frac{\partial^2 z}{\partial x^2} - \frac{\partial^2 z}{\partial x \partial y} - 6\frac{\partial^2 z}{\partial y^2} = 0$ (*ii*) $(D + 2D')(D - 3D')^2z = 0$ (*iii*) $4r - 12s + 9t = 0$.

Sol. (*i*) The given equation in symbolic form is

$$(D^2 - DD' - 6D'^2)z = 0$$

Its A.E. is $m^2 - m - 6 = 0$ or $(m - 3)\,(m + 2) = 0$ $\therefore$ $m = 3, -2$

Hence the complete solution is

$$z = f_1(y + 3x) + f_2(y - 2x).$$

(*ii*) The given equation is

$$(D + 2D')(D - 3D')^2z = 0$$

Its A.E. is $(m + 2)(m - 3)^2 = 0$ which gives $m = -2, 3, 3$

Hence the complete solution is

$$z = f_1(y - 2x) + f_2(y + 3x) + xf_3(y + 3x).$$

(*iii*) The given equation in symbolic form is $(4D^2 - 12DD' + 9D'^2)\,z = 0$

Since $r = \frac{\partial^2 z}{\partial x^2} = D^2 z,\ s = \frac{\partial^2 z}{\partial x \partial y} = DD'z,\ t = \frac{\partial^2 z}{\partial y^2} = D'^2 z$

Its A.E. is $4m^2 - 12m + 9 = 0$ or $(2m - 3)^2 = 0$ $\therefore$ $m = 3/2, 3/2$

Hence the complete solution is

$$z = \phi_1\left(y + \frac{3}{2}x\right) + x\phi_2\left(y + \frac{3}{2}x\right)$$

which may be written as $z = f_1\,(2y + 3x) + xf_2\,(2y + 3x).$

3.12. RULES FOR FINDING THE P.I.

The P.I. of the equation $\phi(D, D')z = F(x, y)$ is given by $\dfrac{1}{\phi(D, D')} F(x, y)$.

I. Short methods to find the P.I.

Analogous to the short methods for finding the P.I. of an ordinary linear differential equation with constant co-efficients, we have short methods for finding the P.I. of homogeneous linear partial differential equations. We give below, without proof, the short methods for the P.I.

(*i*) When $F(x, y) = e^{ax+by}$

$$\text{P.I.} = \frac{1}{\phi(D, D')} e^{ax+by} = \frac{1}{\phi(a, b)} e^{ax+by} \quad (\textit{i.e.}, \text{ put } D = a \text{ and } D' = b) \text{ provided } \phi(a, b) \neq 0.$$

If $\phi(a, b) = 0$, then it is called a case of failure.

(*ii*) When $F(x, y) = \sin(ax + by)$

$$\text{P.I.} = \frac{1}{\phi(D^2, DD', D'^2)} \sin(ax + by) = \frac{1}{\phi(-a^2, -ab, -b^2)} \sin(ax + by)$$

(*i.e.* put $D^2 = -a^2$, $DD' = -ab$, $D'^2 = -b^2$) provided $\phi(-a^2, -ab, -b^2) \neq 0$.

If $\phi(-a^2, -ab, -b^2) = 0$, then it is called a case of failure.

A similar rule holds when $F(x, y) = \cos(ax + by)$.

(*iii*) When $F(x, y) = x^p y^q$, where p, q are positive integers.

$$\text{P.I.} = \frac{1}{\phi(D, D')} x^p y^q = [\phi(D, D')]^{-1} x^p y^q$$

If $p < q$, expand $[\phi(D, D')]^{-1}$ in powers of $\dfrac{D}{D'}$

If $q < p$ expand $[\phi(D, D')]^{-1}$ in powers of $\dfrac{D'}{D}$

Also, we have $\dfrac{1}{D} F(x, y) = \displaystyle\int_{y \text{ constant}} F(x, y)$ and $\dfrac{1}{D'} F(x, y) = \displaystyle\int_{x \text{ constant}} F(x, y)\, dy$

II. General method to find the P.I.

$F(x, y)$ is not always of the forms given before. Moreover, there are cases of failure in the short methods. The general method is applicable to all cases where $F(x, y)$ is not of the forms given before or the short methods fail.

Now, $\phi(D, D')$ can be factorised, in general, into n linear factors.

$$\therefore \quad \text{P.I.} = \frac{1}{\phi(D, D')} F(x, y)$$

$$= \frac{1}{(D - m_1 D')(D - m_2 D') \ldots (D - m_n D')} F(x, y)$$

$$= \frac{1}{D - m_1 D'} \cdot \frac{1}{D - m_2 D'} \cdots\cdots \frac{1}{D - m_n D'} F(x, y)$$

Let us evaluate $\frac{1}{D - mD'} F(x, y)$

Consider the equation, $(D - mD')z = F(x, y)$ or $p - mq = F(x, y)$ [Lagrange's form]

The auxiliary equations are $\frac{dx}{1} = \frac{dy}{-m} = \frac{dz}{F(x, y)}$

From the first two members $dy + mdx = 0$ or $y + mx = c$

From the first and last members, we have

$$dz = F(x, y)dx = F(x, c - mx)dx$$

$\therefore$ $z = \int F(x, c - mx)dx$ or $\frac{1}{D - mD'} F(x, y) = \int F(x, c - mx)\, dx$

where c is replaced by $y + mx$ after integration.

By repeated application of the above rule, the P.I. can be evaluated.

ILLUSTRATIVE EXAMPLES

Example 1. *Solve* $\frac{\partial^3 z}{\partial x^3} - 3\frac{\partial^3 z}{\partial x^2 \partial y} + 4\frac{\partial^3 z}{\partial y^3} = e^{x+2y}$.

Sol. The given equation in symbolic form is $(D^3 - 3D^2D' + 4D'^3)z = e^{x+2y}$

Its A.E. is $m^3 - 3m^2 + 4 = 0$

or $(m + 1)(m^2 - 4m + 4) = 0$ or $(m + 1)(m - 2)^2 = 0$

$\therefore$ $m = -1, 2, 2$

$$\text{C.F.} = f_1(y - x) + f_2(y + 2x) + xf_3(y + 2x)$$

$$\text{P.I.} = \frac{1}{D^3 - 3D^2D' + 4D'^3} e^{x+2y} = \frac{1}{1^3 - 3.1^2.2 + 4.2^3} e^{x+2y} \quad \text{(Putting D = 1, D}' = 2)$$

$$= \frac{1}{27} e^{x+2y}$$

$\therefore$ The complete solution is $z = f_1(y - x) + f_2(y + 2x) + xf_3(y + 2x) + \frac{1}{27} e^{x+2y}$.

Example 2. *Solve* $(D^3 - 7DD'^2 - 6D'^3)z = \sin (x + 2y) + e^{2x+y}$.

Sol. The A.E. is $m^3 - 7m - 6 = 0$

or $(m + 1)(m^2 - m - 6) = 0$ or $(m + 1)(m - 3)(m + 2) = 0$

$\therefore$ $m = -1, 3, -2$

$$\text{C.F.} = f_1(y - x) + f_2(y + 3x) + f_3(y - 2x)$$

$$\text{P.I.} = \frac{1}{D^3 - 7DD'^2 - 6D'^3} [\sin (x + 2y) + e^{2x+y}]$$

$$= \frac{1}{D^3 - 7DD'^2 - 6D'^3} \sin (x + 2y) + \frac{1}{D^3 - 7DD'^2 - 6D'^3} e^{2x+y}$$

$$= \frac{1}{D(D^2) - 7D(D'^2) - 6D'(D'^2)} \sin (x + 2y) + \frac{1}{2^3 - 7(2)(1)^2 - 6.1^3} e^{2x+y}$$

$$= \frac{1}{D(-1^2) - 7D(-2^2) - 6D'(-2^2)} \sin(x + 2y) - \frac{1}{12} e^{2x+y}$$

$$= \frac{1}{27D + 24D'} \sin(x + 2y) - \frac{1}{12} e^{2x+y}$$

$$= \frac{D}{27D^2 + 24DD'} \sin(x + 2y) - \frac{1}{12} e^{2x+y}$$

$$= \frac{D}{27(-1^2) + 24(-1.2)} \sin(x + 2y) - \frac{1}{12} e^{2x+y} = -\frac{1}{75} \cos(x + 2y) - \frac{1}{12} e^{2x+y}$$

∴ The complete solution is

$$z = f_1(y - x) + f_2(y + 3x) + f_3(y - 2x) - \frac{1}{75} \cos(x + 2y) - \frac{1}{12} e^{2x+y}.$$

Example 3. *Solve* $r - 4s + 4t = e^{2x+y}$.

Sol. The equation is $\frac{\partial^2 z}{\partial x^2} - 4\frac{\partial^2 z}{\partial x \partial y} + 4\frac{\partial^2 z}{\partial y^2} = e^{2x+y}$.

Its symbolic form is $(D^2 - 4DD' + 4D'^2)z = e^{2x+y}$

Its A.E. is $m^2 - 4m + 4 = 0$ or $(m - 2)^2 = 0$ ∴ $m = 2, 2$

$$\text{C.F.} = f_1(y + 2x) + x f_2(y + 2x)$$

$$\text{P.I.} = \frac{1}{(D - 2D')^2} e^{2x+y}$$

Putting D = 2 and D′ = 1, the denominator vanishes.

∴ It is a case of failure.

Since $\frac{1}{D - mD'} F(x, y) = \int F(x, c - mx)dx$, where c is to be replaced by $(y + mx)$ after integration.

$$\therefore \quad \text{P.I.} = \frac{1}{D - 2D'} \cdot \frac{1}{D - 2D'} e^{2x+y} = \frac{1}{D - 2D'} \int e^{2x+(c-2x)}\, dx \quad |\text{Here } m = 2$$

$$= \frac{1}{D - 2D'} \int e^c\, dx = \frac{1}{D - 2D'} x e^c = \frac{1}{D - 2D'} x e^{y+2x}$$

$$= \int x\, e^{(c-2x)+2x}\, dx = e^c \int x dx = \frac{1}{2} x^2 e^c = \frac{1}{2} x^2 e^{y+2x}$$

Hence the complete solution is $z = f_1(y + 2x) + x f_2(y + 2x) + \frac{1}{2} x^2 e^{y+2x}$.

Example 4. *Solve:* $(2D^2 - 5DD' + 2D'^2)z = 5 \sin(2x + y)$.

Sol. The A.E. is $2m^2 - 5m + 2 = 0$ or $(2m - 1)(m - 2) = 0$ ∴ $m = \frac{1}{2}, 2$

$$\text{C.F.} = f_1(2y + x) + f_2(y + 2x)$$

$$\text{P.I.} = 5. \frac{1}{2D^2 - 5DD' + 2D'^2} \sin(2x + y)$$

Putting $D^2 = -2^2$, $DD' = -2.1$, $D'^2 = -1^2$

Denominator $= 2(-4) - 5(-2) + 2(-1) = 0$

∴ It is a case of failure.

Now $$\text{P.I.} = 5 \,.\, \frac{1}{(2D - D')(D - 2D')} \sin(2x + y)$$

$$= 5 \,.\, \frac{1}{2D - D'} \cdot \frac{1}{D - 2D'} \sin(2x + y)$$

$$= 5 \,.\, \frac{1}{2D - D'} \int \sin[2x + (c - 2x)]dx \qquad \text{[Here } m = 2\text{]}$$

$$= \frac{5}{2} \cdot \frac{1}{D - \frac{1}{2}D'} \int \sin c \, dx = \frac{5}{2} \cdot \frac{1}{D - \frac{1}{2}D'} x \sin c$$

$$= \frac{5}{2} \cdot \frac{1}{D - \frac{1}{2}D'} x \sin(y + 2x)$$

$$= \frac{5}{2} \int x \sin\left[\left(c - \frac{1}{2}x\right) + 2x\right] dx \qquad \text{[Here } m = 1/2\text{]}$$

$$= \frac{5}{2} \int x \sin\left(c + \frac{3}{2}x\right) dx$$

$$= \frac{5}{2}\left[x \,.\, \frac{-\cos\left(c + \frac{3}{2}x\right)}{\frac{3}{2}} - \int 1 \,.\, \frac{-\cos\left(c + \frac{3}{2}x\right)}{\frac{3}{2}} dx\right]$$

$$= -\frac{5}{3} x \cos\left(c + \frac{3}{2}x\right) + \frac{10}{9} \sin\left(c + \frac{3}{2}x\right)$$

$$= -\frac{5}{3} x \cos\left[\left(y + \frac{1}{2}x\right) + \frac{3}{2}\right] + \frac{10}{9} \sin\left[\left(y + \frac{1}{2}x\right) + \frac{3}{2}x\right]$$

$$= -\frac{5}{3} x \cos(y + 2x) + \frac{10}{9} \sin(y + 2x)$$

Hence the complete solution is

$$z = f_1(2y + x) + f_2(y + 2x) - \frac{5}{3} x \cos(y + 2x) + \frac{10}{9} \sin(y + 2x)$$

or $$z = f_1(2y + x) + F_2(y + 2x) - \frac{5}{3} x \cos(y + 2x)$$

Since $$f_2(y + 2x) + \frac{10}{9} \sin(y + 2x) = f_2(y + 2x) + \phi(y + 2x) = F_2(y + 2x).$$

Note. If we write

$$\text{P.I.} = 5 \,.\, \frac{1}{(D - 2D')(2D - D')} \sin(2x + y) = \frac{5}{2} \cdot \frac{1}{D - 2D'} \cdot \frac{1}{D - \frac{1}{2}D'} \sin(2x + y)$$

$$= \frac{5}{2} \cdot \frac{1}{D - 2D'} \int \sin\left[2x + \left(c - \frac{1}{2}x\right)\right] dx$$

$$= \frac{5}{2} \cdot \frac{1}{D - 2D'} \left[-\frac{2}{3} \cos\left(\frac{3}{2}x + c\right)\right]$$

$$= -\frac{5}{3}\cdot\frac{D}{D-2D'}\cos\left(\frac{3}{2}x+y+\frac{1}{2}x\right) = -\frac{5}{3}\cdot\frac{1}{D-2D'}\cos(y+2x)$$

$$= -\frac{5}{3}\int \cos[(c-2x)+2x]dx = -\frac{5}{3}x\cos c = -\frac{5}{3}x\cos(y+2x)$$

we directly get the result.

Example 5. *Solve: $r - 2s + t = \sin(2x + 3y)$.*

Or

$(D^2 - 2DD' + D'^2)\, z = \sin(2x + 3y).$

Sol. Here the given differential equation is

$$\frac{\partial^2 z}{\partial x^2} - 2\frac{\partial^2 z}{\partial x\,\partial y} + \frac{\partial^2 z}{\partial y^2} = \sin(2x+3y)$$

i.e., $(D^2 - 2DD' + D'^2)\, z = \sin(2x + 3y)$. ...(1)

Its auxiliary equation is

$$m^2 - 2m + 1 = 0 \Rightarrow (m-1)^2 = 0; \quad \therefore \quad m = 1, 1.$$

$\therefore$ C.F. $= \phi_1(y + x) + x\phi_2(y + x)$.

and the particular integral,

$$\text{P.I.} = \frac{1}{D^2 - 2DD' + D'^2}\sin(2x+3y). \qquad [\text{Here } F(a, b) \neq 0]$$

$$= \frac{1}{2^2 - 2.2.3 + 3^2}\{-\sin(2x+3y)\} \qquad \left[\because \iint \sin v\,(dv)^2 = -\sin v\right]$$

$$= -\text{S in}(2x + 3y).$$

Hence the required solution of the given differential equation is

$$z = \phi_1(y + x) + x\phi_2(y + x) - \sin(2x + 3y).$$

Example 6. *Solve: $(D^2 - 6DD' + 9D'^2)\, z = 6x + 2y$.*

Sol. Here the auxiliary equation is

$$m^2 - 6m + 9 = 0 \quad \text{or} \quad (m-3)^2 = 0; \quad \therefore \quad m = 3, 3.$$

$\therefore$ C.F. $= \phi_1(y + 3x) + x\phi_2(y + 3x)$.

Now, the particular integral,

$$\text{P.I.} = \frac{1}{(D-3D')^2}(6x+2y) \qquad [\text{Here } F(a, b) = 0]$$

$$= \frac{x^2}{2!}\cdot(6x+2y) = x^2(3x+y).$$

Hence the required general solution is

$$z = \phi_1(y + 3x) + x\phi_2(y + 3x) + x^2(3x + y).$$

Example 7. *Solve: $(D^2 - 3DD' + 2D'^2)z = \sin(x - 2y)$.*

Sol. Here the auxiliary equation is

$$m^2 - 3m + 2 = 0 \Rightarrow (m-1)(m-2) = 0; \quad \therefore \quad m = 1, 2.$$

$\therefore$ C.F $= \phi_1(y + x) + \phi_2(y + 2x)$.

Now, $\text{P.I.} = \dfrac{1}{D^2 - 3DD' + 2D'^2} \sin(x - 2y)$ [Here $F(a, b) \neq 0$]

$$= \frac{1}{1^2 - 3(1)(-2) + 2(-2)^2}\{-\sin(x - 2y)\} = -\frac{1}{15}\sin(x - 2y).$$

Hence the general solution of the given differential equation is

$$z = \text{C.F.} + \text{P.I.}$$

or $$z = \phi_1(y + x) + \phi_2(y + 2x) - \frac{1}{15}\sin(x - 2y).$$

Example 8. *Solve:* $(D^2 - 5DD' + 4D'^2)\, z = \sin(4x + y)$.

Sol. Hence A. E. is $m^2 - 5m + 4 = 0 \Rightarrow (m - 1)(m - 4) = 0;\ \therefore\ m = 1, 4.$

$\therefore$ $\text{C.F.} = \phi_1(y + x) + \phi_2(y + 4x).$

$$\text{P.I.} = \frac{1}{D^2 - 5DD' + 4D'^2}\sin(4x + y) \qquad [\because\ f(a, b) = 0]$$

$$= \frac{1}{(D - 4D')}\cdot\frac{1}{(D - D')}\sin(4x + y)$$

$$= \frac{1}{(D - 4D')}\cdot\frac{1}{4 - 1}\{-\cos(4x + y)\}$$

$$= \frac{1}{(D - 4D')}\cdot\left\{-\frac{1}{3}\cos(4x + y)\right\} \qquad [\because\ F(a, b) = 0]$$

$$= -\frac{x}{3}\cos(4x + y).$$

$\therefore$ The solution is $z = f_1(y + x) + f_2(y + 4x) - \dfrac{1}{3}x\cos(4x + y).$

Example 9. *Solve:* $(D^2 + 2DD' + D'^2)\, z = 2\cos y - x\sin y$.

Sol. Here the auxiliary equation is

$m^2 + 2m + 1 = 0$ or $(m + 1)^2 = 0;\ \therefore\ m = -1, -1.$

$\therefore$ $\text{C.F.} = \phi_1(y - x) + x\phi_2(y - x)$

$$\text{P.I.} = \frac{1}{(D + D')^2}(2\cos y - x\sin y) = \frac{1}{D + D'}\frac{1}{D + D'}(2\cos y - x\sin y)$$

$$= \frac{1}{D + D'}\int [2\cos(a + x) - x\sin(a + x)]\, dx, \quad \text{where } a = y - x$$

$$= \frac{1}{D + D'}[2\sin(a + x) + x\cos(a + x) - \sin(a + x)]$$

$$= \frac{1}{D + D'}(\sin y + x\cos y)$$

$$= \int [\sin(b + x) + x\cos(b + x)]\, dx, \quad \text{where } b = y - x$$

$$= -\cos(b + x) + x\sin(b + x) + \cos(b + x) = x\sin(b + x) = x\sin y.$$

Hence the required complete solution is

$\therefore$ $$z = \phi_1(y - x) + x\phi_2(y - x) + x\sin y.$$

TEST YOUR KNOWLEDGE

Solve the following equations:

1. $2\dfrac{\partial^2 z}{\partial x^2}+5\dfrac{\partial^2 z}{\partial x\partial y}+2\dfrac{\partial^2 z}{\partial y^2}=0$

2. $\dfrac{\partial^3 z}{\partial x^3}-4\dfrac{\partial^3 z}{\partial x^2\partial y}+4\dfrac{\partial^3 z}{\partial x\partial y^2}=0$

3. $\dfrac{\partial^3 z}{\partial x^3}-7\dfrac{\partial^3 z}{\partial x\partial y^2}+6\dfrac{\partial^3 z}{\partial y^3}=0$

4. $25r-40s+16t=0$

5. $\dfrac{\partial^2 z}{\partial x^2}-7\dfrac{\partial^2 z}{\partial x\partial y}+12\dfrac{\partial^2 z}{\partial y^2}=e^{x-y}$

6. $(D^2+2DD'+D'^2)z=e^{2x+3y}$

7. $\dfrac{\partial^2 z}{\partial x^2}+\dfrac{\partial^2 z}{\partial x\partial y}-6\dfrac{\partial^2 z}{\partial y^2}=\cos(3x+y)$

8. $(D^3-4D^2D'+4DD'^2)z=6\sin(3x+2y)$

9. $\dfrac{\partial^2 z}{\partial x^2}-2\dfrac{\partial^2 z}{\partial x\partial y}+\dfrac{\partial^2 z}{\partial y^2}=\sin x$

10. $\dfrac{\partial^2 y}{\partial t^2}-a^2\dfrac{\partial^2 y}{\partial x^2}=E\sin pt$

11. $\dfrac{\partial^2 z}{\partial x^2}-\dfrac{\partial^2 z}{\partial x\partial y}=\cos x\cos 2y$

12. $\dfrac{\partial^2 z}{\partial x^2}-3\dfrac{\partial^2 z}{\partial x\partial y}+2\dfrac{\partial^2 z}{\partial y^2}=e^{2x+3y}+\sin(x-2y)$

13. $\dfrac{\partial^2 z}{\partial x^2}+3\dfrac{\partial^2 z}{\partial x\partial y}+2\dfrac{\partial^2 z}{\partial y^2}=12xy$

14. $\dfrac{\partial^2 z}{\partial x^2}+2\dfrac{\partial^2 z}{\partial x\partial y}+\dfrac{\partial^2 z}{\partial y^2}=x^2+xy+y^2$

15. $(D^2-6DD'+9D'^2)z=6x+2y$

16. $\dfrac{\partial^3 z}{\partial x^3}-2\dfrac{\partial^3 z}{\partial x^2\partial y}=2e^{2x}+3x^2y$

17. $(D^2-6DD'+9D'^2)z=12x^2+36xy$

18. $(D^2-2DD'+D'^2)z=e^{x+y}$

19. $4r+12s+9t=e^{3x-2y}$

20. $\dfrac{\partial^3 z}{\partial x^2\partial y}-5\dfrac{\partial^3 z}{\partial x\partial y^2}+6\dfrac{\partial^3 z}{\partial y^3}=e^x$

21. $r+2s+t=2(y-x)+\sin(x-y)$

22. $r+s-6t=\cos(2x+y)$

23. $\dfrac{\partial^3 z}{\partial x^3}-4\dfrac{\partial^3 z}{\partial x^2\partial y}+4\dfrac{\partial^3 z}{\partial x\partial y^2}=4\sin(2x+y)$

24. $(D^2-DD'-2D'^2)z=(y-1)e^x$

25. $(D^2+2DD'+D'^2)z=2\cos y-x\sin y$

26. $4r-4s+t=16\log(x+2y)$.

Answers

1. $z=f_1(y-2x)+f_2(2y-x)$
2. $z=f_1(y)+f_2(y+2x)+xf_3(y+2x)$
3. $z=f_1(y+x)+f_2(y+2x)+f_3(y-3x)$
4. $z=f_1(5y+4x)+xf_2(5y+4x)$
5. $z=f_1(y+3x)+f_2(y+4x)+\dfrac{1}{20}e^{x-y}$
6. $z=f_1(y-x)+xf_2(y-x)+\dfrac{1}{25}e^{2x+3y}$
7. $z=f_1(y-3x)+f_2(y+2x)-\dfrac{1}{6}\cos(3x+y)$
8. $z=f_1(y)+f_2(y+2x)+xf_3(y+2x)+2\cos(3x+2y)$
9. $z=f_1(y+x)+xf_2(y+x)-\sin x$
10. $y=f_1(x-at)+f_2(x-at)-\dfrac{E}{p^2}\sin pt$
11. $z=f_1(y)+f_2(y+x)+\dfrac{1}{2}\cos(x+2y)-\dfrac{1}{6}\cos(x-2y)$

12. $z = f_1(y + x) + f_2(y + 2x) + \frac{1}{4} e^{2x+3y} - \frac{1}{15} \sin (x - 2y)$

13. $z = f_1(y - x) + f_2(y - 2x) + 2x^3y - \frac{3}{2} x^4$

14. $z = f_1(y - x) + xf_2(y - x) + \frac{1}{4} (x^4 - 2x^3y + 2x^2y^2)$

15. $z = f_1(y + 3x) + xf_2(y + 3x) + x^2(3x + y)$

16. $z = f_1(y) + xf_2(y) + f_3(y + 2x) + \frac{1}{4} e^{2x} + \frac{1}{60} (3x^5y + x^6)$

17. $z = f_1(y + 3x) + xf_2(y + 3x) + 10x^4 + 6x^3y$

18. $z = f_1(y + x) + xf_2(y + x) + \frac{1}{2} x^2 e^{x+y}$

19. $z = f_1(2y - 3x) + xf_2(2y - 3x) + \frac{1}{8} x^2 e^{3x-2y}$

20. $z = f_1(x) + f_2(y + 2x) + f_3(y + 3x) + ye^x$

21. $z = f_1(y - x) + xf_2(y - x) + x^2y - x^3 + \frac{1}{2} x^2 \sin (x - y)$

22. $z = f_1(y - 3x) + f_2(y + 2x) + \frac{1}{5} x \sin (2x + y)$

23. $z = f_1(y) + f_2(y + 2x) + xf_3(y + 2x) - x^2 \cos (2x + y)$

24. $z = f_1(y - x) + f_2(y + 2x) + ye^x$

25. $z = f_1(y - x) + xf_2(y - x) + x \sin y$

26. $z = f_1(2y + x) + xf_2(2y + x) + 2x^2 \log (x + 2y)$.

3.13. NON-HOMOGENEOUS LINEAR EQUATIONS

In the equation $\phi(D, D')z = F(x, y)$ ...(1)

if the polynomial $\phi(D, D')$ in D, D′ is not homogeneous, then (1) is called a non-homogeneous linear partial differential equation.

Its complete solution is z = C.F. + P.I.

The methods to find the P.I. are exactly the same as those for homogeneous linear equations.

To find the C.F. we resolve $\phi(D, D')$ into linear factors of the form $D - mD' - a$.

Now consider the equation $(D - mD' - a) z = 0$

or $$p - mq = az \quad ...(2)$$

Lagranges's auxiliary equations are $$\frac{dx}{1} = \frac{dy}{-m} = \frac{dz}{az}$$

From the first two members $dy + mdx = 0 \quad \therefore \quad y + mx = b$

From the first and last members $\frac{dz}{z} = adx \quad \therefore \quad \log z = ax + \log c \quad \text{or} \quad z = ce^{ax}$

$\therefore$ The complete solution of (2) is $z = e^{ax} f(y + mx)$

Hence the C.F. of (1), *i.e.*, the complete solution of

$$(D - m_1D' - a_1)(D - m_2D' - a_2)......(D - m_nD' - a_n)z = 0 \text{ is}$$

$$z = e^{a_1x} f_1(y + m_1x) + e^{a_2x} f_2(y + m_2x) +......+ e^{a_nx} f_n(y + m_nx)$$

Also in the case of repeated factors, for example $(D - mD' - a)^3 z = 0$

We have $z = e^{ax} f_1(y + mx) + xe^{ax} f_2(y + mx) + x^2e^{ax} f_3(y + mx)$.

ILLUSTRATIVE EXAMPLES

Example 1. *Solve* $(D^2 - DD' + D' - 1)z = \cos(x + 2y)$.

Sol. Here the A.E. is $D^2 - DD' + D' - 1 = 0$ or $(D + 1)(D - 1) - D'(D - 1) = 0$

or $(D - 1)(D - D' + 1) = 0$ or $(D - 0D' - 1)[D - D' - (-1)] = 0$

$$\therefore \quad \text{C.F.} = e^x f_1(y + 0x) + e^{-x} f_2(y + x) = e^x f_1(y) + e^{-x} f_2(y + x)$$

$$\text{P.I.} = \frac{1}{D^2 - DD' + D' - 1} \cos(x + 2y)$$

$$= \frac{1}{-1^2 - (-1.2) + D' - 1} \cos(x + 2y)$$

[Putting $D^2 = -1^2$ and $DD' = -1.2$]

$$= \frac{1}{D'} \cos(x + 2y) = \int_{x \text{ constant}} \cos(x + 2y)dy = \frac{1}{2} \sin(x + 2y)$$

Hence the complete solution is $z = e^x f_1(y) + e^{-x} f_2(y + x) + \frac{1}{2} \sin(x + 2y)$.

Example 2. *Solve* $(D^2 - D'^2 + 3D' - 3D)z = e^{x+2y+xy}$.

Sol. Here the A.E. is $D^2 - D'^2 + 3D' - 3D = 0$

or $(D - D')(D + D' - 3) = 0$ or $(D - D' - 0)(D + D' - 3) = 0$

$$\therefore \quad \text{C.F.} = f_1(y + x) + e^{3x} f_2(y - x)$$

P.I. corresponding to e^{x+2y}

$$= \frac{1}{(D - D')(D + D' - 3)} e^{x+2y} = \frac{1}{(D + D' - 3)(1 - 2)} e^{x+2y}$$

$$= -\frac{1}{(D + D' - 3)} e^{x+2y} = -e^x \frac{1}{(1 + D' - 3)} e^{2y} = -e^x \frac{1}{D' - 2} e^{2y}$$

$$= -e^x . e^{2y} \frac{1}{D' + 2 - 2} . 1 = -e^{x+2y} \frac{1}{D'}(1) = -e^{x+2y} \int 1\, dy = -ye^{x+2y}$$

P.I. corresponding to xy

$$= \frac{1}{(D - D')(D + D' - 3)} xy = -\frac{1}{3D}\left(1 - \frac{D'}{D}\right)^{-1}\left(1 - \frac{D}{3} - \frac{D'}{3}\right)^{-1} xy$$

$$= -\frac{1}{3D}\left(1 + \frac{D'}{D} + \ldots\ldots\right)\left(1 + \frac{D}{3} + \frac{D'}{3} + \frac{2DD'}{9} \ldots\ldots\right) xy$$

$$= -\frac{1}{3D}\left(1 + \frac{D}{3} + \frac{D'}{3} + \frac{D'}{D} + \frac{D'}{3} + \frac{2DD'}{9} \ldots\ldots\right) xy$$

$$= -\frac{1}{3D}\left(xy + \frac{y}{3} + \frac{x}{3} + \frac{1}{D} x + \frac{x}{3} + \frac{2}{9}\right) = -\frac{1}{3}\left(\frac{x^2 y}{2} + \frac{xy}{3} + \frac{x^2}{2} + \frac{2x}{9}\right)$$

Hence the complete solution is

$$z = f_1(y + x) + e^{3x} f_2(y - x) - ye^{x+2y} - \frac{1}{3}\left(\frac{x^2y}{2} + \frac{xy}{3} + \frac{x^2}{2} + \frac{2x}{9}\right).$$

Example 3. *Solve (D – 3D′ – 2)³z = 6e²ˣ sin (3x + y).*

Sol. Here the A.E. is $(D - 3D' - 2)^3 = 0$

$\therefore$ $$\text{C.F.} = e^{2x} f_1(y + 3x) + xe^{2x} f_2(y + 3x) + x^2e^{2x} f_3(y + 3x)$$

$$\text{P.I.} = \frac{1}{(D - 3D' - 2)^3} 6e^{2x} \sin(3x + y) = 6e^{2x} \frac{1}{(D + 2 - 3D' - 2)^3} \sin(3x + y)$$

$$= 6e^{2x} \frac{1}{(D - 3D')^3} \sin(3x + y) = 6e^{2x} . \frac{x^3}{6} \sin(3x + y) = x^3e^{2x} \sin(3x + y)$$

Hence complete solution is

$$z = e^{2x} f_1(y + 3x) + xe^{2x} f_2(y + 3x) + x^2e^{2x} f_3(y + 3x) + x^3e^{2x} \sin(3x + y).$$

Note 1. When $\phi(a, b) \neq 0$, then $\frac{1}{(bD - aD')^n} \psi(ax + by) = \frac{x^n}{b^n\, n!} \psi(ax + by).$

Note 2. In case $\phi(D, D')$ cannot be resolved into linear factors in D and D′ the equation cannot be integrated as above and in such cases a trial method is used in find solutions.

Example 4. *Solve $(D^3 - 3DD' + D' + 4)z = e^{2x+y}$.*

Sol. Here $D^3 - 3DD' + D' + 4$ cannot be resolved into linear factors in D and D′.

Let $z = Ae^{hx+ky}$

$\therefore$ $$(D^3 - 3DD' + D' + 4)z = A(h^3 - 3hk + k + 4)\, e^{hx+ky}$$

Then $$(D^3 - 3DD' + D' + k + 4)z = 0 \quad \text{iff} \quad h^3 - 3hk + k + 4 = 0$$

$\therefore$ The C.F. is $z = \Sigma Ae^{hx+ky}$, where $h^3 - 3\,hk + k + 4 = 0$

$$\text{P.I.} = \frac{1}{D^3 - 3DD' + D' + 4} e^{2x+y} = \frac{e^{2x+y}}{2^3 - 3(2)(1) + 1 + 4} = \frac{1}{7} e^{2x+y}$$

Hence complete solution is $z = \Sigma\, Ae^{hx+kh} + \frac{1}{7} e^{2x+y}$, where $h^3 - 3hk + k + 4 = 0.$

3.14. EQUATIONS REDUCIBLE TO PARTIAL DIFFERENTIAL EQUATIONS WITH CONSTANT CO-EFFICIENTS

An equation in which the co-efficient of derivative of any order say k is a multiple of the variables of the degree k then it can be reduced to partial differential equation with constant co-efficients in the following way.

Let $$x = e^X, y = e^Y \text{ so that } X = \log x,\ Y = \log y$$

$$\frac{\partial z}{\partial x} = \frac{\partial z}{\partial X} . \frac{\partial X}{\partial x} = \frac{1}{x} \frac{\partial z}{\partial X}$$

or $$x \frac{\partial z}{\partial x} = \frac{\partial z}{\partial X} \quad \therefore \quad x \frac{\partial}{\partial x} = D\left(= \frac{\partial}{\partial X}\right)$$

Now $$x \frac{\partial}{\partial x}\left(x^{k-1} \frac{\partial^{k-1} z}{\partial x^{k-1}}\right) = x^k \frac{\partial^k z}{\partial x^k} + (k - 1)x^{k-1} \frac{\partial^{k-1} z}{\partial x^{k-1}}$$

or $$x^k \frac{\partial^k z}{\partial x^k} = \left(x \frac{\partial}{\partial x} - k + 1\right) x^{k-1} \frac{\partial^{k-1} z}{\partial x^{k-1}}$$

Putting $k = 2, 3, \ldots\ldots$, we get

$$x^2 \frac{\partial^2 z}{\partial x^2} = (D - 1)x \frac{\partial z}{\partial x} = (D - 1)Dz$$

$$x^3 \frac{\partial^3 z}{\partial x^3} = (D - 2)x^3 \frac{\partial^2 z}{\partial x^2} = (D - 2)(D - 1)Dz \text{ etc.}$$

Similarly, $$y\frac{\partial z}{\partial y} = D'z, \quad y^2 \frac{\partial^2 z}{\partial y^2} = (D' - 1)D'z, \quad y^3 \frac{\partial^3 z}{\partial y^3} = (D' - 2)(D' - 1)D'z \text{ etc.}$$

and $$xy \frac{\partial^2 z}{\partial x \partial y} = DD'z \ldots\ldots$$

Substituting in the given equation it reduces to $\psi(D, D')z = V$ which is an equation with constant co-efficients.

Example 5. *Solve:* $x^2 \dfrac{\partial^2 z}{\partial x^2} - 4xy \dfrac{\partial^2 z}{\partial x \partial y} + 4y^2 \dfrac{\partial^2 z}{\partial y^2} + 6y \dfrac{\partial z}{\partial y} = x^3 y^4.$

Sol. Substituting $x = e^X$, $y = e^Y$ and denoting $D = \dfrac{\partial}{\partial X}$, $D' = \dfrac{\partial}{\partial Y}$ the given equation reduces to

$$[D(D - 1) - 4DD' + 4(D' - 1)D' + 6D']z = e^{3X + 4Y}$$

or $$(D - 2D')(D - 2D' - 1)z = e^{3X+4Y}$$

$\therefore$ The C.F. is $$z = f_1(Y + 2X) + e^X f_2(Y + 2X)$$

or $$z = f_1(\log y + 2 \log x) + e^{\log x} f_2(\log y + 2 \log x)$$

$$= f_1(\log (yx^2)) + xf_2(\log (yx^2)) = g_1(yx^2) + xg_2(yx^2)$$

$$\text{P.I.} = \frac{1}{(D - 2D')(D - 2D' - 1)} e^{3X+4Y} = \frac{e^{3X+4Y}}{[3 - 2(4)]\,[3 - 2(4) - 1]} = \frac{1}{30} x^3 y^4$$

$\therefore$ Complete solution is $z = g_1(yx^2) + xg_2(yx^2) + \dfrac{1}{30} x^3 y^4.$

Example 6. *Solve:* $x^2 \dfrac{\partial^2 z}{\partial x^2} - y^2 \dfrac{\partial^2 z}{\partial y^2} = xy.$

Sol. Substituting $x = e^X$, $y = e^Y$; *i.e.*, $X = \log x$,

$Y = \log y$ in the given equation and denoting $\dfrac{\partial}{\partial X} \equiv D$ and $\dfrac{\partial}{\partial Y} = D'$, we get

$$[D (D - 1) - D' (D' - 1)]\, z = e^X e^Y$$

or $[D^2 - D - D'^2 + D']\, z = e^{X + Y}$ or $(D - D')(D + D' - 1)\, z = e^{X + Y}$ ...(1)

$\therefore$ $$\text{C. F.} = \phi_1(Y + X) + e^Y \phi_2(Y - X) = \phi_1(\log y + \log x) + x\phi_2(\log y - \log x)$$

$$= \phi_1 \{\log (xy)\} + x\phi_2 \{\log (y/x)\} = f_1(xy) + xf_2(y/x)$$

Also, $$\text{P. I.} = \frac{1}{(D-D')(D+D'-1)}\, e^{X+Y} = \frac{1}{D-D'}\cdot\frac{1}{(1+1+1)}\, e^{X+Y}$$

$$= \frac{1}{D-D'}\, e^{X+Y} = e^{X+Y}\cdot\frac{1}{(D+1)-(D'+1)}\, 1 = e^{X+Y}\,\frac{1}{D-D'}\, 1$$

$$= e^{X+Y}\,\frac{1}{D}\left(1-\frac{D'}{D}\right)^{-1} 1 = e^{X+Y}\,\frac{1}{D}(1) = Xe^{X+Y} = \log x\,.\,(xy) = xy\log x.$$

Hence the required solution is

$z = f_1(xy) + xf_2(y/x) + xy \log x.$

Example 7. *Solve:* $x^2\dfrac{\partial^2 z}{\partial x^2} - y^2\dfrac{\partial^2 z}{\partial y^2} = x^2 y.$

Sol. Substituting as in Example 1, the given equation becomes

$(D - D')(D + D' - 1)\, z = e^{2X+Y}.$

$\therefore$ $\quad$ C. F. $= f_1(xy) + xf_2(y/x)$

Now, $$\text{P.I.} = \frac{1}{(D-D')(D+D'-1)}\, e^{2X+Y}$$

$$= \frac{1}{(2-1)(2+1-1)}\, e^{2X+Y} = \frac{1}{2}\, e^{2X+Y} = \frac{1}{2}x^2 y.$$

Hence the required solution is

$$z = f_1(xy) + xf_2(y/x) + \frac{1}{2}x^2 y.$$

Example 8. *Solve:* $x^2\dfrac{\partial^2 z}{\partial x^2} - y^2\dfrac{\partial^2 z}{\partial y^2} - y\dfrac{\partial z}{\partial y} - x\dfrac{\partial z}{\partial x} = 0.$

Sol. Substituting $x = e^X$, $y = e^Y$ so that $X = \log x$, $Y = \log y$, and denoting $\dfrac{\partial}{\partial X}$ and $\dfrac{\partial}{\partial Y}$ by D and D′ respectively the given equation reduces to

$$[D(D-1) - D'(D'-1) - D' + D]\, z = 0$$

$\Rightarrow \quad (D^2 - D'^2)\, z = 0 \;\Rightarrow\; (D - D')(D + D')\, z = 0.$

$\therefore$ The solution is

$$z = f_1(Y + X) + f_2(Y - X) = f_1(\log y \log x) + f_2(\log y - \log x)$$

$$= f_1(\log xy) + f_2\left\{\log\left(\frac{y}{x}\right)\right\} = \phi_1(xy) + \phi_2(y/x).$$

Example 9. *Solve:* $x^2\dfrac{\partial^2 z}{\partial x^2} + 2xy\dfrac{\partial^2 z}{\partial x\,\partial y} + y^2\dfrac{\partial^2 z}{\partial y^2} = 0$

Sol. Substituting $x = e^X$, $y = e^Y$ so that $X = \log x$, $Y = \log y$ and denoting $\dfrac{\partial}{\partial X}$ and $\dfrac{\partial}{\partial Y}$ by D and D′ respectively the given equation reduces to

$$[D(D-1) + 2DD' + D'(D'-1)]\, z = 0 \;\Rightarrow\; (D + D')(D + D' - 1)\, z = 0.$$

$\therefore$ The solution is

$$z = f_1\,(\mathrm{Y}-\mathrm{X}) + e^{\mathrm{X}} f_2(\mathrm{Y}-\mathrm{X}) = f_1\,(\log y - \log x) + x f_2(\log y - \log x)$$

$$= f_1\left(\log\frac{y}{x}\right) + x f_2\left(\log\frac{y}{x}\right) = \phi_1\left(\frac{y}{x}\right) + x\phi_2\left(\frac{y}{x}\right).$$

Example 10. *Solve:* $x^2\dfrac{\partial^2 z}{\partial x^2} + 2xy\dfrac{\partial^2 z}{\partial x\partial y} + y^2\dfrac{\partial^2 z}{\partial y^2} - nx\dfrac{\partial z}{\partial x} - ny\dfrac{\partial z}{\partial y} + nz = x^2 + y^2.$

Sol. Substituting $x = e^{\mathrm{X}}$, $y = e^{\mathrm{Y}}$.

If D and D′ denote $\dfrac{\partial}{\partial \mathrm{X}}$ and $\dfrac{\partial}{\partial \mathrm{Y}}$, then given equation reduces to

$$[\mathrm{D}(\mathrm{D}-1) + 2\mathrm{DD}' + \mathrm{D}'(\mathrm{D}'-1) - n\mathrm{D} - n\mathrm{D}' + n]z = e^{2\mathrm{X}} + e^{2\mathrm{Y}}$$

$$\Rightarrow \quad (\mathrm{D}+\mathrm{D}'-1)(\mathrm{D}+\mathrm{D}'-n)\,z = e^{2\mathrm{X}} + e^{2\mathrm{Y}}.$$

$$\therefore \quad \text{C. F.} = e^{\mathrm{X}} f_1\,(\mathrm{Y}-\mathrm{X}) + e^{n\mathrm{X}} f_2(\mathrm{Y}-\mathrm{X})$$

$$= x f_1\,(\log y - \log x) + x^n f_2(\log y - \log x)$$

$$= x f_1\left(\log\frac{y}{x}\right) + x^n f_2\left(\log\frac{y}{x}\right) = x\phi_1\left(\frac{y}{x}\right) + x^n\phi_2\left(\frac{y}{x}\right).$$

$$\text{P.I.} = \frac{1}{(\mathrm{D}+\mathrm{D}'-1)(\mathrm{D}+\mathrm{D}'-n)}\cdot e^{2\mathrm{X}} + \frac{1}{(\mathrm{D}+\mathrm{D}'-1)(\mathrm{D}-\mathrm{D}'-n)}\,e^{2\mathrm{Y}}$$

$$= \frac{e^{2\mathrm{X}}}{(2+0-1)(2+0-n)} + \frac{e^{2\mathrm{Y}}}{(0+2-1)(0+2-n)}$$

$$= \frac{e^{2\mathrm{X}} + e^{2\mathrm{Y}}}{-n} = \frac{x^2+y^2}{2-n}.$$

Hence the solution is

$$z = x\phi_1\left(\frac{y}{x}\right) + x^n\phi_2\left(\frac{y}{x}\right) + \frac{x^2+y^2}{2-n}.$$

TEST YOUR KNOWLEDGE

Solve the following equations:

1. $\mathrm{DD}'(\mathrm{D}+2\mathrm{D}'+1)\,z = 0$
2. $r + 2s + t + 2p + 2q + z = 0$
3. $\mathrm{D}(\mathrm{D}-2\mathrm{D}'-3)z = e^{x+2y}$
4. $(\mathrm{D}^2 + 2\mathrm{DD}' + \mathrm{D}'^2 - 2\mathrm{D} - 2\mathrm{D}')z = \sin(x+2y)$
5. $(\mathrm{D}-\mathrm{D}'-1)(\mathrm{D}-\mathrm{D}'-2)z = e^{2x-y} + x$
6. $(\mathrm{D}+\mathrm{D}'-1)(\mathrm{D}+2\mathrm{D}'-3)z = 4 + 3x + 6y$
7. $(\mathrm{D}^2 - \mathrm{DD}' + \mathrm{D}' - 1)z = \cos(x+2y) + e^y$
8. $(\mathrm{D}^2 - \mathrm{D}')z = 2y - x^2$
9. $x^2 r + 2xys + y^2 t = 0$
10. $x^2 r - 3xys + 2y^2 t + px + 2qy = x + 2y.$

Answers

1. $z = f_1(y) + f_2(-x) + e^{-x} f_3(y-2x)$
2. $z = e^{-x} f_1(y-x) + xe^{-x} f_2(y-x)$
3. $z = f_1(y) + e^{3x} f_2(y+2x) - \dfrac{1}{6}\, e^{x+2y}$

4. $z = f_1(y - x) + e^{2x} f_2(y - x) + \frac{1}{39}[2\cos(x + 2y) - 3\sin(x + 2y)]$

5. $z = e^x f_1(y + x) + e^{2x} f_2(y + x) + \frac{1}{2}e^{2x-y} + \frac{1}{4}(2x + 3)$

6. $z = e^x f_1(y - x) + e^{3x} f_2(y - 2x) + x + 2y + 6$ **7.** $z = e^x f_1(y) + e^{-x} f_2(y + x) + \frac{1}{2}\sin(x + 2y) - xe^y$

8. $z = \sum Ae^{hx+h^2y} + x^2y$ **9.** $z = f_1\left(\frac{y}{x}\right) + xf_2\left(\frac{y}{x}\right)$

10. $z = f_1(xy) + f_2(x^2y) + x + y.$

3.15. NON-LINEAR PARTIAL DIFFERENTIAL EQUATION OF SECOND ORDER

Equation which includes at least one of the partial derivatives r, s, t but none of higher order is called a partial differential equation of second order. We now study a method due to Monge's to solve a non-linear partial differential equation of the form $Rr + Ss + Tt = V$, where R, S, T and V are functions of x, y, z, p and q.

3.16. MONGE'S METHOD OF INTEGRATING Rr + Ss + Tt = V

The given equation is $Rr + Ss + Tt = V$ ...(1)

$$dp = \frac{\partial p}{\partial x}dx + \frac{\partial p}{\partial y}dy = rdx + sdy$$

or
$$r = \frac{dp - sdy}{dx}$$

Similarly $dp = sdx + tdy$ or $t = \dfrac{dq - sdx}{dy}$

Substituting r and t in (1), we get

$$(Rdpdy + Tdqdx - Vdxdy) - s(Rdy^2 - Sdydx + Tdx^2) = 0 \quad ...(2)$$

If some relation between x, y, z, p and q exists such that

$$Rdpdy + Tdqdx - Vdxdy = 0 \quad ...(3)$$

$$Rdy^2 - Sdydx + Tdx^2 = 0 \quad ...(4)$$

then that relation will satisfy (2). Equations (3) and (4) are called **Monge's equations.**

Let Eq. (2) be resolved into two linear factors say

$$dy - \lambda_1 dx = 0 \quad ...(5)$$

$$dy - \lambda_2 dx = 0 \quad ...(6)$$

Now (3) and (5), combined if necessary with $dz = pdx + qdy$, we obtain two integrals $u_1 = c_1$ and $v_1 = c_2$.

Then the relation

$$u_1 = f(v_1) \quad ...(7)$$

is the solution.

Similarly from (3) and (6), we obtain another integral say

$$u_2 = f(v_2) \quad ...(8)$$

Solving equations (7) and (8) for p and q in terms of x and y and using $z = \int pdx + \int qdy$, we get the solution of (1).

Note. If $\lambda_1 = \lambda_2$, then equation (4) becomes perfect square, then we get only one integral which is of the form $Pp + Qq = R$, which can be solved with the help of Lagranges method.

ILLUSTRATIVE EXAMPLES

Example 1. *Solve $r = a^2t$ by Monge's method.*

Sol. The given equation is $r = a^2t$.

Comparing it with the standard equation $R = 1, T = -a^2, S = 0, V = 0$

$\therefore$ Monge's equations are $dpdy - a^2dqdx = 0$ ...(1)

$$dy^2 - a^2dx^2 = 0 \quad ...(2)$$

From equation (2), we have $dy - adx = 0$ ...(3)

$$dy + adx = 0 \quad ...(4)$$

From (1) and (3), $dp(adx) - a^2\, dq\, dx = 0$

or $dp - adq = 0$ ...(5)

From (3) and (5), we obtain $y - ax = c_1$ and $p - aq = c_2$

$\therefore$ $p - aq = f(y - ax)$ ...(6)

Similarly from (4) and (1), we obtain $p + aq = g(y + ax)$ ...(7)

Solving (6) and (7) for p and q, we get

$$p = \frac{1}{2}\{f(y - ax) + g(y + ax)\}$$

and

$$q = \frac{1}{2a}\{g(y + ax) - f(y - ax)\}$$

Substituting in $dz = pdx + qdy$

$$dz = \frac{1}{2}\{f(y - ax) + g(y + ax)\}\, dx + \frac{1}{2a}\{g(y + ax) - f(y - ax)\}\, dy$$

$$= \frac{1}{2a}\{g(y + ax)(dy + adx) - f(y - ax)(dy - adx)\}$$

Integrating, $z = \dfrac{1}{2a}\{G(y + ax) - F(y - ax)\}$ or $z = G_1(y + ax) + F_1(y - ax)$

which is the required solution.

Example 2. *Solve $r - t\cos^2 x + p\tan x = 0$.*

Sol. Comparing the given equation with $Rr + Ss + Tt = V$, we have

$$R = 1, S = 0, t = -\cos^2 x, V = -p\tan x$$

Substituting in Monge's equations, we get

$$dpdy - \cos^2 x\, dqdx + p\tan x\, dxdy = 0 \quad ...(1)$$

and $dy^2 - \cos^2 x\, dx^2 = 0$ ...(2)

Equation (2) gives $dy - \cos x\, dx = 0$ ...(3)

$$dy + \cos x\, dx = 0 \quad ...(4)$$

From (3), we have $y - \sin x = c_1$

Again from (3) and (1), we have

$$dp(\cos x\, dx) - \cos^2 x\, dqdx + p\tan x\, dx \cos x\, dx = 0$$

or $dp - \cos x\, dq + p\tan x\, dx = 0$ or $\sec x\, dp - dq + p\tan x \sec x\, dx = 0$

or $d(p\sec x) - dq = 0$ or $p\sec x - q = c_2$

$\therefore$ $p\sec x - q = f(y - \sin x)$ ...(5)

Similarly from (4) and (1), we have

$$p \sec x + q = g(y + \sin x) \quad ...(6)$$

Solving (5) and (6), we have

$$p = \frac{1}{2} \cos x[f(y - \sin x) + g(y + \sin x)]$$

$$q = \frac{1}{2} [g(y + \sin x) - f(y - \sin x)]$$

Substituting in $dz = pdx + qdy$, we have

$$dz = \frac{1}{2} \cos x[f(y - \sin x) + g(y + \sin x)]\, dx + \frac{1}{2} [g(y + \sin x) - f(y - \sin x)]\, dy$$

$$= \frac{1}{2} [g(y + \sin x)(dy + \cos x\, dx) - f(y - \sin x)(dy - \cos x\, dx)$$

or $$z = \frac{1}{2} [G(y + \sin x) + F(y - \sin x)]$$

which is the required solution.

Example 3. *Solve* $\quad q(q^2 + s) = pt.$

Sol. Comparing with the standard equation

$$R = 0, S = - q, T = p, V = q^3$$

Monge's equations are $pdqdx - q^3\, dxdy = 0$

and $$qdxdy + pdx^2 = 0$$

or $$pdq - q^3\, dy = 0 \quad ...(1)$$

and $$qdy + pdx = 0 \quad ...(2)$$

But $$dz = pdx + qdy = 0 \;\Rightarrow\; z = c_1$$

Combining (1) and (2), we have

$$pdq - q^2(- pdx) = 0 \text{ or } dq + q^2 dx = 0$$

or $$\frac{dq}{q^2} + dx = 0 \text{ or } -\frac{1}{q} + x = c_2$$

or $$\frac{\partial y}{\partial z} - x = - f(z), \text{ where } c_2 = f(z)$$

Integrating, $\quad y = xz - \int f(z)\, dz + c_3 \;$ or $\; y = xz + g(z) + h(z)$

which is the required solution.

Example 4. *Solve:* $pt - qs = q^3$.

Sol. We know that $\quad dq = \frac{\partial q}{\partial x} dx + \frac{\partial q}{\partial y} dy = s\, dx + t\, dy.$

$$\therefore \quad t = \frac{dq - s\,dx}{dy}.$$

Substituting the value of t in the given equation, we have

$$p\frac{dq - s\,dx}{dy} - qs = q^3 \text{ or } (p\, dq - q^3\, dy) - s\,(p\, dx - q\, dy) = 0.$$

∴ The Monge's subsidiary equations are

$$p\,dq - q^3\,dy = 0 \qquad ...(1)$$

and $$p\,dx + q\,dy = 0. \qquad ...(2)$$

From (2), we have

$$dz = 0 \qquad [\because\ dz = p\,dx + q\,dy]$$

which on integration gives

$$z = A. \qquad ...(3)$$

Combining (1) and (2), we have

$$dq + q^2 dx = 0 \quad \text{or} \quad dq/q^2 + dx = 0 \quad \text{or} \quad d\left(-\frac{1}{q} + x\right) = 0$$

which on integration given

$$-\frac{1}{q} + x = B \quad \text{or} \quad -\frac{1}{q} + x = f(z) \quad \text{or} \quad \frac{\partial y}{\partial z} - x = -f(z).$$

Integrating, we get

$$y - xz = -\int_{x = \text{constant}} f(z)\,dz + c$$

$$\therefore \qquad y = xz + f_1(z) + c \quad \text{or} \quad y = xz + f_1(z) + f_2(x),$$

Since c is a function of x which is regarded constant at the time of integration.

TEST YOUR KNOWLEDGE

Solve the following equations by Monge's method:

1. $r = t$
2. $x^2 - 2xs + t + q = 0$
3. $t - r\sec^4 y - 2q\tan y = 0$
4. $y^2 r - 2ys + t = p + 6y$
5. $x^2 r + 2yxs + y^2 t = 0$
6. $(x - y)(xr - xs - ys + yt) = (x + y)(p - q)$
7. $r + (a + b)s + abt = xy.$

Answers

1. $z = F(y - x) + G(y + x)$
2. $z = f(y + \log x) + xg(y + \log x)$
3. $z = f(x - \tan y) + g(x + \tan y)$
4. $z = y^3 - yf(y^2 + 2x) + g(y^2 + 2x)$
5. $z = f_1(y/x) + yg(y/x)$
6. $z = f(xy) + g(x + y)$
7. $Z = \frac{1}{6}x^3y - \frac{1}{24}(a + b)x^4 + f(y - ax) + g(y - bx).$

APPLICATIONS OF PARTIAL DIFFERENTIAL EQUATIONS

3.17. INTRODUCTION

Many physical and engineering problems when formulated in the mathematical language give rise to partial differential equations. Besides these, partial differential equations also play an important role in the theory of Elasticity, Hydraulics etc.

Since the general solution of a partial differential equation in a region R contains arbitrary constants or arbitrary functions, the unique solution of a partial differential equation corresponding to a physical problem will satisfy certain other conditions at the boundary of the region R. These are known as *boundary conditions.* When these conditions are specified for the time $t = 0$, they are known as *initial conditions.* A partial differential equation together with boundary conditions constitutes a *boundary value problem.*

In the applications of ordinary linear differential equations, we first find the general solution and then determine the arbitrary constants from the initial values. But the same method is not applicable to problems involving partial differential equations. Most of the boundary value problems involving linear partial differential equations can be solved by the method of separation of variables. In this method, right from the beginning, we try to find the particular solutions of the partial differential equation which satisfy all or some of the boundary conditions and then adjust them till the remaining conditions are also satisfied. A combination of these particular solutions gives the solution of the problem.

Fourier series is a powerful aid in determining the arbitrary functions.

3.18. METHOD OF SEPARATION OF VARIABLES

In this method, we assume the solution to be the product of two functions, each of which involves only one of the variables. The following examples explain the method.

ILLUSTRATIVE EXAMPLES

Example 1. *Solve the equation* $\dfrac{\partial u}{\partial x} = 2\dfrac{\partial u}{\partial t} + u$, *given* $u(x, 0) = 6e^{-3x}$.

Sol. Here u is a function of x and t.

Let $$u = XT \qquad ...(1)$$

where X is a function of x only and T is a function of t only, be a solution of the given equation.

Then $$\frac{\partial u}{\partial x} = X'T, \quad \frac{\partial u}{\partial t} = XT'$$

Substituting in the given equation, we have

$$X'T = 2XT' + XT \quad \text{or} \quad X'T = (2T' + T)X$$

Separating the variables, we get $$\frac{X'}{X} = \frac{2T' + T}{T} \qquad ...(2)$$

Since x and t are independent variables, as t varies x remains constant, so that the L.H.S. and hence the R.H.S. is constant. Therefore, equation (2) can hold only when each side is equal to the some constant, say k.

$$\therefore \quad \frac{X'}{X} = k \quad \textit{i.e.,} \quad \log X = kx + \log c_1$$

$$\text{or} \quad \log \frac{X}{c_1} = kx \quad \text{or} \quad X = c_1 e^{kx} \qquad ...(3)$$

$$\text{and} \quad \frac{2T' + T}{T} = k \quad \textit{i.e.,} \quad \frac{T'}{T} = \frac{1}{2}(k - 1)$$

or $$\log T = \frac{1}{2}(k-1)t + \log c_2 \quad \text{or} \quad \log \frac{T}{c_2} = \frac{1}{2}(k-1)t$$

or $$T = c_2 \, e^{1/2\,(k-1)t} \qquad \text{...(4)}$$

From (1), (3) and (4), we have $u = u(x, t) = c_1 e^{kx} . \, c_2 e^{1/2\,(k-1)t}$

Since $u(x, 0) = 6e^{-3x}$

$\therefore$ $c_1 c_2 e^{kx} = 6e^{-3x}$ (given)

$\Rightarrow$ $c_1 c_2 = 6$ and $k = -3$

$\therefore$ The unique solution of the given equation is

$$u = 6e^{-3x} . \, e^{-2t} \quad i.e., \quad u = 6e^{-(3x + 2t)}.$$

Example 2. *Solve the equation* $\frac{\partial^2 z}{\partial x^2} - 2\frac{\partial z}{\partial x} + \frac{\partial z}{\partial y} = 0$ *by the method of separation of variables.*

Sol. Here z is a function of x and y

Let $z = XY$ ...(1)

where X is a function of x only and Y is a function of y only, be a solution of the given equation.

Then $$\frac{\partial z}{\partial x} = X'Y, \quad \frac{\partial z}{\partial y} = XY' \quad \text{and} \quad \frac{\partial^2 z}{\partial x^2} = X''Y.$$

Substituting in the given equation, we have

$$X''Y - 2X'Y + XY' = 0 \quad \text{or} \quad (X'' - 2X')Y + XY' = 0$$

Separating the variables, we get $$\frac{X'' - 2X'}{X} = -\frac{Y'}{Y} \qquad \text{...(2)}$$

Since x and y are independent variables, equation (2) can hold only when each side is equal to some constant, say k.

$$\therefore \quad \frac{X'' - 2X'}{X} = k \quad \text{or} \quad X'' - 2X' - kX = 0 \qquad \text{...(3)}$$

and $$-\frac{Y'}{Y} = k \quad \text{or} \quad Y' + kY = 0 \qquad \text{...(4)}$$

These are ordinary differential equations.

For equation (3), the auxiliary equation is $m^2 - 2m - k = 0$, whose roots are $m = 1 \pm \sqrt{1+k}$.

$\therefore$ The solution of equation (3) is $X = c_1 e^{(1+\sqrt{1+k})x} + c_2 e^{(1-\sqrt{1+k})x}$

Also, the solution of (4) is $Y = c_3 e^{-ky}$

Substituting these values of X and Y in (1), the complete solution of the given equation is

$$z = [c_1 e^{(1+\sqrt{1+k})x} + c_2 e^{(1-\sqrt{1+k})x}] . \, c_3 e^{-ky}$$

or $$z = [Ae^{(1+\sqrt{1+k})x} + Be^{(1-\sqrt{1+k})x}]\, e^{-ky}, \quad \text{where } A = c_1 c_3 \text{ and } B = c_2 c_3.$$

[The arbitrary constants A, B and k can be determined so as to satisfy the given boundary conditions.]

Example 3. *Solve the following equation by the method of separation of variables;*

$$4\frac{\partial u}{\partial x}+\frac{\partial u}{\partial y} = 3u, \text{ given } u = 3e^{-y} - e^{-5y} \text{ when } x = 0.$$

Sol. Here u is a function of x and y.

Let $u = XY$ ...(1)

where X is a function of x only and Y is a function of y only, be a solution of the given equation.

Then $\frac{\partial u}{\partial x} = X'Y, \frac{\partial u}{\partial y} = XY'$

Substituting in the given equation, we have

$$4X'Y + XY' = 3XY \quad \text{or} \quad 4X'Y = X(3Y - Y')$$

Separating the variables, we get

$$\frac{4X'}{X} = \frac{3Y - Y'}{Y}$$

Since L.H.S is a function of x only while R.H.S. is a function of y only, the equality can hold only when each side is equal to the same constant, say k.

$$\therefore \quad \frac{4X'}{X} = k \quad \text{or} \quad \frac{X'}{X} = \frac{1}{4}k$$

Integrating $\log X = \frac{1}{4}kx + \log c_1$

$$\log\frac{X}{c_1} = \frac{1}{4}kx \quad \text{or} \quad \frac{X}{c_1} = e^{\frac{1}{4}kx}$$

$$\therefore \quad X = c_1 e^{\frac{1}{4}kx}$$

Also $\frac{3Y - Y'}{Y} = k$ or $3 - \frac{Y'}{Y} = k$

or $\frac{Y'}{Y} = 3 - k$. Integrating $\log Y = (3 - k)y + \log c_2$

or $\log\frac{Y}{c_2} = (3 - k)y$ or $\frac{Y}{c_2} = e^{(3-k)y}$

$$\therefore \quad Y = c_2 e^{(3-k)y}$$

$\therefore$ From (1), $u = c_1 e^{\frac{1}{4}kx} \cdot c_2 e^{(3-k)y} = c\, e^{\frac{1}{4}kx + (3-k)y}$ ...(2)

where $c = c_1c_2$.

Now $u(0, y) = 3e^{-y} - e^{-5y}$, which consists of two terms.

$\Rightarrow$ $u = u(x, y)$ must also consist of two terms.

$\Rightarrow$ k must have two values, say k_1 and k_2. Let the corresponding values of c be c_1 and c_2 respectively.

$\therefore$ $u = c_1 e^{\frac{1}{4}k_1x + (3-k_1)y} + c_2 e^{\frac{1}{4}k_2x + (3-k_2)y}$ ...(3)

Putting $x = 0$, we get

$$u = c_1 e^{(3-k_1)y} + c_2 e^{(3-k_2)y}$$

But $u = 3e^{-y} - e^{-5y}$ when $x = 0$

$$\therefore \quad c_1 e^{(3-k_1)y} + c_2 e^{(3-k_2)y} = 3e^{-y} - e^{-5y}$$

Comparing corresponding terms, we have

$$c_1 = 3, \quad 3 - k_1 = -1$$

$$c_2 = -1, 3 - k_2 = -5$$

or
$$c_1 = 3, c_2 = -1, k_1 = 4, k_2 = 8$$

Putting in (3), the required solution is

$$u = 3e^{x-y} - e^{2x-5y}.$$

Example 4. *Solve:* $2x\dfrac{\partial z}{\partial x} - 3y\dfrac{\partial z}{\partial y} = 0$ *by the method of separation of variables.*

Sol. Here z is a function of x and y.

Let
$$z = XY \qquad \text{...(1)}$$

where X is a function of x only and Y is a function of y only, be a solution of the given equation.

Then
$$\frac{\partial z}{\partial x} = X'Y \quad \text{and} \quad \frac{\partial z}{\partial y} = XY'$$

Substituting in the given equation, we have

$$2x\,X'Y - 3yXY' = 0 \quad \text{or} \quad 2xX'Y = 3yXY'$$

Separating the variables, we get

$$\frac{xX'}{3X} = \frac{yY'}{2Y} = k \text{ (say)}$$

Now
$$\frac{xX'}{3X} = k \Rightarrow \frac{x}{3X}\cdot\frac{dX}{dx} = k \quad \text{or} \quad \frac{dX}{X} = 3k\frac{dx}{x}$$

Integrating
$$\log X = 3k \log x + \log c_1$$

or
$$\log X + \log c_1 x^{3k} \quad \text{or} \quad X = c_1 x^{3k}$$

Also
$$\frac{yY'}{2Y} = k \Rightarrow \frac{y}{2Y}\cdot\frac{dY}{dy} = k \quad \text{or} \quad \frac{dY}{Y} = 2k\frac{dy}{y}$$

Integrating
$$\log Y = 2\,k \log y + \log c_2$$

or
$$\log Y = \log c_2 y^{2k} \quad \text{or} \quad Y = c_2 y^{2k}$$

Substituting the values of X and Y in (1), the complete solution of the given equation is

$$z = (c_1 x^{3k})(c_2 y^{2k})$$

or
$$z = A x^{3k} y^{2k}, \quad \text{where} \quad A = c_1 c_2.$$

Example 5. *Using the method of separation of variables, solve*

$$\frac{\partial^2 z}{\partial x^2} = \frac{\partial z}{\partial y} \text{ given } z = 0 \text{ when } x = 0 \text{ and } \pi\,;\ z = \sin 3x \text{ when } y = 0 \text{ and } 0 < x < \pi.$$

Sol. Here z is a function of x and y.

Let
$$z = XY \qquad \text{...(1)}$$

where X is a function of x only and Y is a function of y only, be a solution of the given equation.

Then
$$\frac{\partial^2 z}{\partial x^2} = X''Y \quad \text{and} \quad \frac{\partial z}{\partial y} = XY'$$

Substituting in the given equation, we have

$$X''Y = XY'$$

Separating the variables, we get

$$\frac{X''}{X} = \frac{Y'}{Y} \qquad ...(2)$$

Since x and y are independent variables, as x varies R.H.S. is a constant and as y varies L.H.S. is a constant. The equality can hold only when each side is equal to the same constant. The given form of the boundary conditions implies that $X = X(x)$ must be of trigonometric form. To satisfy this requirement, the constant of separation in (2) is taken as $-k^2$. Then (2) gives

$$\frac{X''}{X} = -k^2 \qquad ...(3)$$

and

$$\frac{Y'}{Y} = -k^2 \qquad ...(4)$$

From (3), $X'' + k^2X = 0$ or $(D^2 + k^2)X = 0$

whose solution is $X = c_1 \cos kx + c_2 \sin kx$

Also , from (4) $\frac{Y'}{Y} = -k^2$ or $\frac{dY}{dy} = -k^2Y$ or $\frac{dY}{Y} = -k^2 dy$

Integrating $\log Y = -k^2y + \log c_3 \quad \therefore \quad Y = c_3\, e^{-k^2y}$

Substituting the values of X and Y in (1), we get

$$z = (c_1 \cos kx + c_2 \sin kx) \,.\, c_3 e^{-k^2y}$$

$$= (c_1 c_3 \cos kx + c_2c_3 \sin kx)\, e^{-k^2y}$$

or

$$z = (A \cos kx + B \sin kx)\, e^{-k^2y} \qquad ...(5)$$

where $A = c_1c_3$, $B = c_2c_3$.

Now $z = 0$ when $x = 0$ $\therefore$ From (5)

$$0 = Ae^{-k^2y} \quad \text{or} \quad A = 0 \qquad (\because \; e^{-k^2y} \neq 0)$$

$\therefore$ Equation (5) reduces to

$$z = B \sin kx.\, e^{-k^2y} \qquad ...(6)$$

Also $z = 0$ when $x = \pi$ $\therefore$ From (6)

$$0 = B \sin k\pi\, e^{-k^2y} \quad \text{or} \quad B \sin k\pi = 0 \qquad (\because \; e^{-k^2y} \neq 0)$$

$\Rightarrow$ either $B = 0$ or $\sin k\pi = 0$

For non-trivial solution, B must be non-zero.

$\therefore$ $\sin k\pi = 0$ so that k must be an integer.

Since $z = \sin 3x$ when $y = 0$ and $0 < x < \pi$ $\therefore$ From (6)

$$\sin 3x = B \sin kx$$

This is possible only when $B = 1$ and $k = 3$

$\therefore$ From (6), the required solution is

$$z = \sin 3x \,.\, e^{-9y}.$$

TEST YOUR KNOWLEDGE

1. Solve the following equations by the method of separation of variables:

(*i*) $3\frac{\partial u}{\partial x} + 2\frac{\partial u}{\partial y} = 0$, where $u(x, 0) = 4e^{-x}$ (*ii*) $\frac{\partial u}{\partial x} = 4\frac{\partial u}{\partial y}$, where $u(0, y) = 8e^{-3y}$

(*iii*) $4\frac{\partial u}{\partial x} + \frac{\partial u}{\partial y} = 3u$, given $u = 3e^{-y} - e^{-5y}$, when $x = 0$.

2. Find a solution of the equation $\frac{\partial^2 u}{\partial x^2} = \frac{\partial u}{\partial y} + 2u$ in the form $u = f(x)g(y)$. Solve the equation subject to the conditions $u = 0$ and $\frac{\partial u}{\partial x} = 1 + e^{-3y}$, when $x = 0$ for all values of y.

3. Solve the equation $\frac{\partial^2 u}{\partial x \partial t} = e^{-t} \cos x$ given that $u = 0$, when $t = 0$ and $\frac{\partial u}{\partial t} = 0$, when $x = 0$.

4. Obtain the solution of the equation $\frac{\partial^2 z}{\partial x \partial y} = \sin x \sin y$ for which $\frac{\partial z}{\partial y} = -2 \sin y$, when $x = 0$ and $z = 0$, when y is an odd multiple of $\frac{\pi}{2}$.

Answers

1. (*i*) $u = 4e^{\frac{1}{2}(3y - 2x)}$ (*ii*) $u = 8e^{-12x - 3y}$ (*iii*) $u = 3e^{x-y} - e^{2x-5y}$

2. $u = (c_1 e^{\sqrt{kx}} + c_2 e^{-\sqrt{kx}}) \cdot e^{(k-2)y}$; $u = \frac{1}{\sqrt{2}} \sinh \sqrt{2}x + e^{-3y} \sin x$

3. $u = (1 - e^{-t}) \sin x$ 4. $z = \cos y(1 + \cos x)$.

3.19. VIBRATIONS OF A STRETCHED STRING, ONE DIMENSIONAL WAVE EQUATION

$$\frac{\partial^2 \mathbf{y}}{\partial \mathbf{t}^2} = \mathbf{c}^2 \frac{\partial^2 \mathbf{y}}{\partial \mathbf{x}^2}$$

Consider a uniform elastic string of length l stretched tightly between two points O and A, and displaced slightly from its equilibrium position OA. Taking the end O as the origin, OA as the x-axis and a perpendicular line through O as the y-axis, we shall find the displacement y as a function of the distance x and the time t.

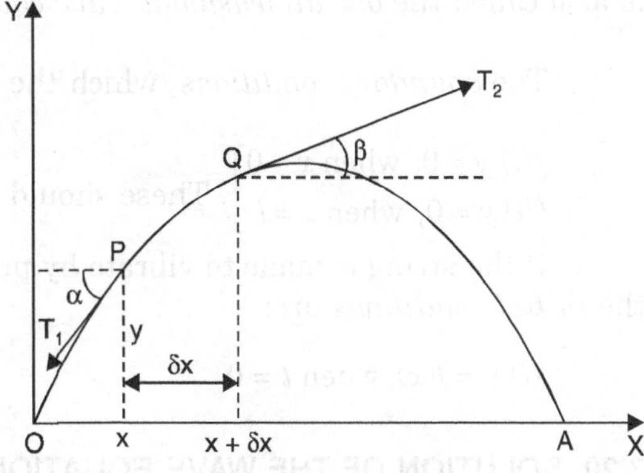

We shall obtain the equation of motion for the string under the following assumptions:

(*i*) The motion takes place entirely in the xy-plane and each particle of the string moves perpendicular to the equilibrium position OA of the string.

(*ii*) The string is perfectly flexible and does not offer resistance to bending.

(*iii*) The tension in the string is so large that the forces due to weight of the string can be neglected.

(*iv*) The displacement y and the slope $\frac{\partial y}{\partial x}$ are small, so that their higher powers can be neglected.

Let m be the mass per unit length of the string. Consider the motion of an element PQ of length δs. Since the string does not offer resistance to bending (by assumption), the tensions T_1 and T_2 at P and Q respectively are tangential to the curve.

Since there is no motion in the horizontal direction, we have

$$T_1 \cos \alpha = T_2 \cos \beta = T \text{ (constant)} \qquad \text{...(1)}$$

Mass of element PQ is $m\delta s$. By Newton's second law of motion, the equation of motion in the vertical direction is

$$m\delta s \frac{\partial^2 y}{\partial t^2} = T_2 \sin \beta - T_1 \sin \alpha$$

or $$\frac{m\delta s}{T} \frac{\partial^2 y}{\partial t^2} = \frac{T_2 \sin \beta}{T_2 \cos \beta} - \frac{T_1 \sin \alpha}{T_1 \cos \alpha} \qquad \text{[By using (1)]}$$

or $$\frac{\partial^2 y}{\partial t^2} = \frac{T}{m\delta s} (\tan \beta - \tan \alpha)$$

or $$\frac{\partial^2 y}{\partial t^2} = \frac{T}{m\delta x} \left[\left(\frac{\partial y}{\partial x} \right)_{x + \delta x} - \left(\frac{\partial y}{\partial x} \right)_x \right]$$

[Since $\delta s = \delta x$ to a first approximation and $\tan \alpha$ and $\tan \beta$ are the slopes of the curve of the string at x and $x + \delta x$]

or $$\frac{\partial^2 y}{\partial t^2} = \frac{T}{m} \left[\frac{\left(\frac{\partial y}{\partial x} \right)_{x + \delta x} - \left(\frac{\partial y}{\partial x} \right)_x}{\delta x} \right] = \frac{T}{m} \frac{\partial^2 y}{\partial x^2}, \text{ as } \delta x \to 0$$

or $$\frac{\partial^2 y}{\partial t^2} = c^2 \frac{\partial^2 y}{\partial x^2}, \quad \text{where } c^2 = \frac{T}{m}$$

This is the partial differential equation giving the transverse vibrations of the string. It is also called the *one dimensional wave equation*.

The *boundary conditions*, which the equation $\frac{\partial^2 y}{\partial t^2} = c^2 \frac{\partial^2 y}{\partial x^2}$ has to satisfy are:

$$\left. \begin{array}{l} (i)\ y = 0, \text{ when } x = 0 \\ (ii)\ y = 0, \text{ when } x = l \end{array} \right] . \text{ These should be satisfied for every value of } t.$$

If the string is made to vibrate by pulling it into a curve $y = f(x)$ and then releasing it, the *initial conditions* are:

(*i*) $y = f(x)$, when $t = 0$ (*ii*) $\frac{\partial y}{\partial t} = 0$, when $t = 0$.

3.20. SOLUTION OF THE WAVE EQUATION

The wave equation is

$$\frac{\partial^2 y}{\partial t^2} = c^2 \frac{\partial^2 y}{\partial x^2} \qquad \text{...(1)}$$

Let $y = XT$ ...(2)

where X is a function of x only and T is a function of t only, be a solution of (1)

Then $\frac{\partial^2 y}{\partial t^2} = XT''$ and $\frac{\partial^2 y}{\partial x^2} = X''T$

Substituting in (1), we have $XT'' = c^2X''T$

Separating the variables, we get $\frac{X''}{X} = \frac{1}{c^2} \cdot \frac{T''}{T}$ ...(3)

Now, the L.H.S. of (3) is a function of x only and the R.H.S. is a function of t only. Since x and t are independent variables, this equation can hold only when both sides reduce to a constant, say k. Then equation (3) leads to the ordinary linear differential equations

$$X''k - X = 0 \quad \text{and} \quad T'' - kc^2T = 0 \quad ...(4)$$

Solving equations (4), we get

(*i*) *When k is positive* and $= p^2$, say

$$X = c_1e^{px} + c_2e^{-px},\ T = c_3e^{cpt} + c_4e^{-cpt}$$

(*ii*) *When k is negative* and $= -p^2$, say

$$X = c_1 \cos px + c_2 \sin px$$

$$T = c_3 \cos cpt + c_4 \sin cpt$$

(*iii*) *When k = 0*

$$X = c_1x + c_2 \quad T = c_3t + c_4$$

Thus, the various possible solutions of the wave equation (1) are :

$$y = (c_1e^{px} + c_2e^{-px})(c_3e^{cpt} + c_4e^{-cpt})$$

$$y = (c_1 \cos px + c_2 \sin px)(c_3 \cos cpt + c_4 \sin cpt)$$

$$y = (c_1x + c_2)(c_3t + c_4)$$

Of these three solutions, we have to choose that solution which is consistent with the physical nature of the problem. Since we are dealing with a problem on vibrations, y must be a periodic function of x and t. Therefore, the solution must involve trigonometric terms.

Accordingly $y = (c_1 \cos px + c_2 \sin px)(c_3 \cos cpt + c_4 \sin cpt)$ ...(5)

is the only suitable solution of the wave equation and it corresponds to $k = -p^2$.

Now, applying boundary conditions that

$$y = 0, \quad \text{when } x = 0$$

and $y = 0$, when $x = l$, we get

$$0 = c_1(c_3 \cos cpt + c_4 \sin cpt) \quad ...(6)$$

and $0 = (c_1 \cos pl + c_2 \sin pl)(c_3 \cos cpt + c_4 \sin cpt)$ ...(7)

From (6), we have $c_1 = 0$ and equation (7) reduces to

$$c_2 \sin pl(c_3 \cos cpt + c_4 \sin cpt) = 0$$

which is satisfied when $\sin pl = 0$ or $pl = n\pi$ or $p = \frac{n\pi}{l}$, where $n = 1, 2, 3, ...$

∴ A solution of the wave equation satisfying the boundary conditions is

$$y = c_2\left(c_3 \cos\frac{n\pi ct}{l} + c_4 \sin\frac{n\pi ct}{l}\right)\sin\frac{n\pi x}{l}$$

$$= \left(a_n \cos \frac{n\pi ct}{l} + b_n \sin \frac{n\pi ct}{l}\right) \sin \frac{n\pi x}{l}$$

on replacing $c_2 c_3$ by a_n and $c_2 c_4$ by b_n.

Adding up the solutions for different values of n, we get

$$y = \sum_{n=1}^{\infty} \left(a_n \cos \frac{n\pi ct}{l} + b_n \sin \frac{n\pi ct}{l}\right) \sin \frac{n\pi x}{l} \qquad \text{...(8)}$$

is also a solution.

Now, applying the initial conditions

$$y = f(x) \quad \text{and} \quad \frac{\partial y}{\partial t} = 0, \quad \text{when } t = 0, \text{ we have}$$

$$f(x) = \sum_{n=1}^{\infty} a_n \sin \frac{n\pi x}{l} \qquad \text{...(9)}$$

and

$$0 = \sum_{n=1}^{\infty} \frac{n\pi c}{l} b_n \sin \frac{n\pi x}{l} \qquad \text{...(10)}$$

Since equation (9) represents Fourier series for $f(x)$, we have

$$a_n = \frac{2}{l} \int_0^l f(x) \sin \frac{n\pi x}{l}\, dx \qquad \text{...(11)}$$

From (10), $b_n = 0$, for all n.

Hence (8) reduces to $y = \sum_{n=1}^{\infty} a_n \cos \frac{n\pi ct}{l} \sin \frac{n\pi x}{l}$ ...(12)

where a_n is given by (11) when $f(x)$ *i.e.*, $y(x, 0)$ is known.

ILLUSTRATIVE EXAMPLES

Example 1. *Obtain the solution of the wave equation*

$$\frac{\partial^2 y}{\partial t^2} = c^2 \frac{\partial^2 y}{\partial x^2}$$

using the method of separation of variables.

Sol. The wave equation is

$$\frac{\partial^2 y}{\partial t^2} = c^2 \frac{\partial^2 y}{\partial x^2} \qquad \text{...(1)}$$

Let $y = \text{XT}$ ...(2)

where X is a function of x only and T is a function of t only be a solution of (1). Then

$$\frac{\partial^2 y}{\partial t^2} = \text{XT}'' \quad \text{and} \quad \frac{\partial^2 y}{\partial x^2} = \text{X}''\text{T}.$$

Substituting in (1), we have

$$\text{XT}'' = c^2\text{X}''\text{T} \quad \text{or} \quad \frac{\text{X}''}{\text{X}} = \frac{1}{c^2} \cdot \frac{\text{T}''}{\text{T}}$$

∵ L.H.S. is a function of x only while R.H.S. is a function of t only, the equality can hold only when each side is a constant say k.

$$\therefore \quad \frac{X''}{X} = \frac{1}{c^2} \cdot \frac{T''}{T} = k \qquad ...(3)$$

Three cases arise according as k is positive, zero or negative.

Case I. Let k be positive and $= p^2$. Then from (3),

$$X'' - p^2X = 0, \quad T'' - c^2p^2T = 0$$

$$X = c_1 e^{px} + c_2e^{-px}, \ T = c_3e^{cpt} + c_4e^{-cpt}$$

∴ From (2), $y = (c_1e^{px} + c_2e^{-px})(c_3e^{cpt} + c_4e^{-cpt})$.

Case II. Let $k = 0$, then from (3)

$$X'' = 0, T'' = 0$$

$$X = c_1x + c_2, \ T = c_3 t + c_4$$

∴ From (2), $y = (c_1x + c_2)(c_3t + c_4)$.

Case III. Let k be negative and $= -p^2$. Then from (3)

$$X'' + p^2X = 0 \quad T'' + p^2c^2T = 0$$

$$X = c_1 \cos px + c_2 \sin px, \ T = c_3 \cos pct + c_4 \sin pct.$$

∴ From (2), $y = (c_1 \cos px + c_2 \sin px)(c_3 \cos pct + c_4 \sin pct)$.

Since it is a problem on vibrating strings, every position of any point on the string is repeated after a fixed interval of time.

⇒ The solution should be a periodic function of time t.

⇒ The solution must involve sines and cosines of t.

Hence, of the three solutions obtained above, the proper solution is the one obtained in Case III.

∴ $y(x, t) = (c_1 \cos px + c_2 \sin px)(c_3 \cos pct + c_4 \sin pct)$.

Note. The boundary conditions which the equation

$$\frac{\partial^2 y}{\partial t^2} = c^2 \frac{\partial^2 y}{\partial x^2} \text{ has to satisfy are:}$$

$$\left.\begin{array}{l}(i)\ y = 0 \text{ when } x = 0 \\ (ii)\ y = 0 \text{ when } x = l\end{array}\right] \text{ for all } t \ge 0$$

If the string is made to vibrate by pulling it into a curve $y = f(x)$ and then releasing it, the initial conditions are

(i) $y = f(x)$ when $t = 0$ (ii) $\frac{\partial y}{\partial t} = 0$ when $t = 0$.

Example 2. *A string is stretched and fastened to two points l apart. Motion is started by displacing the string in the form*

$$y = a \sin \frac{\pi x}{l}$$

from which it is released at time t = 0. Show that the displacement of any point at a distance x from one end at time t is given by

$$y(x, t) = a \sin \frac{\pi x}{l} \cos \frac{\pi ct}{l}.$$

Sol. The vibration of the string is given by

$$\frac{\partial^2 y}{\partial t^2} = c^2 \frac{\partial^2 y}{\partial x^2} \qquad \text{[Wave equation]}$$

Since the vibration of the string is periodic, the solution of is of the form

$$y(x, t) = (c_1 \cos px + c_2 \sin px)(c_3 \cos pct + c_4 \sin pct) \quad ...(2)$$

Since $y = 0$ when $x = 0$

$$\therefore \quad 0 = c_1(c_3 \cos pct + c_4 \sin pct) \Rightarrow c_1 = 0$$

$\therefore$ From (2), we have

$$y(x, t) = c_2 \sin px\,(c_3 \cos pct + c_4 \sin pct)$$

or

$$y(x, t) = \sin px\,(\text{A} \cos pct + \text{B} \sin pct) \quad ...(3)$$

where

$$\text{A} = c_2 c_3 \quad \text{and} \quad \text{B} = c_2 c_4$$

Since $y = 0$ when $x = l$

$$\therefore \quad 0 = \sin pl\,(\text{A} \cos pct + \text{B} \sin pct)$$

$$\Rightarrow \quad \sin pl = 0 \Rightarrow pl = n\pi, n \geq 1$$

or

$$p = \frac{n\pi}{l}$$

$\therefore$ From (3), we have

$$y(x, t) = \sin \frac{n\pi x}{l}\left(\text{A} \cos \frac{n\pi ct}{l} + \text{B} \sin \frac{n\pi ct}{l}\right) \quad ...(4)$$

and

$$\frac{\partial y}{\partial t} = \sin \frac{n\pi x}{l}\left[-\text{A}\frac{n\pi c}{l} \sin \frac{n\pi ct}{l} + \text{B}\frac{n\pi c}{l} \cos \frac{n\pi ct}{l}\right]$$

Since $\frac{\partial y}{\partial t} = 0$ when $t = 0$

$$\therefore \quad 0 = \sin \frac{n\pi x}{l}\left[\text{B}\frac{n\pi c}{l}\right] \Rightarrow \text{B} = 0$$

$\therefore$ From (4), we have

$$y(x, t) = \text{A} \sin \frac{n\pi x}{l} \cos \frac{n\pi ct}{l} \quad ...(5)$$

Also

$$y = a \sin \frac{\pi x}{l} \quad \text{when } t = 0$$

$$\therefore \quad a \sin \frac{\pi x}{l} = \text{A} \sin \frac{n\pi x}{l}$$

$$\Rightarrow \quad \text{A} = a \quad \text{and} \quad n = 1$$

$\therefore$ From (5), we have

$$y(x, t) = a \sin \frac{\pi x}{l} \cos \frac{\pi ct}{l}.$$

Example 3. *A tightly stretched string of length l with fixed ends is initially in equilibrium position. If is set vibrating by giving each point a velocity $v_0 \sin 3\pi x/l$. Find the displacement (x, t).*

Sol. Here the vibrations of the string are governed by one dimensional wave equation. Therefore they are given by

$$\frac{\partial^2 y}{\partial t^2} = c^2 \frac{\partial^2 y}{\partial x^2} \qquad ...(1)$$

The boundary and initial conditions are given by:

The boundary conditions are $y(0, t) = 0$

and $y(l, t) = 0$ ...(2)

The initial conditions are $y(x, 0) = 0$.

$\therefore$ $f(x) = 0$

and $$\left(\frac{\partial y}{\partial t}\right)_{t=0} = v_0 \sin^3 \frac{\pi x}{l} = g(x).$$

Therefore, by § 3.2-2, the subsequent displacement $y(x, t)$ of the string is given by

$$y(x, t) = \sum_{n=1}^{\infty} \left(C_n \cos \frac{n\pi ct}{l} + D_n \sin \frac{n\pi ct}{l} \right) \sin \frac{n\pi x}{l} \qquad ...(3)$$

where $$C_n = \frac{2}{l} \int_0^l f(x) \sin \frac{n\pi x}{l}\, dx = 0 \qquad [\because f(x) = y(x, 0) = 0]$$

and $$\frac{n\pi c}{l} D_n = \frac{2}{l} \int_0^l g(x) \sin \frac{n\pi x}{l} = \frac{2}{l} \int_0^l v_0 \sin^3 \frac{\pi x}{l} \sin \frac{n\pi x}{l}\, dx$$

or $$D_n = \frac{2v_0}{n\pi c} \int_0^l \frac{1}{4} \left(3 \sin \frac{\pi x}{l} - \sin \frac{3\pi x}{l} \right) \sin \frac{n\pi x}{l}\, dx$$

[using $\sin 3\theta = 3 \sin \theta - 4 \sin^3 \theta$]

$$= \frac{v_0}{2n\pi c} \left[\int_0^l 3 \sin \frac{\pi x}{l} \sin \frac{n\pi x}{l}\, dx - \int_0^l \sin \frac{3\pi x}{l} \sin \frac{n\pi x}{l}\, dx \right].$$

Clearly D_n vanishes for all n except when $n = 1$ and $n = 3$.

Now $$D_1 = \frac{3v_0}{2\pi c} \int_0^l \sin^2 \frac{\pi x}{l}\, dx = \frac{3v_0}{4\pi c} \int_0^l \left(1 - \cos \frac{2\pi x}{l} \right) dx = \frac{3lv_0}{4\pi c}$$

and $$D_3 = \frac{-v_0}{6\pi c} \int_0^l \sin^2 \frac{3\pi x}{l}\, dx = \frac{-v_0}{12\pi c} \int_0^l \left(1 - \cos \frac{6\pi x}{l} \right) dx = \frac{-lv_0}{12\pi c}.$$

Thus the solution becomes

$$y(x, t) = D_1 \sin \frac{\pi ct}{l} \sin \frac{\pi x}{l} + D_3 \sin \frac{3\pi ct}{l} \sin \frac{3\pi x}{l}$$

$$= \frac{lv_0}{12\pi c} \left[9 \sin \frac{\pi ct}{l} \sin \frac{\pi x}{l} - \sin \frac{3\pi ct}{l} \sin \frac{3\pi x}{l} \right]$$

which is the required displacement.

Example 4. *A string is stretched between the fixed points (0, 0) and (l, 0) and released at rest from the initial deflection given by*

$$f(x) = \begin{cases} \dfrac{2k}{l}x & \text{when } 0 < x < \dfrac{l}{2} \\ \dfrac{2k}{l}(l-x) & \text{when } \dfrac{l}{2} < x < l \end{cases}$$

Find the deflection of the string at any time t.

Sol. Here the vibration of the string are governed by one dimensional wave equation. Proceeding as in the deflection of the string is given by

$$u(x, t) = \sum_{n=1}^{\infty} C_n \cos\frac{n\pi ct}{l} \sin\frac{n\pi x}{l} \qquad ...(1)$$

where the coeffients C_n are given by

$$C_n = \frac{2}{l}\int_0^l f(x)\sin\frac{n\pi x}{l}\,dx = \frac{2}{l}\left[\int_0^{l/2} f(x)\sin\frac{n\pi x}{l}\,dx + \int_{l/2}^{l} f(x)\sin\frac{n\pi x}{l}\,dx\right]$$

$$= \frac{2}{l}\cdot\left[\int_0^{l/2}\frac{2k}{l}x\sin\frac{n\pi x}{l}\,dx + \int_{l/2}^{l}\frac{2k}{l}(l-x)\sin\frac{n\pi x}{l}\,dx\right]$$

$$= \frac{2}{l}\cdot\frac{2k}{l}\left[\int_0^{l/2} x\,.\sin\frac{n\pi x}{l}\,dx + \int_{l/2}^{l}(l-x)\sin\frac{n\pi x}{l}\,dx\right]$$

$$= \frac{4k}{l^2}\cdot\left[\left\{-x\frac{l}{n\pi}\cos\frac{n\pi x}{l}\right\}_0^{l/2} + \frac{l}{n\pi}\int_0^{l/2}\cos\frac{n\pi x}{l}\,dx\right.$$
$$\left. + \left\{-(l-x)\frac{l}{n\pi}\cos\frac{n\pi x}{l}\right\}_{l/2}^{l} - \frac{l}{n\pi}\int_{l/2}^{l}\cos\frac{n\pi x}{l}\,dx\right]$$

$$= \frac{4k}{l^2}\cdot\left[-\frac{l^2}{2n\pi}\cos\frac{n\pi}{2} + \frac{l}{n\pi}\cdot\frac{l}{n\pi}\left(\sin\frac{n\pi x}{l}\right)_0^{l/2} - 0 + \frac{l}{2}\cdot\frac{l}{n\pi}\cos\frac{n\pi}{2} - \frac{l}{n\pi}\cdot\frac{l}{n\pi}\left(\sin\frac{n\pi x}{l}\right)_{l/2}^{l}\right]$$

$$= \frac{4k}{l^2}\cdot\left[\frac{l^2}{n^2\pi^2}\sin\frac{n\pi}{2} - \frac{l^2}{n^2\pi^2}\left(\sin n\pi - \sin\frac{n\pi}{2}\right)\right]$$

$$= \frac{4k}{l^2}\cdot\left[\frac{2l^2}{n^2\pi^2}\sin\frac{n\pi}{2}\right] = \frac{8k}{n^2\pi^2}\sin\frac{n\pi}{2}.$$

$$\therefore \qquad u(x, t) = \sum_{n=1}^{\infty}\frac{8k}{n^2\pi^2}\sin\frac{n\pi}{2}\cos\frac{n\pi ct}{l}\sin\frac{n\pi x}{l}$$

$$= \frac{8k}{\pi^2}\left[\frac{1}{1^2}\sin\frac{\pi}{l}x\cos\frac{\pi c}{l}t - \frac{1}{3^2}\sin\frac{3\pi}{l}x\cos\frac{3\pi c}{l}t + ...\right] \qquad ...(2)$$

Thus deflection is given by (1) where C_n are given by (2).

Example 5. *A string is stretched and fastened to two points l apart. Motion is started by displacing the string into the form $y = k(lx - x^2)$ from which it is released at time $t = 0$. Find the displacement of any point on the string at a distance x from one end at time t.*

Sol. The vibration of the string is given by

$$\frac{\partial^2 y}{\partial t^2} = c^2 \frac{\partial^2 y}{\partial x^2} \qquad \text{...(1)}$$

Since the vibration of the string is periodic, the solution of (1) is of the form

$$y(x, t) = (c_1 \cos px + c_2 \sin px)(c_3 \cos pct + c_4 \sin pct) \qquad \text{...(2)}$$

Since $y = 0$ where $x = 0$

$$\therefore \quad 0 = c_1 (c_3 \cos pct + c_4 \sin pct) \Rightarrow c_1 = 0$$

$\therefore$ From (2), we have

$$y(x, t) = c_2 \sin px \, (c_3 \cos pct + c_4 \sin pct)$$

or
$$y(x, t) = \sin px \, (\text{A} \cos pct + \text{B} \sin pct) \qquad \text{...(3)}$$

where
$$\text{A} = c_2 c_3 \text{ and } \text{B} = c_2 c_4$$

Since, $y = 0$ when $x = l$

$$\therefore \quad 0 = \sin pl \, (\text{A} \cos pct + \text{B} \sin pct)$$

$$\Rightarrow \quad \sin pl = 0 \text{ or } pl = n\pi$$

or
$$p = \frac{n\pi}{l}, \text{ where } n = 1, 2, 3, \ldots\ldots$$

$\therefore$ A solution of (1) satisfying the boundary conditions is

$$y(x, t) = \left(a_n \cos \frac{n\pi ct}{l} + b_n \sin \frac{n\pi ct}{l}\right) \sin \frac{n\pi x}{l}$$

on replacing A by a_n and B by b_n.

Adding up the solutions for different values of n, we get

$$y(x, t) = \sum_{n=1}^{\infty} \left(a_n \cos \frac{n\pi ct}{l} + b_n \sin \frac{n\pi ct}{l}\right) \sin \frac{n\pi x}{l} \qquad \text{...(4)}$$

is also a solution.

Now applying the initial conditions

$$y = k(lx - x^2) \text{ and } \frac{\partial y}{\partial t} = 0 \text{ when } t = 0, \text{ we have}$$

$$k(lx - x^2) = \sum_{n=1}^{\infty} a_n \sin \frac{n\pi x}{l} \qquad \text{...(5)}$$

and
$$0 = \sum_{n=1}^{\infty} \frac{n\pi c}{l} b_n \sin \frac{n\pi x}{l} \qquad \text{...(6)}$$

Equation (5) represents Fourier series for $k(lx - x^2)$.

$$\therefore \quad a_n = \frac{2}{l} \int_0^l k(lx - x^2) \sin \frac{n\pi x}{l} \, dx$$

$$= \frac{2k}{l}\left[(lx - x^2)\left(-\frac{\cos\frac{n\pi x}{l}}{\frac{n\pi}{l}}\right) - (l-2x)\left(\frac{\sin\frac{n\pi x}{l}}{\left(\frac{n\pi}{l}\right)^2}\right) + (-2)\left[\frac{\cos\frac{n\pi x}{l}}{\left(\frac{n\pi}{l}\right)^3}\right]\right]_0^l$$

$$= \frac{2k}{l}\left(-\frac{2l^3}{n^3\pi^3}\cos n\pi + \frac{2l^3}{n^3\pi^3}\right) = \frac{4kl^2}{n^3\pi^3}\,[1-(-1)^n]$$

$$= \begin{cases} \dfrac{8kl^2}{n^3\pi^3} & \text{when } n \text{ is odd} \\ 0 & \text{when } n \text{ is even} \end{cases}$$

From (6), $b_n = 0$ for all n.

$\therefore$ From (4), we get

$$y(x, t) = \frac{8kl^2}{\pi^3}\sum_{n=1,3,5,\ldots\ldots} \frac{1}{n^3}\cos\frac{n\pi ct}{l}\sin\frac{n\pi x}{l}$$

or
$$y(x, t) = \frac{8kl^2}{\pi^3}\sum_{n=1}^{\infty}\frac{1}{(2n-1)^3}\cos\frac{(2n-1)\pi ct}{l}\sin\frac{(2n-1)\pi x}{l}$$

which is the required solution.

Example 6. *A string of length l is initially at rest in equilibrium position and each of its points is given the velocity*

$$\left(\frac{\partial y}{\partial t}\right)_{t=0} = b\, sin^3\left(\frac{\pi x}{l}\right)$$

Find the displacement y(x, t).

Sol. Proceeding as previous example,

$$y(x, t) = \sum_{n=1}^{\infty}\left(a_n \cos\frac{n\pi ct}{l} + b_n \sin\frac{n\pi ct}{l}\right)\sin\frac{n\pi x}{l}$$

is a solution.

Now applying the initial conditions,

$$y = 0 \quad \text{and} \quad \frac{\partial y}{\partial t} = b\sin^3\left(\frac{\pi x}{l}\right) \text{ when } t = 0\text{, we have}$$

$$0 = \sum_{n=1}^{\infty} a_n \sin\frac{n\pi x}{l} \text{ which gives } a_n = 0$$

and
$$b\sin^3\left(\frac{\pi x}{l}\right) = \sum_{n=1}^{\infty}\frac{n\pi c}{l}\, b_n \sin\frac{n\pi x}{l}$$

$$\Rightarrow \quad b\left[\frac{3}{4}\sin\frac{\pi x}{l} - \frac{1}{4}\sin\frac{3\pi x}{l}\right]$$

$$= \frac{\pi c}{l} b_1 \sin\frac{\pi x}{l} + \frac{2\pi c}{l} b_2 \sin\frac{2\pi x}{l} + \frac{3\pi c}{l} b_3 \sin\frac{3\pi x}{l} + \ldots\ldots$$

$$\left[\because \ \sin 3A = 3\sin A - 4\sin^3 A \ \therefore \ \sin^3 A = \frac{3}{4}\sin A - \frac{1}{4}\sin 3A\right]$$

Equating coefficients of like terms, we have

$$\frac{3b}{4} = \frac{\pi c}{l} b_1, \ b_2 = 0,$$

$$-\frac{b}{4} = \frac{3\pi c}{l} b_3, \ b_4 = b_5 = b_6 = \ldots\ldots = 0$$

$$\Rightarrow \quad b_1 = \frac{3bl}{4c\pi}, \ b_3 = -\frac{bl}{12c\pi}, \ b_2 = b_4 = b_5 = b_6 = \ldots\ldots$$

$\therefore$ From (4), we have

$$y(x, t) = b_1 \sin\frac{\pi ct}{l}\sin\frac{\pi x}{l} + b_3 \sin\frac{3\pi ct}{l}\sin\frac{3\pi x}{l}$$

$$= \frac{3bl}{4c\pi}\sin\frac{\pi ct}{l}\sin\frac{\pi x}{l} - \frac{bl}{12c\pi}\sin\frac{3\pi ct}{l}\sin\frac{3\pi x}{l}$$

$$= \frac{bl}{12c\pi}\left[9\sin\frac{\pi ct}{l}\sin\frac{\pi x}{l} - \sin\frac{3\pi ct}{l}\sin\frac{3\pi x}{l}\right]$$

which is the required displacement.

Example 7. *The points of trisection of a string are pulled aside through the same distance on opposite sides of the position of equilibrium and the string is released from rest. Derive an expression for the displacement of the string at subsequent time and show that the mid-point of the string always remains at rest.*

Sol. Let B and C be the points of trisection of the string OA of length l, say. Initially the string is held in the form OB′C′A, where BB′ = CC′ = a (say)

The equation of OB′ is $\quad y = \frac{3a}{l}x$

The equation of B′C′ is $\quad y - a = -\frac{2a}{\frac{l}{3}}\left(x - \frac{l}{3}\right)$

i.e., $\quad y = \frac{3a}{l}(l - 2x)$

The equation of C′A is $\quad y = \frac{3a}{l}(x - l)$

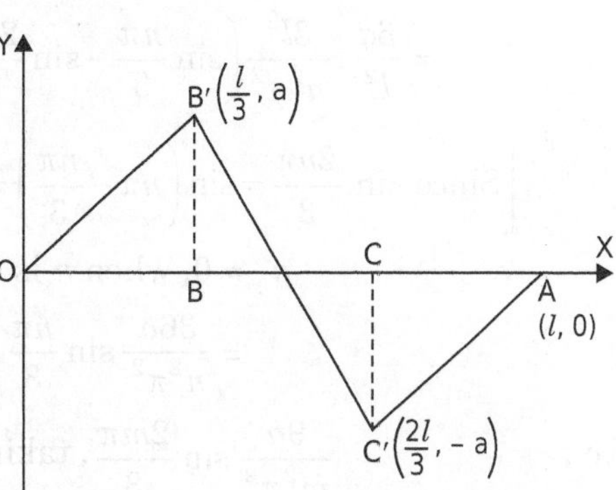

Here the boundary conditions are

$$y(0, t) = y(l, t) = 0$$

and the initial conditions are $y(x, 0) = \begin{cases} \dfrac{3a}{l}x, & 0 \le x \le \dfrac{l}{3} \\ \dfrac{3a}{l}(l - 2x), & \dfrac{l}{3} \le x \le \dfrac{2l}{3} \\ \dfrac{3a}{l}(x - l), & \dfrac{2l}{3} \le x \le l \end{cases}$ and $\dfrac{\partial y}{\partial t} = 0$, when $t = 0$

As proved in we have $y(x, t) = \sum_{n=1}^{\infty} a_n \cos\dfrac{n\pi ct}{l} \sin\dfrac{n\pi x}{l}$

where
$$a_n = \frac{2}{l}\int_0^l y(x, 0) \sin\frac{n\pi x}{l}\, dx$$

$$= \frac{2}{l}\left[\int_0^{l/3} \frac{3ax}{l} \sin\frac{n\pi x}{l} dx + \int_{l/3}^{2l/3} \frac{3a}{l}(l - 2x) \sin\frac{n\pi x}{l} dx + \int_{2l/3}^{l} \frac{3a}{l}(x - l) \sin\frac{n\pi x}{l} dx\right]$$

$$= \frac{6a}{l^2}\left[\left| x\left(-\frac{l}{n\pi}\cos\frac{n\pi x}{l}\right) - 1.\left(-\frac{l^2}{n^2\pi^2}\sin\frac{n\pi x}{l}\right)\right|_0^{l/3}\right.$$
$$+ \left|(l - 2x)\left(-\frac{l}{n\pi}\cos\frac{n\pi x}{l}\right) - (-2)\left(-\frac{l^2}{n^2\pi^2}\sin\frac{n\pi x}{l}\right)\right|_{l/3}^{2l/3}$$
$$\left. + \left|(x - l)\left(-\frac{l}{n\pi}\cos\frac{n\pi x}{l}\right) - 1.\left(-\frac{l^2}{n^2\pi^2}\sin\frac{n\pi x}{l}\right)\right|_{2l/3}^{l}\right]$$

$$= \frac{6a}{l^2}\left[\left(-\frac{l^2}{3n\pi}\cos\frac{n\pi}{3} + \frac{l^2}{n^2\pi^2}\sin\frac{n\pi}{3}\right) + \left(\frac{l^2}{3n\pi}\cos\frac{2n\pi}{3} - \frac{2l^2}{n^2\pi^2}\sin\frac{2n\pi}{3} + \frac{l^2}{3n\pi}\cos\frac{n\pi}{3}\right.\right.$$
$$\left.\left. + \frac{2l^2}{n^2\pi^2}\sin\frac{n\pi}{3}\right) - \left(\frac{l^2}{3n\pi}\cos\frac{2n\pi}{3} + \frac{l^2}{n^2\pi^2}\sin\frac{2n\pi}{3}\right)\right]$$

$$= \frac{6a}{l^2}\cdot\frac{3l^2}{n^2\pi^2}\left(\sin\frac{n\pi}{3} - \sin\frac{2n\pi}{3}\right) = \frac{18a}{n^2\pi^2}\sin\frac{n\pi}{3}\,[1 + (-1)^n]$$

$$\left[\text{Since } \sin\frac{2n\pi}{3} = \sin\left(n\pi - \frac{n\pi}{3}\right) = \sin n\pi\cos\frac{n\pi}{3} - \cos n\pi \sin\frac{n\pi}{3} = -(-1)^n \sin\frac{n\pi}{3}\right]$$

$\therefore$ $a_n = 0$, when n is odd

$= \dfrac{36a}{n^2\pi^2}\sin\dfrac{n\pi}{3}$, when n is even

i.e., $\dfrac{9a}{m^2\pi^2}\sin\dfrac{2m\pi}{3}$, taking $n = 2m$

Hence
$$y(x, t) = \frac{9a}{\pi^2}\sum_{m=1}^{\infty}\frac{1}{m^2}\sin\frac{2m\pi}{3}\cos\frac{2m\pi ct}{l}\sin\frac{2m\pi x}{l}$$

Also $\quad y\left(\frac{l}{2}, t\right) = \frac{9a}{\pi^2} \sum_{m=1}^{\infty} \frac{1}{m^2} \sin \frac{2m\pi}{3} \cos \frac{2m\pi ct}{l} \sin m\pi$

$= 0$, since $\sin m\pi = 0$

$\Rightarrow$ The displacement of the mid-point of the string is zero for all values of t.

Thus the mid-point of the string is always at rest.

Example 8. *A tightly stretched string with fixed end points $x = 0$ and $x = l$ is initially at rest in its equilibrium position. If it is set vibrating by giving to each of its points a velocity $\lambda x(l - x)$, find the displacement of the string at any distance x from one end at any time t.*

Sol. Here the boundary conditions are $y(0, t) = y(l, t) = 0$

As proved above, we have

$$y(x, t) = \sum_{n=1}^{\infty} \left(a_n \cos \frac{n\pi ct}{l} + b_n \sin \frac{n\pi ct}{l} \right) \sin \frac{n\pi x}{l} \quad \text{...(1)}$$

Since the string was at rest initially, $y(x, 0) = 0$

$\therefore$ From (1), $\quad 0 = \sum_{n=1}^{\infty} a_n \sin \frac{n\pi x}{l} \Rightarrow a_n = 0$

$$\therefore \quad y(x, t) = \sum_{n=1}^{\infty} b_n \sin \frac{n\pi ct}{l} \sin \frac{n\pi x}{l} \quad \text{...(2)}$$

and $\quad \frac{\partial y}{\partial t} = \sum_{n=1}^{\infty} \frac{n\pi c}{l} b_n \cos \frac{n\pi ct}{l} \sin \frac{n\pi x}{l} = \frac{\pi c}{l} \sum_{n=1}^{\infty} n b_n \cos \frac{n\pi ct}{l} \sin \frac{n\pi x}{l}$

But $\quad \frac{\partial y}{\partial t} = \lambda x(l - x)$, when $t = 0$

$$\therefore \quad \lambda x(l - x) = \frac{\pi c}{l} \sum_{n=1}^{\infty} n b_n \sin \frac{n\pi x}{l}$$

$$\Rightarrow \quad \frac{\pi c}{l} n b_n = \frac{2}{l} \int_0^l \lambda x(l - x) \sin \frac{n\pi x}{l} \, dx$$

$$= \frac{2\lambda}{l} \left[x(l - x)\left(-\frac{l}{n\pi} \cos \frac{n\pi x}{l} \right) - (l - 2x)\left(-\frac{l^2}{n^2\pi^2} \sin \frac{n\pi x}{l} \right) + (-2)\left(\frac{l^3}{n^3\pi^3} \cos \frac{n\pi x}{l} \right) \right]_0^l$$

$$= \frac{4\lambda l^2}{n^3\pi^3} (1 - \cos n\pi) = \frac{4\lambda l^2}{n^3\pi^3} [1 - (-1)^n]$$

$$= \begin{cases} 0, & \text{when } n \text{ is even} \\ \dfrac{8\lambda l^2}{n^3\pi^3}, & \text{when } n \text{ is odd} \end{cases}$$

i.e., $\quad \frac{8\lambda l^2}{\pi^3 (2m - 1)^3}$, taking $n = 2m - 1$

$$\Rightarrow \qquad b_n = \frac{8\lambda l^3}{c\pi^4 (2m-1)^4}$$

$\therefore$ From (2), the required solution is

$$y(x, t) = \frac{8\lambda l^3}{c\pi^4} \sum_{m=1}^{\infty} \frac{1}{(2m-1)^4} \sin \frac{(2m-1)\pi ct}{l} \sin \frac{(2m-1)\pi x}{l}.$$

Example 9. *Find the temperature u(x, t) in a bar of length l, which is perfectly insulated already whose ends are kept at temperature zero and initial temperature is*

$$f(x) = \begin{cases} x & \text{when } 0 < x < l/2 \\ l - x & \text{when } l/2 < x < l. \end{cases}$$

Sol. Proceeding as previous example, we have

$$u(x, t) = \sum_{n=1}^{\infty} B_n \sin \frac{n\pi x}{l} . e^{-\frac{c^2\pi^2 n^2 t}{l^2}} \qquad \text{...(1)}$$

where
$$B_n = \frac{2}{l} \int_0^l f(x) \sin \frac{n\pi}{l} x \, dx$$

$$= \frac{2}{l} \left[\int_0^{l/2} x \sin \frac{n\pi}{l} x \, dx + \int_{l/2}^{l} (l - x) \sin \frac{n\pi}{l} x \, dx \right].$$

Integrating by parts, we get

$$B_n = \frac{2}{l} \left[\left(-x . \frac{l}{n\pi} \cos \frac{n\pi}{l} x \right)_0^{l/2} + \frac{l}{n\pi} \int_0^{l/2} \cos \frac{n\pi}{l} x \, dx + \left\{ (l - x) \left(-\frac{l}{n\pi} \right) \cos \frac{n\pi x}{l} \right\}_{l/2}^{l} - \frac{l}{n\pi} \int_{l/2}^{l} \cos \frac{n\pi x}{l} dx \right]$$

$$= \frac{2}{l} \left[-\frac{l^2}{2n\pi} \cos \frac{n\pi}{2} + \frac{l^2}{n^2\pi^2} \left(\sin \frac{n\pi}{l} x \right)_0^{l/2} + \frac{l}{2} . \frac{l}{n\pi} \cos \frac{n\pi}{2} - \frac{l^2}{n^2\pi^2} \left(\sin \frac{n\pi}{l} x \right)_{l/2}^{l} \right]$$

$$= \frac{2}{l} \left[\frac{2l^2}{n^2\pi^2} \sin \frac{n\pi}{2} \right] = \frac{4l}{n^2\pi^2} \sin \frac{n\pi}{2}.$$

$\therefore$ $B_n = 0$ when n is even,

$$B_n = \frac{4l}{n^2\pi^2} \text{ (when } n = 1, 5, 9, \ldots)$$

and
$$B_n = -\frac{4l}{n^2\pi^2} \text{ (when } n = 3, 7, 11, \ldots)$$

Hence the required value of $u(x, t)$ is given by

$$u(x, t) = \sum_{n=1}^{\infty} B_n \sin \frac{n\pi}{l} x . e^{-\frac{n^2\pi^2 c^2}{l^2} t}$$

$$= \frac{4l}{\pi^2} \cdot \left[\frac{1}{1^2} \sin \frac{\pi x}{l} . e^{-\frac{c^2\pi^2}{l^2} t} - \frac{1}{3^2} \sin \frac{3\pi x}{l} . e^{-\frac{3^2\pi^2 c^2}{l^2} t} + \ldots \right].$$

TEST YOUR KNOWLEDGE

1. Solve completely the equation $\frac{\partial^2 y}{\partial t^2} = c^2 \frac{\partial^2 y}{\partial x^2}$, representing the vibrations of a string of length l, fixed at both ends, given that $y(0, t) = 0$; $y(l, t) = 0$; $y(x, 0) = f(x)$ and $\frac{\partial y(x, 0)}{\partial t} = 0$, $0 < x < l$.
2. Find the deflection $y(x, t)$ of the vibrating string of length π and ends fixed, corresponding to zero initial velocity and initial deflection $f(x) = k(\sin x - \sin 2x)$, given $c^2 = 1$.
3. Solve the boundary value problem

$$\frac{\partial^2 y}{\partial t^2} = 4 \frac{\partial^2 y}{\partial x^2},\ y(0, t) = y\ (5, t) = 0,\ y(x, 0) = 0,\ \left(\frac{\partial y}{\partial t}\right)_{t=0} = f(x)$$

if (*i*) $f(x) = 5 \sin \pi x$ (*ii*) $f(x) = 3 \sin 2\pi x - 2 \sin 5\pi x$.

4. A tightly stretched string with fixed end points $x = 0$ and $x = l$ is initially in a position given by $y = y_0 \sin^3 \left(\frac{\pi x}{l}\right)$. If it is released from rest from this position, find the displacement $y(x, t)$.
5. A tightly stretched string of length l and fixed at both ends is plucked at $x = \frac{l}{3}$ and assumes initially the shape of a triangle of height h. Find the displacement $y(x, t)$ after the string is released from rest.
6. A tightly stretched flexible string has its ends fixed at $x = 0$ and $x = l$. At time $t = 0$, the string is given a shape defined by $f(x) = \mu x\ (l - x)$, where μ is a constant and then released. Find the displacement of any point x of the string at any time $t > 0$.
7. A string is stretched between the fixed points (0, 0) and (l, 0) and released at rest from the initial deflection given by

$$f(x) = \begin{cases} \frac{2k}{l}\, x, & \text{when } 0 < x < \frac{l}{2} \\ \frac{2k}{l}\,(l - x), & \text{when } \frac{l}{2} < x < l \end{cases}$$

Find the deflection of the string at any time t.

8. The ends of a tightly stretched string of length l are fixed at $x = 0$ and $x = l$. The string is at rest with the point $x = b$ drawn aside through a small distance d and released at time $t = 0$. Show that

$$y(x, t) = \frac{2dl^2}{\pi^2 b(l - b)} \sum_{n=1}^{\infty} \frac{1}{n^2} \sin \frac{n\pi b}{l} \sin \frac{n\pi x}{l} \cos \frac{n\pi ct}{l}.$$

9. A string of length l is initially at rest in equilibrium position and each of its points is given the velocity

$$\left(\frac{\partial y}{\partial t}\right)_{t=0} = b \sin^3 \frac{\pi x}{l}$$

Find the displacement $y(x, t)$.

Answers

1. $y(x, t) = \sum_{n=1}^{\infty} a_n \cos \frac{n\pi ct}{l} \sin \frac{n\pi x}{l}$, where $a_n = \frac{2}{l} \int_0^l f(x). \sin \frac{n\pi x}{l}\, dx$

2. $y(x, t) = k(\cos t \sin x - \cos 2t \sin 2x)$

3. (i) $y = \dfrac{5}{2\pi} \sin \pi x \sin 2\pi t$ (ii) $y = \dfrac{3}{4\pi} \sin 2\pi x \sin 4\pi t - \dfrac{1}{5\pi} \sin 5\pi x \sin 10\pi t$

4. $y(x, t) = \dfrac{y_0}{4}\left(3 \sin \dfrac{\pi x}{l} \cos \dfrac{\pi ct}{l} - \sin \dfrac{3\pi x}{l} \cos \dfrac{3\pi ct}{l}\right)$

5. $y(x, t) = \dfrac{9h}{\pi^2} \sum_{n=1}^{\infty} \dfrac{1}{n^2} \sin \dfrac{n\pi}{3} \sin \dfrac{n\pi x}{l} \cos \dfrac{n\pi ct}{l}$

6. $y = \dfrac{8\mu l^2}{\pi^3} \sum_{n=1}^{\infty} \dfrac{1}{(2n-1)^3} \sin \dfrac{(2n-1)\pi x}{l} \cos \dfrac{(2n-1)\pi ct}{l}$

7. $y(x, t) = \dfrac{8k}{\pi^2} \sum_{n=1}^{\infty} \dfrac{1}{n^2} \sin \dfrac{n\pi}{2} \cos \dfrac{n\pi ct}{l} \sin \dfrac{n\pi x}{l}$

9. $y(x, t) = \dfrac{bl}{12c\pi}\left[9 \sin \dfrac{\pi x}{l} \sin \dfrac{\pi ct}{l} - \sin \dfrac{3\pi x}{l} \sin \dfrac{3\pi ct}{l}\right]$.

3.21. ONE-DIMENSIONAL HEAT FLOW $\dfrac{\partial u}{\partial t} = c^2 \dfrac{\partial^2 u}{\partial x^2}$

Consider the flow of heat by conduction in a uniform bar. It is assumed that the sides of the bar are insulated and the loss of heat from the sides by conduction or radiation is negligible. Take one end of the bar as origin and the direction of flow as the positive x-axis. The temperature u at any point of the bar depends on the distance x of the point from one end and the time t. Also, the temperature of all points of any cross-section is the same.

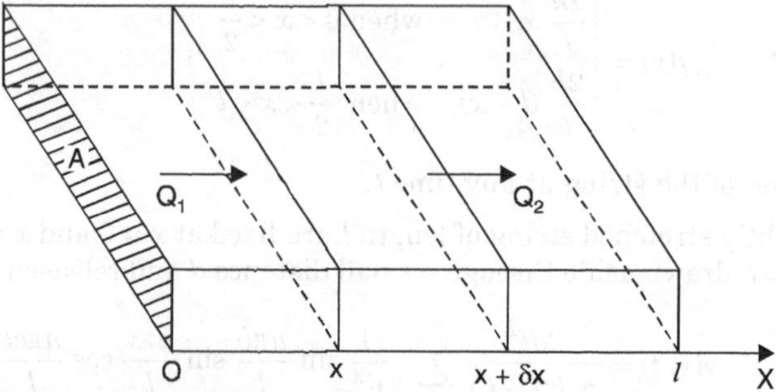

The amount of heat crossing any section of the bar per second depends on the area A of the cross-section, the conductivity K of the material of the bar and the temperature gradient $\dfrac{\partial u}{\partial x}$ *i.e.*, rate of change of temperature w.r.t. distance normal to the area.

∴ Q_1, the quantity of heat flowing into the section at a distance x

$$= -\text{KA}\left(\frac{\partial u}{\partial x}\right)_x \text{ per sec.}$$

(the negative sign on the right is attached because as x increases, u decreases).

Q_2, the quantity of heat flowing out of the section at a distance $x + \delta x$

$$= -\,KA\left(\frac{\partial u}{\partial x}\right)_{x+\delta x} \text{ per sec.}$$

Hence the amount of heat retained by the slab with thickness δx is

$$Q_1 - Q_2 = KA\left[\left(\frac{\partial u}{\partial x}\right)_{x+\delta x} - \left(\frac{\partial u}{\partial x}\right)_x\right] \text{ per sec} \quad ...(1)$$

But the rate of increase of heat in the slab $= S\rho A\,\delta x\,\dfrac{\partial u}{\partial x}$ ...(2)

where S is the specific heat and ρ, the density of the material.

$$\therefore \quad \text{From (1) and (2), } S\rho A\,\delta x\frac{\partial u}{\partial x} = KA\left[\left(\frac{\partial u}{\partial x}\right)_{x+\delta x} - \left(\frac{\partial u}{\partial x}\right)_x\right]$$

or
$$S\rho\frac{\partial u}{\partial t} = K\left[\frac{\left(\frac{\partial u}{\partial x}\right)_{x+\delta x} - \left(\frac{\partial u}{\partial x}\right)_x}{\delta x}\right]$$

Taking the limit as $\delta x \to 0$, we have

$$S\rho\frac{\partial u}{\partial t} = K\frac{\partial^2 u}{\partial x^2} \quad \text{or} \quad \frac{\partial u}{\partial t} = \frac{K}{S\rho}\frac{\partial^2 u}{\partial x^2}$$

or
$$\frac{\partial u}{\partial t} = c^2\frac{\partial^2 u}{\partial x^2}, \text{ where } c^2 = \frac{K}{S\rho}$$

is known as diffusivity of the material of the bar.

3.22. SOLUTION OF THE HEAT EQUATION

The heat equation is
$$\frac{\partial u}{\partial t} = c^2\frac{\partial^2 u}{\partial x^2} \quad ...(1)$$

Let
$$u = XT \quad ...(2)$$

where X is a function of x only and T is a function of t only be a solution of (1).

Then
$$\frac{\partial u}{\partial t} = XT' \quad \text{and} \quad \frac{\partial^2 u}{\partial x^2} = X''T$$

Substituting in (1), we have $XT' = c^2X''T$

Separating the variables, we get
$$\frac{X''}{X} = \frac{1}{c^2}\frac{T'}{T} \quad ...(3)$$

Now, the L.H.S. of (3) is a function of x only and the R.H.S. is a function of t only. Since x and t are independent variables, this equation can hold only when both sides reduce to a constant, say k. Then equation (3) leads to the ordinary differential equations

$$\frac{d^2X}{dx^2} - kX = 0 \quad \text{and} \quad \frac{dT}{dt} - kc^2T = 0 \qquad ...(4)$$

Solving equations (4), we get

(*i*) *When k is positive* and $= p^2$, say

$$X = c_1e^{px} + c_2e^{-px}, \; T = c_3e^{c^2p^2t}$$

(*ii*) *When k is negative* and $= -p^2$, say

$$X = c_1 \cos px + c_2 \sin px, \; T = c_3e^{-c^2p^2t}$$

(*iii*) *When k = 0*

$$X = c_1x + c_2, \; T = c_3.$$

Thus the various possible solutions of the heat equation (1) are :

$$u = (c_1e^{px} + c_2e^{-px}).c_3e^{c^2p^2t}$$

$$u = (c_1 \cos px + c_2 \sin px) \,.\, c_3e^{-c^2p^2t}$$

$$u = (c_1x + c_2)c_3.$$

Of these three solutions, we have to choose that solution which is consistent with the physical nature of the problem. Since u decreases as time t increases, the only suitable solution of the heat equation is

$$u = (c_1 \cos px + c_2 \sin px)\, e^{-c^2p^2t}.$$

ILLUSTRATIVE EXAMPLES

Example 1. *A rod of length l with insulated sides is initially at a uniform temperature* u_0. *Its ends are suddenly cooled to 0°C and are kept at that temperature. Find the temperature function u(x, t).*

Sol. The temperature function $u(x, t)$ satisfies the differential equation

$$\frac{\partial u}{\partial t} = c^2 \frac{\partial^2 u}{\partial x^2}$$

$$u(x, t) = (c_1 \cos px + c_2 \sin px)e^{-c^2p^2t} \qquad ...(1)$$

Since the ends $x = 0$ and $x = l$ are cooled to 0°C and kept at that temperature throughout, the boundary conditions are $u(0, t) = u(l, t) = 0$ for all t

Also $u(x, 0) = u_0$ is the initial condition.

Since $u(0, t) = 0$, we have from (1), $0 = c_1\, e^{-c^2p^2t} \;\Rightarrow\; c_1 = 0$

$\therefore$ From (1),
$$u(x, t) = c_2 \sin px.\, e^{-c^2p^2t} \qquad ...(2)$$

Since $u(l, t) = 0$, we have from (2), $0 = c_2 \sin pl \,.\, e^{-c^2p^2t}$

$$\Rightarrow \quad \sin pl = 0 \;\Rightarrow\; pl = n\pi$$

$$\therefore \quad p = \frac{n\pi}{l}, \; n \text{ being an integer}$$

Solution (2) reduces to $u(x, t) = b_n \sin \frac{n\pi x}{l} . e^{\frac{-c^2n^2\pi^2t}{l^2}}$ on replacing c_2 by b_n.

The most general solution is obtained by adding all such solutions for $n = 1, 2, 3, ...$

$$\therefore \quad u(x, t) = \sum_{n=1}^{\infty} b_n \sin \frac{n\pi x}{l} . e^{\frac{-c^2n^2\pi^2t}{l^2}} \quad ...(3)$$

Since $u(x, 0) = u_0$, we have $u_0 = \sum_{n=1}^{\infty} b_n \sin \frac{n\pi x}{l}$

which is half-range sine series for u_0.

$$\therefore \quad b_n = \frac{2}{l}\int_0^l u_0 \sin \frac{n\pi x}{l}\, dx = \begin{cases} 0, & \text{when } n \text{ is even} \\ \frac{4u_0}{n\pi}, & \text{when } n \text{ is odd} \end{cases}$$

Hence the temperature function

$$u(x, t) = \frac{4u_0}{\pi} \sum_{n=1,3,5,......} \frac{1}{n} \sin \frac{n\pi x}{l} e^{-\frac{c^2n^2\pi^2t}{l^2}}$$

$$= \frac{4u_0}{\pi} \sum_{n=1}^{\infty} \frac{1}{2n-1} \sin \frac{(2n-1)\pi x}{l} e^{-\frac{c^2(2n-1)^2\pi^2t}{l^2}}.$$

Example 2. *(a) An insulated rod of length l has its ends A and B maintained at 0°C and 100°C respectively until steady state conditions prevail. If B is suddenly reduced to 0°C and maintained at 0°C, find the temperature at a distance x from A at time t.*

(b) Find also the temperature if the change consists of raising the temperature of A to 20° C and reducing that of B to 80° C.

Sol. (*a*) The temperature function $u(x, t)$ satisfies the differential equation

$$\frac{\partial u}{\partial t} = c^2 \frac{\partial^2 u}{\partial x^2} \quad ...(1)$$

Prior to the temperature change at the end B, when $t = 0$, the heat flow was independent of time (steady state condition). When the temperature u depends only upon x and not on t, (1) reduces to $\frac{\partial^2 u}{\partial x^2} = 0$

Its general solution is $\quad u = ax + b \quad ...(2)$

where a, b are arbitrary constants.

Since $u = 0$ for $x = 0$ and $u = 100$ for $x = l$, we get from (2), $b = 0$ and $a = \frac{100}{l}$

$\therefore$ The *initial condition* is expressed by $\quad u(x, 0) = \frac{100}{l} x$

Also the *boundary conditions* for the subsequent flow are $u(0, t) = 0$ and $u(l, t) = 0$ for all values of t.

The most general solution of (1) satisfying the boundary conditions is

$$u(x, t) = \sum_{n=1}^{\infty} b_n \sin\frac{n\pi x}{l} \cdot e^{\frac{-c^2 n^2 \pi^2 t}{l^2}} \quad \text{...(3)}$$

Since $u(x, 0) = \dfrac{100}{l} x$, we have $\dfrac{100}{l} x = \displaystyle\sum_{n=1}^{\infty} b_n \sin\frac{n\pi x}{l}$

which is half-range sine series for $\dfrac{100}{l} x$.

$$\therefore \qquad b_n = \frac{2}{l}\int_0^l \frac{100x}{l} \sin\frac{n\pi x}{l}\, dx$$

$$= \frac{200}{l^2}\left[x\left(-\frac{\cos\frac{n\pi x}{l}}{\frac{n\pi}{l}} \right) - (1)\left(-\frac{\sin\frac{n\pi x}{l}}{\left(\frac{n\pi}{l}\right)^2} \right) \right]_0^l$$

$$= \frac{200}{l^2}\left[-\frac{l^2}{n\pi}\cos n\pi \right] = -\frac{200}{n\pi}(-1)^n = \frac{200}{n\pi}(-1)^{n+1}$$

Hence the temperature function

$$u(x, t) = \frac{200}{\pi}\sum_{n=1}^{\infty} \frac{(-1)^{n+1}}{n} \sin\frac{n\pi x}{l} \cdot e^{-\frac{c^2 n^2 \pi^2 t}{l^2}}$$

(*b*) Here the *initial condition* remains the same as in part (*a*), *i.e.*, $u(x, 0) = \dfrac{100}{l} x$ and the *boundary conditions* are $u(0, t) = 20$ for all values of t.

$u(l, t) = 80$ for all values of t.

Since the boundary values are non-zero, we modify the procedure. We break up the temperature function $u(x, t)$ into two parts as

$$u(x, t) = u_s(x) + u_t(x, t) \quad \text{...(1)}$$

where $u_s(x)$ is a solution of (1) involving x only and satisfying the boundary conditions ; $u_t(x, t)$ is then a function defined by (4). Thus $u_s(x)$ is a steady state solution of the form (2) and $u_t(x, t)$ may be regarded as a transient part of the solution which decreases with increase of time.

Since $u_s(0) = 20$ and $u_s(l) = 80$, we have from (2)

$$20 = b \quad \text{and} \quad 80 = al + b$$

$$b = 20 \quad \text{and} \quad a = \frac{60}{l}$$

and
$$u_s(x) = \frac{60}{l} x + 20 \quad \text{...(5)}$$

Putting $x = 0$ in (4), we have $u_t(0, t) = u(0, t) - u_s(0) = 20 - 20 = 0$ ...(6)

Putting $x = l$ in (4), we have $u_t(l, t) = u(l, t) - u_s(l) = 80 - 80 = 0$ ...(7)

Also $$u_t(x, 0) = u(x, 0) - u_s(x)$$

$$= \frac{100}{l}x - \left(\frac{60}{l}x + 20\right) = \frac{40}{l}x - 20 \qquad \text{...(8)}$$

Hence (6) and (7) give the boundary conditions and (7) gives the initial condition relative to the transient solution $u_t(x, t)$. Since the boundary values given by (6) and (7) are both zero, therefore as in part (*a*), we have

$$u_t(x, t) = \sum_{n=1}^{\infty} b_n \sin\frac{n\pi x}{l} e^{-\frac{c^2n^2\pi^2t}{l^2}} \qquad \text{...(9)}$$

where b_n is given by $$b_n = \frac{2}{l}\int_0^l \left(\frac{40}{l}x - 20\right)\sin\frac{n\pi x}{l}\,dx = \frac{-40}{n\pi}(1 + \cos n\pi)$$

$$= \begin{cases} 0, & \text{when } n \text{ is odd} \\ -\frac{80}{n\pi}, & \text{when } n \text{ is even} \end{cases}$$

Hence (9) becomes

$$u_t(x, t) = -\frac{80}{\pi}\sum_{n=2, 4, 6, \ldots\ldots}^{\infty} \frac{1}{n}\sin\frac{n\pi x}{l} e^{-\frac{c^2n^2\pi^2t}{l^2}}$$

$$= -\frac{40}{\pi}\sum_{m=1}^{\infty}\frac{1}{m}\sin\frac{2m\pi x}{l} e^{-\frac{4c^2m^2\pi^2t}{l^2}} \qquad \text{...(10)} \quad (\text{taking } n = 2m)$$

Combining (5) and (10), the required solution is

$$u(x, t) = 20 + \frac{60}{l}x - \frac{40}{\pi}\sum_{m=1}^{\infty}\frac{1}{m}\sin\frac{2m\pi x}{l} e^{-\frac{4c^2m^2\pi^2t}{l^2}}.$$

Example 3. *The ends A and B of a rod 20 cm long have the temperatures at 30°C and 80°C until steady state prevails. The temperatures of the ends are changed to 40°C and 60°C respectively. Find the temperature distribution in the rod at time t.*

Sol. The boundary conditions of the problem are

$$u(0, t) = 40 \text{ and } u(20, t) = 60 \qquad \text{...(1)}$$

To derive the initial conditions, we observe that prior to the temperature changes at the ends, when $t = 0$, steady state conditions prevailed and the temperature at every point of the rod was independent of time.

i.e., $$\frac{\partial u}{\partial t} = 0$$

The equation of heat conduction

$$\frac{\partial u}{\partial t} = c^2\frac{\partial^2 u}{\partial x^2} \qquad \text{...(2)}$$

reduces to $$\frac{\partial^2 u}{\partial x^2} = 0$$

Its general solution is

$$u = ax + b \qquad ...(3)$$

where a and b are arbitrary constants.

Since $u = 30$ when $x = 0$ and $u = 80$ when $x = 20$

we have $a = \frac{5}{2}$ and $b = 30$

∴ The initial condition is given by

$$u(x, 0) = \frac{5}{2}x + 30 \qquad ...(4)$$

It is now required to find $u(x, t)$ satisfying the differential equation (2) subject to the boundary conditions (1) and initial condition (4).

Since the boundary values are non-zero, we modify the procedure. We break up the temperature function $u(x, t)$ into two parts as

$$u(x, t) = u_s(x) + u_t(x, t) \qquad ...(5)$$

where $u_s(x)$ is a solution of (2) involving x only and satisfying the boundary conditions (1) and $u_t(x, t)$ is then defined by (5). Thus $u_s(x)$ is a steady state solution while $u_t(x, t)$ may be regarded as a transient part of the solution which decreases as t increases.

Now $u_s(0) = 40$ and $u_s(20) = 60$

From (3), $a = 1$ and $b = 40$

so that $u_s(x) = x + 40 \qquad ...(6)$

Consequently, $u_t(x, t) = u(x, t) - u_s(x)$

$= u(x, t) - x - 40$

Using the boundary and initial conditions (1) and (4), we get

$$\left.\begin{aligned} u_t(0, t) &= u(0, t) - 40 = 40 - 40 = 0 \\ u_t(20, t) &= u(20, t) - 20 - 40 = 60 - 60 = 0 \end{aligned}\right\} \qquad ...(7)$$

$$u_t(x, 0) = u(x, 0) - x - 40 = \frac{5}{2}x + 30 - x - 40 = \frac{3}{2}x - 10 \qquad ...(8)$$

Thus relations (7) and (8) serve as boundary and initial conditions for the transient part $u_t(x, t)$.

Now since the boundary values expressed by (7) are both zero, $u_t(x, t)$ is given by

$$u_t(x, t) = \sum_{n=1}^{\infty} b_n \sin\frac{n\pi x}{l} e^{-\frac{c^2n^2\pi^2 t}{l^2}}$$

$$= \sum_{n=1}^{\infty} b_n \sin\frac{n\pi x}{20} e^{-\frac{c^2n^2\pi^2 t}{400}} \qquad [\because\ l = 20]$$

...(9)

Using the initial condition (8), we get

$$\frac{3}{2}x - 10 = \sum_{n=1}^{\infty} b_n \sin\frac{n\pi x}{20}, \quad 0 < x < 20$$

which is half-range sine series for $\frac{3}{2}x - 10$.

$$\therefore \qquad b_n = \frac{2}{20}\int_0^{20}\left(\frac{3}{2}x - 10\right)\sin\frac{n\pi x}{20}\,dx$$

$$= \frac{1}{10}\left[\left(\frac{3}{2}x - 10\right)\left(-\frac{\cos\frac{n\pi x}{20}}{\frac{n\pi}{20}}\right) - \left(\frac{3}{2}\right)\left(-\frac{\sin\frac{n\pi x}{20}}{\left(\frac{n\pi}{20}\right)^2}\right)\right]_0^{20}$$

$$= \frac{1}{10}\left[-\frac{400}{n\pi}\cos n\pi - \frac{200}{n\pi}\right]$$

$$= -\frac{20}{n\pi}\,[2\cos n\pi + 1] = -\frac{20}{n\pi}\,[2(-1)^n + 1]$$

Substituting for b_n in (9), we have

$$u_t(x, t) = -\frac{20}{\pi}\sum_{n=1}^{\infty}\left(\frac{2(-1)^n + 1}{n}\right)\sin\frac{n\pi x}{20}\,e^{\frac{-c^2n^2\pi^2t}{400}}$$

Hence the general solution of (2) is given by

$$u(x, t) = x + 40 - \frac{20}{\pi}\sum_{n=1}^{\infty}\left(\frac{2(-1)^n + 1}{n}\right)\sin\frac{n\pi x}{20}\,e^{-\frac{c^2n^2\pi^2t}{400}}.$$

TEST YOUR KNOWLEDGE

1. Determine the solution of one-dimensional heat equation $\frac{\partial u}{\partial t} = c^2\frac{\partial^2 u}{\partial x^2}$, where the boundary conditions are $u(0, t) = 0$, $u(l, t) = 0$ $(t > 0)$ and the initial condition $u(x, 0) = x$, l being the length of the bar.
2. A bar with insulated sides is initially at temperature 0°C throughout. The end $x = 0$ is kept at 0°C and heat is suddenly applied at the end $x = l$ so that $\frac{\partial u}{\partial x} = A$ for $x = l$, where A is a constant. Find the temperature function $u(x, t)$.
3. A homogeneous rod of conducting material of length 100 cm has its ends kept as zero temperature and the temperature initially is $u(x, 0) = \begin{cases} x, & 0 \le x \le 50 \\ 100 - x, & 50 \le x \le 100 \end{cases}$

   Find the temperature $u(x, t)$ at any time.
4. Find the temperature $u(x, t)$ in a slab whose ends $x = 0$ and $x = L$ are kept at zero temperature and whose initial temperature $f(x)$ is given by

$$f(x) = \begin{cases} k, & \text{when } 0 < x < \tfrac{1}{2}L \\ 0, & \text{when } \tfrac{1}{2}L < x < L \end{cases}.$$

5. Solve the equation $\frac{\partial u}{\partial t} = \frac{\partial^2 u}{\partial x^2}$ with boundary conditions $u(x, 0) = 3\sin n\pi x$, $u(0, t) = 0$, $u(1, t) = 0$, where $0 < x < 1$, $t > 0$.

6. Find the solution of $\frac{\partial V}{\partial t} = k\frac{\partial^2 V}{\partial x^2}$, having given that $V = V_0 \sin nt$ when $x = 0$ for all values of t and $V = 0$ when x is very large.

7. Show that the solution of the differential equation $\frac{\partial u}{\partial t} = k\frac{\partial^2 u}{\partial x^2}$, subject to the conditions

(*i*) u not infinite for $t \to \infty$, (*ii*) $\frac{\partial u}{\partial x} = 0$ for $x = 0$ and $x = l$,

(*iii*) $u = lx - x^2$ for $t = 0$, between $x = 0$ and $x = l$, is $u = \frac{1}{6}l^2 - \frac{l^2}{\pi^2}\sum_{n=1}^{\infty}\frac{1}{n^2}\cos\frac{2n\pi x}{l}\, e^{-4n^2\pi^2 kt/l^2}$.

8. Solve $\frac{\partial \theta}{\partial t} = k\frac{\partial^2 \theta}{\partial x^2}$, such that

(*i*) θ is finite, when $t \to \infty$, (*ii*) $\frac{\partial \theta}{\partial x} = 0$ when $x = 0$ and $\theta = 0$ when $x = l$ for all t,

(*iii*) $\theta = \theta_0$, when $t = 0$ for all values of x between 0 and l.

9. Find a solution of the heat conduction equation $\frac{\partial u}{\partial t} = \alpha\frac{\partial^2 u}{\partial x^2}$ such that

(*i*) u is finite when $t \to \infty$, (*ii*) $u = 100$ when $x = 0$ or π for all values of t,

(*iii*) $u = 0$ when $t = 0$ for all values of x between 0 and π.

(Here the initially ice-cold rod has its ends in boiling water.)

10. A bar 10 cm long, with insulated sides, has its ends A and B maintained at temperatures 50°C and 100°C respectively, until steady-state conditions prevail. The temperature at A is suddenly raised to 90°C and at the same time that at B is lowered to 60°C. Find the temperature distribution in the bar at time t.

11. The ends A and B of a rod 20 cm long have the temperature at 30°C and 80°C until steady state prevails. The temperatures of the ends are changed to 40°C and 60°C respectively. Find the temperature distribution in the rod at time t.

Answers

1. $u(x, t) = -\frac{2l}{\pi}\sum_{n=1}^{\infty}\frac{\cos n\pi}{n}\sin\frac{n\pi x}{l}\, e^{(-c^2n^2\pi^2 t)/l^2}$

2. $u(x, t) = Ax + \frac{8Al}{\pi^2}\sum_{m=1}^{\infty}\frac{(-1)^m}{(2m-1)^2}\sin\frac{(2m-1)\pi x}{l}\, e^{[-(2m-1)c^2\pi^2 t]/4l^2}$

3. $u(x, t) = \frac{400}{\pi^2}\sum_{m=0}^{\infty}\frac{(-1)^m}{(2m+1)^2}\sin\frac{(2m+1)\pi x}{100}\, e^{-[(2m+1)c\pi/100]^2 t}$

4. $u(x, t) = \frac{4k}{\pi}\sum_{n=1}^{\infty}\frac{1}{n}\sin^2\left(\frac{n\pi}{4}\right)\sin\frac{n\pi x}{L}\, e^{(-c^2n^2\pi^2)t/L^2}$

5. $u(x, t) = 3\sum_{n=1}^{\infty}\sin n\pi x\; e^{-n^2\pi^2 t}$

6. $V = V_0 e^{-\sqrt{\left(\frac{n}{2k}\right)}x}\sin\left[nt - \sqrt{\left(\frac{n}{2k}\right)}x\right]$

8. $\theta = \frac{4\theta_0}{\pi}\left[e^{-\left(\frac{\pi}{2l}\right)^2 kt}\cos\frac{\pi x}{2l} - \frac{1}{3}e^{-\left(\frac{3\pi}{2l}\right)^2 kt}\cos\frac{3\pi x}{2l} + \frac{1}{5}e^{-\left(\frac{5\pi}{2l}\right)^2 kt}\cos\frac{5\pi x}{2l} - \ldots\ldots\right]$

9. $u = 100 - \frac{400}{\pi}\sum_{m=1}^{\infty}\frac{\sin(2m-1)x}{2m-1}e^{-(2m-1)^2\alpha t}$

10. $u(x, t) = 90 - 3x - \frac{80}{\pi}\sum_{n=1}^{\infty}\frac{1}{n}\sin\frac{n\pi x}{5}e^{-c^2n^2\pi^2t/25}$

11. $u(x, t) = 40 + x - \frac{20}{\pi}\sum_{n=1}^{\infty}\frac{1}{n}(2\cos n\pi + 1)\sin\frac{n\pi x}{20}e^{-\frac{c^2n^2\pi^2t}{400}}$.

3.23. TWO-DIMENSIONAL HEAT FLOW

Consider the flow of heat in a metal plate, in the XOY plane. If the temperature at any point is independent of the z-coordinate and depends on x, y and t only, then the flow is called two dimensional and the heat flow lies in the plane XOY only and is zero along the normal to the plane XOY.

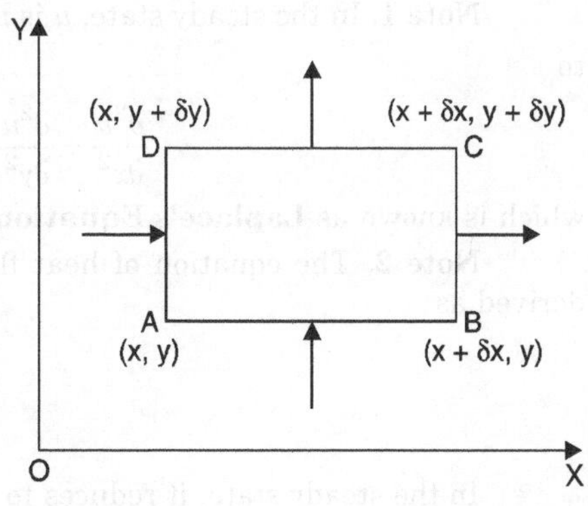

Take a rectangular element of the plate with sides δx and δy and thickness α. As discussed in the one-dimensional heat flow along a bar, the quantity of heat that enters the plate per second from the sides AB and AD is given by

$$-k\alpha\,\delta x\left(\frac{\partial u}{\partial y}\right)_y \quad \text{and} \quad -k\alpha\,\delta y\left(\frac{\partial u}{\partial x}\right)_x$$

respectively and that which flows out through the sides CD and BC per second is

$$-k\alpha\,\delta x\left(\frac{\partial u}{\partial y}\right)_{y+\delta y} \quad \text{and} \quad -k\alpha\,\delta y\left(\frac{\partial u}{\partial x}\right)_{x+\delta x} \quad \text{respectively.}$$

Therefore, the total gain of heat by the rectangular plate ABCD per second

$$= -k\alpha\delta x\left(\frac{\partial u}{\partial y}\right)_y - k\alpha\delta y\left(\frac{\partial u}{\partial x}\right)_x + k\alpha\delta x\left(\frac{\partial u}{\partial y}\right)_{y+\delta y} + k\alpha\delta y\left(\frac{\partial u}{\partial x}\right)_{x+\delta x}$$

$$= k\alpha\delta x\delta y\left[\frac{\left(\frac{\partial u}{\partial x}\right)_{x+\delta x} - \left(\frac{\partial u}{\partial x}\right)_x}{\delta x} + \frac{\left(\frac{\partial u}{\partial y}\right)_{y+\delta y} - \left(\frac{\partial u}{\partial y}\right)_y}{\delta y}\right] \quad \ldots(1)$$

The rate of gain of heat by the plate is also given by

$$s\rho\delta x\delta y\frac{\partial u}{\partial t} \quad \ldots(2)$$

where s = specific heat and ρ = density of the metal plate.

Equating (1) and (2), we have

$$k\alpha\delta x\delta y\left[\frac{\left(\frac{\partial u}{\partial x}\right)_{x+\delta x}-\left(\frac{\partial u}{\partial x}\right)_{x}}{\delta x}+\frac{\left(\frac{\partial u}{\partial y}\right)_{y+\delta y}-\left(\frac{\partial u}{\partial y}\right)_{y}}{\delta y}\right]=s\rho\delta x\delta y\frac{\partial u}{\partial t}$$

Dividing both sides by $\alpha\delta x\delta y$ and taking the limit as $\delta x\to 0$, $\delta y\to 0$, we get

$$k\left(\frac{\partial^2 u}{\partial x^2}+\frac{\partial^2 u}{\partial y^2}\right)=s\rho\frac{\partial u}{\partial t}$$

or $$c^2\left(\frac{\partial^2 u}{\partial x^2}+\frac{\partial^2 u}{\partial y^2}\right)=\frac{\partial u}{\partial t},\text{ where } c^2=\frac{k}{s\rho} \quad ...(3)$$

Equation (3) gives the temperature distribution of the plate in the transient state.

Note 1. In the steady state, u is independent of t, so that $\frac{\partial u}{\partial t}=0$ and the above equation reduces to

$$\frac{\partial^2 u}{\partial x^2}+\frac{\partial^2 u}{\partial y^2}=0 \quad ...(4)$$

which is known as **Laplace's Equation in two dimensions.**

Note 2. The equation of heat flow in a solid (Three-dimensional heat flow) can similarly be derived as

$$c^2\left(\frac{\partial^2 u}{\partial x^2}+\frac{\partial^2 u}{\partial y^2}+\frac{\partial^2 u}{\partial z^2}\right)=\frac{\partial u}{\partial t}$$

In the steady state, it reduces to $$\frac{\partial^2 u}{\partial x^2}+\frac{\partial^2 u}{\partial y^2}+\frac{\partial^2 u}{\partial z^2}=0$$

which is **Laplace's Equation in three-dimensions.**

3.24. SOLUTION OF LAPLACE'S EQUATION IN TWO DIMENSIONS

Laplace's equation in two dimensions is

$$\frac{\partial^2 u}{\partial x^2}+\frac{\partial^2 u}{\partial y^2}=0 \quad ...(1)$$

Let $$u = \text{XY} \quad ...(2)$$

where X is a function of x only and Y is a function of y only, be a solution of (1).

Then $$\frac{\partial^2 u}{\partial x^2}=\text{X}''\text{Y} \quad\text{and}\quad \frac{\partial^2 u}{\partial y^2}=\text{XY}''$$

Substituting in (1), we have $\text{X}''\text{Y}+\text{XY}''=0$ or $\frac{\text{X}''}{\text{X}}=-\frac{\text{Y}''}{\text{Y}}$ ...(3)

Now the L.H.S. of (3) is a function of x only and the R.H.S. is a function of y only. Since x and y are independent variables, this equation can hold only when both sides reduce to a constant, say k. Then equation (3) leads to the ordinary differential equations

$$\frac{d^2\text{X}}{dx^2}-k\text{X}=0 \quad\text{and}\quad \frac{d^2\text{Y}}{dy^2}+k\text{Y}=0 \quad ...(4)$$

Solving equations (4), we get

(*i*) *When k is positive* and = p^2, say

$$X = c_1e^{px} + c_2e^{-px},\ Y = c_3 \cos py + c_4 \sin py$$

(*ii*) *When k is negative* and = $-p^2$, say

$$X = c_1 \cos px + c_2 \sin px,\ Y = c_3e^{py} + c_4e^{-py}$$

(*iii*) *When k = 0*

$$X = c_1x + c_2,\ Y = c_3y + c_4$$

Thus the various possible solutions of Laplace's equation (1) are:

$$u = (c_1e^{px} + c_2e^{-px})(c_3 \cos py + c_4 \sin py)$$

$$u = (c_1 \cos px + c_2 \sin px)(c_3e^{py} + c_4e^{-py})$$

$$u = (c_1x + c_2)(c_3y + c_4)$$

Of these three solutions, we have to choose that solution which is consistent with the physical nature of the problem and the given boundary conditions.

ILLUSTRATIVE EXAMPLES

Example 1. *Solve* $\dfrac{\partial^2 u}{\partial x^2} + \dfrac{\partial^2 u}{\partial y^2} = 0$ *which satisfies the conditions:*

$$u(0, y) = u(l, y) = u(x, 0) = 0 \text{ and } u(x, a) = \sin \frac{n\pi x}{l}.$$

Sol. The given equation is $\dfrac{\partial^2 u}{\partial x^2} + \dfrac{\partial^2 u}{\partial y^2} = 0$ ...(1)

The three possible solutions of eqn. (1) are

$u = (c_1e^{px} + c_2e^{-px})(c_3 \cos py + c_4 \sin py)$ ...(2)

$u = (c_1 \cos px + c_2 \sin px)(c_3e^{py} + c_4e^{-py})$ ...(3)

$u = (c_1x + c_2)(c_3y + c_4)$ ...(4)

Y
u = sin $\frac{n\pi x}{l}$
a
u = 0
• P(x, y)
u = 0
O
u = 0
l
X

Keeping in view the given boundary conditions, the only possible solution is (3)

$\therefore$ $u(x, y) = (c_1 \cos px + c_2 \sin px)(c_3e^{py} + c_4e^{-py})$ ...(5)

Since $u(0, y) = 0$ $\quad\therefore\quad 0 = c_1(c_3e^{py} + c_4e^{-py})$

$\Rightarrow$ $c_1 = 0$

Equation (5) reduces to $u(x, y) = c_2 \sin px(c_3e^{py} + c_4e^{-py})$ ...(6)

Since $u(l, y) = 0$ $\quad\therefore\quad 0 = c_2 \sin pl(c_3e^{py} + c_4e^{-py})$

$\Rightarrow$ $\sin pl = 0$ $\quad$ *i.e.,* $\quad pl = n\pi$ or $p = \dfrac{n\pi}{l}$, *n* being an integer

Also $u(x, 0) = 0$ $\quad\therefore\quad 0 = c_2 \sin \dfrac{n\pi x}{l}(c_3 + c_4)$

$\Rightarrow$ $c_3 + c_4 = 0$ $\quad\therefore\quad c_4 = -c_3$

$\therefore$ Equation (6) becomes $u(x, y) = c_2 \sin \dfrac{n\pi x}{l}\left(c_3e^{\frac{n\pi y}{l}} - c_3e^{\frac{-n\pi y}{l}}\right)$

Replacing c_2c_3 by b_n, we have $u(x, y) = b_n \sin \frac{n\pi x}{l}\left(e^{\frac{n\pi y}{l}} - e^{\frac{-n\pi y}{l}}\right)$

$$= 2b_n \sin \frac{n\pi x}{l} \sinh \frac{n\pi y}{l} \qquad ...(7)$$

Putting $y = a$, we have $u(x, a) = \sin \frac{n\pi x}{l} = 2b_n \sin \frac{n\pi x}{l} \sinh \frac{n\pi a}{l}$

$$\Rightarrow \quad 2b_n \sinh \frac{n\pi a}{l} = 1 \quad \text{or} \quad b_n = \frac{1}{2\sinh \frac{n\pi a}{l}}$$

Hence (7) reduces to $u(x, y) = \sin \frac{n\pi x}{l}\left[\frac{\sinh \frac{n\pi y}{l}}{\sinh \frac{n\pi a}{l}}\right]$

which is the required solution of (1).

Example 2. *A rectangular plate with insulated surface is 10 cm wide and so long compared to its width that it may be considered infinite in length without introducing an appreciable error. If the temperature of the short edge y = 0 is given by*

$$u = 20x \quad \text{for} \quad 0 \le x \le 5$$

and

$$u = 20(10 - x) \quad \text{for} \quad 5 \le x \le 10$$

and the two long edges x = 0, x = 10 as well as the other short edge are kept at 0°C, prove that the temperature u at any point (x, y) is given by

$$u = \frac{800}{\pi^2}\sum_{1}^{\infty} \frac{(-1)^{n+1}}{(2n-1)^2} \sin \frac{(2n-1)\pi x}{10} . e^{-\frac{(2n-1)\pi y}{10}}.$$

Sol. The temperature $u(x, y)$ at any point P(x, y) satisfies the equation

$$\frac{\partial^2 u}{\partial x^2} + \frac{\partial^2 u}{\partial y^2} = 0. \qquad ...(1)$$

The boundary conditions are $u(0, y) = 0$

for all values of $y \ge 0$ ...(2)

$u(10, y) = 0$

for all values of $y \ge 0$ ...(3)

$u(x, \infty) = 0$

in $0 \le x \le 10$ ...(4)

$$u(x, 0) = \begin{cases} 20x, & 0 \le x \le 5 \\ 20(10 - x), & 5 \le x \le 10 \end{cases} \qquad ...(5)$$

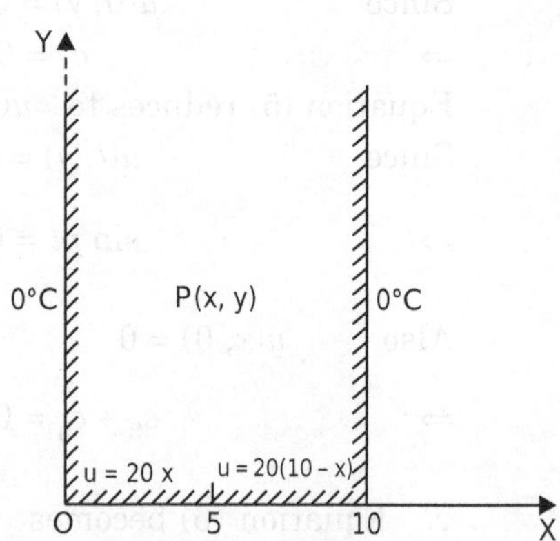

Now the three possible solutions of (1) are

$u = (c_1e^{px} + c_2e^{-px})(c_3 \cos py + c_4 \sin py)$ ...(6)

$u = (c_1 \cos px + c_2 \sin px)(c_3e^{py} + c_4e^{-py})$ ...(7)

$u = (c_1x + c_2)(c_3y + c_4)$ ...(8)

The solution (6) cannot satisfy the condition (2) since, we get $u \neq 0$ for $x = 0$, for all values of y. The solution (8) cannot satisfy the condition (4). Thus the only possible solution is (7).

$$\therefore \quad u(x, y) = (c_1 \cos px + c_2 \sin px)(c_3 e^{py} + c_4 e^{-py}) \qquad ...(9)$$

Since $u(0, y) = 0 \quad \therefore \quad 0 = c_1(c_3\, e^{py} + c_4\, e^{-py}) \Rightarrow c_1 = 0$

$\therefore$ Equation (9) reduces to $u(x, y) = c_2 \sin px(c_3\, e^{py} + c_4\, e^{-py})$ ...(10)

Since $u(10, y) = 0 \quad \therefore \quad 0 = c_2 \sin 10p\, (c_3\, e^{py} + c_4\, e^{-py})$

$\Rightarrow \quad \sin 10p = 0 \quad$ *i.e.*, $10p = n\pi$ or $p = \dfrac{n\pi}{10}$, n being an integer.

Also $u(x, \infty) = 0 \quad \therefore \quad c_3 = 0$

Hence from (10) a solution satisfying (2), (3) and (4) is $u(x, y) = c_2 c_4 \sin \dfrac{n\pi x}{10}\, e^{-\frac{n\pi y}{10}}$

Replacing c_2c_4 by b_n, the most general solution is

$$u(x, y) = \sum_{n=1}^{\infty} b_n \sin \frac{n\pi x}{10} . e^{-\frac{n\pi y}{10}} \qquad ...(11)$$

Putting $y = 0$, we have $u(x, 0) = \displaystyle\sum_{n=1}^{\infty} b_n \sin \frac{n\pi x}{10}$

$$\therefore\ b_n = \frac{2}{10} \int_0^{10} u(x, 0) \sin \frac{n\pi x}{10}\, dx$$

$$= \frac{1}{5}\left[\int_0^5 20x \sin \frac{n\pi x}{10}\, dx + \int_5^{10} 20(10 - x) \sin \frac{n\pi x}{10}\, dx\right]$$

$$= 4\left[x\left(\frac{-\cos \frac{n\pi x}{10}}{\frac{n\pi}{10}}\right) - (1)\left(-\frac{\sin \frac{n\pi x}{10}}{\left(\frac{n\pi}{10}\right)^2}\right)\right]_0^5 + 4\left[(10 - x)\left(-\frac{\cos \frac{n\pi x}{10}}{\frac{n\pi}{10}}\right) - (-1)\left(-\frac{\sin \frac{n\pi x}{10}}{\left(\frac{n\pi}{10}\right)^2}\right)\right]_5^{10}$$

$$= 4\left[-\frac{50}{n\pi} \cos \frac{n\pi}{2} + \left(\frac{10}{n\pi}\right)^2 \sin \frac{n\pi}{2}\right] + 4\left[\frac{50}{n\pi} \cos \frac{n\pi}{2} + \left(\frac{10}{n\pi}\right)^2 \sin \frac{n\pi}{2}\right] = \frac{800}{n^2\pi^2} \sin \frac{n\pi}{2}$$

$$= \begin{cases} \dfrac{800}{n^2\pi^2}(-1)^{\frac{n-1}{2}}, & \text{when } n \text{ is odd} \\ 0, & \text{when } n \text{ is even} \end{cases}$$

Hence from (11), the required solution is

$$u(x, y) = \frac{800}{\pi^2} \sum_{n = 1, 3, 5, \ldots\ldots} \frac{(-1)^{\frac{n-1}{2}}}{n^2} \sin \frac{n\pi x}{10}\, e^{-\frac{n\pi y}{10}} \qquad \text{(Replacing } n \text{ by } 2n - 1\text{)}$$

$$= \frac{800}{\pi^2} \sum_{n=1}^{\infty} \frac{(-1)^{n-1}}{(2n-1)^2} \sin \frac{(2n-1)\pi x}{10} e^{-\frac{(2n-1)\pi y}{10}}$$

or $$u = \frac{800}{\pi^2} \sum_{1}^{\infty} \frac{(-1)^{n+1}}{(2n-1)^2} \sin \frac{(2n-1)\pi x}{10} e^{-\frac{(2n-1)\pi y}{10}}$$

$$\mid \because \ (-1)^{n+1} = (-1)^{n-1} \cdot (-1)^2 = (-1)^{n-1}.$$

Example 3. *An infinitely long plane uniform plate is bounded by two parallel edges and an end at right angles to them. The breadth is π; this end is maintained at temperature u_0 at all points and the other edges are at zero temperature. Determine the temperature at any point of the plate in the steady state.*

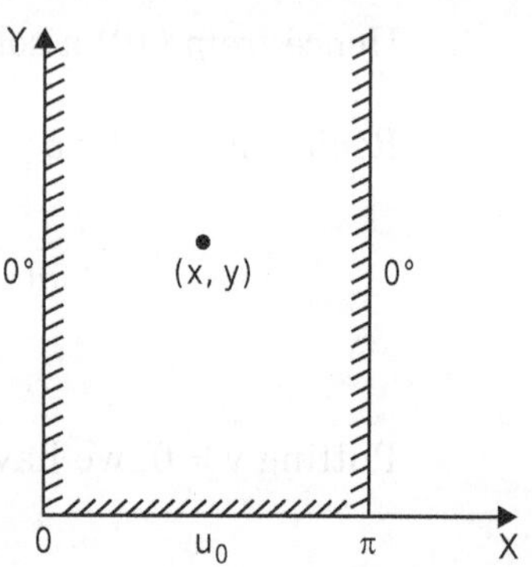

Sol. In the steady state, the temperature $u(x, y)$ at any point (x, y) is given by

$$\frac{\partial^2 u}{\partial x^2} + \frac{\partial^2 u}{\partial y^2} = 0 \qquad ...(1)$$

Let $$u = XY \qquad ...(2)$$

where X is a function of x only and Y is a function of y only, be a solution of (1)

Then $$\frac{\partial^2 u}{\partial x^2} = X''Y. \frac{\partial^2 u}{\partial y^2} = XY''$$

Substituting in (1), we have

$$X''Y + XY'' = 0 \quad \text{or} \quad \frac{X''}{X} = -\frac{Y''}{Y}$$

$\because$ L.H.S. is a function of x only while R.H.S. is a function of y only, the equality holds only when each side is a constant, say k.

$$\therefore \qquad \frac{X''}{X} = -\frac{Y''}{Y} = k \qquad ...(3)$$

Three cases arise according as k is positive, zero or negative.

Case I. Let k be positive and $= p^2$. Then from (3),

$$X'' - p^2X = 0 \qquad ...(4) \qquad Y'' + p^2Y = 0 \qquad ...(5)$$

To solve (4), A.E. is $m^2 - p^2 = 0$ so that $m = \pm p$

$$\therefore \qquad X = c_1 e^{px} + c_2 e^{-px}$$

To solve (5), A.E. is $m^2 + p^2 = 0$ so that $m = \pm ip$

$$\therefore \qquad Y = c_3 \cos py + c_4 \sin py$$

$\therefore$ From (2), $$u = (c_1 e^{px} + c_2 e^{-px})(c_3 \cos py + c_4 \sin py).$$

Case II. Let $k = 0$, then from (3),

$$X'' = 0, Y'' = 0$$

$$\Rightarrow \qquad X' = c_1, Y' = c_3$$

$$\Rightarrow \qquad X = c_1x + c_2, \quad Y = c_3y + c_4$$

$\therefore$ From (2), $$u = (c_1x + c_2)(c_3y + c_4).$$

Case III. Let k be negative and $= -p^2$. Then from (3),

$$X'' + p^2X = 0, \quad Y'' = p^2Y = 0$$

Proceeding as in case I,

$$X = c_1 \cos px + c_2 \sin px,\ Y = c_3 e^{py} + c_4 e^{-py}$$

From (2), $u = (c_1 \cos px + c_2 \sin px)(c_3 e^{py} + c_4 e^{-py})$

Of the three solutions obtained above, we have to choose the one which satisfies the given boundary conditions.

The given boundary conditions are

(i) $u(0, y) = 0$ for all y.

(ii) $u(\pi, y) = 0$ for all y.

(iii) $u(x, \infty) = 0$ for all $0 < x < \pi$.

(iv) $u(x, 0) = u_0$ for all $0 < x < \pi$.

Putting $x = 0$ in the solution of Case I, we have

$$u(0, y) = (c_1 + c_2)(c_3 \cos py + c_4 \sin py)$$

or $$0 = (c_1 + c_2)(c_3 \cos py + c_4 \sin py)$$

This equation is satisfied only when $c_1 = c_2 = 0$

This leads to the trivial solution $u = 0$

Thus the solution of Case I is not consistent with the given boundary conditions.

As $y \to \infty$ in the solution of Case II, we have $u \to \infty$

But $u(x, \infty) = 0$ for $0 < x < \pi$

∴ The solution of Case II is also not consistent with the given boundary conditions.

∴ The solution of Case III is the suitable solution.

Thus $$u(x, y) = (c_1 \cos px + c_2 \sin px)(c_3 e^{py} + c_4 e^{-py}) \quad ...(6)$$

Putting $x = 0$ in (6), $u(0, y) = c_1 (c_3 e^{py} + c_4 e^{-py})$

$\Rightarrow$ $0 = c_1(c_3 e^{py} + c_4 e^{-py})$ [First boundary condition]

∴ $c_1 = 0$

Thus $u(x, y) = c_2 \sin px(c_3 e^{py} + c_4 e^{-py})$

Now $u \to 0$ as $y \to \infty$ [Third boundary condition] only when $c_3 = 0$.

Thus $u(x, y) = c_2 c_4 \sin px \,.\, e^{-py}$

Putting $x = \pi$, $u(\pi, y) = c_2 c_4 \sin p\pi \,.\, e^{-py}$

$\Rightarrow$ $0 = c_2 c_4 \sin p\pi e^{-py}$ [Second boundary condition]

Since $c_2 \neq 0$, $c_4 \neq 0$, $\sin p\pi = 0$

$\Rightarrow$ p is an integer. Let $p = n$.

Thus $u(x, y) = c_2 c_4 \sin nx\, e^{-ny} = b_n \sin nx\, e^{-ny}$ (where $b_n = c_2 c_4$)

The general solution is obtained by adding all such solutions for $n = 1, 2, 3, ...$

∴ The general solution is given by

$$u(x, y) = \sum_{n=1}^{\infty} b_n \sin nx . e^{-ny} \quad ...(7)$$

Putting $y = 0$ in (7), $u(x, 0) = \sum_{n=1}^{\infty} b_n \sin nx$

$$\Rightarrow \quad u_0 = \sum_{n=1}^{\infty} b_n \sin nx \quad \text{[Fourth boundary condition]}$$

This is a Fourier sine series, where b_n is given by

$$b_n = \frac{2}{\pi}\int_0^{\pi} u_0 \sin nx\, dx = \frac{2u_0}{\pi}\left[-\frac{\cos nx}{n}\right]_0^{\pi}$$

$$= \frac{2u_0}{n\pi}[1 - \cos n\pi] = \frac{2u_0}{n\pi}\ [1 - (-1)^n]$$

$$= \begin{cases} 0 & \text{when } n \text{ is even} \\ \dfrac{4u_0}{n\pi} & \text{when } n \text{ is odd} \end{cases}$$

$$= \frac{4u_0}{(2n-1)\pi} \qquad \text{(replacing } n \text{ by } 2n - 1)$$

Hence from (7), we have

$$u(x, y) = \frac{4u_0}{\pi}\sum_{n=1}^{\infty}\frac{1}{2n-1}\sin(2n-1)x \, . \, e^{-(2n-1)y}.$$

Example 4. *Solve* $\dfrac{\partial^2 u}{\partial x^2} + \dfrac{\partial^2 u}{\partial y^2} = 0$ *which satisfies the conditions:*

$$u(0, y) = u(1, y) = u(x, 0) = 0$$

and $$u(x, a) = \sin\frac{n\pi x}{l}.$$

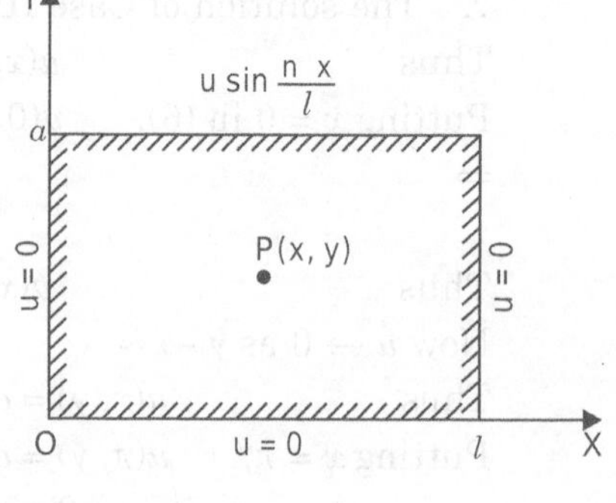

Sol. The given equation is

$$\frac{\partial^2 u}{\partial x^2} + \frac{\partial^2 u}{\partial y^2} = 0 \qquad ...(1)$$

The three possible solutions of (1) are

$$u = (c_1 e^{px} + c_2 e^{-px})(c_3 \cos py + c_4 \sin py) \qquad ...(2)$$

$$u = (c_1 x + c_2)(x_3 y + x_4) \qquad ...(3)$$

$$u = (c_1 \cos px + c_2 \sin px)(c_3 e^{py} + c_4 e^{-py}) \qquad ...(4)$$

Keeping in view the given boundary conditions, the only possible solution is (4).

$$\therefore \quad u(x, y) = (c_1 \cos px + c_2 \sin px)(c_3 e^{py} + c_4 e^{-py}) \qquad ...(5)$$

Since $u(0, y) = 0 \quad \therefore \quad 0 = c_1(c_3 e^{py} + c_4 e^{-py})$

$\Rightarrow \quad c_1 = 0.$

Equation (5) reduces to $u(x, y) = c_2 \sin px\,(c_3 e^{py} + c_4 e^{-py})$ ...(6)

Since $u(l, y) = 0 \quad \therefore \quad 0 = c_2 \sin pl\,(c_3 e^{py} + c_4 e^{-py})$

$\Rightarrow \quad \sin pl = 0 \qquad (\because \ c_2 \neq 0)$

$\Rightarrow \quad pl = n\pi \quad \text{or} \quad p = \dfrac{n\pi}{l}, n \geq 1$

Also $u(x, 0) = 0 \quad \therefore \quad 0 = c_2 \sin\dfrac{n\pi x}{l}(c_3 + c_4)$

$\Rightarrow \quad c_3 + c_4 = 0 \quad \therefore \quad c_4 = -c_3$

$\therefore$ Equation (6) becomes

$$u(x, y) = c_2 \sin \frac{n\pi x}{l} \left(c_3 e^{\frac{n\pi x}{l}} - c_3 e^{-\frac{n\pi x}{l}} \right)$$

Replacing c_2c_3 by b_n, we have

$$u(x, y) = b_n \sin \frac{n\pi x}{l} \left(e^{\frac{n\pi x}{l}} - e^{\frac{n\pi x}{l}} \right)$$

$$= 2b_n \sin \frac{n\pi x}{l} \sinh \frac{n\pi y}{l} \quad \text{...(7)}$$

Since $\qquad u(x, a) = \sin \dfrac{n\pi x}{l}$

$$\therefore \qquad \sin \frac{n\pi x}{l} = 2b_n \sin \frac{n\pi x}{l} \sinh \frac{n\pi a}{l}$$

$$\Rightarrow \qquad b_n = \frac{1}{2 \sinh \frac{n\pi a}{l}}$$

Hence equations (7) reduces to

$$u(x, y) = \sin \frac{n\pi x}{l} \left(\frac{\sinh \frac{n\pi y}{l}}{\sinh \frac{n\pi a}{l}} \right)$$

which is the required solution of (1).

TEST YOUR KNOWLEDGE

1. A long rectangular plate of width a cm with insulated surface has its temperature v equal to zero on both the long sides and one of the short sides so that $v(0, y) = 0$, $v(a, y) = 0$, $v(x, \infty) = 0$, $v(x, 0) = kx$. Show that the steady-state temperature within the plate is

$$v(x, y) = \frac{2ak}{\pi} \sum_{n=1}^{\infty} \frac{(-1)^{n+1}}{n} e^{-n\pi y/a} \sin \frac{n\pi x}{a}.$$

2. A rectangular plate with insulated surfaces is 8 cm wide and so long compared to its width that it may be considered infinite in length without introducing an appreciable error. If the temperature along one short edge $y = 0$ is given by

$$u(x, 0) = 100 \sin \frac{\pi x}{8}, \; 0 < x < 8$$

while the two long edges $x = 0$ and $x = 8$ as well as the other short edge are kept at 0°C, show that the steady-state temperature at any point of the plate is given by $u(x, y) = 100\, e^{-\pi y/8} \sin \dfrac{\pi x}{8}$.

3. Solve $\dfrac{\partial^2 u}{\partial x^2} + \dfrac{\partial^2 u}{\partial y^2} = 0$ for $0 < x < \pi$, $0 < y < \pi$, given that $u(0, y) = u(\pi, y) = u(x, \pi) = 0$, $u(x, 0) = \sin^2 x$.

4. Solve $\frac{\partial^2 u}{\partial x^2} + \frac{\partial^2 u}{\partial y^2} = 0$ within the rectangle $0 \le x \le a, 0 \le y \le b$ given that

$$u(0, y) = u(a, y) = u(x, b) = 0 \quad \text{and} \quad u(x, 0) = x(a - x).$$

5. A square plate is bounded by the lines $x = 0, y = 0, x = 20$ and $y = 20$. Its faces are insulated. The temperature along the upper horizontal edge is given by $u(x, 20) = x(20 - x)$, when $0 < x < 20$, while other three edges are kept at 0°C. Find the steady state temperature in the plate.

6. An infinitely long plane uniform plate is bounded by two parallel edges and an end at right angles to them. The breadth is π. This end is maintained at a temperature u_0 at all points and the other edges are at zero temperature. Determine the temperature at any point of the plate in the steady-state.

7. A rectangular plate has sides a and b. Let the side of length a be taken along OX and that of length b along OY and the other sides along $x = a$ and $y = b$. The sides $x = 0, x = a, y = b$ are insulated and the edge $y = 0$ is kept at temperature $u_0 \cos \frac{\pi x}{a}$. Find the steady-state temperature at any point (x, y).

8. The temperature u is maintained at 0°C along three edges of a square plate of length 100 cm and the fourth edge is maintained at 100°C until steady-state conditions prevail. Find an expression for the temperature u at any point (x, y). Hence show that the temperature at the centre of the plate

$$= \frac{200}{\pi}\left[\frac{1}{\cosh \frac{\pi}{2}} - \frac{1}{3\cosh \frac{3\pi}{2}} + \frac{1}{5\cosh \frac{5\pi}{2}} - \ldots\ldots\right].$$

Answers

3. $u(x, y) = -\frac{8}{\pi} \sum_{n=1,3,5,\ldots}^{\infty} \frac{\sin nx \sinh n(\pi - y)}{n(n^2 - 4)\sinh n\pi}$

4. $u(x, y) = \frac{8a^2}{\pi^3} \sum_{n=0}^{\infty} \frac{1}{(2n+1)^3} \sin \frac{(2n+1)\pi x}{a} \times \frac{\sinh \frac{(2n+1)\pi}{a}(b - y)}{\sinh \frac{(2n+1)\pi b}{a}}$

5. $u(x, y) = \frac{3200}{\pi^3} \sum_{n=1}^{\infty} \frac{\sin \frac{(2n-1)\pi x}{20} \sinh \frac{(n-1)\pi y}{20}}{(2n-1)^3 \sinh (2n-1)\pi}$

6. $u(x, y) = \frac{4u_0}{\pi}\left[e^{-y} \sin x + \frac{1}{3} e^{-3y} \sin 3x + \frac{1}{5} e^{-5y} \sin 5x + \ldots\ldots\right]$

7. $u(x, y) = u_0 \cos \frac{\pi x}{a} \cosh \frac{\pi}{a}(b - y) \operatorname{sech} \frac{\pi b}{a}$.

3.25. VIBRATING MEMBRANE—TWO-DIMENSIONAL WAVE EQUATION

We shall now obtain the equation for the vibrations of a tightly stretched membrane (such as the membrane of a drum). We shall assume that the membrane is uniform and the tension in it per unit length is the same at every point in all directions. Let T be the tension per unit length and m be the mass of the membrane per unit area.

Consider the forces on an element $\delta x \delta y$ of the membrane. Due to its displacement u, perpendicular to the xy-plane, the forces $T\delta y$ (tangential to the membrane) on its opposite edges of length δy act at angles α and β to the horizontal. So their vertical component

$= (T\delta y) \sin \beta - (T\delta y) \sin \alpha = T\delta y(\tan \beta - \tan \alpha)$ approximately, since α and β are small

$$= T\delta y \left[\left(\frac{\partial u}{\partial x} \right)_{x + \delta x} - \left(\frac{\partial u}{\partial x} \right)_x \right] = T\delta y \delta x \left[\frac{\left(\frac{\partial u}{\partial x} \right)_{x + \delta x} - \left(\frac{\partial u}{\partial x} \right)_x}{\delta x} \right]$$

$= T\delta x \delta y \dfrac{\partial^2 u}{\partial x^2}$ upto a first order of approximation.

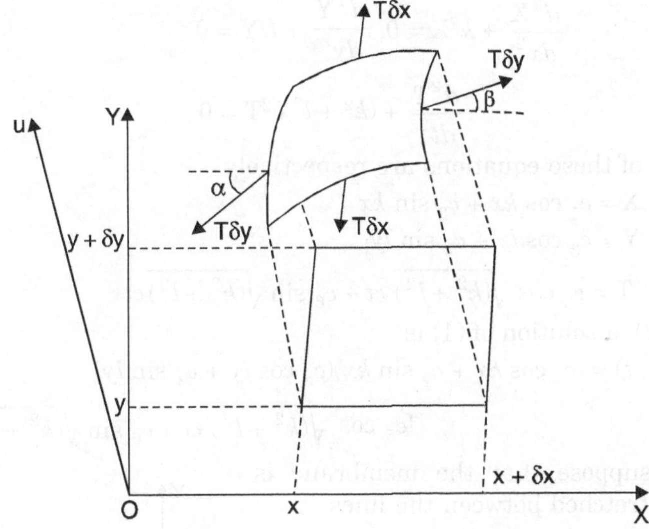

Similarly, the forces $T\delta x$ acting on the edges of length δx have the vertical component $T\delta x \delta y \dfrac{\partial^2 u}{\partial y^2}$.

Hence the equation of motion of the element is

$$(m\delta x \delta y) \frac{\partial^2 u}{\partial t^2} = T \left(\frac{\partial^2 u}{\partial x^2} + \frac{\partial^2 u}{\partial y^2} \right) \delta x \delta y$$

or
$$\frac{\partial^2 u}{\partial t^2} = c^2 \left(\frac{\partial^2 u}{\partial x^2} + \frac{\partial^2 u}{\partial y^2} \right), \quad \text{where } c^2 = \frac{T}{m}.$$

This is the wave equation in two dimensions.

3.26. SOLUTION OF THE TWO-DIMENSIONAL WAVE EQUATION

The two-dimensional wave equation is

$$\frac{\partial^2 u}{\partial t^2} = c^2 \left(\frac{\partial^2 u}{\partial x^2} + \frac{\partial^2 u}{\partial y^2} \right) \quad \text{...(1)}$$

Let $u = \text{XYT}$ ...(2)

where X is a function of x only, Y is a function of y only and T is a function of t only, be a solution of (1).

Then $\dfrac{\partial^2 u}{\partial t^2} = \text{XYT}''$, $\dfrac{\partial^2 u}{\partial x^2} = \text{X}''\text{YT}$ and $\dfrac{\partial^2 u}{\partial y^2} = \text{XY}''\text{T}$

Substituting in (1), we have $\dfrac{1}{c^2}\text{XYT}'' = \text{X}''\text{YT} + \text{XY}''\text{T}$

Dividing by XYT, we have $\dfrac{1}{c^2}\dfrac{\text{T}''}{\text{T}} = \dfrac{\text{X}''}{\text{X}} + \dfrac{\text{Y}''}{\text{Y}}$ ...(3)

This will be true only when *each member is a constant.* Choosing the constants suitably, we have

$$\frac{d^2\text{X}}{dx^2} + k^2\text{X} = 0, \quad \frac{d^2\text{Y}}{dy^2} + l^2\text{Y} = 0$$

and
$$\frac{d^2\text{T}}{dt^2} + (k^2 + l^2)c^2\text{T} = 0$$

The solutions of these equations are respectively

$$\text{X} = c_1 \cos kx + c_2 \sin kx$$
$$\text{Y} = c_3 \cos ly + c_4 \sin ly$$

and
$$\text{T} = c_5 \cos \sqrt{(k^2 + l^2)}\, ct + c_6 \sin \sqrt{(k^2 + l^2)}\, ct$$

Hence from (2), a solution of (1) is

$$u(x, y, t) = (c_1 \cos kx + c_2 \sin kx)(c_3 \cos ly + c_4 \sin ly)$$
$$[c_5 \cos \sqrt{(k^2 + l^2)}\, ct + c_6 \sin \sqrt{(k^2 + l^2)}\, ct] \quad ...(4)$$

Now let us suppose that the membrane is rectangular and is stretched between the lines

$$x = 0, x = a, y = 0, y = b.$$

Then the boundary conditions are :

(*i*) $u = 0$, when $x = 0$

(*ii*) $u = 0$, when $x = a$

(*iii*) $u = 0$, when $y = 0$

(*iv*) $u = 0$, when $y = b$

for all t.

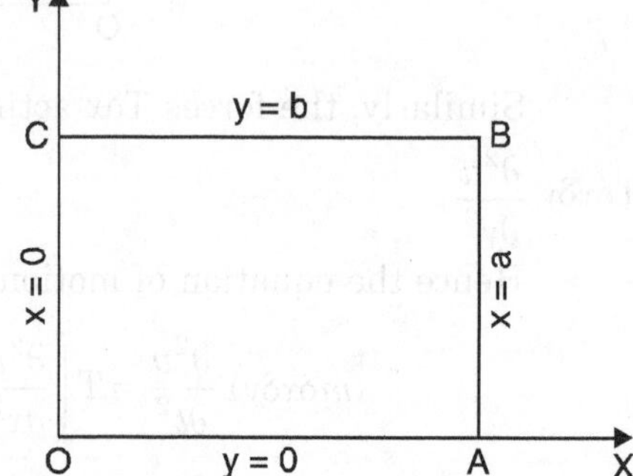

Applying the condition (*i*), we have

$$0 = c_1(c_3 \cos ly + c_4 \sin ly)[c_5 \cos \sqrt{(k^2 + l^2)}\, ct + c_6 \sin \sqrt{(k^2 + l^2)}\, ct]$$

i.e., $c_1 = 0$

Putting $c_1 = 0$ in (3) and applying the condition (*ii*), we have $\sin ka = 0$ or $k = \dfrac{m\pi}{a}$, where m is an integer.

Similarly, applying the conditions (*iii*) and (*iv*), we get

$$c_3 = 0 \quad \text{or} \quad l = \frac{n\pi}{b}, \text{ where } n \text{ is an integer.}$$

Therefore, the solution (3) becomes

$$u(x, y, t) = c_2c_4 \sin\frac{m\pi x}{a} \sin\frac{n\pi y}{b} (c_5 \cos pt + c_6 \sin pt)$$

where $$p = \pi c\sqrt{\frac{m^2}{a^2}+\frac{n^2}{b^2}}$$

Replacing the arbitrary constants suitably, we can write the general solution as

$$u(x, y, t) = \sum_{m=1}^{\infty}\sum_{n=1}^{\infty} \sin\frac{m\pi x}{a} \sin\frac{n\pi y}{b} (A_{mn} \cos pt + B_{mn} \sin pt) \qquad ...(5)$$

Now suppose the membrane starts from rest from the initial position $u = f(x, y)$ *i.e.*, $u(x, y, 0) = f(x, y)$.

Then applying the condition: $\frac{\partial u}{\partial t} = 0$ when $t = 0$, we get $B_{mn} = 0$.

Also using the condition: $u = f(x, y)$ when $t = 0$, we get

$$f(x, y) = \sum_{m=1}^{\infty}\sum_{n=1}^{\infty} A_{mn} \sin\frac{m\pi x}{a} \sin\frac{n\pi y}{b} \qquad ...(6)$$

This is a double Fourier series. Multiplying both sides by $\sin\frac{m\pi x}{a} \cdot \sin\frac{n\pi y}{b}$ and integrating from $x = 0$ to $x = a$ and $y = 0$ to $y = b$, every term on the right except one becomes zero. Thus, we get

$$\int_0^a\int_0^b f(x, y) \sin\frac{m\pi x}{a} \sin\frac{n\pi y}{b}\, dy\, dx = \frac{ab}{4} A_{mn}$$

i.e., $$A_{mn} = \frac{4}{ab}\int_0^a\int_0^b f(x, y) \sin\frac{m\pi x}{b} \sin\frac{n\pi y}{a}\, dy\, dx \qquad ...(7)$$

Hence, from (5), the required solution is

$$u(x, y, t) = \sum_{m=1}^{\infty}\sum_{n=1}^{\infty} A_{mn} \sin\frac{m\pi x}{a} \sin\frac{n\pi y}{b} \cos pt$$

where A_{mn} is given by (7) and $p = \pi c\sqrt{\frac{m^2}{a^2}+\frac{n^2}{b^2}}$.

Example. *A tightly stretched unit square membrane starts vibrating from rest and its initial displacement is k sin 2πx sin πy. Show that the deflection at any instant is k sin 2πx sin πy cos* ($\sqrt{5}\,\pi\, ct$).

Sol. Here we have to solve the equation $\frac{\partial^2 u}{\partial t^2} = c^2\left(\frac{\partial^2 u}{\partial x^2}+\frac{\partial^2 u}{\partial y^2}\right)$

with boundary conditions $u(0, y, t) = u(l, y, t) = u(x, 0, t) = u(x, 1, t) = 0$

and the initial conditions $u(x, y, 0) = f(x, y) = k \sin 2\pi x \sin \pi y$

$$\frac{\partial u}{\partial t} = 0 \text{ when } t = 0$$

$$u(x, y, t) = \sum_{m=1}^{\infty} \sum_{n=1}^{\infty} A_{mn} \sin m\pi x \sin n\pi y \cos pt \quad ...(1)$$

Since $\quad a = b = 1 \quad \text{where } p = \pi c \sqrt{m^2 + n^2}$

and
$$A_{mn} = \frac{4}{1 \times 1} \int_0^1 \int_0^1 k \sin 2\pi x \sin \pi y \sin m\pi x \sin n\pi y \, dydx$$

$$= 4k \int_0^1 \sin m\pi x \sin 2\pi x \, dx \int_0^1 \sin n\pi y \sin \pi y \, dy$$

$$= 0 \text{ for } m \neq 2 \text{ or } n \neq 1$$

$\therefore \quad A_{21} = 4k \left(\frac{1}{2}\right)\left(\frac{1}{2}\right) = k \quad \text{and} \quad p = \pi c \sqrt{(2)^2 + (1)^2} = \sqrt{5}\, \pi c$

Hence solution (1) reduces to $u(x, y, t) = k \sin 2\pi x \sin \pi y \cos (\sqrt{5}\, \pi ct)$.

TEST YOUR KNOWLEDGE

1. Find the deflection $u(x, y, t)$ of a square membrane with $a = b = c = 1$, if the initial velocity is zero and the initial deflection $f(x, y) = k \sin \pi x \sin 2\pi y$.
2. Find the deflection $u(x, y, t)$ of a rectangular membrane $(0 < x < 1, 0 < y < 2)$ whose boundary is fixed, given that it starts from rest and $u(x, y, 0) = xy(1 - x)(2 - y)$.
3. Find the deflection $u(x, y, t)$ of a rectangular membrane $(0 < x < a, 0 < y < b)$ whose boundary is fixed, given that it starts from rest and $u(x, y, 0) = xy(a^2 - x^2)(b^2 - y^2)$.
4. Find the deflection $u(x, y, t)$ of the rectangular membrane $(0 < x < a, 0 < y < b)$ whose boundary is fixed, given that the initial velocity is zero and the initial deflection $f(x, y) = k\,(ax - x^2)(by - y^2)$.

Answers

1. $u(x, y, t) = k \sin \pi x \sin 2\pi y \cos (\sqrt{5}\, \pi t)$

2. $u(x, y, t) = \sum_{m=1}^{\infty} \sum_{n=1}^{\infty} A_{mn} \sin m\pi x \sin \frac{n\pi y}{2} \cos pt$, where $A_{mn} = \frac{256}{m^3 n^3 \pi^6}$ both m, n odd

   and $p = \pi c \sqrt{m^2 + \frac{n^2}{4}}$

3. $u(x, y, t) = \sum_{m=1}^{\infty} \sum_{n=1}^{\infty} A_{mn} \sin \frac{m\pi x}{a} \sin \frac{n\pi y}{b} \cos pt$, where $A_{mn} = \frac{144 a^3 b^3}{m^3 n^3 \pi^6}$ and $p = \pi c \sqrt{\frac{m^2}{a^2} + \frac{n^2}{b^2}}$.

4. $u(x, y, t) = \sum_{m=1}^{\infty} \sum_{n=1}^{\infty} A_{mn} \sin \frac{m\pi x}{a} \sin \frac{n\pi y}{b} \cos pt$, where $A_{mn} = \frac{64 a^2 b^2 k}{m^3 n^3 \pi^6}$ both m, n odd

   and $$p = \pi c \sqrt{\frac{m^2}{a^2} + \frac{n^2}{b^2}}.$$

3.27. TRANSMISSION LINE EQUATIONS

Consider the flow of electricity in an insulated cable. Let V be the potential and I the current at time t at a point P of the cable at a distance x from a given point O. Let V + δV and I + δI be the potential and the current at the point Q$(x + \delta x)$ at the same time.

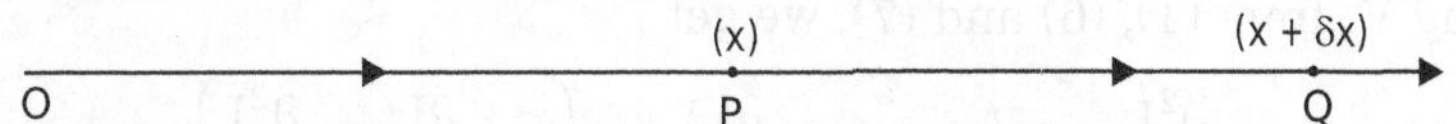

Let R, L, C, G be respectively the resistance, inductance, capacitance and leakance to the ground per unit length of the cable, each assumed to be constant.

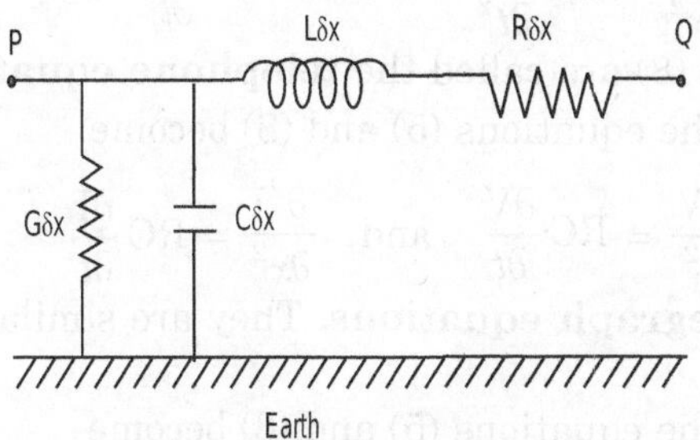

Since the potential drop across the segment PQ = potential drop due to resistance + potential drop due to inductance

$$\therefore \qquad - \delta V = (R\delta x)I + (L\delta x)\frac{\partial I}{\partial t}$$

Dividing by δx and proceeding to the limits as $\delta x \to 0$, we have

$$-\frac{\partial V}{\partial x} = RI + L\frac{\partial I}{\partial t} \qquad ...(1)$$

Similarly, the decrease in current in crossing the segment PQ = decrease in current due to capacitance and leakance

$$\therefore \qquad - \delta I = (G\delta x)V + (C\delta x)\frac{\partial V}{\partial t}$$

Dividing by δx and proceeding to the limits as $\delta x \to 0$, we have

$$-\frac{\partial I}{\partial x} = GV + C\frac{\partial V}{\partial t} \qquad ...(2)$$

Differentiating (1) and (2) partially w.r.t. x and t respectively, we have

$$-\frac{\partial^2 V}{\partial x^2} = R\frac{\partial I}{\partial x} + L\frac{\partial^2 I}{\partial x \partial t} \qquad ...(3)$$

and

$$-\frac{\partial^2 I}{\partial x \partial t} = G\frac{\partial V}{\partial t} + C\frac{\partial^2 V}{\partial t^2} \qquad ...(4)$$

Eliminating I from (2), (3) and (4), we get

$$-\frac{\partial^2 V}{\partial x^2} = R\left(-GV - C\frac{\partial V}{\partial t}\right) + L\left(-G\frac{\partial V}{\partial t} - C\frac{\partial^2 V}{\partial t^2}\right)$$

or

$$\frac{\partial^2 V}{\partial x^2} = LC\frac{\partial^2 V}{\partial t^2} + (RC + LG)\frac{\partial V}{\partial t} + RGV \qquad ...(5)$$

Now differentiating (1) and (2) w.r.t. t and x respectively, we have

$$-\frac{\partial^2 V}{\partial x \partial t} = R\frac{\partial I}{\partial t} + L\frac{\partial^2 I}{\partial t^2} \qquad ...(6)$$

and

$$-\frac{\partial^2 I}{\partial x^2} = G\frac{\partial V}{\partial x} + C\frac{\partial^2 V}{\partial x \partial t} \qquad ...(7)$$

Eliminating V from (1), (6) and (7), we get

$$-\frac{\partial^2 I}{\partial x^2} = G\left(-RI - L\frac{\partial I}{\partial t}\right) + C\left(-R\frac{\partial I}{\partial t} - L\frac{\partial^2 I}{\partial t^2}\right)$$

or
$$\frac{\partial^2 I}{\partial x^2} = LC\frac{\partial^2 I}{\partial t^2} + (LG + RC)\frac{\partial I}{\partial t} + RGI \quad \text{...(8)}$$

The equations (5) and (8) are called the **telephone equations.**

Cor. 1. If L = G = 0, the equations (5) and (8) become

$$\frac{\partial^2 V}{\partial x^2} = RC\frac{\partial V}{\partial t} \quad \text{and} \quad \frac{\partial^2 I}{\partial x^2} = RC\frac{\partial I}{\partial t}$$

which are known as the **telegraph equations.** They are similar to the equations in one dimensional heat flow.

Cor. 2. If R = G = 0, the equations (5) and (8) become

$$\frac{\partial^2 V}{\partial x^2} = LC\frac{\partial^2 V}{\partial t^2} \quad \text{and} \quad \frac{\partial^2 I}{\partial x^2} = LC\frac{\partial^2 I}{\partial t^2}$$

which are known as the **radio equations.** They are similar to the wave equation.

ILLUSTRATIVE EXAMPLES

Example 1. *Find the current i and voltage e in a line of length l, t seconds after the ends are suddenly grounded, given that*

$$i(x, 0) = i_0,\ e(x, 0) = e_0 \sin\frac{\pi x}{l}$$

Also R and G are negligible.

Sol. Since R and G are negligible, we use Radio equations

viz.,
$$\frac{\partial e}{\partial x} = -L\frac{\partial i}{\partial t} \quad \text{...(1)}$$

and
$$\frac{\partial i}{\partial x} = -C\frac{\partial e}{\partial t} \quad \text{...(2)}$$

For elimination of i, differentiating (1) partially w.r.t. x and (2) partially w.r.t. t, we have

$$\frac{\partial^2 e}{\partial x^2} = -L\frac{\partial^2 i}{\partial x \partial t} \quad \text{and} \quad \frac{\partial^2 i}{\partial t \partial x} = -C\frac{\partial^2 e}{\partial t^2}$$

whence
$$\frac{\partial^2 e}{\partial x^2} = LC\frac{\partial^2 e}{\partial t^2} \quad \text{...(3)}$$

The initial conditions are $i(x, 0) = i_0,\ e(x, 0) = e_0 \sin\frac{\pi x}{l}$ ...(4)

Since the ends are suddenly grounded, the boundary conditions are

$$e(0, t) = e(l, t) = 0 \quad \text{...(5)}$$

Also $i = i_0$ (constant) when $t = 0$

$$\therefore \quad \frac{\partial i}{\partial x} = 0 \text{ which gives } \frac{\partial e}{\partial t} = 0 \text{ when } t = 0 \qquad ...(6)$$

$$\left(\because \frac{\partial i}{\partial x} = -C\frac{\partial e}{\partial t}\right)$$

Now let $e = XT$ be a solution of (3) where X is a function of x only and T is a function of t only.

$$\frac{\partial^2 e}{\partial x^2} = X''T \quad \text{and} \quad \frac{\partial^2 e}{\partial t^2} = XT''$$

$\therefore$ From (3), $\quad X''T = LCXT''$

Separating the variables $\quad \dfrac{X''}{X} = LC\dfrac{T''}{T} = -p^2$ (say)

This leads to the ordinary differential equations

$$\frac{d^2X}{dx^2} + p^2X = 0 \quad \text{and} \quad \frac{d^2T}{dt^2} + \frac{p^2}{LC}T = 0$$

$$\therefore \quad X = c_1 \cos px + c_2 \sin px$$

$$T = c_3 \cos\frac{pt}{\sqrt{LC}} + c_4 \sin\frac{pt}{\sqrt{LC}}$$

$$\Rightarrow \quad e = XT = (c_1 \cos px + c_2 \sin px)\left(c_3 \cos\frac{pt}{\sqrt{LC}} + c_4 \sin\frac{pt}{\sqrt{LC}}\right) \qquad ...(7)$$

Applying the boundary conditions (5), we have

$$c_1 = 0 \text{ and } p = \frac{n\pi}{l}, \ n \text{ being an integer}$$

$\therefore$ Equation (7) becomes

$$e = c_2 \sin\frac{n\pi x}{l}\left(c_3 \cos\frac{n\pi t}{l\sqrt{LC}} + c_4 \sin\frac{n\pi t}{l\sqrt{LC}}\right)$$

or

$$e = \sin\frac{n\pi x}{l}\left(A\cos\frac{n\pi t}{l\sqrt{LC}} + B\sin\frac{n\pi t}{l\sqrt{LC}}\right) \qquad ...(8)$$

where $A = c_2c_3$ and $B = c_2c_4$

$$\frac{\partial e}{\partial t} = \sin\frac{n\pi x}{l}\left(-\frac{An\pi}{l\sqrt{LC}}\sin\frac{n\pi t}{l\sqrt{LC}} + \frac{Bn\pi}{l\sqrt{LC}}\cos\frac{n\pi t}{l\sqrt{LC}}\right)$$

Since $\quad \dfrac{\partial e}{\partial t} = 0$ when $t = 0$ $\qquad$ [from (6)]

We get $\quad B = 0$

$\therefore$ From (8), $\quad e = A\sin\dfrac{n\pi x}{l}\cos\dfrac{n\pi t}{l\sqrt{LC}}$

By superposition, $\quad e = \displaystyle\sum_{n=1}^{\infty} A_n \sin\frac{n\pi x}{l}\cos\frac{n\pi t}{l\sqrt{LC}}$ is also a solution.

But $\quad e = e_0 \sin\dfrac{\pi x}{l}$ when $t = 0$ $\qquad$ [from (4)]

$$\therefore \qquad e_0 \sin \frac{\pi x}{l} = \sum_{n=1}^{\infty} A_n \sin \frac{n\pi x}{l}$$

$$\Rightarrow \qquad A_1 = e_0 \quad \text{and} \quad A_2 = A_3 = \ldots\ldots = 0$$

Hence
$$e = e_0 \sin \frac{\pi x}{l} \cos \frac{\pi t}{l\sqrt{LC}}$$

Now
$$-L\frac{\partial i}{\partial t} = \frac{\partial e}{\partial x} \qquad \text{[from (1)]}$$

$$\therefore \qquad \frac{\partial i}{\partial t} = -\frac{1}{L} \cdot \frac{e_0 \pi}{l} \cos \frac{\pi x}{l} \cos \frac{\pi t}{l\sqrt{LC}}$$

Integrating w.r.t. t, regarding x as constant

$$i = -\frac{e_0 \pi}{Ll} \cos \frac{\pi x}{l} \cdot \frac{l\sqrt{LC}}{\pi} \sin \frac{\pi t}{l\sqrt{LC}} + f(x) \qquad \ldots(9)$$

where $f(x)$ is an arbitrary constant function.

Since $i = i_0$ when $t = 0$, we have $i_0 = 0 + f(x)$ or $f(x) = i_0$

$\therefore$ From (9), we have
$$i = i_0 - e_0 \sqrt{\frac{C}{L}} \cos \frac{\pi x}{l} \sin \frac{\pi t}{l\sqrt{LC}}.$$

Example 2. *A transmission line 1000 km long is initially under steady-state conditions with potential 1300 volts at the sending end (x = 0) and 1200 volts at the receiving end (x = 1000). The terminal end of the line is suddenly grounded, but the potential at the source is kept at 1300 volts. Assuming the inductance and leakance to be negligible, find the potential E(x, t).*

Sol. Since L and G are negligible, we use the telegraph equation

$$\frac{\partial^2 E}{\partial x^2} = RC \frac{\partial E}{\partial t} \quad \text{or} \quad \frac{\partial E}{\partial t} = \frac{1}{RC} \cdot \frac{\partial^2 E}{\partial x^2} \qquad \ldots(1)$$

Here E_s = initial steady voltage satisfying $\frac{\partial^2 E}{\partial x^2} = 0$

$$= 1300 - \frac{1300 - 1200}{1000} x = 1300 - 0.1x = E(x, 0) \qquad \ldots(2)$$

E_s' = steady voltage (after grounding the terminal end) when steady conditions are ultimately reached

$$= 1300 - \frac{1300 - 0}{1000} x = 1300 - 1.3x$$

$\therefore$ $E(x, t) = E_s' + E_t(x, t)$, where $E_t(x, t)$ is the transient part

$$= 1300 - 1.3x + \sum_{n=1}^{\infty} b_n \sin \frac{n\pi x}{l} \cdot e^{-\frac{n^2\pi^2 t}{l^2 RC}} \qquad \ldots(3)$$

where $l = 1000$ km.

Putting $t = 0$ in (3), we have

$$E(x, 0) = 1300 - 1.3x + \sum_{n=1}^{\infty} b_n \sin \frac{n\pi x}{l}$$

or $$1300 - 0.1x = 1300 - 1.3x + \sum_{n=1}^{\infty} b_n \sin \frac{n\pi x}{l} \qquad \text{[Using (2)]}$$

or $$1.2x = \sum_{n=1}^{\infty} b_n \sin \frac{n\pi x}{l}$$

where $b_n = \frac{2}{l} \int_0^l 1.2x \sin \frac{n\pi x}{l}\, dx$

$$= \frac{2.4}{l}\left[x\left(-\frac{\cos \frac{n\pi x}{l}}{\frac{n\pi}{l}} \right) - (1)\left(-\frac{\sin \frac{n\pi x}{l}}{\left(\frac{n\pi}{l}\right)^2} \right) \right]_0^l$$

$$= \frac{2.4}{l} \cdot \frac{l}{n\pi} [-l \cos n\pi]$$

$$= -\frac{2.4 \times 1000}{n\pi} \times (-1)^n \qquad [\because \; l = 1000]$$

$$= \frac{2400}{\pi} \cdot \frac{(-1)^{n+1}}{n}$$

Hence $$E(x, t) = 1300 - 1.3x + \frac{2400}{\pi} \sum_{n=1}^{\infty} \frac{(-1)^{n+1}}{n} \sin \frac{n\pi x}{l}\, e^{-\frac{n^2\pi^2 t}{l^2 RC}}$$

Example 3. *Neglecting R and G, find then e.m.f $v(x, t)$ in a line of length l, t seconds after the ends were suddenly grounded, given that $i(x, 0) = i_0$ and $v(x, 0) = e_1 \sin \frac{\pi x}{l} + e_5 \sin \frac{5\pi x}{l}$.*

Sol. Since R and G are negligible, we use the Radio equation $\frac{\partial^2 v}{\partial x^2} = LC \frac{\partial^2 v}{\partial t^2}$. ...(*i*)

Since the ends are suddenly grounded, we have the boundary conditions

$v(0, t) = 0, v(l, t) = 0.$ ...(*ii*)

Also the initial conditions are $i(x, 0) = i_0$

and $v(x, 0) = e_1 \sin \pi x/l + e_5 \sin 5\pi x/l.$ ...(*iii*)

$\therefore$ $\frac{\partial i}{\partial x} - c\frac{\partial v}{\partial t}$ gives $\frac{\partial v}{\partial t}(x, 0) = 0.$ ...(*iv*)

Let $v = X(x)\, T(x)$ be the solution of (*i*).

$\therefore$ (*i*) gives $X''T = LCXT''$

$$\frac{X''}{X} = LC\frac{T''}{T} = k^2 \text{ (say).}$$

$\therefore$ $X'' + h^2X = 0$ and $T'' + (k^2/LC)\, T = 0.$

Solving these equations, we get

$$v = (c_1 \cos kx + c_2 \sin kx)\left(c_3 \cos \frac{k}{\sqrt{LC}} . t + c_4 \sin \frac{k}{\sqrt{LC}} t \right)$$

Using the boundary conditions (*ii*), we get

$$c_1 = 0 \quad \text{and} \quad k = n\pi/l.$$

$$\therefore \qquad v = \sin\frac{n\pi x}{l}\left(a_n \cos\frac{n\pi}{l\sqrt{\text{LC}}}t + b_n \sin\frac{n\pi}{l\sqrt{\text{LC}}}t\right).$$

Using the initial condition (*iv*), we get $b_n = 0$.

$$\therefore \qquad v = a_n \sin\frac{n\pi x}{l}\cos\frac{n\pi}{l\sqrt{\text{LC}}}t.$$

Thus the most general solution of (*i*) is

$$v = \sum_{n=1}^{\infty} a_n \sin\frac{n\pi x}{l}\cos\frac{n\pi t}{l\sqrt{\text{LC}}}.$$

Finally by the initial condition (*iii*), we have

$$e_1 \sin\frac{\pi x}{l} + e_5 \sin\frac{5\pi x}{l} = \Sigma a_n \sin\frac{n\pi x}{l}$$

$\therefore$ $a_1 = e_1$ and $a_5 = e_5$ while all other a's are zero.

Hence $$v = e_1 \sin\frac{\pi x}{l}\cos\frac{\pi t}{l\sqrt{\text{LC}}} + e_5 \sin\frac{5\pi x}{l}\cos\frac{5\pi t}{l\sqrt{\text{LC}}}$$

which is the required solution.

Example 4. *A telephone line 3000 km. long has a resistance of 4 ohms/km. and a capacitance of 5 × 10^{-7} farad/km. Initially both the ends are grounded so that the line is unchanged. At time t = 0, a constant e.m.f. E is applied to one end, while the other end is left grounded. Assuming the inductance and leakance to be negligible, show that the steady state current of the grounded end at the end of 1 sec. is 5.3%.*

Sol. Since L = 0, G = 0, we use the telegraph equation

$$\frac{\partial^2 v}{\partial x^2} = \text{RC}\frac{\partial v}{\partial t}.$$

Let $v = \text{X}(x)\,\text{T}(t)$ be its solution so that

$$\text{TX}'' = \text{RCXT}' \quad \text{or} \quad \frac{\text{X}''}{\text{X}} = \text{RC}\frac{\text{T}'}{\text{T}} = -k^2 \text{ (say)}.$$

$$\therefore \qquad \text{X}'' + k^2\text{X} = 0 \quad \text{and} \quad \text{T}' + (k^2/\text{RC})\,\text{T} = 0.$$

Solving these equations, we get

$$\text{X} = c_1 \cos kx + c_2 \sin kx,\ \text{T} = c_3 e^{-k^2 t/\text{RC}}$$

giving $$v = (c_1 \cos kx + c_2 \sin kx)\, c_3 e^{-k^2 t/\text{RC}}. \qquad \text{...}(i)$$

When $t = 0$; $v = 0$ at $x = 0$ and $v = 0$ at $x = l$.

$$\therefore \qquad 0 = c_1 c_3;\ 0 = (c_1 \cos kl + c_2 \sin kl)c_3$$

i.e., $c_1 c_3 = 0$ and $kl = n\pi$ (n an integer).

Putting these in (*i*) and adding a linear term, we have

$$v = a_0 x + b_0 + \sum_{n=1}^{\infty} \sin\frac{n\pi x}{l} e^{-n^2\pi^2 t/\text{RC}t^2}. \qquad \text{...}(ii)$$

The end conditions of the problem are

$$v = 0 \text{ at } x = 0 \quad \text{and} \quad v = \text{E at } x = l.$$

Using these, (*ii*) gives $v_0 = 0$ and $a_0 = \text{E}/l$.

Then (*ii*) becomes $$v = \frac{E}{l}x + \Sigma b_n \sin\frac{n\pi x}{l}e^{-n^2\pi^2 t/\text{R}Cl^2}.$$

Also $v = 0$ when $t = 0$, we get $-\text{E}x/l = \Sigma b_n \sin n\pi x/l$.

This requires the expansion of $(-\text{E}x/l)$ as a half-range sine series in $(0, l)$.

$$\therefore \quad b_n = \frac{2}{l}\int_0^l \left(\frac{-\text{E}x}{l}\right)\sin\left(\frac{n\pi x}{l}\right)dx$$

$$= \frac{2}{l}\left[\left(\frac{-\text{E}x}{l}\right)\left(-\frac{l}{n\pi}\cos\frac{n\pi x}{l}\right) - \left(\frac{-\text{E}}{l}\right)\left(-\frac{l^2}{n^2\pi^2}\sin\frac{n\pi x}{l}\right)\right]_0^l$$

$$= \frac{2}{l}\left(\frac{\text{E}l}{n\pi}\cos n\pi\right) = \frac{2\text{E}}{n\pi}(-1)^n.$$

Thus $$v = \frac{Ex}{l} + \frac{2E}{\pi}\sum_{n=1}^{\infty}\sin\frac{n\pi x}{l}e^{-n^2\pi^2 t/\text{R}Cl^2}. \qquad ...(iii)$$

Also when L = 0, $$\frac{-\partial v}{\partial x} = \text{R}i$$

i.e., $$i = -\frac{1}{\text{R}}\frac{\partial v}{\partial x} = -\frac{\text{E}}{l\text{R}} - \frac{2\text{E}}{l\text{R}}\sum_{n=1}^{\infty}(-1)^n\cos\frac{n\pi x}{l}e^{-n^2\pi^2 t/\text{R}Cl^2}.$$

At the grounded end ($x = 0$), the current is

$$i = -\frac{\text{E}}{l\text{R}} - \frac{2\text{E}}{l\text{R}}\sum_{n=1}^{\infty}(-1)^n e^{-n^2\pi^2 t/\text{R}Cl^2}.$$

When $t = 1$ sec, $$i = -\frac{\text{E}}{l\text{R}}(1 - 2e^{\pi^2/\text{R}Cl^2} + 2e^{-4\pi^2/\text{R}Cl^2} - ...) \qquad ...(iv)$$

Since $$\frac{\pi^2}{\text{R}Cl^2} = \frac{(3.14)^2}{4(5\times 10^{-7})(3000)^2} = 0.548.$$

$$\therefore \quad e^{-\pi^2/\text{R}Cl^2} = e^{-0.578} = 0.578.$$

When $t \to \infty$, $i \to -\text{E}/l\text{R}$.

Hence from (*iv*), we have

$$i = -\frac{\text{E}}{l\text{R}}\{1 - 2(0.578) + 2(0.578)^4 - 2(0.578)^9 + ...\}$$

$$= -\frac{\text{E}}{l\text{R}}\{1 - 1.156 + 0.223 - 0.014 + ...\}$$

$$= i_\infty(0.053) = 5.3\% \text{ of } i_\infty.$$

TEST YOUR KNOWLEDGE

1. A steady voltage distribution of 20 volts at the sending end and 12 volts at the receiving end is maintained in a telephone wire of length l. At time $t = 0$, the receiving end is grounded. Find the voltage and current t sec later. Neglect leakance and inductance.

2. Obtain the solution of the radio equation, $\frac{\partial^2 e}{\partial x^2} = \text{LC}\frac{\partial^2 e}{\partial t^2}$ when a periodic *e.m.f.* $\text{E}_0 \cos pt$ is applied at the end $x = 0$ of the line.

3. Show that a transmission line with negligible resistance and leakage propagates waves of current and potential with a velocity equal to $\frac{i}{\sqrt{\text{LC}}}$, where L is the self-inductance and C is the capacitance.

4. A line of length l is initially uncharged, so that $i(x, 0) = 0$. $v(x, 0) = 0$, $\frac{\partial v}{\partial t} = 0$ when $t = 0$. At $t = 0$, one end $x = l$ is suddenly connected with a constant potential E while the other end is grounded. Neglecting R and G. Show that

$$v(x, t) = \frac{\text{E}x}{l} + \frac{2\text{E}}{\pi}\sum_{n=1}^{\infty}\frac{(-1)^n}{n}\sin\frac{n\pi x}{l}\cos\left(\frac{n\pi t}{l\sqrt{\text{LC}}}\right).$$

5. Solve the telephone equation $\frac{\partial^2 \text{V}}{\partial x^2} = \text{LC}\frac{\partial^2 \text{V}}{\partial t^2} + \text{RC}\frac{\partial \text{V}}{\partial t}$ given that $\text{V} = \text{V}_0 \sin qt$ when $x = 0$ and $\text{V} = 0$ when $x \to \infty$, assuming that $\frac{q\text{L}}{\text{R}}$ is large compared with unity.

[**Hint.** Assume $\text{V} = \text{A}e^{px+iqt}$.]

Answers

1. $$v(x, t) = \frac{20(l - x)}{l} + \frac{24}{\pi}\sum_{n=1}^{\infty}\frac{(-1)^{n+1}}{n}\sin\frac{n\pi x}{l}\, e^{-n^2\pi^2 t/\text{RC}l^2}$$

$$i(x, t) = \frac{20}{l\text{R}} + \frac{24}{l\text{R}}\sum_{n=1}^{\infty}(-1)^n\cos\frac{n\pi x}{l}\, e^{-n^2\pi^2 t/\text{RC}l^2}$$

2. $e(x, t) = \text{E}_0 \cos p(t - x\sqrt{\text{LC}})$

5. $\text{V}(x, t) = \text{V}_0\, e^{-\frac{\text{R}}{2}\sqrt{\frac{\text{C}}{\text{L}}}\,x} \sin q(t - \sqrt{\text{LC}}\, x)$.

3.28. LAPLACE'S EQUATION

We know that the two dimensional heat flow equation in steady state reduces to

$$\frac{\partial^2 u}{\partial x^2} + \frac{\partial^2 u}{\partial y^2} = 0 \qquad ...(1)$$

which is *Laplace's equation in two dimensions.*

Also the three dimensional heat flow equation in steady state reduces to

$$\nabla^2 u = \frac{\partial^2 u}{\partial x^2} + \frac{\partial^2 u}{\partial y^2} + \frac{\partial^2 u}{\partial z^2} = 0 \qquad ...(2)$$

which is *Laplace's equation in three dimensions.*

Laplace's equation has wide applications in physics and engineering. The theory of its solutions is called the *potential theory* and its solutions are called *harmonic functions.*

The solution of Laplace's equation, subject to certain boundary conditions, is simplified by a proper choice of coordinate system.

If the problem involves *rectangular boundaries*, we prefer to take Laplace's equation in *cartesian coordinates* given by (1) and (2).

If the problem involves circular boundaries, we prefer to take Laplace's equation in polar coordinates given by

$$\frac{\partial^2 u}{\partial r^2} + \frac{1}{r}\frac{\partial u}{\partial r} + \frac{1}{r^2}\frac{\partial^2 u}{\partial \theta^2} = 0 \qquad ...(3)$$

This equation can be obtained from (1) by putting $x = r \cos \theta$, $y = r \sin \theta$, thus changing the independent variables from (x, y) to (r, θ).

If the problem involves *cylindrical boundaries*, we prefer to take Laplace's equation in *cylindrical coordinates* given by

$$\frac{\partial^2 u}{\partial r^2} + \frac{1}{r}\frac{\partial u}{\partial r} + \frac{1}{r^2}\frac{\partial^2 u}{\partial \theta^2} + \frac{\partial^2 u}{\partial z^2} = 0 \qquad ...(4)$$

This equation can be obtained from (2) by putting $x = r \cos \theta$, $y = r \sin \theta$, $z = z$, thus changing the independent variables (x, y, z) to (r, θ, z).

If the problem involves *spherical boundaries*, we prefer to take Laplace's equation in *spherical coordinates* given by

$$\frac{\partial^2 u}{\partial r^2} + \frac{2}{r}\frac{\partial u}{\partial r} + \frac{1}{r^2}\frac{\partial^2 u}{\partial \theta^2} + \frac{\cot\theta}{r}\frac{\partial u}{\partial \theta} + \frac{1}{r^2 \sin^2\theta}\frac{\partial^2 u}{\partial \phi^2} = 0 \qquad ...(5)$$

This equation can be obtained from (2) by putting,

$$x = r \sin \theta \cos \phi,\ y = r \sin \theta \sin \phi,\ z = r \cos \theta,$$

thus changing the independent variables (x, y, z) to (r, θ, ϕ).

3.29. SOLUTIONS OF LAPLACE'S EQUATION

(*a*) Solution of Laplace's Equation in Two-dimensional Cartesian Form

We have already discussed the solution of $\dfrac{\partial^2 u}{\partial x^2} + \dfrac{\partial^2 u}{\partial y^2} = 0$

(*b*) Solution of Laplace's Equation in Polar Coordinates

Laplace's equation in polar coordinates is $r^2 \dfrac{\partial^2 u}{\partial r^2} + r\dfrac{\partial u}{\partial r} + \dfrac{\partial^2 u}{\partial \theta^2} = 0$ ...(1)

Let $u(r, \theta) = R(r)F(\theta)$ or simply $u = RF$ ...(2)

where R is a function of r only and F is a function of θ only, be a solution of (1).

Substituting it in (1), we get $r^2R''F + rR'F + RF'' = 0$

or $(r^2R'' + rR')F + RF'' = 0$

Separating the variables, $\dfrac{r^2R'' + rR'}{R} = -\dfrac{F''}{F}$ = constant = k (say)

Thus, we get ordinary differential equations

$$r^2 \frac{d^2R}{dr^2} + r\frac{dR}{dr} - kR = 0 \qquad ...(3)$$

and $$\frac{d^2F}{d\theta^2} + kF = 0 \quad ...(4)$$

Now (3) is a homogeneous linear differential equation.

Putting $r = e^z$, (3) reduces to $\dfrac{d^2R}{dz^2} - kR = 0$ ...(5)

Solving (5) and (4), we get

(*i*) *When k is positive* and $= p^2$, say

$$R = c_1e^{pz} + c_2e^{-pz} = c_1r^p + c_2r^{-p},$$
$$F = c_3 \cos p\theta + c_4 \sin p\theta.$$

(*ii*) *When k is negative* and $= -p^2$, say

$$R = c_1 \cos pz + c_2 \sin pz = c_1 \cos (p \log r) + c_2 \sin (p \log r)$$
$$F = c_3e^{p\theta} + c_4e^{-p\theta}$$

(*iii*) *When* $k = 0$

$$R = c_1z + c_2 = c_1 \log r + c_2$$
$$F = c_3\theta + c_4$$

Thus the three possible solutions of (1) are

$$u = (c_1r^p + c_2r^{-p})(c_3 \cos p\theta + c_4 \sin p\theta) \quad ...(6)$$
$$u = [c_1 \cos (p \log r) + c_2 \sin (p \log r)]\,(c_3\,e^{p\theta} + c_4\,e^{-p\theta}) \quad ...(7)$$
$$u = (c_1 \log r + c_2)(c_3\theta + c_4) \quad ...(8)$$

Of these solutions, we choose the one which is consistent with the physical nature of the problem.

Note. Usually we require a solution extending up to the origin.

Since u must be finite at the origin, we reject solutions (7) and (8). Also from (6), $c_2 = 0$.

∴ In this case, the solution may be written as

$$u = (A \cos p\theta + B \sin p\theta)r^p$$

The general solution will consist of a sum of similar terms with different (arbitrary) values of A, B and p.

(*c*) Solution of Laplace's Equation in Three-dimensional Cartesian Form

Laplace's equation in three-dimensional cartesian form is

$$\frac{\partial^2 u}{\partial x^2} + \frac{\partial^2 u}{\partial y^2} + \frac{\partial^2 u}{\partial z^2} = 0 \quad ...(1)$$

Let $u\,(x, y, z) = X(x)\,Y(y)\,Z(z)$ or simply $u = XYZ$ ...(2)

be a solution of (1).

Substituting it in (1), we get $X''YZ + XY''Z + XYZ'' = 0$

Dividing by XYZ, we get $$\frac{X''}{X} + \frac{Y''}{Y} + \frac{Z''}{Z} = 0$$

or $$\frac{1}{X}\cdot\frac{d^2X}{dx^2} + \frac{1}{Y}\cdot\frac{d^2Y}{dy^2} + \frac{1}{Z}\cdot\frac{d^2Z}{dz^2} = 0 \quad ...(3)$$

which is of the form $F_1(x) + F_2(y) + F_3(z) = 0$.

Since x, y, z are independent, this is possible only when F_1, F_2, F_3 are constants. Assuming these constants to be k^2, l^2 and $-(k^2 + l^2)$ respectively, (3) gives rise to the following equations:

$$\frac{d^2X}{dx^2} - k^2X = 0, \quad \frac{d^2Y}{dy^2} - l^2Y = 0, \quad \frac{d^2Z}{dz^2} + (k^2 + l^2)Z = 0$$

Their solutions are $\quad X = c_1e^{kx} + c_2e^{-kx}, \quad Y = c_3e^{ly} + c_4e^{-ly}$

$$Z = c_5 \cos \sqrt{(k^2 + l^2)}\, z + c_6 \sin \sqrt{(k^2 + l^2)}\, z$$

Hence a solution of (1) is

$$u = (c_1e^{kx} + c_2e^{-kx})(c_3e^{ly} + c_4e^{-ly}) \left[c_5 \cos \sqrt{(k^2 + l^2)}\, z + c_6 \sin \sqrt{(k^2 + l^2)}\, z\right]$$

Since the three constants could have been taken as $-k^2$, $-l^2$ and $k^2 + l^2$, an alternative solution of (1) is

$$u = (c_1 \cos kx + c_2 \sin kx)(c_3 \cos ly + c_4 \sin ly)\left[c_5e^{\sqrt{k^2+l^2}\,z} + c_6e^{-\sqrt{k^2+l^2}\,z}\right]$$

The choice of the constants and hence the general solution depends on the given initial and boundary conditions.

(*d*) **Solution of Laplace's Equation in Cylindrical Coordinates**

Laplace's equation in cylindrical coordinates is

$$\frac{\partial^2 u}{\partial r^2} + \frac{1}{r}\frac{\partial u}{\partial r} + \frac{1}{r^2}\frac{\partial^2 u}{\partial \theta^2} + \frac{\partial^2 u}{\partial z^2} = 0 \quad \text{...(1)}$$

Let $\quad u(r, \theta, z) = R(r)F(\theta)Z(z)$

or $\quad$ simply $\quad u = RFZ \quad$ ...(2)

be a solution of (1).

Substituting it in (1), we get $\quad R''FZ + \frac{1}{r}R'FZ + \frac{1}{r^2}RF''Z + RFZ'' = 0$

Dividing by RFZ, we get

$$\frac{1}{R}\left(\frac{d^2R}{dr^2} + \frac{1}{r}\frac{dR}{dr}\right) + \frac{1}{r^2F}\frac{d^2F}{d\theta^2} + \frac{1}{z}\frac{d^2Z}{dz^2} = 0 \quad \text{...(3)}$$

Assuming $\quad \frac{d^2F}{d\theta^2} = -n^2F$ and $\frac{d^2Z}{dz^2} = k^2Z \quad$ ...(4)

Equation (3) reduces to $\quad \frac{1}{R}\left(\frac{d^2R}{dr^2} + \frac{1}{r}\frac{dR}{dr}\right) - \frac{n^2}{r^2} + k^2 = 0$

or

$$r^2\frac{d^2R}{dr^2} + r\frac{dR}{dr} + (k^2r^2 - n^2)R = 0$$

This is Bessel's equation. Its solution is $R = c_1J_n(kr) + c_2Y_n(kr)$

The solutions of equations (3) are $\quad F = c_3 \cos n\theta + c_4 \sin n\theta, \quad Z = c_5e^{kz} + c_6e^{-kz}$

Hence a solution of (1) is $\quad u = [c_1J_n(kr) + c_2Y_n(kr)]\,(c_3 \cos n\theta + c_4 \sin n\theta)(c_5e^{kz} + c_6e^{-kz})$ which is known as a *cylindrical harmonic*.

(e) Solution of Laplace's Equation in Spherical Coordinates

Laplace's equation in spherical coordinates is

$$\frac{\partial^2 u}{\partial r^2}+\frac{2}{r}\frac{\partial u}{\partial r}+\frac{1}{r^2}\frac{\partial^2 u}{\partial \theta^2}+\frac{\cot\theta}{r^2}\frac{\partial u}{\partial \theta}+\frac{1}{r^2\sin^2\theta}\frac{\partial^2 u}{\partial \phi^2}=0 \qquad ...(1)$$

Let $u(r, \theta, \phi) = \mathrm{R}(r)\mathrm{G}(\theta)\mathrm{H}(\phi)$

or simply $u = \mathrm{RGH}$ ...(2)

be a solution of (1).

Substituting it in (1), we get

$$\mathrm{R}''\,\mathrm{GH}+\frac{2}{r}\mathrm{R}'\mathrm{GH}+\frac{1}{r^2}\mathrm{RG}''\mathrm{H}+\frac{\cot\theta}{r^2}\mathrm{RG}'\mathrm{H}+\frac{1}{r^2\sin^2\theta}\mathrm{RGH}''=0$$

Dividing by RGH, we get

$$\frac{1}{\mathrm{R}}\left(r^2\frac{d^2\mathrm{R}}{dr^2}+2r\frac{d\mathrm{R}}{dr}\right)+\frac{1}{\mathrm{G}}\left(\frac{d^2\mathrm{G}}{d\theta^2}+\cot\theta\frac{d\mathrm{G}}{d\theta}\right)+\frac{1}{\mathrm{H}\sin^2\theta}\frac{d^2\mathrm{H}}{d\phi^2}=0 \qquad ...(3)$$

Putting
$$\frac{1}{\mathrm{R}}\left(r^2\frac{d^2\mathrm{R}}{dr^2}+2r\frac{d\mathrm{R}}{dr}\right)=n(n+1) \qquad ...(4)$$

and
$$\frac{1}{\mathrm{H}}\frac{d^2\mathrm{H}}{d\phi^2}=-m^2 \qquad ...(5)$$

Equation (3) reduces to $\dfrac{d^2\mathrm{G}}{d\theta^2}+\cot\theta\dfrac{d\mathrm{G}}{d\theta}+[n(n+1)-m^2\operatorname{cosec}^2\theta]\mathrm{G}=0$

This is associated Legendre's equation and its solution is

$$\mathrm{G}=c_1\,\mathrm{P}_n^{\ m}(\cos\theta)+c_2\,\mathrm{Q}_n^{\ m}(\cos\theta)$$

The solution of (5) is $\mathrm{H}=c_3\cos m\phi+c_4\sin m\phi$

To solve (4), assume that $\mathrm{R}=r^k$ so that

$$k(k-1)+2k=n(n+1) \quad \text{or} \quad (k^2-n^2)+(k-n)=0$$

or
$$(k-n)(k+n+1)=0 \quad \therefore \quad k=n \quad \text{or} \quad -n-1$$

Thus $\mathrm{R}=c_5r^n+c_6r^{-n-1}$

Hence the general solution of (1) is

$$u=\sum_{n=0}^{\infty}\sum_{m=0}^{\infty}[c_1\mathrm{P}_n^{\ m}(\cos\theta)+c_2\mathrm{Q}_n^{\ m}(\cos\theta)]\,(c_3\cos m\phi+c_4\sin m\phi)(c_5r^n+c_6r^{-n-1})$$

Any solution of (1) is known as a *spherical harmonic*.

Example 1. *The diameter of a semi-circular plate of radius a is kept at 0°C and the temperature at the semi-circular boundary is T°C. Show that the steady state temperature in the plate is given by*

$$u(r,\theta)=\frac{4T}{\pi}\sum_{n=1}^{\infty}\frac{1}{2n-1}\left(\frac{r}{a}\right)^{2n-1}\sin(2n-1)\theta.$$

Sol. Take the centre of the circle as the pole and the bounding diameter as the initial line. Let the steady state temperature at any point $\mathrm{P}(r, \theta)$ be $u(r, \theta)$, so that u satisfies the equation

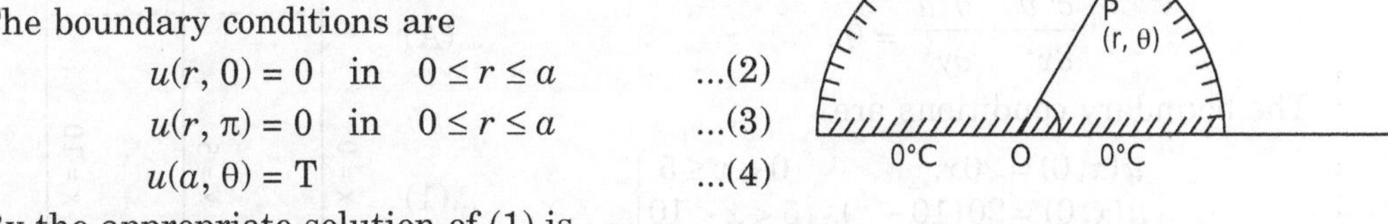

$$r^2 \frac{\partial^2 u}{\partial r^2} + r\frac{\partial u}{\partial r} + \frac{\partial^2 u}{\partial \theta^2} = 0 \qquad ...(1)$$

The boundary conditions are

$$u(r, 0) = 0 \quad \text{in} \quad 0 \le r \le a \qquad ...(2)$$

$$u(r, \pi) = 0 \quad \text{in} \quad 0 \le r \le a \qquad ...(3)$$

and $$u(a, \theta) = \mathrm{T} \qquad ...(4)$$

By the appropriate solution of (1) is

$$u(r, \theta) = (c_1 r^p + c_2 r^{-p})(c_3 \cos p\theta + c_4 \sin p\theta) \qquad ...(5)$$

Applying condition (2), $u(r, 0) = (c_1 r^p + c_2 r^{-p})c_3 = 0$

$\therefore$ $c_3 = 0$ and equation (5) becomes

$$u\,(r, \theta) = (c_1 r^p + c_2 r^{-p})c_4 \sin p\theta \qquad ...(6)$$

Applying condition (3), $u(r, \pi) = (c_1 r^p + c_2 r^{-p})c_4 \sin p\pi = 0$

or $\sin p\pi = 0 \quad \therefore \quad p = n$, where n is an integer.

Hence equation (6) reduces to $u(r, \theta) = (c_1 r^n + c_2 r^{-n})c_4 \sin n\theta$ ...(7)

Since $u = 0$ when $r = 0$, $\therefore \quad c_2 = 0$

and equation (7) becomes $u(r, \theta) = b_n r^n \sin n\theta$, where $b_n = c_1 c_4$

$\therefore$ The most general solution of (1) is of the form

$$u(r, \theta) = \sum_{n=1}^{\infty} b_n r^n \sin n\theta \qquad ...(8)$$

Applying condition (4), $u(a, \theta) = \sum_{n=1}^{\infty} b_n a^n \sin n\theta = \mathrm{T}$

or $$\mathrm{T} = \sum_{n=1}^{\infty} \mathrm{B}_n \sin n\theta, \quad \text{where } \mathrm{B}_n = b_n a^n$$

and $$\mathrm{B}_n = \frac{2}{\pi}\int_o^{\pi} \mathrm{T} \sin n\theta \; d\theta = \frac{2\mathrm{T}}{n\pi}(1 - \cos n\pi)$$

$$\therefore \quad b_n = \frac{\mathrm{B}_n}{a^n} = \frac{2\mathrm{T}}{n\pi a^n}(1 - \cos n\pi) = \begin{cases} 0, & \text{if } n \text{ is even} \\ \dfrac{4\mathrm{T}}{n\pi a^n}, & \text{if } n \text{ is odd} \end{cases}$$

Hence from (8), we have

$$u(r, \theta) = \frac{4\mathrm{T}}{\pi}\left[\frac{(r/a)}{1}\sin\theta + \frac{(r/a)^3}{3}\sin 3\theta + \frac{(r/a)^5}{5}\sin 5\theta + \ldots\right]$$

$$= \frac{4\mathrm{T}}{\pi}\sum_{n=1}^{\infty}\frac{1}{2n-1}\left(\frac{r}{a}\right)^{2n-1}\sin\,(2n-1)\,\theta.$$

Example 2. *A rectangular plate with insulated surfaces is 10 cms wide and so long compared to its width that it may be considered infinite in length without introducing an appreciable error. If the temperature along the short edge y = 0 is given by*

$$u(x, 0) = 20x, \qquad 0 \le x \le 5$$

$$u(x, 0) = 20\,(10 - x), \qquad 5 < x < 10.$$

While the two long edges x = 0 and x = 10, as well as the other short edge are kept at 0°C. Find the steady state temperature at any point (x, y) of the plate.

Sol. Here the temperature distribution is governed by two dimensional Laplace's equation

$$\frac{\partial^2 u}{\partial x^2} + \frac{\partial^2 u}{\partial y^2} = 0. \quad \text{...(A)}$$

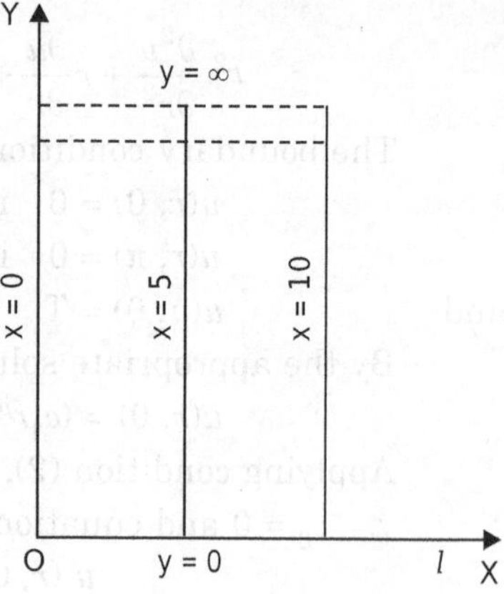

The boundary conditions are

$$\left.\begin{aligned} u(x, 0) &= 20x, & 0 < x \le 5 \\ u(x, 0) &= 20(10 - x), & 5 < x < 10 \end{aligned}\right\} \quad \text{...(1)}$$

$$u(0, y) = 0 \quad \text{...(2)}$$

$$u(10, y) = 0 \quad \text{...(3)}$$

$$u = 0 \text{ as } y \to \infty. \quad \text{...(4)}$$

Now proceeding, the solution of (A) is given by

$$u(x, y) = \sum_{n=1}^{\infty} B_n \sin \frac{n\pi x}{10} e^{-n\pi y/10} \qquad [\because \ l = 10 \text{ here}] \quad \text{...(5)}$$

where $$B_n = \frac{2}{10} \int_0^{10} f(x) \sin \frac{n\pi x}{10} \, dx \qquad [\because \ l = 10]$$

where $u(x, 0) = f(x)$ defined by (1).

$$\therefore \quad B_n = \frac{2}{10} \left[\int_0^5 20x \sin \frac{n\pi x}{10} dx + \int_5^{10} 20(10 - x) \sin \frac{n\pi x}{10} dx \right]$$

$$= \frac{20}{5} \left[x \left(-\cos \frac{n\pi x}{10} \right) \frac{10}{n\pi} - \left(-\sin \frac{n\pi x}{10} \right) \left(\frac{10}{n\pi} \right)^2 \right]_0^5$$

$$+ \frac{20}{5} \left[(10 - x) \left(-\cos \frac{n\pi x}{10} \right) \frac{10}{n\pi} - (-1) \left(-\sin \frac{n\pi x}{10} \right) \left(\frac{10}{n\pi} \right)^2 \right]_5^{10}$$

$$= \frac{40}{n\pi} \left[-5 \cos \frac{1}{2} n\pi + \frac{10}{n\pi} \sin \frac{1}{2} n\pi \right] + \frac{40}{n\pi} \left[5 \cos \frac{1}{2} n\pi + \frac{10}{n\pi} \sin \frac{1}{2} n\pi \right]$$

$$= \frac{800}{n\pi} \sin \frac{1}{2} \pi$$

$$= \begin{cases} \dfrac{800}{n\pi} & \text{when } n \text{ is odd of the form } 4m + 1 \\ \dfrac{-800}{n\pi} & \text{when } n \text{ is of the form } 4m + 3 \\ 0 & \text{when } n \text{ is of the form } 2m. \end{cases}$$

Hence the required temperature is given by

$$u = \frac{800}{\pi^2} \left[e^{-\pi y/10} \sin \frac{\pi x}{10} - \frac{1}{3} e^{-3\pi y/10} \sin \frac{3\pi x}{10} + \ldots \right].$$

4 Complex Variable

4.1. INTRODUCTION

A complex number z is an ordered pair (x, y) of real numbers and is written as

$$z = x + iy, \quad \text{where } i = \sqrt{-1}.$$

The real numbers x and y are called the real and imaginary parts of z. In the Argand's diagram, the complex number z is represented by the point $P(x, y)$. If (r, θ) are the polar coordinates of P, then $r = \sqrt{x^2 + y^2}$ is called the modulus of z and is denoted by $|z|$. Also $\theta = \tan^{-1} \dfrac{y}{x}$ is called the argument of z and is denoted by arg. z. Every non-zero complex number z can be expressed as

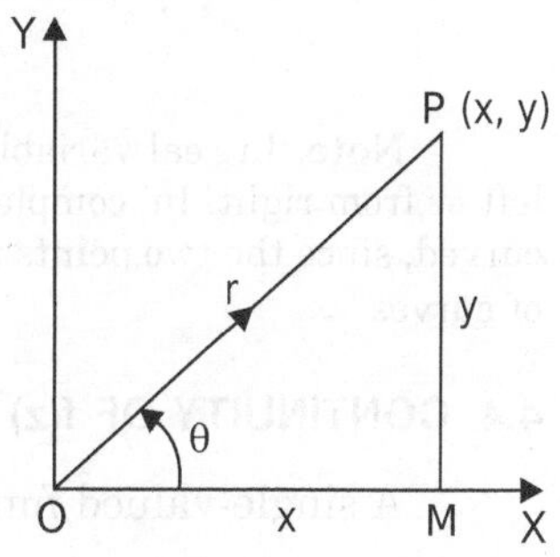

$$z = r(\cos\theta + i\sin\theta) = re^{i\theta}$$

If $z = x + iy$, then the complex number $x - iy$ is called the conjugate of the complex number z and is denoted by $\bar{z}$.

Clearly, $\quad |\bar{z}| = |z|, |z|^2 = z\bar{z},$

$$R(z) = \frac{z + \bar{z}}{2}, \quad I(z) = \frac{z - \bar{z}}{2i}.$$

4.2. FUNCTION OF A COMPLEX VARIABLE

If x and y are real variables, then $z = x + iy$ is called a **complex variable.** If corresponding to each value of a complex variable $z(= x + iy)$ in a given region R, there correspond one or more values of another complex variable $w\ (= u + iv)$, then w is called a function of the complex variable z and is denoted by

$$w = f(z) = u + iv$$

For example, if $\quad w = z^2, \quad$ where $z = x + iy$ and $w = f(z) = u + iv$

then $\quad u + iv = (x + iy)^2 = (x^2 - y^2) + i(2xy)$

$\Rightarrow \quad u = x^2 - y^2 \quad$ and $\quad v = 2xy$

Thus u and v, the real and imaginary parts of w, are functions of the real variables x and y.

$\therefore \quad w = f(z) = u(x, y) + iv(x, y)$

If to each value of z, there corresponds one and only one value of w, then w is called a *single-valued function* of z. If to each value of z, there correspond more than one values of w, then w is called a *multi-valued function of z*.

To represent $w = f(z)$ graphically, we take two Argand diagrams: one to represent the point z and the other to represent w. The former diagram is called the *xoy*-plane or the z-plane and the latter *uov*-plane or the w-plane.

4.3. LIMIT OF f(z)

A function $f(z)$ tends to the limit l as z tends to z_0 *along any path,* if to each positive arbitrary number ε, however small, there corresponds a positive number δ, such that

$$| f(z) - l | < \varepsilon \quad \text{whenever} \quad 0 < | z - z_0 | < \delta$$

i.e., $\qquad l - \varepsilon < f(z) < l + \varepsilon$

whenever $\qquad z_0 - \delta < z < z_0 + \delta, z \neq z_0$

and we write $\qquad \underset{z \to z_0}{\text{Lt}} f(z) = l.$

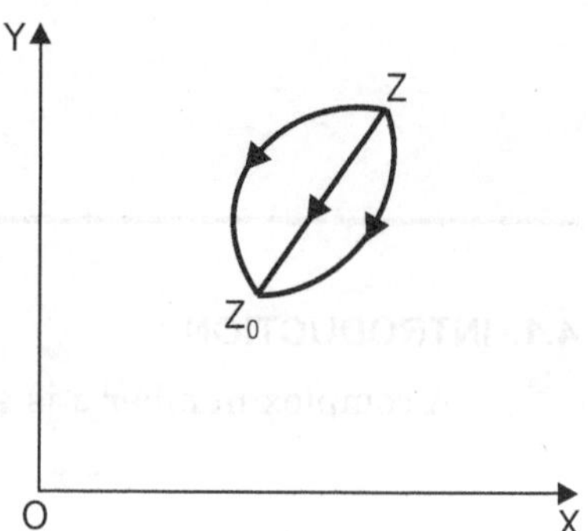

Note. In real variables, $x \to x_0$ implies that x approaches x_0 along the number line, either from left or from right. In complex variables, $z \to z_0$ implies that z approaches z_0 along any path, straight or curved, since the two points representing z and z_0 in a complex plane can be joined by an infinite number of curves.

4.4. CONTINUITY OF f(z)

A single-valued function $f(z)$ is said to be continuous at a point $z = z_0$ if $\underset{z \to z_0}{\text{Lt}} f(z) = f(z_0)$.

A function $f(z)$ is said to be continuous in a region R of the z-plane if it is continuous at every point of the region.

4.5. DERIVATIVE OF f(z)

Let $w = f(z)$ be a single-valued function of the variable $z(= x + iy)$, then the derivative or differential co-efficient of $w = f(z)$ is defined as

$$\frac{dw}{dz} = f'(z) = \underset{\delta z \to 0}{\text{Lt}} \frac{f(z + \delta z) - f(z)}{\delta z}$$

provided the limit exists, independent of the manner in which $\delta z \to 0$.

4.6. ANALYTIC FUNCTION

If a single-valued function $f(z)$ possesses a unique derivative at every point of a region R, then $f(z)$ is called an **analytic function** or a **regular function** or a **holomorphic function** of z in R.

A point where the function ceases to be analytic is called a **singular point.**

4.7. NECESSARY AND SUFFICIENT CONDITIONS FOR f(z) TO BE ANALYTIC

The necessary and sufficient conditions for the function

$$w = f(z) = u(x, y) + iv(x, y)$$

to be analytic in a region R, are

(*i*) $\dfrac{\partial u}{\partial x}, \dfrac{\partial u}{\partial y}, \dfrac{\partial v}{\partial x}, \dfrac{\partial v}{\partial y}$ *are continuous functions of x and y in the region R.*

(*ii*) $\frac{\partial u}{\partial x} = \frac{\partial v}{\partial y}, \frac{\partial u}{\partial y} = -\frac{\partial v}{\partial x}$.

The conditions in (*ii*) are known as **Cauchy-Riemann equations** or briefly **C-R equations.**

Proof. (*a*) **Necessary Condition.** Let $w = f(z) = u(x, y) + iv(x, y)$ be analytic in a region R, then $\frac{dw}{dz} = f'(z)$ exists uniquely at every point of that region.

Let δx and δy be the increments in x and y respectively. Let δu, δv and δz be the corresponding increments in u, v and z respectively. Then,

$$f'(z) = \underset{\delta z \to 0}{\text{Lt}} \frac{f(z + \delta z) - f(z)}{\delta z} = \underset{\delta z \to 0}{\text{Lt}} \frac{(u + \delta u) + i(v + \delta v) - (u + iv)}{\delta z}$$

$$= \underset{\delta z \to 0}{\text{Lt}} \left(\frac{\delta u}{\delta z} + i\frac{\delta v}{\delta z}\right) \quad \text{...(1)}$$

Since the function $w = f(z)$ is analytic in the region R, the limit (1) must exist independent of the manner in which $\delta z \to 0$, *i.e.*, along whichever path δx and $\delta y \to 0$.

First, let $\delta z \to 0$ along a line parallel to x-axis so that $\delta y = 0$ and $\delta z = \delta x$.

[since $z = x + iy$, $z + \delta z = (x + \delta x) + i(y + \delta y)$ and $\delta z = \delta x + i\delta y$]

$$\therefore \text{ From (1), } f'(z) = \underset{\delta x \to 0}{\text{Lt}} \left(\frac{\delta u}{\delta x} + i\frac{\delta v}{\delta x}\right) = \frac{\partial u}{\partial x} + i\frac{\partial v}{\partial x} \quad \text{...(2)}$$

Now, let $\delta z \to 0$ along a line parallel to y-axis so that $\delta x = 0$ and $\delta z = i\,\delta y$.

$$\therefore \text{ From (1), } f'(z) = \underset{\delta y \to 0}{\text{Lt}} \left(\frac{\delta u}{i\delta y} + i\frac{\delta v}{i\delta y}\right) = \frac{1}{i}\frac{\partial u}{\partial y} + \frac{\partial v}{\partial y}$$

$$= \frac{\partial v}{\partial y} - i\frac{\partial u}{\partial y} \quad \text{...(3)} \quad \left|\ \because \frac{1}{i} = -i\right.$$

From (2) and (3), we have $\frac{\partial u}{\partial x} + i\frac{\partial v}{\partial x} = \frac{\partial v}{\partial y} - i\frac{\partial u}{\partial y}$

Equating the real and imaginary parts, $\frac{\partial u}{\partial x} = \frac{\partial v}{\partial y}$ and $\frac{\partial u}{\partial y} = -\frac{\partial v}{\partial x}$.

Hence the necessary condition for f(z) to be analytic is that the C-R equations must be satisfied.

(*b*) **Sufficient Condition.** Let $f(z) = u + iv$ be a single-valued function possessing partial derivatives $\frac{\partial u}{\partial x}, \frac{\partial u}{\partial y}, \frac{\partial v}{\partial x}, \frac{\partial v}{\partial y}$ at each point of a region R and satisfying C-R equations.

i.e., $$\frac{\partial u}{\partial x} = \frac{\partial v}{\partial y} \quad \text{and} \quad \frac{\partial u}{\partial y} = -\frac{\partial v}{\partial x}.$$

We shall show that $f(z)$ is analytic, *i.e.*, $f'(z)$ exists at every point of the region R.

By Taylor's theorem for functions of two variables, we have, on omitting second and higher degree terms of δx and δy.

$$f(z + \delta z) = u(x + \delta x, y + \delta y) + iv(x + \delta x, y + \delta y)$$

$$= \left[u(x, y) + \left(\frac{\partial u}{\partial x}\delta x + \frac{\partial u}{\partial y}\delta y\right)\right] + i\left[v(x, y) + \left(\frac{\partial v}{\partial x}\delta x + \frac{\partial v}{\partial y}\delta y\right)\right]$$

$$= [u(x, y) + iv(x, y)] + \left(\frac{\partial u}{\partial x} + i\frac{\partial v}{\partial x}\right)\delta x + \left(\frac{\partial u}{\partial y} + i\frac{\partial v}{\partial y}\right)\delta y$$

$$= f(z) + \left(\frac{\partial u}{\partial x} + i\frac{\partial v}{\partial x}\right)\delta x + \left(\frac{\partial u}{\partial y} + i\frac{\partial v}{\partial y}\right)\delta y$$

or
$$f(z + \delta z) - f(z) = \left(\frac{\partial u}{\partial x} + i\frac{\partial v}{\partial x}\right)\delta x + \left(\frac{\partial u}{\partial y} + i\frac{\partial v}{\partial y}\right)\delta y$$

$$= \left(\frac{\partial u}{\partial x} + i\frac{\partial v}{\partial x}\right)\delta x + \left(-\frac{\partial v}{\partial x} + i\frac{\partial u}{\partial x}\right)\delta y \qquad \mid \text{ Using C-R equations}$$

$$= \left(\frac{\partial u}{\partial x} + i\frac{\partial v}{\partial x}\right)\delta x + \left(\frac{\partial u}{\partial x} + i\frac{\partial v}{\partial x}\right)i\delta y \qquad \mid \because -1 = i^2$$

$$= \left(\frac{\partial u}{\partial x} + i\frac{\partial v}{\partial x}\right)(\delta x + i\delta y) = \left(\frac{\partial u}{\partial x} + i\frac{\partial v}{\partial x}\right)\delta z \qquad \mid \because \delta x + i\delta y = \delta z$$

$$\Rightarrow \qquad \frac{f(z + \delta z) - f(z)}{\delta z} = \frac{\partial u}{\partial x} + i\frac{\partial v}{\partial x}$$

$$\therefore \qquad f'(z) = \underset{\delta z \to 0}{\text{Lt}} \frac{f(z + \delta z) - f(z)}{\delta z} = \frac{\partial u}{\partial x} + i\frac{\partial v}{\partial x}$$

Thus $f'(z)$ exists, because $\dfrac{\partial u}{\partial x}, \dfrac{\partial v}{\partial x}$ exist.

Hence $f(z)$ is analytic.

Note 1. The real and imaginary parts of an analytic function are called **conjugate functions.** Thus, if $f(z) = u(x, y) + iv(x, y)$ is an analytic function, then $u(x, y)$ and $v(x, y)$ are conjugate functions. The relation between two conjugate functions is given by C-R equations.

Note 2. When a function $f(z)$ is known to be analytic, it can be differentiated in the ordinary way as if z is a real variable.

Thus, $\qquad f(z) = z^2 \quad \Rightarrow \quad f'(z) = 2z$

$\qquad f(z) = \sin z \quad \Rightarrow \quad f'(z) = \cos z$ etc.

4.8. CAUCHY-RIEMANN EQUATIONS IN POLAR COORDINATES

Let (r, θ) be the polar coordinates of the point whose cartesian coordinates are (x, y), then

$$x = r\cos\theta, y = r\sin\theta,$$

$$z = x + iy = r(\cos\theta + i\sin\theta) = re^{i\theta}$$

$$\therefore \qquad u + iv = f(z) = f(re^{i\theta}) \qquad \text{...(1)}$$

Differentiating (1) partially w.r.t. r, we have

$$\frac{\partial u}{\partial r} + i\frac{\partial v}{\partial r} = f'(re^{i\theta}) \cdot e^{i\theta} \qquad \text{...(2)}$$

Differentiating (1) partially w.r.t. θ, we have

$$\frac{\partial u}{\partial \theta} + i\frac{\partial v}{\partial \theta} = f'(re^{i\theta}) \,.\, ire^{i\theta} = ir\left(\frac{\partial u}{\partial r} + i\frac{\partial v}{\partial r}\right) \qquad | \text{ Using (2)}$$

$$= -r\frac{\partial v}{\partial r} + ir\frac{\partial u}{\partial r}$$

Equating real and imaginary parts, we get

$$\frac{\partial u}{\partial \theta} = -r\frac{\partial v}{\partial r} \quad \text{and} \quad \frac{\partial v}{\partial \theta} = r\frac{\partial u}{\partial r}$$

or $$\frac{\partial u}{\partial r} = \frac{1}{r}\frac{\partial v}{\partial \theta} \quad \text{and} \quad \frac{\partial v}{\partial r} = -\frac{1}{r}\frac{\partial u}{\partial \theta}$$ which is the polar form of C-R equations.

4.9. HARMONIC FUNCTIONS

Any solution of the Laplace's equation, $\frac{\partial^2 \phi}{\partial x^2} + \frac{\partial^2 \phi}{\partial y^2} = 0$ is called a harmonic function.

Let $f(z) = u + iv$ be analytic in some region of the z-plane, then u and v satisfy C-R equations.

$$\therefore \qquad \frac{\partial u}{\partial x} = \frac{\partial v}{\partial y} \qquad ...(1)$$

and $$\frac{\partial u}{\partial y} = -\frac{\partial v}{\partial x} \qquad ...(2)$$

Differentiating (1) partially w.r.t. x and (2) w.r.t. y, we get

$$\frac{\partial^2 u}{\partial x^2} = \frac{\partial^2 v}{\partial x \partial y} \qquad ...(3)$$

and $$\frac{\partial^2 u}{\partial y^2} = -\frac{\partial^2 v}{\partial y \partial x} \qquad ...(4)$$

Assuming $\frac{\partial^2 v}{\partial x \partial y} = \frac{\partial^2 v}{\partial y \partial x}$ and adding (3) and (4), we get

$$\frac{\partial^2 u}{\partial x^2} + \frac{\partial^2 u}{\partial y^2} = 0 \qquad ...(5)$$

Now, differentiating (1) partially w.r.t. y and (2) w.r.t. x, we get

$$\frac{\partial^2 u}{\partial y \partial x} = \frac{\partial^2 v}{\partial y^2} \qquad ...(6)$$

and $$\frac{\partial^2 u}{\partial x \partial y} = -\frac{\partial^2 v}{\partial x^2} \qquad ...(7)$$

Assuming $\frac{\partial^2 u}{\partial y \partial x} = \frac{\partial^2 u}{\partial x \partial y}$ and subtracting (7) from (6), we get

$$\frac{\partial^2 v}{\partial x^2} + \frac{\partial^2 v}{\partial y^2} = 0 \qquad ...(8)$$

Equations (5) and (8) show that the real and imaginary parts u and v of an analytic function satisfy the Laplace's equation.

Hence u and v are known as harmonic functions.

4.10. ORTHOGONAL SYSTEM

Every analytic function $f(z) = u + iv$ defines two families of curves $u(x, y) = c_1$ and $v(x, y) = c_2$, which form an orthogonal system.

Consider the two families of curves

$$u(x, y) = c_1 \qquad \text{...(1)}$$

and

$$v(x, y) = c_2 \qquad \text{...(2)}$$

Differentiating (1) w.r.t. x, we get

$$\frac{\partial u}{\partial x} + \frac{\partial u}{\partial y} \cdot \frac{dy}{dx} = 0 \quad \text{or} \quad \frac{dy}{dx} = -\frac{\frac{\partial u}{\partial x}}{\frac{\partial u}{\partial y}} = m_1 \text{ (say)}$$

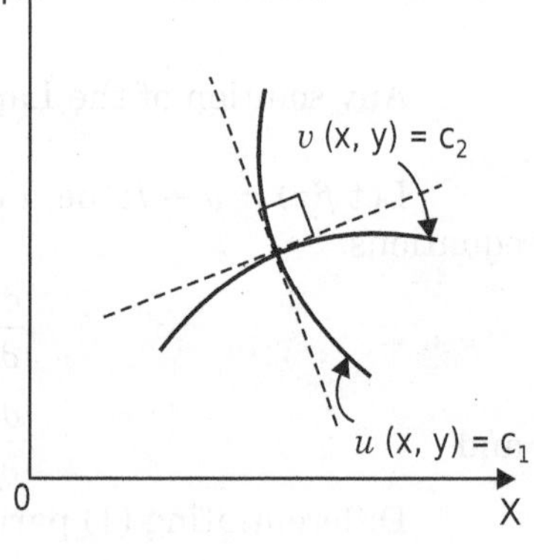

Similarly, from (2), we get $\quad \dfrac{dy}{dx} = -\dfrac{\frac{\partial v}{\partial x}}{\frac{\partial v}{\partial y}} = m_2$ (say)

$$\therefore \qquad m_1 m_2 = \frac{\frac{\partial u}{\partial x} \cdot \frac{\partial v}{\partial x}}{\frac{\partial u}{\partial y} \cdot \frac{\partial v}{\partial y}} \qquad \text{...(3)}$$

Since $f(z)$ is analytic, u and v satisfy C-R equations

i.e.,

$$\frac{\partial u}{\partial x} = \frac{\partial v}{\partial y} \quad \text{and} \quad \frac{\partial u}{\partial y} = -\frac{\partial v}{\partial x}$$

$\therefore$ From (3),

$$m_1 m_2 = \frac{\frac{\partial v}{\partial y} \cdot \frac{\partial v}{\partial x}}{-\frac{\partial v}{\partial x} \cdot \frac{\partial v}{\partial y}} = -1$$

Thus the product of the slopes of the curves (1) and (2) is – 1. Hence the curves intersect at right angles, *i.e.,* they form an orthogonal system.

4.11. APPLICATION OF ANALYTIC FUNCTIONS TO FLOW PROBLEMS

Since the real and imaginary parts of an analytic function satisfy the Laplace's equation in two variables, these conjugate functions provide solutions to a number of field and flow problems.

For example, consider the two dimensional irrotational motion of an incompressible fluid, in planes parallel to xy-plane.

Let V be the velocity of a fluid particle, then it can be expressed as

$$V = v_x \hat{i} + v_y \hat{j} \qquad \text{...(1)}$$

Since the motion is irrotational, there exists a scalar function $\phi(x, y)$, such that

$$V = \nabla\phi(x, y) = \frac{\partial\phi}{\partial x}\hat{i} + \frac{\partial\phi}{\partial y}\hat{j} \qquad ...(2)$$

From (1) and (2), we have $v_x = \frac{\partial\phi}{\partial x}$ and $v_y = \frac{\partial\phi}{\partial y}$ ...(3)

The scalar function $\phi(x, y)$, which gives the velocity components, is called the **velocity potential function** or simply the **velocity potential.**

Also the fluid being incompressible, div V = 0

$$\Rightarrow \qquad \left(\hat{i}\frac{\partial}{\partial x} + \hat{j}\frac{\partial}{\partial y}\right)(v_x\hat{i} + v_y\hat{j}) = 0$$

$$\Rightarrow \qquad \frac{\partial v_x}{\partial x} + \frac{\partial v_y}{\partial y} = 0 \qquad ...(4)$$

Substituting the values of v_x and v_y from (3) in (4), we get

$$\frac{\partial}{\partial x}\left(\frac{\partial\phi}{\partial x}\right) + \frac{\partial}{\partial y}\left(\frac{\partial\phi}{\partial y}\right) = 0 \quad \text{or} \quad \frac{\partial^2\phi}{\partial x^2} + \frac{\partial^2\phi}{\partial y^2} = 0$$

Thus the function ϕ is harmonic and can be treated as real part of an analytic function

$$w = f(z) = \phi(x, y) + i\,\psi(x, y)$$

For interpretation of conjugate function $\psi(x, y)$, the slope at any point of the curve $\psi(x, y) = c'$ is given by

$$\frac{dy}{dx} = -\frac{\dfrac{\partial\psi}{\partial x}}{\dfrac{\partial\psi}{\partial y}} = \frac{\dfrac{\partial\phi}{\partial y}}{\dfrac{\partial\phi}{\partial x}} \qquad \text{[By C-R equations]}$$

$$= \frac{v_y}{v_x} \qquad \text{[By (3)]}$$

This shows that the resultant velocity $\sqrt{v_x^2 + v_y^2}$ of the fluid particle is along the tangent to the curve $\psi(x, y) = c'$ *i.e.,* the fluid particles move along this curve. Such curves are known as *stream lines* and $\psi(x, y)$ is called the *stream function*. The curves represented by $\phi(x, y) = c$ are called *equipotential lines*.

Since $\phi(x, y)$ and $\psi(x, y)$ are conjugate functions of analytic function $w = f(z)$, the equipotential lines $\phi(x, y) = c$ and the stream lines $\phi(x, y) = c'$, intersect each other orthogonally.

Now, $$\frac{dw}{dz} = \frac{\partial\phi}{\partial x} + i\frac{\partial\psi}{\partial x}$$

$$= \frac{\partial\phi}{\partial x} - i\frac{\partial\phi}{\partial y} \qquad \text{[By C-R equations]}$$

$$= v_x - iv_y \qquad \text{[By (3)]}$$

$\therefore$ The magnitude of resultant velocity $= \left|\frac{dw}{dz}\right| = \sqrt{v_x^2 + v_y^2}$

The function $w = f(z)$ which fully represents the flow pattern is called the *complex potential*.

In the study of electrostatics and gravitational fields, the curves $\phi(x, y) = c$ and $\psi(x, y) = c'$ are called *equipotential lines* and *lines of force* respectively. In heat flow problems, the curves $\phi(x, y) = c$ and $\psi(x, y) = c'$ are known as *isothermals* and *heat flow lines* respectively.

ILLUSTRATIVE EXAMPLES

Example 1. *Show that the function $f(z) = \sqrt{|xy|}$ is not regular at the origin, although Cauchy-Riemann equations are satisfied.*

Sol. Let $\quad f(z) = u(x, y) + iv(x, y) = \sqrt{|xy|}$, then $u(x, y) = \sqrt{|xy|}$, $v(x, y) = 0$

At the origin (0, 0), we have

$$\frac{\partial u}{\partial x} = \underset{x \to 0}{\text{Lt}} \frac{u(x, 0) - u(0, 0)}{x} = \underset{x \to 0}{\text{Lt}} \frac{0 - 0}{x} = 0$$

$$\frac{\partial u}{\partial y} = \underset{y \to 0}{\text{Lt}} \frac{u(0, y) - u(0, 0)}{y} = \underset{y \to 0}{\text{Lt}} \frac{0 - 0}{y} = 0$$

$$\frac{\partial v}{\partial x} = \underset{x \to 0}{\text{Lt}} \frac{v(x, 0) - v(0, 0)}{x} = \underset{x \to 0}{\text{Lt}} \frac{0 - 0}{x} = 0$$

$$\frac{\partial v}{\partial y} = \underset{y \to 0}{\text{Lt}} \frac{v(0, y) - v(0, 0)}{y} = \underset{y \to 0}{\text{Lt}} \frac{0 - 0}{y} = 0$$

Clearly, $\quad \dfrac{\partial u}{\partial x} = \dfrac{\partial v}{\partial y}, \dfrac{\partial u}{\partial y} = -\dfrac{\partial v}{\partial x}$

Hence C-R equations are satisfied at the origin.

Now $\quad f'(0) = \underset{z \to 0}{\text{Lt}} \dfrac{f(z) - f(0)}{z} = \underset{z \to 0}{\text{Lt}} \dfrac{\sqrt{|xy|} - 0}{x + iy}$

If $z \to 0$ along the line $y = mx$, we get

$$f'(0) = \underset{x \to 0}{\text{Lt}} \frac{\sqrt{|mx^2|}}{x(1 + im)} = \underset{x \to 0}{\text{Lt}} \frac{\sqrt{|m|}}{1 + im}$$

Now this limit is not unique since it depends on m. Therefore, $f'(0)$ does not exist.

Hence the function $f(z)$ is not regular at the origin.

Example 2. *Prove that the function $f(z)$ defined by*

$$f(z) = \frac{x^3(1 + i) - y^3(1 - i)}{x^2 + y^2}, \; z \neq 0 \text{ and } f(0) = 0$$

is continuous and the Cauchy-Riemann equations are satisfied at the origin, yet $f'(0)$ does not exist.

Sol. Here, $\quad f(z) = \dfrac{(x^3 - y^3) + i(x^3 + y^3)}{x^2 + y^2}, \; z \neq 0$

Let $\quad f(z) = u + iv = \dfrac{x^3 - y^3}{x^2 + y^2} + i\,\dfrac{x^3 + y^3}{x^2 + y^2},$

then $$u = \frac{x^3 - y^3}{x^2 + y^2},\ v = \frac{x^3 + y^3}{x^2 + y^2}$$

Since $$z \neq 0 \ \Rightarrow\ x \neq 0, y \neq 0$$

$\therefore$ u and v are rational functions of x and y with non-zero denominators. Thus, u, v and hence $f(z)$ *are continuous functions when* $z \neq 0$. To test them for continuity at $z = 0$, on changing u, v to polar coordinates by putting $x = r\cos\theta$, $y = r\sin\theta$, we get

$$u = r(\cos^3\theta - \sin^3\theta) \text{ and } v = r(\cos^3\theta + \sin^3\theta)$$

When $z \to 0$, $r \to 0$

$$\therefore \quad \underset{z\to 0}{\text{Lt}}\ u = \underset{r\to 0}{\text{Lt}}\ r(\cos^3\theta - \sin^3\theta) = 0$$

Similarly, $$\underset{z\to 0}{\text{Lt}}\ v = 0$$

$$\therefore \quad \underset{r\to 0}{\text{Lt}}\ f(z) = 0 = f(0)$$

$\Rightarrow$ $f(z)$ is continuous at $z = 0$.

Hence $f(z)$ is continuous for all values of z.

At the origin (0, 0), we have

$$\frac{\partial u}{\partial x} = \underset{x\to 0}{\text{Lt}}\ \frac{u(x,0) - u(0,0)}{x} = \underset{x\to 0}{\text{Lt}}\ \frac{x - 0}{x} = 1$$

$$\frac{\partial u}{\partial y} = \underset{y\to 0}{\text{Lt}}\ \frac{u(0,y) - u(0,0)}{y} = \underset{y\to 0}{\text{Lt}}\ \frac{-y - 0}{y} = -1$$

$$\frac{\partial v}{\partial x} = \underset{x\to 0}{\text{Lt}}\ \frac{v(x,0) - v(0,0)}{x} = \underset{x\to 0}{\text{Lt}}\ \frac{x - 0}{x} = 1$$

$$\frac{\partial v}{\partial y} = \underset{y\to 0}{\text{Lt}}\ \frac{v(0,y) - v(0,0)}{y} = \underset{y\to 0}{\text{Lt}}\ \frac{y - 0}{y} = 1$$

$$\therefore \quad \frac{\partial u}{\partial x} = \frac{\partial v}{\partial y} \text{ and } \frac{\partial u}{\partial y} = -\frac{\partial v}{\partial x}$$

Hence C-R equations are satisfied at the origin.

Now, $$f'(0) = \underset{z\to 0}{\text{Lt}}\ \frac{f(z) - f(0)}{z} = \underset{z\to 0}{\text{Lt}}\ \frac{(x^3 - y^3) + i(x^3 + y^3) - 0}{(x^2 + y^2)(x + iy)}$$

Let $z \to 0$ along the line $y = x$, then

$$f'(0) = \underset{x\to 0}{\text{Lt}}\ \frac{0 + 2ix^3}{2x^3(1+i)} = \frac{i}{1+i} = \frac{i(1-i)}{2} = \frac{1+i}{2} \quad \text{...(1)}$$

Also, let $z \to 0$ along the x-axis (*i.e.*, $y = 0$), then

$$f'(0) = \underset{x\to 0}{\text{Lt}}\ \frac{x^3 + ix^3}{x^3} = 1 + i \quad \text{...(2)}$$

Since the limits (1) and (2) are different, $f'(0)$ does not exist.

Example 3. *Prove that the function sinh z is analytic and find its derivative.*

Sol. Here $$f(z) = u + iv = \sinh z = \sinh(x + iy) = \sinh x \cos y + i \cosh x \sin y$$

$$\therefore \quad u = \sinh x \cos y \quad \text{and} \quad v = \cosh x \sin y$$

$$\frac{\partial u}{\partial x} = \cosh x \cos y, \ \frac{\partial u}{\partial y} = -\sinh x \sin y$$

$$\frac{\partial v}{\partial x} = \sinh x \sin y, \ \frac{\partial v}{\partial y} = \cosh x \cos y$$

$$\therefore \quad \frac{\partial u}{\partial x} = \frac{\partial v}{\partial y} \text{ and } \frac{\partial u}{\partial y} = -\frac{\partial v}{\partial x}$$

Thus C-R equations are satisfied.

Since sinh x, cosh x, sin y and cos y are continuous functions, $\frac{\partial u}{\partial x}, \frac{\partial u}{\partial y}, \frac{\partial v}{\partial x}$ and $\frac{\partial v}{\partial y}$ are also continuous functions satisfying C-R equations.

Hence $f(z)$ is analytic everywhere.

Now $\quad f'(z) = \frac{\partial u}{\partial x} + i\frac{\partial v}{\partial x}$

$$= \cosh x \cos y + i \sinh x \sin y = \cosh(x + iy) = \cosh z.$$

Example 4. *Find the analytic function $f(z) = u + iv$ of which the real part is $u = e^x (x \cos y - y \sin y)$.*

Sol. We have $dv = \frac{\partial v}{\partial x} dx + \frac{\partial v}{\partial y} dy$

$$= -e^x(-x \sin y - y \cos y - \sin y)\,dx + e^z(x \cos y - y \sin y + \cos y)\,dy$$

Integrating, we get

$$v = \int_{y \text{ const.}} e^x(x \sin y + \sin y + y \cos y)\,dx$$

$$+ \int (\text{those terms which do not contain } x)\,dx + c$$

$$= \sin y \,.\int xe^x\,dx + (y \cos y + \sin y)\int e^x\,dx + \int 0dy + c$$

$$= [(x-1)\sin y + y \cos y + \sin y]\,e^x + c$$

$$= (x \sin y + y \sin y)\,e^x + c, \text{ where } c \text{ is a constant.}$$

$$\therefore \quad f(z) = u + iv = e^x(x \cos y - y \sin y) + i\,[e^x(x \sin y + y \cos y) + c]$$

$$= xe^x(\cos y + i \sin y) + iy\,e^x(\cos y + i \sin y) + ic$$

$$= (x + iy)\,e^x \,.\, e^{iy} + c' = ze^z + c'.$$

Example 5. *Find the analytic function, whose real part is* $\dfrac{\sin 2x}{\cosh 2y - \cos 2x}$.

Sol. Let $\quad f(z) = u + iv$ where $u = \dfrac{\sin 2x}{\cosh 2y - \cos 2x}$

Now $\quad f'(z) = \frac{\partial u}{\partial x} + i\frac{\partial v}{\partial x} = \frac{\partial u}{\partial x} - i\frac{\partial u}{\partial y}$ [by C-R equations]

$$= \frac{(\cosh 2y - \cos 2x)\, 2\cos 2x - \sin 2x\,(2\sin 2x)}{(\cosh 2y - \cos 2x)^2} - i\, \frac{\sin 2x\,(-2\sinh 2y)}{(\cosh 2y - \cos 2x)^2}$$

$$= \frac{2\cos 2x \cosh 2y - 2}{(\cosh 2y - \cos 2x)^2} + i\, \frac{2\sin 2x \sinh 2y}{(\cosh 2y - \cos 2x)^2}$$

By Milne-Thomson's method, we can express $f'(z)$ in terms of z by putting $x = z$ and $y = 0$.

$$\therefore \qquad f'(z) = \frac{2\cos 2z - 2}{(1 - \cos 2z)^2} + i(0) = \frac{-2}{1 - \cos 2z}$$

$$= \frac{-2}{2\sin^2 z} = -\operatorname{cosec}^2 z.$$

Now integrating w.r.t. z, we get

$$f(z) = \cot z + ic$$

[We take the constant of integration as imaginary since u does not contain by constant]

Example 6. *Determine the analytic function whose real part is* e^{2x} *(x cos 2y – y sin 2y).*

Sol. Let $f(z) = u + iv$ be the analytic function,

where $u = e^{2x}(x\cos 2y - y\sin 2y)$

$$\therefore \qquad \frac{\partial u}{\partial x} = 2e^{2x}(x\cos 2y - y\sin 2y) + e^{2x}\cos 2y$$

$$= e^{2x}(2x\cos 2y - 2y\sin 2y + \cos 2y) \qquad \text{...(1)}$$

$$\frac{\partial u}{\partial y} = e^{2x}(-2x\sin 2y - \sin 2y - 2y\cos 2y)$$

$$= -e^{2x}(2x\sin 2y + \sin 2y + 2y\cos 2y) \qquad \text{...(2)}$$

Since $f(z)$ is analytic, u and v must satisfy C-R equations

$$\frac{\partial u}{\partial x} = \frac{\partial v}{\partial y} \quad \text{and} \quad \frac{\partial u}{\partial y} = -\frac{\partial v}{\partial x}$$

Now $$\frac{\partial v}{\partial y} = \frac{\partial u}{\partial x} = e^{2x}(2x\cos 2y - 2y\sin 2y + \cos 2y)$$

Integrating w.r.t. y, treating x as constant, we get

$$v = e^{2x}\left[2x\,.\frac{\sin 2y}{2} - \left\{2y\left(-\frac{\cos 2y}{2}\right) - (2)\left(-\frac{\sin 2y}{4}\right)\right\} + \frac{\sin 2y}{2}\right] + \phi(x)$$

$$= e^{2x}(x\sin 2y + y\cos 2y) + \phi(x) \qquad \text{...(3)}$$

where $\phi(x)$ is an arbitrary function of x.

$$\therefore \qquad \frac{\partial v}{\partial x} = 2e^{2x}(x\sin 2y + y\cos 2y) + e^{2x}(\sin 2y) + \phi'(x)$$

$$= e^{2x}(2x\sin 2y + \sin 2y + 2y\cos 2y) + \phi'(x)$$

But $\quad \dfrac{\partial v}{\partial x} = -\dfrac{\partial u}{\partial y} = e^{2x}(2x\sin 2y + \sin 2y + 2y\cos 2y)$ [From (2)]

$\therefore \quad \phi'(x) = 0$

$\Rightarrow \quad \phi(x) = c$, an arbitrary constant.

$\therefore$ From (3), $\quad v = e^{2x}(x\sin 2y + y\cos 2y) + c$

$$f(z) = u + iv = e^{2x}(x\cos 2y - y\sin 2y) + ie^{2x}(x\sin 2y + y\cos 2y) + ic$$
$$= e^{2x}[(x + iy)\cos 2y + i(x + iy)\sin 2y] + c'$$
$$= (x + iy)\,e^{2x}(\cos 2y + i\sin 2y) + c'$$
$$= ze^{2z}\,.\,e^{2iy} + c' = ze^{2(x+iy)} + c' = ze^{2z} + c'.$$

(Milne Thomson's Method)

This method determines the analytic function $f(z)$ when u or v is given.

Since $\quad z = x + iy,\ \bar{z} = x - iy \quad$ so that $x = \dfrac{z + \bar{z}}{2},\ y = \dfrac{z - \bar{z}}{2i}$

Let $\quad f(z) = u(x, y) + iv(x, y) \quad \ldots(1)$

$$= u\left(\frac{z+\bar{z}}{2}, \frac{z-\bar{z}}{2i}\right) + iv\left(\frac{z+\bar{z}}{2}, \frac{z-\bar{z}}{2i}\right)$$

Considering this as an identity in the two independent variables z, $\bar{z}$ and putting $\bar{z} = z$, we get

$$f(z) = u(z, 0) + iv(z, 0)$$

which is the same as (1) if we replace x by z and y by 0.

Now $\quad f(z) = u + iv$

$\Rightarrow \quad f'(z) = \dfrac{\partial u}{\partial x} + i\dfrac{\partial v}{\partial x} = \dfrac{\partial u}{\partial x} - i\dfrac{\partial u}{\partial y}$ [C-R equations]

$$= e^{2x}(2x\cos 2y - 2y\sin 2y + \cos 2y) + ie^{2x}(2x\sin 2y + \sin 2y + 2y\cos 2y)$$

On replacing x by z and y by 0 on R.H.S., we get $f'(z) = e^{2z}(2z + 1)$

Integrating w.r.t. z, we have $\quad f(z) = (2z + 1)\dfrac{e^{2z}}{2} - 2\,.\,\dfrac{e^{2z}}{4} + c = ze^{2z} + c.$

Example 7. *Show that the function $u = x^3 - 3xy^2$ is harmonic and find the corresponding analytic function of this as the real part.*

Sol. Here $\quad u = x^3 - 3xy^2.$

$\therefore \quad \dfrac{\partial u}{\partial x} = 3x^2 - 3y^2 = \phi_1(x, y)$, say $\quad \ldots(1)$

$\dfrac{\partial u}{\partial y} = -6xy = \phi_2(x, y)$, say. $\quad \ldots(2)$

Again, $\quad \dfrac{\partial^2 u}{\partial x^2} = 6x\,.\,\dfrac{\partial^2 u}{\partial y^2} = -6x.$

$$\therefore \quad \frac{\partial^2 u}{\partial x^2} + \frac{\partial^2 u}{\partial y^2} = 0, \text{ so that } u \text{ is a harmonic function.}$$

Putting $x = z, y = 0$ in (1) and (2),

$$\phi_1(z, 0) = 3z^2, \phi_2(z, 0) = 0.$$

Hence by Milne-Thomson's method.

$$f(z) = \int [\phi_1(z, 0) - i\phi_2(z, 0)]\, dz + c = \int 3z^2\, dz + c.$$

$$f(z) = z^3 + c.$$

Example 8. *Determine the analytic function $w = u + iv$, if $v = \log(x^2 + y^2) + x - 2y$.*

Sol. Here $v = \log(x^2 + y^2) + x - 2y$

$$\therefore \quad \frac{\partial v}{\partial x} = \frac{2x}{x^2 + y^2} + 1$$

$$\frac{\partial v}{\partial y} = \frac{2y}{x^2 + y^2} - 2$$

Since $w = u + iv$

$$\frac{dw}{dz} = \frac{\partial u}{\partial x} + i\frac{\partial v}{\partial x} = \frac{\partial v}{\partial y} + i\frac{\partial v}{\partial x} \qquad \text{[C-R equations]}$$

$$= \left(\frac{2y}{x^2 + y^2} - 2\right) + i\left(\frac{2x}{x^2 + y^2} + 1\right)$$

Replacing x by z and y by 0, we get $\frac{dw}{dz} = -2 + i\left(\frac{2}{z} + 1\right) = (i - 2) + \frac{2i}{z}$

Integrating w.r.t.z, we have $w = (i - 2)z + 2i \log z + c.$

Example 9. *Show that the function $u = e^{-2xy} \sin(x^2 - y^2)$ is harmonic. Find the conjugate function v and express $u + iv$ as an analytic function of z.*

Sol. Here $u = e^{-2xy} \sin(x^2 - y^2)$

$$\therefore \quad \frac{\partial u}{\partial x} = -2y\, e^{-2xy} \sin(x^2 - y^2) + 2x\, e^{-2xy} \cos(x^2 - y^2)$$

$$\frac{\partial^2 u}{\partial x^2} = 4y^2 e^{-2xy} \sin(x^2 - y^2) - 4xy\, e^{-2xy} \cos(x^2 - y^2) + 2e^{-2xy} \cos(x^2 - y^2)$$
$$- 4xy\, e^{-2xy} \cos(x^2 - y^2) - 4x^2 e^{-2xy} \sin(x^2 - y^2) \qquad ...(1)$$

$$\frac{\partial u}{\partial y} = -2x\, e^{-2xy} \sin(x^2 - y^2) - 2y\, e^{-2xy} \cos(x^2 - y^2)$$

$$\frac{\partial^2 u}{\partial y^2} = 4x^2 e^{-2xy} \sin(x^2 - y^2) + 4xy\, e^{-2xy} \cos(x^2 - y^2) - 2e^{-2xy} \cos(x^2 - y^2)$$
$$+ 4xy\, e^{-2xy} \cos(x^2 - y^2) - 4y^2 e^{-2xy} \sin(x^2 - y^2) \qquad ...(2)$$

Adding (1) and (2), we get

$$\frac{\partial^2 u}{\partial x^2} + \frac{\partial^2 u}{\partial y^2} = 0 \qquad \text{which proves that } u \text{ is harmonic.}$$

Now, let $f(z) = u + iv$

then $$f'(z) = \frac{\partial u}{\partial x} + i\frac{\partial v}{\partial x} = \frac{\partial u}{\partial x} - i\frac{\partial u}{\partial y} \qquad \text{[C-R equations]}$$

$$= [-2y\, e^{-2xy} \sin(x^2 - y^2) + 2x\, e^{-2xy} \cos(x^2 - y^2)] + i[2x\, e^{-2xy} \sin(x^2 - y^2) + 2y\, e^{-2xy} \cos(x^2 - y^2)]$$

Replacing x by z and y by 0, we get

$$f'(z) = 2z \cos z^2 + 2iz \sin z^2 = 2z(\cos z^2 + i \sin z^2) = 2ze^{iz^2}$$

Integrating w.r.t. z, we have

$$f(z) = -ie^{iz^2} + C \text{ (C = A + } i\text{B is a complex constant)}$$

which expresses $u + iv$ as an analytic function of z.

Since $$u + iv = -ie^{iz^2} + C = -ie^{i(x+iy)^2} + C$$

$$= -ie^{i(x^2-y^2+2ixy)} + C = -ie^{-2xy} . e^{i(x^2-y^2)} + C$$

$$= -ie^{-2xy}[\cos(x^2 - y^2) + i \sin(x^2 - y^2)] + C$$

$$= e^{-2xy} \sin(x^2 - y^2) + i[-e^{-2xy} \cos(x^2 - y^2)] + C$$

$\therefore$ $$v = -e^{-2xy} \cos(x^2 - y^2) + B.$$

Example 10. *Prove the function $u = x^3 - 3xy^2 + 3x^2 - 3y^2 + 1$ satisfies Laplace's equation and determine corresponding analytic function $u + iv$.*

Sol. Here $u = x^3 - 3xy^2 + 3x^2 - 3y^2 + 1$.

$\therefore$ $$\frac{\partial u}{\partial x} = 3x^2 - 3y^2 + 6x = \phi_1(x, y), \text{ say} \qquad ...(1)$$

$$\frac{\partial u}{\partial y} = -6xy - 6y = \phi_2(x, y) \text{ say.} \qquad ...(2)$$

Also, $$\frac{\partial^2 u}{\partial x^2} = 6x + 6 \quad \text{and} \quad \frac{\partial^2 u}{\partial y^2} = -6x - 6.$$

$\therefore$ $\frac{\partial^2 u}{\partial x^2} + \frac{\partial^2 u}{\partial y^2} = 0$, so that u satisfies Laplace's equation.

Now putting $x = z$, $y = 0$ (1) and (2), we get

$$\phi_1(z, 0) = 3z^2 + 6z, \ \phi_2(z, 0) = 0.$$

By Milne-Thomson's method, we have

$$f(z) = \int [\phi_1(z, 0) - i\phi_2(z, 0)]\, dz + c$$

$$= \int [3z^2 + 6z - i . 0]\, dz + c = \int (3z^2 + 6z) + c$$

$\therefore$ $$f(z) = z^3 + 3z^2 + c.$$

Example 11. *Show that $u = \frac{1}{2} \log(x^2 + y^2)$ is harmonic and find it harmonic conjugate.*

Sol. Here $u = \frac{1}{2} \log(x^2 + y^2)$. We then have

$$\frac{\partial u}{\partial x} = \frac{x}{x^2 + y^2}, \quad \frac{\partial u}{\partial y} = \frac{y}{x^2 + y^2}$$

$$\frac{\partial^2 u}{\partial x^2} = \frac{y^2 - x^2}{(x^2 + y^2)^2}, \quad \frac{\partial^2 u}{\partial y^2} = \frac{x^2 - y^2}{(x^2 + y^2)^2}$$

Clearly u satisfies Laplace's equation $\frac{\partial^2 u}{\partial x^2} + \frac{\partial^2 u}{\partial y^2} = 0$, and hence u is harmonic.

Let the conjugate of u be v. Then

$$dv = \frac{\partial v}{\partial x}\, dx + \frac{\partial v}{\partial y}\, dy = -\frac{\partial u}{\partial y}\, dx + \frac{\partial u}{\partial x}\, dy,$$ by virtus of Cauchy-Riemann equations,

or
$$dv = -\frac{y}{x^2 + y^2}\, dx + \frac{x}{x^2 + y^2}\, dy = \text{M}dx + \text{N}dy.$$

Since $\frac{\partial \text{M}}{dy} = \frac{y^2 - x^2}{(x^2 + y^2)^2} = \frac{\partial \text{N}}{\partial x}$, the above equation is exact and hence on integration, we get

$$v = \int_{y \text{ const.}} \frac{-y}{x^2 + y^2}\, dx + \int (\text{Terms of } N \text{ containing only } y \text{ or constant})\, dy + c$$

$$= -y \int_{y \text{ cosnt.}} \frac{-dx}{x^2 + y^2} + \int 0\, dy + c = -y \cdot \frac{1}{y} \tan^{-1}\left(\frac{x}{y}\right) + c$$

$$= \tan^{-1}\left(-\frac{x}{y}\right) + c = \frac{\pi}{2} - \cot^{-1}\left(-\frac{x}{y}\right) + c$$

$$\therefore \quad v = \cot^{-1}\left(\frac{x}{y}\right) + c' = \tan^{-1}\left(\frac{y}{x}\right) + c'.$$

Example 12. *If $w = f(z) = u + iv$ is analytic function and $u - v = e^x (\cos y - \sin y)$. Find w in terms of z.*

Sol. We have $\quad u - v = e^x (\cos y - \sin y).$

$$\therefore \quad \frac{\partial u}{\partial x} - \frac{\partial v}{\partial x} = e^x (\cos y - \sin y) \qquad \text{...(1)}$$

and
$$\frac{\partial u}{\partial y} - \frac{\partial v}{\partial y} = -e^x (\sin y + \cos y)$$

or
$$-\frac{\partial v}{\partial x} - \frac{\partial u}{\partial x} = -e^x (\sin y + \cos y) \qquad \text{[Using Cauchy-Riemann equations]}$$

or
$$\frac{\partial v}{\partial x} + \frac{\partial v}{\partial x} = e^x (\sin y + \cos y). \qquad \text{...(2)}$$

Solving (1) and (2), we get

$$\frac{\partial u}{\partial x} = e^x \cos y = \phi_1 (x, y), \text{ say}$$

$$\frac{\partial v}{\partial x} = e^x \sin y = \phi_2 (x, y), \text{ say.}$$

and so $$f'(z) = \frac{\partial u}{\partial x} + i \frac{\partial v}{\partial x} = \phi_1 (x, y) + i\phi_2 (x, y)$$

$\therefore$ $$f(z) = \int [\phi_1 (z, 0) + i\phi_2 (z, 0)]\, dz + c, \text{ where } c \text{ is an arbitrary constant}$$

or $$f(z) = \int e^z\, dz + c = e^z + c.$$

A useful device. We have

$$f(z) = u + iv \text{ so that } i\, f(z) = i\, u - v.$$

$\therefore$ $$(1 + i)\, f(z) = (u - v) + i\, (u + v).$$

Taking $u - v = \text{U}$, $u + v = \text{V}$, $(1 + i)\, f = \text{F}(z)$, we obtain

$$\text{F}(z) = \text{U} + i\text{V}.$$

Since $f(z) = u + iv$ is analytic, $\text{F}(z) = \text{U} + i\text{V}$ is analytic too.

By (1), $$\frac{d\text{U}}{dx} = \frac{du}{dx} - \frac{dv}{dx} = e^x (\cos y - \sin y) = \phi_1 (x, y)$$

and by (2), $$-\frac{d\text{V}}{dx} = \frac{d\text{U}}{dy} = -\, e^x (\sin y + \cos y) = \phi_2 (x, y).$$

By Milne-Thomson's mehtod, we have

$$\text{F}(z) = \int [\phi_1 (z, 0) - i\phi_2 (z, 0)]\, dz + c$$

or $$(1 + i)\, f(z) = \int (e^z + ie^z)\, dz + c$$

or $$f(z) = e^z + c_1, \text{ where } c_1 = c/(1 + i).$$

Example 13. *If* $u = (x - 1)^3 - 3xy^2 + 3y^2$, *determine* v *so that* $u + iv$ *is a regular function of* $x + iy$.

Sol. Here $u = (x - 1)^3 - 3xy^2 + 3y^2$, so that

$$\frac{\partial u}{\partial x} = 3(x - 1)^2 - 3y^2 \quad \text{and} \quad \frac{\partial u}{\partial y} = -\, 6xy + 6y.$$

By Cauchy-Rieman equations, we have

$$\frac{\partial v}{\partial x} = -\frac{\partial u}{\partial y} = 6xy - 6y.$$

Integrating with respects to x, we get

$$v = 3x^2y - 6yx + f(y) \qquad \text{...(1)}$$

$$\therefore \qquad \frac{\partial v}{\partial y} = 3x^2 - 6x + f'(y) \qquad ...(2)$$

$$\text{Also} \qquad \frac{\partial v}{\partial y} = \frac{\partial u}{\partial x} = 3\,(x-1)^2 - 3y^2 \qquad ...(3)$$

[Using Cauchy-Riemann equations]

From (2) and (3), we obtain

$$3x^2 - 6x + f'(y) = 3x^2 - 6x + 3 - 3y^2$$

or
$$f'(y) = 3 - 3y^2.$$

Integrating with respect to y, we get

$$f(y) = 3y - y^3 + c.$$

Substituting this value of $f(y)$ in (1), we get

$$v = 3x^2y - 6yx + 3y - y^3 + c.$$

Example 14. *Show that the function $u = \sin x.\cosh + 2\cos x.\sinh y + x^2 - y^2 + 4xy$ is a harmonic function and determine the corresponding analytic funtion $f(z) = u + iv$.*

Sol. We have given that

$$u = \sin x\,.\cosh y + 2\cos x\,.\sinh y + x^2 - y^2 + 4xy.$$

$$\therefore \qquad \frac{\partial u}{\partial x} = \cos x.\cosh y - 2\sin x.\sinh y + 2x + 4y = \phi_1(x, y) \qquad ...(1)$$

$$\frac{\partial u}{\partial y} = \sin x \sinh y + 2\cos x \cosh y - 2y + 4x = \phi_2(x, y) \qquad ...(2)$$

$$\frac{\partial^2 u}{\partial x^2} = -\sin x \cosh y - 2\cos x \sinh y + 2$$

$$\frac{\partial^2 u}{\partial y^2} = \sin x \cosh y + 2\cos x \sinh y - 2$$

$$\therefore \qquad \frac{\partial^2 u}{\partial x^2} + \frac{\partial^2 u}{\partial y^2} = 0.$$

Consquently u satisfies Laplace's equation so that it is a harmonic function.

Putting $x = z$, $y = 0$ in (1) and (2), we get

$$\phi_1(z, 0) = \cos z + 2z,\ \phi_2(z, 0) = 2\cos z + 4z.$$

By Milne-Thomson's method,

$$f(z) = \int [\phi_1(z, 0) - i\phi_2(z, 0)]\, dz + c = \int [\cos z + 2z - 2i\,\{\cos z + 2z\}]\, dz + c$$

$$= \sin z + z^2 - 2i\,\{\sin z + z^2\} + c$$

or
$$f(z) = (1 - 2i)\,(\sin z + z^2) + c.$$

Example 15. *If $u - v = (x - y)\,(x^2 + 4xy + y^2)$ and $f(z) = u + iv$ is an analytic function of $z = x + iy$, find $f(z)$ in terms of z.*

Sol. Here $f(z) = u + iv$ so $if(z) = iu - v$.

$$\therefore \qquad (1 + i)\,f(z) = (u - v) + i(u + v) = \text{U} + i\text{V}, \text{ say}.$$

Here $\quad U = u - v = (x - y)(x^2 + 4xy + y^2)$

$$\frac{\partial U}{\partial x} = \frac{\partial u}{\partial x} - \frac{\partial v}{\partial x} = x^2 + 4xy + y^2 + (x - y)(2x + 4y)$$

$$= 3x^2 + 6xy - 3y^2 = \phi_1(x, y), \text{ say} \quad \text{...(1)}$$

and
$$\frac{\partial U}{\partial y} = \frac{\partial u}{\partial y} - \frac{\partial v}{\partial y} = -(x^2 + 4xy + y^2) + (x - y)(4x + 2y)$$

$$= 3x^2 - 6xy - 3y^2 = \phi_2(x, y), \text{ say.} \quad \text{...(2)}$$

Putting $x = z$, $y = 0$ in (1) and (2), we get

$$\phi_1(z, 0) = 3z^2, \; \phi_2(z, 0) = 3z^2.$$

By Milne-Thomson's method,

$$(1 + i) f(z) = \int [\phi_1(z, 0) - i\phi_1(z, 0)] \, dz + c$$

$$= \int (3z^2 - 3iz^2) \, dz + c = (1 - i) z^3 + c$$

$$\therefore \quad f(z) = \frac{1-i}{1+i} z^3 + \frac{c}{1+i}$$

$$f(z) = -iz^3 + \alpha, \text{ where } \alpha = \frac{c}{1+i}$$

Example 16. *An electrostatic field in the xy-plane is given by the potential function $\phi = 3x^2y - y^3$, find the stream function.*

Sol. Let $\psi(x, y)$ be a stream function.

By C-R equations
$$\frac{\partial \psi}{\partial x} = -\frac{\partial \phi}{\partial y} = -3x^2 + 3y^2 \quad \text{...(1)}$$

$$\frac{\partial \psi}{\partial y} = \frac{\partial \phi}{\partial x} = 6xy \quad \text{...(2)}$$

Integrating (1) w.r.t. x, treating y as constant, we get

$$\psi = -x^3 + 3xy^2 + F(y)$$

so that
$$\frac{\partial \psi}{\partial y} = 6xy + F'(y) \quad \text{...(3)}$$

From (2) and (3), $\quad 6xy + F'(y) = 6xy \quad$ or $\quad F'(y) = 0 \quad \therefore \quad F(y) = c$

Hence $\quad \psi = -x^3 + 3xy^2 + c.$

Example 17. *If $f(z)$ is a regular function of z, prove that*

$$\left(\frac{\partial^2}{\partial x^2} + \frac{\partial^2}{\partial y^2}\right) |f(z)|^2 = 4 \, |f'(z)|^2.$$

Sol. Let $f(z) = u + iv$ so that $|f(z)| = \sqrt{u^2 + v^2}$

or
$$|f(z)|^2 = u^2 + v^2 = \phi(x, y) \text{ (say)}$$

$$\therefore \qquad \frac{\partial \phi}{\partial x} = 2u\frac{\partial u}{\partial x} + 2v\frac{\partial v}{\partial x}$$

$$\frac{\partial^2 \phi}{\partial x^2} = 2\left[u\frac{\partial^2 u}{\partial x^2} + \left(\frac{\partial u}{\partial x}\right)^2 + v\frac{\partial^2 v}{\partial x^2} + \left(\frac{\partial v}{\partial x}\right)^2\right]$$

Similarly,
$$\frac{\partial^2 \phi}{\partial y^2} = 2\left[u\frac{\partial^2 u}{\partial y^2} + \left(\frac{\partial u}{\partial y}\right)^2 + v\frac{\partial^2 v}{\partial y^2} + \left(\frac{\partial v}{\partial y}\right)^2\right]$$

Adding, we get

$$\frac{\partial^2 \phi}{\partial x^2} + \frac{\partial^2 \phi}{\partial y^2} = 2\left[u\left(\frac{\partial^2 u}{\partial x^2} + \frac{\partial^2 u}{\partial y^2}\right) + \left(\frac{\partial u}{\partial x}\right)^2 + \left(\frac{\partial u}{\partial y}\right)^2 + v\left(\frac{\partial^2 v}{\partial x^2} + \frac{\partial^2 v}{\partial y^2}\right) + \left(\frac{\partial v}{\partial x}\right)^2 + \left(\frac{\partial v}{\partial y}\right)^2\right] \qquad \ldots(1)$$

Since $f(z) = u + iv$ is a regular function of z, u and v satisfy C-R equations and Laplace's equation.

$$\therefore \qquad \frac{\partial u}{\partial x} = \frac{\partial v}{\partial y}, \frac{\partial u}{\partial y} = -\frac{\partial v}{\partial x} \quad \text{and} \quad \frac{\partial^2 u}{\partial x^2} + \frac{\partial^2 u}{\partial y^2} = 0 = \frac{\partial^2 v}{\partial x^2} + \frac{\partial^2 v}{\partial y^2}$$

$\therefore$ From (1), we get

$$\frac{\partial^2 \phi}{\partial x^2} + \frac{\partial^2 \phi}{\partial y^2} = 2\left[0 + \left(\frac{\partial u}{\partial x}\right)^2 + \left(\frac{\partial v}{\partial x}\right)^2 + 0 + \left(\frac{\partial v}{\partial x}\right)^2 + \left(\frac{\partial u}{\partial x}\right)^2\right]$$

$$= 4\left[\left(\frac{\partial u}{\partial x}\right)^2 + \left(\frac{\partial v}{\partial x}\right)^2\right] \qquad \ldots(2)$$

Now
$$f(z) = u + iv$$

$$\therefore \qquad f'(z) = \frac{\partial u}{\partial x} + i\frac{\partial v}{\partial x} \text{ and } |f'(z)|^2 = \left(\frac{\partial u}{\partial x}\right)^2 + \left(\frac{\partial v}{\partial x}\right)^2$$

From (2), we get

$$\left(\frac{\partial^2}{\partial x^2} + \frac{\partial^2}{\partial y^2}\right)\phi = 4|f'(z)|^2 \text{ or } \left(\frac{\partial^2}{\partial x^2} + \frac{\partial^2}{\partial y^2}\right)|f(z)|^2 = 4\,|f'(z)|^2.$$

Example 18. *If $w = f(z)$ is a regular function of z such that $f'(z) \neq 0$, prove that*

$$\left(\frac{\partial^2}{\partial x^2} + \frac{\partial^2}{\partial y^2}\right) \log |f'(z)| = 0.$$

If $|f(z)|$ is the product of a function of x and a function of y, show that $f'(z) = e^{az^2 + \beta z + y}$ where α is real β, Υ are complex constants.

Sol. Recall
$$\frac{\partial^2}{\partial x^2} + \frac{\partial^2}{\partial y^2} = 4\frac{\partial^2}{\partial z\, \partial \bar{z}}$$

$$\therefore \quad \left(\frac{\partial^2}{\partial x^2}+\frac{\partial^2}{\partial y^2}\right) |\log f'(z)| = 4\,\frac{\partial^2}{\partial z\,\partial \bar{z}}\log |f'(z)| = 2\,\frac{\partial^2}{\partial z\,\partial \bar{z}}\log |f'(z)|^2$$

$$= 2\,\frac{\partial^2}{\partial z\,\partial \bar{z}}\log f'(z) f'(\bar{z}) \qquad [\because \ |z|^2 = z\bar{z}]$$

$$= 2\left[\frac{\partial^2}{\partial z \partial \bar{z}}\log f'(z) + \frac{\partial^2}{\partial z\,\partial \bar{z}}\log f'(\bar{z})\right]$$

$$= 2\left[\frac{\partial}{\partial z}\frac{f''(z)}{f'(z)} + \frac{\partial}{\partial z}\frac{f''(z)}{f'(z)}\right] = 2\,[0 + 0] = 0.$$

Second part. Let $|f'(z)| = \phi(x)\,\Psi(y)$. Then since

$$\left(\frac{\partial^2}{\partial x^2}+\frac{\partial^2}{\partial y^2}\right)\log|f'(z)| = 0, \text{ we have}$$

$$\left(\frac{\partial^2}{\partial x^2}+\frac{\partial^2}{\partial y^2}\right)[\log \phi(x) + \log \Psi(y)] = 0$$

i.e.,
$$\frac{d^2}{dx^2}\log \phi(x) + \frac{d^2}{dy^2}\log \Psi(y) = 0$$

or
$$\frac{d^2}{dx^2}\log \phi(x) = -\frac{d^2}{dy^2}\log \Psi(y) = 2c, \text{ say, where } c \text{ is a real constant.}$$

[$\because$ L.H.S. and R.H.S. both are independent of each other]

Now $\dfrac{d^2 \log \phi(x)}{dx^2} = 2c$, gives on integration $\dfrac{d \log \phi(x)}{dx} = 2cx + d$.

Again, integrating, $\log \phi(x) = cx^2 + dx + e$.

Similarly, $\quad -\log \Psi(y) = cy^2 + d'y + e'$, where d, e, d', e' are real constants.

Hence $\quad |f'(z)| = \exp[c(c^2 - y^2) + dx + d'y + e + e']$ ...(1)

Now $|\exp(\alpha z^2 + \beta z + \gamma)|$

$$= |\exp\{\alpha(x+iy)^2 + (a+bi)(x+iy) + (p+iq)\}$$

| where $\beta = a + ib,\ r = p + iq$

$$= |\exp[\alpha(x^2 - y^2) + ax - by + p].\exp[i(2\alpha xy + bx + ay + q)]|$$

$$= \exp[\alpha(x^2 - y^2) + ax - by + p], \qquad [\because \ |e^{A+iB}| = e^A]$$

which is of the same form as (1).

Hence we can write

$$f'(z) = \exp(\alpha z^2 + \beta z + \gamma).$$

Example 19. *If f(z) is an analytic function of z, prove that*

$$\left(\frac{\partial^2}{\partial x^2}+\frac{\partial^2}{\partial y^2}\right)|\,\mathrm{R}\,f(z)|^2 = 2\,|f'(z)|^2.$$

Sol. Let $f(z) = u + iv$. Then $R\,f(z) = u$.

Now $$\frac{\partial u^2}{\partial x} = 2u\,\frac{\partial u}{\partial x}.$$

Again differentiating, we get

$$\frac{\partial^2 u^2}{\partial x^2} = 2\left[\left(\frac{\partial u}{\partial x}\right)^2 + u\,\frac{\partial^2 u}{\partial x^2}\right].$$

Similarly, $$\frac{\partial^2 u^2}{\partial y^2} = 2\left[\left(\frac{\partial u}{\partial x}\right)^2 + u\frac{\partial^2 u}{\partial y^2}\right].$$

Adding,we get

$$\frac{\partial^2 u^2}{\partial x^2} + \frac{\partial^2 u^2}{\partial y^2} = 2\left[\left(\frac{\partial u}{\partial x}\right)^2 + \left(\frac{\partial u}{\partial y}\right)^2 + u\left(\frac{\partial^2 u}{\partial x^2} + \frac{\partial^2 u}{\partial y^2}\right)\right].$$

But u staisfies Laplace's equation, and so

$$\frac{\partial^2 u^2}{\partial x^2} + \frac{\partial^2 u^2}{\partial y^2} = 2\left[\left(\frac{\partial u}{\partial x}\right)^2 + \left(\frac{\partial u}{\partial y}\right)^2\right] = 2[u_x^2 + v_x^2] \qquad [\because\ u_y = -v_x]$$

Also $$f'(z) = \frac{dw}{dz} = \frac{\partial w}{\partial x} = \frac{\partial u}{\partial x} + i\frac{\partial v}{\partial x}.$$

$$\therefore \qquad |f'(z)|^2 = \left(\frac{\partial u}{\partial x}\right)^2 + \left(\frac{\partial v}{\partial x}\right)^2.$$

Hence $$\frac{\partial^2 u^2}{\partial x^2} + \frac{\partial^2 u^2}{\partial y^2} = 2|f'(z)|^2 \text{ or } \left(\frac{\partial^2 u^2}{\partial x^2} + \frac{\partial^2 u^2}{\partial y^2}\right)|R\,f'(z)|^2 = 2|f'(z)|^2$$

Example 20. *Determine the analytic function $f(z) = u + iv$, if*

$$u - v = \frac{\cos x + \sin x - e^{-y}}{2(\cos x - \cosh y)} \text{ and } f\left(\frac{\pi}{2}\right) = 0$$

Sol. Given $f(z) = u + iv$ so that $i\,f(z) = iu - v$.

$\therefore \qquad (1 + i)\,f(z) = (u - v) + i\,(u + v).$

Letting $(1 + i)\,f(z) = F(z)$, $u - v = U$, $u + v = V$, we obtain

$F(z) = U + iV.$

It is given that $$U = u - v = \frac{\cos x + \sin x - e^{-y}}{2(\cos x - \cosh y)}.$$

Now $$\frac{\partial U}{\partial x} = \frac{(-\sin x + \cos x)(\cos x - \cosh y) + \sin x(\cos x + \sin x - e^{-y})}{2(\cos x - \cosh y)^2} = \phi_1(x, y),$$ say

and $$\frac{\partial U}{\partial y} = \frac{e^{-y}(\cos x - \cosh y) - \sinh y(\cos x + \sin x - e^{-y})}{2\,(\cos x - \cosh y)^2} = \phi_2(x, y), \text{ say.}$$

Putting $x = z, y = 0$, we get

$$\phi_1(z, 0) = \frac{(-\sin z + \cos z)(\cos z - 1) + \sin z(\cos z + \sin z - 1)}{2(\cos z - 1)^2}$$

$$= \frac{\cos z(\cos z - 1) + \sin^2 z}{2(\cos z - 1)^2} = \frac{1 - \cos z}{2(\cos z - 1)^2} = \frac{-1}{2(\cos z - 1)}$$

and $\phi_2(z, 0) = \frac{\cos z - 1}{2(\cos z - 1)^2} = \frac{1}{2(\cos z - 1)}$

By Milne Thomson's method,

$$(1 + i) f(z) = F(z) = \int [\phi_1(z, 0) - i\phi_2(z, 0]\, dz + c$$

$$= -\left(\frac{1+i}{2}\right)\int \frac{dz}{\cos z - 1} + c = \left(\frac{1+i}{4}\right)\int \operatorname{cosec}^2\left(\frac{z}{2}\right) dz + c$$

$$= -\left(\frac{1+i}{2}\right)\cot\left(\frac{z}{2}\right) + c.$$

$$\therefore \qquad f(z) = -\frac{1}{2}\cot\left(\frac{z}{2}\right) + \alpha.$$

Subjecting this relation to the condition $f\left(\frac{\pi}{2}\right) = 0$, we get

$$0 = -\frac{1}{2}\cot\frac{\pi}{2} + \alpha = \alpha - \frac{1}{2} \quad \text{or} \quad \alpha = \frac{1}{2}.$$

$$\therefore \qquad f(z) = -\frac{1}{2}\cot\left(\frac{z}{2}\right) + \frac{1}{2} \quad \text{or} \quad 2f(z) = 1 - \cot\left(\frac{z}{2}\right).$$

Example 21. *If $u = x^2 - y^2$, $v = -\frac{y}{(x^2 + y^2)}$, that show that both u and v satisfy Laplace's equation, but $u + iv$ is not an analytic function of z.*

Sol. (*i*) To show that $u + iv$ is not an analytic function of z, it satisfies to show that u and v do not satisfy the equations $u_x = v_y$, $u_y = -v_x$.

Now $$\frac{\partial u}{\partial x} = 2x, \quad \frac{\partial v}{\partial y} = \frac{-(x^2 + y^2) - 2y(-y)}{(x^2 - y^2)^2} = \frac{-(x^2 - y^2)}{(x^2 + y^2)^2}$$

$$\frac{\partial u}{\partial y} = -2y, \quad \frac{\partial v}{\partial x} = -y \cdot \frac{-2x}{(x^2 + y^2)^2} = \frac{2xy}{(x^2 + y^2)^2}.$$

Evidently, $u_x \neq v_y$, $u_y = -v_x$.

It follows that $u + iv$ is not an analytic function of z.

(*ii*) To show that u and v both satisfy Laplace's equations, we have to show that

$$\frac{\partial^2 u}{\partial x^2} + \frac{\partial^2 u}{\partial y^2} = 0, \frac{\partial^2 v}{\partial x^2} + \frac{\partial^2 v}{\partial y^2} = 0.$$

Now $\dfrac{\partial^2 u}{\partial x^2} = 2, \dfrac{\partial^2 u}{\partial x^2} = -2.$

$$\therefore \quad \frac{\partial^2 u}{\partial x^2} + \frac{\partial^2 u}{\partial y^2} = 0$$

Again, $$\frac{\partial^2 v}{\partial x^2} = 2y\left[\frac{(x^2+y^2)^2 - 2(x^2+y^2)2x.x}{(x^2+y^2)^4}\right] = \frac{2y\,(y^2-3x^2)}{(x^2+y^2)^3}$$

and $$\frac{\partial^2 v}{\partial x^2} = \frac{2y\,(y^2+y^2)^2 - (y^2+y^2).2.(x^2+y^2)\,2y}{(x^2+y^2)^4} = \frac{2y\,(3x^2-y^2)}{(x^2+y^2)^3}.$$

$$\therefore \quad \frac{\partial^2 v}{\partial x^2} + \frac{\partial^2 v}{\partial y^2} = 0.$$

Example 22. *Show that the function e^x (cos y + isin y) is holomorphic and find its derivative.*

Sol. Let $f(z) = u + iv = e^x \cos y + ie^x \sin y$. We then have

$$\frac{\partial u}{\partial x} = e^x \cos y, \quad \frac{\partial u}{\partial y} = -\,e^x \sin y$$

$$\frac{\partial v}{\partial x} = e^x \sin y, \frac{\partial v}{\partial y} = e^x \cos y.$$

$$\therefore \quad \frac{\partial u}{\partial x} = \frac{\partial v}{\partial y} \text{ and } \frac{\partial u}{\partial y} = -\frac{\partial v}{\partial x},$$

So that u, v satisfy Cauchy-Riemann equations. It can be easily shown that u and v are harmonic functions.

Hence $f'(z)$ is holomorphic.

Now $$f'(z) = \frac{\partial u}{\partial x} + i\frac{\partial v}{\partial x} = e^x \cos y + ie^x \sin y = e^x(\cos y + i \sin y) = e^{x+iy} = e^z.$$

Thus the derivative is identical with the given function.

TEST YOUR KNOWLEDGE

1. Determine a, b, c, d so that the function $f(z) = (x^2 + axy + by^2) + i(cx^2 + dxy + y^2)$ is analytic.
2. Determine p such that the function $f(z) = \dfrac{1}{2}\log(x^2+y^2) + i\tan^{-1}\dfrac{px}{y}$ be an analytic function.
3. Show that the polar form of Cauchy-Riemann equations are $\dfrac{\partial u}{\partial r} = \dfrac{1}{r}\dfrac{\partial v}{\partial \theta}, \dfrac{\partial v}{\partial r} = -\dfrac{1}{r}\dfrac{\partial u}{\partial \theta}$.

   Deduce that $\dfrac{\partial^2 u}{\partial r^2} + \dfrac{1}{r}\dfrac{\partial u}{\partial r} + \dfrac{1}{r^2}\dfrac{\partial^2 u}{\partial \theta^2} = 0.$
4. If $f(z) = \dfrac{x^3 y(y-ix)}{x^6+y^2}$, $z \neq 0$, $f(0) = 0$, prove that $\dfrac{f(z)-f(0)}{z} \to 0$ as $z \to 0$ along any radius vector but not as $z \to 0$ in any manner.

5. Show that the function $f(z)$ defined by $f(z) = \dfrac{x^3 y^5 (x + iy)}{x^6 + y^{10}}$, $z \neq 0$, $f(0) = 0$, is not analytic at the origin even though it satisfies Cauchy-Riemann equations at the origin.

6. Determine which of the following functions are analytic:

(i) e^z (ii) $\sin z$ (iii) $\cosh z$

(iv) $\dfrac{1}{z}$ (v) $\dfrac{x + iy}{x^2 + y^2}$ (vi) $\dfrac{1}{2} \log (x^2 + y^2) + i \tan^{-1} \dfrac{y}{x}$.

7. Show that $| z |^2$ is not analytic at any point.

8. Show that $u + iv = \dfrac{x - iy}{x - iy + a}$ where $a \neq 0$, is not an analytic function of $z = x + iy$ whereas $u - iv$ is such a function.

9. Determine the analytic function whose real part is

(i) $x^3 - 3xy^2 + 3x^2 - 3y^2 + 1$ (ii) $\log \sqrt{x^2 + y^2}$

(iii) $e^x (x \cos y - y \sin y)$ (iv) $\cos x \cosh y$

(v) $e^{-x} (x \sin y - y \cos y)$ (vi) $\dfrac{\sin 2x}{\cosh 2y - \cos 2x}$.

10. Find the regular function whose imaginary part is

(i) $\dfrac{x - y}{x^2 + y^2}$ (ii) $\cos x \cosh y$ (iii) $\sinh x \cos y$.

11. Prove that $u = x^2 - y^2 - 2xy - 2x + 3y$ is harmonic. Find a function v such that $f(z) = u + iv$ is analytic. Also express $f(z)$ in terms of z.

12. An electrostatic field in the xy-plane is given by the potential function $\phi = x^2 - y^2$, find the stream function.

13. If $w = \phi + i\psi$ represents the complex potential for an electric field and $\psi = x^2 - y^2 + \dfrac{x}{x^2 + y^2}$, determine the function ϕ.

14. If the potential function is $\log (x^2 + y^2)$, find the flux function and the complex potential function.

15. In a two dimensional fluid flow, the stream function is $\psi = \tan^{-1}\left(\dfrac{y}{x}\right)$, find the velocity potential ϕ.

16. If $f(z) = u + iv$ is an analytic function, find $f(z)$ if

(i) $u - v = e^x (\cos y - \sin y)$ (ii) $u + v = \dfrac{x}{x^2 + y^2}$, when $f(1) = 1$

(iii) $u - v = \dfrac{\cos x + \sin x - e^{-y}}{2 \cos x - e^y - e^{-y}}$, when $f\left(\dfrac{\pi}{2}\right) = 0$.

17. If $f(z)$ is an analytic function of z, prove that

(i) $\left(\dfrac{\partial^2}{\partial x^2} + \dfrac{\partial^2}{\partial y^2}\right) \log | f'(z) | = 0$ (ii) $\left(\dfrac{\partial^2}{\partial x^2} + \dfrac{\partial^2}{\partial y^2}\right) | \text{R} f(z) |^2 = 2 | f'(z) |^2$.

Answers

1. $a = 2, b = -1, c = -1, d = 2$ 2. $p = -1$

6. (i), (ii), (iii), (iv) except when $z = 0$, (v) except when $z = 0$

9. (i) $z^3 + 3z^2 + 1 + c$ (ii) $\log z + c$ (iii) $z e^z + c$

(iv) $\cos z + c$ (v) $ize^{-z} + 6i$ (vi) $\cot z + c$

10. (*i*) $\frac{1+i}{z} + c$ (*ii*) $\sin x \sinh y + c$ (*iii*) $-\cosh x \sin y + c$

11. $v = x^2 - y^2 + 2xy - 2y - 3x + 6,\ f(z) = (1+i)z^2 - (2+3i)z + i6$

12. $\psi = 2xy + c$ **13.** $-2xy + \frac{x}{x^2+y^2} + c$ **14.** $2\tan^{-1}\left(\frac{y}{x}\right),\ 2\log z + c$

15. $\frac{1}{2}\log(x^2+y^2) + c$ **16.** (*i*) $e^z + c$ (*ii*) $\frac{1}{1-i}\left(\frac{1}{z} - 1\right)$

(*iii*) $\frac{1}{2}\left(1 - \cot\frac{z}{2}\right)$.

4.12. TRANSFORMATION OR MAPPING

We know that the real function $y = f(x)$ can be represented graphically by a curve in the xy-plane. Also, the real function $z = f(x, y)$ can be represented by a surface in three dimensional space. However, this method of graphical representation fails in the case of complex functions because a complex function $w = f(z)$ *i.e.*, $u + iv = f(x + iy)$ involves four real variables, two independent variables x, y and two dependent variables u, v. Thus a four dimensional region is required to represent it graphically in the cartesian fashion. As it is not possible, we choose, two complex planes and call them z-plane and w-plane. In the z-plane, we plot the point $z = x + iy$ and in the w-plane, we plot the corresponding point $w = u + iv$. Thus the function $w = f(z)$ defines a correspondence between points of these two planes. If the point z describes some curve C in the z-plane, the point w will move along a corresponding curve C′ in the w-plane, since to each (x, y) there corresponds a point (u, v). *The function $w = f(z)$ thus defines a mapping or transformation of the z-plane into the w-plane.*

For example, consider the transformation $w = z + (1 - i)$. Let us determine the region D′ of the w-plane corresponding to the rectangular region D in the z-plane bounded by $x = 0$, $y = 0$, $x = 1$ and $y = 2$.

Since $w = z + (1 - i)$, we have

$$u + iv = (x + iy) + (1 - i) = (x + 1) + i(y - 1)$$

Thus $u = x + 1$ and $v = y - 1$

Hence the lines $x = 0$, $y = 0$, $x = 1$ and $y = 2$ in the z-plane are mapped onto the lines $u = 1$, $v = -1$, $u = 2$ and $v = 1$ in the w-plane. The regions D and D′ are shown shaded in the figure.

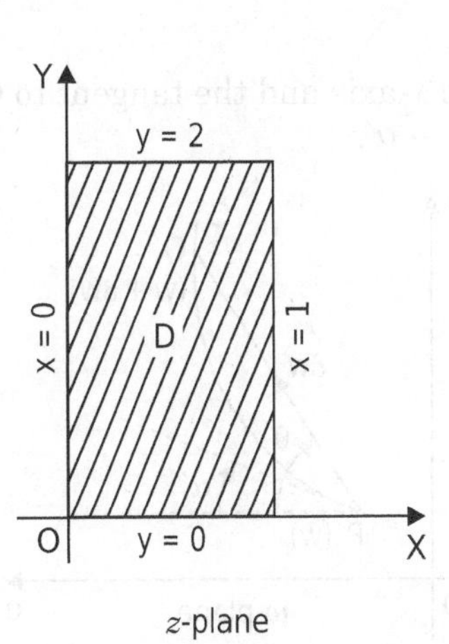

z-plane

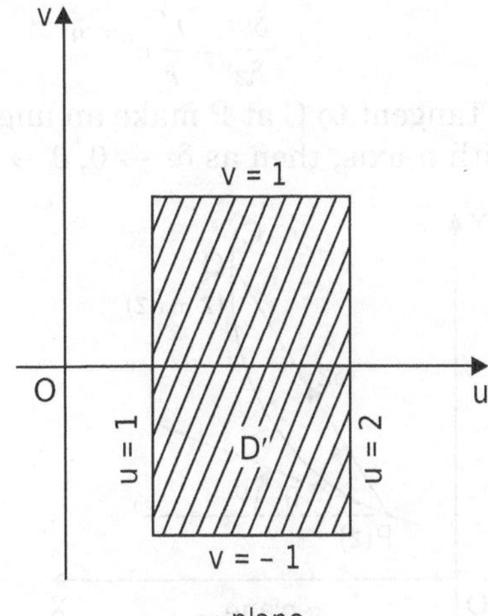

w-plane

4.13. CONFORMAL TRANSFORMATION

Suppose two curves C_1, C_2 in the z-plane intersect at the point P and the corresponding curves C_1', C_2' in the w-plane intersect at P′ under the transformation $w = f(z)$. *If the angle of intersection of the curves at P is the same as the angle of intersection of the curves at P′, both in magnitude and sense, then the transformation is said to be conformal at P.*

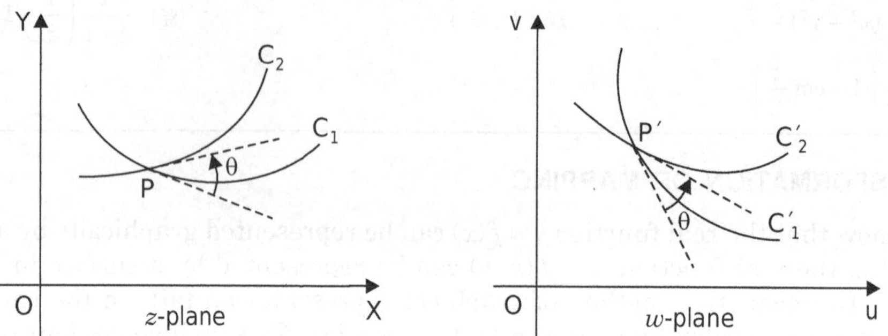

Definition. A transformation which preserves angles both in magnitude and sense between every pair of curves through a point is said to be conformal at the point.

The conditions under which the transformation $w = f(z)$ is conformal are given by the following theorem.

4.14. THEOREM

If f(z) is analytic and $f'(z) \neq 0$ in a region R of the z-plane, then the mapping $w = f(z)$ is conformal at all points of R.

Proof. Let P(z) be a point in the region R of the z-plane and P′(w) the corresponding point in the region R′ of the w-plane. Suppose P moves on a curve C and P′ moves on the corresponding curve C′. Let Q($z + \delta z$) be a neighbouring point on C and Q′($w + \delta w$) the corresponding point on C′ so that $\vec{PQ} = \delta z$ and $\vec{P'Q'} = \delta w$.

Then δz is a complex number whose modulus r is the length PQ and amplitude θ is the angle which PQ makes with the x-axis.

$\therefore$ $$\delta z = re^{i\theta}$$

Similarly, $\delta w = r'e^{i\theta'}$, where r' is the modulus and θ' is the amplitude of δw.

$\therefore$ $$\frac{\delta w}{\delta z} = \frac{r'}{r}e^{i(\theta'-\theta)}$$

Let the tangent to C at P make an angle α with x-axis and the tangent to C′ at P′ make an angle α' with u-axis, then as $\delta z \to 0$, $\theta \to \alpha$ and $\theta' \to \alpha'$.

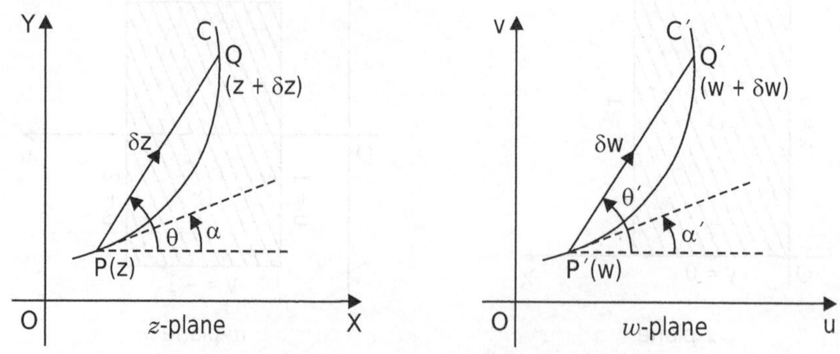

$$\therefore \qquad f'(z) = \frac{dw}{dz} = \underset{\delta z \to 0}{\text{Lt}} \frac{\delta w}{\delta z} = \underset{\delta z \to 0}{\text{Lt}} \left(\frac{r'}{r}\right) e^{i(\theta' - \theta)} \qquad \ldots(1)$$

Since $f'(z) \neq 0$, let $f'(z) = \rho e^{i\phi}$, then $\rho = | f'(z) |$ and ϕ = amplitude of $f'(z)$.

$$\therefore \text{ From (1),} \qquad \rho e^{i\phi} = \underset{\delta z \to 0}{\text{Lt}} \left(\frac{r'}{r}\right) e^{i(\theta' - \theta)}$$

Thus
$$\rho = \underset{\delta z \to 0}{\text{Lt}} \frac{r'}{r} \qquad \ldots(2)$$

and
$$\phi = \underset{\delta z \to 0}{\text{Lt}} (\theta' - \theta) = \alpha' - \alpha \qquad \ldots(3)$$

Now let C_1 be another curve through P in the z-plane and C_1' the corresponding curve through P′ in the w-plane. If the tangent to C_1 at P makes an angle β with x-axis and the tangent to C_1' at P′ makes an angle β' with u-axis, then as in (3),

$$\phi = \beta' - \beta \qquad \ldots(4)$$

From (3) and (4), $\alpha' - \alpha = \beta' - \beta$ or $\beta - \alpha = \beta' - \alpha' = \gamma$

Thus angle between the curves before and after the mapping is preserved in magnitude and sense. Hence the mapping by the analytic function $w = f(z)$ is conformal at each point where $f'(z) \neq 0$.

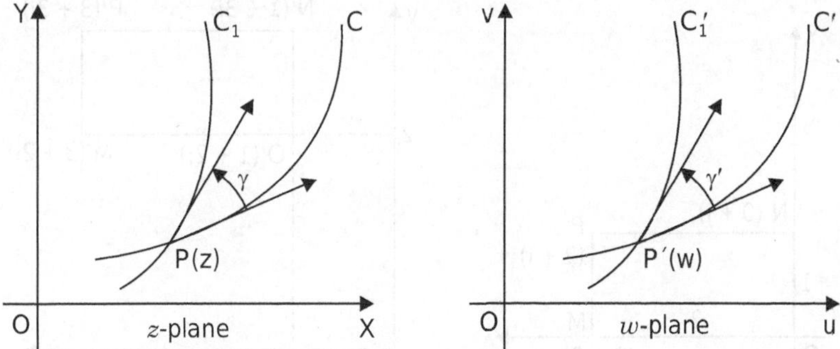

Note 1. A point at which $f'(z) = 0$ is called a **critical point** of the transformation.

Note 2. From (2), $\qquad \rho = \underset{\delta z \to 0}{\text{Lt}} \dfrac{r'}{r}$

It follows that under the conformal transformation $w = f(z)$, the lengths of arcs through P are magnified in the ratio $\rho : 1$, where $\rho = | f'(z) |$. Thus an infinitesimal length in the z-plane is magnified by the factor $| f'(z) |$ in the w-plane and consequently infinitesimal areas in the z-plane are magnified by the factor $| f'(z) |^2$ in the w-plane.

Note 3. From (3), $\alpha' = \alpha + \phi$ shows that the tangent to the curve C at P is rotated through an angle ϕ = amp $[f'(z)]$ under the given transformation.

Note 4. A harmonic function remains harmonic under a conformal transformation.

Example. *For the conformal transformation $w = z^2$, show that*

(i) the co-efficient of magnification at $z = 1 + i$ is $2\sqrt{2}$.

(ii) the angle of rotation at $z = 1 + i$ is $\dfrac{\pi}{4}$.

Sol. Here $f(z) = w = z^2$

$\therefore$ $f'(z) = 2z$ and $f'(1 + i) = 2 + 2i$

$\therefore$ The co-efficient of magnification at $z = 1 + i$ is $|f'(1 + i)| = \sqrt{4+4} = 2\sqrt{2}$

The angle of rotation at $z = 1 + i$ is amp. $[f'(1 + i)] = \tan^{-1}\frac{2}{2} = \frac{\pi}{4}$.

4.15. SOME STANDARD TRANSFORMATIONS

1. Translation: w = z + c, where c is a complex constant.

Let $z = x + iy$, $c = a + ib$ and $w = u + iv$

then the transformation becomes $u + iv = (x + iy) + (a + ib) = (x + a) + i(y + b)$

so that $u = x + a$ and $v = y + b$

Thus the transformation is a mere translation of the axes and preserves the shape and size.

For example, the rectangle OMPN in z-plane is transformed to rectangle O′M′P′N′ in the w-plane under the transformation $w = z + (1 + 2i)$.

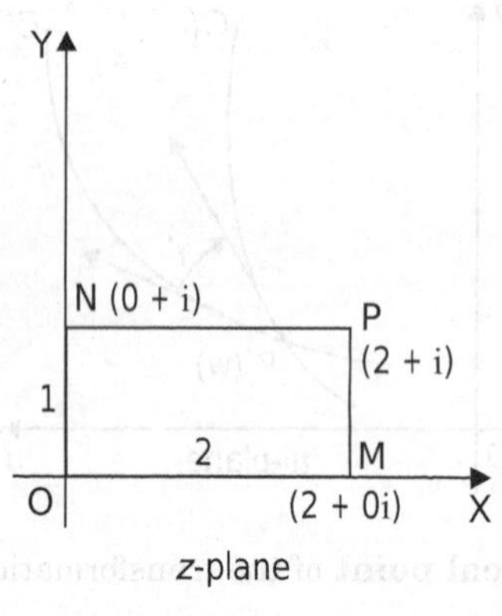

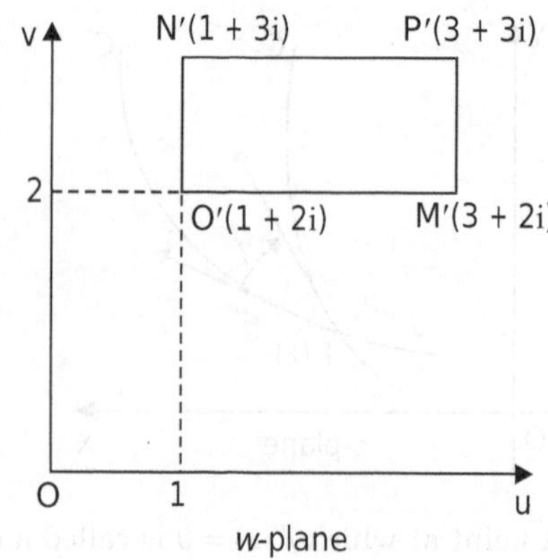

2. Bilinear Transformation:

A transformation of the form $$w = \frac{az+b}{cz+d} \quad ...(1)$$
where a, b, c, d are complex constants and $ad - bc \neq 0$ is called a **bilinear** or **Mobius transformation.**

The transformation given by (1) is conformal, since
$$\frac{dw}{dz} = \frac{ad-bc}{(cz+d)^2} \neq 0$$

The inverse mapping of (1) is $z = \dfrac{-dw+b}{cw-a}$ ...(2)

which is also a bilinear transformation.

The transformation (1) can be written as
$$cwz + wd - az - b = 0$$
which is linear both in w and z and hence the name bilinear transformation.

From (1), we observe that each point in the z-plane except the point $z = -\frac{d}{c}$ maps into a unique point in the w-plane. Similarly, from (2), we observe that each point in the w-plane except the point $w = \frac{a}{c}$ maps into a unique point in the z-plane. Considering the two exceptional points as points at infinity in respective planes, we can say that there is one to one correspondence between all points in the two planes.

Every bilinear transformation $w = \frac{az+b}{cz+d}$, $ad - bc \neq 0$ is the combination of basic transformations

(*i*) translation: $w = z + c$ (*ii*) rotation and magnification: $w = cz$

(*iii*) inversion: $w = \frac{1}{z}$

By actual division, we have $w = \frac{a}{c} + \frac{bc-ad}{c^2} \cdot \frac{1}{z + \frac{d}{c}}$

Taking $w_1 = z + \frac{d}{c}$, $w_2 = \frac{1}{w_1}$, $w_3 = \frac{bc-ad}{c^2} w_2$, we get $w = \frac{a}{c} + w_3$

Thus, by these transformations, we successively pass from z-plane to w_1-plane, from w_1-plane to w_2-plane, from w_2-plane to w_3-plane and finally from w_3-plane to w-plane.

Since each of these auxiliary transformations maps circles into circles, hence a *bilinear transformation also maps circles into circles.*

Note 1. *The cross-ratio is invariant under a bilinear transformation.*

Thus if w_1, w_2, w_3 and w_4 are the respective images of four distinct points z_1, z_2, z_3 and z_4, then

$$\frac{(w_1 - w_3)(w_2 - w_4)}{(w_1 - w_4)(w_2 - w_3)} = \frac{(z_1 - z_3)(z_2 - z_4)}{(z_1 - z_4)(z_2 - z_3)}.$$

Note 2. In the bilinear transformation $w = \frac{az+b}{cz+d}$, $ad - bc \neq 0$ dividing the numerator and denominator of the right hand side by one of the four constants, we observe that there are only three independent constants. Hence *three independent conditions are required to determine a bilinear transformation.*

ILLUSTRATIVE EXAMPLES

Example 1. *Consider the transformation $w = z\, e^{i\pi/4}$ and determine the region in the w-plane corresponding to the triangular region bounded by the lines $x = 0$, $y = 0$ and $x + y = 1$ in the z-plane.*

Sol. The given transformation is

$$w = u + iv = \left(\cos\frac{\pi}{4} + i \sin\frac{\pi}{4}\right)(x + iy)$$

$$= \frac{1}{\sqrt{2}}(1 + i)(x + iy) = \frac{1}{\sqrt{2}}[(x - y) + i(x + y)]$$

$$\therefore \quad u = \frac{1}{\sqrt{2}}(x - y) \quad \text{and} \quad v = \frac{1}{\sqrt{2}}(x + y)$$

The line $x = 0$ maps into $u = -\frac{1}{\sqrt{2}}y$, $v = \frac{1}{\sqrt{2}}y$ or into $v = -u$.

The line $y = 0$ maps into $u = \frac{1}{\sqrt{2}}x$, $u = \frac{1}{\sqrt{2}}x$

or into $v = u$

The line $x + y = 1$ maps into $v = \frac{1}{\sqrt{2}}$.

The two regions are shown shaded in the figure.

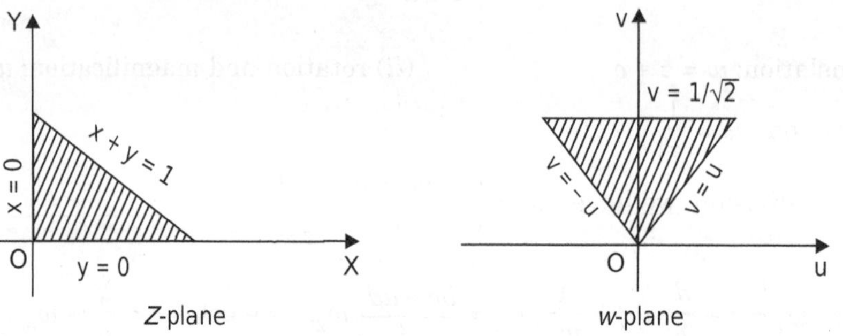

Example 2. *Under the transformation $w = \frac{1}{z}$, find the image of $| z - 2i | = 2$.*

Sol. The given transformation is

$$w = \frac{1}{z} \quad \text{or} \quad z = \frac{1}{w} \quad \text{or} \quad x + iy = \frac{1}{u + iv} = \frac{u - iv}{u^2 + v^2}$$

so that
$$x = \frac{u}{u^2 + v^2} \quad \text{and} \quad y = -\frac{v}{u^2 + v^2} \qquad ...(1)$$

The given curve is $| z - 2i | = 2$ or $| x + i(y - 2) | = 2$

or $x^2 + (y - 2)^2 = 4$ or $x^2 + y^2 - 4y = 0$ ...(2)

which is a circle in the z-plane with centre (0, 2) and radius 2.

Substituting the values of x and y from (1) in (2), we get

$$\frac{u^2}{(u^2 + v^2)^2} + \frac{v^2}{(u^2 + v^2)^2} + \frac{4v}{u^2 + v^2} = 0 \quad \text{or} \quad \frac{u^2 + v^2}{(u^2 + v^2)^2} + \frac{4v}{u^2 + v^2} = 0$$

or $1 + 4v = 0$, a straight line which is the required image of the given curve.

Example 3. *Find the image of the infinite strip $\frac{1}{4} \le y \le \frac{1}{2}$ under the transformation $w = \frac{1}{z}$. Also show the regions graphically.*

Sol. The given transformation is

$$w = \frac{1}{z} \quad \text{or} \quad z = \frac{1}{w} \quad \text{or} \quad x + iy = \frac{1}{u + iv} = \frac{u - iv}{u^2 + v^2}$$

so that
$$x = \frac{u}{u^2 + v^2} \quad \text{and} \quad y = \frac{-v}{u^2 + v^2}$$

If $\quad y = \frac{1}{4}$, then $\quad \frac{-v}{u^2+v^2} = \frac{1}{4}$

or $\quad u^2 + v^2 + 4v = 0 \quad$ or $\quad u^2 + (v+2)^2 = 4$

If $\; y = \frac{1}{2}$, then $\quad \frac{-v}{u^2+v^2} = \frac{1}{2} \quad$ or $\quad u^2 + v^2 + 2v = 0 \quad$ or $\quad u^2 + (v+1)^2 = 1$

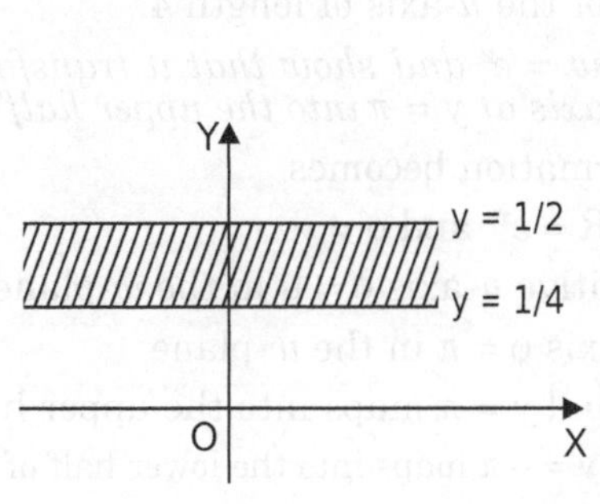

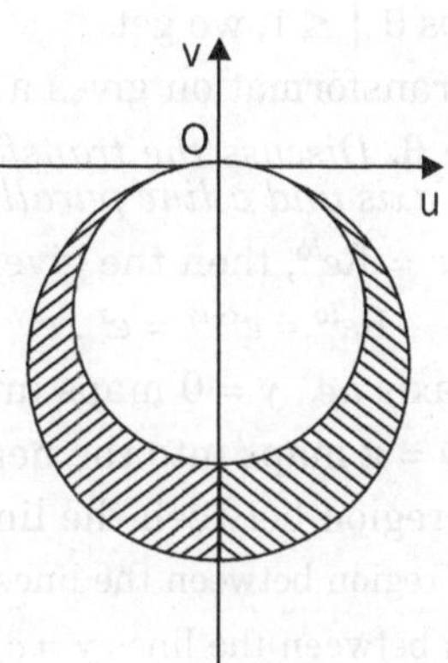

Hence the infinite strip $\frac{1}{4} \le y \le \frac{1}{2}$ is transformed into the region between the two circles

$$u^2 + (v+2)^2 = 4, \text{ centre } (0, -2), \text{ radius } 2$$

and $\quad u^2 + (v+1)^2 = 1$, centre $(0, -1)$, radius 1.

Example 4. *Show that the transformation* $w = \frac{2z+3}{z-4}$ *maps the circle* $x^2 + y^2 - 4x = 0$ *onto the straight line* $4u + 3 = 0$.

Sol. The given transformation is $\qquad w = \frac{2z+3}{z-4}$

The inverse transformation is $\qquad z = \frac{4w+3}{w-2} \qquad$ ...(1)

Now the equation $x^2 + y^2 - 4x = 0$ can be written as $z\bar{z} - 2(z + \bar{z}) = 0$

Substituting for z and $\bar{z}$ from (1), we get

$$\frac{4w+3}{w-2} \cdot \frac{4\bar{w}+3}{\bar{w}-2} - 2\left(\frac{4w+3}{w-2} + \frac{4\bar{w}+3}{\bar{w}-2}\right) = 0$$

or $\quad 16w\bar{w} + 12w + 12\bar{w} + 9 - 2(4w\bar{w} + 3\bar{w} - 8w - 6 + 4w\bar{w} + 3w - 8\bar{w} - 6) = 0$

or $\quad 22(w + \bar{w}) + 33 = 0 \quad$ or $\quad 22(2u) + 33 = 0 \quad$ or $\quad 4u + 3 = 0.$

Example 5. *Show that the transformation* $w = z + \frac{1}{z}$ *maps the circle* $|z| = c$ *into the ellipse* $u = \left(c + \frac{1}{c}\right)\cos\theta,\ v = \left(c - \frac{1}{c}\right)\sin\theta$. *Discuss the case when* $c = 1$.

Sol. A point on the circle $|z| = c$ can be written as $z = ce^{i\theta}$

$\therefore \qquad w = z + \frac{1}{z}$ becomes $u + iv = ce^{i\theta} + \frac{1}{c}e^{-i\theta}$

or $$u + iv = c(\cos\theta + i\sin\theta) + \frac{1}{c}(\cos\theta - i\sin\theta) = \left(c + \frac{1}{c}\right)\cos\theta + i\left(c - \frac{1}{c}\right)\sin\theta$$

$\therefore$ $$u = \left(c + \frac{1}{c}\right)\cos\theta \quad \text{and} \quad v = \left(c - \frac{1}{c}\right)\sin\theta$$

which are the parametric equations of an ellipse.

When $c = 1$, we have $u = 2\cos\theta$ and $v = 0$

Since $|\cos\theta| \le 1$, we get $-2 \le u \le 2$ and $v = 0$

$\therefore$ The transformation gives a segment of the u-axis of length 4.

Example 6. *Discuss the transformation $w = e^z$ and show that it transforms the region between the real axis and a line parallel to real axis at $y = \pi$ into the upper half of the w-plane.*

Sol. Let $w = Re^{i\phi}$, then the given transformation becomes

$$Re^{i\phi} = e^{x+iy} = e^x \cdot e^{iy} \text{ so that } R = e^x \text{ and } \phi = y$$

The real axis *i.e.*, $y = 0$ maps into the positive u-axis $\phi = 0$ in the w-plane.

The line $y = \pi$ maps into the negative u-axis $\phi = \pi$ in the w-plane.

Thus the region between the lines $y = 0$ and $y = \pi$ maps into the upper half of w-plane.

Note. The region between the lines $y = 0$ and $y = -\pi$ maps into the lower half of w-plane.

The region between the lines $y = c$ and $y = c + 2\pi$ maps into the whole of the w-plane, since e^z is periodic with period $2\pi i$.

The imaginary axis $x = 0$ maps into a unit circle $R = e^0 = 1$ in the w-plane.

Example 7. *Discuss the transformation $w = \sin z$.*

Sol. The given transformation is

$$w = u + iv = \sin(x + iy)$$

or $$u + iv = \sin x \cosh y + i \cos x \sinh y$$

so that $$u = \sin x \cosh y$$

and $$v = \cos x \sinh y \qquad \text{...(1)}$$

Eliminating y from equation (1), we get

$$\frac{u^2}{\sin^2 x} - \frac{v^2}{\cos^2 x} = 1 \qquad [\because \cosh^2 y - \sinh^2 y = 1]$$

Thus the straight lines $x = c$ in the z-plane are mapped into confocal hyperbolas in the w-plane.

Eliminating x from equation (1), we get $\dfrac{u^2}{\cosh^2 y} + \dfrac{v^2}{\sinh^2 y} = 1$

Thus the straight lines $y = c$ in the z-plane are mapped into confocal ellipses.

The lines $x = 0$ and $y = 0$ map into the lines $u = 0$ and $v = 0$ respectively in the w-plane.

Example 8. *Find the bilinear transformation which maps the points $z = 1, i, -1$ into the points $w = i, 0, -i$.*

Hence find the image of $|z| < 1$.

Sol. Let the required bilinear transformation be $w = \dfrac{az + b}{cz + d}$ ...(1)

Substituting the corresponding values of w and z in (1), we get

$$i = \frac{a + b}{c + d}, \; 0 = \frac{ai + b}{ci + d}, \quad -i = \frac{-a + b}{-c + d}$$

These equations can be written as

$$(a + b) - i(c + d) = 0 \quad ...(2)$$

$$b + ia = 0 \quad ...(3)$$

and $$(-a + b) + i(-c + d) = 0 \quad ...(4)$$

From (3), $$b = -ia$$

Adding (2) and (4), $$2b - 2ic = 0 \quad \text{or} \quad c = \frac{b}{i} = -a \quad \text{[Using (3)]}$$

Subtracting (4) from (2), $$2a - 2id = 0 \quad \text{or} \quad d = \frac{a}{i} = -ia$$

Substituting for b, c, d in (1), we get $$w = \frac{az - ia}{-az - ia} \quad \text{or} \quad w = \frac{i - z}{i + z} \quad ...(5)$$

which is the required bilinear transformation.

Now, from (5), $$z = i\,\frac{1 - w}{1 + w}$$

$\therefore$ $|z| < 1$ is mapped into the region

$$\left| i\,\frac{1 - w}{1 + w} \right| < 1 \quad \text{or} \quad \frac{|i||1 - w|}{|1 + w|} < 1$$

or $$|1 - w| < |1 + w| \quad \because |i| = 1$$

or $$|1 - u - iv| < |1 + u + iv| \quad \text{or} \quad (1 - u)^2 + v^2 < (1 + u)^2 + v^2$$

or $$u > 0.$$

Hence the interior of the circle $x^2 + y^2 = 1$ in the z-plane is mapped into the entire half of the w-plane to the right of the imaginary axis.

TEST YOUR KNOWLEDGE

1. What is the region of the w-plane into which the rectangular region in the z-plane bounded by the lines $x = 0, y = 0, x = 1$ and $y = 2$ is mapped under the transformation $w = z + (2 - i)$?
2. Find the image of the triangle with vertices at $i, 1 + i, 1 - i$ in the z-plane under the transformation $w = 3z + 4 - 2i$.
3. Determine the region in the w-plane in which the rectangle bounded by the lines $x = 0, y = 0, x = 2$ and $y = 3$ is mapped under the transformation $w = \sqrt{2}\, e^{i\pi/4} . z$.
4. Consider the transformation $w = 2z$ and determine the region D′ of the w-plane into which the triangular region D enclosed by the lines $x = 0, y = 0, x + y = 1$ in the z-plane is mapped under this transformation.
5. Find the image of the semi-infinite strip $x > 0, 0 < y < 2$ under the transformation $w = iz + 1$. Show the region graphically.
6. Find the image of the circle $|z| = 2$ under the transformation $w = z + 3 + 2i$.
7. Find the image of the following curves under the mapping $w = \frac{1}{z}$
   (*i*) the line $y - x + 1 = 0$ (*ii*) the circle $|z - 3| = 5$.
8. Show that the image of the hyperbola $x^2 - y^2 = 1$ under the transformation $w = \frac{1}{z}$ is the lemniscate $\rho^2 = \cos 2\phi$, where $w = \rho e^{i\phi}$.

9. Show that the map of the real axis of the z-plane on the w-plane by the transformation $w = \frac{1}{z+i}$ is a circle and find its centre and radius.

10. Prove that $w = \frac{z}{i-z}$ maps the upper half of the z-plane into the upper half of the w-plane.

11. Under the transformation $w = \frac{z-i}{1-iz}$, find the map of the circle $|z| = 1$ in the w-plane.

12. Show that the transformation $w = i\frac{1-z}{1+z}$ transforms the circle $|z| = 1$ into the real axis of w-plane and the interior of the circle $|z| < 1$ into the upper half of the w-plane.

13. Show that under the transformation $w = \frac{z-i}{z+i}$, real axis in the z-plane is mapped into the circle $|w| = 1$. What portion of the z-plane corresponds to the interior of the circle?

14. Under the transformation $w = z^2$, obtain the map in the w-plane of the square with vertices (0, 0), (2, 0), (2, 2), (0, 2) in the z-plane.

15. Find the images of the straight lines $x = 0, y = 0, x = 1$ and $y = 1$ under the transformation $w = z^2$.

16. Determine the region of the w-plane into which the triangle formed by $x = 1, y = 1$ and $x + y = 1$ is mapped under the transformation $w = z^2$.

17. Show that the transformation $w = \sqrt{z}$ maps the domain in the z-plane to the right of the line $x = a$ into the interior of a hyperbola in the w-plane.

18. Show that the map of the circle $|z| = 2$ under the transformation $w + 2i = z + \frac{1}{z}$ is an ellipse, find its axes and centre.

19. Show that the transformation $w = z + \frac{1}{z}$ converts the straight line $\arg z = \alpha\left(|\alpha| < \frac{\pi}{2}\right)$ into a branch of the hyperbola of eccentricity $\sec \alpha$.

20. Show that the transformation $w = z + \frac{a^2 - b^2}{4z}$ transforms the circle $|z| = \frac{1}{2}(a + b)$ in the z-plane into an ellipse of semi-axes a, b in the w-plane.

21. If $w = z + \frac{a^2}{z}$, prove that, when z describes the circle $x^2 + y^2 = a^2$, w describes a straight line and find its length. Also prove that if z describes the circle $x^2 + y^2 = b^2$, where $b > a$, w describes an ellipse.

22. Show that the transformation $w = \cosh z$ maps the lines parallel to x-axis and lines parallel to y-axis into confocal central conics. What physical problems can we study with the help of this transformation?

23. Find the bilinear transformation which maps:

(*a*) the points $z = 1, -i, -1$ into the points $w = i, 0, -i$.

(*b*) the points $z = 1, i, -1$ into the points $w = 0, 1, \infty$.

(*c*) the points $z = 0, -i, -1$ into the points $w = i, 1, 0$. Find the image of the line $y = mx$ under this transformation.

Answers

1. Rectangular region bounded by the lines $u = 2, v = -1, u = 3, v = 1$.

2. A triangle with vertices $(4, -1)$, $(7, 1)$ and $(7, -3)$.

3. Rectangular region bounded by the lines $v = -u, v = u, u + v = 4, v - u = 6$.

4. A triangular region bounded by the lines $u = 0, v = 0, u + v = 2$.

5. $-1 < u < 1, v > 0$

6. $(u-3)^2 + (v-2)^2 = 4$

7. (*i*) $u^2 + v^2 - u - v = 0$ (*ii*) $\left| w + \frac{3}{16} \right| = \frac{5}{16}$

9. $\left(0, -\frac{1}{2}\right); \frac{1}{2}$

11. $v = 0$

13. $y > 0$

14. $(0, 0), (4, 0), (0, 8), (-4, 0)$; an isosceles triangle

15. Negative real axis, positive real axis, $v^2 = 4(1-u)$, $v^2 = 4(1+u)$

16. Region bounded by the parabolas $v^2 = 4(1 \pm u)$, $u^2 = 1 - 2v$

18. $\frac{5}{2}, \frac{3}{2}; (0, 0)$

23. (*a*) $w = \frac{z+i}{z-i}$ (*b*) $w = \frac{i(1-z)}{1+z}$ (*c*) $w = \frac{i(1+z)}{1-z}$; $m(u^2+v^2) + 2u - m = 0$.

4.16. COMPLEX INTEGRATION

Let $f(z)$ be a continuous function of the complex variable $z = x + iy$ defined at all points of a curve C having end points A and B. Divide the curve C into n parts at the points

$$A = P_0(z_0), P_1(z_1), \ldots\ldots, P_i(z_i), \ldots\ldots, P_n(z_n) = B.$$

Let $\delta z_i = z_i - z_{i-1}$ and ξ_i be any point on the arc $P_{i-1}P_i$. Then the limit of the sum

$$\sum_{i=1}^{n} f(\xi_i)\,\delta z_i$$

as $n \to \infty$ and each $\delta z_i \to 0$, if it exists, is called the **line integral of f(z)** along the curve C. It is denoted by

$$\int_C f(z)\,dz$$

In case the points P_0 and P_n coincide so that C is a closed curve, then this integral is called **contour integral** and is denoted by $\oint_C f(z)\,dz$.

If $f(z) = u(x, y) + iv(x, y)$, then since $dz = dx + idy$, we have

$$\int_C f(z)\,dz = \int_C (u + iv)(dx + idy)$$

$$= \int_C (udx - vdy) + i\int_C (vdx + udy)$$

which shows that the evaluation of the line integral of a complex function can be reduced to the evaluation of two line integrals of real functions.

Moreover, the value of the integral depends on the path of integration unless the integrand is analytic. When the same path of integration is used in each integral, then

$$\int_a^b f(z)\,dz = -\int_b^a f(z)\,dz$$

If C is a point on the arc joining a and b, then

$$\int_a^b f(z)\,dz = \int_a^c f(z)\,d(z) + \int_c^b f(z)\,dz.$$

ILLUSTRATIVE EXAMPLES

Example 1. *Evaluate* $\int_0^{1+i} (x - y + ix^2)\,dz$

(a) along the straight line from z = 0 to z = 1 + i

(b) along the real axis from z = 0 to z = 1 and then along a line parallel to imaginary axis from z = 1 to z = 1 + i.

(c) along the imaginary axis from z = 0 to z = i and then along a line parallel to real axis from z = i to z = 1 + i.

Sol. (*a*) Along the straight line OP joining O(z = 0) and P(z = 1 + i), $y = x$, $dy = dx$ and x varies from 0 to 1.

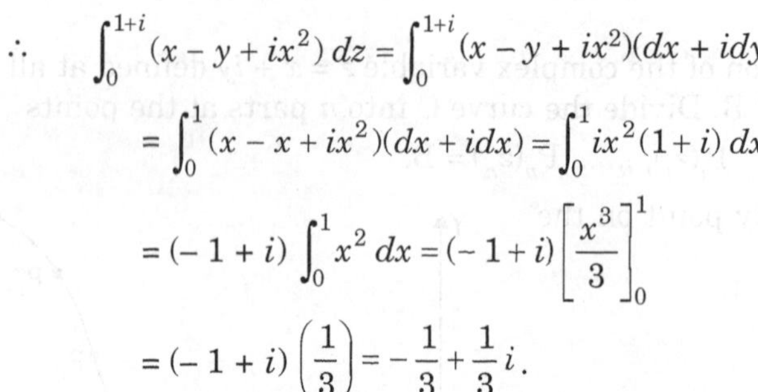

$$\therefore \quad \int_0^{1+i} (x - y + ix^2)\,dz = \int_0^{1+i} (x - y + ix^2)(dx + i\,dy)$$

$$= \int_0^1 (x - x + ix^2)(dx + i\,dx) = \int_0^1 ix^2(1+i)\,dx$$

$$= (-1 + i)\int_0^1 x^2\,dx = (-1+i)\left[\frac{x^3}{3}\right]_0^1$$

$$= (-1 + i)\left(\frac{1}{3}\right) = -\frac{1}{3} + \frac{1}{3}i.$$

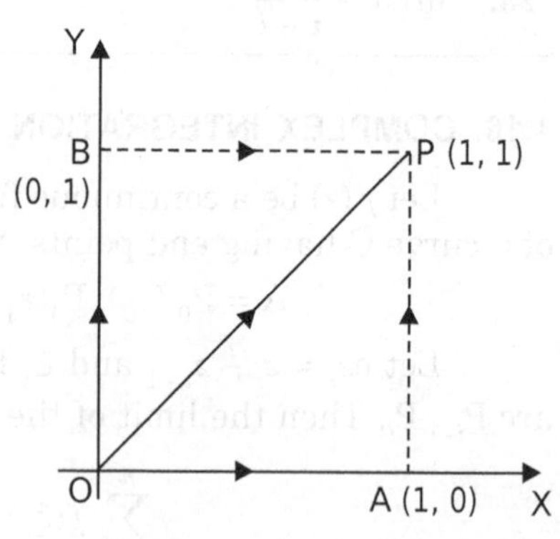

(*b*) Along the path OAP where A is $z = 1$

$$\int_0^{1+i} (x - y + ix^2)\,dz = \int_{OA} (x - y + ix^2)\,dz + \int_{AP} (x - y + ix^2)\,dz \qquad \text{...(1)}$$

Now along OA, $y = 0$, $dz = dx$ and x varies from 0 to 1.

$$\therefore \quad \int_{OA} (x - y + ix^2)\,dz = \int_0^1 (x + ix^2)\,dx = \left[\frac{x^2}{2} + i\frac{x^3}{3}\right]_0^1 = \frac{1}{2} + \frac{1}{3}i$$

Also, along AP, $x = 1$, $dz = i\,dy$ and y varies from 0 to 1

$$\therefore \quad \int_{AP} (x - y + ix^2)\,dz = \int_0^1 (1 - y + i)\,i\,dy = \left[(-1 + i)y - i\frac{y^2}{2}\right]_0^1$$

$$= -1 + i - \frac{1}{2}i = -1 + \frac{1}{2}i$$

Hence, from (1),

$$\int_0^{1+i} (x - y + ix^2)\,dz = \left(\frac{1}{2} + \frac{1}{3}i\right) + \left(-1 + \frac{1}{2}i\right) = -\frac{1}{2} + \frac{5}{6}i.$$

(c) Along the path OBP where B is $z = i$

$$\int_0^{1+i} (x - y + ix^2)\, dz = \int_{OB} (x - y + ix^2)\, dz + \int_{BP} (x - y + ix^2)\, dz \qquad ...(2)$$

Now along OB, $x = 0$, $dz = idy$ and y varies from 0 to 1

$$\therefore \quad \int_{OB} (x - y + ix^2)\, dz = \int_0^1 (-y)\, idy = -i\left[\frac{y^2}{2}\right]_0^1 = -\frac{1}{2}i$$

Also, along BP, $y = 1$, $dz = dx$ and x varies from 0 to 1

$$\therefore \quad \int_{BP} (x - y + ix^2)\, dz = \int_0^1 (x - 1 + ix^2)\, dx = \left[\frac{x^2}{2} - x + i\frac{x^3}{3}\right]_0^1 = -\frac{1}{2} + \frac{1}{3}i$$

Hence, from (2), $\displaystyle\int_0^{1+i} (x - y + ix^2)\, dz = -\frac{1}{2}i + \left(-\frac{1}{2} + \frac{1}{3}i\right) = -\frac{1}{2} - \frac{1}{6}i.$

Note. The values of the integral are different along the three different paths.

Example 2. *Evaluate* $\displaystyle\int_0^{1+i} (x^2 - iy)\, dz$ *along the paths*

(a) $y = x$ *(b)* $y = x^2$.

Sol. (a) Along the line $y = x$,

$dy = dx$ so that $dz = dx + idx = (1 + i)\, dx$

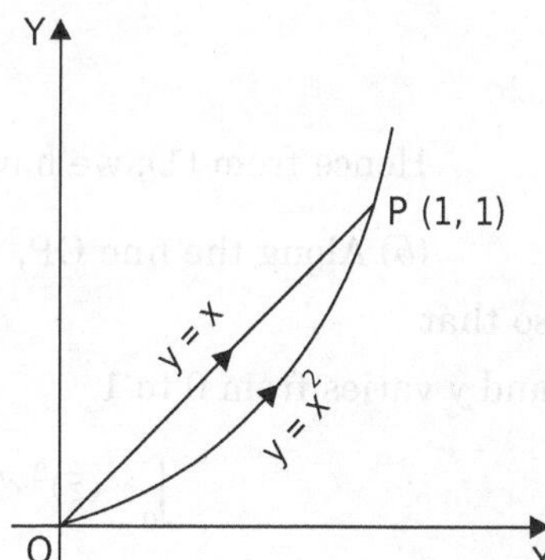

$$\therefore \quad \int_0^{1+i} (x^2 - iy)\, dz = \int_0^1 (x^2 - ix)(1 + i)\, dx$$

$$= (1 + i)\left[\frac{x^3}{3} - i\frac{x^2}{2}\right]_0^1 = (1 + i)\left(\frac{1}{3} - \frac{1}{2}i\right)$$

$$= \frac{(1 + i)(2 - 3i)}{6} = \frac{5}{6} - \frac{1}{6}i.$$

(b) Along the parabola $y = x^2$, $dy = 2x\, dx$ so that

$$dz = dx + 2ix\, dx = (1 + 2ix)\, dx$$

and x varies from 0 to 1.

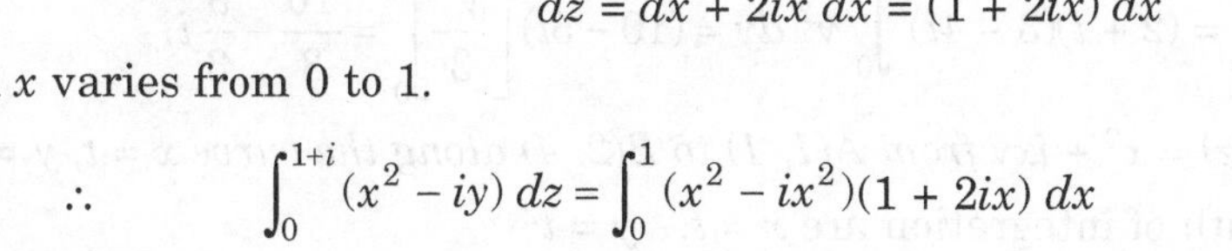

$$\therefore \quad \int_0^{1+i} (x^2 - iy)\, dz = \int_0^1 (x^2 - ix^2)(1 + 2ix)\, dx$$

$$= (1 - i)\int_0^1 x^2(1 + 2ix)\, dx = (1 - i)\left[\frac{x^3}{3} + i\frac{x^4}{2}\right]_0^1$$

$$= (1 - i)\left(\frac{1}{3} + \frac{1}{2}i\right) = \frac{(1 - i)(2 + 3i)}{6} = \frac{5}{6} + \frac{1}{6}i.$$

Example 3. *Evaluate* $\displaystyle\int_0^{2+i} (\bar{z})^2\, dz$*, along*

(a) the real axis to 2 and then vertically to $2 + i$.

(b) along the line $2y = x$.

Sol. $(\bar{z})^2 = (x - iy)^2 = (x^2 - y^2) - 2ixy$

(*a*) Along the path OAP where A is (2, 0) and P is (2, 1).

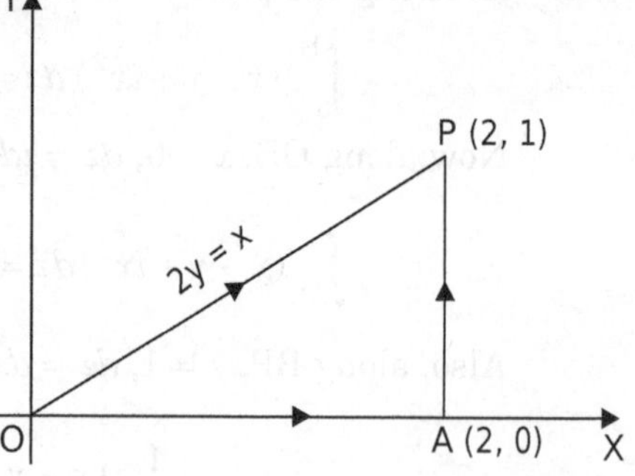

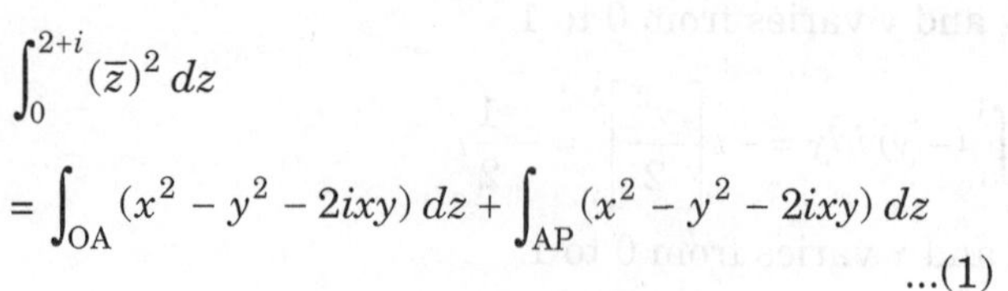

$$\int_0^{2+i} (\bar{z})^2\, dz$$

$$= \int_{OA} (x^2 - y^2 - 2ixy)\, dz + \int_{AP} (x^2 - y^2 - 2ixy)\, dz \quad ...(1)$$

Now, along OA, $y = 0$, $dz = dx$ and x varies from 0 to 2

$$\therefore \quad \int_{OA} (x^2 - y^2 - 2ixy)\, dz = \int_0^2 x^2\, dx = \left[\frac{x^3}{3}\right]_0^2 = \frac{8}{3}$$

Also, along AP, $x = 2$, $dz = idy$ and y varies from 0 to 1

$$\therefore \quad \int_{AP} (x^2 - y^2 - 2ixy)\, dz = \int_0^1 (4 - y^2 - 4iy)\, idy$$

$$= \left[4iy - i\frac{y^3}{3} + 2y^2\right]_0^1 = 4i - \frac{1}{3}i + 2 = 2 + \frac{11}{3}i$$

Hence from (1), we have $\displaystyle\int_0^{2+i} (\bar{z})^2\, dz = \frac{8}{3} + 2 + \frac{11}{3}i = \frac{14}{3} + \frac{11}{3}i.$

(*b*) Along the line OP, $2y = x$, $dx = 2dy$

so that $dz = 2dy + i\, dy = (2 + i)\, dy$

and y varies from 0 to 1.

$$\therefore \quad \int_0^{2+i} (\bar{z})^2\, dz = \int_0^{2+i} (x^2 - y^2 - 2ixy)\, dz = \int_0^1 (4y^2 - y^2 - 4iy^2)(2 + i)\, dy$$

$$= (2 + i)(3 - 4i) \int_0^1 y^2\, dy = (10 - 5i)\left[\frac{y^3}{3}\right]_0^1 = \frac{10}{3} - \frac{5}{3}i.$$

Example 4. *Integrate $f(z) = x^2 + ixy$ from A(1, 1) to B(2, 4) along the curve $x = t$, $y = t^2$.*

Sol. Equations of the path of integration are $x = t$, $y = t^2$

$$\therefore \quad dx = dt, \quad dy = 2t\, dt$$

At A(1, 1), $t = 1$ and at B(2, 4), $t = 2$

$$\therefore \quad \int_A^B f(z)\, dz = \int_A^B (x^2 + ixy)(dx + idy) = \int_1^2 (t^2 + it^3)(dt + 2it\, dt)$$

$$= \int_1^2 (t^2 - 2t^4)\, dt + i\int_1^2 3t^3\, dt = \left[\frac{t^3}{3} - \frac{2t^5}{5}\right]_1^2 + i\left[\frac{3t^4}{4}\right]_1^2$$

$$= \left(\frac{8}{3} - \frac{64}{5}\right) - \left(\frac{1}{3} - \frac{2}{5}\right) + i\left(12 - \frac{3}{4}\right) = -\frac{151}{5} + \frac{45}{4}i.$$

Example 5. *Prove that*

(*i*) $\oint_C \frac{dz}{z-a} = 2\pi i$

(*ii*) $\oint_C (z-a)^n\, dz = 0$ *[n is an integer* $\neq -1$*] where C is the circle* $|\ z-a\ | = r$.

Sol. The equation of the circle C is

$$|\ z-a\ | = r \quad \text{or} \quad z - a = re^{i\theta}$$

where θ varies from 0 to 2π as z describes C once in the anti-clockwise direction. $dz = ire^{i\theta}\, d\theta$.

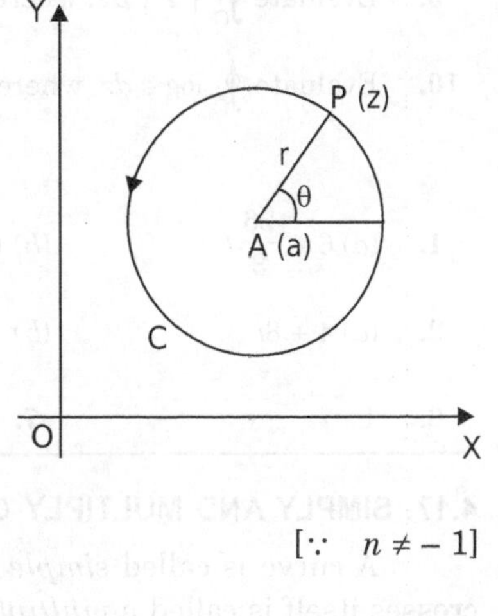

(*i*) $$\oint_C \frac{dz}{z-a} = \int_0^{2\pi} \frac{ire^{i\theta}\, d\theta}{re^{i\theta}} = i\int_0^{2\pi} d\theta = 2\pi i$$

(*ii*) $$\oint_C (z-a)^n\, dz = \int_0^{2\pi} r^n e^{ni\theta}\, .\, ire^{i\theta}\, d\theta$$

$$= ir^{n+1}\int_0^{2\pi} e^{i(n+1)\theta}\, d\theta$$

$$= ir^{n+1}\left[\frac{e^{i(n+1)\theta}}{i(n+1)}\right]_0^{2\pi} \qquad [\because \ n \neq -1]$$

$$= \frac{r^{n+1}}{n+1}\,[e^{i2(n+1)\pi} - 1]$$

$$= 0. \qquad [\because \ e^{i2(n+1)\pi} = \cos 2(n+1)\pi + i\sin 2(n+1)\pi = 1 + i0 = 1]$$

TEST YOUR KNOWLEDGE

1. Evaluate $\int_0^{3+i} z^2\, dz$, along

   (*a*) the line $y = \frac{x}{3}$ (*b*) the real axis to 3 and then vertically to $3 + i$

   (*c*) the parabola $x = 3y^2$.

2. Evaluate $\int_{1-i}^{2+i} (2x + iy + 1)\, dz$, along

   (*a*) the straight line joining $(1 - i)$ to $(2 + i)$. (*b*) the curve $x = t + 1$, $y = 2t^2 - 1$.

3. Evaluate $\int_0^{4+2i} \bar{z}\, dz$, along the curve given by $z = t^2 + it$.

4. Evaluate $\oint_C |z|^2\, dz$, around the square with vertices at (0, 0), (1, 0), (1, 1) and (0, 1).

5. Show that $\oint_C (z+1)\, dz = 0$, where C is the boundary of the square whose vertices are at the points $z = 0$, $z = 1$, $z = 1 + i$ and $z = i$.

6. Evaluate $\oint_C (y - x - 3x^2 i)\, dz$, where C is the straight line from $z = 0$ to $z = 1 + i$.

7. Evaluate $\oint_C (z - z^2)\, dz$, where C is the upper half of the circle $|z| = 1$.

8. Prove that $\int_C \frac{1}{z}\, dz = -\pi i$ or πi according as C is the semi-circular arc $|z| = 1$ from $z = -1$ to $z = 1$ above or below the real axis.

9. Evaluate $\oint_C |z|\, dz$, where C is the left half of the unit circle $|z| = 1$ from $z = -i$ to $z = i$.

10. Evaluate $\oint_C \log z\, dz$, where C is the unit circle $|z| = 1$.

Answers

1. (a) $6 + \frac{26}{3}i$ (b) $6 + \frac{26}{3}i$ (c) $6 + \frac{26}{3}i$

2. (a) $4 + 8i$ (b) $4 + \frac{25}{3}i$ **3.** $10 - \frac{8}{3}i$ **4.** $-1 + i$

6. $1 - i$ **7.** $\frac{2}{3}$ **9.** $2i$ **10.** 0.

4.17. SIMPLY AND MULTIPLY CONNECTED REGIONS

A curve is called *simple closed curve* if it does not cross itself (Fig. 1). A curve which crosses itself is called a *multiple curve* (Fig. 2).

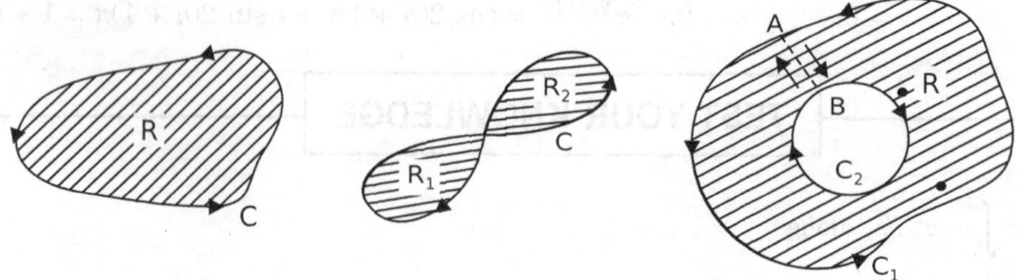

Fig.1 Fig. 2 Fig. 3

A region is called *simply connected* if every closed curve in the region encloses points of the region only, *i.e.*, every closed curve lying in it can be contracted indefinitely without passing out of it. A region which is not simply connected is called a *multiply connected* region. In plain terms, a simply connected region is one which has no holes. Figure 3 shows a multiply connected region R enclosed between two separate curves C_1 and C_2. (There can be more than two separate curves). We can convert a multiply connected region into a simply connected one, by giving it one or more cuts (*e.g.*, along the dotted line AB).

4.18. CAUCHY'S INTEGRAL THEOREM

Statement. *If f(z) is an analytic function and f ′(z) is continuous at each point within and on a simple closed curve C, then*

$$\oint_C f(z)\, dz = 0.$$

Proof. Let R be the region bounded by the curve C.

Let $f(z) = u(x, y) + iv(x, y)$, then

$$\oint_C f(z)\,dz = \oint_C (u + iv)(dx + idy)$$

$$= \oint_C (udx - vdy) + i \oint_C (vdx + udy) \qquad \ldots(1)$$

Since $f'(z)$ is continuous, the partial derivatives $\dfrac{\partial u}{\partial x}, \dfrac{\partial u}{\partial y}, \dfrac{\partial v}{\partial x}, \dfrac{\partial v}{\partial y}$ are also continuous in R. Hence, by Green's Theorem, we have

$$\oint_C f(z)\,dz = \iint_R \left(-\frac{\partial v}{\partial x} - \frac{\partial u}{\partial y}\right) dx\,dy + i \iint_R \left(\frac{\partial u}{\partial x} - \frac{\partial v}{\partial y}\right) dx\,dy \qquad \ldots(2)$$

Now $f(z)$ being analytic at each point of the region R, by Cauchy-Riemann equations, we have

$$\frac{\partial u}{\partial x} = \frac{\partial v}{\partial y} \quad \text{and} \quad \frac{\partial u}{\partial y} = -\frac{\partial v}{\partial x}$$

Thus, the integrals of the two double integrals in (2) vanish.

Hence, $\oint_C f(z)\,dz = 0.$

Cor. 1. *If $f(z)$ is analytic in a region R and P and Q are two points in R, then $\int_P^Q f(z)\,dz$ is independent of the path joining P and Q and lying entirely in R.*

Let PAQ and PBQ be any two paths joining P and Q.

By Cauchy's theorem,

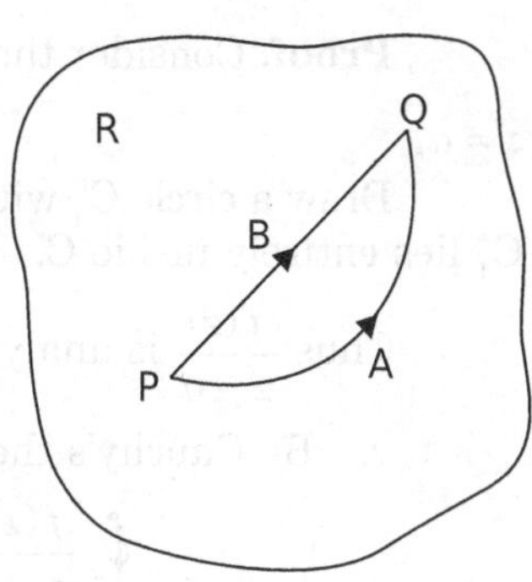

$$\int_{PAQBP} f(z)\,dz = 0$$

$$\Rightarrow \quad \int_{PAQ} f(z)\,dz + \int_{QBP} f(z)\,dz = 0$$

$$\Rightarrow \quad \int_{PAQ} f(z)\,dz - \int_{PBQ} f(z)\,dz = 0$$

Hence $$\int_{PAQ} f(z)\,dz = \int_{PBQ} f(z)\,dz.$$

Cor. 2. *If $f(z)$ is analytic in the region bounded by two simple closed curves C_1 and C_2, then $\oint_C f(z)\,dz = \oint_{C_1} f(z)\,dz$*

Let AB be a cross-cut joining the curves C and C_1, then the doubly connected region becomes simply connected.

By Cauchy's theorem, $\int_{APQABDEBA} f(z)\,dz = 0$

$$\Rightarrow \quad \int_{APQA} f(z)\,dz + \int_{AB} f(z)\,dz + \int_{BDEB} f(z)\,dz + \int_{BA} f(z)\,dz = 0$$

$$\Rightarrow \quad \oint_C f(z)\,dz + \int_{AB} f(z)\,dz - \int_{BEDB} f(z)\,dz - \int_{AB} f(z)\,dz = 0$$

$$\Rightarrow \quad \oint_C f(z)dz - \oint_{C_1} f(z)\,dz = 0$$

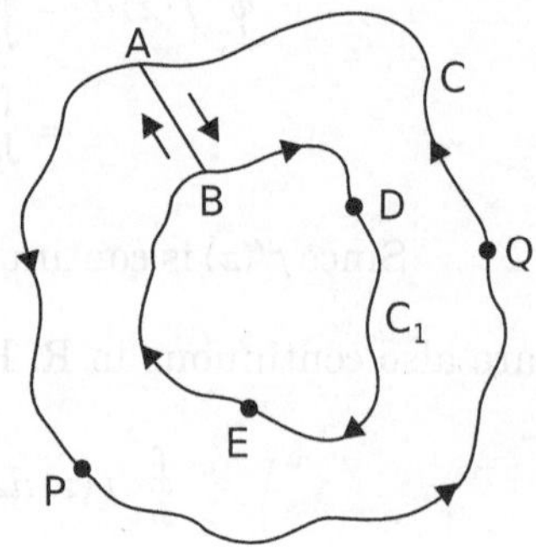

(Integrals around a closed curve are taken positive when the curve is traversed in counter-clockwise direction)

Hence, $\oint_C f(z)\,dz = \oint_{C_1} f(z)\,dz$

The theorem can be extended.

If a closed curve C contains non-intersecting closed curves C_1, C_2, ..., C_n, then by introducing cross-cuts, it can be shown that

$$\oint_C f(z)\,dz = \oint_{C_1} f(z)\,dz + \oint_{C_2} f(z)\,dz + ... + \oint_{C_n} f(z)\,dz.$$

4.19. CAUCHY'S INTEGRAL FORMULA

Statement. *If f(z) is analytic within and on a closed curve C and a is any point within C, then* $f(a) = \dfrac{1}{2\pi i}\oint_C \dfrac{f(z)}{z-a}\,dz.$

Proof. Consider the function $\dfrac{f(z)}{z-a}$, which is analytic at every point within C except at $z = a$.

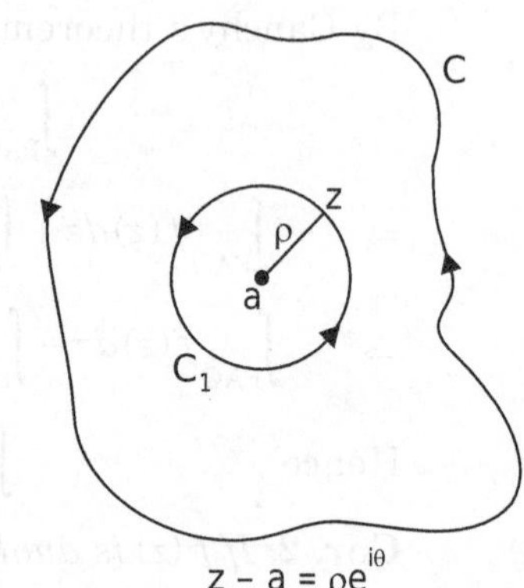

Draw a circle C_1 with a as centre and radius ρ such that C_1 lies entirely inside C.

Thus $\dfrac{f(z)}{z-a}$ is analytic in the region between C and C_1.

∴ By Cauchy's theorem, we have

$$\oint_C \frac{f(z)}{z-a}\,dz = \oint_{C_1} \frac{f(z)}{z-a}\,dz \qquad ...(1)$$

Now, the equation of circle C_1 is

$$|\,z-a\,| = \rho \quad \text{or} \quad z - a = \rho e^{i\theta}$$

so that $$dz = i\rho e^{i\theta}\,d\theta$$

$$\oint_{C_1} \frac{f(z)}{z-a}\,dz = \int_0^{2\pi} \frac{f(a+\rho e^{i\theta})}{\rho e^{i\theta}}\,.\,i\rho e^{i\theta}\,d\theta = i\int_0^{2\pi} f(a+\rho e^{i\theta})\,d\theta$$

Hence, by (1), we have

$$\oint_C \frac{f(z)}{z-a}\,dz = i\int_0^{2\pi} f(a+\rho e^{i\theta})\,d\theta \qquad ...(2)$$

In the limiting form, as the circle C_1 shrinks to the point a, *i.e.*, $\gamma \to 0$, then from (2),

$$\oint_C \frac{f(z)}{z-a}\,dz = i\int_0^{2\pi} f(a)\,d\theta = i\,f(a)\int_0^{2\pi} d\theta = 2\pi i\,f(a)$$

Hence, $$f(a) = \frac{1}{2\pi i}\oint_C \frac{f(z)}{z-a}\,dz$$

which is the required Cauchy's integral formula.

Cor. By Cauchy's integral formula, we have

$$f(a) = \frac{1}{2\pi i}\oint_C \frac{f(z)}{z-a} \qquad \text{...(1)}$$

where a is any point within C, it may be treated as a parameter.

Differentiating both sides of (1) w.r.t. a,

$$f'(a) = \frac{1}{2\pi i}\oint_C \frac{\partial}{\partial a}\left[\frac{f(z)}{z-a}\right] dz = \frac{1}{2\pi i}\oint_C \frac{f(z)}{(z-a)^2}\,dz$$

Similarly, $$f''(a) = \frac{2\,!}{2\pi i}\oint_C \frac{f(z)}{(z-a)^3}\,dz$$

$$f'''(a) = \frac{3\,!}{2\pi i}\oint_C \frac{f(z)}{(z-a)^4}\,dz$$

and in general, $$f^n(a) = \frac{n\,!}{2\pi i}\oint_C \frac{f(z)}{(z-a)^{n+1}}\,dz.$$

Thus, *if a function of a complex variable has a first derivative in a simply connected region, all its higher derivatives exist in that region. This property is not exhibited by the functions of real variables.*

ILLUSTRATIVE EXAMPLES

Example 1. *Evaluate* $\oint_C (x^2 - y^2 + 2ixy)\,dz$, *where C is the contour* $|\,z\,| = 1$.

Sol. $f(z) = x^2 - y^2 + 2ixy = (x + iy)^2 = z^2$ is analytic everywhere within and on $|\,z\,| = 1$.

∴ By Cauchy's integral theorem,

$$\oint_C f(z)\,dz = 0.$$

Example 2. *Evaluate* $\oint_C (3z^2 + 4z + 1)\,dz$, *where C is the arc of the cycloid* $x = a(\theta - \sin\theta)$, $y = a(1 - \cos\theta)$ *between* $(0, 0)$ *and* $(2\pi a, 0)$.

Sol. Here, $f(z) = 3z^2 + 4z + 1$ is analytic everywhere so that the integral is independent of the path of integration and depends only on the end points $z_1 = 0 + i0$ and $z_2 = 2\pi a + i0$.

$$\therefore \quad \int_C (3z^2 + 4z + 1)\,dz = \int_0^{2\pi a} (3z^2 + 4z + 1)\,dz = \Big[z^3 + 2z^2 + z\Big]_0^{2\pi a}$$

$$= 8\pi^3 a^3 + 8\pi^2 a^2 + 2\pi a$$

$$= 2\pi a\,(4\pi^2 a^2 + 4\pi a + 1).$$

Example 3. *Evaluate* $\oint_C \frac{e^{-z}}{z+1} dz$, *where C is the circle*

(*a*) $|z| = 2$ (*b*) $|z| = \frac{1}{2}$.

Sol. $f(z) = e^{-z}$ is an analytic function.

(*a*) The point $a = -1$ lies inside the circle $|z| = 2$.

$\therefore$ By Cauchy's integral formula,

$$\frac{1}{2\pi i}\oint_C \frac{f(z)}{z-(-1)} dz = f(-1) \Rightarrow \oint_C \frac{e^{-z}}{z+1} dz = 2\pi i e.$$

(*b*) The point $a = -1$ lies outside the circle $|z| = \frac{1}{2}$.

$\therefore$ The function $\frac{e^{-z}}{z+1}$ is analytic within and on C.

By Cauchy's integral theorem, we have $\oint_C \frac{e^{-z}}{z+1} dz = 0$.

Example 4. *Evaluate* $\oint_C \frac{\cos \pi z^2}{(z-1)(z-2)} dz$, *where C is the circle* $|z| = 3$.

Sol. The integrand has singularities where

$$(z-1)(z-2) = 0 \quad i.e., \quad \text{at } z = 1 \quad \text{and} \quad z = 2.$$

Both these points lie within the circle $|z| = 3$

$$f(z) = \cos \pi z^2 \text{ is an analytic function.}$$

Also $$\frac{1}{(z-1)(z-2)} = -\frac{1}{z-1} + \frac{1}{z-2}$$

$$\therefore \quad \oint_C \frac{\cos \pi z^2}{(z-1)(z-2)} dz = -\oint_C \frac{\cos \pi z^2}{z-1} dz + \oint_C \frac{\cos \pi z^2}{z-2} dz \qquad ...(1)$$

$$= -2\pi i\, f(1) + 2\pi i\, f(2) \quad \text{where} \quad f(z) = \cos \pi z^2$$

$$= -2\pi i \cos \pi + 2\pi i \cos 4\pi = 2\pi i + 2\pi i = 4\pi i.$$

Example 5. *Evaluate* $\oint_C \frac{3z^2 + z}{z^2 - 1} dz$, *where C is the circle* $|z - 1| = 1$.

Sol. The integrand has singularities where $z^2 - 1 = 0$ *i.e.,* at $z = 1$ and $z = -1$.

The circle $|z - 1| = 1$ has centre at $z = 1$ and radius 1 and includes the point $z = 1$.

$f(z) = 3z^2 + z$ is an analytic function.

Also $$\frac{1}{z^2-1} = \frac{1}{(z-1)(z+1)} = \frac{1}{2}\left(\frac{1}{z-1} - \frac{1}{z+1}\right)$$

$$\therefore \quad \oint_C \frac{3z^2+z}{z^2-1} dz = \frac{1}{2}\oint_C \frac{3z^2+z}{z-1} dz - \frac{1}{2}\oint_C \frac{3z^2+z}{z+1} dz \qquad ...(1)$$

By Cauchy's integral formula,

$$\oint_C \frac{3z^2+z}{z-1}dz = 2\pi i\, f(1), \quad \text{where } f(z) = 3z^2 + z$$

$$= 8\pi i$$

By Cauchy's integral theorem,

$$\oint_C \frac{3z^2+z}{z+1}dz = 0$$

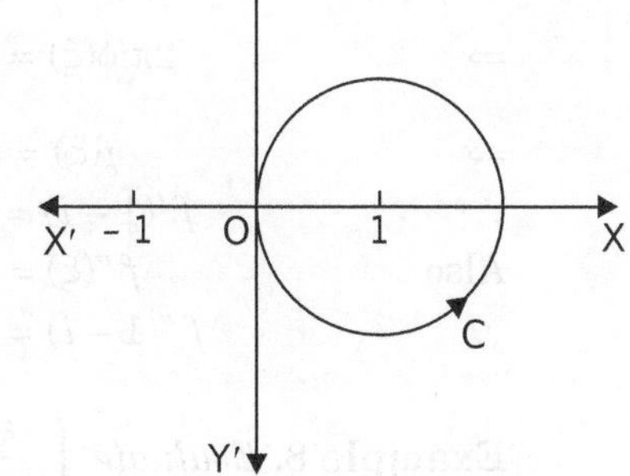

$\therefore$ From (1), we have

$$\oint_C \frac{3z^2+z}{z^2-1}dz = 4\pi i.$$

Example 6. *Use Cauchy's integral formula to evaluate*

$$\oint_C \frac{e^{2z}}{(z+1)^4}dz, \text{ where C is the circle } |\ z\ | = 2.$$

Sol. The integrand has a singularity at $z = -1$ which lies within the circle $|\ z\ | = 2$.

Now $\qquad f^n(a) = \dfrac{n!}{2\pi i}\displaystyle\oint_C \frac{f(z)}{(z-a)^{n+1}}dz$ ...(1)

Here $\quad a = -1, \quad n + 1 = 4 \quad i.e., \quad n = 3, f(z) = e^{2z}$

$\therefore \qquad f^n(z) = 2^n e^{2z}, f^3(a) = f^3(-1) = 2^3 e^{-2}$

From (1), we have $f^3(-1) = \dfrac{3!}{2\pi i}\displaystyle\oint_C \frac{e^{2z}}{(z+1)^4}dz \Rightarrow 8e^{-2} = \dfrac{3}{\pi i}\displaystyle\oint_C \frac{e^{2z}}{(z+1)^4}dz$

Hence $\qquad \displaystyle\oint_C \frac{e^{2z}}{(z+1)^4}dz = \frac{8\pi i}{3e^2}.$

Example 7. *If* $f(\xi) = \displaystyle\int_C \frac{3z^2+7z+1}{z-\xi}dz$, *where C is the circle* $x^2 + y^2 = 4$, *find the values of* $f(3)$, $f'(1-i)$ *and* $f''(1-i)$.

Sol. The given circle C is $x^2 + y^2 = 4$ or $|\ z\ | = 2$.

The point $z = 3$ lies outside the circle $|\ z\ | = 2$ while $z = 1 - i$ and $z = 1 + i$ lie inside the circle.

Now $f(3) = \displaystyle\oint_C \frac{3z^2+7z+1}{z-3}dz$ and $\dfrac{3z^2+7z+1}{z-3}$ is analytic within and on C.

$\therefore$ By Cauchy's integral theorem, we have

$$\oint_C \frac{3z^2+7z+1}{z-3}dz = 0 \Rightarrow f(3) = 0.$$

Now let $\phi(z) = 3z^2 + 7z + 1$ which is analytic everywhere.

$\therefore$ By Cauchy's integral formula, we have

$$\phi(\xi) = \frac{1}{2\pi i}\oint_C \frac{\phi(z)}{z-\xi}\,dz, \quad \text{where } \xi \text{ is a point within C.}$$

$$\Rightarrow \quad 2\pi i\phi(\xi) = \oint_C \frac{3z^2+7z+1}{z-\xi}\,dz = f(\xi)$$

$$\Rightarrow \quad f(\xi) = 2\pi i(3\xi^2+7\xi+1) \Rightarrow f'(\xi) = 2\pi i(6\xi+7)$$

$$\therefore \quad f'(1-i) = 2\pi i[6(1-i)+7] = 2\pi i(13-6i) = 2\pi(6+13i)$$

Also $\quad f''(\xi) = 12\pi i$

$$\therefore \quad f''(1-i) = 12\pi i.$$

Example 8. *Evaluate* $\int_C \frac{z^2-z+1}{z-1}\,dz$*, where C is the circle*

(i) $|z| = 1,$ *(ii)* $|z| = \frac{1}{2}$

Sol. *(i)* Let $f(z) = \frac{z^2-z+1}{z-1}$. Here $a = 1$.

Since $f(z)$ is analytic within and on circle C : $|z| = 1$ lies on C.

$\therefore$ By Cauchy's integral formula

$$f(a) = \frac{1}{2\pi i}\int_C \frac{f(z)}{z-a}\,dz$$

or $$1 = \frac{1}{2\pi i}\int_C \frac{z^2-z+1}{z+1}\,dz \quad \text{or} \quad \int_C \frac{z^2-z+1}{z-1}\,dz = 2\pi i.$$

(ii) In this case, $a = 1$ lies outside the circle $|z| = \frac{1}{2}$

Hence $f(z) = \frac{z^2-z+1}{z-1}$ is analytic everywhere within C.

$\therefore$ By Cauchy's theorem, we have

$$\int_C \frac{z^2-z+1}{z-1}\,dz = 0.$$

Example 9. *If C is a closed contour containing the origin inside it, prove that*

$$\frac{a^n}{n!} = \frac{1}{2\pi i}\int_C \frac{e^{az}\,dz}{z^{n+1}}.$$

Sol. By Cauchy's integral formula:

$$f^{(n)}(z_0) = \frac{n!}{2\pi i}\int_C \frac{f(z)}{(z-z_0)^{n+1}}\,dz$$

$$\therefore \quad f^{(n)}(0) = \frac{n!}{2\pi i}\int_C \frac{f(z)}{z^{n+1}}\,dz.$$

Take $f(z) = e^{az}$ so that $f^{(n)}(z_0) = a^n e^{az}$, we obtain

$$f^{(n)}(0) = a^n = \frac{n!}{2\pi i}\int_C \frac{e^{az}\,dz}{z^{n+1}}$$

or
$$\frac{a^2}{n!} = \frac{1}{2\pi i}\int_C \frac{e^{az}}{z^{n+1}}\,dz.$$

Example 10. *Using Cauchy's integral formula, evaluate* $\frac{e^{2z}}{(z-1)(z-2)}\,dz$, *where C is the circle* $|z| = 3$.

Sol. Let $f(z) = e^{2z}$. Clearly, $f(z)$ is analytic within the circle $C : |z| = 3$ and the two singular points $a = 1$ and $a = 2$ lie inside C.

∴ By Cauchy's integral formula, we have

$$\int_C \frac{e^{2z}}{(z-1)(z-2)}\,dz = \int_C e^{2z}\left(\frac{1}{z-2} - \frac{1}{z-1}\right)dz = \int_C \frac{e^{2z}}{z-2}\,dz - \int_C \frac{e^{2z}}{z-1}\,dz$$

$$= 2\pi i e^4 - 2\pi i e^2 \qquad \left[\because f(a) = \frac{1}{2\pi i}\int_C \frac{f(z)}{z-a}\,dz\right]$$

$$= 2\pi i\,(e^4 - e^2).$$

Example 11. *Using Cauchy's integral formula, prove that*

$$\int_C \frac{e^{2z}}{(z+1)^4}\,dz = \frac{8\pi e^{-2}}{3}i,$$

where C is the Circle $|z| = 2$.

Sol. Let $f(z) = e^{2z}$. Clearly, $f(z)$ is analytic inside the circle $C : |z| = 2$. Also $z = -1$ lies inside C.

∴ By Cauchy's integral formula

$$f^{(n)}(a) = \frac{n!}{2\pi i}\int_C \frac{f(z)}{(z-a)^{n+1}}\,dz. \qquad ...(1)$$

Put $a = -1$, $n = 3$ in (1), we obtain

$$f'''(-1) = \frac{3!}{2\pi i}\int_C \frac{f(z)}{(z+1)^4}\,dz.$$

Take $f(z) = e^{2z}$, so that $f^{(n)}(z) = 2^n e^{2z}$

$$f'''(-1) = 2^3 e^{-2} = 8e^{-2}.$$

Hence
$$8e^{-2} = \frac{3!}{2\pi i}\int_C \frac{e^{2z}}{(z+1)^4}\,dz \quad \text{or} \quad \int_C \frac{e^{2z}}{(z+1)^4}\,dz = \frac{8\pi . e^{-2}}{3}i.$$

Example 12. *Use Cauchy's integral formula to calculate* $\int_C \frac{3z^2 + 7z + 1}{z+1}\,dz$, *where C is* $|z| = \frac{1}{2}$.

Sol. In this case, $a = -1$ lies outside the circle $|z| = \frac{1}{2}$. Hence $f(z) = \frac{3z^2 + 7z + 1}{z + 1}$ is analytic everywhere within the circle C: $|z| = \frac{1}{2}$.

$\therefore$ By Cauchy's theorem, we have

$$\int_C f(z)\, dz = 0 \quad i.e., \quad \int_C \frac{3z^2 + 7z + 1}{z + 1}\, dz = 0.$$

Example 13. *Use Cauchy's integral formula to calculate:*

$$\int_C \frac{\cos \pi z}{z^2 - 1}\, dz \text{ around a rectangle with vertices } 2 \pm i,\ -2 \pm i.$$

Sol. Let $f(z) = \cos \pi z$. Clearly, $f(z)$ is analytic in the region bounded by the given rectangle and the two singular points $a = 1$ and $a = -1$ lie inside the given rectangle.

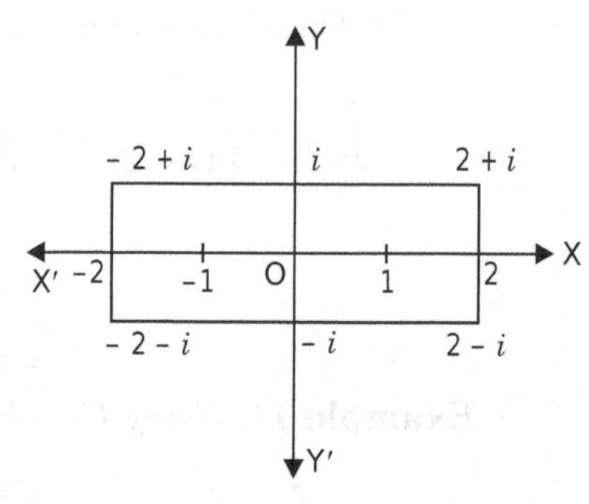

$$\therefore \int_C \frac{\cos \pi z}{z^2 - 1}\, dz = \int_C \frac{\cos \pi z}{(z-1)(z+1)}\, dz$$

$$= \frac{1}{2}\int_C \left(\frac{1}{z-1} - \frac{1}{z+1}\right) \cos \pi z\, dz$$

$$= \frac{1}{2}\int_C \frac{\cos \pi z}{z-1}\, dz - \frac{1}{2}\int_C \frac{\cos \pi z}{z+1}\, dz$$

$$= \frac{1}{2}[2\pi i \cos \pi(1)] - \frac{1}{2}[2\pi i \cos \pi(-1)]$$

$$\left[\text{By Cauchy's integral formula: } f(a) = \frac{1}{2\pi i}\int_C \frac{f(z)}{z-a}\, dz\right]$$

$$= 0.$$

Example 14. *Evaluate by Cauchy's integral formula* $\int_C \frac{dz}{z(z - \pi i)}$ *where C is* $|z + 3i| = 1$.

Sol. We have

$$\int_C \frac{dz}{z(z+\pi i)} = \frac{1}{\pi i}\left[\int_C \frac{dz}{z} - \int_C \frac{dz}{z + \pi i}\right].$$

Also observe that points $z = 0$, $z = -\pi i$ are inside C. Take $f(z) = 1$ and so $f(0) = 1$, $f(-\pi i) = 1$.

Now, by Cauchy's integral formula, we have

$$\int_C \frac{dz}{z} = 2\pi i f(0) = 2\pi i \quad \text{and} \quad \int_C \frac{dz}{z + \pi i} = 2\pi i\, f(-\pi i) = 2\pi i$$

$$\therefore \quad \int_C \frac{dz}{z(z+\pi i)} = \frac{2\pi i - 2\pi i}{\pi i} = 0.$$

Example 15. *Use Cauchy's integral formula to calculate:*

$$\int_C \frac{z^2+1}{z(2z+1)}\,dz, \text{ where C is } |z| = 1.$$

Sol. Let $f(z) = z^2 + 1$. Clearly, $f(z)$ is analytic in the region bounded by the circle C: $|z| = 1$ and the two singular points $a = 0$ and $a = -\frac{1}{2}$ lie inside C.

$$\therefore \quad \int_C \frac{z^2+1}{z(2z+1)}\,dz = \int_C \left(\frac{1}{z} - \frac{2}{2z+1}\right)(z^2+1)\,dz = \int_C \frac{z^2+1}{z}\,dz - 2\int_C \frac{z^2+1}{2z+1}\,dz$$

$$= \int_C \frac{z^2+1}{z}\,dz - \int_C \frac{z^2+1}{z+(1/2)}\,dz$$

$$= 2\pi i\, f(0) - 2\pi i\, f\left(-\frac{1}{2}\right)$$

$$= 2\pi i - 2\pi i \times \frac{5}{4} = -\frac{3\pi i}{2}.$$

Example 16. *Use Cauchy's integral formula to evaluate:*

$$\int_C \frac{\sin^2 z}{(z-\pi/6)^3}\,dz, \text{ where C is the circle } |z| = 1.$$

Sol. Let $f(z) = \sin^2 z$. Clearly, $f(z)$ is analytic inside the circle C : $|z| = 1$ and the point $a = \pi/6$ (= 0.5 approx.) lies within C.

$\therefore$ By Cauchy's integral formula $f''(a) = \dfrac{2!}{2\pi i}\displaystyle\int_C \frac{f(z)}{(z-a)^3}\,dz$, we get

$$\int_C \frac{\sin^2 z}{(z-\pi/6)^3}\,dz = \pi i \left[\frac{d^2}{dz^2}(\sin^2 z)\right]_{z=\pi/6}$$

$$= \pi i\,[2\cos 2z]_{z=\pi/6} = 2\pi i \cos\frac{\pi}{3} = \pi i.$$

Example 17. *Use Cauchy's integral formula to calculate:*

$$\int_C \frac{e^z}{(z^2+\pi^2)^2}\,dz, \text{ where C is } |z| = 4.$$

Sol. Clearly, $\dfrac{e^z}{(z^2+\pi^2)} = \dfrac{e^z}{(z+\pi i)^2(z-\pi i)^2}$ is not analytic at $z = \pm \pi i$.

However both $z = \pm \pi i$ lie within the circle $|z| = 4$.

Now splitting into partial fractions, we obtain

$$\frac{1}{(z+\pi i)^2(z-\pi i)^2} = \frac{A}{z+\pi i} + \frac{B}{(z+\pi i)^2} + \frac{C}{z-\pi i} + \frac{D}{(z-\pi i)^2}$$

where $\quad A = \dfrac{7}{2\pi^3 i}, C = -\dfrac{7}{2\pi^3 i}, B = D = -\dfrac{1}{4\pi^2}.$

$$\therefore \quad \int_C \frac{e^z}{(z^2+\pi^2)^2}\,dz = \frac{7}{2\pi^3 i}\left\{\int_C \frac{e^z}{z+\pi i}\,dz - \int_C \frac{e^z}{z-\pi i}\,dz\right\}$$

$$-\frac{1}{4\pi^2}\left\{\int_C \frac{e^z}{(z+\pi i)^2}\,dz + \int \frac{e^z}{(z-\pi i)^2}\,dz\right\}$$

$$= \frac{7}{2\pi^3 i}\,[2\pi i\, f(-\pi i) - 2\pi i\, f(\pi i)] - \frac{1}{4\pi^2}\,[2\pi i\, f'(-\pi i) = 2\pi i\, f'(\pi i)]$$

[By Cauchy's integral formula where $f(z) = e^z$]

$$= \frac{7}{\pi^2}[e^{-\pi i} - e^{\pi i}] - \frac{i}{2\pi}[e^{-\pi i} + e^{\pi i}]$$

$$= \frac{14i}{\pi^2}\left(\frac{e^{\pi i} - e^{\pi i}}{2i}\right) - \frac{i}{\pi}\left(\frac{e^{\pi i} + e^{-\pi i}}{2}\right)$$

$$= -\frac{14i}{\pi^2}\sin\pi - \frac{i}{\pi}\cos\pi = \frac{i}{\pi}.$$

Example 18. *Evaluate, using Cauchy's integral formulae:*

(*i*) $\displaystyle\int_C \frac{e^z}{(z+1)^2}\,dz$, *where C is* $|z-1| = 3$.

(*ii*) $\displaystyle\int_C \frac{\log z}{(z-1)^3}\,dz$, *where C is* $|z| = \frac{1}{2}$.

Sol. (*i*) The centre of the circle C: $|z-1| = 3$ is $z = 1$ and its radius is 3. Clearly, $\dfrac{e^z}{(z+1)^2}$ is not analytic at $z = -1$. However $z = -1$ lies within the circle C.

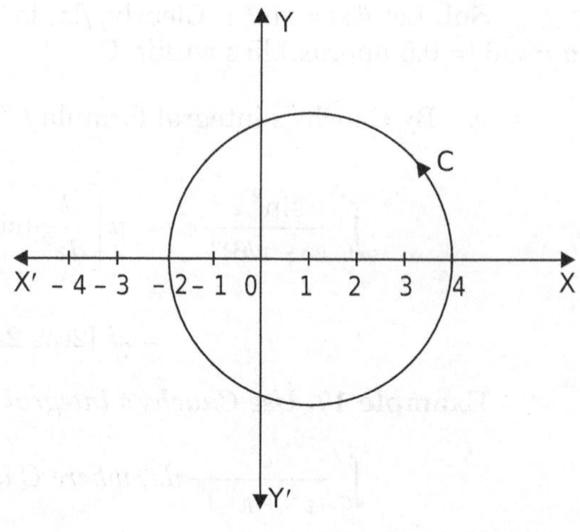

$\therefore$ By Cauchy's integral formula

$$f'(-1) = \frac{1}{2\pi i}\int_C \frac{e^z}{(z+1)^2}\,dz, \text{ where } f(z) = e^z$$

i.e.,
$$e^{-1} = \frac{1}{2\pi i}\int_C \frac{e^z}{(z+1)^2}\,dz$$

or
$$\int_C \frac{e^z}{(z+1)^2}\,dz = 2\pi i e^{-1}.$$

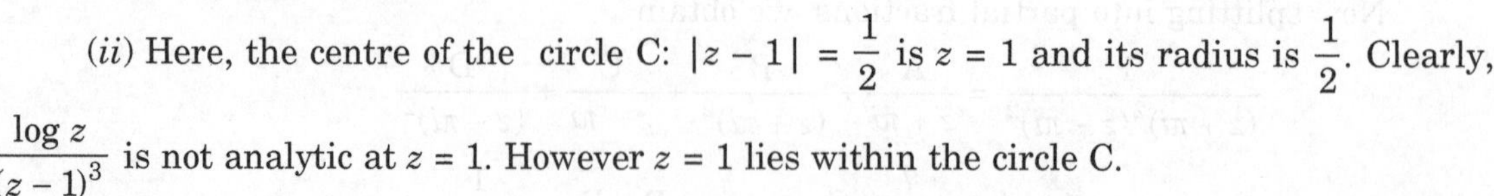

(*ii*) Here, the centre of the circle C: $|z-1| = \frac{1}{2}$ is $z = 1$ and its radius is $\frac{1}{2}$. Clearly, $\dfrac{\log z}{(z-1)^3}$ is not analytic at $z = 1$. However $z = 1$ lies within the circle C.

∴ By Cauchy's integral formula, we have

$$f''(1) = \frac{2!}{2\pi i} \int_C \frac{\log z}{(z-1)^3}\, dz, \text{ where } f(z) = \log z$$

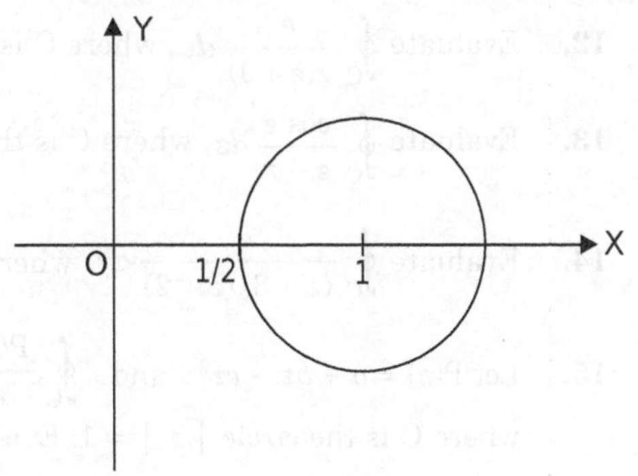

i.e., $$\left[\frac{d^2}{dz^2} \log z\right]_{z=1} = \frac{1}{\pi i} \int_C \frac{\log z}{(z-1)^3}\, dz$$

i.e., $$\left[-\frac{1}{z^2}\right]_{z=1} = \frac{1}{\pi i} \int_C \frac{\log z}{(z-1)^3}\, dz$$

or $$\int_C \frac{\log z}{(z-1)^3}\, dz = -\pi i.$$

TEST YOUR KNOWLEDGE

1. Evaluate $\oint_C \frac{z^2+5}{z-3}\, dz$, where C is the circle

(*a*) $| z | = 4$ (*b*) $| z | = 1$.

2. Evaluate $\int_C \frac{e^z}{z-2}\, dz$, where C is the circle

(*a*) $| z | = 3$ (*b*) $| z | = 1$.

3. Evaluate $\oint_C \frac{3z^2+7z+1}{z+1}\, dz$, where C is the circle

(*a*) $| z | = 1.5$ (*b*) $| z + i | = 1$.

4. Evaluate $\oint_C \frac{\sin 3z}{z+\frac{\pi}{2}}\, dz$, where C is the circle $| z | = 5$.

5. Evaluate $\int_C \frac{z^3+z+1}{z^2-3z+2}\, dz$, where C is the ellipse $4x^2 + 9y^2 = 1$.

6. Evaluate $\oint_C \frac{\cos \pi z}{z^2-1}\, dz$ around a rectangle with vertices

(*a*) $2 \pm i, -2 \pm i$ (*b*) $-i, 2-i, 2+i, i$.

7. Evaluate $\oint_C \frac{\sin^2 z}{\left(z-\frac{\pi}{6}\right)^3}\, dz$, where C is the circle $| z | = 1$.

8. Evaluate $\oint_C \frac{z\, dz}{(z-1)(z-3)}$, where C is the circle

(*a*) $| z | = 3$ (*b*) $| z | = 3/2$.

9. Evaluate $\oint_C \frac{z+4}{z^2+2z+5}\, dz$, where C is the circle $| z | = 1$.

10. Evaluate $\oint_C \frac{\sin \pi z^2 + \cos \pi z^2}{(z-1)(z-2)}\, dz$, where C is the circle $| z | = 3$.

11. Evaluate $\oint_C \frac{e^z}{(z-1)(z-4)}\, dz$, where C is the circle $| z | = 2$.

12. Evaluate $\oint_C \frac{e^z}{z(z+1)} dz$, where C is the circle $|z| = \frac{1}{4}$.

13. Evaluate $\oint_C \frac{\cos z}{z - \pi} dz$, where C is the circle $|z - 1| = 3$.

14. Evaluate $\oint_C \frac{e^z}{(z+3)(z+2)} dz$, where C is the circle $|z - 1| = \frac{1}{2}$.

15. Let $P(z) = a + bz + cz^2$ and $\oint_C \frac{P(z)}{z} dz = \oint_C \frac{P(z)}{z^2} dz = \oint_C \frac{P(z)}{z^3} dz = 2\pi i$ where C is the circle $|z| = 1$. Evaluate P(z).

16. Evaluate $\oint_C \frac{e^{-z}}{z^2} dz$, where C is the circle $|z| = 1$.

17. Evaluate $\oint_C \frac{e^{-2z}\, dz}{(z+1)^3}$, where C is the circle $|z| = 2$.

18. If $f(\xi) = \oint_C \frac{4z^2 + z + 5}{z - \xi} dz$, where C is the ellipse $\frac{x^2}{4} + \frac{y^2}{9} = 1$, find $f(1), f(i), f'(-1)$ and $f''(-i)$.

Answers

1. (*a*) $28\pi i$ (*b*) 0 **2.** (*a*) $2\pi i e^2$ (*b*) 0 **3.** (*a*) $-6\pi i$ (*b*) 0 **4.** $2\pi i$

5. 0 **6.** (*a*) 0 (*b*) $-\pi i$ **7.** πi **8.** (*a*) $2\pi i$ (*b*) $-\pi i$

9. 0 **10.** $4\pi i$ **11.** $-\frac{2}{3}\pi i e$ **12.** $2\pi i$

13. $-2\pi i$ **14.** 0 **15.** $1 + z + z^2$ **16.** $-2\pi i$

17. $4\pi i e^2$ **18.** $20\pi i, 2\pi(i - 1), -14\pi i, 16\pi i$.

4.20. SERIES OF COMPLEX TERMS

A series of the form $(a_1 + ib_1) + (a_2 + ib_2) + \ldots + (a_n + ib_n) + \ldots$ where $a_1, b_1, a_2, b_2, \ldots\ldots$ are real numbers, is called a series of complex terms and can be expressed as

$$\sum_{n=1}^{\infty} a_n + i \sum_{n=1}^{\infty} b_n \qquad \ldots(1)$$

The series (1) is convergent if Σa_n and Σb_n both are convergent.

If the series Σa_n and Σb_n converge to a and b respectively, then the series (1) is said to converge to $a + ib$.

The necessary (not sufficient) condition for convergence of a series (1) is $\underset{n \to \infty}{\text{Lt}} (a_n + ib_n) = 0$.

The series (1) is said to be **absolutely convergent** if the series $|a_1 + ib_1| + |a_2 + ib_2| + \ldots\ldots + |a_n + ib_n| + \ldots\ldots$ is convergent.

Since $|a_n| \le |a_n + ib_n|$ and $|b_n| \le |a_n + ib_n|$, we conclude that an absolutely convergent series is convergent. The converse is not true.

If the series of functions $u_1(z) + u_2(z) + ... + u_n(z) + ... = \sum_{n=1}^{\infty} u_n(z)$ ...(2)

converges to S(z) and $S_n(z)$ is its nth partial sum, then the series (2) is said to be **uniformly convergent** in a region R, if given any positive number ε, there exists a positive number N, depending on ε but not on z, such that for every z in R,

$$| S(z) - S_n(z) | < \varepsilon \text{ for all } n > N.$$

A series of the form

$$a_0 + a_1(z-a) + a_2(z-a)^2 + ... + a_n(z-a)^n + ... = \sum_{n=0}^{\infty} a_n (z-a)^n \quad ...(3)$$

is called a **power series** in $(z - a)$.

If the series (3) is convergent for all values of z satisfying $| z - a | < R$, *i.e.*, for all values of z lying inside the circle with centre at $z = a$ and radius R, then this circle is called the **circle of convergence** and R is called the **radius of convergence.**

The power series (3) is uniformly convergent in a region R if there exists a series of positive terms ΣM_n such that $| a_n(z-a)^n | \le M_n$ for all values of z in the region R.

This is called **Weierstrass M-test** for uniform convergence of a power series.

4.21. TAYLOR'S SERIES

If f(z) is analytic inside a circle C with centre at a, then for all z inside C,

$$f(z) = f(a) + (z-a) f'(a) + \frac{(z-a)^2}{2!} f''(a) + ... + \frac{(z-a)^n}{n!} f^n(a) + ...$$

Proof. Let z be any point inside the circle C. Draw a circle C_1 with centre at a and radius smaller than that of C such that z is an interior point of C_1.

Let w be any point on C_1, then

$$| z - a | < | z - w | \quad i.e., \left| \frac{z-a}{z-w} \right| < 1$$

C
w
z
C_1
a

Now, $$\frac{1}{w-z} = \frac{1}{(w-a)-(z-a)} = \frac{1}{w-a}\left[1 - \frac{z-a}{w-a}\right]^{-1}$$

Expanding R.H.S. by binomial theorem as $\left| \frac{z-a}{w-a} \right| < 1$, we get

$$\frac{1}{w-z} = \frac{1}{w-a}\left[1 + \frac{z-a}{w-a} + \left(\frac{z-a}{w-a}\right)^2 + ... + \left(\frac{z-a}{w-a}\right)^n + ...\right] \quad ...(1)$$

This series converges uniformly since $\left| \frac{z-a}{w-a} \right| < 1$. Multiplying both sides of (1) by $\frac{1}{2\pi i} f(w)$ and integrating term by term w.r.t. w, over C_1, we get

$$\frac{1}{2\pi i}\oint_{C_1}\frac{f(w)}{w-z}\,dw = \frac{1}{2\pi i}\oint_{C_1}\frac{f(w)}{w-a}\,dw + \frac{z-a}{2\pi i}\oint_{C_1}\frac{f(w)}{(w-a)^2}\,dw + \frac{(z-a)^2}{2\pi i}\oint_{C_1}\frac{f(w)}{(w-a)^3}\,dw + \ldots\ldots + \frac{(z-a)^n}{2\pi i}\oint_{C_1}\frac{f(w)}{(w-a)^{n+1}}\,dw + \ldots \quad \ldots(2)$$

Since $f(w)$ is analytic on and inside C_1, by Cauchy's formula,

$$\frac{1}{2\pi i}\oint_{C_1}\frac{f(w)}{w-z}\,dw = f(z)$$

Also $\quad f^n(a) = \frac{n!}{2\pi i}\oint_{C_1}\frac{f(w)}{(w-a)^{n+1}}\,dw \quad$ *i.e.*, $\quad \frac{1}{2\pi i}\oint_{C_1}\frac{f(w)}{(w-a)^{n+1}}\,dw = \frac{f^n(a)}{n!}$

$\therefore$ From (2), we have

$$f(z) = f(a) + (z-a)f'(a) + \frac{(z-a)^2}{2!}f''(a) + \ldots + \frac{(z-a)^n}{n!}f^n(a) + \ldots \quad \ldots(3)$$

which is the required *Taylor's series* for $f(z)$ about $z = a$.

Cor. 1. Putting $z = a + h$ in (3), we get

$$f(a+h) = f(a) + hf'(a) + \frac{h^2}{2!}f''(a) + \ldots\ldots + \frac{h^n}{n!}f^n(a) + \ldots$$

Cor. 2. If $a = 0$, the series (3) becomes

$$f(z) = f(0) + zf'(0) + \frac{z^2}{2!}f''(0) + \ldots\ldots + \frac{z^n}{n!}f^n(0) + \ldots$$

This series is called **Maclaurin's series.**

4.22. LAURENT'S SERIES

If $f(z)$ is analytic inside and on the boundary of the ring shaped region R bounded by two concentric circles C_1 and C_2 of radii r_1 and r_2 ($r_1 > r_2$) respectively having centre at a, then for all z in R,

$$f(z) = a_0 + a_1(z-a) + a_2(z-a)^2 + \ldots + a_{-1}(z-a)^{-1} + a_{-2}(z-a)^{-2} + \ldots$$

where $\quad a_n = \frac{1}{2\pi i}\oint_{C_1}\frac{f(w)}{(w-a)^{n+1}}\,dw;\ n = 0, 1, 2, \ldots$

and $\quad a_{-n} = \frac{1}{2\pi i}\oint_{C_2}\frac{f(w)}{(w-a)^{-n+1}}\,dw;\ n = 1, 2, 3, \ldots$

Proof. Let z be any point in the region R, then by Cauchy's integral formula for double connected region, we have

$$f(z) = \frac{1}{2\pi i}\oint_{C_1}\frac{f(w)}{w-z}\,dw - \frac{1}{2\pi i}\oint_{C_2}\frac{f(w)}{w-z}\,dw \quad \ldots(1)$$

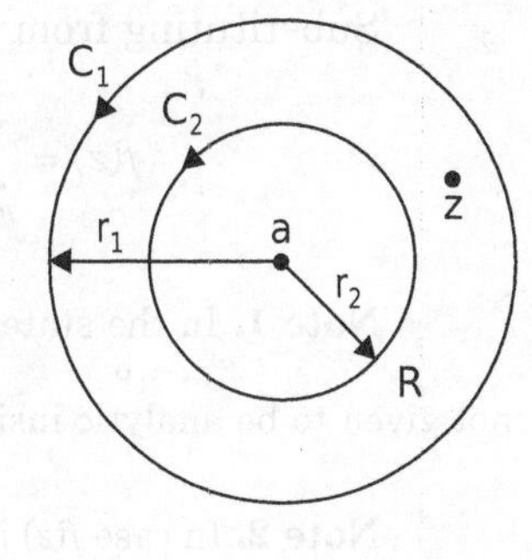

For the first integral in (1), w lies on C_2

$$\therefore \quad |z-a| < |w-a| \quad i.e., \left|\frac{z-a}{w-a}\right| < 1$$

Now
$$\frac{1}{w-a} = \frac{1}{(w-a)-(z-a)} = \frac{1}{w-a}\left(1-\frac{z-a}{w-a}\right)^{-1}$$

$$= \frac{1}{w-a}\left[1+\frac{z-a}{w-a}+\left(\frac{z-a}{w-a}\right)^2+\ldots\ldots\right]$$

Multiplying both sides by $\frac{1}{2\pi i} f(w)$ and integrating term by term w.r.t. w, along the circle C_1, we get

$$\frac{1}{2\pi i}\oint_{C_1}\frac{f(w)}{w-z}\,dw = \frac{1}{2\pi i}\oint_{C_1}\frac{f(w)}{w-a}\,dw + \frac{z-a}{2\pi i}\oint_{C_1}\frac{f(w)}{(w-a)^2}\,dw + \frac{(z-a)^2}{2\pi i}\oint_{C_1}\frac{f(w)}{(w-a)^3}\,dw + \ldots\ldots$$

$$= a_0 + a_1(z-a) + a_2(z-a)^2 + \ldots\ldots$$

$$= \sum_{n=0}^{\infty} a_n (z-a)^n \quad \ldots(2) \quad \left[\because\ a_n = \frac{1}{2\pi i}\oint_{C_1}\frac{f(w)}{(w-a)^{n+1}}\,dw, n = 0, 1, 2, \ldots\ldots\right]$$

For the second integral in (1), w lies on C_2

$$\therefore \quad |w-a| < |z-a| \quad i.e., \left|\frac{w-a}{z-a}\right| < 1$$

Now
$$\frac{1}{w-z} = \frac{1}{(w-a)-(z-a)} = -\frac{1}{z-a}\left(1-\frac{w-a}{z-a}\right)^{-1}$$

$$= -\frac{1}{z-a}\left[1+\frac{w-a}{z-a}+\left(\frac{w-a}{z-a}\right)^2+\ldots\ldots\right]$$

Multiplying both sides by $\frac{1}{2\pi i} f(w)$ and integrating term by term w.r.t. w, along the circle C_2, we get

$$-\frac{1}{2\pi i}\oint_{C_2}\frac{f(w)}{w-z}\,dw = \frac{1}{z-a}\cdot\frac{1}{2\pi i}\oint_{C_2} f(w)\,dw + \frac{1}{(z-a)^2}\cdot\frac{1}{2\pi i}\oint_{C_2}(w-a)f(w)\,dw + \frac{1}{(z-a)^3}\cdot\frac{1}{2\pi i}\oint_{C_2}(w-a)^2 f(w)\,dw + \ldots\ldots$$

$$= a_{-1}(z-a)^{-1} + a_{-2}(z-a)^{-2} + a_{-3}(z-a)^{-3} + \ldots\ldots$$

$$= \sum_{n=1}^{\infty} a_{-n}(z-a)^{-n} \quad \ldots(3)$$

$$\left[\because\ a_{-n} = \frac{1}{2\pi i}\oint_{C_2}\frac{f(w)}{(w-a)^{-n+1}}\,dw, n = 1, 2, 3, \ldots\ldots\right]$$

Substituting from (2) and (3) in (1), we get

$$f(z) = \sum_{n=0}^{\infty} a_n (z-a)^n + \sum_{n=1}^{\infty} a_{-n}(z-a)^{-n}$$

Note 1. In the statement of Laurent's series, $a_n = \dfrac{1}{2\pi i}\displaystyle\oint_{C_1} \dfrac{f(w)}{(w-a)^{n+1}}\,dw \neq \dfrac{f^n(a)}{n!}$ because $f(z)$ is not given to be analytic inside C_1.

Note 2. In case $f(z)$ is analytic inside C_1, then $a_{-n} = 0$ and $a_n = \dfrac{1}{2\pi i}\displaystyle\oint_{C_1} \dfrac{f(w)}{(w-a)^{n+1}}\,dw = \dfrac{f^n(a)}{n!}$ and Laurent's series reduces to Taylor's series.

Note 3. If C is any simple closed curve which lies in the ring-shaped region R and encloses the circle C_1, then

$$\oint_{C_1} \frac{f(w)}{(w-a)^{n+1}}\,dw = \oint_{C} \frac{f(w)}{(w-a)^{n+1}}\,dw$$

and $$\oint_{C_2} \frac{f(w)}{(w-a)^{-n+1}}\,dw = \oint_{C} \frac{f(w)}{(w-a)^{-n+1}}\,dw$$

$\therefore$ Laurent's series can be written as

$$f(z) = \sum_{n=-\infty}^{\infty} a_n (z-a)^n, \quad \text{where } a_n = \frac{1}{2\pi i}\oint_{C} \frac{f(w)}{(w-a)^{n+1}}\,dw.$$

Note 4. The process of finding the co-efficient a_n by complex integration is complicated. In practice, we expand the function $f(z)$ by binomial theorem or by some other method to obtain Taylor's or Laurent's series.

ILLUSTRATIVE EXAMPLES

Example 1. *Expand* $\dfrac{1}{z^2 - 3z + 2}$ *in the region*

(*a*) $|z| < 1$ (*b*) $1 < |z| < 2$

(*c*) $|z| > 2$ (*d*) $0 < |z-1| < 1$.

Sol. Here $f(z) = \dfrac{1}{z^2 - 3z + 2} = \dfrac{1}{(z-1)(z-2)} = \dfrac{1}{z-2} - \dfrac{1}{z-1}$ (Partial Fractions)

(*a*) If $|z| < 1$, then $\left|\dfrac{z}{2}\right| < 1$

$$\therefore \quad f(z) = \frac{1}{-2\left(1-\dfrac{z}{2}\right)} + \frac{1}{1-z}$$

[Arranged suitably to make the binomial expansions valid]

$$= -\frac{1}{2}\left(1-\frac{z}{2}\right)^{-1} + (1-z)^{-1}$$

$$= -\frac{1}{2}\left(1+\frac{z}{2}+\frac{z^2}{4}+\frac{z^3}{8}+......\right)+(1+z+z^2+z^3+......)$$

(By binomial theorem)

$$= \frac{1}{2}+\frac{3z}{4}+\frac{7}{8}z^2+\frac{15}{16}z^3+......$$

(*b*) If $1 < |z| < 2$, then $\left|\frac{1}{z}\right| < 1$ and $\left|\frac{z}{2}\right| < 1$

$$\therefore \quad f(z) = \frac{1}{-2\left(1-\frac{z}{2}\right)}-\frac{1}{z\left(1-\frac{1}{z}\right)} = -\frac{1}{2}\left(1-\frac{z}{2}\right)^{-1}-\frac{1}{z}\left(1-\frac{1}{z}\right)^{-1}$$

$$= -\frac{1}{2}\left(1+\frac{z}{2}+\frac{z^2}{4}+\frac{z^3}{8}+......\right)-\frac{1}{z}\left(1+\frac{1}{z}+\frac{1}{z^2}+\frac{1}{z^3}+......\right)$$

(By binomial theorem)

$$= -\frac{1}{2}\left(1+\frac{z}{2}+\frac{z^2}{4}+......\right)-(z^{-1}+z^{-2}+z^{-3}+......)$$

(*c*) If $|z| > 2$, then $\left|\frac{2}{z}\right| < 1$ and $\left|\frac{1}{z}\right| < 1$

$$\therefore \quad f(z) = \frac{1}{z\left(1-\frac{2}{z}\right)}-\frac{1}{z\left(1-\frac{1}{z}\right)} = \frac{1}{z}\left(1-\frac{2}{z}\right)^{-1}-\frac{1}{z}\left(1-\frac{1}{z}\right)^{-1}$$

$$= \frac{1}{z}\left(1+\frac{2}{z}+\frac{4}{z^2}+\frac{8}{z^3}+......\right)-\frac{1}{z}\left(1+\frac{1}{z}+\frac{1}{z^2}+\frac{1}{z^3}+......\right)$$

(By binomial theorem)

$$= z^{-2}+3z^{-3}+7z^{-4}+......$$

(*d*) For $0 < |z-1| < 1$, we have

$$f(z) = \frac{1}{(z-1)-1}-\frac{1}{z-1} = -\frac{1}{1-(z-1)}-\frac{1}{z-1} = -(z-1)^{-1}-[1-(z-1)]^{-1}$$

$$= -(z-1)^{-1}-[1+(z-1)+(z-1)^2+(z-1)^3+......].$$

Example 2. *Show that when* $|z+1| < 1$, $z^{-2} = 1+\sum_{n=1}^{\infty}(n+1)(z+1)^n$.

Sol. $$f(z) = z^{-2} = \frac{1}{z^2} = \frac{1}{[(z+1)-1]^2} = \frac{1}{[1-(z+1)]^2} = [1-(z+1)]^{-2}$$

$$= 1+2(z+1)+3(z+1)^2+4(z+1)^3+......$$

[By binomial theorem, since $|z+1| < 1$]

$$= 1+\sum_{n=1}^{\infty}(n+1)(z+1)^n.$$

Example 3. *Expand cos z in a Taylor's series about* $z = \frac{\pi}{4}$.

Sol. Here $f(z) = \cos z, f'(z) = -\sin z, f''(z) = -\cos z, f'''(z) = \sin z$, ...

$$\therefore \quad f\left(\frac{\pi}{4}\right)=\frac{1}{\sqrt{2}},\ f'\left(\frac{\pi}{4}\right)=-\frac{1}{\sqrt{2}},\ f''\left(\frac{\pi}{4}\right)=-\frac{1}{\sqrt{2}},\ f'''\left(\frac{\pi}{4}\right)=\frac{1}{\sqrt{2}},\ ......$$

Hence $\cos z = f(z)$

$$= f\left(\frac{\pi}{4}\right)+\left(z-\frac{\pi}{4}\right)f'\left(\frac{\pi}{4}\right)+\frac{\left(z-\frac{\pi}{4}\right)^2}{2!}f''\left(\frac{\pi}{4}\right)+\frac{\left(z-\frac{\pi}{4}\right)^3}{3!}f'''\left(\frac{\pi}{4}\right)+ ......$$

$$= \frac{1}{\sqrt{2}}\left[1-\left(z-\frac{\pi}{4}\right)-\frac{1}{2!}\left(z-\frac{\pi}{4}\right)^2+\frac{1}{3!}\left(z-\frac{\pi}{4}\right)^3+ ......\right].$$

Example 4. *Expand the function* $\frac{\sin z}{z-\pi}$ *about* $z = \pi$.

Sol. Putting $z - \pi = t$, we have

$$\frac{\sin z}{z-\pi}=\frac{\sin(\pi+t)}{t}=\frac{-\sin t}{t}$$

$$=-\frac{1}{t}\left(t-\frac{t^3}{3!}+\frac{t^5}{5!} ......\right)=-1+\frac{t^2}{3!}-\frac{t^4}{5!}+ ...... =-1+\frac{(z-\pi)^2}{3!}-\frac{(z-\pi)^4}{5!}+ ......$$

Example 5. *Expand* $f(z) = \frac{z}{(z+1)(z+2)}$ *about* $z = -2$.

Sol. To expand $f(z)$ about $z = -2$, *i.e.,* in powers of $z + 2$, we put $z + 2 = t$.

$$\therefore \quad f(z)=\frac{z}{(z+1)(z+2)}=\frac{t-2}{(t-1)t}=\frac{2-t}{t(1-t)}=\frac{2-t}{t}(1-t)^{-1}$$

$$=\frac{2-t}{t}(1+t+t^2+t^3+ ......) \qquad \text{for } 0 < |t| < 1$$

$$=\frac{1}{t}(2+t+t^2+t^3+ ......)=\frac{2}{t}+1+t+t^2+ ....$$

$$=\frac{2}{z+2}+1+(z+2)+(z+2)^2+ ...... \qquad \text{for } 0 < |z+2| < 1$$

which is the required Laurent's series.

Example 6. *Expand in the series the function* $f(z) = \frac{1}{(z-1)(z-2)}$ *in the regions*

(*a*) $0 < |z| < 1$, (*b*) $1 < |z| < 2$,

(*c*) $|z| > 2$, (*d*) $0 < |z - 1| < 1$.

Sol. Resolving into partial fractions, we obtain

$$f(z)=\frac{1}{(z-1)(z-2)}=\frac{1}{z-2}-\frac{1}{z-1}. \qquad ...(1)$$

(*a*) For $0 < |z| < 1$, we have $\frac{|z|}{2} < 1$.

$$f(z) = (1-z)^{-1} - \frac{1}{2}\left(1-\frac{z}{2}\right)^{-1} = \sum_{n=0}^{\infty} z^n - \frac{1}{2}\sum_{n=0}^{\infty}\left(\frac{z}{2}\right)^n = \sum_{n=0}^{\infty}\left(1-\frac{1}{2^{n+1}}\right)z^n,$$

By Binomial expansion
which is a series of positive power of z, and hence it is a Taylor's expansion for $f(z)$ within the circle $|z| = 1$.

(*b*) For $1 < |z| < 2$, we write $f(z)$ in such a way that the binomial expansion is valid in the given range so that

$$\frac{1}{|z|} < 1, \frac{|z|}{2} < 1.$$

$$\therefore \quad f(z) = -\frac{1}{2}\left(1-\frac{z}{2}\right)^{-1} - \frac{1}{z}\left(1-\frac{1}{z}\right)^{-1} = -\frac{1}{2}\sum_{n=0}^{\infty}\left(\frac{z}{2}\right)^n - \frac{1}{z}\sum_{n=0}^{\infty}\left(\frac{1}{z}\right)^n$$

$$= -\frac{1}{2}\sum_{n=0}^{\infty}\left(\frac{z}{2}\right)^n - \frac{1}{z}\sum_{n=0}^{\infty} z^{-n},$$

which being a series of positive and negative powers of z is Laurent's expansion in the annulus $1 < |z| < 2$.

(*c*) For $|z| > 2$, we write $f(z)$ in such a way that the binomial expansion is valid for the given range so that

$$\frac{2}{|z|} < 1 \text{ and } \frac{1}{|z|} < \frac{1}{2} < 1.$$

$$\therefore \quad f(z) = \frac{1}{z}\left(1-\frac{2}{z}\right)^{-1} - \frac{1}{z}\left(1-\frac{1}{z}\right)^{-1} = \frac{1}{z}\sum_{n=1}^{\infty}\left(\frac{2}{z}\right)^n - \frac{1}{z}\sum_{n=1}^{\infty}\left(\frac{1}{z}\right)^n = \sum_{n=0}^{\infty}(2^n-1)\frac{1}{z^{n+1}},$$

which being a series in negative powers of z is Laurent's expansion in the annulus $2 < |z| < R$ where R is large.

(*d*) For $0 < |z-1| < 1$, we write (*i*) as

$$f(z) = \frac{1}{(z-1)-1} - \frac{1}{z-1} = -(z-1)^{-1} - [1-(z-1)]^{-1}$$

$$= -(z-1)^{-1} - [1 + (z-1) + (z-1)^2 + (z-1)^3 + \ldots].$$

Example 7. *Expand* $f(z) = \frac{1}{(z+1)(z+3)}$ *in a Laurent's series valid for the regions*

(*i*) $|z| < 1$, (*ii*) $1 < |z| < 3$, (*iii*) $|z| > 3$, (*iv*) $0 < |z+1| < 2$.

Sol. Resolving into partial fractions, we obtain

$$f(z) = \frac{1}{2(z+1)} - \frac{1}{2(z+3)}.$$

(*i*) For $|z| < 1$, we have $\frac{|z|}{3} < 1$.

$$\therefore \quad f(z) = \frac{1}{2}(1+z)^{-1} - \frac{1}{6}\left(2+\frac{z}{3}\right)^{-1}$$

$$= \frac{1}{2}[1 - z + z^2 - z^3 + \ldots] - \frac{1}{6}\left[1 - \frac{z}{3} + \left(\frac{z}{3}\right)^2 - \left(\frac{z}{3}\right)^3 + \ldots\right]$$

$$= \left(\frac{1}{2} - \frac{1}{6}\right) - \left(\frac{1}{2} - \frac{1}{18}\right)z + \left(\frac{1}{2} - \frac{1}{54}\right)z^2 - \left(\frac{1}{2} - \frac{1}{162}\right)z^3 + \ldots$$

$$= \frac{1}{3} - \frac{4}{9}z + \frac{13}{27}z^2 - \frac{40}{21}z^3 + \ldots$$

(*ii*) For $|z| > 1$, we have $\frac{1}{|z|} < 1$.

$$\therefore \qquad \frac{1}{2(z+1)} = \frac{1}{2z\left(1 + \frac{1}{z}\right)} = \frac{1}{2z}\left(1 + \frac{1}{z}\right)^{-1}$$

$$= \frac{1}{2z}\left(1 - \frac{1}{z} + \frac{1}{z^2} - \frac{1}{z^3} + \ldots\right) = \frac{1}{2z} - \frac{1}{2z^2} + \frac{1}{2z^3} - \frac{1}{2z^4} + \ldots$$

For $|z| < 3$, we have $\frac{|z|}{3} < 1$.

$$\therefore \qquad \frac{1}{2(z+3)} = \frac{1}{6\left(1 + \frac{z}{3}\right)} = \frac{1}{6}\left(1 + \frac{z}{3}\right)^{-1}$$

$$= \frac{1}{6}\left[1 - \frac{z}{3} + \left(\frac{z}{3}\right)^2 - \left(\frac{z}{3}\right)^3 + \left(\frac{z}{3}\right)^4 - \ldots\right] = \frac{1}{6} - \frac{z}{18} + \frac{z^2}{54} - \frac{z^3}{162} + \frac{z^4}{1536} - \ldots$$

Hence, for $1 < |z| < 3$, the Laurent's series for $f(z)$ is given by

$$f(z) = \ldots - \frac{1}{2z^4} + \frac{1}{2z^3} - \frac{1}{2z^2} + \frac{1}{2z} - \frac{1}{6} + \frac{z}{18} - \frac{z^2}{54} + \frac{z^3}{162} - \frac{z^4}{1536} + \ldots.$$

(*iii*) For $|z| > 3$, we have $\frac{1}{|z|} < 1$ and $\frac{3}{|z|} < 1$

$$\therefore \quad f(z) = \frac{1}{2z}\left(1 + \frac{1}{z}\right)^{-1} - \frac{1}{2z}\left(1 + \frac{3}{z}\right)^{-1}$$

$$= \frac{1}{2z}\left[1 - \frac{1}{z} + \left(\frac{1}{z}\right)^2 - \left(\frac{1}{z}\right)^3 + \left(\frac{1}{z}\right)^4 - \ldots\right] - \frac{1}{2z}\left[1 - \frac{3}{z} + \left(\frac{3}{z}\right)^2 - \left(\frac{3}{z}\right)^3 + \left(\frac{3}{z}\right)^4 - \ldots\right]$$

$$= \frac{1}{z^2} - \frac{4}{z^3} + \frac{40}{z^5} + \ldots.$$

(*iv*) Let $z + 1 = u$. Then $0 < |z + 1| < 2$ implies $0 < |u| < 2$ so that $\frac{|u|}{2} < 1$.

$$\therefore \qquad f(z) = \frac{1}{(z+1)(z+3)} = \frac{1}{u(u+2)} = \frac{1}{2u}\left(1 + \frac{u}{2}\right)^{-1}$$

$$= \frac{1}{2u}\left[1 - \frac{u}{2} + \left(\frac{u}{2}\right)^2 - \left(\frac{u}{2}\right)^3 + \dots\right] = \frac{1}{2z} - \frac{1}{4} + \frac{u}{8} - \frac{u^2}{16} + \dots$$

$$= \frac{1}{2(z+1)} - \frac{1}{4} + \frac{z+1}{8} - \frac{(z+1)^2}{16} + \dots.$$

Example 8. *Find the laurent's expansion of* $f(z) = \dfrac{7z-2}{(z+1)z(z-2)}$ *in the region* $1 < z + 1 < 3$.

Sol. The given function is $f(z) = \dfrac{7z-2}{(z+1)z(z-2)}$

Writing $z + 1 = u$, we have

$$f(z) = \frac{7(u-1)-2}{u(u-1)(u-1-2)} = \frac{7u-9}{u(u-1)(u-3)}$$

$$= -\frac{3}{u} + \frac{1}{u-1} + \frac{2}{u-3} \qquad \text{[Splitting in to partial fraction]}$$

$$= -\frac{3}{u} + \frac{1}{u(1-1/u)} - \frac{2}{3(1-u/3)}$$

$$= -\frac{3}{u} + \frac{1}{u}\left(1 - \frac{1}{u}\right)^{-1} - \frac{2}{3}\left(1 - \frac{u}{3}\right)^{-1}.$$

Since $1 < u < 3$ or $\dfrac{1}{u} < 1$ and $\dfrac{u}{3} < 1$, expanding by binomial theorem, we have

$$f(z) = -\frac{3}{u} + \frac{1}{u}\left(1 + \frac{1}{u} + \frac{1}{u^2} + \frac{1}{u^3} + \dots \infty\right) - \frac{2}{3}\left(1 + \frac{u}{3} + \frac{u^2}{3^2} + \frac{u^3}{3^3} + \dots \infty\right)$$

$$= \left(-\frac{2}{u} + \frac{1}{u^2} + \frac{1}{u^3} + \dots \infty\right) - \frac{2}{3}\left(1 + \frac{u}{3} + \frac{u^2}{3^2} + \frac{u^3}{3^2} + \dots \infty\right).$$

Hence $$f(z) = \left[-\frac{2}{z+1} + \frac{1}{(z+1)^2} + \frac{1}{(z+1)^3} + \dots \infty\right]$$

$$- \frac{2}{3}\left[1 + \frac{z+1}{3} + \frac{(z+1)^2}{3^2} + \frac{(z+1)^3}{3^3} + \dots \infty\right]$$

which is valid in the region $1 < z + 1 < 3$.

Example 9. *Find the Taylor's and Laurent's series expansion of the function* $\dfrac{z^2-1}{(z+2)(z+3)}$ *about* $z = 0$ *in the regions*

(*i*) $|z| < 2$, (*ii*) $2 < |z| < 3$, (*iii*) $|z| > 3$.

Sol. We have

$$f(z) = \frac{z^2-1}{(z+2)(z+3)} = 1 - \frac{5z+7}{(z+2)(z+3)} = 1 + \frac{3}{z+2} - \frac{8}{z+3}$$

(*i*) For $|z| < 2$, we have

$$f(z) = 1+\frac{3}{2}\left(1+\frac{z}{2}\right)^{-1} - \frac{8}{3}\left(1+\frac{z}{3}\right)^{-1}$$

$$= 1+\frac{3}{2}\left[1-\frac{z}{2}+\frac{z^2}{2^2}-\frac{z^3}{2^3}+\ldots\right]-\frac{8}{3}\left[1-\frac{z}{3}+\frac{z^2}{3^2}-\frac{z^3}{3^3}+\ldots\right],$$

Binomial expansion valid for $|z| < 2$

$$= 1+\frac{3}{2}\sum_{n=0}^{\infty}(-1)^n\frac{z^n}{2^n}-\frac{8}{3}\sum_{n=0}^{\infty}(-1)^n\frac{z^n}{3^n}$$

$$= 1+\sum_{n=0}^{\infty}\left[(-1)^n.\left\{\frac{3}{2^{n+1}}-\frac{8}{3^{n+1}}\right\}z^n\right]$$

which is a 'Taylor's series valid for $|z| < 2$.

(*ii*) For $2 < |z| < 3$, we have $\frac{2}{|z|} < 1$ and $\frac{|z|}{3} < 1$.

$$\therefore \qquad f(z) = 1+\frac{3}{2}\left(1+\frac{2}{z}\right)^{-1} - \frac{8}{3}\left(1+\frac{z}{3}\right)^{-1}$$

$$= 1+\frac{3}{2}\left[1-\frac{2}{z}+\frac{2^2}{z^2}-\frac{2^3}{z^3}+\ldots\right]-\frac{8}{3}\left[1-\frac{z}{3}+\frac{z^2}{3^2}-\frac{z^3}{3^3}+\ldots\right]$$

$$= 1+\frac{3}{2}\sum_{n=0}^{\infty}(-1)^n.\frac{2^n}{z^n}-\frac{8}{3}\sum_{n=0}^{\infty}(-1)^n.\frac{z^n}{3^n}.$$

(*iii*) For $|z| > 3$, we have $\frac{2}{|z|} < 1$ and $\frac{3}{|z|} < 1$.

$$\therefore \qquad f(z) = 1+\frac{3}{z}\left(1+\frac{2}{z}\right)^{-1} - \frac{8}{z}\left(1+\frac{3}{z}\right)^{-1}$$

$$= 1+\frac{3}{z}\left[1-\frac{2}{z}+\frac{2^2}{z^2}-\frac{2^3}{z^3}+\ldots\right]-\frac{8}{z}\left[1-\frac{3}{z}+\frac{3^2}{z^2}-\frac{3^3}{z^3}+\ldots\right]$$

$$= 1+\frac{3}{z}\sum_{n=0}^{\infty}(-1)^n\frac{2^n}{z^n}-\frac{8}{z}\sum_{n=0}^{\infty}(-1)^n\frac{3^n}{z^n}.$$

Example 10. *Obtain the expression for* $\frac{z^2-4}{(z+1)(z+4)}$ *Which are vaild when*

(*i*) $|z| < 1$. (*ii*) $1 < |z| < 4$. (*iii*) $|z| > 4$.

Sol. Let $f(z) = \frac{z^2-4}{z^2+5z+4} = 1-\frac{5z-1}{(z+4)(z+1)} = 1-\frac{1}{1+z}-\frac{4}{4+z}.$

(*i*) For $|z| < 1$, we have

$$f(z) = 1 - (1+z)^{-1} - \left(1+\frac{z}{4}\right)^{-1}$$

$$= 1 - [1 - z + z^2 - \ldots + (-1)^n z^n + \ldots] - \left[1 - \frac{z}{4} + \left(\frac{z}{4}\right)^2 - \ldots + (-1)^n \left(\frac{z}{4}\right)^n + \ldots\right]$$

$$= -1 + [z - z^2 + \ldots + (-1)^{n+1} z^n + \ldots] + \left[\left(\frac{z}{4}\right) - \left(\frac{z}{4}\right)^2 + \ldots + (-1)^{n+1}\left(\frac{z}{4}\right)^n + \ldots\right]$$

$$= -1 + \sum_{n=0}^{\infty} (-1)^{n+1}[1 + 4^{-n}]\, z^n,$$

which is the Maclaurin's series of $f(z)$ valid in $|z| < 1$.

(*ii*) For $1 < |z| < 4$, we have $\frac{1}{|z|} < 1, \frac{|z|}{4} < 1.$

Now $f(z)$ is expressible as

$$f(z) = 1 - \frac{1}{z}\left(1+\frac{1}{z}\right)^{-1} - \left(1+\frac{z}{4}\right)^{-1}$$

$$= 1 - \frac{1}{z}\left[1 - \frac{1}{z} + \frac{1}{z^2} - \ldots\right] - \left[1 - \frac{z}{4} + \left(\frac{z}{4}\right)^2 - \ldots\right]$$

$$= \left[-\frac{1}{z} + \frac{1}{z^2} - \frac{1}{z^3} + \ldots\right] - \left[-\frac{z}{4} + \left(\frac{z}{4}\right)^2 - \left(\frac{z}{4}\right)^3 + \ldots\right]$$

$$= \sum_{n=1}^{\infty} (-1)^n \left[\frac{1}{z^n} - \left(\frac{z}{4}\right)^n\right],$$

which being a series of positive and negative powers of z is the Laurent's expansion of $f(z)$ valid in $1 < |z| < 4$.

(*iii*) For $|z| > 4$, we have $\frac{4}{|z|} < 1.$

Now $f(z)$ is expressible as

$$f(z) = 1 - \frac{1}{z}\left(1+\frac{1}{z}\right)^{-1} - \frac{4}{z}\left(1+\frac{4}{z}\right)^{-1}$$

$$= 1 - \frac{1}{z}\left[1 - \frac{1}{z} + \frac{1}{z^2} - \ldots\right] - \frac{4}{z}\left[1 - \frac{4}{z} + \left(\frac{4}{z}\right)^2 - \ldots\right]$$

$$= 1 - \frac{1}{z}\sum_{n=0}^{\infty} (-1)^n \frac{1}{z^n} - \frac{4}{z}\sum_{n=0}^{\infty} (-1)^n \left(\frac{4}{z}\right)^n$$

$$= 1 - \sum_{n=0}^{\infty} (-1)^n \left[\frac{1}{z^{n+1}} + \left(\frac{4}{z}\right)^{n+1} \right]$$

$$= 1 + \sum_{n=0}^{\infty} \frac{(-1)^{n+1}}{z^{n+1}} (1 + 4^{n+1}) = 1 + \sum_{n=1}^{\infty} (-1)^n (1 + 4^n) . \frac{1}{z^n}$$

TEST YOUR KNOWLEDGE

Expand the following functions as a Taylor's series:

1. $\log (1 + z)$ about $z = 0$. **2.** $\sin z$ about $z = \frac{\pi}{4}$. **3.** $\frac{z}{(z+1)(z+2)}$ about $z = 2$.

Expand the following functions in Laurent's series:

4. $\frac{1}{z-2}$, for $|z| > 2$. **5.** $\frac{1}{z^2 - 4z + 3}$, for $1 < |z| < 3$. **6.** $\frac{1}{z(z-1)(z-2)}$, for $|z| > 2$.

7. $\frac{z^2 - 1}{(z+2)(z+3)}$, for $|z| > 3$. **8.** $\frac{1 - \cos z}{z^3}$, about $z = 0$. **9.** $\frac{e^z}{(z-1)^2}$, about $z = 1$.

10. Expand $f(z) = \frac{(z-2)(z+2)}{(z+1)(z+4)}$ in the region:

(*a*) $|z| < 1$ (*b*) $1 < |z| < 4$ (*c*) $|z| > 4$.

Answers

1. $z - \frac{z^2}{2} + \frac{z^3}{3} - \frac{z^4}{4} + \ldots\ldots$ **2.** $\frac{1}{\sqrt{2}} \left[1 + \left(z - \frac{\pi}{4}\right) - \frac{1}{2!}\left(z - \frac{\pi}{4}\right)^2 - \frac{1}{3!}\left(z - \frac{\pi}{4}\right)^3 + \ldots\ldots \right]$

3. $\left(\frac{1}{2} - \frac{1}{3}\right) - \left(\frac{1}{2^3} - \frac{1}{3^2}\right)(z-2) + \left(\frac{1}{2^5} - \frac{1}{3^3}\right)(z-2)^2 - \ldots\ldots$

4. $\frac{1}{z} + \frac{2}{z^2} + \frac{2^2}{z^3} + \frac{2^3}{z^4} + \ldots\ldots$ **5.** $-\frac{1}{6} - \frac{1}{6} \sum_{n=1}^{\infty} \frac{z^n}{3^n} - \frac{1}{2} \sum_{n=1}^{\infty} z^{-n}$

6. $z^{-3} + 3z^{-4} + 7z^{-5} + \ldots\ldots + (2^{n-1} - 1)z^{-(n+1)} + \ldots\ldots$

7. $1 + \frac{3}{z} \sum_{n=0}^{\infty} \left(-\frac{2}{z}\right)^n - \frac{8}{z} \sum_{n=0}^{\infty} \left(-\frac{3}{z}\right)^n$ **8.** $-\sum_{n=2}^{\infty} \frac{z^{2n-5}}{(2n-2)!}$

9. $e \left[(z-1)^{-2} + (z-1)^{-1} + \frac{1}{2!} + \frac{1}{3!}(z-1) + \frac{1}{4!}(z-1)^2 + \ldots\ldots \right]$

10. (*a*) $-1 + \sum_{n=1}^{\infty} (-1)^{n+1} (1 + 4^{-n}) z^n$ (*b*) $\sum_{n=0}^{\infty} \frac{(-1)^{n+1}}{4^n} z^n + \sum_{n=1}^{\infty} (-1)^n z^{-n}$

(*c*) $1 + \sum_{n=1}^{\infty} (-1)^n (1 + 4^n) z^{-n}$.

4.23. SINGULAR POINTS, RESIDUES

A point at which a function $f(z)$ ceases to be analytic is called a *singular point* or *singularity* of $f(z)$.

For example, $z = 2$ is a singular point of $f(z) = \dfrac{1}{z-2}$.

A singular point $z = a$ of a function $f(z)$ is called an *isolated singular point* if there exists a circle with centre a which contains no other singular point of $f(z)$.

For example, $z = -1, 1$ are two isolated singular points of the function $f(z) = \dfrac{z}{z^2-1}$.

The function $f(z) = \dfrac{1}{\sin(\pi z)}$ has an infinite number of isolated singular points at $z = \pm 1, \pm 2, ...$

When $z = a$ is an isolated singular point of $f(z)$, we can expand $f(z)$ in a Laurent's series about $z = a$.

Thus $$f(z) = \sum_{n=0}^{\infty} a_n (z-a)^n + \sum_{n=1}^{\infty} a_{-n}(z-a)^{-n}$$

$$= \sum_{n=0}^{\infty} a_n (z-a)^n + [a_{-1}(z-a)^{-1} + a_{-2}(z-a)^{-2} + ... + a_{-m}(z-a)^{-m} + ...] \qquad ...(1)$$

If the series with negative powers has a finite number of terms, say m, so that $a_{-m} \neq 0$; $a_{-m-1} = a_{-m-2} = ... = 0$, then $z = a$ is called a *pole of order m*. A pole of order one is called a *simple pole.*

If the series with negative powers does not terminate, *i.e.*, has an infinite number of terms, then $z = a$ is called an *essential singular point* of $f(z)$.

The co-efficient of $(z-a)^{-1}$ in the expansion of $f(z)$ about an isolated singularity $z = a$ is called the *residue* of $f(z)$ at $z = a$. Thus from equation (1), the residue of $f(z)$ at $z = a$ is a_{-1} and may be denoted by

$$a_{-1} = \text{Res }\{f(z), a\}$$

Since $$a_{-1} = \frac{1}{2\pi i}\oint_C f(z)\,dz$$

$$\therefore \quad \oint_C f(z)\,dz = 2\pi i a_{-1} = 2\pi i \text{ Res }\{f(z), a\}.$$

4.24. RESIDUE THEOREM

If f(z) is analytic at all points inside and on a simple closed curve C, except at a finite number of isolated singular points within C, then

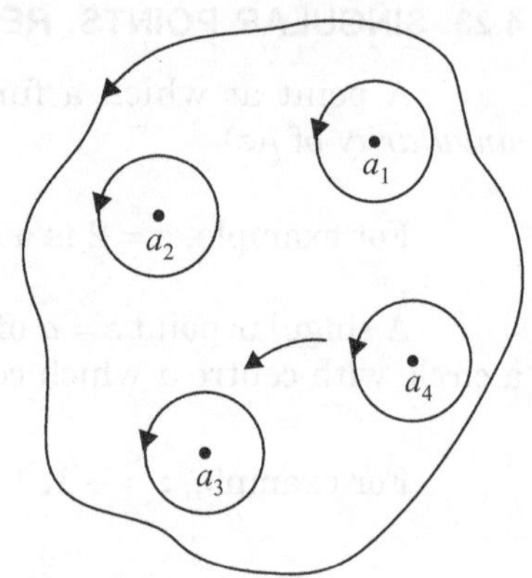

$\oint_C f(z)\,dz = 2\pi i$ *(sum of residues at singular points within C).*

Proof. Around each of the isolated singular points $a_1, a_2, ..., a_n$ draw small non-intersecting circles $C_1, C_2, ..., C_n$ lying wholly inside C, with centres at $z = a_1, a_2, ..., a_n$ respectively.

Since $f(z)$ is analytic in the multiply connected region bounded by $C, C_1, C_2, ..., C_n$, we have by Cauchy's theorem for multiply connected region.

$$\oint_C f(z)\,dz = \oint_{C_1} f(z)\,dz + \oint_{C_2} f(z)\,dz + ... + \oint_{C_n} f(z)\,dz$$

$$= 2\pi i\,[\text{Res}\,\{(z), a_1\} + \text{Res}\,\{f(z), a_2\} + ... + \text{Res}\,\{f(z), a_n\}]$$

$$= 2\pi i\,(\text{sum of residues at } a_1, a_2, ..., a_n).$$

4.25. CALCULATION OF RESIDUES

(1) *If $f(z)$ has a simple pole (i.e., pole of order 1) at $z = a$, then* $\text{Res}\,\{f(z), a\} = \underset{z \to a}{\text{Lt}}\,(z - a)\,f(z)$.

Since $z = a$ is a pole of order 1, the Laurent's series becomes

$$f(z) = a_0 + a_1(z - a) + a_2(z - a)^2 + ... + a_{-1}(z - a)^{-1}.$$

Multiplying both sides by $(z - a)$, we get

$$(z - a)\,f(z) = a_0(z - a) + a_1(z - a)^2 + a_2(z - a)^3 + ... + a_{-1}$$

$$\therefore \qquad \underset{z \to a}{\text{Lt}}\,(z - a)\,f(z) = a_{-1} = \text{Res}\,\{f(z), a\}$$

(2) *If $f(z)$ has a pole of order m at $z = a$, then*

$$\textbf{Res}\,\{\mathbf{f(z), a}\} = \frac{1}{(\mathbf{m - 1})!}\,\underset{\mathbf{z \to a}}{\textbf{Lt}}\,\frac{\mathbf{d^{m-1}}}{\mathbf{dz^{m-1}}}\,[\mathbf{(z - a)^m\,f(z)}]$$

Since $z = a$ is a pole of order m, the Laurent's series becomes

$$f(z) = a_0 + a_1(z - a) + a_2(z - a)^2 + ... + a_{-1}(z - a)^{-1} + a_{-2}(z - a)^{-2} + ... + a_{-m}(z - a)^{-m}$$

Multiplying both sides by $(z - a)^m$, we get

$$(z - a)^m f(z) = a_0(z - a)^m + a_1(z - a)^{m+1} + a_2(z - a)^{m+2} + ... + a_{-1}(z - a)^{m-1} + a_{-2}(z - a)^{m-2} + ... + a_{-m}$$

Differentiating both sides $(m - 1)$ times w.r.t. z and taking the limit as $z \to a$, we get

$$\underset{z \to a}{\text{Lt}}\,\frac{d^{m-1}}{dz^{m-1}}\,[(z - a)^m f(z)] = a_{-1} = (m - 1)\,!$$

or

$$\frac{1}{(m - 1)\,!}\,\underset{z \to a}{\text{Lt}}\,\frac{d^{m-1}}{dz^{m-1}}\,[(z - a)^m f(z)] = a_{-1} = \text{Res}\,\{f(z), a\}.$$

ILLUSTRATIVE EXAMPLES

Example 1. *Determine the poles of the function* $f(z) = \dfrac{z^2}{(z-1)^2(z+2)}$ *and the residue at each pole.*

Sol. The function $f(z)$ has a pole of order 2 at $z = 1$ and a simple pole at $z = -2$.

Residue of $f(z)$ at $z = 1$ is

$$\underset{z\to 1}{\text{Lt}} \frac{d}{dz}[(z-1)^2 f(z)] = \underset{z\to 1}{\text{Lt}} \frac{d}{dz}\left(\frac{z^2}{z+2}\right)$$

$$= \underset{z\to 1}{\text{Lt}} \frac{(z+2).2z - z^2.1}{(z+2)^2} = \underset{z\to 1}{\text{Lt}} \frac{z^2+4z}{(z+2)^2} = \frac{5}{9}$$

Residue of $f(z)$ at $z = -2$ is

$$= \underset{z\to -2}{\text{Lt}} [(z+2) f(z)] = \underset{z\to -2}{\text{Lt}} \frac{z^2}{(z-1)^2} = \frac{4}{9}.$$

Example 2. *Find the sum of the residues of the function* $f(z) = \dfrac{\sin z}{z\cos z}$ *at its poles inside the circle* $|z| = 2$.

Sol. The function $f(z)$ has simple poles at $z = 0, \pm\frac{\pi}{2}, \pm\frac{3\pi}{2}, \ldots\ldots$ of these, $z = 0, \pm\frac{\pi}{2}$ lie inside $|z| = 2$.

Residue of $f(z)$ at $z = 0$ is $\underset{z\to 0}{\text{Lt}} [z.f(z)] = \underset{z\to 0}{\text{Lt}} \dfrac{\sin z}{\cos z} = 0$

Residue of $f(z)$ at $z = \frac{\pi}{2}$ is

$$\underset{z\to\frac{\pi}{2}}{\text{Lt}} \left(z-\frac{\pi}{2}\right) f(z) = \underset{z\to\frac{\pi}{2}}{\text{Lt}} \frac{\left(z-\frac{\pi}{2}\right)\sin z}{z\cos z} \qquad \Big|\text{ Form } \frac{0}{0}$$

$$= \underset{z\to\frac{\pi}{2}}{\text{Lt}} \frac{\left(z-\frac{\pi}{2}\right)\cos z + \sin z}{\cos z - z\sin z} \qquad \text{[By L' Hospital's Rule]}$$

$$= \frac{1}{-\frac{\pi}{2}} = -\frac{2}{\pi}$$

Similarly, residue of $f(z)$ at $z = -\frac{\pi}{2}$ is $\frac{2}{\pi}$.

$\therefore$ Sum of residues $= 0 - \frac{2}{\pi} + \frac{2}{\pi} = 0.$

Example 3. *Evaluate* $\oint_C \frac{e^z}{(z+1)^2}\,dz$*, where C is the circle* $|z-3| = 3$.

Sol. Here $f(z) = \frac{e^z}{(z+1)^2}$ has only one singular point $z = -1$ which is a pole of order 2 and it lies inside the circle $|z-1| = 3$.

Residue of $f(z)$ at $z = -1$ is $\underset{z \to -1}{\text{Lt}} \frac{d}{dz}[(z+1)^2 f(z)] = \underset{z \to -1}{\text{Lt}} \frac{d}{dz}(e^z) = \underset{z \to -1}{\text{Lt}}\, e^z = e^{-1}$

$\therefore$ By residue theorem, we have $\oint_C \frac{e^z}{(z+1)^2}\,dz = 2\pi i(e^{-1}) = \frac{2\pi i}{e}$.

Example 4. *Evaluate* $\oint_C \frac{2z-1}{z(z+1)(z-3)}\,dz$*, where C is the circle* $|z| = 2$.

Sol. Here $f(z) = \frac{2z-1}{z(z+1)(z-3)}$ has three simple poles at $z = 0, -1, 3$ of which only $z = 0, -1$ lie inside the circle $|z| = 2$.

Residue of $f(z)$ at $z = 0$ is $\underset{z \to 0}{\text{Lt}}\, z f(z) = \underset{z \to 0}{\text{Lt}} \frac{2z-1}{(z+1)(z-3)} = \frac{1}{3}$

Residue of $f(z)$ at $z = -1$ is $\underset{z \to -1}{\text{Lt}}\,(z+1) f(z) = \underset{z \to -1}{\text{Lt}} \frac{2z-1}{z(z-3)} = -\frac{3}{4}$

$\therefore$ By residue theorem,

$$\oint_C \frac{2z-1}{z(z+1)(z-3)}\,dz = 2\pi i \text{ (sum of residues)}$$

$$= 2\pi i\left(\frac{1}{3} - \frac{3}{4}\right) = -\frac{5\pi i}{6}.$$

Example 5. *Evaluate* $\oint_C \frac{z-3}{z^2+2z+5}\,dz$*, where C is the circle*

(i) $|z| = 1$ *(ii)* $|z+1-i| = 2$ *(iii)* $|z+1+i| = 2$.

Sol. The poles of $f(z) = \frac{z-3}{z^2+2z+5}$ are given by

$$z^2 + 2z + 5 = 0 \quad i.e., \quad z = \frac{-2 \pm \sqrt{4-20}}{2} = -1 \pm 2i$$

(i) Both the poles lie outside the circle $|z| = 1$.

$\therefore$ By Cauchy's theorem, we have $\oint_C \frac{z-3}{z^2+2z+5}\,dz = 0$

(ii) Only the pole $z = -1 + 2i$ lies inside the circle $|z+1-i| = 2$.

Residue of $f(z)$ at $z = -1 + 2i$ is

$$\underset{z \to -1+2i}{\text{Lt}}\,(z+1-2i) f(z) = \underset{z \to \alpha}{\text{Lt}} \frac{(z-\alpha)(z-3)}{z^2+2z+5}, \text{ where } \alpha = -1+2i \qquad \Big|\ \text{Form } \frac{0}{0}$$

$$= \underset{z \to \alpha}{\text{Lt}} \frac{(z-\alpha)+(z-3)}{2z+2} \qquad |\ \text{By L' Hospital's Rule}$$

$$= \frac{\alpha - 3}{2\alpha + 2} = \frac{-1 + 2i - 3}{-2 + 4i + 2} = \frac{i - 2}{2i}$$

$\therefore$ By residue theorem, $\oint_C \frac{z - 3}{z^2 + 2z + 5} dz = 2\pi i \left(\frac{i - 2}{2i}\right) = \pi(i - 2).$

(*iii*) Only the pole $z = -1 - 2i$ lies inside the circle $|z + 1 + i| = 2$

Residue of $f(z)$ at $z = -1 - 2i$ is

$$\underset{z \to -1-2i}{\text{Lt}} (z + 1 + 2i) f(z) = \underset{z \to \beta}{\text{Lt}} \frac{(z - \beta)(z - 3)}{z^2 + 2z + 5}, \text{ where } \beta = -1 - 2i \qquad \Big| \text{ Form } \frac{0}{0}$$

$$= \underset{z \to \beta}{\text{Lt}} \frac{(z - \beta)(z - 3)}{2z + 2} \qquad \Big| \text{ By L' Hospital's Rule}$$

$$= \frac{\beta - 3}{2\beta + 2} = \frac{-4 - 2i}{-4i} = \frac{i + 2}{2i}$$

$\therefore$ By residue theorem, $\oint_C \frac{z - 3}{z^2 + 2z + 5} dz = 2\pi i \left(\frac{i + 2}{2i}\right) = \pi(i + 2).$

Example 6. *Evaluate* $\oint_C \frac{e^z}{\cos \pi z}$, *where C is the unit circle* $|z| = 1$.

Sol. Here $f(z) = \frac{e^z}{\cos \pi z}$ has simple poles at $z = \pm \frac{1}{2}, \pm \frac{3}{2}, \pm \frac{5}{2}, \ldots\ldots$, of which only $z = \pm \frac{1}{2}$ lie inside the circle $|z| = 1$.

Residue of $f(z)$ at $z = \frac{1}{2}$ is

$$\underset{z \to \frac{1}{2}}{\text{Lt}} (z - \tfrac{1}{2}) f(z) = \underset{z \to \frac{1}{2}}{\text{Lt}} \frac{(z - \frac{1}{2}) e^z}{\cos \pi z} \qquad \Big| \text{ Form } \frac{0}{0}$$

$$= \underset{z \to \frac{1}{2}}{\text{Lt}} \frac{(z - \frac{1}{2}) e^z + e^z}{-\pi \sin \pi z} \qquad \Big| \text{ By L' Hospital's Rule}$$

$$= \frac{e^{1/2}}{-\pi}.$$

Similarly, residue of $f(z)$ at $z = -\frac{1}{2}$ is $\frac{e^{-1/2}}{\pi}$.

$\therefore$ By residue theorem, $\oint_C \frac{e^z}{\cos \pi z} dz = 2\pi i$ (sum of residues)

$$= 2\pi i \left(-\frac{e^{1/2}}{\pi} + \frac{e^{-1/2}}{\pi}\right) = -4i \left(\frac{e^{1/2} - e^{-1/2}}{2}\right) = -4i \sinh \tfrac{1}{2}.$$

Example 7. *Find the residue of* $\frac{1}{(z^2+a^2)^3}$ *at* $z = ia$.

Sol. Let $$f(z) = \frac{1}{(z+ia)^3\,(z-ia)^3} = \frac{\phi(z)}{(z-ia)^3}$$

where $$\phi(z) = \frac{1}{(z+ia)^3}$$

Hene $z = ia$ is a pole of order three and

$$\text{Res}_{z = ia}\, f(z) = \frac{\phi''(ia)}{2!}$$

Now $\phi''(z) = \frac{12}{(z+ia)^5}$ so that $\phi''(ia) = \frac{12}{32\,ia^5}$

$$\therefore\quad \text{Res}_{z = ia}\quad f(z) = \frac{1}{2!}\cdot\frac{12}{32ia^5} = \frac{3}{16\,ia^5}$$

Example 8. *Find the residue of* $f(z) = \frac{e^z}{z\,\sin(mz)}$ *at the origin.*

Sol. Here both the factors z and sin (mz) of the denominator of $f(z)$ vanish at $z = 0$, and that $f(z)$ has a double pole at $z = 0$. Now expanding both the numerator and the denominator, we have

$$f(z) = \frac{1 + z + \frac{z^2}{2!} + \ldots}{mz^2\left(1 - \frac{m^2z^2}{3!} + \frac{m^4z^4}{5!} - \ldots\right)}$$

Hence for all sufficiently small z with $|z| > 0$, we have

$$f(z) = \frac{1}{mz^2}\left(1 + z + \frac{z^2}{2!} + \ldots\right)\left(1 + \frac{m^2z^2}{3!} - \ldots\right) = \frac{1}{m}\cdot\frac{1}{z^2} + \frac{1}{m}\cdot\frac{1}{z} + \ldots.$$

Thus, the residue of $f(z)$ at the pole (at the origin)

= coefficient of $\frac{1}{z}$ in the Laurent's expansion of $f(z)$

$= \frac{1}{m}$.

Example 9. (*i*) *Show that* $\int_0^{2\pi} \frac{d\theta}{a + b\cos\theta} = \int_0^{2\pi} \frac{d\theta}{a + b\sin\theta} = \frac{2\pi}{\sqrt{a^2 - b^2}}$, $a > b > 0$.

(*ii*) *Show that* $\int_0^{2\pi} \frac{d\theta}{1 + a\cos\theta} = \frac{2\pi}{\sqrt{1 - a^2}}$, $a^2 < 1$.

Sol. (*i*) Let C be the unit circle $|z| = 1$ and

$$I = \int_0^{2\pi} \frac{d\theta}{a + b\cos\theta},\ a > b > 0.$$

Putting $z = e^{i\theta}$, so that $\frac{dz}{iz} = d\theta$, we get

$$I = \frac{1}{i}\int_C \frac{dz}{z[a + \frac{1}{2}b(z + 1/z)]} = \frac{2}{i}\int_C \frac{dz}{bz^2 + 2az + b} = \frac{2}{i}\int_C f(z)\,dz.$$

Now the poles of $f(z)$ are given by the roots of $bz^2 + 2az + b = 0$

or $$z = \frac{-2a \pm \sqrt{4a^2 - 4b^2}}{2b} = \frac{-a \pm \sqrt{a^2 - b^2}}{b}$$

Let $$\alpha = \frac{-a + \sqrt{a^2 - b^2}}{b},\ \beta = \frac{-a - \sqrt{a^2 - b^2}}{b}.$$

Since $a > b > 0$, we have $|\beta| > 1$.

But $$|\alpha\beta| = |\alpha|\,|\beta| = 1,$$

it follows that $|\alpha| < 1$. Hence $z = \alpha$ is the only simple pole inside C.

Also $$f(z) = \frac{1}{b(z - \alpha)(z - \beta)}$$

$$\therefore\quad \text{Res}_{z = \alpha} f(z) = \lim_{z \to \alpha} (z - \alpha) \cdot \frac{2}{bi(z - \alpha)(z - \beta)}$$

$$= \lim_{z \to \alpha} \frac{2}{bi(z - \beta)} = \frac{2}{bi(\alpha - \beta)} = \frac{2}{bi \cdot \frac{2\sqrt{a^2 - b^2}}{b}} = \frac{1}{i\sqrt{a^2 - b^2}}$$

Hence, $$\int_0^{2\pi} \frac{d\theta}{a + b\cos\theta} = 2\pi i \cdot \frac{1}{i\sqrt{a^2 - b^2}} = \frac{2\pi}{\sqrt{a^2 - b^2}}$$

(*ii*) Similarly, $\int_0^{2\pi} \frac{d\theta}{a + b\sin\theta}$ can be evaluated. This is left as an exercise for the reader.

Example 10. *Evaluate* $\int_0^{\pi} \frac{a\,d\theta}{a^2 + \sin^2\theta}$, $a > 0$.

Sol. Let $$I = \int_0^{\pi} \frac{a\,d\theta}{a^2 + \sin^2\theta}$$

Then $$I = \int_0^{\pi} \frac{2a\,d\theta}{2a^2 + 2\sin^2\theta} = \int_0^{\pi} \frac{2a\,d\theta}{2a^2 + 1 - \cos 2\theta}$$

$$= \int_0^{2\pi} \frac{a\,dt}{2a^2 + 1 - \cos t}, \text{ putting } 2\theta = t \text{ so that } 2d\theta = dt$$

$$= \int_0^{2\pi} \frac{a\,dt}{2a^2 + 1 - \frac{1}{2}(e^{it} + e^{-it})}$$

Let C denote the unit circle $|z| = 1$. Now writing $z = e^{it}$, so that $dz = i\, e^{it}\, dt$ or $\frac{dz}{iz} = dt$, we get

$$I = \int_C \frac{2a}{2(2a^2 + 1) - (z + 1/z)} \cdot \frac{dz}{iz}$$

$$= \frac{2a}{i} \int_C \frac{dz}{2(2a^2 + 1)z - z^2 - 1} = 2ai \int_C \frac{dz}{z^2 - 2(2a^2 + 1)\, z + 1}$$

$$= \int_C f(z)\, dz, \text{ where } f(z) = \frac{2ai}{z^2 - z(2a^2 + 1)\, z + 1}$$

Now poles of $f(z)$ are given by

$$z^2 - 2(2a^2 + 1)\, z + 1 = 0$$

or $$z = 2a^2 + 1 \pm 2a\sqrt{a^2 + 1}$$

Let $$\alpha = (2a^2 + 1) + 2a\sqrt{a^2 + 1},\ \beta = (2a^2 + 1) - 2a\sqrt{a^2 + 1}.$$

Clearly $|\alpha| > 1$ and since $|\alpha\beta| = |\alpha|\,|\beta| = 1$, we have $|\beta| < 1$. Hence $f(z)$ has only one simple pole $z = \beta$ lying inside C.

$$\text{Residue (at } z = \beta) = \lim_{z \to \beta} (z - \beta) \frac{2ai}{(z - \alpha)(z - \beta)} = \frac{2ai}{\beta - \alpha} = \frac{2ai}{(-4a)\sqrt{a^2 + 1}} = \frac{1}{2i\sqrt{a^2 + 1}}$$

By Cauchy's residue theorem.

$$\therefore \quad \int_0^{\pi} \frac{a\,d\theta}{a^2 + \sin^2\theta} = 2\pi i \frac{1}{2i\sqrt{a^2 + 1}} = \frac{\pi}{\sqrt{a^2 + 1}}$$

Example 11. *Evaluate* $\int_0^{2\pi} \frac{a\,d\theta}{(a + b\cos\theta)^2}$ $(a > b > 0)$.

Sol. Putting $z = e^{i\theta}$ so that $\frac{dz}{iz} = d\theta$, we get

$$I = \int_0^{2\pi} \frac{d\theta}{(a + b\cos\theta)^2} = \frac{1}{i} \int_C \frac{dz}{z[a + (b/z)(z + 1/z)]^2}$$

$$= -4i \int_C \frac{z\,dz}{(bz^2 + 2az + b)^2} = -\frac{4i}{b^2} \int_C \frac{z\,dz}{(z^2 (2a/b)\, z + 1)^2}$$

$$= -\frac{4i}{b^2}\int_C f(z)\,dz$$

where C is the unit circle $|z| = 1$,

Poles of order two of $f(z)$ are given by the roots of

$$z^2 + \frac{2a}{b}z + 1 = 0$$

$$z = \frac{1}{2}\left[-\frac{2a}{b} \pm \sqrt{\left(\frac{4a^2}{b^2} - 4\right)}\right] = \frac{-a \pm \sqrt{a^2 - b^2}}{b}$$

Let $$\alpha = \frac{-a + \sqrt{a^2 - b^2}}{b},\ \beta = \frac{-a - \sqrt{a^2 - b^2}}{b}$$

Since $a > b > 0$, $|\beta| < 1$ and that $\alpha\beta = 1$, it follows $|\alpha| < 1$. Hence α is the only pole of order two inside C.

Now $$-\frac{4i}{b^2}f(z) = -\frac{4iz}{b^2(z-\alpha)^2(z-\beta)^2} = \frac{\phi(z)}{(z-\alpha)^2},$$

where $$\phi(z) = -\frac{4iz}{b^2(z-\beta)^2}.$$

Residue at the double pole $z = \alpha$ is

$$\phi'(\alpha) = -\frac{4i}{b^2}\left[\frac{(z-\beta)^2 - 2z(z-\beta)}{(z-\beta)^4}\right]_{z=\alpha} = -\frac{4i}{b^2}\left[\frac{(\alpha-\beta) - 2\alpha}{(\alpha-\beta)^3}\right] = \frac{4i}{b^2}\cdot\frac{(\alpha+\beta)}{(\alpha-\beta)^3}$$

$$= \frac{4i}{b^2}\cdot\frac{(-2a/b)}{(8/b^3)(a^2-b^2)^{3/2}} = -\frac{ai}{(a^2-b^2)^{3/2}}.$$

Hence by Cauchy's residue theorem

$$\int_0^{2\pi}\frac{d\theta}{(a+b\cos\theta)^2} = 2\pi i\left[-\frac{ai}{(a^2-b^2)^{3/2}}\right] = \frac{2\pi a}{(a^2-b^2)^{3/2}}.$$

Example 12. *Show that* $\int_0^{\pi}\frac{\cos 2\theta\,d\theta}{1 - 2a\cos\theta + a^2} = \frac{\pi a^2}{1-a^2}\,(a^2 < 1).$

Sol. Let $$\text{I} = \int_0^{\pi}\frac{\cos 2\theta\,d\theta}{1 - 2a\cos\theta + a^2}\,(a^2 < 1)$$

then $$\text{I} = \int_0^{\pi}\frac{\cos 2\theta\,(1 + 2a\cos\theta + a^2)}{(1+a^2)^2 - 4a^2\cos^2\theta}\,d\theta$$

$$= \int_0^{\pi}\frac{(1+a^2)\cos 2\theta}{(1+a^2)^2 - 2a^2(1+\cos 2\theta)}\,d\theta + 2a\int_0^{\pi}\frac{\cos 2\theta\cos\theta}{(1+a^2)^2 - 4a^2\cos^2\theta}\,d\theta$$

$$= \int_0^{\pi} \frac{(1+a^2)\cos 2\theta}{(1+a^2)^2 - 2a^2(1+\cos 2\theta)}\, d\theta$$

Since the second integral vanishes by the property of definite integrals

$$= \frac{1}{2}\int_0^{2\pi} \frac{(1+a^2)\cos\phi\, d\phi}{1+a^4 - 2a^2\cos\phi}, \text{ putting } 2\theta = \phi$$

$$= \frac{1}{2i}\int_C \frac{(1+a^2)\frac{1}{2}\left(z+\frac{1}{z}\right)}{1+a^4 - a^2\left(z+\frac{1}{z}\right)} \cdot \frac{dz}{z}, \text{ Putting } z = e^{i\phi} \text{ where C is the circle } |z| = 1$$

$$= \frac{(1+a^2)}{4i}\int_C \frac{(z^2+1)}{[-a^2 z^2 + (1+a^4)z - a^2]}\, dz$$

$$= \frac{(1+a^2)\, i}{4a^2}\int_C \frac{(z^2+1)}{z[z^2 - (a^2+1/a^2)z + 1]}\, dz$$

$$= \frac{(1+a^2)\, i}{4a^2}\int_C \frac{(z^2+1)\, dz}{z(z-a^2)(z-1/a^2)} = \int_C f(z)\, dz.$$

Now $f(z)$ has simple poles at $z = 0, a^2, 1/a^2$ of which the poles at $z = 0, a^2$ lie inside the circle since $a^2 < 1$.

$$\text{Res}_{z=0} f(z) + \text{Res}_{z=a^2} f(z) = \lim_{z\to 0} z\,.\, f(z) + \lim_{z\to a^2}(z-a^2) f(z)$$

$$= \lim_{z\to 0}\frac{(1+a^2)\, i}{4a^2}\cdot\frac{(z^2+1)}{(z+a^2)(z-1/a^2)} + \lim_{z\to a^2}\frac{(1+a^2)\, i}{4a^2}\cdot\frac{(z^2+1)}{z(z-1/a^2)}$$

$$= \frac{(1+a^2)\, i}{4a^2}\left[1 + \frac{a^4+1}{a^2(a^2-1/a^2)}\right] = \frac{(1+a^2)\, i}{4a^2}\left[\frac{2a^4}{a^4-1}\right] = \frac{a^2 i}{2(a^2-1)}$$

Hence $\displaystyle\int_0^{\pi} \frac{\cos 2\theta}{1-2a\cos\theta + a^2}\, d\theta = 2\pi i \times$ (sum of residues at poles inside C)

$$= 2\pi i\,.\,\frac{a^2 i}{2(a^2-1)} = \frac{\pi a^2}{1-a^2}$$

Example 13. *Apply calculus of residues to prove that* $\displaystyle\int_0^{2\pi}\frac{\cos 2\theta\, d\theta}{5+4\cos\theta} = \frac{\pi}{6}.$

Sol. Let C denote the unit circle $|z| = 1$

and $$I = \int_0^{2\pi}\frac{\cos 2\theta\, d\theta}{5+4\cos\theta}. \text{ Then}$$

$$I = \text{real part of } \int_0^{2\pi}\frac{e^{2i\theta}}{(5+4\cos\theta)}\, d\theta$$

$$= \int_C \frac{z^2}{[5 + 2(z + 1/z)]} \cdot \frac{dz}{iz}, \text{ putting } z = e^{i\theta}$$

$$= \frac{1}{i} \int_C \frac{z^2 dz}{2z^2 + 5z + 2} = \frac{1}{i} \int_C \frac{z^2\, dz}{(2z + 1)(z + 2)}$$

$$= \frac{1}{2i} \int_C \frac{z^2\, dz}{(z + \frac{1}{2})(z + 2)} = \int_C f(z)\, dz.$$

Now $f(z)$ has simple poles at $z = -\frac{1}{2}$ and $z = -2$ of which only $z = -\frac{1}{2}$ lies inside C.

$$\text{Res}_{z = -\frac{1}{2}} f(z) = \lim_{z \to -\frac{1}{2}} (z + \frac{1}{2}) \cdot \frac{z^2}{2i(z + \frac{1}{2})(z + 2)} = \frac{1}{12i}$$

$$\therefore \qquad I = \int_0^{2\pi} \frac{e^{2i\theta}\, d\theta}{5 + 4\cos\theta} = 2\pi i \cdot \frac{1}{12i} = \frac{\pi}{6}$$

Now equating real parts on both sides, we get $I = \pi/6$.

TEST YOUR KNOWLEDGE

Determine the poles of the following functions and the residue at each pole:

1. $\frac{2z + 1}{z^2 - z - 2}$
2. $\frac{z + 1}{z^2(z - 2)}$
3. $\frac{z^2}{(z - 1)(z - 2)^2}$
4. $\frac{1 - e^{2z}}{z^4}$
5. $\frac{e^z}{z^2 + \pi^2}$
6. $\frac{z}{\cos z}$
7. $\frac{z^2 - 2z}{(z + 1)^2 (z^2 + 4)}$.

Find the residues at z = 0 of the following functions:

8. $z \cos \frac{1}{z}$
9. $\text{cosec}^2 z$
10. $\frac{1 + e^z}{\sin z + z \cos z}$.

Evaluate the following integrals:

11. $\oint_C \frac{z^2 + 2z - 2}{z - 4}\, dz$, where C is a closed curve containing the point $z = 4$ in its interior.
12. $\oint_C \frac{1 - 2z}{z(z - 1)(z - 2)}\, dz$, where C is the circle $|z| = 1.5$.
13. $\oint_C \frac{z}{(z - 1)(z - 2)^2}\, dz$, where C is the circle $|z - 2| = \frac{1}{2}$.
14. $\oint_C \frac{12z - 7}{(z - 1)^2 (2z + 3)}\, dz$, where C is the circle $|z| = 2$.
15. $\oint_C \frac{\sin \pi z^2 + \cos \pi z^2}{(z - 1)^2 (z - 2)}\, dz$, where C is the circle $|z| = 3$.

16. $\oint_C \dfrac{z \sec z}{(1-z)^2}\, dz$, where C is the circle $| z | = 2$.

17. $\oint_C \dfrac{dz}{\sinh z}$, where C is the circle $| z | = 4$.

18. $\oint_C \tan z\, dz$, where C is the circle $| z | = 2$.

Answers

1. $z = -1, 2; \frac{1}{3}, \frac{5}{3}$ **2.** $z = 0, 2 ; -\frac{3}{4}, \frac{3}{4}$ **3.** $z = 1, 2 ; 1, 0$ **4.** $z = 0; -\frac{4}{3}$

5. $z = \pm \pi i; \pm \frac{i}{2\pi}$ **6.** $z = (2n + 1)\frac{\pi}{2}$; $(-1)^{n+1} (2n + 1)\frac{\pi}{2}$, where n is an integer.

7. $z = -1, \pm 2i; -\frac{14}{25}, \frac{7 \pm i}{25}$ **8.** $-\frac{1}{2}$ **9.** 0

10. 1 **11.** $44\pi i$ **12.** $3\pi i$

13. $-2\pi i$ **14.** 0 **15.** $4\pi i(\pi + 1)$

16. $2\pi i \sec 1(1 + \tan 1)$ **17.** $-2\pi i$ **18.** $-4\pi i$.

4.26. APPLICATION OF RESIDUES TO EVALUATE REAL INTEGRALS

The residue theorem provides a simple and elegant method for evaluating many important definite integrals of real variables. Some of these are illustrated below.

(1) Integrals of the type $\int_0^{2\pi} F(\cos\theta, \sin\theta)\, d\theta$,

where F(cos θ, sin θ) is a rational function of cos θ and sin θ.

Such integrals can be reduced to complex line integrals by the substitution $z = e^{i\theta}$, so that

$$dz = ie^{i\theta}\, d\theta, \text{ i.e., } d\theta = \frac{dz}{iz}.$$

Also
$$\cos\theta = \frac{e^{i\theta} + e^{-i\theta}}{2} = \frac{1}{2}\left(z + \frac{1}{z}\right)$$

$$\sin\theta = \frac{e^{i\theta} + e^{-i\theta}}{2i} = \frac{1}{2i}\left(z - \frac{1}{z}\right).$$

As θ varies from 0 to 2π, z moves once round the unit circle in the anti-clockwise direction.

$$\therefore \quad \int_0^{2\pi} F(\cos\theta, \sin\theta)\, d\theta = \oint_C F\left(\frac{z + z^{-1}}{2}, \frac{z - z^{-1}}{2i}\right)\frac{dz}{iz}$$

where C is the unit circle $| z | = 1$.

The integral on the right can be evaluated by using the residue theorem.

ILLUSTRATIVE EXAMPLES

Example 1. *Evaluate* $\int_0^{2\pi} \frac{d\theta}{2+\cos\theta}$.

Sol. Put $z = e^{i\theta}$ so that $\cos\theta = \frac{1}{2}\left(z+\frac{1}{z}\right)$ and $d\theta = \frac{dz}{iz}$

Then $$\int_0^{2\pi} \frac{d\theta}{2+\cos\theta} = \oint_C \frac{1}{2+\frac{1}{2}\left(z+\frac{1}{z}\right)}\frac{dz}{iz} = \frac{2}{i}\oint_C \frac{1}{z^2+4z+1}\,dz$$

where C is the circle $|z| = 1$.

The poles of the integrand are the roots of $z^2 + 4z + 1 = 0$, which are $z = -2 \pm \sqrt{3}$.

Of the two poles, only $z = -2 + \sqrt{3}$ lies inside the circle C.

Residue at $z = -2 + \sqrt{3}$ is

$$= \underset{z\to\alpha}{\text{Lt}}\ (z-\alpha)\frac{1}{z^2+4z+1}, \quad \text{where } \alpha = -2+\sqrt{3} \qquad \left|\ \text{Form } \frac{0}{0}\right.$$

$$= \underset{z\to\alpha}{\text{Lt}}\ \frac{1}{2z+4} = \frac{1}{2\alpha+4} = \frac{1}{2\sqrt{3}}$$

$\therefore$ By residue theorem, $\oint_C \frac{dz}{z^2+4z+1} = 2\pi i\left(\frac{1}{2\sqrt{3}}\right) = \frac{\pi i}{\sqrt{3}}$

Hence $$\int_0^{2\pi} \frac{d\theta}{2+\cos\theta} = \frac{2}{i}\oint_C \frac{dz}{z^2+4z+1} = \frac{2}{i}\left(\frac{\pi i}{\sqrt{3}}\right) = \frac{2\pi}{\sqrt{3}}.$$

Example 2. *Evaluate* $\int_0^{2\pi} \frac{d\theta}{1-2a\sin\theta+a^2},\ 0 < a < 1.$

Sol. Put $z = e^{i\theta}$ so that $\sin\theta = \frac{1}{2i}\left(z-\frac{1}{z}\right)$ and $d\theta = \frac{dz}{iz}$.

Then, $$\int_0^{2\pi} \frac{d\theta}{1-2a\sin\theta+a^2} = \oint_C \frac{1}{1-2a\,.\,\frac{1}{2i}\left(z-\frac{1}{z}\right)+a^2}\frac{dz}{iz}$$

$$= -\oint_C \frac{dz}{az^2 - i(a^2+1)z - a} = -\oint_C \frac{dz}{(az-i)(z-ia)}$$

where C is the circle $|z| = 1$.

The integrand has simple poles at $z = \frac{i}{a}$ and $z = ia$ of which only $z = ia$ lies inside the circle C. $[\because\ 0 < a < 1]$

Residue at $z = ia$ is $\underset{z\to ia}{\text{Lt}}\ (z-ia)\frac{1}{(az-i)(z-ia)} = \underset{z\to ia}{\text{Lt}}\ \frac{1}{az-i} = \frac{1}{i(a^2-1)}$

$\therefore$ By residue theorem,

$$\oint_C \frac{dz}{(az-i)(z-ia)} = 2\pi i\,.\,\frac{1}{i(a^2-1)} = \frac{2\pi}{a^2-1}$$

Hence $\displaystyle\int_0^{2\pi} \frac{d\theta}{1-2a\sin\theta+a^2} = -\oint_C \frac{dz}{(az-i)(z-ia)} = \frac{2\pi}{1-a^2}.$

Example 3. *Evaluate* $\displaystyle\int_0^{\pi} \frac{d\theta}{a+b\cos\theta}$, *where* $a > |\,b\,|$.

Sol. Since $\displaystyle\int_0^{2a} f(x)\,dx = 2\int_0^{a} f(x)\,dx$, if $f(2a-x) = f(x)$

$$\therefore \quad \int_0^{2\pi} \frac{d\theta}{a+b\cos\theta} = 2\int_0^{\pi} \frac{d\theta}{a+b\cos\theta} \qquad [\because \cos(2\pi-\theta) = \cos\theta]$$

or
$$\int_0^{\pi} \frac{d\theta}{a+b\cos\theta} = \frac{1}{2}\int_0^{2\pi} \frac{d\theta}{a+b\cos\theta} \qquad \text{...(1)}$$

Putting $z = e^{i\theta}$, so that $\cos\theta = \dfrac{1}{2}\left(z+\dfrac{1}{z}\right)$ and $d\theta = \dfrac{dz}{iz}$, we have

$$\int_0^{2\pi} \frac{d\theta}{a+b\cos\theta} = \oint_C \frac{1}{a+\frac{b}{2}\left(z+\frac{1}{z}\right)}\,\frac{dz}{iz} = \oint_C \frac{2dz}{i(bz^2+2az+b)}$$

where C is the circle $|\,z\,| = 1$.

Proceeding as in example 3, residue at $z = \alpha$ is

$$= \underset{z\to\alpha}{\text{Lt}}\,(z-\alpha)\,.\,\frac{2}{ib(z-\alpha)(z-\beta)} = \underset{z\to\alpha}{\text{Lt}}\,\frac{2}{ib(z-\beta)} = \frac{2}{ib(\alpha-\beta)}$$

$$= \frac{2}{ib}\,.\,\frac{b}{2\sqrt{a^2-b^2}} = \frac{1}{i\sqrt{a^2-b^2}}$$

$\therefore$ By residue theorem,

$$\oint_C \frac{2dz}{i(bz^2+2az+b)} = 2\pi i\,.\,\frac{1}{i\sqrt{a^2-b^2}} = \frac{2\pi}{\sqrt{a^2-b^2}}$$

Hence from (1), $\displaystyle\int_0^{\pi} \frac{d\theta}{a+b\cos\theta} = \frac{\pi}{\sqrt{a^2-b^2}}.$

(2) **Integrals of the type** $\displaystyle\int_{-\infty}^{\infty} \frac{f(x)}{F(x)}\,dx$**, where f(x) and F(x) are polynomials in x such that** $\dfrac{x\,f(x)}{F(x)} \to 0$ **as** $x \to \infty$ **and F(x) has no zeros on the real axis.**

Consider the integral $\displaystyle\oint_C \frac{f(z)}{F(z)}\,dz$

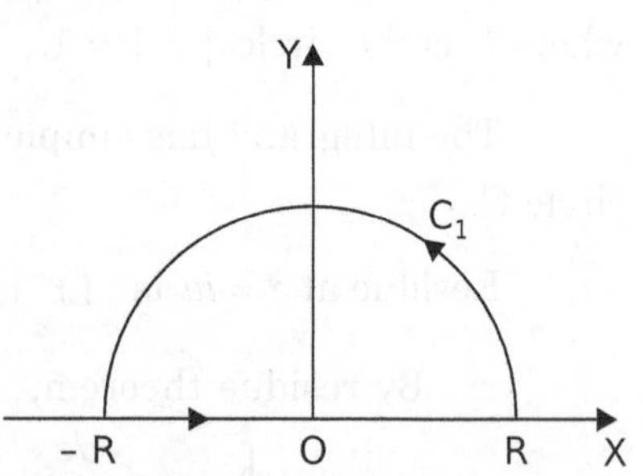

over the closed contour C consisting of the real axis from – R to R and the semi-circle C_1 of radius R in the upper half plane.

We take R large enough so that all the poles of $\dfrac{f(z)}{F(z)}$ in the upper half plane lie within C.

By residue theorem, we have

$$\oint_C \frac{f(z)}{F(z)}\,dz = 2\pi i\left[\text{sum of the residues of } \frac{f(z)}{F(z)} \text{ in the upper half plane}\right]$$

or $$\int_{C_1} \frac{f(z)}{F(z)}\,dz + \int_{-R}^{R} \frac{f(x)}{F(x)}\,dx = 2\pi i\left[\text{sum of the residues of } \frac{f(z)}{F(z)} \text{ in the upper half plane}\right] \quad ...(1)$$

($\because$ on the real axis, $z = x$)

If we put $z = Re^{i\theta}$ in the first integral on the left side, then R is constant on C_1 and as z moves along C_1, θ varies from 0 to π.

$$\therefore \quad \int_{C_1} \frac{f(z)}{F(z)}\,dz = \int_0^{\pi} \frac{f(Re^{i\theta})}{F(Re^{i\theta})}\, Re^{i\theta}\, i d\theta$$

For large R, $\left|\int_0^{\pi} \frac{f(Re^{i\theta})}{F(Re^{i\theta})}\, Re^{i\theta}\, i d\theta\right| \to 0$ $\qquad\Big|$ is of the order $\frac{Rf(R)}{F(R)}$

$$\therefore \quad \int_0^{\pi} \frac{f(Re^{i\theta})}{F(Re^{i\theta})}\, Re^{i\theta}\, i d\theta \to 0 \text{ when } R \to \infty$$

Hence from (1), we have $\displaystyle\int_{-\infty}^{\infty} \frac{f(x)}{F(x)}\,dx = 2\pi i$

$$\left[\text{Sum of the residues of } \frac{f(z)}{F(z)} \text{ in the upper half plane}\right].$$

Example 4. *Evaluate* $\displaystyle\int_{-\infty}^{\infty} \frac{x^2}{(x^2+a^2)(x^2+b^2)}\,dx$ *(a > 0, b > 0).*

Sol. The poles of $\phi(z) = \dfrac{z^2}{(z^2+a^2)(z^2+b^2)}$ are $z = \pm\, ia, \pm\, ib$.

Of these, $z = ia$ and $z = ib$ lie in the upper half of the z-plane.

Residue of $\phi(z)$ at $z = ia$ is

$$\underset{z\to ia}{\text{Lt}}\ (z - ia)\,.\,\frac{z^2}{(z^2+a^2)(z^2+b^2)} = \underset{z\to ia}{\text{Lt}}\ \frac{z^2}{(z+ia)(z^2+b^2)} = \frac{-a^2}{2ia(-a^2+b^2)} = \frac{a}{2i(a^2-b^2)}$$

Residue of $\phi(z)$ at $z = ib$ is

$$\underset{z\to ib}{\text{Lt}}\ (z - ib)\,.\,\frac{z^2}{(z^2+a^2)(z^2+b^2)} = \underset{z\to ib}{\text{Lt}}\ \frac{z^2}{(z^2+a^2)(z+ib)} = \frac{-b^2}{(-b^2+a^2)(2ib)} = \frac{-b}{2i(a^2-b^2)}$$

$$\therefore \quad \int_{-\infty}^{\infty} \frac{x^2\,dx}{(x^2+a^2)(x^2+b^2)} = 2\pi i\ [\text{sum of residues of } \phi(z) \text{ in the upper half of } z\text{-plane}]$$

$$= 2\pi i\left[\frac{a}{2i(a^2-b^2)} - \frac{b}{2i(a^2-b^2)}\right] = \pi\left[\frac{a-b}{a^2-b^2}\right] = \frac{\pi}{a+b}.$$

Example 5. *Evaluate* $\displaystyle\int_0^{\infty} \frac{dx}{x^4+1}.$

Sol. The poles of $\phi(z) = \dfrac{1}{z^4+1}$ are obtained by solving $z^4 + 1 = 0$.

Now $z^4 + 1 = 0.$

$\Rightarrow$ $z = (-1)^{1/4} = (\cos \pi + i \sin \pi)^{1/4} = [\cos (2n\pi + \pi) + i \sin (2n\pi + \pi)]^{1/4}$

$= \cos \frac{(2n+1)\pi}{4} + i \sin \frac{(2n+1)\pi}{4}$ (By De Moivre's theorem)

where $n = 0, 1, 2, 3.$

When $n = 0$, $z = \cos \frac{\pi}{4} + i \sin \frac{\pi}{4} = \frac{1}{\sqrt{2}} + i \frac{1}{\sqrt{2}}$

When $n = 1$, $z = \cos \frac{3\pi}{4} + i \sin \frac{3\pi}{4} = -\frac{1}{\sqrt{2}} + i \frac{1}{\sqrt{2}}$

When $n = 2$, $z = \cos \frac{5\pi}{4} + i \sin \frac{5\pi}{4} = -\frac{1}{\sqrt{2}} - i \frac{1}{\sqrt{2}}$

When $n = 3$, $z = \cos \frac{7\pi}{4} + i \sin \frac{7\pi}{4} = \frac{1}{\sqrt{2}} - i \frac{1}{\sqrt{2}}$

Of these, only the poles corresponding to $n = 0, 1$, *viz*, $z = e^{i\frac{\pi}{4}}$ and $z = e^{3i\frac{\pi}{4}}$ lie in the upper half of z-plane.

Residue of $\phi(z)$ at $z = e^{i\frac{\pi}{4}}$ is $\lim_{z \to e^{i\frac{\pi}{4}}} \frac{z - e^{i\frac{\pi}{4}}}{z^4 + 1}$ | Form $\frac{0}{0}$

$= \lim_{z \to e^{i\frac{\pi}{4}}} \frac{1}{4z^3}$ | By L' Hospital's rule

$= \frac{1}{4e^{3i\frac{\pi}{4}}} = \frac{1}{4} e^{-3i\frac{\pi}{4}}$

Similarly, residue of $\phi(z)$ at $z = e^{3i\frac{\pi}{4}}$ is $\frac{1}{4} e^{-9i\frac{\pi}{4}}$

$$\therefore \int_0^\infty \frac{dx}{x^4+1} = \frac{1}{2} \int_{-\infty}^\infty \frac{dx}{x^4+1}$$

$= \frac{1}{2} \cdot 2\pi i$ [Sum of residues of $\phi(z)$ in the upper half of z-plane]

$$= \pi i \left[\frac{1}{4} e^{-3i\frac{\pi}{4}} + \frac{1}{4} e^{-9i\frac{\pi}{4}} \right] = \frac{1}{4} \pi i \left[\left(\cos \frac{3\pi}{4} - i \sin \frac{3\pi}{4} \right) + \left(\cos \frac{9\pi}{4} - i \sin \frac{9\pi}{4} \right) \right]$$

$$= \frac{1}{4} \pi i \left[-\frac{1}{\sqrt{2}} - \frac{i}{\sqrt{2}} + \frac{1}{\sqrt{2}} - \frac{i}{\sqrt{2}} \right] = \frac{1}{4} \pi i (-i\sqrt{2}) = \frac{\pi\sqrt{2}}{4}.$$

Note. The above method can also be applied to some cases where $f(x)$ contains trigonometric functions also.

(3) Integrals of the type $\int_{-\infty}^{\infty} \frac{f(x)}{F(x)} dx$, when F(x) has zeros on the real axis.

In such cases we proceed in a manner similar to above, except that the singularities on the real axis are enclosed in small semi-circles to avoid their inclusion in C, *i.e.,* the contour C is indented at these singularities.

Example 6. *Evaluate* $\int_0^{\infty} \frac{\sin mx}{x} dx, m > 0.$

Sol. Since sin mx is the imaginary part of e^{imx}, we consider the function

$$\phi(z) = \frac{e^{imz}}{z}.$$

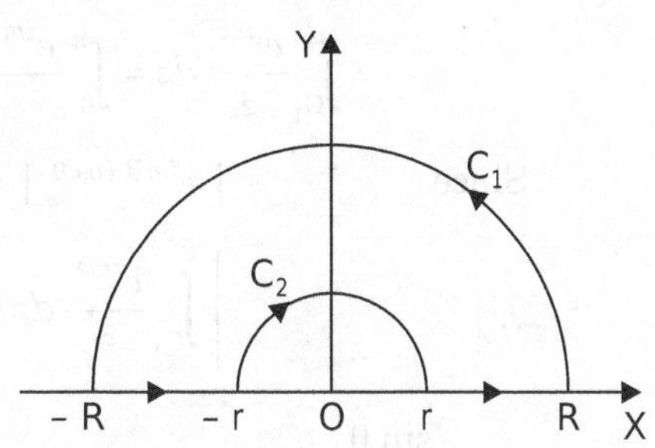

This has a simple pole at $z = 0$, which lies on the real axis. Enclose this singularity by a small semi-circle $C_2 : | z | = r$. Evaluate the function $\phi(z)$ over the contour C shown in the figure consisting of parts of the real axis from – R to – r and r to R, the small semi-circle C_2 and the large semi-circle C_1. Since the function has no singularity within this contour, by Cauchy's theorem, we have

$$\oint_C \phi(z)\, dz = 0$$

$$\Rightarrow \quad \int_{-R}^{-r} \phi(x)\, dx + \int_{C_2} \phi(z)\, dz + \int_r^R \phi(x)\, dx + \int_{C_1} \phi(z)\, dz = 0$$

$$\Rightarrow \quad \int_{-R}^{-r} \frac{e^{imx}}{x} dx + \int_{C_2} \frac{e^{imz}}{z} dz + \int_r^R \frac{e^{imx}}{x} dx + \int_{C_1} \frac{e^{imz}}{z} dz = 0 \qquad ...(1)$$

Substituting – x for x in the first integral and combining it with the third integral, we get

$$\int_r^R \frac{e^{imx} - e^{-imx}}{x} dx + \int_{C_2} \frac{e^{imz}}{z} dz + \int_{C_1} \frac{e^{imz}}{z} dz = 0$$

or
$$2i \int_r^R \frac{\sin mx}{x} dx + \int_{C_2} \frac{e^{imz}}{z} dz + \int_{C_1} \frac{e^{imz}}{z} dz = 0 \qquad ...(2)$$

Now $$\int_{C_2} \frac{e^{imz}}{z} dz = \int_{C_2} \frac{1}{z} dz + \int_{C_2} \frac{e^{imz} - 1}{z} dz \qquad ...(3)$$

On C_2, $z = re^{i\theta}$

$$\therefore \quad \int_{C_2} \frac{1}{z} dz = \int_{\pi}^{0} \frac{re^{i\theta} i\, d\theta}{re^{i\theta}} = - \int_0^{\pi} i d\theta = - i\pi$$

Also $\left| \int_{C_2} \frac{e^{imz} - 1}{z} dz \right| \le M \int_{C_2} \frac{|dz|}{|z|} = \pi M$

where M is the maximum value on C_2 of $| e^{imz} - 1 | = | e^{imr(\cos\theta + i \sin\theta)} - 1 |$

Clearly, $M \to 0$ as $r \to 0$

$\therefore$ From (3), $\int_{C_2} \frac{e^{imz}}{z} dz = -i\pi$ ...(4)

Putting $z = Re^{i\theta}$ in the integral over C_1, we get

$$\int_{C_1} \frac{e^{imz}}{z} dz = \int_0^\pi \frac{e^{imR(\cos\theta + i\sin\theta)}}{Re^{i\theta}} Re^{i\theta}\, i\, d\theta = i \int_0^\pi e^{imR\cos\theta} . e^{-mR\sin\theta}\, d\theta$$

Since $| e^{imR\cos\theta} | \le 1$

$$\therefore \quad \left| \int_{C_1} \frac{e^{imz}}{z} dz \right| \le \int_0^\pi e^{-mR\sin\theta} d\theta = 2\int_0^{\pi/2} e^{-mR\sin\theta} d\theta$$

Also $\frac{\sin\theta}{\theta}$ continually decreases from 1 to $\frac{2}{\pi}$ as θ increases from 0 to $\frac{\pi}{2}$.

$\therefore$ For $0 \le \theta \le \frac{\pi}{2}$, $\frac{\sin\theta}{\theta} \ge \frac{2}{\pi}$ or $\sin\theta \ge \frac{2\theta}{\pi}$

$$\therefore \quad \left| \int \frac{e^{imz}}{z} dz \right| \le 2\int_0^{\pi/2} e^{-2mR\theta/\pi} d\theta = \left[-\frac{\pi}{mR} e^{-2mR\theta/\pi} \right]_0^{\pi/2} = \frac{\pi}{mR}(1 - e^{-mR})$$

As $R \to \infty$, $\frac{\pi}{mR}(1 - e^{-mR}) \to 0$

$$\therefore \quad \int_{C_1} \frac{e^{imz}}{z} dz = 0$$

Hence from (2), on taking the limit as $r \to 0$ and $R \to \infty$, we get

$$2i\int_0^\infty \frac{\sin mx}{x} dx - i\pi = 0 \quad \text{or} \quad \int_0^\infty \frac{\sin mx}{x} dx = \frac{\pi}{2}.$$

Example 7. *If $a > b > 0$, prove that* $\int_{-\infty}^\infty \frac{\cos x\, dx}{(x^2 + a^2)(x^2 + b^2)} = \frac{\pi}{a^2 - b^2}\left(\frac{e^{-b}}{b} - \frac{e^{-a}}{a} \right)$.

Sol. Consider $\int_C \frac{e^{iz}}{(z^2 + a^2)(z^2 + b^2)} dz = \int_C f(z) dz$ where z is a closed contour as discussed in Ex. 2.

By Cauchy's residue theorem, we get

$$\int_C f(z)\,dz = \int_{-R}^{R} f(x)\,dx + \int_{C_R} f(z)\,dz = 2\pi i\, \Sigma R^+ . \qquad ...(1)$$

Evidently $\lim\limits_{|z|\to\infty} \dfrac{1}{(z^2+a^2)(z^2+b^2)} = 0$.

Hence, by Jordan's lemma,

$$\lim_{R\to\infty} \int_{C_R} \frac{e^{iz}\,dz}{(z^2+a^2)(z^2+b^2)} = \lim_{R\to\infty}\int_{C_R} f(z)\,dz = 0. \qquad ...(2)$$

Proceeding to the limit, when $R \to \infty$, we have from (1) and (2),

$$\int_{-\infty}^{\infty} f(x)\,dx = 2\pi i\, \Sigma R^+ .$$

Now $f(z)$ has simple poles at $z = \pm\, ai$ and $z = \pm\, bi$ of which $z = ai$ and $z = bi$ lie in the upper half plane.

$$\text{Res}_{z=ai} f(z) = \lim_{z\to ai} (z-ai)\, f(z) = \lim_{z\to ai} \frac{e^{iz}}{(z+ai)(z^2+b^2)} = \frac{e^{-a}}{2ai(b^2-a^2)}$$

Similarly, $\text{Res}_{z=bi} f(z) = \dfrac{e^{-b}}{2bi(a^2-b^2)}$

Sum of residues $= \dfrac{1}{2i(a^2-b^2)}\left[\dfrac{e^{-b}}{b} - \dfrac{e^{-a}}{a}\right]$.

Hence from (3), $\displaystyle\int_{-\infty}^{\infty} f(x)\,dx = \frac{\pi}{(a^2-b^2)}\left[\frac{e^{-b}}{b} - \frac{e^{-a}}{a}\right]$

or $\displaystyle\int_0^{\infty} f(x)\,dx = \frac{\pi}{2(a^2-b^2)}\left[\frac{e^{-b}}{b} - \frac{e^{-a}}{a}\right]$.

Example 8. *Show by contour integration,* $\displaystyle\int_0^{\infty} \frac{\sin x}{x(x^2+a^2)}\,dx = \frac{\pi}{2a^2}(1-e^{-a}),\ (a>0).$

Sol. Consider $\displaystyle\int_C \frac{e^{iz}\,dz}{z(z^2+a^2)} = \int_C f(z)\,dz$,

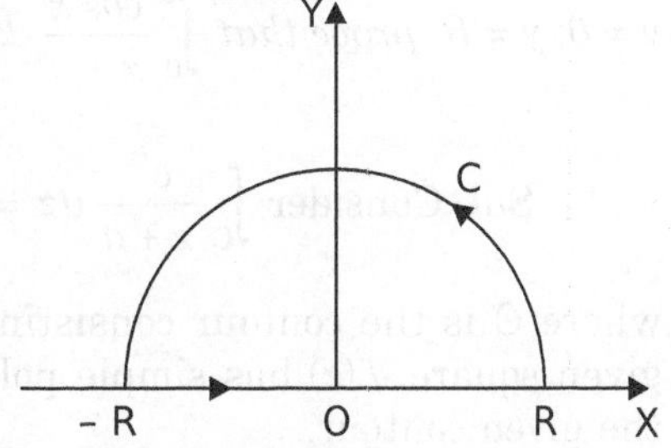

where C is the closed contour consisting of

(*i*) the real axis from ρ ro R,

(*ii*) a large semi-circle Γ in the upper half plane given by $|z| = R$,

(*iii*) the real axis $-R$ to $-\rho$, and

(*iv*) a small semi-circle γ given by $|z| = \rho$.

Now $f(z)$ has simple poles at $z = 0, \pm ai$ of which only $z = ai$ lies within C. The pole $z = 0$ is avoided by indentation.

$$\text{Res}_{z=ai} f(z) = \lim_{z \to ai} (z - ai) \frac{e^{iz}}{z(z-ai)(z-ai)} = \frac{e^{-a}}{-2a^2}$$

Hence by Cauchy's residue theorem, we get

$$\int_C f(z)\,dz = \int_\rho^R f(z)\,dz + \int_\Gamma f(z)\,dz + \int_{-R}^{-\rho} f(x)\,dx + \int_\gamma f(z)\,dz$$

$$= 2\pi i \,.\, \frac{e^{-a}}{-2a^2} = -\frac{\pi i}{a^2} e^{-a}. \qquad \ldots(1)$$

Since, $\dfrac{1}{z(z^2+a^2)} \to 0$ as $|z| \to \infty$, and hence by Jordan's lemma, $\lim_{R \to \infty} \int_\Gamma f(z)\,dz = 0$.

Also, since $\quad \lim_{z \to 0} z\, f(z) = \lim_{z \to 0} \dfrac{e^{iz}}{z^2 + a^2} = \dfrac{1}{a^2}$.

Hence $\quad \displaystyle\int_\gamma f(z)\,dz = \frac{i}{a^2}(0 - \pi) = -\frac{\pi i}{a^2}$

Hence as $\rho \to 0$ and $R \to \infty$, we get from (1),

$$\int_0^\infty f(x)\,dx + \int_{-\infty}^0 f(x)\,dx - \frac{\pi i}{a^2} = -\frac{\pi i}{a^2} e^{-a}$$

or

$$\int_{-\infty}^\infty \frac{e^{ix}}{x(x^2+a^2)}\,dx = \frac{\pi i}{a^2}(1 - e^{-a})$$

Equating imaginary parts, we get $\displaystyle\int_{-\infty}^\infty \frac{\sin x}{x(x^2+a^2)}\,dx = \frac{\pi}{a^2}(1 - e^{-a})$.

Example 9. *Integrating* $\dfrac{e^{iz}}{z+a}$ *along the boundary of the square defined by x = 0, x = R, y = 0, y = R, prove that* $\displaystyle\int_0^\infty \frac{\cos x}{x+a}\,dx = \int_0^\infty \frac{xe^{-ax}}{1+x^2}\,dx,\ a > 0$ *and* $\displaystyle\int_0^\infty \frac{\sin x}{x+a}\,dx = \int_0^\infty \frac{e^{-ax}}{1+x^2}\,dx.$

Sol. Consider $\displaystyle\int_C \frac{e^{iz}}{z+a}\,dz = \int_C f(z)\,dz$,

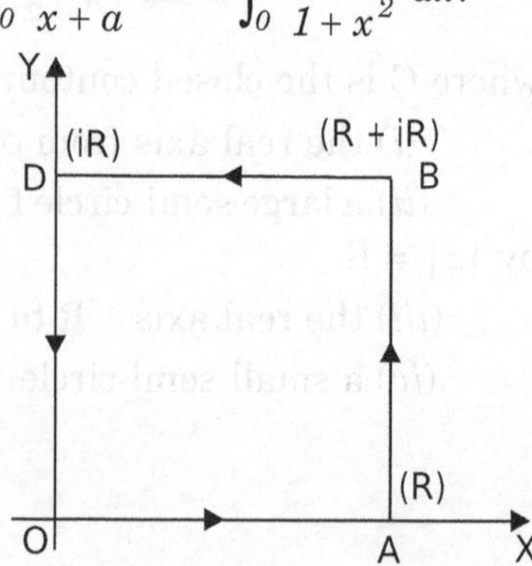

where C is the contour consisting of the perimeter of the given square. $f(z)$ has simple pole at $z = -a$ lying outside the given contour.

Hence by Cauchy's theorem, we have

$$\int_C f(z)\,dz = \left(\int_{AB} + \int_{BD} + \int_{DO} + \int_{OA}\right) f(z)\,dz = 0. \qquad ...(1)$$

On AB, $z = R + iy$, $dz = idy$ and y goes from 0 to R, so

$$\left|\int_{AB} f(z)\,dz\right| \le \int_{AB} |f(z)||dz| = \int_{AB} \frac{|e^{iz}|}{|z+a|}|dz| \le \int_{AB} \frac{|e^{iz}|}{|z|-a}|dz|$$

$$= \int_0^R \frac{|e^{i(R+iy)}|}{|R+iy|-a}|i\,dy| \le \int_0^R \frac{e^{-y}}{|R|-|iy|-a}\,dy$$

$$= \int_0^R \frac{e^{-y}}{R-y-a}\,dy = -\int_{R-a}^{-a} \frac{e^{u+a-R}}{u}\,du, \text{ putting } R-y-a=u$$

$$= \frac{e^a}{e^R}\int_{-a}^{R-a} \frac{e^u}{u}\,du \to 0 \text{ as } R \to \infty.$$

Similarly, $\int_{BD} f(z)\,dz = 0$ when $R \to \infty$.

On DO, $z = iy$, $dz = idy$ and on OA, $z = x$, $dz = dx$.

Hence making $R \to \infty$, we get from (1),

$$\int_\infty^0 \frac{e^{i(iy)}}{iy+a}\,i\,dy + \int_0^\infty \frac{e^{ix}}{x+a}\,dx = 0$$

or

$$\int_0^\infty \frac{e^{ix}}{x+a}\,dx = i\int_0^\infty \frac{e^{-y}}{iy+a}\,dy = i\int_0^\infty \frac{(i+x)\,e^{-ax}\,a\,dx}{a\,(1+ix)\,(i+x)}, \text{ putting } y = ax$$

$$= \int_0^\infty \frac{(i+x)\,e^{-ax}}{1+x^2}\,dx.$$

Equating real and imaginary parts, we have

$$\int_0^\infty \frac{\cos x}{x+a}\,dx = \int_0^\infty \frac{xe^{-ax}}{1+x^2}\,dx$$

and

$$\int_0^\infty \frac{\sin x}{x+a}\,dx = \int_0^\infty \frac{e^{-ax}}{1+x^2}\,dx.$$

Example 10. *Integrating* e^{-z^2} *round the rectangle with vertices at* $\pm R$, $\pm R + ia$, *prove that* $\int_{-\infty}^{\infty} e^{-(x+ai)^2}\,dx = \int_{-\infty}^{\infty} e^{-x^2}\,dx = \sqrt{\pi}$.

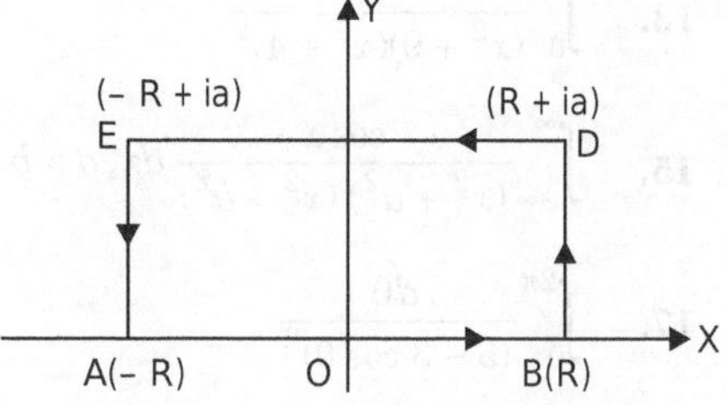

Sol. Consider $\int_C e^{-z^2}\,dz = \int_C f(z)\,dz$, where C is the perimeter of the given rectangle.

Since $f(z)$ is regular on the within C, we have by Cauchy's theorem,

$$\int_C f(z)\,dz \left(\int_{AB} + \int_{BD} + \int_{DE} + \int_{EA}\right) f(z)\,dz = 0$$

or $\int_{-R}^{R} f(x)\,dx + \int_0^a f(R+iy)\,i\,dy + \int_R^{-R} f(x+ia)\,dx + \int_a^0 f(-R+iy)\,i\,dy = 0.$ ...(1)

Now $\left|\int_0^a f(R+iy)\,i\,dy\right| \le \int_0^a \left| e^{-(R+iy)^2}\,i\,dy\right|$

$$\le \int_0^a e^{y^2-R^2}\,dy \to 0 \text{ as } R \to \infty.$$

Similarly, $\int_0^a f(-R+iy)\,i\,dy \to 0$ as $R \to \infty$.

Hence when $R \to \infty$, we get from (1),

$$\int_{-\infty}^{\infty} e^{-x^2}\,dx - \int_{-\infty}^{\infty} e^{-(x+ia)^2}\,dx = 0$$

or $$\int_{-\infty}^{\infty} e^{-(x+ai)^2}\,dx = \int_{-\infty}^{\infty} e^{-x^2}\,dx = 2\int_0^{\infty} e^{-x^2}\,dx = 2\cdot\frac{\sqrt{\pi}}{2} = \sqrt{\pi}.$$

TEST YOUR KNOWLEDGE

Evaluate the following integrals by contour integration:

1. $\int_0^{2\pi} \frac{d\theta}{5-3\cos\theta}$

2. $\int_0^{2\pi} \frac{d\theta}{a+b\sin\theta}$, where $a > |b|$

3. $\int_0^{\pi} \frac{\cos 3\theta}{5-4\cos\theta}\,d\theta$

4. $\int_0^{2\pi} \frac{\cos 2\theta}{5+4\cos\theta}\,d\theta$

5. $\int_0^{2\pi} \frac{d\theta}{1-2p\cos\theta+p^2}$, $0<p<1$

6. $\int_0^{2\pi} \frac{\cos 2\theta\,d\theta}{1-2p\cos\theta+p^2}$, $0<p<1$

7. $\int_0^{\pi} \frac{a\,d\theta}{a^2+\sin^2\theta}$

8. $\int_0^{\infty} \frac{dx}{(x^2+1)^2}$

9. $\int_{-\infty}^{\infty} \frac{dx}{x^4+1}$

10. $\int_0^{\infty} \frac{dx}{x^6+1}$

11. $\int_{-\infty}^{\infty} \frac{dx}{(x^2+a^2)(x^2+b^2)}$

12. $\int_{-\infty}^{\infty} \frac{x^2-x+2}{x^4+10x^2+9}\,dx$

13. $\int_0^{\infty} \frac{x^2\,dx}{(x^2+9)(x^2+4)^2}$

14. $\int_0^{\infty} \frac{x\sin x}{x^2+a^2}\,dx$, $a>0$

15. $\int_{-\infty}^{\infty} \frac{\cos x}{(x^2+a^2)(x^2+b^2)}\,dx$, $a>b>0$

16. $\int_0^{\infty} \frac{\sin mx}{x(x^2+a^2)}\,dx$, $m>0, a>0$

17. $\int_0^{2\pi} \frac{d\theta}{(5-3\cos\theta)^2}$

18. $\int_{-\infty}^{\infty} \frac{x^2}{(x^2+1)(x^2+4)}\,dx$

19. $\int_0^{\infty} \frac{\cos x}{x}\,dx$

20. $\int_{-\infty}^{\infty} \frac{\sin x}{x^2+4x+5}\,dx.$

Answers

1. $\dfrac{\pi}{2}$ **2.** $\dfrac{2\pi}{\sqrt{a^2-b^2}}$ **3.** $\dfrac{\pi}{24}$ **4.** $\dfrac{\pi}{6}$

5. $\dfrac{2\pi}{1-p^2}$ **6.** $\dfrac{2\pi p^2}{1-p^2}$ **7.** $\dfrac{\pi}{\sqrt{1+a^2}}$ **8.** $\dfrac{\pi}{4}$

9. $\dfrac{\pi}{\sqrt{2}}$ **10.** $\dfrac{\pi}{3}$ **11.** $\dfrac{\pi}{ab\,(a+b)}$ **12.** $\dfrac{5\pi}{12}$

13. $\dfrac{\pi}{200}$ **14.** $\dfrac{1}{2}\pi e^{-a}$ **15.** $\dfrac{\pi}{a^2-b^2}\left(\dfrac{e^{-b}}{b}-\dfrac{e^{-a}}{a}\right)$ **16.** $\dfrac{\pi}{2a^2}(1-e^{-ma})$

17. $\dfrac{5\pi}{32}$ **18.** $\dfrac{\pi}{3}$ **19.** 0 **20.** $\dfrac{-\pi\sin 2}{e}$.

5

Statistics and Probability

5.1. INTRODUCTION

Statistics is as old as the human society itself.

It is difficult to imagine any face of our life untouched by numerical data. The modern society is essentially data-oriented. It is, therefore, essential to know how to extract useful information from such data. This is the primary objective of statistics. Statistics concerns itself with the collection, presentation and drawing of inferences from numerical data which vary.

In singular sense, Statistics is used to describe the principles and methods which are employed in collection, presentation, analysis and interpretation of data. These devices help to simplify the complex data and make it possible for a common man to understand it without much difficulty. Human mind is unable to assimilate complicated data at a stretch. Statistical methods make these figures intelligible and readily understandable.

In plural sense, Statistics is considered as a numerical description of quantitative aspect of things.

Definition *Statistics is the science which deals with methods of collecting, classifying, presenting, comparing and interpreting numerical data collected to throw light on any sphere of enquiry.*

5.2. VARIABLE (OR VARIATE)

A quantity which can vary from one individual to another is called a **variable** or **variate,** *e.g.*, heights, weights, ages, wages of persons, rainfall records of cities etc.

Quantities which can take any numerical value within a certain range are called **continuous variables** *e.g.*, as the child grows, his/her height takes all possible values from 50 cm to 100 cm.

Quantities which are incapable of taking all possible values are called **discrete** or **discontinuous variables** *e.g.*, the number of children a man can have are positive integers 1, 2, 3 etc. (no value between any two consecutive integers).

5.3. FREQUENCY DISTRIBUTIONS

Consider the marks obtained by 60 students of B.A./B.Sc. Part III class in a college test in Mathematics according to their roll numbers:

38, 11, 40, 0, 26, 15, 5, 45, 7, 32, 2, 18, 42, 8, 31, 27, 4, 12, 35, 15, 0, 7, 28, 46, 9, 16, 29, 34, 10, 7, 5, 1, 17, 22, 35, 8, 36, 47, 11, 30, 19, 0, 16, 14, 16, 18, 41, 38, 2, 17, 42, 45, 48, 28, 7, 21, 8, 28, 5, 20.

The data does not give any useful information. It is rather confusing to mind. These are called **raw data** or **ungrouped data.**

We would like to bring out certain salient features of this data. If we express the data in ascending or descending order of magnitude, this does not reduce the bulk of the data. We condense the data into **classes** or **groups** as below:

(*i*) Determine the **range** of the data *i.e.*, difference between the largest and smallest numbers occurring in the data.

Here range = 48 – 0 = 48.

(*ii*) Decide upon the number of classes or groups into which raw data is to be grouped. There are no hard and fast rules for this. The insight of the experimentor determines this number. However, the number of classes should not be less than 5 and more than 30. With less number of classes accuracy is lost and with more number of classes the computations become tedious.

Let us take the number of classes = 7 here.

(*iii*) Divide the range by the desired number of classes to determine the approximate *width* or *size of class interval.* If the quotient is a fraction, take the next integer. In the above example, size of class interval is $\frac{48}{7}$ or 7.

As far as possible, classes should be of the same size.

(*iv*) Using the size of the interval, set up the class limits, making sure that the minimum and the maximum numbers occurring in the data are included in some class. As far as possible, open end classes ($a < x < b$) should be avoided since they create difficulty in analysis and interpretation. Boundaries of each class are selected in such a way that there is no ambiguity as to which class a particular item of the data belongs.

(*v*) The observations corresponding to the common point of two classes should always be included in the higher class, *e.g.,* if 20 is an element of the data and 10–20 and 20–30 are two classes, then 20 is to be put in the class 20–30 and not 10–20. That is to say every class should be regarded as open to the right.

(*vi*) Take each item from the data, one at a time, and place a *tally mark* (/) opposite the class to which it belongs. Tally marks are recorded in bunches of five. Having occurred four times, the fifth occurrence is represented by putting a cross-tally (\ or /) on the first four tallies ($\bcancel{||||}$ or $\cancel{||||}$). This technique facilitates the counting of the tally marks at the end.

(*vii*) *The count of tally marks in a particular class provides us with the frequency in that class*. The word 'frequency' is derived from 'how frequently' a variable occurs.

(*viii*) Marks are called the variable (x) and the number of students in a class is known as the frequency (f) or class frequency of the variable.

(*ix*) The total of all frequencies must equal the number of observations in the raw data.

(*x*) The table displaying the manner in which frequencies are distributed over various classes is called *frequency table.*

(*xi*) We are often interested in knowing, at a glance, the number of observations less than a particular value. This is done by finding cumulative frequency. The *cumulative frequency* corresponding to a class is the sum of frequencies of that class and of all classes prior to that class.

(*xii*) The table displaying the manner in which cumulative frequencies are distributed is called *cumulative frequency table.*

Using the above steps, we have the following cumulative frequency table for the example under consideration.

Class interval (marks x)	*Tally marks (number of students)*	*Frequency (f)*	*Cumulative Frequency*
0–7	𝍸 𝍸	10	10
7–14	𝍸 𝍸 ||	12	22
14–21	𝍸 𝍸 ||	12	34
21–28	||||	4	38
28–35	𝍸 |||	8	46
35–42	𝍸 ||	7	53
42–49	𝍸 ||	7	60
Total		60	

ILLUSTRATIVE EXAMPLES

Example 1. *The weights in grams of 50 apples picked at random from a consignment are as follows:*

106, 107, 76, 82, 109, 107, 115, 93, 187, 195, 123, 125, 111, 92, 86, 70, 126, 68, 130, 129, 139, 119, 115, 128, 100, 186, 84, 99, 113, 204, 111, 141, 136, 123, 90, 115, 98, 110, 78, 90, 107, 81, 131, 75, 84, 104, 110, 80, 118, 82.

Form the grouped frequency table by dividing the variate range into intervals of equal width, each corresponding to 20 gms in such a way that the mid-value of the first class corresponds to 70 gms.

Sol. Mid-value of first class = 70, Width of each class = 20 (Given)

∴ The first class interval is (70 – 10) – (70 + 10) *i.e.*, 60 – 80.

Weight in grams	*No. of apples*	*Frequency*
60–80	𝍸	5
80–100	𝍸 𝍸 |||	13
100–120	𝍸 𝍸 𝍸 ||	17
120–140	𝍸 𝍸	10
140–160	|	1
160–180		0
180–200	|||	3
200–220	|	1
Total		50

Example 2. *Form an ordinary frequency table from the following table:*

Marks	*No. of Students*	*Marks*	*No. of Students*
Above 0	*40*	*Above 30*	*18*
Above 10	*30*	*Above 40*	*12*
Above 20	*25*	*Above 50*	*0*

Sol.

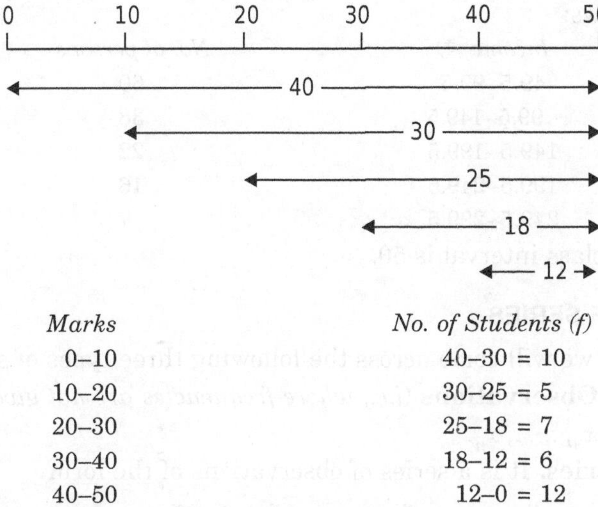

Marks	*No. of Students (f)*
0–10	40–30 = 10
10–20	30–25 = 5
20–30	25–18 = 7
30–40	18–12 = 6
40–50	12–0 = 12

5.4. 'EXCLUSIVE' AND 'INCLUSIVE' CLASS-INTERVALS

Class-intervals of the type $\{x : a \leq x < b\} = [a, b)$ are called 'exclusive' since they exclude the upper limit of the class. The following data are classified on this basis.

Income (₹)	*No. of persons*
50–100	88
100–150	70
150–200	52
200–250	30
250–300	23

In this method, the upper limit of one class is the lower limit of the next class. In this example, there are 88 persons whose income is from ₹ 50 to ₹ 99.99. A person whose income is ₹ 100 is included in the class ₹ 100—₹ 150.

Class-intervals of the type $\{x : a \leq x \leq b\} = [a, b]$ are called 'inclusive' since they include the upper limit of the class. The following data are classified on this basis.

Income (₹)	*No. of persons*
50–99	60
100–149	38
150–199	22
200–249	16
250–299	7

However, to ensure continuity and to get correct class-limits, exclusive method of classification should be adopted. To convert inclusive class-intervals into exclusive, we have to make an adjustment.

Adjustment. Find the difference between the lower-limit of the second class and the upper limit of the first class. Divide it by 2. Subtract the value so obtained from all the lower limits and add the value to all upper limits.

In the above example, the adjustment factor is $\dfrac{100-99}{2} = 0.5$. The adjusted classes would then be as follows:

Income (₹)	*No. of persons*
49.5–99.5	60
99.5–149.5	38
149.5–199.5	22
199.5–249.5	16
249.5–299.5	7

The size of the class interval is 50.

5.5. THREE TYPES OF SERIES

In this chapter, we will come across the following three types of series:

(*a*) **Individual Observations** (*i.e., where frequencies are not given*).

Form $x : x_1, x_2, x_3, ..., x_n$.

(*b*) **Discrete Series.** It is a series of observations of the form

$$\begin{array}{lllll} x: & x_1, & x_2, & x_3, ..., x_n \\ f: & f_1, & f_2, & f_3, ..., f_n \end{array}$$

(*c*) **Continuous Series.** It is a series of observations of the form

$$\begin{array}{lllll} \text{Class Interval:} & a_1 - a_2 & a_2 - a_3 & ...... & a_n - a_{n+1} \\ f: & f_1 & f_2 & ...... & f_n \end{array}$$

For the purpose of further calculations in statistical work, the mid-point of each class is taken to represent the class.

Thus, if m_i is the mid-point of the ith class, then $m_i = \dfrac{a_i + a_{i+1}}{2}$ and the above series takes the form

$$\begin{array}{lllll} \textit{Mid-value} \quad m: & m_1, & m_2, & m_3, ..., m_n \\ \textit{Frequency} \quad f: & f_1, & f_2, & f_3, ..., f_n. \end{array}$$

The mid-value of the ith class may also be denoted by x_i. Thus, a continuous series is reduced to the form of a discrete series.

5.6. GRAPHICAL REPRESENTATION

A frequency distribution when represented by means of a graph makes the unwieldy data intelligible. A better perspective can be had by representing the frequency distribution graphically since graphs, if drawn attractively, are eye-catching and leave a more lasting impression on the mind of the observer. Graphs are a good visual aid. But graphs do not give accurate measurements of the variable as are given by the tables. Another disadvantage is that by taking different scales, the facts may be misrepresented.

Some important types of graphs are given below:

(A) Histogram

In drawing the histogram of a given grouped frequency distribution:

(*a*) Mark off along the *x*-axis all the class intervals on a suitable scale. (If class-intervals are equal, then each = 1 cm is quite suitable).

(*b*) Mark frequencies along *y*-axis on a suitable scale.

(*c*) It must not be assumed that scale for both the axes will be same. We can have different scales for the two axes. The determination of scale depends upon our convenience and the type and nature of data. The scale or scales should be so chosen as to fit the size of graph-paper and to hold all the figures of the data.

(*d*) Construct rectangles with the class-intervals as bases and heights proportional (if the class intervals are equal) to frequencies.

A diagram with all these rectangles is called a **histogram.**

Example 3. *The weights (in grams) of 40 oranges picked at random from a basket are as follows:*

45, 55, 30, 110, 75, 100, 40, 60, 65, 40, 100, 75, 70, 60, 70, 95, 85, 80, 35, 45, 40, 50, 60, 65, 55, 45, 90, 85, 75, 85, 75, 70, 110, 100, 80, 70, 55, 30, 70.

Represent the data by means of a histogram.

Sol. Range = max. (110) – min. (30) = 80

Let number of class intervals = 7

Width of class interval = $\left(\frac{80}{7} \text{ or}\right)$ 12.

Wts. of oranges (in gms.)	*No. of oranges*	*Frequency*
30–42	卌 II	7
42–54	IIII	4
54–66	卌 III	8
66–78	卌 IIII	9
78–90	卌	5
90–102	卌	5
102–114	II	2
Total		40

The histogram of the above frequency distribution is given below:

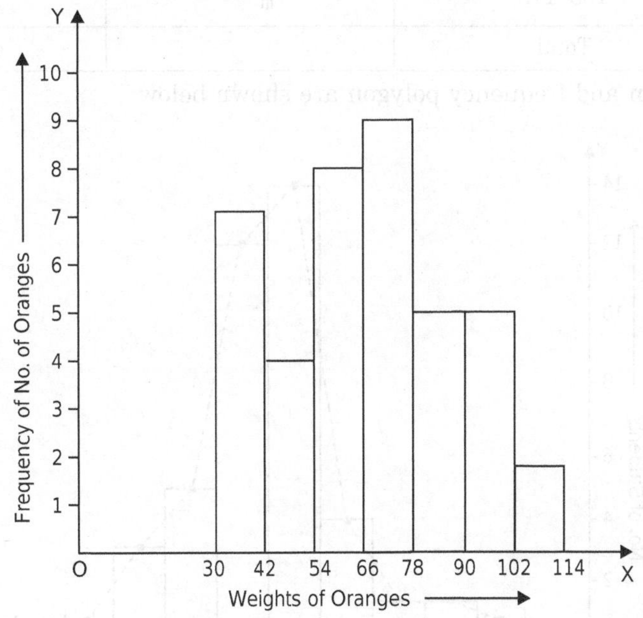

(B) Frequency Polygon

For a grouped frequency distribution, with equal class-intervals, a frequency polygon is obtained by joining the middle points of the upper sides (tops) of the adjacent rectangles of the histogram by means of straight lines. To complete the polygon, the mid-points at each end are joined to the immediately lower and higher mid-points at zero frequency *i.e.*, on the x-axis.

Example 4. *The following table gives the weights (to the nearest pound) of 40 male students at a university. Construct a frequency distribution with 7 classes and draw the histogram and frequency polygon.*

138, 164, 150, 132, 144, 125, 149, 157, 146, 158, 140, 147, 136, 148, 152, 144, 168, 126, 138, 176, 163, 119, 154, 165, 146, 173, 142, 147, 135, 140, 135, 102, 145, 135, 142, 150, 156, 145, 128.

Sol. Range of raw data = max. (176) – min. (102) = 74

Number of classes = 7

$\therefore$ Width of class interval = $\left(\dfrac{74}{7} \text{ or}\right)$ 11.

Weight (to the nearest pound)	*Tally marks*	*Frequency*
102–113	|	1
113–124	|	1
124–135	||||	4
135–146	卌 卌 ||||	14
146–157	卌 卌 ||	12
157–168	卌	5
168–179	|||	3
Total		40

The histogram and frequency polygon are shown below:

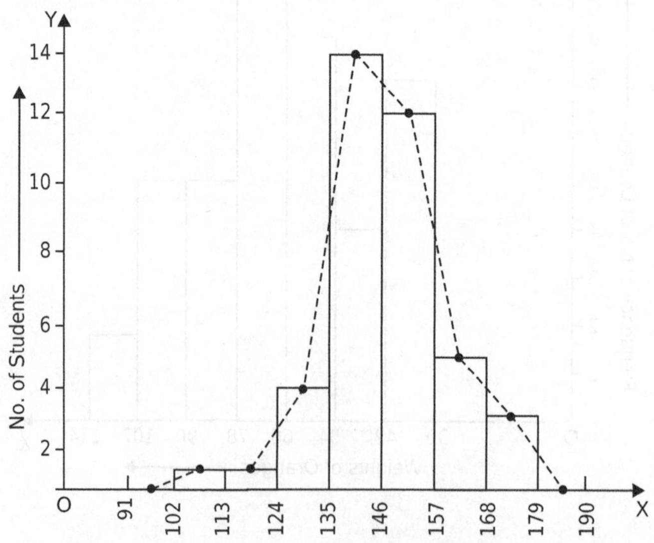

(*Histogram:* rectangles; *Frequency polygon:* shown dotted).

(C) Cumulative Frequency Curve or the Ogive

The curve obtained by plotting cumulative frequency is called a cumulative frequency curve or an ogive (pronounced ojive). There are two types of ogives.

(*i*) **Less than ogive.** Plot the points with the *upper limits* of the classes as abscissae and the corresponding less than cumulative frequency as ordinates. Join the points by a free hand smooth curve to get the less than ogive. It is a rising curve. (An ogive usually means less than ogive).

(*ii*) **More than ogive.** Plot the points with the *lower limits* of the classes as abscissae and the corresponding more than cumulative frequency as ordinates. Join the points by a free-hand smooth curve to get the more than ogive. It is a falling curve.

Consider the following frequency distribution:

Marks	*No. of students*	*Marks*	*No. of students*
10–20	4	40–50	20
20–30	6	50–60	18
30–40	10	60–70	2

Let us convert it first into a 'less than c.f.' distribution and then into a 'more than c.f.' distribution.

Marks less than	*No. of students*	*Marks more than*	*No. of students*
20	4	10	60
30	(+ 6 =) 10	20	(– 4 =) 56
40	(+ 10 =) 20	30	(– 6 =) 50
50	(+ 20 =) 40	40	(– 10 =) 40
60	(+ 18 =) 58	50	(– 20 =) 20
70	(+ 2 =) 60	60	(– 18 =) 2
		70	(– 2 =) 0

5.7. MEASURES OF CENTRAL TENDENCY

Tabulation arranges facts in a logical order and helps their understanding and comparison. But often, the groups tabulated are still too large for their characteristics to be readily grasped. What is desired is a numerical expression which summarises the characteristic of the group. Measures of central tendency or measures of location (also popularly called Averages) serve this purpose.

A figure which is used to represent a whole series should neither have the lowest value nor the highest in the series, but a value somewhere between these two limits, possibly in the centre, where most of the items of the series cluster. Such figures are called *Measures of Central Tendency* (or Averages).

According to Professor Bowley, "Averages are Statistical Constants which enable us to comprehend in a single effort the significance of the whole."

There are five types of averages in common use:

1. Arithmetic Average or Mean
2. Median
3. Mode
4. Geometric Mean
5. Harmonic Mean

We shall take them one by one.

5.7.1. Arithmetic Mean

In the case of Individual Observations (i.e., where frequency is not given):

1. Direct Method. If $x : x_1, x_2, ..., x_n$ then A.M. $\bar{x}$ is given by

$$\bar{x} = \frac{x_1 + x_2 + ... + x_n}{n} = \frac{1}{n}\Sigma x.$$

2. Short Cut Method. (*Shift of origin*) Shifting the origin to an *arbitrary point a*, the formula

$$\bar{x} = \frac{1}{n}\Sigma x \quad \text{becomes} \quad \bar{x} - a = \frac{1}{n}\Sigma(x - a)$$

or, $$\bar{x} = a + \frac{1}{n}\Sigma d_x \quad \text{where} \quad d_x = x - a$$

Here, a = arbitrary number, called *Assumed Mean*

$\Sigma d_x = \Sigma(x - a) = (x_1 - a) + (x_2 - a) + ... + (x_n - a)$

= sum of the deviations of the variate x from a

n = number of observations.

In the case of Discrete Series:

1. Direct Method. If the frequency distribution is

$x : x_1, x_2, ..., x_n$

$f : f_1, f_2, ..., f_n$, then

$$\bar{x} = \frac{f_1x_1 + f_2x_2 + ... + f_nx_n}{f_1 + f_2 + ... + f_n} = \frac{\Sigma fx}{N} \quad \text{where } N = f_1 + f_2 + ... + f_n = \Sigma f$$

2. Short Cut Method. (*Shift of origin*) Shifting the origin to an arbitrary point 'a', the formula

$$\bar{x} = \frac{1}{N}\Sigma fx \quad \text{becomes} \quad \bar{x} - a = \frac{1}{N}\Sigma f(x - a)$$

or $$\bar{x} = a + \frac{1}{N}\Sigma fd_x, \quad \text{where} \quad d_x = x - a$$

Thus $$\bar{x} = a + \frac{\mathbf{1}}{\mathbf{N}}\boldsymbol{\Sigma} \mathbf{fd_x}, \quad \text{where} \quad a = \text{assumed mean}$$

$\Sigma fd_x = \Sigma f(x - a)$

$= f_1(x_1 - a) + f_2(x_2 - a) + ... + f_n(x_n - a)$

= sum of the products of f and the deviation of the corresponding variate x from a.

$N = f_1 + f_2 + ... + f_n = \Sigma f.$

Note. If the frequencies are given in terms of class intervals, the mid-values of class intervals are considered as x and then the above formulae are applied.

In the case of Continuous Series having equal class intervals, say of width h, we use a different formula (Shift of origin and change of scale ; Step Deviation Method).

Let $$u = \frac{x - a}{h} \text{ then } x = a + hu$$

$\therefore$ $$\Sigma fx = \Sigma f(a + hu) = a\Sigma f + h\Sigma fu$$

Dividing both sides by N = Σf, we get

$$\frac{\Sigma fx}{N} = a + \frac{h\Sigma fu}{N} \quad \text{or} \quad \bar{x} = a + \mathbf{h}\frac{\Sigma \mathbf{fu}}{\mathbf{N}} \quad \text{where } u = \frac{x-a}{h}.$$

Weighted Arithmetic Mean. If the variate-values are not of equal importance, we may attach to them 'weights' $w_1, w_2, \ldots\ldots, w_n$ as measures of their importance.

The weighted mean $\bar{x}_w$ is defined as $\bar{x}_w = \dfrac{w_1x_1 + w_2x_2 + \ldots\ldots + w_nx_n}{w_1 + w_2 + \ldots\ldots + w_n} = \dfrac{\Sigma wx}{\Sigma w}$ (*i.e.*, write w for f).

Example 5. *Find the mean from the following data:*

Marks	*No. of students*	*Marks*	*No. of students*
Below 10	*5*	*Below 60*	*60*
Below 20	*9*	*Below 70*	*70*
Below 30	*17*	*Below 80*	*78*
Below 40	*29*	*Below 90*	*83*
Below 50	*45*	*Below 100*	*85*

Sol. The frequency distribution table can be written as:

Marks	*Mid values* (x)	f	$x - 55$	$u = \frac{x-55}{10}$	fu
0–10	5	5	– 50	– 5	– 25
10–20	15	4	– 40	– 4	– 16
20–30	25	8	– 30	– 3	– 24
30–40	35	12	– 20	– 2	– 24
40–50	45	16	– 10	– 1	– 16
50–60	55	15	0	0	0
60–70	65	10	10	1	10
70–80	75	8	20	2	16
80–90	85	5	30	3	15
90–100	95	2	40	4	8
		N = Σf = 85			Σfu = – 56

$$\text{Mean } \bar{x} = a + h\frac{\Sigma fu}{N} = 55 + 10 \times \left(\frac{-56}{86}\right) \qquad [\text{Here } a = 55,\ h = 10]$$

$$= 55 - \frac{112}{17} = 48.41 \text{ marks.}$$

Example 6. *The mean of 200 items was 50. Later on it was discovered that two items were misread as 92 and 8 instead of 192 and 88. Find out the correct mean.*

Sol. Here incorrect value of $\bar{x} = 50,\ n = 200$

Since $\bar{x} = \dfrac{\Sigma x}{n} \quad \therefore \quad \Sigma x = n\bar{x}$

Using incorrect value of $\bar{x}$,

Incorrect $\Sigma x = 200 \times 50 = 10000$

$\therefore$ Corrected value of $\Sigma x = 10000 - (92 + 8) + (192 + 88) = 10180$

$$\text{Correct mean} = \frac{\text{Corrected } \Sigma x}{n} = \frac{10180}{200} = 50.9.$$

Properties of Arithmetic Mean

Property I. *The algebraic sum of the deviations of all the variates from their arithmetic mean is zero.*

Proof. Let d_x be the deviation of the variate x from the mean $\bar{x}$, then $d_x = x - \bar{x}$

$$\therefore \quad \Sigma f d_x = \Sigma f(x - \bar{x}) = \Sigma fx - \bar{x}\Sigma f$$

$$= N\bar{x} - N\bar{x} = 0 \qquad \left| \because \bar{x} = \frac{\Sigma fx}{N}, \text{ where } N = \Sigma f. \right.$$

Property II. *The sum of the squares of the deviations of a set of values is minimum when taken about mean.*

Proof. Let the frequency distribution be x_i/f_i, $i = 1, 2, \ldots\ldots, n$. Let z be the sum of the squares of the deviations of the given values from an arbitrary point a (say).

$$\Rightarrow \text{ Let } \quad z = \sum_{i=1}^{n} f(x - a)^2.$$

We have to show that z is minimum when $a = \bar{x}$.

$$z \text{ will be minimum when } \frac{dz}{da} = 0 \text{ and } \frac{d^2z}{da^2} > 0$$

$$\text{Now} \quad \frac{dz}{da} = \sum_{i=1}^{n} 2f(x - a).(-1) = -2\sum_{i=1}^{n} f(x - a)$$

$$\therefore \quad \frac{dz}{da} = 0 \Rightarrow -2\Sigma f(x - a) = 0$$

$$\Rightarrow \Sigma fx - a\Sigma f = 0$$

$$\Rightarrow N\bar{x} - aN = 0 \qquad \left[\because \bar{x} = \frac{\Sigma fx}{N}, \Sigma f = N\right]$$

$$\Rightarrow \bar{x} - a = 0 \qquad (\because N = \Sigma f \neq 0)$$

$$\Rightarrow a = \bar{x}$$

$$\text{Also} \quad \frac{d^2z}{da^2} = -2\sum_{i=1}^{n} f(-1) = 2\Sigma f = 2N > 0$$

Hence z is minimum when $a = \bar{x}$.

Property III. (*Mean of the composite series*)

If $\bar{x}_i$ $(i = 1, 2, \ldots\ldots, k)$ be the arithmetic means of k distributions with respective frequencies n_i $(i = 1, 2, \ldots\ldots, k)$, then mean $\bar{x}$ of the whole distribution obtained by combining the k distributions is given by

$$\bar{x} = \frac{n_1\bar{x}_1 + n_2\bar{x}_2 + \ldots + n_k\bar{x}_k}{n_1 + n_2 + \ldots + n_k} = \frac{\sum_i n_i\bar{x}_i}{\sum_i n_i}$$

Proof. Let $x_{11}, x_{12}, x_{13}, \ldots\ldots, x_{1n_1}$ be the variables of the first distribution, $x_{21}, x_{22}, \ldots\ldots, x_{2n_2}$ be the variables of the second distribution, and so on. Then by definition

$$\left.\begin{aligned} \bar{x}_1 &= \frac{1}{n_1}(x_{11} + x_{12} + \ldots\ldots + x_{1n_1}) \\ \bar{x}_2 &= \frac{1}{n_2}(x_{21} + x_{22} + \ldots\ldots + x_{2n_2}) \\ &\ldots\ldots\ldots\ldots\ldots\ldots\ldots\ldots\ldots\ldots \\ \bar{x}_k &= \frac{1}{n_k}(x_{k_1} + x_{k_2} + \ldots\ldots + x_{kn_k}) \end{aligned}\right\} \qquad \ldots(A)$$

The mean $\bar{x}$ of the whole distribution of size $(n_1 + n_2 + \ldots\ldots + n_k)$ is given by

$$\bar{x} = \frac{(x_{11} + x_{12} + \ldots\ldots + x_{1n_1}) + (x_{21} + x_{22} + \ldots\ldots + x_{2n_2}) + \ldots\ldots + (x_{k_1} + x_{k_2} + \ldots\ldots + x_{kn_k})}{n_1 + n_2 + \ldots\ldots + n_k}$$

$$= \frac{n_1\bar{x}_1 + n_2\bar{x}_2 + \ldots\ldots + n_k\bar{x}_k}{n_1 + n_2 + \ldots\ldots + n_k} = \frac{\sum\limits_i n_i\bar{x}_i}{\sum\limits_i n_i}.$$

Example 7. *The mean annual salary paid to all employees of a company was ₹ 5000. The mean annual salaries paid to male and female employees were ₹ 5200 and ₹ 4200 respectively. Determine the percentage of males and females employed by the company.*

Sol. Let p_1 and p_2 represent percentage of males and females respectively.

Then $p_1 + p_2 = 100$ ...(1)

Mean annual salary of all employees $(\bar{x})$ = ₹ 5000

Mean annual salary of all males $(\bar{x}_1)$ = ₹ 5200

Mean annual salary of all females $(\bar{x}_2)$ = ₹ 4200

Using $\bar{x} = \dfrac{p_1\bar{x}_1 + p_2\bar{x}_2}{p_1 + p_2}$, we get $5000 = \dfrac{5200p_1 + 4200p_2}{100}$

or $52p_1 + 42p_2 = 5000$ or $26p_1 + 21p_2 = 2500$ or $26p_1 + 21(100 - p_1) = 2500$ [Using (1)]

or $5p_1 = 2500 - 2100 = 400$ $\therefore$ $p_1 = 80$ and $p_2 = 100 - 80 = 20$

Hence the percentage of males and females is 80 and 20 respectively.

5.7.2. Median

1. Median is the central value of the variable **when the values are arranged in ascending or descending order of magnitude.** When the observations are arranged in the order of their size, the median is the value of that item which has equal number of observations on either side. Median divides the distribution into two equal parts. Median is, thus, a potential average.

For the computation of median, it is necessary that the items be arranged in ascending or descending order.

2. For an ungrouped frequency distribution, if the n values of the variate are arranged in ascending or descending order of magnitude.

(*a*) *When n is odd,* the middle value *i.e.,* $\left(\frac{n+1}{2}\right)^{\text{th}}$ value gives the median.

(*b*) *When n is even,* there are two middle values $\left(\frac{n}{2}\right)^{\text{th}}$ and $\left(\frac{n}{2}+1\right)^{\text{th}}$.

The arithmetic mean of these two values gives the median.

3. For discrete frequency distribution, median is obtained by considering cumulative frequencies. Find $\frac{N+1}{2}$ where $N = \Sigma f_i$. Find cumulative frequency just $\geq \frac{N+1}{2}$. The corresponding value of x is the median.

4. For a grouped frequency distribution, the median is given by the formula,

$$\textbf{Median} = l + \frac{\mathbf{h}}{\mathbf{f}}\left(\frac{\mathbf{N}}{\mathbf{2}} - \mathbf{C}\right)$$

where, l = lower limit of median class, where median class is the class corresponding to cumulative frequency just $\geq \frac{N}{2}$

h = width of the median class ; f = frequency of the median class

$N = \Sigma f$; C = cumulative frequency of the class **preceding** the median class.

5. Partition values. These are the values of the variate which divide the total frequency into a number of equal parts. Median being that value of the variate which divides the total frequency into two equal parts.

Quartiles. Quartiles are those values of the variate which divide the total frequency into four equal parts. When the lower half before the median is divided into two equal parts, the value of the dividing variate is called Lower Quartile and is denoted by Q_1. The value of the variate dividing the upper half into two equal parts is called the Upper Quartile and is denoted by Q_3. (Q_2 being the median). The formulae for computation are

$$Q_1 = l + \frac{h}{f}\left(\frac{N}{4} - C\right) \; ; \; Q_3 = l + \frac{h}{f}\left(\frac{3N}{4} - C\right)$$

Example 8. *Below are given the marks obtained by a batch of 20 students in a certain class in Mathematics and Physics:*

Roll Nos.:	1	2	3	4	5	6	7	8	9	10
Marks in Maths.:	53	54	52	32	30	60	47	46	35	28
Marks in Physics:	58	55	25	32	26	85	44	80	33	72
Roll Nos.:	11	12	13	14	15	16	17	18	19	20
Marks in Maths.:	25	42	33	48	72	51	45	33	65	29
Marks in Physics:	10	42	15	46	50	64	39	38	30	36

In which subject is the level of knowledge of the students higher?

Sol. To find out the subject in which the level of knowledge of the students is higher, we find out the medians of both the series. The subject for which the median value is higher will be the subject in which the level of knowledge of the students is higher. Let us arrange the marks in ascending order of magnitude.

S. No.	*Marks in Maths*	*Marks in Physics*	*S. No.*	*Marks in Maths*	*Marks in Physics*
1	25	10	11	46	42
2	28	15	12	47	44
3	29	25	13	48	46
4	30	26	14	51	50
5	32	30	15	52	55
6	33	32	16	53	58
7	33	33	17	54	64
8	35	36	18	60	72
9	42	38	19	65	80
10	45	39	20	72	85

Number of items in each case = 20 (even)

Median marks in Mathematics

$$= \text{A.M. of sizes of } \left(\frac{20}{2}\right)^{\text{th}} \text{ and } \left(\frac{20}{2}+1\right)^{\text{th}} \text{ items}$$

$$= \text{A.M. of sizes of 10th and 11th items} = \frac{45+46}{2} = 45.5.$$

Median marks in Physics $= \text{A.M. of sizes of 10th and 11th items} = \dfrac{39+42}{2} = 40.5.$

Since the median marks in Mathematics are greater than the median marks in Physics, the level of knowledge in Mathematics is higher.

Example 9. *Obtain the median for the following frequency distribution*:

x :	*1*	*2*	*3*	*4*	*5*	*6*	*7*	*8*	*9*
f:	*8*	*10*	*11*	*16*	*20*	*25*	*15*	*9*	*6*

Sol. The cumulative frequency distribution table is given below:

x	f	*c.f.*
1	8	8
2	10	18
3	11	29
4	16	45
5	20	65
6	25	90
7	15	105
8	9	114
9	6	120

Here N = 120 $\therefore \quad \dfrac{N+1}{2} = 60.5$

Cumulative frequency just greater than $\frac{N+1}{2}$ is 65 and the value of x corresponding to *c.f.* 65 is 5. Hence median is 5.

Example 10. *Find the median, lower and upper quartiles from the following table:*

Marks	*No. of students*	*Marks*	*No. of students*
Below 10	*15*	*Below 50*	*94*
Below 20	*35*	*Below 60*	*127*
Below 30	*60*	*Below 70*	*198*
Below 40	*84*	*Below 80*	*249*

Sol. From the above table, we reconstruct the *c.f.* table with class intervals.

Marks	*No. of students (f)*	*c.f.*
0–10	15	15
10–20	20	35
20–30	25	60
30–40	24	84
40–50	10	94
50–60	33	127
60–70	71	198
70–80	51	249

Here $N = 249$

(*i*) **Calculation of Median**

$\frac{N}{2} = 124.5$ ∴ median class is 50–60, $l = 50$; $h = 10, f = 33, C = 94$

$$\therefore \quad \text{Median} = l + \frac{h}{f}\left(\frac{N}{2} - C\right) = 50 + \frac{10}{33}(124.5 - 94)$$

$$= 50 + \frac{305}{33} = 50 + 9.24 = 59.24 \text{ marks.}$$

(*ii*) **Calculation of lower quartile Q_1**

$\frac{N}{4} = 62.25$ ∴ lower quartile class is 30–40, $l = 30$

$h = 10, f = 24, C = 60$

$$\therefore \quad Q_1 = l + \frac{h}{f}\left(\frac{N}{4} - C\right) = 30 + \frac{10}{24}(62.25 - 60)$$

$$= 30 + \frac{22.5}{24} = 30 + .94 = 30.94 \text{ marks.}$$

(*iii*) **Calculation of upper quartile Q_3**

$\frac{3N}{4} = \frac{747}{4} = 186.75$ ∴ upper quartile class is 60–70

$l = 60, h = 10, f = 71, C = 127$

$$\therefore \quad Q_3 = l + \frac{h}{f}\left(\frac{3N}{4} - C\right) = 60 + \frac{10}{71}(186.75 - 127)$$

$$= 60 + \frac{597.5}{71} = 60 + 8.41 = 68.41 \text{ marks.}$$

5.7.3. Mode

1. ***Mode.*** Mode is the value which occurs most frequently in a set of observations and around which the other items of the set cluster densely. It is the point of maximum frequency or the point of greatest density. In other words, the mode or modal value of the distribution is that value of the variate for which frequency is maximum.

2. ***Calculation of Mode***

(*a*) In the case of discrete frequency distribution, mode is the value of x corresponding to maximum frequency.

But in any one (or more) of the following cases:

(*i*) if the maximum frequency is repeated.

(*ii*) if the maximum frequency occurs in the very beginning or at the end of the distribution.

(*iii*) if there are irregularities in the distribution, the value of mode is determined by the *method of grouping* (illustrated in examples below).

(*b*) In the case of continuous frequency distribution, mode is given by the formula:

$$\text{Mode} = l + \frac{f_m - f_1}{2f_m - f_1 - f_2} \times h$$

where l is the lower limit, h is the width and f_m is the frequency of the model class, f_1 and f_2 are the frequencies of the classes preceding and succeeding the modal class respectively.

While applying the above formula, it is necessary to see that the class-intervals are of the same size. **If they are unequal, they should first be made equal** on the assumption that the frequencies are equally distributed throughout the class.

In case $f_m - f_1 < 0$ or $2f_m - f_1 - f_2 = 0$, use the formula

$$\text{Mode} = l + \frac{\Delta_1}{\Delta_1 + \Delta_2} \times h$$

where $\Delta_1 = | f_m - f_1 |$ and $\Delta_2 = | f_m - f_2 |$.

(*c*) **For a symmetrical distribution, mean, median and mode coincide.**

(*d*) Where mode is **ill-defined** *i.e.,* where the method of grouping also, fails, its value can be ascertained by the formula

Mode = 3 Median – 2 Mean

This measure is called the **empirical mode.**

Example 11. *Calculate the mode from the following frequency distribution*:

Size (x):	4	5	6	7	8	9	10	11	12	13
Frequency (f):	2	5	8	9	12	14	14	15	11	13

Sol. Method of Grouping:

Size (x)	*Frequency*					
	I	*II*	*III*	*IV*	*V*	*VI*
4	2	7		15		
5	5		13		22	
6	8	17				29
7	9		21	35		
8	12	26			**40**	
9	14		**28**			**43**
10	14	**29**		**40**		
11	15		26		39	
12	11	24				
13	13					

Explanation:

In column I, *original* frequencies are written.

In column II, frequencies of column I are combined *two by two*.

In column III, *leave the first frequency of column I* and combine the others *two by two*.

In column IV, frequencies of column I are combined *three by three*.

In column V, *leave the first frequency of column I* and combine the others *three by three*.

In column VI, *leave the first two frequencies in column I* and combine the others *three by three*.

In all these columns, the maximum frequency is written in bold black type.

Note. All operations are done on column I.

Now we frame another table in which against every maximum item of columns I to VI, we write down the corresponding size or sizes. The size (x) which occurs maximum number of times is the mode.

Columns	Size of item having max. frequency
I	11
II	10, 11
III	9, 10
IV	10, 11, 12
V	8, 9, 10
VI	9, 10, 11

Since the item 10 occurs maximum number of times (*i.e.*, 5 times), hence the mode is 10.

Example 12. *Find the mode of the following*:

Marks:	*1–5*	*6–10*	*11–15*	*16–20*	*21–25*
No. of candidates:	*7*	*10*	*16*	*32*	*24*
Marks:	*26–30*	*31–35*	*36–40*	*41–45*	
No. of candidates:	*18*	*10*	*5*	*1*	

Sol. Here the greatest frequency 32 lies in the class 16–20. Hence modal class is 16–20. But the actual limits of this class are 15.5–20.5.

$$l = 15.5, f_m = 32, f_1 = 16, f_2 = 24, h = 5$$

$$\therefore \quad \text{Mode} = l + \frac{f_m - f_1}{2f_m - f_1 - f_2} \times h = 15.5 + \frac{32 - 16}{64 - 16 - 24} \times 5$$

$$= 15.5 + \frac{16}{24} \times 5 = 15.5 + \frac{10}{3} = 18.83 \text{ marks.}$$

5.7.4. Geometric Mean

Geometric Mean. (*a*) Geometric Mean (G.M.) of n individual observations $x_1, x_2, \ldots\ldots, x_n$ $(x_i \neq 0)$ is the nth root of their product.

Thus $$G = (x_1, x_2, \ldots, x_n)^{1/n}$$

Taking logarithms of both sides $\log G = \frac{1}{n}(\log x_1 + \log x_2 + \ldots + \log x_n) = \frac{1}{n}\sum_{i=1}^{n} \log x_i$

$$\therefore \quad G = \text{antilog}\left[\frac{1}{n}\sum_{i=1}^{n} \log x_i\right]$$

(*b*) If $x_1, x_2, \ldots\ldots, x_n$ occur $f_1, f_2, \ldots\ldots, f_n$ times respectively and $N = \sum_{i=1}^{n} f_i$, then G.M. is given by

$$G = (x_1^{f_1} x_2^{f_2} \ldots\ldots x_n^{f_n})^{1/N}$$

Taking logarithms of both sides

$$\log G = \frac{1}{N}(f_1 \log x_1 + f_2 \log x_2 + \ldots + f_n \log x_n) = \frac{1}{N}\sum_{i=1}^{n} f_i \log x_i$$

$$G = \text{antilog}\left[\frac{1}{N}\sum_{i=1}^{n} f_i \log x_i\right]$$

(*c*) In the case of continuous frequency distribution, x is taken to be the value corresponding to the mid-points of the class-intervals.

Example 13. *Compute the geometric mean from the following data*:

Marks	*No. of students*
0–10	*10*
10–20	*5*
20–30	*8*
30–40	*7*
40–50	*20*

Sol.

Marks	*Mid-values* (*x*)	*No. of Students* (*f*)	*log x*	*f log x*
0–10	5	10	0.6990	6.9900
10–20	15	5	1.1761	5.8805
20–30	25	8	1.3979	11.1832
30–40	35	7	1.5441	10.8087
40–50	45	20	1.6532	33.0640
		50		67.9264

$$\log G = \frac{1}{N} \Sigma f \log x = \frac{67.9264}{50} = 1.3585$$

$$G = \text{antilog } 1.3585 = 22.83.$$

5.7.5. Harmonic Mean

Harmonic Mean. Harmonic mean of a number of observations is the reciprocal of the arithmetic mean of the reciprocals of the given values. Thus, the harmonic mean H of n observations $x_1, x_2, ..., x_n$ is

$$H = \frac{1}{\frac{1}{n}\sum_{i=1}^{n} \frac{1}{x_i}} = \frac{n}{\frac{1}{x_1} + \frac{1}{x_2} + ... + \frac{1}{x_n}}.$$

If $x_1, x_2, ..., x_n$ (none of them being zero) have the frequencies $f_1, f_2, ..., f_n$ respectively, then harmonic mean is given by

$$H = \frac{1}{\frac{1}{N}\sum_{i=1}^{n} \frac{f_i}{x_i}} = \frac{N}{\frac{f_1}{x_1} + \frac{f_2}{x_2} + ... + \frac{f_n}{x_n}}, \; N = \sum_{i=1}^{n} f_i$$

In the case of class-intervals, x is taken to be the mid-value of the class-interval.

Example 14. *Find out the harmonic mean of the following data*:

Marks (out of 150)	*No. of students*
10	*2*
20	*3*
40	*6*
60	*5*
120	*4*

Sol.

x	f	$\frac{1}{x}$	$\frac{f}{x}$
10	2	.100	.200
20	3	.050	.150
40	6	.025	.150
60	5	.017	.085
120	4	.008	.032
	20		.617

$$\text{H.M.} = \frac{N}{\Sigma \frac{f}{x}} = \frac{20}{.617} = 32.4.$$

Example 15. *An aeroplane flies along the four sides of a square at speeds of 100, 200, 300 and 400 km/hr. respectively. What is the average speed of the aeroplane in its flight around the square?*

Sol. When *equal distances* are covered with unequal speeds, the harmonic mean is the proper average.

$$\therefore \quad \text{Average speed} = \frac{4}{\frac{1}{100}+\frac{1}{200}+\frac{1}{300}+\frac{1}{400}} = 192 \text{ km/hr.}$$

TEST YOUR KNOWLEDGE

1. The minimum temperature in (°C) for Delhi for the month of July, 1983 as reported by the Meterological Department is given below. Construct a frequency distribution table for it.
30.3, 30.0, 25.8, 26.5, 24.2, 25.2, 28.0, 28.0, 29.5, 27.8, 30.0, 31.1, 27.2, 25.9, 27.6, 24.5, 24.4, 27.0, 28.1, 26.0, 25.4, 28.0, 26.9, 25.7, 27.2, 25.5, 26.6, 28.5, 28.0, 27.7, 24.0.
2. The following are the monthly rents (in rupees) of 40 shops. Tabulate the data by grouping in intervals of ₹ 8.
38, 42, 49, 37, 82, 37, 75, 62, 54, 79, 84, 75, 63, 44, 74, 44, 36, 69, 54, 48, 74, 47, 52, 57, 62, 67, 72, 77, 82, 51, 31, 38, 43, 75, 67, 77, 47, 64, 84, 81.
3. Draw a histogram representing the following frequency distribution:

Monthly Wages (in ₹)	*Number of Workers*
15	2
20	20
25	26
30	16
35	9
40	4
45	3

[**Hint.** Mid-values of class intervals of size 5 are given.]

4. Represent the following distribution by a (*i*) histogram and (*ii*) frequency polygon.

Scores	*Frequency*
90–99	2
80–89	12
70–79	22
60–69	20

50–59	14
40–49	3
30–39	1

5. Represent the following distribution by an ogive:

Marks	*No. of students*	*Marks*	*No. of students*
0–10	5	50–60	4
10–20	13	60–70	1
20–30	12	70–80	3
30–40	11	80–90	1
40–50	8	90–100	2

6. Compute the arithmetic mean for the following data:

Height (in cm):	219	216	213	210	207	204	201	198	195
No. of persons:	2	4	6	10	11	7	5	4	1

7. Find the average marks of students from the following data:

Marks	*No. of students*	*Marks*	*No. of students*
Above 0	80	Above 60	28
Above 10	77	Above 70	16
Above 20	72	Above 80	10
Above 30	65	Above 90	8
Above 40	55	Above 100	0
Above 50	43		

8. Two hundred persons were interviewed by a public opinion polling agency. The frequency distribution gives the ages of persons interviewed.

Age Group (years)	*Frequency*	*Age Group (years)*	*Frequency*
80–89	2	40–49	56
70–79	2	30–39	40
60–69	6	20–29	42
50–59	20	10–19	32

Calculate the arithmetic mean of the data.

9. Calculate the arithmetic mean from the following data:

Class interval	*Frequency*	*Class interval*	*Frequency*
0–1	8	15–25	11
1–3	8	25–28	10
3—5	10	28—30	9
5–10	12	30–45	8
10–15	18	45–60	6

10. Find the class intervals if the arithmetic mean of the following distribution is 33 and assumed mean is 35.

Step deviation (u):	– 3	– 2	– 1	0	1	2
Frequency (f):	5	10	25	30	20	10

11. The average height of a group of 25 children was calculated to be 78.4 cm. It was later discovered that one value was misread as 69 cm instead of the correct value 96 cm. Calculate the correct average.

12. A candidate obtains the following percentage in an examination : English 60, Hindi 75, Mathematics 63, Physics 59 and Chemistry 55. Find the weighted mean if weights 2, 1, 5, 5, 3 are allotted to the subjects.

13. From the following data calculate the missing frequency:

No. of tablets	*No. of persons cured*	*No. of tablets*	*No. of persons cured*
4–8	11	24–28	9
8–12	13	28–32	17
12–16	16	32–36	6
16–20	14	36–40	4
20–24	?		

The average number of tablets to cure a person is 20.

14. The frequencies of values 0, 1, 2, ..., n of a variable are given by

$q^n, {}^nC_1q^{n-1}p, {}^nC_2q^{n-2}p^2, ..., p^n$ where $p + q = 1$. Show that the mean is np.

15. The mean marks obtained by 300 students in the subject of Statistics are 45. The mean of the top 100 of them was found to be 70 and the mean of the last 100 was known to be 20. What is the mean of the remaining 100 students?

Answers

6. 207.54 cm **7.** 51.75 marks **8.** 35.8 years **9.** 17.36
10. 0–10, 10–20, 20–30, 30–40, 40–50, 50–60 **11.** 79.48 cm **12.** 60.63% marks
13. 14 **15.** 45

5.8. DISPERSION

A measure of central tendency by itself can exhibit only one of the important characteristics of distribution. It can represent a series only 'as best as a single figure can'. It is inadequate to give us a complete idea of the distribution. It must be supported and supplemented by some other measures. One such measure is **Dispersion.**

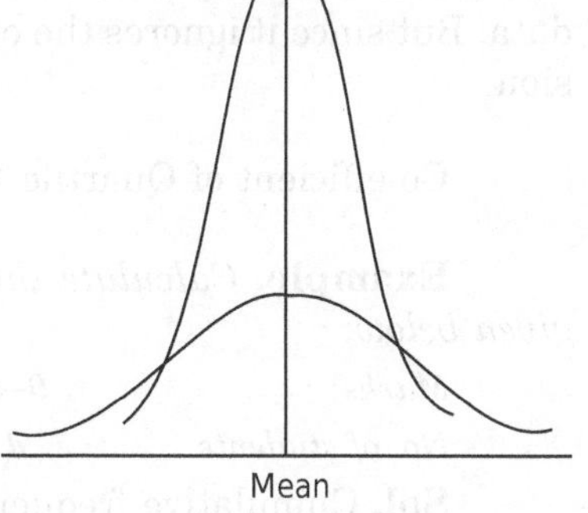

Two or more frequency distributions may have exactly identical averages but even then they may differ markedly in several ways. Further analysis is, therefore, essential to account for these differences. Consider the following example:

Distribution A:	75	85	95	105	115	125
Distribution B:	10	20	30	70	180	290

A.M. of each distribution is $\frac{600}{6} = 100$. In distribution A, the values of the variate differ from 100 but the difference is small. In distribution B, the items are widely scattered and lie far from the mean. Although the A.M. is the same, yet the two distributions widely differ from each other in their formation.

Therefore, while studying a distribution, it is equally important to know how the variates are clustered around or scattered away from the point of central tendency. Such variation is called *dispersion* or *spread* or *scatter* or *variability*. Thus, *dispersion is the extent to which the values are dispersed about the central value.*

5.9. MEASURES OF DISPERSION

The following are the measure of dispersion:

(*a*) Range (*b*) Quartile deviation or Semi-inter-quartile range

(*c*) Average (or mean) deviation (*d*) Standard deviation.

(*a*) **Range.** Range is the difference between the extreme values of the variate.

$$\text{Range} = \text{L} - \text{S}, \text{ where L = Largest and S = Smallest}$$

$$\text{Co-efficient of Range} = \frac{\text{L} - \text{S}}{\text{L} + \text{S}}.$$

It is easily understood and computed. But it suffers from the drawback that it depends exclusively on the two extreme values. It is not a reliable measure of dispersion.

(*b*) **Quartile Deviation.** The difference between the upper and lower quartiles *i.e.*, $Q_3 - Q_1$ is known as the *inter quartile range* and half of it *i.e.*, $\frac{1}{2}(Q_3 - Q_1)$, is called the *semi-inter-quartile range* or the *quartile deviation*.

$$\text{Quartile Deviation} = \frac{1}{2}(Q_3 - Q_1).$$

It is definitely a better measure of dispersion than range as it makes use of 50% of the data. But since it ignores the other 50% of the data, it is also not a reliable measure of dispersion.

$$\text{Co-efficient of Quartile Deviation} = \frac{Q_3 - Q_1}{Q_3 + Q_1}.$$

Example. *Calculate the quartile deviation of the marks of 39 students in Statistics given below*:

Marks:	0–5	5—10	10—15	15–20	20–25	25–30
No. of students:	4	6	8	12	7	2

Sol. Cumulative frequency table is given below:

Marks	*No. of students (f)*	*c.f.*
0–5	4	4
5–10	6	10
10–15	8	18
15–20	12	30
20–25	7	37
25–30	2	39

Here $N = \Sigma f = 39$; $\frac{N}{4} = 9.75$ $\therefore$ Class of Q_1 is 5–10

$$Q_1 = l + \frac{h}{f}\left(\frac{N}{4} - C\right) = 5 + \frac{5}{6}(9.75 - 4) = 5 + \frac{5 \times 5.75}{6} = 9.79$$

$\frac{3N}{4} = 29.25$ $\therefore$ Class of Q_3 is 15–20

$$Q_3 = l + \frac{h}{f}\left(\frac{3N}{4} - C\right) = 15 + \frac{5}{12}(29.25 - 18) = 15 + \frac{5 \times 11.25}{12} = 19.69$$

Quartile deviation $= \frac{1}{2}(Q_3 - Q_1) = \frac{1}{2}(19.69 - 9.79) = \frac{1}{2} \times 9.90 = 4.95$ marks.

(*c*) **Average Deviation or Mean Deviation.** If $x_1, x_2, x_3, \ldots\ldots\ldots, x_n$ occur $f_1, f_2, f_3, \ldots\ldots, f_n$ times respectively and $N = \sum_{i=1}^{n} f_i$, the mean deviation from the average A (usually mean or median) is given by

$$\textit{Mean deviation} = \frac{1}{N}\sum_{i=1}^{n} f_i \mid x_i - A \mid,$$

where $\mid x_i - A \mid$ represents the modulus or the absolute value of the deviation $(x_i - A)$.

Since the mean deviation is based on all the values of the variate, it is a better measure of dispersion than range or quartile deviation. But some artificially is created due to the ignoring of the signs of the deviations $(x_i - A)$. This renders it useless for further mathematical treatment.

$$\text{Co-efficient of Mean Deviation} = \frac{\text{Mean Deviation}}{\text{Average from which it is calculated}}.$$

Example. *Find the mean deviation from the median of the following frequency distribution:*

Marks:	*0–10*	*10–20*	*20–30*	*30–40*	*40–50*
No. of students:	5	8	15	16	6

Sol.

Mid-value	*f*	*c.f.*	$\lvert x - M_d \rvert$	$f\lvert x - M_d \rvert$
5	5	5	23	115
15	8	13	13	104
25	15	28	3	45
35	16	44	7	112
45	6	50	17	102
	50			478

$\frac{N}{2} = 25 \quad \therefore$ Median class corresponds to *c.f.* 28 *i.e.*, median class is 20–30

$$\text{Median } M_d = l + \frac{h}{f}\left(\frac{N}{2} - C\right) = 20 + \frac{10}{15}(25 - 13) = 20 + 8 = 28$$

$$\text{Mean deviation from median} = \frac{1}{N}\Sigma f \mid x - M_d \mid = \frac{478}{50} = 9.56 \text{ marks.}$$

(*d*) **Standard Deviation.** *Root mean Square Deviation.* The root-mean square deviation denoted by *s*, is defined as the positive square root of the mean of the squares of the deviations from an arbitrary origin A. Thus

$$s = +\sqrt{\frac{1}{N}\Sigma f_i(x_i - A)^2}$$

When the deviations are taken from the mean $\bar{x}$, the root-mean square deviation is called the *standard deviation* and is denoted by the Greek letter σ. Thus

$$\sigma = \sqrt{\frac{1}{N}\Sigma f_i(x_i - \bar{x})^2}\ .$$

Note. The square of the standard deviation σ^2 is called *variance*.

Short-cut methods for calculating Standard Deviation (σ).

(*i*) **Direct Method**

$$\sigma = \sqrt{\frac{1}{N}\Sigma f_i(x_i - \bar{x})^2}$$

$$\Rightarrow \qquad \sigma^2 = \frac{1}{N}\Sigma f_i({x_i}^2 - 2x_i\,\bar{x} + \bar{x}^2) = \frac{1}{N}\Sigma f_i{x_i}^2 - 2\bar{x}\,.\frac{1}{N}\Sigma f_i x_i + \bar{x}^2\,.\frac{1}{N}\Sigma f_i$$

(taking the constants $\bar{x}$, $\bar{x}^2$ outside the summation sign)

$$= \frac{1}{N}\Sigma f_i{x_i}^2 - 2\bar{x}\,.\,\bar{x} + \bar{x}^2\,.\frac{1}{N}\,.\,N = \frac{1}{N}\Sigma f_i{x_i}^2 - \bar{x}^2$$

$$\therefore \qquad \sigma = \sqrt{\frac{1}{N}\Sigma f_i{x_i}^2 - \bar{x}^2} = \sqrt{\frac{1}{N}\Sigma f_i{x_i}^2 - \left(\frac{1}{N}\Sigma f_i x_i\right)^2}\ .$$

(*ii*) **Change of Origin**

Let the origin be shifted to an arbitrary point '*a*'. Let $d = x - a$ denote the deviation of variate x from the new origin

$$d = x - a \ \Rightarrow\ \bar{d} = \bar{x} - a$$

$$\therefore \qquad d - \bar{d} = x - \bar{x}$$

$$\sigma_x = \sqrt{\frac{1}{N}\Sigma f(x - \bar{x})^2} = \sqrt{\frac{1}{N}\Sigma f(d - \bar{d})^2} = \sigma_d$$

∴ S.D. remains unchanged by shift of origin.

$$\sigma_x = \sigma_d = \sqrt{\frac{1}{N}\Sigma fd^2 - \left(\frac{1}{N}\Sigma fd\right)^2}\ .$$

Note. In the case of series of individual observations, if the mean is a whole number, take $a = \bar{x}$. In the case of discrete series, when the values of x are not equidistant, take '*a*' somewhere in the middle of the x-series.

(*iii*) **Shift of Origin and Change of Scale** (Step Deviation Method)

Let the origin be shifted to an arbitrary point '*a*'. Let the new scale be $\frac{1}{h}$ times the original scale.

Let $u = \dfrac{x-a}{h}$ then $hu = x - a \Rightarrow h\bar{u} = \bar{x} - a \quad \therefore \quad h(u - \bar{u}) = x - \bar{x}$

$$\sigma_x = \sqrt{\frac{1}{N}\Sigma f(x-\bar{x})^2} = \sqrt{\frac{1}{N}\Sigma f h^2 (u-\bar{u})^2} = h\sqrt{\frac{1}{N}\Sigma f(u-\bar{u})^2} = h\sigma_u$$

which is independent of a but not h. Hence S.D. is independent of change of origin but not of change of scale.

$$\sigma_x = h\sigma_u = h\sqrt{\frac{1}{N}\Sigma fu^2 - \left(\frac{1}{N}\Sigma fu\right)^2}$$

Note. In the case of Discrete Series, when the values of x are equidistant at intervals of h or in the case of continuous series having equal class intervals of width h, use Step Deviation Method.

Relation between σ and s

By definition, we have

$$s^2 = \frac{1}{N}\Sigma f_i(x_i - a)^2 = \frac{1}{N}\Sigma f_i(x_i - \bar{x} + \bar{x} - a)^2$$

$$= \frac{1}{N}\Sigma f_i(x_i - \bar{x} + d)^2 \text{ where } d = \bar{x} - a$$

$$= \frac{1}{N}\Sigma f_i[(x_i - \bar{x})^2 + d^2 + 2d(x_i - \bar{x})]$$

$$= \frac{1}{N}\Sigma f_i(x_i - \bar{x})^2 + \frac{d^2}{N}\Sigma f_i + \frac{2d}{N}\Sigma f_i(x_i - \bar{x}) = \sigma^2 + \frac{d^2}{N}.N + \frac{2d}{N}(0)$$

$[\because \Sigma f_i(x_i - \bar{x})$ = algebraic sum of the deviations from mean = 0]

$$= \sigma^2 + d^2$$

Hence $s^2 = \sigma^2 + d^2 \quad \because d^2 \geq 0 \quad \therefore s^2 \geq \sigma^2$

Clearly s^2 is least when $d = 0$, *i.e.*, $\bar{x} = a$

∴ Mean square deviation (s^2) and consequently root-mean square deviation (s) is least when the deviations are measured from the mean.

Hence *standard deviation is the least possible root-mean square deviation.*

5.10. RELATIONS BETWEEN MEASURES OF DISPERSION

Mean Deviation $= \dfrac{4}{5}$ (standard deviation) $= \dfrac{4}{5}\sigma$

Semi-interquartile range $= \dfrac{2}{3}$ (standard deviation) $= \dfrac{2}{3}\sigma$.

5.11. CO-EFFICIENT OF DISPERSION

Whenever we want to compare the variability of two series which differ widely in their averages or which are measured in different units, we calculate the co-efficients of dispersion, which being ratios, are numbers independent of the units of measurement. The co-efficients of dispersion (C.D.) based on different measures of dispersion are as follows:

(*a*) C.D. based on range: $= \dfrac{x_{max} - x_{min}}{x_{max} + x_{min}}$.

(*b*) Based on quartile deviation: C.D. $= \dfrac{Q_3 - Q_1}{Q_3 + Q_1}$

(*c*) Based on mean deviation: $\text{C.D.} = \dfrac{\text{mean deviation}}{\text{average from which it is calculated}}$

(*d*) Based on standard deviation: $\text{C.D.} = \dfrac{\text{S.D.}}{\text{Mean}} = \dfrac{\sigma}{\overline{x}}$

Co-efficient of variation. It is the percentage variation in the mean, standard deviation being considered as the total variation in the mean.

$$\text{C.V.} = \frac{\sigma}{\overline{x}} \times 100.$$

ILLUSTRATIVE EXAMPLES

Example 1. *Find the mean and standard deviation of the following:*

Series	*Frequency*	*Series*	*Frequency*
15–20	*2*	*45–50*	*20*
20–25	*5*	*50–55*	*17*
25–30	*8*	*55–60*	*16*
30–35	*11*	*60–65*	*13*
35–40	*15*	*65–70*	*11*
40–45	*20*	*70–75*	*5*

Sol.

Mid-values x	*f*	$u = \dfrac{x - 47.5}{5}$	*fu*	fu^2
17.5	2	− 6	− 12	72
22.5	5	− 5	− 25	125
27.5	8	− 4	− 32	128
32.5	11	− 3	− 33	99
37.5	15	− 2	− 30	60
42.5	20	− 1	− 20	20
47.5	20	0	0	0
52.5	17	1	17	17
57.5	16	2	32	64
62.5	13	3	39	117
67.5	11	4	44	176
72.5	5	5	25	125
	N = 143		5	1003

$$\overline{x} = a + h \,.\, \frac{\Sigma fu}{\text{N}} = 47.5 + 5 \times \frac{5}{143} = 47.7$$

$$\sigma_x = h\sigma_u = h\sqrt{\frac{1}{\text{N}}\Sigma fu^2 - \left(\frac{\Sigma fu}{\text{N}}\right)^2} = 5\sqrt{\frac{1003}{143} - \left(\frac{5}{143}\right)^2} = 5 \times 2.65 = 13.25.$$

Example 2. *Goals scored by two teams A and B in a football season were as follows:*

No. of goals scored in a match	*No. of matches*	
	A	*B*
0	*27*	*17*
1	*9*	*9*
2	*8*	*6*
3	*5*	*5*
4	*4*	*3*

Find out which team is more consistent.

Sol. *Calculation of co-efficient of variation for the team A:*

No. of goals scored (*x*)	*No. of matches* (*f*)	$d_x = x - 2$	fd_x	fd_x^2
0	27	– 2	– 54	108
1	9	– 1	– 9	9
2	8	0	0	0
3	5	1	5	1
4	4	2	8	56
	N = 53		– 50	138

$$\bar{x} = a + \frac{\Sigma fd_x}{N} = 2 + \frac{-50}{53} = 2 - 0.94 = 1.06$$

$$\sigma = \sqrt{\frac{1}{N}\Sigma fd_x^2 - \left(\frac{\Sigma fd_x}{N}\right)^2} = \sqrt{\frac{138}{53} - \left(\frac{-50}{53}\right)^2} = 1.31$$

Co-efficient of variation for the team A = $\frac{\sigma}{\bar{x}} \times 100 = \frac{1.31 \times 100}{1.06} = 123.6$

Calculation of co-efficient of variation for the team B:

No. of goals scored (*x*)	*No. of matches* (*f*)	$d_x = x - 2$	fd_x	fd_x^2
0	17	– 2	– 34	68
1	9	– 1	– 9	9
2	6	0	0	0
3	5	1	5	5
4	3	2	6	12
	N = 40		– 32	94

$$\bar{x} = a + \frac{\Sigma fd_x}{N} = 2 - \frac{32}{40} = 2 - .8 = 1.2$$

$$\sigma = \sqrt{\frac{1}{N}\Sigma fd_x^2 - \left(\frac{\Sigma fd_x}{N}\right)^2} = \sqrt{\frac{94}{40} - \left(\frac{-32}{40}\right)^2} = 1.3$$

Co-efficient of variation for the team B = $\frac{\sigma}{\bar{x}} \times 100 = \frac{1.3 \times 100}{1.2} = 108.3$

Since the co-efficient of variation is less for the team B, hence team B is more consistent.

Example 3. *Calculate the mean and the standard deviation of the series:*

Marks		*No. of students*
More than	*0*	*100*
More than	*10*	*90*
More than	*20*	*75*
More than	*30*	*50*
More than	*40*	*25*
More than	*50*	*15*
More than	*60*	*5*
More than	*70*	*0*

Sol.

Marks	*Mid-values* (x)	f	$u = \frac{x - 35}{10}$	fu	fu^2
0–10	5	10	– 3	– 30	90
10–20	15	15	– 2	– 30	60
20–30	25	25	– 1	– 25	25
30–40	35	25	0	0	0
40–50	45	10	1	10	10
50–60	55	10	2	20	40
60–70	65	5	3	15	45
		N = 100		– 40	270

$$\text{Mean } \bar{x} = a + h\frac{\Sigma fu}{N} = 35 + 10\left(\frac{-40}{100}\right) = 31$$

$$\text{Standard deviation} \quad \sigma = h\sqrt{\frac{1}{N}\Sigma fu^2 - \left(\frac{\Sigma fu}{N}\right)^2}$$

$$= 10\sqrt{\frac{270}{100} - \left(\frac{-40}{100}\right)^2} = 10\sqrt{2.7 - 0.16}$$

$$= 10 \times 1.59 = 15.9 \text{ marks.}$$

Example 4. *Calculate the mean and standard deviation for the following:*

Size of item:	*6*	*7*	*8*	*9*	*10*	*11*	*12*
Frequency:	*3*	*6*	*9*	*13*	*8*	*5*	*4*

Sol.

Size x	f	$u = x - 9$	fu	fu^2
6	3	-3	-9	27
7	6	-2	-12	24
8	9	-1	-9	9
9	13	0	0	0
10	8	1	8	8
11	5	2	10	20
12	4	3	12	36
	N = 48		0	124

$$\text{Mean } \bar{x} = a + \frac{\Sigma fu}{N} = 9$$

$$\text{Standard deviation} \quad \sigma = \sqrt{\frac{1}{N}\Sigma fu^2 - \left(\frac{\Sigma fu}{N}\right)^2} = \sqrt{\frac{124}{48}} = \sqrt{2.5833} = 1.607.$$

Example 5. *A student obtained the mean and standard deviation of 100 observations as 40 and 5 respectively. It was later discovered that he had wrongly copied down an observation as 50 instead of 40. Calculate the correct mean and standard deviation.*

Sol. Here $n = 100$, incorrect mean $\bar{x} = 40$

Using $\bar{x} = \dfrac{\Sigma x}{n}$ incorrect $\Sigma x = n\bar{x} = 4000$

Correct $\Sigma x = 4000 - 50 + 40 = 3990$

Correct mean $= \dfrac{\text{Correct } \Sigma x}{n} = \dfrac{3990}{100} = 39.9$

Incorrect variance $\sigma^2 = (5)^2 = 25$

Using $\sigma^2 = \dfrac{1}{n}\Sigma x^2 - (\bar{x})^2$

$$25 = \frac{1}{100}\Sigma x^2 - (40)^2$$

$\Rightarrow$ $2500 = \Sigma x^2 - 160000 \Rightarrow \Sigma x^2 = 162500$

$\therefore$ Incorrect $\Sigma x^2 = 162500$

Correct $\Sigma x^2 = 162500 - (50)^2 + (40)^2 = 161600$

$\therefore$ Correct $\sigma^2 = \dfrac{1}{n}(\text{correct } \Sigma x^2) - (\text{correct } \bar{x})^2$

$$= \frac{161600}{100} - (39.9)^2 = 23.99$$

$\therefore$ Correct S.D. $= \sqrt{23.99} = 4.9$.

Example 6. *An original frequency table with mean 11 and variance 9.9 was lost but the following table derived from it was found. Construct the original table.*

Step deviation:	-2	-1	0	1	2
Frequency:	1	6	7	4	2

Sol.

Step deviation (u)	f	fu	fu^2
2	1	-2	4
1	6	-6	6
0	7	0	0
1	4	4	4
2	2	4	8
	N = 20	$\Sigma fu = 0$	$\Sigma fu^2 = 22$

$$\sigma = h\sqrt{\frac{1}{N}\Sigma fu^2 - \left(\frac{\Sigma fu}{N}\right)^2}$$

$\Rightarrow$ Variance $\quad \sigma^2 = h^2\left[\frac{1}{N}\Sigma fu^2 - \left(\frac{\Sigma fu}{N}\right)^2\right]$

$\Rightarrow \quad 9.9 = h^2\left[\frac{22}{20} - 0\right] \Rightarrow h^2 = 9 \quad \therefore \quad h = 3$

Now $\quad$ mean $\bar{x} = a + h\,\frac{\Sigma fu}{N}$

$\Rightarrow \quad 11 = a + 3 \times (0) \quad \Rightarrow a = 11$

Also $\quad u = \frac{x-a}{h} \quad$ Taking $u = -2$

$$-2 = \frac{x-11}{3} \quad \Rightarrow x = 5.$$

$\therefore$ Mid-value of first class-interval is 5. Since the size of each class-interval is $h = 3$, the class intervals are

3.5–6.5, 6.5–9.5, 9.5–12.5, 12.5–15.5, 15.5–18.5

Hence the original table is

Class-intervals:	3.5–6.5	6.5–9.5	9.5–12.5	12.5–15.5	15.5–18.5
f:	1	6	7	4	2

5.12. METHOD OF GROUP AVERAGES

Let the straight line $y = a + bx$ ...(1) fit the set of n observations (x_i, y_i) $i = 1, 2, ..., n$ very closely.

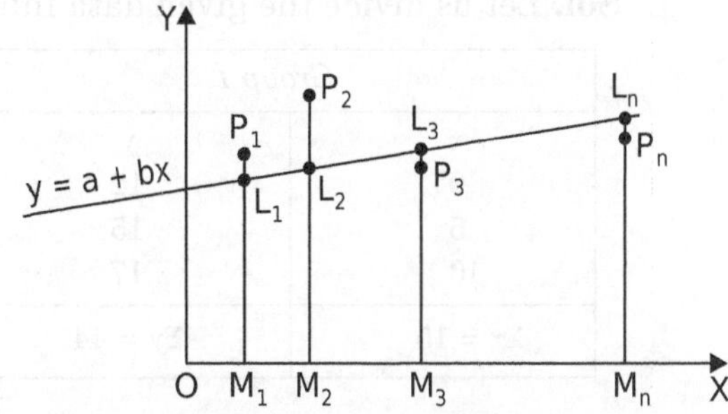

Now from the figure when $x = x_1$, the observed value of $y = y_1 = P_1M_1$ and expected value from (1) is $y = a + bx_1 = L_1M_1$.

Then d_1 = observed value of L_1 – expected value at L_1

$= y_1 - (a + bx_1) = P_1M_1 - L_1M_1 = P_1L_1$

which is called the **residual** or **error** at x_1.

Similarly the residuals for remaining observations are $d_i = y_i - (a + bx_i) = P_iL_i$; $i = 2, 3, \ldots, n$. From the figure it is clear that some of these residuals may be positive and others are negative. **The method of group averages is based on the assumption that the sum of the residuals is zero.** *i.e.*, $\sum_{i=1}^{n} d_i = 0$.

To find the constants a and b in (1) we require two equations connecting them. Therefore we divide the given data into two groups:

(*i*) **The first group containing** k **observations** $(x_1, y_1), (x_2, y_2), \ldots, (x_k, y_k)$ and

(*ii*) **the second group containing the remaining** $n - k$ **observations** $(x_{k+1}, y_{k+1}), (x_{k+2}, y_{k+2}), \ldots, (x_n, y_n)$. Assuming that **the sum of the residuals in each group is zero**, we get

$$\sum_{i=1}^{k} d_i = \sum_{i=1}^{k} [y_i - (a + bx_i)] = 0 \text{ and } \sum_{i=k+1}^{n} d_i = \sum_{i=k+1}^{n} [y_i - (a + bx_i)] = 0$$

which gives on simplification $\bar{y}_1 = a + b\bar{x}_1$ ...(2)

and $\bar{y}_2 = a + b\bar{x}_2$ ...(3)

where $\bar{x}_1 = \frac{1}{k}\sum_{i=1}^{k} x_i, \bar{y}_1 = \frac{1}{k}\sum_{i=1}^{k} y_i$, are the averages of x's, y's of the first group and

$\bar{x}_2 = \frac{1}{n-k}\sum_{i=k+1}^{n} x_i, \bar{y}_2 = \frac{1}{n-k}\sum_{i=k+1}^{n} y_i$, are the averages of x's, y's of the second group.

This shows that the equations (2) and (3) are obtained from (1) by writing the averages of x's, y's of the two groups in the places of x and y. Solving (2) and (3), we get a and b and hence the equation (1).

Note. Since grouping of observations can be done in different ways, the values of a and b are not unique. This is the main defect of this method. To overcome this, we generally divide the data in two groups so that both contain approximately the equal number of observations.

Example 7. *The following table gives corresponding values of x and y. Obtain an equation of the form y = a + bx using the method of group averages*

x:	*0*	*5*	*10*	*15*	*20*	*25*
y:	*12*	*15*	*17*	*22*	*24*	*30.*

Sol. Let us divide the given data into two groups as follows:

Group I		*Group II*	
x	y	x	y
0	12	15	22
5	15	20	24
10	17	25	30
$\Sigma x = 15$	$\Sigma y = 44$	$\Sigma x = 60$	$\Sigma y = 76$

$\therefore$ the averages of group I are $\bar{x}_1 = \dfrac{15}{3} = 5;\ \bar{y}_1 = \dfrac{44}{3} = 14.667$

and for group II are $\bar{x}_2 = \dfrac{60}{3} = 20;\ \bar{y}_2 = \dfrac{76}{3} = 25.333.$

Substituting the averages of x's, y's of the two groups in the required line $y = a + bx$, we get

$$\bar{y}_1 = a + b\bar{x}_1 \text{ i.e., } 14.667 = a + 5b$$

and
$$\bar{y}_2 = a + b\bar{x}_2 \text{ i.e., } 25.333 = a + 20b$$

Solving these two equations, we get

$$a = 11.1117 \text{ and } b = 0.7111 \text{ approximately.}$$

Hence the required line is $y = 11.1117 + 0.7111x$.

Example 8. *The head of water H(ft) and the quantity of water Q(ft^3) flowing per second are related by the law $Q = CH^n$. Find the best values of C and n by the method of group averages for the following data:*

H:	*1.2*	*1.4*	*1.6*	*1.8*	*2.0*	*2.4*	*2.6*
Q:	*4.2*	*6.1*	*8.5*	*11.5*	*14.9*	*23.5*	*27.1*

Sol. The given law can be written in linear form as $y = a + nx$ ...(1) where $y = \log_{10}$ Q, $a = \log_{10}$ C and $x = \log_{10}$ H.

Now considering the first four observations as one group and remaining three observations as second group, we have the following table:

Group I		*Group II*	
$x = \log_{10} H$	$y = \log_{10} Q$	$x = \log_{10} H$	$y = \log_{10} Q$
0.07918	0.62325	0.30103	1.17319
0.14163	0.78533	0.38021	1.37107
0.20412	0.92942	0.41497	1.43297
0.25527	1.06070		
$\Sigma x = 0.6802$	$\Sigma y = 3.3987$	$\Sigma x = 1.09621$	$\Sigma y = 3.97723$

$\therefore$ The averages of group I are

$$\bar{x}_1 = \frac{0.6802}{4} = 0.17005;\ \bar{y}_1 = \frac{3.3987}{4} = 0.84968$$

and for group II are
$$\bar{x}_2 = \frac{1.09621}{3} = 0.3654;\ \bar{y}_2 = \frac{3.97723}{3} = 1.32574$$

Substituting the averages of x's, y's of the two groups in (1), we get

$$0.84968 = a + 0.17005n$$

and $$1.32574 = a + 0.3654n$$

Solving these equations, we get

$$a = 2.4369593 \; ; \; n = -9.3342 \text{ (approximately)}$$

But $C = \text{antilog}_{10}\, a = 273.50124$

which are the best values as per the method.

Note. By taking three observations in first group and four observations in second group, we get different values of C and n.

5.13. EQUATIONS INVOLVING THREE CONSTANTS

Sometimes it is required of us to fit the observed data to the curves of the form (*i*) $y = a + bx + cx^2$ (*ii*) $y = ax^b + c$ and (*iii*) $y = ae^{bx} + c$ which involve three constants a, b and c. In such cases we reduce them to a linear form by reducing the number of constants from three to two. We will see below how these curves can be fitted.

(*i*) **The curve of the form** $y = a + bx + cx^2$

Let (x_1, y_1) be a particular point on the curve, satisfying the data

Then $$y_1 = a + bx_1 + cx_1^2; \quad \therefore \quad y - y_1 = b(x - x_1) + c(x^2 - x_1^2)$$

or $$\frac{y - y_1}{x - x_1} = b + c(x + x_1) \quad \text{or} \quad Y = b + cX \qquad ...(1)$$

where $$Y = \frac{y - y_1}{x - x_1}, \quad \text{and} \quad X = x + x_1$$

Now equation (1) is linear in Y and X and can be easily fitted.

Example 9. *Using the method of averages, fit a parabola* $y = a + bx + cx^2$ *to the following data:*

x:	*20*	*40*	*60*	*80*	*100*	*120*
y:	*5.5*	*9.1*	*14.9*	*22.8*	*33.3*	*46*

Sol. Let the point (20, 5.5) lies on the parabola

$$y = a + bx + cx^2 \qquad ...(1)$$

$$\therefore \qquad 5.5 = a + 20b + 400c \qquad ...(2)$$

(1) – (2) gives, $$\frac{y - 5.5}{x - 20} = b + c(x + 20) \quad \text{or} \quad Y = b + cX \qquad ...(3)$$

where $$Y = \frac{y - 5.5}{x - 20}, \; X = x + 20$$

Choosing first three observations as one group and last three observations as another group, we make the following table for the values of X and Y.

Group I				Group II			
x	y	$X = x + 20$	$Y = \frac{y - 5.5}{x - 20}$	x	y	$X = x + 20$	$Y = \frac{y - 5.5}{x - 20}$
20	5.5	—	—	80	22.8	100	0.288
40	9.1	60	0.18	100	33.3	120	0.348
60	14.9	80	0.235	120	46.0	140	0.405
		$\Sigma X = 140$	$\Sigma Y = 0.415$			$\Sigma X = 360$	$\Sigma Y = 1.041$
$\overline{X}_1 = \frac{140}{2} = 70$; $\overline{Y}_1 = \frac{0.415}{2} = 0.2075$				$\overline{X}_2 = \frac{360}{3} = 120$; $\overline{Y}_2 = \frac{1.041}{3} = 0.347$			

Now the average equations for the two groups are

$$\overline{Y}_1 = b + c\overline{X}_1 \quad i.e., \quad 0.2075 = b + c(70)$$

and $$\overline{Y}_2 = b + c\overline{X}_2 \quad i.e., \quad 0.347 = b + c(120)$$

by solving these two equations $b = 0.0122;\ c = 0.00279$

Substituting in (3), we get $\frac{y - 5.5}{x - 20} = 0.0122 + 0.00279(x + 20)$

or $$y = 0.00279x^2 + 0.0122x + 4.14$$

which is the required curve.

(ii) **The curve of the form** $y = ax^b + c$

It can be written as $$y - c = ax^b \qquad ...(1)$$

or $$Y = A + bX \qquad ...(2)$$

where $Y = \log_{10} (y - c)$

$A = \log_{10} a$ and $X = \log_{10} x$. Now (2) is linear in X and Y.

Let (x_1, y_1), (x_2, y_2) and (x_3, y_3) be three particular points on the curve (1) such that x_1, x_2, x_3 are in geometric progression *i.e.*, $x_2^2 = x_1x_3$.

$$\therefore \quad y_1 - c = ax_1^b\ ;\ y_2 - c = ax_2^b \text{ and } y_3 - c = ax_3^b$$

$$\therefore \quad (y_1 - c)(y_3 - c) = a^2(x_1x_3)^b = (ax_2^b)^2 = (y_2 - c)^2$$

and on simplification $c = \dfrac{y_1y_3 - y_2^2}{y_1 + y_3 - 2y_2}$.

Using this, values for Y can be determined and hence the required curve.

Example 10. *In magnetic arc at constant arc length the voltage V consumed by the arc is observed for values of the current i*

i:	*0.5*	*1*	*2*	*4*	*8*	*12*
V:	*160*	*120*	*94*	*75*	*62*	*56*

If V and i are connected by a relation of the form $V = ai^b + c$, *find a, b and c and hence the curve.*

Sol. Take $i_1 = 1$, $i_2 = 2$ and $i_3 = 4$ which are in G.P. to find c. Then $V_1 = 120$, $V_2 = 94$ and $V_3 = 75$.

$$\therefore \qquad c = \frac{V_1V_3 - V_2^2}{V_1 + V_3 - 2V_2} = \frac{(120)(75) - (94)^2}{120 + 75 - 2(94)} = 23.4$$

$$\therefore \qquad V = ai^b + 23.4 \quad \text{or} \quad V - 23.4 = ai^b \qquad ...(1)$$

or $$Y = A + bX \qquad ...(2)$$

where $Y = \log_{10} (V - 23.4)$, $A = \log_{10} a$ and $X = \log_{10} i$

Choosing first three observations as one group and other three as second group, we have the following table:

Group I				Group II			
i	V	$X = \log_{10} i$	$Y = \log_{10} (V - 23.4)$	i	V	$X = \log_{10} i$	$Y = \log_{10} (V - 23.4)$
0.5	160	$\bar{1}.6990$	2.1354	4	75	0.6021	1.7126
1	120	0.0000	1.9850	8	62	0.9031	1.5866
2	94	0.3010	1.8488	12	56	1.0792	1.5132
		$\Sigma X = 0.0000$	$\Sigma Y = 5.9692$			$\Sigma X = 2.5844$	$\Sigma Y = 4.8124$
$\overline{X}_1 = \frac{\Sigma X}{3} = 0$; $\overline{Y}_1 = \frac{\Sigma Y}{3} = 1.9897$				$\overline{X}_2 = \frac{\Sigma X}{3} = 0.8615$; $\overline{Y}_2 = \frac{\Sigma Y}{3} = 1.6041$			

Substituting the averages of X's, Y's of two groups in (2), we get

$$\overline{Y}_1 = A + b\overline{X}_1 \Rightarrow 1.9897 = A + b(0) \quad \therefore \quad A = 1.9897$$

$$\overline{Y}_2 = A + b\overline{X}_2 \Rightarrow 1.6041 = A + 0.8615b \quad \therefore \quad b = -0.4476$$

$\therefore$ $a = \text{antilog}_{10} A = 97.66$, $b = -0.45$, $c = 23.4$ approximately

and the curve is $V = 97.66i^{-0.45} + 23.4$.

(*iii*) **The curve of the form** $y = ae^{bx} + c$

which can be written as $y - c = ae^{bx}$ ...(1)

and also in linear form as $Y = A + Bx$

where $Y = \log_{10} (y - c)$, $A = \log_{10} a$ and $B = b \log_{10} e$.

Let (x_1, y_1), (x_2, y_2) and (x_3, y_3) be three particular points on the curve such that x_1, x_2 and x_3 are in arithmetic progression *i.e.*, $x_1 + x_3 = 2x_2$.

Substituting these points in (1), we have

$$y_1 - c = ae^{bx_1}, \; y_2 - c = ae^{bx_2} \quad \text{and} \quad y_3 - c = ae^{bx_3}$$

$$\therefore \qquad (y_1 - c)(y_3 - c) = a^2e^{b(x_1 + x_2)} = a^2e^{b(2x_2)} = a^2(e^{bx_2})^2 = (y_2 - c)^2$$

which gives on simplification $c = \dfrac{y_1y_3 - y_2^2}{y_1 + y_3 - 2y_2}$.

Using this, values of Y can be determined and hence the required curve.

Note. The same procedure can be adopted to the curve of the form $y = ab^x + c$

Example 11. *The variables s and t are connected by the relation $s = a + be^{nt}$ and their corresponding values are given in the following table:*

t:	1	2	6	8	11
s:	12.71	12.46	11.65	11.34	10.99

Find a, b and n and hence the curve.

Sol. The given curve can be written as

$$s - a = be^{nt} \quad \text{...(1)}$$

and in linear form as

$$Y = B + Nt \quad \text{...(2)}$$

where $Y = \log_{10}(s - a)$, $B = \log_{10} b$ and $N = n \log_{10} e$.

Take $t_1 = 1, t_2 = 6$ and $t_3 = 11$ such that $t_1 + t_3 = 2t_2$ the corresponding values of s are $s_1 = 12.71, s_2 = 11.65$ and $s_3 = 10.99$.

$$\therefore \quad a = \frac{s_1 s_3 - s_2^2}{s_1 + s_3 - 2s_3} = \frac{(12.71)(10.99) - (11.65)^2}{12.71 + 10.99 - 2(11.65)} = 9.901$$

Now choosing first three observations as one group and other two observations as second group, we have the following table:

Group I			*Group II*		
t	s	$Y = \log_{10}(s - 9.901)$	t	s	$Y = \log_{10}(s - 9.901)$
1	12.71	0.4486	8	11.34	0.1581
2	12.46	0.4081	11	10.99	0.0370
6	11.65	0.2428			
$\Sigma t = 9$		$\Sigma Y = 1.0995$	$\Sigma t = 19$		$\Sigma Y = 0.1951$
$\bar{t}_1 = 3; \bar{Y}_1 = 0.3665$			$\bar{t}_2 = 9.5; \bar{Y}_2 = 0.0976$		

Substituting the averages of t's and Y's of two groups in (2), we get

$$0.3665 = B + 3N \text{ and } 0.0976 = B + 9.5\,N$$

On solving, $B = 0.4907$; $N = -0.0414$

Now $b = \text{antilog}_{10} B = 3.097$ and $n = \dfrac{N}{\log_{10} e} = -0.0953$

Hence the required curve is $s = 9.901 - 3.097e^{-0.0953t}$.

TEST YOUR KNOWLEDGE

1. Fit a curve of the form $y = a + bx$ to the following data by the method of group averages:

x:	1	2	3	4	5	6	7	8	9	10
y:	52.5	58.7	65	70.2	75.4	81.1	87.2	95.5	102.2	108.4

2. Fit a curve of the form $y = ax^b$ to the following data by the method of group averages:

x:	10	20	30	40	50	60	70	80
y:	1.06	1.33	1.52	1.68	1.81	1.91	2.01	2.11

3. Using method of group averages fit a curve of the form $y = ab^x$ to the following data:

x:	2	4	6	8	10	12
y:	7.32	8.24	9.20	10.19	11.01	12.05

4. Fit a curve of the form $y = a + bx + cx^2$ to the following data by the method of group averages:

x:	2	2.5	3	3.5	4	4.5	5	5.5	6
y:	18	17.8	17.5	17	15.8	14.8	13.3	11.7	9

5. Fit a curve of the form $y = ae^{bx} + c$ to the following data by the method of averages:

x:	0	1	2	3	5	7	10	15	20
y:	52.2	48.8	46.0	43.5	39.7	36.5	33.0	28.7	28.0

6. Fit a curve of the form $y = a + bx^c$ to the following data by the method of group averages:

x:	1	2	4	6	10	16
y:	15	45	165	364	1004	2564

7. Fit a curve of the type $y = a + bc^x$ to the following data by the method of group averages:

x:	0	1	2	3	4	5	6	7	8
y:	2.4	3.2	3.7	5.1	7.8	13.2	23.6	44.8	87

Answers

1. $y = 46.048 + 6.104x$ **2.** $y = 0.4851x^{0.3354}$ **3.** $y = (6.7468)(1.0505)^x$

4. $y = 1.58 + 2.1x - 0.5x^2$ **5.** $y = 29.54e^{(-0.1)x} + 21.98$ **6.** $y = 5 + 10x^2$

7. $y = 2.26 + (0.3)(2.07)^x$.

5.14. PRINCIPLE OF LEAST SQUARES

So far we have studied the curve fitting by (*i*) Graphical method and (*ii*) Method of group averages. The values of the constants in the first method depend upon the choice of the points on the line and in the second method on grouping the observations. Hence both the methods fail to give a unique curve of fit. So we require some other method to get the best values of the constants. Principle of least squares provides unique set of values to the constants and hence suggests a curve of best fit to the given data.

Let (x_i, y_i), $i = 1, 2, \ldots\ldots, n$ be the n sets of observations and let $y = f(x)$ ...(1) be the relation suggested between x and y.

When $x = x_i$ the observed value of y_i is $y_i = P_iM_i$.

Expected value $= L_iM_i = f(x_i)$.

$\therefore$ Residual (or error) at $x = x_i$ is $e_i = y_i - f(x_i)$

It is obvious that some of e_i's may be + ve or – ve. Thus by giving equal weightage to each residuals consider

$$E = \sum_{i=1}^{n} e_i^2 = \sum_{i=1}^{n} [y_i - f(x_i)]^2 \qquad ...(2)$$

Now if E = 0 then $y_i = f(x_i)\ \forall\ i$ *i.e.*, all the points lie on the curve. Otherwise the minimum of E results the best fitting curve to the data. Thus the curve of best fit is that for which the sum of squares of the residuals is minimum using this principle we shall fit the following curves:

(*i*) **Fitting a straight line** $y = ax + b$

(*ii*) **Fitting a parabola** $y = ax^2 + bx + c$

(*iii*) **Fitting the exponential curve** $y = ae^{bx}$ and

(*iv*) **Fitting the curve** $y = ax^b$.

(*i*) **Fitting a straight line.** Let (x_i, y_i), $i = 1, 2, ..., n$ be n sets of observations of related data and $y = ax + b$ ...(1) be the straight line to be fitted. The residual at $x = x_i$ is

$$e_i = y_i - f(x_i) = y_i - (ax_i + b), \; i = 1, 2, ..., n$$

$$\therefore \quad E = \sum_{i=1}^{n} e_i^2 = \sum_{i=1}^{n} [y_i - (ax_i + b)]^2$$

By the principle of least squares, E is minimum

$$\therefore \quad \frac{\partial E}{\partial a} = 0 \quad \text{and} \quad \frac{\partial E}{\partial b} = 0$$

$$\therefore \quad \sum_{i=1}^{n} 2(y_i - ax_i - b)(-x_i) = 0 \quad \text{or} \quad \sum_{i=1}^{n} (y_i x_i - ax_i^2 - bx_i) = 0$$

and
$$\sum_{i=1}^{n} 2(y_i - ax_i - b)(-1) = 0 \quad \text{or} \quad \sum_{i=1}^{n} (y_i - ax_i - b) = 0$$

$$\therefore \quad \sum_{i=1}^{n} y_i x_i = a \sum_{i=1}^{n} x_i^2 + b \sum_{i=1}^{n} x_i \qquad ...(2)$$

and
$$\sum_{i=1}^{n} y_i = a \sum_{i=1}^{n} x_i + nb \qquad ...(3)$$

Since x_i, y_i are known, equations (2) and (3) give two equations in a and b. Solving these the best values for a and b can be known and hence equation (1).

Note. (1) Dropping the suffix i, we can write (2), (3) as $\mathbf{\Sigma xy = a\Sigma x^2 + b\Sigma x}$ and $\mathbf{\Sigma y = a\Sigma x + nb}$. These are known as normal equations to the curve $y = ax + b$.

(2) If the numbers in the given data are large, we can change the origin and scale with the substitutions $X = \dfrac{x - A}{h}$, $Y = \dfrac{y - B}{k}$ for the sake of convenience and ease in calculation work.

Example. *By the method of least squares, find the straight line that best fits the following data:*

x:	*1*	*2*	*3*	*4*	*5*
y:	*14*	*27*	*40*	*55*	*68*

Sol. Let the straight line of best fit be $y = ax + b$ ...(1)

The normal equations are $\Sigma xy = a\Sigma x^2 + b\Sigma x$ ...(2)

and $\Sigma y = a\Sigma x + 5b$ ...(3)

The values of Σx, Σy, Σx^2 and Σxy are calculated as shown below:

x	y	x^2	xy
1	14	1	14
2	27	4	54
3	40	9	120
4	55	16	220
5	68	25	340
$\Sigma x = 15$	$\Sigma y = 204$	$\Sigma x^2 = 55$	$\Sigma xy = 748$

∴ The equations (2) and (3) becomes

$$748 = 55a + 15b \quad \text{and} \quad 204 = 15a + 5b$$

Solving these, we have $a = 13.6$ and $b = 0$

Putting these values in (1), we get the line of best fit as $y = 13.6x$.

Aliter. Let $X = x - 3$, $Y = y - 40$ and let the straight line in new variables be

$$Y = AX + B \qquad ...(1)$$

Now the normal equations are $\Sigma XY = A\Sigma X^2 + B\Sigma X$ ...(2)

and $$\Sigma Y = A\Sigma X + 5B \qquad ...(3)$$

x	y	$X = x - 3$	$Y = y - 40$	X^2	XY
1	14	– 2	– 26	4	52
2	27	– 1	– 13	1	13
3	40	0	0	0	0
4	55	1	15	1	15
5	68	2	28	4	56
		$\Sigma X = 0$	$\Sigma Y = 4$	$\Sigma X^2 = 10$	$\Sigma XY = 136$

Substituting $\Sigma X = 0$, $\Sigma Y = 4$, $\Sigma XY = 136$ and $\Sigma X^2 = 10$ in (2) and (3), we get

$$136 = 10A \quad \text{and} \quad 4 = 5B \quad \text{or} \quad A = 13.6 \quad \text{and} \quad B = 0.8$$

∴ From (1); $Y = 13.6X + 0.8$ *i.e.*, $(y - 40) = 13.6(x - 3) + 0.8$

or $y = 13.6x$ which is the same as seen before.

(ii) **Fitting a parabola.** Let $y = ax^2 + bx + c$ ...(1) be a parabola to be fitted for the data (x_i, y_i), $i = 1, 2, ..., n$. As explained in Art. 2.26 the residual at $x = x_i$ is

$$e_i = y_i - f(x_i) = y_i - (ax_i^2 + bx_i + c)$$

$$E = \sum_{i=1}^{n} e_i^2 = \sum_{i=1}^{n} [y_i - (ax_i^2 + bx_i + c)]^2$$

Now E should be minimum for the best values of a, b and c (Principle of least squares)

$$\therefore \quad \frac{\partial E}{\partial a} = 0, \quad \frac{\partial E}{\partial b} = 0 \quad \text{and} \quad \frac{\partial E}{\partial c} = 0$$

i.e.,

$$\sum_{i=1}^{n} 2[y_i - (ax_i^2 + bx_i + c)](-x_i^2) = 0$$

$$\sum_{i=1}^{n} 2[y_i - (ax_i^2 + bx_i + c)](-x_i) = 0$$

and

$$\sum_{i=1}^{n} 2[y_i - (ax_i^2 + bx_i + c)](-1) = 0$$

which results on simplification after dropping the suffix as

$$\mathbf{\Sigma x^2 y = a\Sigma x^4 + b\Sigma x^3 + c\Sigma x^2}$$

$$\mathbf{\Sigma xy = a\Sigma x^3 + b\Sigma x^2 + c\Sigma x}$$

and $$\Sigma y = a\Sigma x^2 + b\Sigma x + nc$$

where $\Sigma = \sum_{i=1}^{n}$. These are called the normal equations of the parabola (1). Solving these, we get the best values of the constants a, b and c and hence best fit for (1).

Example. *Fit a parabola $y = ax^2 + bx + c$ in least square sense to the data:*

x:	10	12	15	23	20
y:	14	17	23	25	21

Sol. The normal equations to the curve are

$$\left.\begin{aligned} \Sigma y &= a\Sigma x^2 + b\Sigma x + 5c \\ \Sigma xy &= a\Sigma x^3 + b\Sigma x^2 + c\Sigma x \\ \text{and} \quad \Sigma x^2 y &= a\Sigma x^4 + b\Sigma x^3 + c\Sigma x^2 \end{aligned}\right\} \quad ...(1)$$

The values of Σx, Σx^2, ... etc., are calculated by means of the following table:

x	y	x^2	x^3	x^4	xy	x^2y
10	14	100	1000	10000	140	1400
12	17	144	1728	20736	204	2448
15	23	225	3375	50625	345	5175
23	25	529	12167	279841	575	13225
20	21	400	8000	160000	420	8400
$\Sigma x = 80$	$\Sigma y = 100$	$\Sigma x^2 = 1398$	$\Sigma x^3 = 26270$	$\Sigma x^4 = 521202$	$\Sigma xy = 1684$	$\Sigma x^2y = 30648$

Substituting the obtained values from the table in normal equations (1), we have

$$100 = 1398a + 80b + 5c$$

$$1684 = 26270a + 1398b + 80c$$

$$30648 = 521202a + 26270b + 1398c$$

On solving $a = -0.07$, $b = 3.03$, $c = -8.89$

$\therefore$ The required equation is $y = -0.07x^2 + 3.03x - 8.89$

Aliter. Let $X = x - 16$ and $Y = y - 20$ and the parabola of fit be $Y = AX^2 + BX + C$ in the new variables, whose normal equations are

$$\Sigma Y = A\Sigma X^2 + B\Sigma X + 5C$$

$$\Sigma XY = A\Sigma X^3 + B\Sigma X^2 + C\Sigma X$$

$$\Sigma X^2Y = A\Sigma X^4 + B\Sigma X^3 + C\Sigma X^2$$

The values of ΣX, etc., are calculated as below:

x	y	X	Y	X^2	X^3	X^4	XY	X^2Y
10	14	− 6	− 6	36	− 216	1296	36	− 216
12	17	− 4	− 3	16	− 64	256	12	− 48
15	23	− 1	3	1	1	1	− 3	3
23	25	7	5	49	343	2401	35	245
20	21	4	1	16	64	256	4	16
		$\Sigma X = 0$	$\Sigma Y = 0$	$\Sigma X^2 = 118$	$\Sigma X^3 = 128$	$\Sigma X^4 = 4210$	$\Sigma XY = 84$	$\Sigma X^2Y = 0$

$\therefore$ The normal equations become

$$0 = 118A + 5C$$
$$84 = 128A + 118B$$
$$0 = 4210A + 128B + 118C$$

On solving A = – 0.07, B = 0.79 and C = 1.67

$$\therefore \quad Y = -0.07X^2 + 0.79X + 1.67$$

or $$(Y - 20) = -0.07(x - 16)^2 + 0.79(x - 16) + 1.67$$

which simplifies to $$y = -0.07x^2 + 3.03x - 8.89 \text{ as observed earlier.}$$

(*iii*) **Fitting an exponential curve.** $y = ae^{bx}$

Taking logarithms on both sides, we get

$$\log_{10} y = \log_{10} a + bx \log_{10} e$$

i.e., $$Y = A + Bx \quad ...(1)$$

where $Y = \log_{10} y$, $A = \log_{10} a$ and $B = b \log_{10} e$

The normal equations for (1) are $\Sigma Y = nA + B\Sigma x$ and $\Sigma xY = A\Sigma x + B\Sigma x^2$

Solving these, we get A and B.

Then a = antilog A and $b = \dfrac{B}{\log_{10} e}$.

(*iv*) **Fitting the curve.** $y = ax^b$

Taking logarithms on both sides, we get

$$\log_{10} y = \log_{10} a + b \log_{10} x$$

i.e., $$Y = A + bX \quad ...(1)$$

where $Y = \log_{10} y$, $A = \log_{10} a$ and $X = \log_{10} x$.

The normal equations to (1) are $\Sigma Y = nA + b\Sigma X$ and $\Sigma XY = A\Sigma X + b\Sigma X^2$

which give A and b on solving and a = antilog A.

ILLUSTRATIVE EXAMPLES

Example 1. *Find the curve of best fit of the type $y = ae^{bx}$ to the following data by the method of least squares:*

x:	*1*	*5*	*7*	*9*	*12*
y:	*10*	*15*	*12*	*15*	*21*

Sol. The curve to be fitted is $y = ae^{bx}$

or $$Y = A + Bx \quad \text{where } Y = \log_{10} y, A = \log_{10} a, \text{ and } B = b \log_{10} e$$

$\therefore$ The normal equations are $\Sigma Y = 5A + B\Sigma x$ and $\Sigma xY = A\Sigma x + B\Sigma x^2$

x	y	$Y = \log_{10} y$	x^2	xY
1	10	1.0000	1	1
5	15	1.1761	25	5.8805
7	12	1.0792	49	7.5544
9	15	1.1761	81	10.5849
12	21	1.3222	144	15.8664
$\Sigma x = 34$		$\Sigma Y = 5.7536$	$\Sigma x^2 = 300$	$\Sigma xY = 40.8862$

Substituting the values of Σx, etc. calculated by means of above table in the normal equations.

We get 5.7536 = 5A + 34B and 40.8862 = 34A + 300B

On solving A = 0.9766; B = 0.02561

$\therefore \quad a = \text{antilog}_{10} \text{ A} = 9.4754; \; b = \dfrac{\text{B}}{\log_{10} e} = 0.059$

Hence the required curve is $y = 9.4754e^{0.059x}$.

Example 2. *Obtain a relation of the form $y = ab^x$ for the following data by the method of least squares:*

x	2	3	4	5	6
y	8.3	15.4	33.1	65.2	127.4

Sol. The curve to be fitted is $y = ab^x$

or $\quad \text{Y} = \text{A} + \text{B}x \quad$ where $\text{A} = \log_{10} a$, $\text{B} = \log_{10} b$ and $\text{Y} = \log_{10} y$.

$\therefore$ The normal equations are $\Sigma\text{Y} = 5\text{A} + \text{B}\Sigma x$ and $\Sigma x\text{Y} = \text{A}\Sigma x + \text{B}\Sigma x^2$.

x	y	$Y = \log_{10} y$	x^2	xY
2	8.3	0.9191	4	1.8382
3	15.4	1.1872	9	3.5616
4	33.1	1.5198	16	6.0792
5	65.2	1.8142	25	9.0710
6	127.4	2.1052	36	12.6312
$\Sigma x = 20$		$\Sigma\text{Y} = 7.5455$	$\Sigma x^2 = 90$	$\Sigma x\text{Y} = 33.1812$

Substituting the values of Σx, etc. from the above table in normal equations, we get

$$7.5455 = 5\text{A} + 20\text{B} \text{ and } 33.1812 = 20\text{A} + 90\text{B}.$$

On solving A = 0.31 and B = 0.3

$\therefore \quad a = \text{antilog A} = 2.04$ and $b = \text{antilog B} = 1.995$.

Hence the required curve is $y = 2.04(1.995)^x$.

TEST YOUR KNOWLEDGE

1. A simply supported beam carries a concentrated load P(lb) at its mid-point. Corresponding to various values of P, the maximum deflection Y (in) is measured. The data are given below. Find a law of the type $\text{Y} = a + b\text{P}$

P	100	120	140	160	180	200
Y	0.45	0.55	0.60	0.70	0.80	0.85

2. In the following table y is the weight of potassium bromide which will dissolve in 100 grams of water at temperature x^0. Find a linear law between x and y

$x°$ (c)	0	10	20	30	40	50	60	70
y (gm)	53.5	59.5	65.2	70.6	75.5	80.2	85.5	90

3. By the method of least squares, find the curve $y = ax + bx^2$ that best fits the following data:

x	1	2	3	4	5
y	1.8	5.1	8.9	14.1	19.8

4. Find the parabola of the form $y = a + bx + cx^2$ which fits most closely with the observations

x	– 3	– 2	– 1	0	1	2	3
y	4.63	2.11	0.67	0.09	0.63	2.15	4.58

5. By the method of least squares, fit a second degree curve $y = a + bx + cx^2$ to the following data:

x	1	2	3	4	5	6	7	8	9
y	2	6	7	8	10	11	11	10	9

6. Fit a parabola $y = a + bx + cx^2$ to the following data:

x	2	4	6	8	10
y	3.07	12.85	31.47	57.38	91.29

7. Fit an equation of the form $y = ae^{bx}$ to the following data by the method of least squares:

x	1	2	3	4
y	1.65	2.7	4.5	7.35

8. The voltage V across a capacitor at time t seconds is given by the following table. Use the principle of least squares to fit a curve of the form $V = ae^{kt}$ to the data:

t	0	2	4	6	8
V	150	63	28	12	5.6

9. Fit a curve of the form $y = ae^{bx}$ to the following data:

x	0	2	4
y	5.102	10	31.62

10. Fit a curve of the form $y = ax^b$ to the data given below:

x	1	2	3	4	5
y	7.1	27.8	62.1	110	161

11. Fit a curve of the form $y = ab^x$ in least square sence to the data given below:

x	2	3	4	5	6
y	144	172.8	207.4	248.8	298.5

Answers

1. Y = 0.048 + 0.0041P	**2.** $y = 54.35 + 0.5184x$	**3.** $y = 1.37x + 0.53x^2$
4. $y = 1.243 - 0.004x + 0.22x^2$	**5.** $y = -1 + 3.55x - 0.27x^2$	**6.** $y = 0.34 - 0.78x + 0.99x^2$
7. $y = e^{0.5x}$	**8.** $V = 146.3\, e^{-0.4118t}$	**9.** $y = 4.642e^{0.46x}$
10. $y = 7.173x^{1.952}$	**11.** $y = 99.86(1.2)^x$.	

5.15. METHOD OF MOMENTS

Let (x_i, y_i), $i = 1, 2, \ldots\ldots, n$ be the set of n observations of related data so that the x's are equally spaced.

i.e., $$x_2 - x_1 = x_3 - x_2 = \ldots\ldots = x_n - x_{n-1} = h \text{ (say)}.$$

We define the moments of the observed values of y as follows:

The first moment $= \mu_1 = \Sigma yh = h\Sigma y$

The second moment $= \mu_2 = h\Sigma xy$

The third moment $= \mu_3 = h\Sigma x^2 y$ and so on.

Let $y = f(x)$ be a curve fitting the data. Then the moments of the expected values of y are

The first moment $= \gamma_1 = \int y\, dx = \int f(x)\, dx$

The second moment $= \gamma_2 = \int xy\, dx = \int x f(x)\, dx$

The third moment $= \gamma_3 = \int x^2 y\, dx = \int x^2 f(x)\, dx$ and so on.

This method is based on the assumption that the moments of the observed values of y are respectively equal to the moments of the expected values of y

i.e., $$\mu_1 = \gamma_1\, ;\ \mu_2 = \gamma_2,\ \mu_3 = \gamma_3 \qquad \ldots(1)$$

and so on.

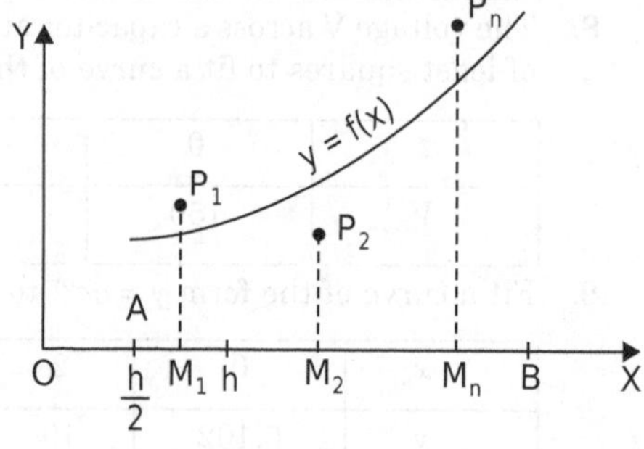

μ_is are calculated from the tabulated values of x and y while γ_is are computed as follows:

In the figure the ordinate $P_1(x = x_1)$ can be taken as the value of y at the mid-point of the interval $\left(x_1 - \frac{h}{2}, x_1 + \frac{h}{2}\right)$. Similarly, $P_n(x = x_n)$ can be taken as the value of y at the mid-point of the interval $\left(x_n - \frac{h}{2}, x_n + \frac{h}{2}\right)$. Let A and B be the points on OX such that $OA = x_1 - \frac{h}{2}$ and $OB = x_n + \frac{h}{2}$.

Now $$\gamma_1 = \int y\, dx = \int f(x)\, dx = \int_{x_1 - \frac{h}{2}}^{x_n + \frac{h}{2}} f(x)\, dx$$

$$\gamma_2 = \int x f(x)\, dx = \int_{x_1 - \frac{h}{2}}^{x_n + \frac{h}{2}} x f(x)\, dx,\ \gamma_3 = \int x^2 f(x)\, dx = \int_{x_1 - \frac{h}{2}}^{x_n + \frac{h}{2}} x^2 f(x)\, dx \text{ and so on.}$$

Now using the equations (1) [called observation equations], we can solve for the unknowns involved in $f(x)$.

Note. If $y = a + bx$ then it involves only two unknowns hence the observation equations are $\mu_1 = \gamma_1$ and $\mu_2 = \gamma_2$. For parabola $y = a + bx + cx^2$, the observation equations are $\mu_1 = \gamma_1$, $\mu_2 = \gamma_2$ and $\mu_3 = \gamma_3$ since it involves three unknowns. In general number of unknowns = number of observation equations.

Example. *Fit a straight line and a parabola to the following data:*

x:	*1*	*2*	*3*	*4*	*5*	*6*
y:	*4*	*8*	*10*	*12*	*16*	*20*

Sol. (*i*) **Fitting a straight line.** Let the equation be $y = a + bx$ ...(1)

This involves only two unknowns a and b.

$\therefore$ Observation equations are $\mu_1 = \gamma_1$ ...(2) and $\mu_2 = \gamma_2$ ...(3) Here $h = 1$,

$$\mu_1 = h\Sigma y = 1(4 + 8 + 10 + 12 + 16 + 20) = 70$$

$$\mu_2 = h\Sigma xy = 1(4 + 16 + 30 + 48 + 80 + 120) = 298$$

The limits of integration are $\left[1-\frac{1}{2}, 6+\frac{1}{2}\right]$ *i.e.*, [0.5, 6.5]

$$\therefore \quad \gamma_1 = \int_{0.5}^{6.5} y\,dx = \int_{0.5}^{6.5} (a + bx)\,dx = \left[ax + \frac{bx^2}{2}\right]_{0.5}^{6.5} = 6a + 21b$$

$$\gamma_2 = \int_{0.5}^{6.5} xy\,dx = \int_{0.5}^{6.5} (ax + bx^2)\,dx = \left[\frac{ax^2}{2} + \frac{bx^3}{3}\right]_{0.5}^{6.5} = 21a + 91.5b$$

$\therefore$ From the observation equations (2) and (3), we have

$$70 = 6a + 21b \text{ and } 298 = 21a + 91.5b$$

Solving these equations, we get $a = 1.3611$ and $b = 2.9444$.

Hence the required equation is $y = 1.3611 + 2.9444x$.

(*ii*) **Fitting a parabola.** Let the equation be $y = a + bx + cx^2$ ...(1) involving three unknowns a, b and c.

$\therefore$ $\mu_1 = \gamma_1$, $\mu_2 = \gamma_2$ and $\mu_3 = \gamma_3$ are the observation equations. Here $h = 1$.

$$\mu_1 = h\Sigma y = 70,\ \mu_2 = h\Sigma xy = 298 \text{ and}$$

$$\mu_3 = h\Sigma x^2 y = 1 \,.\, [4 + 32 + 90 + 192 + 400 + 720] = 1438$$

Limits of integration [0.5, 6.5]

$$\gamma_1 = \int_{0.5}^{6.5} (a + bx + cx^2)\,dx = \left[ax + \frac{bx^2}{2} + \frac{cx^3}{3}\right]_{0.5}^{6.5} = 6a + 21b + 91.5c$$

$$\gamma_2 = \int_{0.5}^{6.5} x(a + bx + cx^2)\,dx = \left[a\frac{x^2}{2} + b\frac{x^3}{3} + c\frac{x^4}{4}\right]_{0.5}^{6.5} = 21a + 91.5b + 446.25c$$

$$\gamma_3 = \int_{0.5}^{6.5} x^2\,(a + bx + cx^2)\,dx = 91.5a + 446.25b + 2320.575c$$

Thus the observation equations are

$$70 = 6a + 21b + 91.5c$$
$$298 = 21a + 91.5b + 446.25c$$
$$1438 = 91.5a + 446.25b + 2320.575c$$

Solving these equations, we get $a = 1.2540$, $b = 3.0255$ and $c = -0.0116$

Hence the parabola is $y = 1.254 + 3.0255x - 0.0116x^2$.

TEST YOUR KNOWLEDGE

1. Use the method of moments to fit a straight line to the data given below:

x	1	3	5	7	9
y	1.5	2.8	4.0	4.7	6.0

2. Fit a parabola of the form $y = ax^2 + bx + c$ to the data:

x	1	2	3	4
y	1.7	1.8	2.3	3.2

by the method of moments.

Answers

1. $y = 1.1845 + 0.5231x$ 2. $y = 0.074x^2 + 0.063x + 1.53$.

5.16. MOMENTS

The rth *moment* of a variable x *about any point* A is denoted by μ_r' and is defined as

$$\mu_r' = \frac{1}{N} \Sigma f(x - A)^r \quad \text{where} \quad N = \Sigma f$$

The rth moment of a variable x about the mean M is denoted by μ_r and is defined as

$$\mu_r = \frac{1}{N} \Sigma f(x - M)^r$$

In particular $\mu_0' = \frac{1}{N} \Sigma f(x - A)^0 = \frac{1}{N} \Sigma f = \frac{1}{N} . N = 1$

Similarly, $\mu_0 = 1$

$$\mu_1 = \frac{1}{N} \Sigma f(x - M) = 0$$

| Being the algebraic sum of the deviations from the mean

$$\mu_2 = \frac{1}{N} \Sigma f(x - M)^2 = \sigma^2, \text{ by definition.}$$

The *results* $\mu_0 = 1$, $\mu_1 = 0$, $\mu_2 = \sigma^2$ are of fundamental importance and should be committed to memory.

5.17. RELATION BETWEEN MOMENTS ABOUT MEAN IN TERMS OF MOMENTS ABOUT ANY POINT AND VICE VERSA

By definition, $\mu_r' = \frac{1}{N} \Sigma f(x - A)^r$ where A is any point

$= \frac{1}{N} \Sigma fd^r$ where $d = x - A$ ...(*i*)

Putting $r = 1$, $\mu_1' = \frac{1}{N} \Sigma fd$

$\therefore \quad M = A + \frac{1}{N} \Sigma fd = A + \mu_1'$

or $\quad \mu_1' = M - A$ ...(*ii*)

Now $\quad \mu_r = \frac{1}{N} \Sigma f(x - M)^r$

$= \frac{1}{N} \Sigma f(x - A + A - M)^r = \frac{1}{N} \Sigma f(d - \mu_1')^r$ | Using (*ii*)

$= \frac{1}{N} \Sigma f\,[d^r - {}^rC_1 d^{r-1}\mu_1' + {}^rC_2 d^{r-2}\mu_1'^2 - {}^rC_3 d^{r-3}\mu_1'^3 + \ldots\ldots + (-1)^r . \mu_1'^r]$

$= \frac{1}{N} \Sigma fd^r - {}^rC_1\mu_1' . \frac{1}{N} \Sigma fd^{r-1} + {}^rC_2\mu_1'^2 \frac{1}{N} \Sigma fd^{r-2}$

$- {}^rC_3\mu'^3 \frac{1}{N} \Sigma fd^{r-3} + \ldots\ldots + (-1)^r\mu_1'^r . \frac{1}{N} \Sigma f$

$= \mu_r' - {}^rC_1\mu_{r-1}' \mu_1' + {}^rC_2\mu_{r-2}' \mu_1'^2 - {}^rC_3\mu_{r-3}' \mu_1'^3 + \ldots\ldots + (-1)^r\mu_1'^r$

| Using (*i*)

In particular, putting $r = 2, 3, 4$, we get

$\mu_2 = \mu_2' - 2\mu_1'^2 + \mu_0' \mu_1'^2 = \mu_2' - \mu_1'^2$ | $\because \mu_0' = 1$

$\mu_3 = \mu_3' - 3\mu_2' \mu_1' + 3\mu_2'^3 - \mu_0' \mu_1'^3 = \mu_3' - 3\mu_2' \mu_1' + 2\mu_1'^3$

$\mu_4 = \mu_4' - 4\mu_3' \mu_1' + 6\mu_2' \mu_1'^2 - 4\mu_1' \mu_1'^3 + \mu_0' \mu_1'^4$

$= \mu_4' - 4\mu_3' \mu_1' + 6\mu_2' \mu_1'^2 - 3\mu_1'^4$

Hence $\quad \mu_1 = 0$

$\mu_2 = \mu_2' - \mu_1'^2$

$\mu_3 = \mu_3' - 3\mu_2' \mu_1' + 2\mu_3'^3$ $\quad (\mu_1' = M - A)$

$\mu_4 = \mu_4' - 4\mu_3' \mu_1' + 6\mu_2' \mu_1'^2 - 3\mu_1'^4$

Conversely, $\mu_r = \frac{1}{N} \Sigma f(x - M)^r = \frac{1}{N} \Sigma fd^r$ where $d = x - M$ ...(*iii*)

Now $\quad \mu_r' = \frac{1}{N} \Sigma f(x - A)^r = \frac{1}{N} \Sigma f(x - M + M - A)^r = \frac{1}{N} \Sigma f(d + \mu_1')^r$

| Using (*ii*)

$= \frac{1}{N} \Sigma f(d^r + {}^rC_1 d^{r-1} \mu_1' + {}^rC_2 d^{r-2} \mu_1'^2 + {}^rC_3 d^{r-3} \mu_1'^3 + \ldots\ldots + \mu_1'^r)$

$$= \frac{1}{N}\Sigma fd^r + {}^rC_1\mu_1' \cdot \frac{1}{N}\Sigma fd^{r-1} + {}^rC_2\mu_1'^2 \frac{1}{N}\Sigma fd^{r-2} + \ldots\ldots + \mu_1'^r \cdot \frac{1}{N}\Sigma f$$

$$= \mu_r + {}^rC_1\mu_{r-1}\,\mu_1' + {}^rC_2\mu_{r-2}\,\mu_1'^2 + \ldots\ldots + \mu_1'^r \qquad | \text{ Using } (iii)$$

In particular, putting $r = 2, 3, 4$ and noting that $\mu_1 = 0$, $\mu_0 = 1$, we get

$$\mu_2' = \mu_2 + 2\mu_1\mu_1' + \mu_0\,\mu_1'^2 = \mu_2 + \mu_1'^2$$

$$\mu_3' = \mu_3 + 3\mu_2\mu_1' + 3\mu_1\mu_1'^2 + \mu_0\mu_1'^3 = \mu_3 + 3\mu_2\mu_1' + \mu_1'^3$$

$$\mu_4' = \mu_4 + 4\mu_3\mu_1' + 6\mu_2\mu_1'^2 + 4\mu_1\mu_1'^3 + \mu_0\mu_1'^4 = \mu_4 + 4\mu_3\mu_1' + 6\mu_2\mu_1'^2 + \mu_1'^4.$$

5.18. EFFECT OF CHANGE OF ORIGIN AND SCALE ON MOMENTS

Let $\quad u = \dfrac{x - A}{h} \quad i.e., \quad x = A + hu$

$\therefore \quad \bar{x} = A + h\bar{u}$, where bar denotes the mean of the respective variable

$\therefore \quad x - \bar{x} = h(u - \bar{u})$

$$\mu_r' = \frac{1}{N}\Sigma f(x - A)^r = \frac{1}{N}\Sigma fh^r u^r = h^r \cdot \frac{1}{N}\Sigma fu^r$$

Also $\quad \mu_r = \dfrac{1}{N}\Sigma f(x - \bar{x})^r = \dfrac{1}{N}\Sigma fh^r(u - \bar{u})^r = h^r \cdot \dfrac{1}{N}\Sigma f(u - \bar{u})^r$

Hence *the* r^{th} *moment of the variable* x *is* h^r *times the corresponding moment of the variable* u.

Example. *Calculate the first four moments of the following distribution about the mean and hence find* β_1 *and* β_2*:*

x:	*0*	*1*	*2*	*3*	*4*	*5*	*6*	*7*	*8*
f:	*1*	*8*	*28*	*56*	*70*	*56*	*28*	*8*	*1*

Sol. Let us first calculate moments about $x = 4$.

$$\mu_r' = \frac{1}{N}\Sigma f(x-4)^r = \frac{1}{N}\Sigma fd^r \quad \text{where } d = x - 4$$

x	f	$d = x - 4$	fd	fd^2	fd^3	fd^4
0	1	– 4	– 4	16	– 64	256
1	8	– 3	– 24	72	– 216	648
2	28	– 2	– 56	112	– 224	448
3	56	– 1	– 56	56	– 56	56
4	70	0	0	0	0	0
5	56	1	56	56	56	56
6	28	2	56	112	224	448
7	8	3	24	72	216	648
8	1	4	4	16	64	256
	N = 256		0	512	0	2816

$\mu_1' = \dfrac{1}{N}\Sigma fd = 0; \qquad \mu_2' = \dfrac{1}{N}\Sigma fd^2 = \dfrac{512}{256} = 2$

$\mu_3' = \dfrac{1}{N}\Sigma fd^3 = 0; \qquad \mu_4' = \dfrac{1}{N}\Sigma fd^4 = \dfrac{2816}{256} = 11$

Moments about mean are

$\mu_1 = 0$ (*always*); $\quad \mu_2 = \mu_2' - \mu_1'^2 = 2$

$\mu_3 = \mu_3' - 3\mu_2'\mu_1' + 2\mu_1'^3 = 0$; $\quad \mu_4 = \mu_4' - 4\mu_3'\mu_1' + 6\mu_2'\mu_1'^2 - 3\mu_1'^4 = 11$

$$\beta_1 = \frac{\mu_3^2}{\mu_2^3} = 0; \qquad \beta_2 = \frac{\mu_4}{\mu_2^2} = \frac{11}{4} = 2.75.$$

TEST YOUR KNOWLEDGE

1. Calculate the quartile deviation of the marks of 63 students in Physics given below:

Marks	*No. of students*	*Marks*	*No. of students*
0–10	5	50–60	7
10–20	7	60–70	3
20–30	10	70–80	2
30–40	16	80–90	2
40–50	11	90–100	0

2. Find the mean deviation from the mean of the following distribution:

Class:	0–6	6–12	12–18	18–24	24–30
Frequency:	8	10	12	9	5

3. Compute mean deviation from median of the following distribution:

Marks:	0–10	10–20	20–30	30–40	40–50
No. of students:	5	10	20	5	10

4. Compute the standard deviation for the following data relating to marks obtained by 15 students:

12, 21, 21, 23, 27, 28, 30, 34, 37, 39, 39, 39, 40, 49, 54.

5. Calculate mean and standard deviation for the following distribution:

x:	56	63	70	77	84	91	98
f:	3	6	14	16	13	6	2

6. Calculate the mean and standard deviation for the following:

Size of item:	6	7	8	9	10	11	12
Frequency:	3	6	9	13	8	5	4

7. The following table shows the marks obtained by 100 candidates in an examination. Calculate the mean, median and standard deviation:

Marks obtained:	1–10	11–20	21–30	31–40	41–50	51–60
No. of candidates:	3	16	26	31	16	8

8. Calculate the mean and standard deviation of the following frequency distribution:

Weekly wages in ₹	*No. of men*
4.5–12.5	4
12.5–20.5	24
20.5–28.5	21
28.5–36.5	18
36.5–44.5	5
44.5–52.5	3
52.5–60.5	5
60.5–68.5	8
68.5–76.5	2

9. (*i*) The mean of five items of an observation is 4 and the variance is 5.2. If three of the items are 1, 2 and 6, then find the other two.

(*ii*) Show that the variance of the first n positive integers is $\frac{1}{12}(n^2 - 1)$.

Answers

1. 12.32 marks	**2.** 6.3	**3.** 9	**4.** 10.9 marks
5. 76.53, 9.87	**6.** 9, 1.61	**7.** 32, 32.6, 12.4	
8. ₹ 31.35, ₹ 16.64	**9.** (*i*) 4, 7.		

5.19. CORRELATION

In a bivariate distribution, if the change in one variable affects a change in the other variable, the variables are said to be *correlated.*

If the two variables deviate in the same direction *i.e.*, if the increase (or decrease) in one results in a corresponding increase (or decrease) in the other, correlation is said to be *direct or positive*.

e.g., the correlation between income and expenditure is positive.

If the two variables deviate in opposite direction *i.e.*, if the increase (or decrease) in one results in a corresponding decrease (or increase) in the other, correlation is said to be *inverse or negative*.

e.g., the correlation between volume and the pressure of a perfect gas or the correlation between price and demand is negative.

Correlation is said to be *perfect* if the deviation in one variable is followed by a corresponding **proportional deviation** in the other.

5.20. SCATTER OR DOT DIAGRAMS

It is the simplest method of the diagrammatic representation of bivariate data. Let (x_i, y_i) $i = 1, 2, 3, \ldots, n$ be a bivariate distribution. Let the values of the variables x and y be plotted along the x-axis and y-axis on a suitable scale. Then corresponding to every ordered pair, there corresponds a point or dot in the xy-plane. The diagram of dots so obtained is called a *dot or scatter diagram.*

If the dots are vary close to each other and the number of observations is not vary large, a fairly good correlation is expected. If the dots are widely scattered, a poor correlation is expected.

5.21. KARL PEARSON'S CO-EFFICIENT OF CORRELATION (OR PRODUCT MOMENT CORRELATION CO-EFFICIENT)

Correlation co-efficient between two variables x and y, usually denoted by $r(x, y)$ or r_{xy} is a numerical measure of linear relationship between them and is defined as

$$r_{xy} = \frac{\Sigma(x_i - \bar{x})(y_1 - \bar{y})}{\sqrt{\Sigma(x_i - \bar{x})^2 \, \Sigma(y_i - \bar{y})^2}} = \frac{\frac{1}{n}\Sigma(x_i - \bar{x})(y_1 - \bar{y})}{\sqrt{\frac{1}{n}\Sigma(x_i - \bar{x})^2 \cdot \frac{1}{n}\Sigma(y_i - \bar{y})^2}} = \frac{\frac{1}{n}\Sigma(x_i - \bar{x})(y_i - \bar{y})}{\sigma_x \sigma_y}.$$

Note. Correlation co-efficient is independent of change of origin and scale.

Let us define two new variables u and v as

$$u = \frac{x-a}{h}, \; v = \frac{y-b}{k} \quad \text{where } a, b, h, k \text{ are constants, then } r_{xy} = r_{uv}.$$

5.22. COMPUTATION OF CORRELATION CO-EFFICIENT

We know that $r_{xy} = \dfrac{\frac{1}{n}\Sigma(x_i - \bar{x})(y_i - \bar{y})}{\sigma_x \sigma_y}$

Now $\dfrac{1}{n}\Sigma(x_i - \bar{x})(y_i - \bar{y}) = \dfrac{1}{n}\Sigma(x_i y_i - x_i\bar{y} - y_i\bar{x} + \bar{x}\,\bar{y})$

$$= \frac{1}{n}\Sigma x_i y_i - \bar{y}\,.\,\frac{1}{n}\Sigma x_i - \bar{x}\,.\,\frac{1}{n}\Sigma y_i + \frac{1}{n}(n\bar{x}\,\bar{y})$$

$$= \frac{1}{n}\Sigma x_i y_i - \bar{y}\,.\,\bar{x} - \bar{x}\,.\,\bar{y} + \bar{x}\,.\,\bar{y} = \frac{1}{n}\Sigma x_i y_i - \bar{x}\,.\,\bar{y}$$

$$\sigma_x^2 = \frac{1}{n}\Sigma(x_i - \bar{x})^2 = \frac{1}{n}\Sigma(x_i^2 - 2x_i\bar{x} + \bar{x}^2)$$

$$= \frac{1}{n}\Sigma x_i^2 - 2\bar{x}\,.\,\frac{1}{n}\Sigma x_i + \frac{1}{n}n\bar{x}^2 = \frac{1}{n}\Sigma x_i^2 - 2\bar{x}\,.\,\bar{x} + \bar{x}^2 = \frac{1}{n}\Sigma x_i^2 - \bar{x}^2$$

Similarly, $\quad \sigma_y^2 = \dfrac{1}{n}\Sigma y_i^2 - \bar{y}^2$

$$\therefore \quad r_{xy} = \frac{\frac{1}{n}\Sigma x_i y_i - \bar{x}\,\bar{y}}{\sqrt{\left(\frac{1}{n}\Sigma x_i^2 - \bar{x}^2\right)\left(\frac{1}{n}\Sigma y_i^2 - \bar{y}^2\right)}}$$

If $\quad u = \dfrac{x-a}{h}, v = \dfrac{y-b}{k}$ then $r_{xy} = r_{uv} = \dfrac{\frac{1}{n}\Sigma u_i v_i - \bar{u}\,\bar{v}}{\sqrt{\left(\frac{1}{n}\Sigma u_i^2 - \bar{u}^2\right)\left(\frac{1}{n}\Sigma v_i^2 - \bar{v}^2\right)}}$.

ILLUSTRATIVE EXAMPLES

Example 1. *Ten students got the following percentage of marks in Principles of Economics and Statistics:*

Roll Nos.:	1	2	3	4	5	6	7	8	9	10
Marks in Economics:	78	36	98	25	75	82	90	62	65	39
Marks in Statistics:	84	51	91	60	68	62	86	58	53	47

Calculate the co-efficient of correlation.

Sol. Let the marks in the two subjects be denoted by x and y respectively.

x	y	$u = x - 65$	$v = y - 66$	u^2	v^2	uv
78	84	13	18	169	324	234
36	51	– 29	– 15	841	225	435
98	91	33	25	1089	625	825
25	60	– 40	– 6	1600	36	240
75	68	10	2	100	4	20
82	62	17	– 4	289	16	– 68
90	86	25	20	625	400	500
62	58	– 3	– 8	9	64	24
65	53	0	– 13	0	169	0
39	47	– 26	– 19	676	361	494
Total		0	0	5398	2224	2734

$$\bar{u} = \frac{1}{n}\Sigma u_i = 0,\ \bar{v} = \frac{1}{n}\Sigma v_i = 0$$

$$r_{uv} = \frac{\frac{1}{n}\Sigma u_i v_i - \bar{u}\,\bar{v}}{\sqrt{\left(\frac{1}{n}\Sigma u_i^2 - \bar{u}^2\right)\left(\frac{1}{n}\Sigma v_i^2 - \bar{v}^2\right)}} = \frac{\frac{1}{10}(2734)}{\sqrt{\frac{1}{10}(5398).\frac{1}{10}(2224)}}$$

$$= \frac{2734}{\sqrt{5398 \times 2224}} = 0.787$$

Hence $\qquad r_{xy} = r_{uv} = 0.787.$

Example 2. *Find the co-efficient of correlation for the following table:*

x:	*10*	*14*	*18*	*22*	*26*	*30*
y:	*18*	*12*	*24*	*6*	*30*	*36*

Sol. Let $\qquad u = \dfrac{x-22}{4},\quad v = \dfrac{y-24}{6}.$

x	y	u	v	u^2	v^2	uv
10	18	– 3	– 1	9	1	3
14	12	– 2	– 2	4	4	4
18	24	– 1	0	1	0	0
22	6	0	– 3	0	9	0
26	30	1	1	1	1	1
30	36	2	2	4	4	4
Total		– 3	– 3	19	19	12

$$\bar{u} = \frac{1}{n}\Sigma u_i = \frac{1}{6}(-3) = -\frac{1}{2};\quad \bar{v} = \frac{1}{n}\Sigma v_i = \frac{1}{6}(-3) = -\frac{1}{2}$$

$$r_{uv} = \frac{\frac{1}{n}\Sigma u_i v_i - \bar{u}\,\bar{v}}{\sqrt{\left(\frac{1}{n}\Sigma u_i^2 - \bar{u}^2\right)\left(\frac{1}{n}\Sigma v_i^2 - \bar{v}^2\right)}} = \frac{\frac{1}{6}(12) - \frac{1}{4}}{\sqrt{\left[\frac{1}{6}(19) - \frac{1}{4}\right]\left[\frac{1}{6}(19) - \frac{1}{4}\right]}} = 0.6$$

Hence $\qquad r_{xy} = r_{uv} = 0.6.$

Example 3. *A computer while calculating correlation co-efficient between two variables X and Y from 25 pairs of observations obtained the following results:*

$$n = 25, \quad \Sigma X = 125, \quad \Sigma X^2 = 650,$$
$$\Sigma Y = 100, \quad \Sigma Y^2 = 460, \quad \Sigma XY = 508.$$

It was, however, later discovered at the time of checking that he had copied down two pairs as

X	Y
6	14
8	6

while the correct values were

X	Y
8	12
6	8

Obtain the correct value of correlation co-efficient.

Sol. Corrected $\Sigma X = 125 - 6 - 8 + 8 + 6 = 125$

Corrected $\Sigma Y = 100 - 14 - 6 + 12 + 8 = 100$

Corrected $\Sigma X^2 = 650 - 6^2 - 8^2 + 8^2 + 6^2 = 650$

Corrected $\Sigma Y^2 = 460 - 14^2 - 6^2 + 12^2 + 8^2 = 436$

Corrected $\Sigma XY = 508 - 6 \times 14 - 8 \times 6 + 8 \times 12 + 6 \times 8 = 520$

(Subtract the incorrect values and add the corresponding correct values)

$$\overline{X} = \frac{1}{n}\Sigma X = \frac{1}{25} \times 125 = 5; \quad \overline{Y} = \frac{1}{n}\Sigma Y = \frac{1}{25} \times 100 = 4$$

Corrected $$r_{xy} = \frac{\frac{1}{n}\Sigma XY - \overline{X}\,\overline{Y}}{\sqrt{\left(\frac{1}{n}\Sigma X^2 - \overline{X}^2\right)\left(\frac{1}{n}\Sigma Y^2 - \overline{Y}^2\right)}}$$

$$= \frac{\frac{1}{25} \times 520 - 5 \times 4}{\sqrt{\left(\frac{1}{25} \times 650 - 25\right)\left(\frac{1}{25} \times 436 - 16\right)}} = \frac{\frac{4}{5}}{\sqrt{(1)\left(\frac{36}{25}\right)}} = \frac{4}{5} \times \frac{5}{6} = \frac{2}{3} = 0.67.$$

Example 4. *If $z = ax + by$ and r is the correlation co-efficient between x and y, show that*

$$\sigma_z^2 = a^2\sigma_x^2 + b^2\sigma_y^2 + 2abr\,\sigma_x\,\sigma_y.$$

Sol. $z = ax + by$

$\Rightarrow \quad \bar{z} = a\bar{x} + b\bar{y}, \qquad z_i = ax_i + by_i$

$$z_i - \bar{z} = a(x_i - \bar{x}) + b(y_i - \bar{y})$$

Now $$\sigma_z^2 = \frac{1}{n}\Sigma(z_i - \bar{z})^2 = \frac{1}{n}\Sigma[a(x_i - \bar{x}) + b(y_i - \bar{y})]^2$$

$$= \frac{1}{n}\Sigma[a^2(x_i - \bar{x})^2 + b^2(y_i - \bar{y})^2 + 2ab(x_i - \bar{x})(y_i - \bar{y})]$$

$$= a^2.\frac{1}{n}\Sigma(x_i - \bar{x})^2 + b^2.\frac{1}{n}\Sigma(y_i - \bar{y})^2 + 2ab.\frac{1}{n}\Sigma(x_i - \bar{x})(y_i - \bar{y})$$

$$= a^2\sigma_x^2 + b^2\sigma_y^2 + 2abr\,\sigma_x\sigma_y \qquad \left| \because r = \frac{\frac{1}{n}\Sigma(x_i - \bar{x})(y_i - \bar{y})}{\sigma_x\sigma_y} \right.$$

5.23. CALCULATION OF CO-EFFICIENT OF CORRELATION FOR A BIVARIATE FREQUENCY DISTRIBUTION

If the bivariate data on x and y is presented on a two way correlation table and f is the frequency of a particular rectangle in the correlation table, then

$$r_{xy} = \frac{\Sigma fxy - \dfrac{1}{n}\Sigma fx\,\Sigma fy}{\sqrt{\left[\Sigma fx^2 - \dfrac{1}{n}(\Sigma fx)^2\right]\left[\Sigma fy^2 - \dfrac{1}{n}(\Sigma fy)^2\right]}}$$

Since change of origin and scale do not affect the co-efficient of correlation,

$\therefore$ $r_{xy} = r_{uv}$ where the new variables u, v are properly chosen.

Example 5. *The following table gives according to age the frequency of marks obtained by 100 students in an intelligence test:*

Age (in years) → / Marks ↓	18	19	20	21	Total
10–20	4	2	2		8
20–30	5	4	6	4	19
30–40	6	8	10	11	35
40–50	4	4	6	8	22
50–60		2	4	4	10
60–70		2	3	1	6
Total	19	22	31	28	100

Calculate the co-efficient of correlation between age and intelligence.

Sol. Let age and intelligence be denoted by x and y respectively.

Mid value	x / y	18	19	20	21	f	u	fu	fu^2	fuv
15	10–20	4	2	2		8	– 3	24	72	30
25	20–30	5	4	6	4	19	– 2	– 38	76	20
35	30–40	6	8	10	11	35	– 1	– 35	35	9
45	40–50	4	4	6	8	22	0	0	0	0
55	50–60		2	4	4	10	1	10	10	2
65	60–70		2	3	1	6	2	12	24	– 2
	f	19	22	31	28	100	Totals	**– 75**	**217**	**59**
	v	2	– 1	0	1	Totals				
	fv	– 38	– 22	0	28	**– 32**				
	fv^2	76	22	0	28	**126**				
	fuv	56	16	0	13	**59**				

Let us define two new variables u and v as $u = \dfrac{y-45}{10}$, $v = x - 20$

$$r_{xy} = r_{uv} = \frac{\Sigma fuv - \frac{1}{n}\Sigma fu\,\Sigma fv}{\sqrt{\left[\Sigma fu^2 - \frac{1}{n}(\Sigma fu)^2\right]\left[\Sigma fv^2 - \frac{1}{n}(\Sigma fv)^2\right]}}$$

$$= \frac{59 - \frac{1}{100}(-75)(-32)}{\sqrt{\left[217 - \frac{1}{100}(-75)^2\right]\left[126 - \frac{1}{100}(-32)^2\right]}} = \frac{59-24}{\sqrt{\frac{643}{4} \times \frac{2894}{25}}} = 0.25.$$

5.24. RANK CORRELATION

Sometimes we have to deal with problems in which data cannot be quantitatively measured but qualitative assessment is possible.

Let a group of n individuals be arranged in order of merit or proficiency in possession of two characteristics A and B. The ranks in the two characteristics are, in general, different. For example, if A stands for intelligence and B for beauty, it is not necessary that the most intelligent individual may be the most beautiful and *vice versa.* Thus an individual who is ranked at the top for the characteristic A *may be* ranked at the bottom for the characteristic B. Let (x_i, y_i), $i = 1, 2, ..., n$ be the ranks of the n individuals in the group for the characteristics A and B respectively. Pearsonian co-efficient of correlation between the ranks x_i's and y_i's is called the *rank correlation co-efficient* between the characteristics A and B for that group of individuals.

Thus rank correlation co-efficient

$$r = \frac{\Sigma(x_i - \bar{x})(y_i - \bar{y})}{\sqrt{\Sigma(x_i - \bar{x})^2\,\Sigma(y_i - \bar{y})^2}} = \frac{\frac{1}{n}\Sigma(x_i - \bar{x})(y_i - \bar{y})}{\sigma_x \sigma_y} \qquad ...(1)$$

Now x_i's and y_i's are merely the permutations of n numbers from 1 to n. *Assuming that no two individuals are bracketed or tied in either classification i.e., $(x_i, y_i) \neq (x_j, y_j)$ for $i \neq j$, both x and y take all integral values from 1 to n.*

$$\therefore \qquad \bar{x} = \bar{y} = \frac{1}{n}(1 + 2 + 3 + ... + n) = \frac{1}{n} \cdot \frac{n(n+1)}{2} = \frac{n+1}{2}$$

$$\Sigma x_i = 1 + 2 + 3 + ... + n = \frac{n(n+1)}{2} = \Sigma y_i$$

$$\Sigma x_i^2 = 1^2 + 2^2 + ... + n^2 = \frac{n(n+1)(2n+1)}{6} = \Sigma y_i^2$$

If d_i denotes the difference in ranks of the ith individual, then

$$d_i = x_i - y_i = (x_i - \bar{x}) - (y_i - \bar{y}) \qquad [\because \ \bar{x} = \bar{y}]$$

$$\frac{1}{n}\Sigma d_i^2 = \frac{1}{n}\Sigma[(x_i - \bar{x}) - (y_i - \bar{y})]^2$$

$$= \frac{1}{n}\Sigma(x_i - \bar{x})^2 + \frac{1}{n}\Sigma(y_i - \bar{y})^2 - 2.\frac{1}{n}\Sigma(x_i - \bar{x})(y_i - \bar{y})$$

$$= \sigma_x^2 + \sigma_y^2 - 2r\sigma_x\sigma_y \qquad ...(2) \quad \text{[Using (1)]}$$

But $\qquad \sigma_x^2 = \frac{1}{n}\,\Sigma x_i^2 - \bar{x}^2 = \frac{1}{n}\,\Sigma y_i^{\,2} - \bar{y}^2 = \sigma_y^2$

$\therefore$ From (2), $\frac{1}{n}\,\Sigma d_i^2 = 2\sigma_x^2 - 2r\sigma_x^2 = 2(1-r)\,\sigma_x^2 = 2(1-r)\left[\frac{1}{n}\,\Sigma x_i^2 - \bar{x}^2\right]$

$$= 2(1-r)\left[\frac{1}{n}\cdot\frac{n(n+1)(2n+1)}{6} - \frac{(n+1)^2}{4}\right]$$

$$= (1-r)\,(n+1)\left[\frac{4n+2-3n-3}{6}\right] = \frac{(1-r)(n^2-1)}{6} \quad \text{or} \quad 1-r = \frac{6\Sigma d_i^2}{n(n^2-1)}$$

Hence $\qquad r = 1 - \frac{6\Sigma d_i^2}{n(n^2-1)}.$

Note. This is called *Spearman's Formula for* Rank Correlation.

$$\Sigma d_i = \Sigma(x_i - y_i) = \Sigma x_i - \Sigma y_i = 0$$

always. This serves as a check on calculations.

Example 6. *The marks secured by recruits in the selection test (X) and in the proficiency test (Y) are given below*:

Serial No:	*1*	*2*	*3*	*4*	*5*	*6*	*7*	*8*	*9*
X:	*10*	*15*	*12*	*17*	*13*	*16*	*24*	*14*	*22*
Y:	*30*	*42*	*45*	*46*	*33*	*34*	*40*	*35*	*39*

Calculate the rank correlation co-efficient.

Sol. *Here the marks are given. Therefore, first of all, write down ranks.* In each series, the item with the largest size is ranked 1, next largest 2 and so on.

X	10	15	12	17	13	16	24	14	22	Total
Y	30	42	45	46	33	34	40	35	39	
Ranks in X (*x*)	9	5	8	3	7	4	1	6	2	
Ranks in Y (*y*)	9	3	2	1	8	7	4	6	5	
$d = x - y$	0	2	6	2	− 1	− 3	− 3	0	− 3	0
d^2	0	4	36	4	1	9	9	0	9	72

$\therefore \qquad r = 1 - \frac{6\Sigma d^2}{n(n^2-1)} = 1 - \frac{6 \times 72}{9 \times 80} = 1 - 0.6 = 0.4 \quad$ Here $n = 9$.

5.25. REPEATED RANKS

If any two or more individuals have same rank or the same value in the series of marks, then the above formula fails and requires an adjustment. In such cases, each individual is given an average rank. This common average rank is the average of the ranks which these individuals would have assumed if they were slightly different from each other. Thus, if two individual are ranked equal at the sixth place, they would have assumed the 6th and 7th ranks if they were ranked slightly different. Their common rank $= \frac{6+7}{2} = 6.5$. If three individuals are ranked equal at fourth place, they would have assumed the 4th, 5th and 6th ranks if they were ranked slightly different. Their common rank = $\frac{4+5+6}{3}$ = 5.

Adjustment. Add $\frac{1}{12}m(m^2 - 1)$ to Σd^2 where m stands for the number of times an item is repeated.

☞**This adjustment factor is to be added for each repeated item.**

Thus
$$r = 1 - \frac{6\left\{\Sigma d^2 + \frac{1}{12}m(m^2-1) + \frac{1}{12}m(m^2-1) + \ldots\ldots\right\}}{n(n^2-1)}.$$

Example 7. *Obtain the rank correlation co-efficient for the following data:*

X: 68 64 75 50 64 80 75 40 55 64
Y: 62 58 68 45 81 60 68 48 50 70.

Sol. Here, marks are given, so write down the ranks.

X	68	64	75	50	64	80	75	40	55	64	Total
Y	62	58	68	45	81	60	68	48	50	70	
Ranks in X (x)	4	6	2.5	9	6	1	2.5	10	8	6	
Ranks in Y (y)	5	7	3.5	10	1	6	3.5	9	8	2	
$d = x - y$	–1	–1	–1	–1	5	–5	–1	1	0	4	0
d^2	1	1	1	1	25	25	1	1	0	16	72

In the X-series, the value 75 occurs twice. Had these values been slightly different, they would have been given the ranks 2 and 3. Therefore, the common rank given to them is $\frac{2+3}{2}$ = 2.5. The value 64 occurs thrice. Had these values been slightly different, they would have been given the ranks 5, 6 and 7. Therefore the common rank given to them is $\frac{5+6+7}{3}$ = 6. Similarly, in the Y-series, the value 68 occurs twice. Had these values been slightly different, they would have been given the ranks 3 and 4. Therefore, the common rank given to them is $\frac{3+4}{2}$ = 3.5.

Thus, m has the values 2, 3, 2.

$$\therefore \quad r = 1 - \frac{6\left\{\Sigma d^2 + \frac{1}{12}m(m^2-1) + \frac{1}{12}m(m^2-1) + \ldots\ldots\right\}}{n(n^2-1)}$$

$$= 1 - \frac{6\left[72 + \frac{1}{12}\{2(2^2-1)\} + \frac{1}{12}\{3(3^2-1)\} + \frac{1}{12}\{2(2^2-1)\}\right]}{10(10^2-1)}$$

$$= 1 - \frac{6 \times 75}{990} = \frac{6}{11} = 0.545.$$

5.26. REGRESSION

Regression is the estimation or prediction of unknown values of one variable from known values of another variable.

After establishing the fact of correlation between two variables, it is natural curiosity to know the extent to which one variable varies in response to a given variation in the other variable *i.e.*, one is interested to know the nature of relationship between the two variables.

Regression measures the nature and extent of correlation.

5.27. LINEAR REGRESSION

If two variates x and y are correlated *i.e.*, there exists an association or relationship between them, then the scatter diagram will be more or less concentrated round a curve. This curve is called the *curve of regression* and the relationship is said to be expressed by means of *curvilinear regression.* In the particular case, when the curve is a straight line, it is called a *line of regression* and the regression is said to be *linear*.

A line of regression is the straight line which gives the best fit in the least square sense to the given frequency.

If the line of regression is so chosen that the sum of squares of deviation parallel to the axis of y is minimised [*See Fig. (a)*], it is called *the line of regression of y on x* and it gives *the best estimate of y for any given value of x.*

If the line of regression is so chosen that the sum of squares of deviations parallel to the axis of x is minimised [*See Fig. (b)*], it is called *the line of regression of x on y* and it gives *the best estimate of x for any given value of y.*

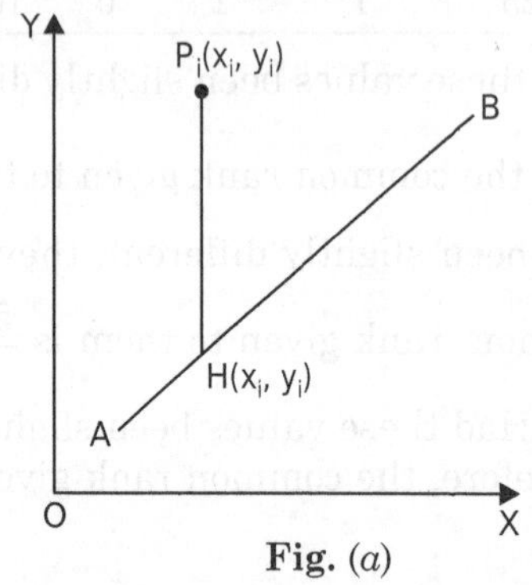

Fig. (*a*)

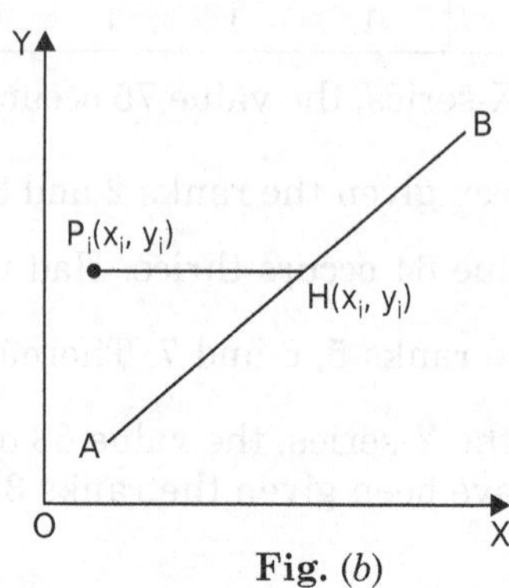

Fig. (*b*)

5.28. LINES OF REGRESSION

Let the equation of line of regression of y on x be

$$y = a + bx \qquad \text{...(1)}$$

Then $$\bar{y} = a + b\bar{x} \qquad \text{...(2)}$$

Subtracting (2) from (1), we have

$$y - \bar{y} = b(x - \bar{x}) \qquad \text{...(3)}$$

The normal equations are
$$\Sigma y = na + b\Sigma x$$
$$\Sigma xy = a\Sigma x + b\Sigma x^2 \qquad \text{...(4)}$$

Shifting the origin to $(\bar{x}, \bar{y})$, (4) becomes

$$\Sigma(x-\bar{x})(y-\bar{y}) = a\Sigma(x-\bar{x}) + b\Sigma(x-\bar{x})^2 \quad \text{...(5)}$$

Since $\dfrac{\Sigma(x-\bar{x})(y-\bar{y})}{n\,\sigma_x\sigma_y} = r \quad \therefore \quad \Sigma(x-\bar{x}) = 0; \quad$ and $\quad \dfrac{1}{n}\Sigma(x-\bar{x})^2 = \sigma_x^2$

$\therefore$ From (5), $\quad nr\sigma_x\sigma_y = a.0 + b.n\sigma_x^2 \quad \Rightarrow \quad b = \dfrac{r\sigma_y}{\sigma_x}$

Hence, from (3), the line of regression of y on x is $\quad y - \bar{y} = r\dfrac{\sigma_y}{\sigma_x}(x-\bar{x})$

Similarly, the line of regression of x on y is $\quad x - \bar{x} = r\dfrac{\sigma_x}{\sigma_y}(y-\bar{y})$

$\dfrac{r\sigma_y}{\sigma_x}$ is called the regression co-efficient of y on x and is denoted by b_{yx}.

$\dfrac{r\sigma_x}{\sigma_y}$ is called the regression co-efficient of x on y and is denoted by b_{xy}.

Note. If $r = 0$, the two lines of regression become $y = \bar{y}$ and $x = \bar{x}$ which are two straight lines parallel to X and Y axes respectively and passing through their means $\bar{y}$ and $\bar{x}$. They are mutually perpendicular.

If $r = \pm 1$, the two lines of regression will coincide.

5.29. PROPERTIES OF REGRESSION CO-EFFICIENTS

Property I. *Correlation co-efficient is the geometric mean between the regression co-efficients.*

Proof. The co-efficients of regression are $\dfrac{r\sigma_y}{\sigma_x}$ and $\dfrac{r\sigma_x}{\sigma_y}$.

G.M. between them $= \sqrt{\dfrac{r\sigma_y}{\sigma_x}\times\dfrac{r\sigma_x}{\sigma_y}} = \sqrt{r^2} = r =$ co-efficient of correlation.

Property II. *If one of the regression co-efficients is greater than unity, the other must be less than unity.*

Proof. The two regression co-efficients are $b_{yx} = \dfrac{r\sigma_y}{\sigma_x}$ and $b_{xy} = \dfrac{r\sigma_x}{\sigma_y}$.

Let $b_{yx} > 1$, then $\dfrac{1}{b_{yx}} < 1$ ...(1)

Since $\quad b_{yx}\,.\,b_{xy} = r^2 \le 1 \quad (\because \; -1 \le r \le 1) \quad \therefore \quad b_{xy} \le \dfrac{1}{b_{yx}} < 1.$ | Using (1)

Similarly, if $b_{xy} > 1$, then $b_{yx} < 1$.

Property III. *Arithmetic mean of regression co-efficients is greater than the correlation co-efficient.*

Proof. We have to prove that $\dfrac{b_{yx}+b_{xy}}{2} > r$ or $\dfrac{\dfrac{r\sigma_y}{\sigma_x}+\dfrac{r\sigma_x}{\sigma_y}}{2} > r$

or $\sigma_y^2 + \sigma_x^2 > 2\sigma_x\sigma_y$ or $(\sigma_x - \sigma_y)^2 > 0$ which is true.

Property IV. *Regression co-efficients are independent of the origin but not of scale.*

Proof. Let $u = \dfrac{x-a}{h}$, $v = \dfrac{y-b}{k}$ where a, b, h and k are constants

$$b_{yx} = \frac{r\sigma_y}{\sigma_x} = r \cdot \frac{k\sigma_v}{h\sigma_u} = \frac{k}{h}\left(\frac{r\sigma_v}{\sigma_u}\right) = \frac{k}{h}\, b_{vu}$$

Similarly, $b_{xy} = \dfrac{h}{k}\, b_{uv}$.

Thus, b_{yx} and b_{xy} are both independent of a and b but not of h and k.

Property V. *The correlation co-efficient and the two regression co-efficients have same sign.*

Proof. Regression co-efficient of y on $x = b_{yx} = r\dfrac{\sigma_y}{\sigma_x}$

Regression co-efficient of x on $y = b_{xy} = r\dfrac{\sigma_x}{\sigma_y}$

Since σ_x and σ_y are both positive, b_{yx}, b_{xy} and r have same sign.

5.30. ANGLE BETWEEN TWO LINES OF REGRESSION

If θ is the acute angle between the two regression lines in the case of two variables x and y, show that

$$\tan\theta = \frac{1-r^2}{r}\cdot\frac{\sigma_x\sigma_y}{\sigma_x^2+\sigma_y^2} \quad \textit{where } r, \sigma_x, \sigma_y \textit{ have their usual meanings.}$$

Explain the significance of the formula when r = 0 and r = ± 1.

Proof. Equations to the lines of regression of y on x and x on y are

$$y - \bar{y} = \frac{r\sigma_y}{\sigma_x}(x - \bar{x}) \quad \text{and} \quad x - \bar{x} = \frac{r\sigma_x}{\sigma_y}(y - \bar{y})$$

Their slopes are $m_1 = \dfrac{r\sigma_y}{\sigma_x}$ and $m_2 = \dfrac{\sigma_y}{r\sigma_x}$.

$$\therefore \quad \tan\theta = \pm\frac{m_2 - m_1}{1 + m_2 m_1} = \pm\frac{\dfrac{\sigma_y}{r\sigma_x} - \dfrac{r\sigma_y}{\sigma_x}}{1 + \dfrac{\sigma_y^2}{\sigma_x^2}}$$

$$= \pm\frac{1-r^2}{r}\cdot\frac{\sigma_y}{\sigma_x}\cdot\frac{\sigma_x^2}{\sigma_x^2+\sigma_y^2} = \pm\frac{1-r^2}{r}\cdot\frac{\sigma_x\sigma_y}{\sigma_x^2+\sigma_y^2}$$

Since $r^2 \le 1$ and σ_x, σ_y are positive.

$\therefore$ +ve sign gives the acute angle between the lines.

Hence $$\tan\theta = \frac{1-r^2}{r} \cdot \frac{\sigma_x \sigma_y}{\sigma_x^2 + \sigma_y^2}$$

when $r = 0$, $\theta = \dfrac{\pi}{2}$

$\therefore$ The two lines of regression are perpendicular to each other.

Hence the estimated value of y is the same for all values of x and *vice versa*.

when $r = \pm 1$, $\tan\theta = 0$ so that, $\theta = 0$ or π.

Hence the lines of regression coincide and there is perfect correlation between the two variates x and y.

Note. $$\frac{r\sigma_x}{\sigma_y} = \frac{\frac{1}{n}\Sigma xy - \bar{x}\,\bar{y}}{\sigma_x \sigma_y} \cdot \frac{\sigma_x}{\sigma_y} = \frac{\frac{1}{n}\Sigma xy - \bar{x}\,\bar{y}}{\sigma_y^2} = \frac{\frac{1}{n}\Sigma xy - \bar{x}\,\bar{y}}{\frac{1}{n}\Sigma y^2 - \bar{y}^2}$$

Similarly, $$\frac{r\sigma_y}{\sigma_x} = \frac{\frac{1}{n}\Sigma xy - \bar{x}\,\bar{y}}{\frac{1}{n}\Sigma x^2 - \bar{x}^2}.$$

ILLUSTRATIVE EXAMPLES

Example 1. *Calculate the co-efficient of correlation and obtain the least square regression line of y on x for the following data:*

x:	1	2	3	4	5	6	7	8	9
y:	9	8	10	12	11	13	14	16	15

Also obtain an estimate of y which should correspond on the average to x = 6.2.

Sol.

x	y	$u = x - 5$	$v = y - 12$	u^2	v^2	uv
1	9	– 4	– 3	16	9	12
2	8	– 3	– 4	9	16	12
3	10	– 2	– 2	4	4	4
4	12	– 1	0	1	0	0
5	11	0	– 1	0	1	0
6	13	1	1	1	1	1
7	14	2	2	4	4	4
8	16	3	4	9	16	12
9	15	4	3	16	9	12
Total		0	0	60	60	57

$$r_{xy} = r_{uv} = \frac{\frac{1}{n}\Sigma uv - \overline{u}\,\overline{v}}{\sqrt{\left(\frac{1}{n}\Sigma \overline{u}^2 - \overline{u}^2\right)\left(\frac{1}{n}\Sigma v^2 - \overline{v}^2\right)}} = \frac{\frac{1}{9}(57) - 0}{\sqrt{\left[\frac{1}{9}(60) - 0\right]\left[\frac{1}{9}(60) - 0\right]}}$$

$$= \frac{19}{20} = 0.95$$

$$\frac{r\sigma_y}{\sigma_x} = \frac{r\sigma_v}{\sigma_u} = \frac{\frac{1}{n}\Sigma uv - \overline{u}\,\overline{v}}{\frac{1}{n}\Sigma u^2 - \overline{u}^2} = \frac{\frac{1}{9}(57) - 0}{\frac{1}{9}(60) - 0} = \frac{19}{20} = 0.95$$

Also $\overline{x} = 5 + \frac{1}{9}\Sigma u = 5, \overline{y} = 12 + \frac{1}{9}\Sigma v = 12$

Equation of the line of regression of y on x is

$$y - \overline{y} = \frac{r\sigma_y}{\sigma_x}(x - \overline{x})$$

or $$y - 12 = 0.95\,(x - 5)$$

or $$y = 0.95x + 7.25$$

When $x = 6.2$, estimated value of $y = 0.95 \times 6.2 + 7.25 = 5.89 + 7.25 = 13.14$.

Example 2. *In a partially destroyed laboratory record of an analysis of a correlation data, the following results only are eligible:*

Variance of $x = 9$

Regression equations: $8x - 10y + 66 = 0$, $40x - 18y = 214$.

What were (a) the mean values of x and y, (b) the standard deviation of y and (c) the co-efficient of correlation between x and y.

Sol. (*i*) **Since both the lines of regression pass through the point** $(\overline{x}, \overline{y})$ therefore, we have

$$8\overline{x} - 10\overline{y} + 66 = 0 \quad \text{...(1)}$$

$$40\overline{x} - 18\overline{y} - 214 = 0 \quad \text{...(2)}$$

Multiplying (1) by 5, $40\overline{x} - 50\overline{y} + 330 = 0$ ...(3)

Substracting (3) from (2), $32\overline{y} - 544 = 0 \quad \therefore \quad \overline{y} = 17$

$\therefore$ From (1), $8\overline{x} - 170 + 66 = 0$ or $8\overline{x} = 104 \quad \therefore \quad \overline{x} = 13$

Hence $\overline{x} = 13, \quad \overline{y} = 17$ ...(*a*)

(*ii*) Variance of $x = \sigma_x^2 = 9$ (given)

$\therefore$ $\sigma_x = 3$

The equations of lines of regression can be written as

$$y = .8x + 6.6 \quad \text{and} \quad x = .45y + 5.35$$

$\therefore$ The regression co-efficient of y on x is $\frac{r\sigma_y}{\sigma_x} = .8$ ...(4)

The regression co-efficient of x on y is $\dfrac{r\sigma_x}{\sigma_y} = .45$ ...(5)

Multiplying (4) and (5), $r^2 = .8 \times .45 = .36 \quad \therefore \quad r = 0.6$ ...(b)

(+ve sign with sq. root is taken because regression co-efficients are +ve).

From (4), $\sigma_y = \dfrac{.8\sigma_x}{r} = \dfrac{.8 \times 3}{0.6} = 4.$ ...(c)

TEST YOUR KNOWLEDGE

1. (*a*) Calculate the correlation co-efficient for the following heights in inches of fathers (X) and their sons (Y).

X:	65	66	67	67	68	69	70	72
Y:	67	68	65	68	72	72	69	71

(*b*) Find the correlation co-efficient between x and y from the given data:

x:	78	89	97	69	59	79	68	57
y:	125	137	156	112	107	138	123	108

(*c*) Find the correlation co-efficient from the following data:

x:	92	89	87	86	83	77	71	63	53	50
y:	86	88	91	77	68	85	52	82	37	57

2. Calculate the co-efficient of correlation for the following ages of husbands and wives:

Husbands's age x:	23	27	28	28	29	30	31	33	35	36
Wife's age y:	18	20	22	27	21	29	27	29	28	29

3. Establish the formula $\sigma^2_{x-y} = \sigma_x^2 + \sigma_y^2 - 2r\sigma_x\sigma_y$

where r is the correlation co-efficient between x and y.

4. (*a*) Calculate the co-efficient of correlation for the following table:

y \ x	*16–18*	*18–20*	*20–22*	*22–24*
10–20	2	1	1	
20–30	3	2	3	2
30–40	3	4	5	6
40–50	2	2	3	4
50–60		1	2	2
60–70		1	2	1

(*b*) Find the correlation between x (marks in Mathematics) and y (marks in Engineering Drawing) given in the following data:

y \ x	*10—40*	*40—70*	*70—100*	*Total*
0–30	5	20	—	25
30–60	—	28	2	30
60–90	—	32	13	45
Total	5	80	15	100

5. Ten students got the following percentage of marks in Chemistry and Physics:

Students:	1	2	3	4	5	6	7	8	9	10
Marks in Chemistry:	78	36	98	25	75	82	90	62	65	39
Marks in Physics:	84	51	91	60	68	62	86	58	63	47

Calculate the rank correlation co-efficient.

6. Ten competitors in a musical test were ranked by the three judges x, y and z in the following order:

Ranks by x:	1	6	5	10	3	2	4	9	7	8
Ranks by y:	3	5	8	4	7	10	2	1	6	9
Ranks by z:	6	4	9	8	1	2	3	10	5	7

Using rank correlation method, discuss which pair of judges has the nearest approach to common likings in music.

7. A sample of 12 fathers and their eldest sons gave the following data about their heights in inches:

Father:	65	63	67	64	68	62	70	66	68	67	69	71
Son:	68	66	68	65	69	66	68	65	71	67	68	70

Calculate the co-efficient of rank correlation.

8. If $r = 0$, show that the two lines of regression are paralled to the axes.

9. If the two regression co-efficients are 0.8 and 0.2, what would be the value of co-efficient of correlation.

10. (*a*) Find the correlation co-efficient and the equations of regression lines for the following values of x and y:

x:	1	2	3	4	5
y:	2	5	3	8	7

(*b*) Find the correlation co-efficient between x and y for the given values. Find also the two regression lines.

x:	1	2	3	4	5	6	7	8	9	10
y:	10	12	16	28	25	36	41	49	40	50

11. The two regression equations of the variables x and y are $x = 19.13 - 0.87y$ and $y = 11.64 - 0.50x$. Find (*i*) mean of x's, (*ii*) mean of y's and (*iii*) the correlation co-efficient between x and y.

12. Two random variables have the regression lines with equations $3x + 2y = 26$ and $6x + y = 31$. Find the mean values and the correlation co-efficient between x and y.

13. In a partially destroyed laboratory data, only the equations giving the two lines of regression of y on x and x on y are available and are respectively, $7x - 16y + 9 = 0$, $5y - 4x - 3 = 0$.

Calculate the co-efficient of correlation, $\bar{x}$ and $\bar{y}$.

Answers

1. (*a*) 0.603 (*b*) 0.96 (*c*) 0.7291 **2.** 0.82 **4.** (*a*) 0.28 (*b*) 0.4517

5. 0.84 **6.** x and z **7.** 0.722

9. 0.4

10. (*a*) $r = 0.8$; $y = 1.3x + 1.1$; $x = 0.5y + 0.5$ (*b*) $r = 0.96$; $y = 4.69x + 4.9$; $x = 0.2y - 0.64$

11. (*i*) 15.79 (*ii*) 3.74 (*iii*) – 0.6595

12. $\bar{x} = 4, \bar{y} = 7; r = -0.5$ **13.** $r = 0.7395$; $\bar{x} = -0.1034$; $\bar{y} = 0.5172$.

5.31. THEORY OF PROBABILITY

Here we define and explain certain terms which are used frequently.

(*a*) **Trial and event.** Let an experiment be repeated under essentially the same conditions and let it result in any one of the several possible outcomes. Then, the experiment is called a *trial* and the possible outcomes are known as *events* or *cases*.

For example: (*i*) Tossing of a coin is a trial and the turning up of head or tail is an event.

(*ii*) Throwing a die is a trial and getting 1 or 2 or 3 or 4 or 5 or 6 is an event.

(*b*) **Exhaustive events.** The total number of all possible outcomes in any trial is known as *exhaustive events* or *exhaustive cases.*

For example: (*i*) In tossing a coin, there are two exhaustive cases, head and tail.

(*ii*) In throwing a dice, there are 6 exhaustive cases, for any one of the six faces may turn up.

(*iii*) In throwing two dice, the exhaustive cases are $6 \times 6 = 6^2$ for any of the 6 numbers from 1 to 6 on one die can be associated with any of the 6 numbers on the other die.

In general, in throwing n dice, the exhaustive cases are 6^n.

(*c*) **Favourable events or cases.** The cases which entail the happening of an event are said to be *favourable* to the event. It is the total number of possible outcomes in which the specified event happens.

For example: (*i*) In throwing a die, the number of cases favourable to the appearance of a multiple of 3 are two *viz.* 3 and 6 while the number of cases favourable to the appearance of an even number are three, *viz.*, 2, 4 and 6.

(*ii*)In a throw of two dice, the number of cases favourable to getting a sum 6 is 5, *viz.*, (1, 5); (5, 1); (2, 4); (4, 2); (3, 3).

(*d*) **Mutually exclusive events.** Events are said to be *mutually exclusive* or *incompatible* if the happening of any one of them precludes (*i.e.,* rules out) the happening of all others, *i.e.*, if no two or more than two of them can happen simultaneously in the same trial.

For example : (*i*) In tossing a coin, the events head and tail are mutually exclusive, since if the outcome is head, the possibility of getting tail in the same trial is ruled out.

(*ii*) In throwing a die, all the six faces numbered, 1, 2, 3, 4, 5, 6 are mutually exclusive since any outcome rules out the possibility of getting any other.

(*e*) **Equally likely events.** Events are said to be *equally likely* if there is no reason to expect any one in preference to any other.

For example: (*i*) When a card is drawn from a well shuffled pack, any card may appear in the draw so that the 52 different cases are equally likely.

(*ii*) In throwing a die, all the six faces are equally likely to come.

(*f*) **Independent and dependent events.** Two or more events are said to be *independent* if the happening or non-happening of any one does not depend (or is not affected) by the happening or non-happening of any other. Otherwise they are said to be *dependent*.

For example: If a card is drawn from a pack of well shuffled cards and replaced before drawing the second card, the result of the second draw is independent of the first draw. However, if the first card drawn is not replaced, then, the second draw is dependent on the first draw.

5.32. (a) MATHEMATICAL (OR CLASSICAL) DEFINITION OF PROBABILITY

If a trial results in n *exhaustive, mutually exclusive* and *equally likely cases* and m of them are favourable to the happening of an event E, then the probability of happening of E is given by

$$p \text{ or } P(E) = \frac{\text{Favourable number of cases}}{\text{Exhaustive number of cases}} = \frac{m}{n}.$$

Note 1. Since the number of cases favourable to happening of E is m and the exhaustive number of cases is n, therefore, the number of cases unfavourable to happening of E are $n - m$.

Note 2. The probability that the event E will not happen is given by

$$q \text{ or } P(\overline{E}) = \frac{\text{Unfavourable number of cases}}{\text{Exhaustive number of cases}} = \frac{n-m}{n} = 1 - \frac{m}{n} = 1 - p$$

$$p + q = 1 \quad i.e., \quad P(E) + P(\overline{E}) = 1$$

Obviously, p and q are non-negative and cannot exceed unity, *i.e.*, $0 \le p \le 1, 0 \le q \le 1$.

Note 3. If P(E) = 1, E is called a *certain event i.e.*, the chance of its happening is cent-percent. If P(E) = 0, then E is an *impossible event*.

Note 4. If n cases are favourable to E and m cases are favourable to $\overline{E}$ (*i.e.*, unfavourable to E), then exhaustive number of cases = $n + m$.

$$P(E) = \frac{n}{n+m} \quad \text{and} \quad P(\overline{E}) = \frac{m}{n+m}$$

We say that "odds in *favour of E*" are $n : m$ and "odds *against* E" are $m : n$.

5.32. (b) STATISTICAL (OR EMPIRICAL) DEFINITION OF PROBABILITY

If in n trials, an event E happens m times, then the probability of happening of E is given by

$$p = P(E) = \underset{n \to \infty}{\text{Lt}} \frac{m}{n}.$$

ILLUSTRATIVE EXAMPLES

Example 1. *A bag contains 7 white, 6 red and 5 black balls. Two balls are drawn at random. Find the probability that they will both be white.*

Sol. Total number of balls = 7 + 6 + 5 = 18.

Out of 18 balls, 2 can be drawn in $^{18}C_2$ ways.

$\therefore$ Exhaustive number of cases = $^{18}C_2 = \dfrac{18 \times 17}{2 \times 1} = 153$

Out of 7 white balls, 2 can be drawn in $^7C_2 = \dfrac{7 \times 6}{2 \times 1} = 21$ ways.

$\therefore$ Favourable number of cases = 21

$$\text{Probability} = \frac{21}{153} = \frac{7}{51}.$$

Example 2. *Four cards are drawn from a pack of cards. Find the probability that (i) all are diamonds, (ii) there is one card of each suit, and (iii) there are two spades and two hearts.*

Sol. 4 cards can be drawn from a pack of 52 cards in $^{52}C_4$ ways.

$\therefore$ Exhaustive number of cases = $^{52}C_4 = \dfrac{52 \times 51 \times 50 \times 49}{4 \times 3 \times 2 \times 1} = 270725.$

(*i*) There are 13 diamonds in the pack and 4 can be drawn out of them in $^{13}C_4$ ways.

$\therefore$ Favourable number of cases = $^{13}C_4 = \dfrac{13 \times 12 \times 11 \times 10}{4 \times 3 \times 2 \times 1} = 715.$

$$\text{Required probability} = \frac{715}{270725} = \frac{143}{54145} = \frac{11}{4165}.$$

(*ii*) There are 4 suits, each containing 13 cards.

$\therefore$ Favourable number of cases = $^{13}C_1 \times {}^{13}C_1 \times {}^{13}C_1 \times {}^{13}C_1 = 13 \times 13 \times 13 \times 13.$

$$\text{Required probability} = \frac{13 \times 13 \times 13 \times 13}{270725} = \frac{2197}{20825}.$$

(*iii*) 2 spades out of 13 can be drawn in $^{13}C_2$ ways.

2 hearts out of 13 can be drawn in $^{13}C_2$ ways.

$\therefore$ Favourable number of cases = $^{13}C_2 \times {}^{13}C_2 = 78 \times 78$

$$\text{Required probability} = \frac{78 \times 78}{270725} = \frac{468}{20825}.$$

Example 3. *A bag contains 50 tickets numbered 1, 2, 3, ..., 50, of which five are drawn at random and arranged in ascending order of magnitude* ($x_1 < x_2 < x_3 < x_4 < x_5$). *What is the probability that* $x_3 = 30$?

Sol. Exhaustive number of cases $^{50}C_5$.

If $x_3 = 30$, then the two tickets with numbers x_1 and x_2 must come out of 29 tickets numbered 1 to 29 and this can be done in $^{29}C_2$ ways. The other two tickets with numbers x_4 and x_5 must come out of the 20 tickets number 31 to 50 and this can be done in $^{20}C_2$ ways.

$\therefore$ Favourable number of cases = $^{29}C_2 \times {}^{20}C_2$.

$$\text{Required probability} = \frac{^{29}C_2 \times {}^{20}C_2}{^{50}C_5} = \frac{551}{15134}.$$

5.33. RANDOM EXPERIMENT

Occurrences which can be repeated a number of times, essentially under the same conditions, and whose result cannot be predicted before hand are known as **random experiments.**

For example, rolling of a die, tossing a coin, taking out balls from an urn.

Sample Space. Out of the several possible outcomes of a random experiment, one and only one can take place in a trial. The set of all these possible outcomes is called the **sample space** for the particular experiment and is denoted by S.

For example, if a coin is tossed, the possible outcomes are H (Head) and T (Tail).

Thus $\quad S = \{H, T\}$.

Sample Point. The elements of S, the sample space, are called **sample points.**

For example, if a coin is tossed and H and T denote 'Head' and 'Tail' respectively, then S = {H, T}.

The two sample points are H and T.

Finite Sample Space. If the number of sample points in a sample space is finite, we call it a **finite sample space.** (In this chapter, we shall deal with finite sample spaces only).

Event. Every subset of S, the sample space, is called an event.

Since $S \subset S$, S itself is an event; called a **certain event.**

Also, $\phi \subset S$, the null set is also an event, called an **impossible event.**

If $e \in S$, then e is called an **elementary event.** Every elementary event contains only one sample point.

5.34. AXIOMS

(*i*) With each event E (*i.e.*, a sample point) is associated a real number between 0 and 1, called the probability of that event and is denoted by P(E). Thus $0 \le P(E) \le 1$.

(*ii*) The sum of the probabilities of all simple (elementary) events constituting the sample space is 1. Thus P(S) = 1.

(*iii*) The probability of a compound event (*i.e.*, an event made up of two or more sample events) is the sum of the probabilities of the simple events comprising the compound event.

Thus, if there are n *equally likely* possible outcomes of a random experiment, then the sample space S contains n sample points and the probability associated with each sample point is $\frac{1}{n}$. [By Axiom (*ii*)]

Now, if an event E consists of m sample points, then the probability of E is

$$P(E) = \frac{1}{n} + \frac{1}{n} + \ldots\ldots m \text{ times} = \frac{m}{n}$$

$$= \frac{\text{Number of sample points in E}}{\text{Number of sample points in S}}.$$

This closely agrees with the classical definition of probability.

5.35. PROBABILITY OF THE IMPOSSIBLE EVENT IS ZERO, *i.e.*, P (ϕ) = 0

Impossible event contains no sample point. As such, the sample space S and the impossible event ϕ are *mutually exclusive.*

$$\Rightarrow \quad S \cup \phi = S \quad \Rightarrow \quad P(S \cup \phi) = P(S)$$

$$\Rightarrow \quad P(S) + P(\phi) = P(S) \quad \Rightarrow \quad P(\phi) = 0.$$

5.36. PROBABILITY OF THE COMPLEMENTARY EVENT $\overline{A}$ OF A IS GIVEN BY P ($\overline{A}$) = 1 – P(A)

$\overline{A}$ and A are disjoint events. Also $A \cup \overline{A} = S$

$$\therefore \quad P(A \cup \overline{A}) = P(S)$$

$$\Rightarrow \quad P(A) + P(\overline{A}) = 1 \text{ Hence } P(\overline{A}) = 1 - P(A).$$

5.37. FOR ANY TWO EVENTS A AND B, P($\overline{A} \cap$ B) = P(B) – P(A $\cap$ B)

$$\overline{A} \cap B = \{p : p \in B \text{ and } p \notin A\}$$

Now $\overline{A} \cap B$ and $A \cap B$ are disjoint sets and

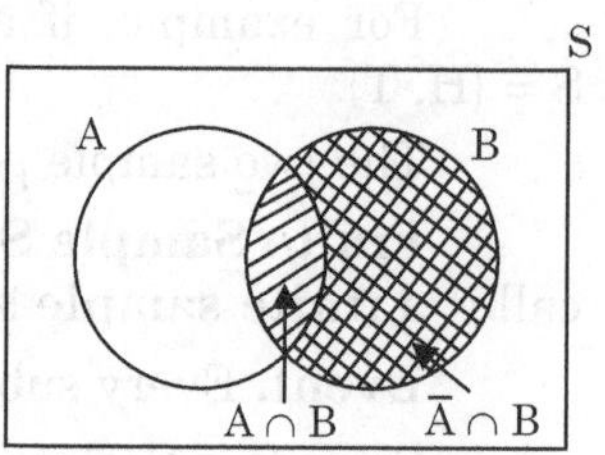

$$(\overline{A} \cap B) \cup (A \cap B) = B$$

$$\Rightarrow \quad P[(\overline{A} \cap B) \cup (A \cap B)] = P(B)$$

$$\Rightarrow \quad P(\overline{A} \cap B) + P(A \cap B) = P(B)$$

$$\Rightarrow \quad P(\overline{A} \cap B) = P(B) - P(A \cap B).$$

Note. Similarly, it can be proved that $P(A \cap \overline{B}) = P(A) - P(A \cap B)$.

5.38. IF B $\subset$ A, THEN

(*i*) **P(A $\cap$ $\overline{B}$) = P(A) – P(B)** (*ii*) **P(B) $\leq$ P(A)**

Proof. When $B \subset A$, B and $A \cap \overline{B}$ are disjoint and their union is A.

$\Rightarrow \quad B \cup (A \cap \overline{B}) = A$

$\Rightarrow \quad P[B \cup (A \cap \overline{B})] = P(A)$

$\Rightarrow \quad P(B) + P(A \cap \overline{B}) = P(A)$

$\Rightarrow \quad P(A \cap \overline{B}) = P(A) - P(B) \quad ...(1)$

Now, if E is any event,

then $0 \le P(E) \le 1$, *i.e.*, $P(E) \ge 0$

$\therefore \quad P(A \cap \overline{B}) \ge 0 \Rightarrow P(A) - P(B) \ge 0$ [Using (1)]

$\Rightarrow \quad P(B) \le P(A).$

S, A, B, $A \cap \overline{B}$

5.39. P(A ∩ B) ≤ P(A) AND P(A ∩ B) ≤ P(B)

Proof. We have $B \subset A \Rightarrow P(B) \le P(A)$

Since $(A \cap B) \subset A$ and $(A \cap B) \subset B$

$\therefore \quad P(A \cap B) \le P(A)$ and $P(A \cap B) \le P(B)$.

5.40. ADDITION THEOREM OF PROBABILITIES (OR THEOREM OF TOTAL PROBABILITY)

Statement. *If A and B are any two events, then*

$$P(A \cup B) = P(A) + P(B) - P(A \cap B)$$

i.e.,

$$P(A \text{ or } B) = P(A) + P(B) - P(A \text{ and } B).$$

S, A, B, $A \cap \overline{B}$, $A \cap B$, $\overline{A} \cap B$

Proof. A and $\overline{A} \cap B$ are *disjoint sets* and their union is $A \cup B$.

$\Rightarrow \quad A \cup B = A \cup (\overline{A} \cap B)$

$\Rightarrow \quad P(A \cup B) = P[A \cup (\overline{A} \cap B)] = P(A) + P(\overline{A} \cap B)$

$= P(A) + [P(\overline{A} \cap B) + P(A \cap B) - P(A \cap B)]$

$= P(A) + P[(\overline{A} \cap B) \cup (A \cap B)] - P[(A \cap B)]$

[$\because$ $\overline{A} \cap B$ and $A \cap B$ are disjoint]

$= P(A) + P(B) - P(A \cap B)$ [$\because$ $(\overline{A} \cap B) \cup (A \cap B) = B$]

$P(A \cup B) = P(A) + P(B) - P(A \cap B).$

Note 1. If A and B are two mutually disjoint events, then $A \cap B = \phi$, so that $P(A \cap B) = P(\phi) = 0$.

$\therefore \quad P(A \cup B) = P(A) + P(B).$

Note 2. $P(A \cup B)$ is also written as $P(A + B)$. Thus, for mutually disjoint events A and B,

$$P(A + B) = P(A) + P(B).$$

$P(A \cap B)$ is also written as $P(AB)$.

5.41. IF A, B AND C ARE ANY THREE EVENTS, THEN

$$\mathbf{P(A \cup B \cup C) = P(A) + P(B) + P(C) - P(A \cap B) - P(B \cap C) - P(C \cap A) + P(A \cap B \cap C)}$$

Or

$$\mathbf{P(A + B + C) = P(A) + P(B) + P(C) - P(AB) - P(BC) - P(CA) + P(ABC)}$$

Proof. Using the above Art. for two events, we have

$P(A \cup B \cup C) = P[(A \cup B) \cup C]$

$= P(A \cup B) + P(C) - P[(A \cup B) \cap C]$

$= [P(A) + P(B) - P(A \cap B)] + P(C) - P[(A \cap C) \cup (B \cap C)]$

[By distributive law]

$= P(A) + P(B) + P(C) - P(A \cap B) - [P(A \cap C)]$

$+ P(B \cap C) - P\{(A \cap C) \cap (B \cap C)\}$

$$= P(A) + P(B) + P(C) - P(A \cap B) - P(A \cap C) - P(B \cap C) + P(A \cap B \cap C)$$

$$[\because \ (A \cap C) \cap (B \cap C) = A \cap B \cap C]$$

$$= P(A) + P(B) + P(C) - P(A \cap B) - P(B \cap C) - P(C \cap A) + P(A \cap B \cap C)$$

$$[\because \ A \cap C = C \cap A]$$

or $\quad P(A + B + C) = P(A) + P(B) + P(C) - P(AB) - P(BC) - P(CA) + P(ABC).$

5.42. IF $A_1, A_2, ..., A_n$ ARE n MUTUALLY EXCLUSIVE EVENTS, THEN THE PROBABILITY OF THE HAPPENING OF ONE OF THEM IS

$$\mathbf{P(A_1 \cup A_2 \cup ...... \cup A_n) = P(A_1 + A_2 + ... + A_n) = P(A_1) + P(A_2) + ... + P(A_n)}$$

Proof. Let N be the total number of mutually exclusive, exhaustive and equally likely cases of which m_1 are favourable to A_1, m_2 are favourable to A_2 and so on.

$$\left.\begin{array}{l} \text{Probability of occurrence of event } A_1 = P(A_1) = \dfrac{m_1}{N} \\ \text{Probability of occurrence of event } A_2 = P(A_2) = \dfrac{m_2}{N} \\ \ldots\ldots\ldots\ldots\ldots\ldots\ldots\ldots\ldots\ldots\ldots\ldots \\ \text{Probability of occurrence of event } A_n = P(A_n) = \dfrac{m_n}{N} \end{array}\right\} \quad ...(1)$$

The events being mutually exclusive and equally likely, the number of cases favourable to the event

A_1 or A_2 or ... or A_n is $m_1 + m_2 + ... + m_n$.

$\therefore$ Probability of occurrence of one of the events $A_1, A_2, ..., A_n$ is $P(A_1 + A_2 + ... + A_n)$

$$= \frac{m_1 + m_2 + ... + m_n}{N} = \frac{m_1}{N} + \frac{m_2}{N} + ... + \frac{m_n}{N}$$

$$= P(A_1) + P(A_2) + ... + P(A_n) \qquad | \text{ Using (1)}$$

Example 4. *In a given race, the odds in favour of four horses A, B, C, D are 1 : 3, 1 : 4, 1 : 5, 1 : 6 respectively. Assuming that a dead heat is impossible; find the chance that one of them wins the race.*

Sol. Let p_1, p_2, p_3, p_4 be the probabilities of winning of the horses A, B, C, D respectively.

Since a *dead heat* (in which all the four horses cover *same distance in same time*) is not possible, the events are mutually exclusive.

Odds infavour of A are 1 : 3 $\therefore \quad p_1 = \dfrac{1}{1+3} = \dfrac{1}{4}$

Similarly, $\qquad p_2 = \dfrac{1}{5}, \ p_3 = \dfrac{1}{6}, \ p_4 = \dfrac{1}{7}.$

If p is the chance that one of them wins, then

$$p = p_1 + p_2 + p_3 + p_4 = \frac{1}{4} + \frac{1}{5} + \frac{1}{6} + \frac{1}{7} = \frac{319}{420}.$$

Example 5. *A card is drawn from a well-shuffled pack of playing cards. What is the probability that it is either a spade or an ace?*

Sol. Let A = the event of drawing a spade

and B = the event of drawing an ace

A and B are *not* mutually exclusive.

AB = the event of drawing the ace of spades

$$P(A) = \frac{13}{52},\ P(B) = \frac{4}{52},\ P(AB) = \frac{1}{52}$$

$$\therefore \quad P(A + B) = P(A) + P(B) - P(AB) = \frac{13}{52} + \frac{4}{52} - \frac{1}{52} = \frac{16}{52} = \frac{4}{13}.$$

5.43. CONDITIONAL PROBABILITY

The probability of the happening of an event E_1 when another event E_2 is known to have already happened is called *Conditional Probability* and is denoted by $P(E_1/E_2)$.

Mutually Independent Events. An event E_1 is said to be independent of an event E_2 if

$$P(E_1/E_2) = P(E_1).$$

i.e., if the probability of happening of E_1 is independent of the happening of E_2.

5.44. MULTIPLICATIVE LAW OF PROBABILITY (or Theorem of Compound Probability)

The probability of simultaneous occurrence of two events is equal to the probability of one of the events multiplied by the conditional probability of the other, i.e., for two events A and B,

$$P(A \cap B) = P(A) \times P(B/A)$$

where P(B/A) represents the conditional probability of occurrence of B when the event A has already happened.

Proof. Suppose a trial results in n exhaustive, mutually exclusive and equally likely outcomes, m of them being favourable to the happening of the event A.

$\therefore$ Probability of happening of the event A = $P(A) = \dfrac{m}{n}$ ...(1)

Out of m outcomes favourable to the happening of A, let m_1 be favourable to the happening of the event B.

$\therefore$ Conditional probability of B, given that A has happened = $P(B/A) = \dfrac{m_1}{m}$ ...(2)

Now out of n exhaustive, mutually exclusive and equally likely outcomes, m_1 are favourable to the happening of '*A and B*'.

$\therefore$ Probability of simultaneous occurrence of A and B

$$= P(A \cap B) = \frac{m_1}{n} = \frac{m_1}{m} \times \frac{m}{n} = \frac{m}{n} \times \frac{m_1}{m}$$

$$= P(A) \times P(B/A) \qquad \text{[Using (1) and (2)]}$$

Hence $P(A \cap B) = P(A) \times P(B/A)$.

Note. $P(A \cap B)$ is also written as $P(AB)$.

Thus $P(AB) = P(A) \times P(B/A)$.

Cor. 1. Interchanging A and B

$$P(BA) = P(B) \times P(A/B)$$

or $$P(AB) = P(B) \times P(A/B) \qquad [\because\ B \cap A = A \cap B]$$

Cor. 2. If A and B are independent events, then P(B/A) = P(B)

$\therefore$ P(AB) = P(A) × P(B).

Generalisation. If $A_1, A_2, ..., A_n$ are n independent events, then

$$P(A_1 A_2 ... A_n) = P(A_1) \times P(A_2) \times ...... \times P(A_n).$$

Cor. 3. If p is the chance that an event will happen in one trial then the chance that it will happen in a succession of r trials is

$$p.p ...... p\ (r \text{ times}) = p^r.$$

Cor. 4. If $p_1, p_2, ..., p_n$ are the probabilities that certain events happen, then the probabilities of their non-happening are $1 - p_1, 1 - p_2, ..., 1 - p_n$ and, therefore, the probability of all of these failing is

$$(1 - p_1)(1 - p_2)......(1 - p_n).$$

Hence the chance in which at least one of these events much happen is

$$1 - (1 - p_1)(1 - p_2) ...... (1 - p_n).$$

Example 6. *A problem in mechanics is given to three students A, B, C whose chances of solving it are* $\frac{1}{2}, \frac{1}{3}, \frac{1}{4}$ *respectively. What is the probability that the problem will be solved?*

Sol. The probabilities of A, B, C solving the problem are $\frac{1}{2}, \frac{1}{3}, \frac{1}{4}$.

The probabilities of A, B, C not solving the problem are $1 - \frac{1}{2}, 1 - \frac{1}{3}, 1 - \frac{1}{4}$ *i.e.,* $\frac{1}{2}, \frac{2}{3}, \frac{3}{4}$.

$\therefore$ The probability that the problem is not solved by any of them $= \frac{1}{2} \times \frac{2}{3} \times \frac{3}{4} = \frac{1}{4}$.

Hence the probability that the problem is solved by at least one of them $= 1 - \frac{1}{4} = \frac{3}{4}$.

Example 7. *The odds that a book will be favourably reviewed by three independent critics are 5 to 2, 4 to 3 and 3 to 4 respectively. What is the probability that, of the three reviews, a majority will be favourable?*

Sol. Let the three critics be A, B, C. The probabilities p_1, p_2, p_3 of the book being favourably reviewed by A, B, C are $\frac{5}{7}, \frac{4}{7}, \frac{3}{7}$ respectively.

$\therefore$ The probabilities that the book is unfavourably reviewed by A, B, C are

$$1 - \frac{5}{7} = \frac{2}{7}, 1 - \frac{4}{7} = \frac{3}{7}, 1 - \frac{3}{7} = \frac{4}{7}.$$

A majority will be favourable if the reviews of at least two are favourable.

(*i*) If A, B, C all review favourably, the probability is

$$\frac{5}{7} \times \frac{4}{7} \times \frac{3}{7} = \frac{60}{343} \qquad | \ p_1 p_2 p_3$$

(*ii*) If A, B review favourably and C reviews unfavourably, the probability is

$$\frac{5}{7} \times \frac{4}{7} \times \frac{4}{7} = \frac{80}{343} \qquad | \ p_1 p_2 (1 - p_3)$$

(*iii*) If A, C review favourably and B reviews unfavourably, probability is

$$\frac{5}{7} \times \frac{3}{7} \times \frac{3}{7} = \frac{45}{343} \qquad | \ p_1 (1 - p_2) p_3$$

(*iv*) If B, C review favourably and A reviews unfavourably, the probability is

$$\frac{2}{7}\times\frac{4}{7}\times\frac{3}{7}=\frac{24}{343} \qquad | \ (1-p_1)p_2p_3$$

Hence the probability that a majority will be favourable is

$$\frac{60}{343}+\frac{80}{343}+\frac{45}{343}+\frac{24}{343}=\frac{209}{343}.$$

Example 8. *A can hit a target 4 times in 5 shots; B 3 times in 4 shots; C twice in 3 shots. They fire a volley. What is the probability that at least two shots hit?*

Sol. Probability of A's hitting the target $=\frac{4}{5}$

Probability of B's hitting the target $=\frac{3}{4}$

Probability of C's hitting the target $=\frac{2}{3}$.

For at least two hits, we may have

(*i*) A, B, C all hit the target, the probability for which is

$$\frac{4}{5}\times\frac{3}{4}\times\frac{2}{3}=\frac{24}{60}.$$

(*ii*) A, B hit the target and C misses it, the probability for which is

$$\frac{4}{5}\times\frac{3}{4}\times\left(1-\frac{2}{3}\right)=\frac{4}{5}\times\frac{3}{4}\times\frac{1}{3}=\frac{12}{60}.$$

(*iii*) A, C hit the target and B misses it, the probability for which is

$$\frac{4}{5}\times\left(1-\frac{3}{4}\right)\times\frac{2}{3}=\frac{4}{5}\times\frac{1}{4}\times\frac{2}{3}=\frac{8}{60}.$$

(*iv*) B, C hit the target and A misses it, the probability for which is

$$\left(1-\frac{4}{5}\right)\times\frac{3}{4}\times\frac{2}{3}=\frac{1}{5}\times\frac{3}{4}\times\frac{2}{3}=\frac{6}{60}.$$

Since these are mutually exclusive events, required probability

$$=\frac{24}{60}+\frac{12}{60}+\frac{8}{60}+\frac{6}{60}=\frac{50}{60}=\frac{5}{6}.$$

Example 9. *A has 2 shares in a lottery in which there are 3 prizes and 5 blanks; B has 3 shares in a lottery in which there are 4 prizes and 6 blanks. Show that A's chance of success is to B's as 27 : 35.*

Sol. A can draw two tickets (out of 3 + 5 = 8) in $^8C_3 = 28$ ways.

A will get the blanks in $^5C_2 = 10$ ways. ∴ A can win a prize in 28 – 10 = 18 ways

Hence A's chance of success $=\frac{18}{28}=\frac{9}{14}$

B can draw 3 tickets in $^{10}C_3 = 120$ ways; B will get all blanks in $^6C_3 = 20$ ways.

∴ B can win a prize in 120 – 20 = 100 ways.

Hence B's chance of success $=\frac{100}{120}=\frac{5}{6}$.

∴ A's chance : B's chance $=\frac{9}{14}:\frac{5}{6}$ = 27 : 35.

Example 10. *A and B throw alternately with a single die, A having the first throw. The person who first throws ace is to win. What are their respective chances of winning?*

Sol. The chance of throwing an ace with a single die $= \frac{1}{6}$

The chance of not throwing an ace with a single die $= 1 - \frac{1}{6} = \frac{5}{6}$.

If A is to win, he should throw an ace in the first or third or fifth, ..., throws.

If B is to win, he should throw an ace in the second or fourth or sixth, ..., throws.

The chances that an ace is thrown in the first, second, third, ..., throws are

$$\frac{1}{6}, \frac{5}{6}.\frac{1}{6}, \frac{5}{6}.\frac{5}{6}.\frac{1}{6}, \frac{5}{6}.\frac{5}{6}.\frac{5}{6}.\frac{1}{6} \text{ ......} \quad \text{or} \quad \frac{1}{6}, \frac{5}{6}.\frac{1}{6}, \left(\frac{5}{6}\right)^2.\frac{1}{6}, \left(\frac{5}{6}\right)^3.\frac{1}{6}, \text{......}$$

$$\therefore \quad \text{A's chance} = \frac{1}{6} + \left(\frac{5}{6}\right)^2.\frac{1}{6} + \left(\frac{5}{6}\right)^4.\frac{1}{6} + \text{......} = \frac{\frac{1}{6}}{1-\left(\frac{5}{6}\right)^2} = \frac{6}{11}$$

Sum of an infinite G.P. $= \frac{a}{1-r}$

$$\text{B's chance} = 1 - \frac{6}{11} = \frac{5}{11}.$$

Example 11. *An urn contains 10 white and 3 black balls, while another urn contains 3 white and 5 black balls. Two balls are drawn from the first urn and put into the second urn and then a ball is drawn from the latter. What is the probability that it is a white ball?*

Sol. The two balls drawn from the first urn may be

(*i*) both white (*ii*) both black (*iii*) one white and one black.

Let these events be denoted by A, B, C respectively.

$$P(A) = \frac{{}^{10}C_2}{{}^{13}C_2} = \frac{10 \times 9}{13 \times 12} = \frac{15}{26}; \qquad P(B) = \frac{{}^{3}C_2}{{}^{13}C_2} = \frac{3 \times 2}{13 \times 12} = \frac{1}{26}$$

$$P(C) = \frac{{}^{10}C_1 \times {}^{3}C_1}{{}^{13}C_2} = \frac{10 \times 3}{\frac{13 \times 12}{2 \times 1}} = \frac{10}{26}$$

When two balls are transferred from first urn to second urn, the second urn will contain.

(*i*) 5 white and 5 black balls (*ii*) 3 white and 7 black balls

(*iii*) 4 white and 6 black balls.

Let W denote the event of drawing a white ball from the second urn in the three cases (*i*), (*ii*) and (*iii*).

Now $P(W/A) = \frac{5}{10}, \quad P(W/B) = \frac{3}{10}, \quad P(W/C) = \frac{4}{10}$

$\therefore$ Reqd. probability $= P(A) . P(W/A) + P(B) . P(W/B) + P(C) . P(W/C)$

$$= \frac{15}{26}.\frac{5}{10} + \frac{1}{26}.\frac{3}{10} + \frac{10}{26}.\frac{4}{10} = \frac{75 + 3 + 40}{260} = \frac{118}{260} = \frac{59}{130}.$$

TEST YOUR KNOWLEDGE

1. In a class of 10 students, 4 are boys and the rest are girls. Find the probability that a student selected will be a girl.
2. What is the chance that a (*i*) non-leap year (*ii*) leap year should have fifty three Sundays?
3. A card is drawn from an ordinary pack and a gambler bets that it is a spade or an ace. What are the odds against his winning the bet?
4. An integer is chosen at random from the first two hundred positive integers. What is the probability that the integer chosen is divisible by 6 or 8?
5. Six cards are drawn at random from a pack of 52 cards. What is the probability that 3 will be red and 3 black?
6. From a set of raffle tickets numbered 1 to 100, three are drawn at random. What is the probability that all are odd numbered?
7. (*a*) If from a lottery of 30 tickets, marked, 1, 2, 3, ..., 30, four tickets be drawn, what is the chance that those marked 1 and 2 are among them?

   (*b*) An urn contains 5 red and 10 black balls. Eight of them are placed in another urn. What is the chance that the latter then contains 2 red and 6 black balls?
8. A party of n persons sit at a round table. Find the odds against two specified individuals sitting next to each other.
9. A five-figured number is formed by the digits 0, 1, 2, 3, 4 (without repetition). Find the probability that the number formed is divisible by 4.
10. Three newspapers A, B, C are published in a city and a survey of readers indicates the following: 20% read A, 16% read B, 14% read C, 8% read both A and B, 5% read both A and C, 4% read both B and C, 2% read all the three.

    For a person chosen at random, find the probability that he reads none of the papers.
11. A problem in Statistics is given to five students. Their chances of solving it are $\frac{1}{2}, \frac{1}{3}, \frac{1}{4}, \frac{1}{4}$ and $\frac{1}{5}$. What is the probability that the problem will be solved?
12. A can hit a target 5 times in 6 shots, B 4 times in 5 shots and C 3 times in 4 shots. They fire a volley. What is the probability that two shots, at least hit the target?
13. Three groups of children contain respectively 3 girls and 1 boy; 2 girls and 2 boys; 1 girl and 3 boys. One child is selected at random from each group. Show that the chance that the three selected consist of 1 girl and 2 boys is $\frac{13}{32}$.
14. Four persons are chosen at random from a group containing 3 men, 2 women and 4 children. Show that the chance that exactly two of them will be children is $\frac{5}{21}$.
15. A bag contains 10 balls, two of which are red, three blue and five black. Three balls are drawn at random from the bag. What is the probability that

    (*i*) the three balls are of different colours, (*ii*) two balls are of the same colour,

    (*iii*) the balls are all of the same colour.

Answers

1. $\frac{3}{5}$ **2.** (*i*) $\frac{1}{7}$ (*ii*) $\frac{2}{7}$ **3.** 9 : 4 **4.** $\frac{1}{4}$

5. $\frac{13000}{39151}$ **6.** $\frac{4}{33}$ **7.** (*a*) $\frac{2}{145}$, (*b*) $\frac{140}{429}$ **8.** $(n - 3) : 2$

9. $\frac{5}{16}$ **10.** $\frac{13}{20}$ **11.** $\frac{17}{20}$ **12.** $\frac{107}{120}$

15. *(i)* $\frac{1}{4}$ *(ii)* $\frac{79}{120}$ *(iii)* $\frac{11}{120}$.

5.43. BAYE'S THEOREM

If $E_1, E_2, ..., E_n$ are mutually exclusive and exhaustive events with $P(E_i) \neq 0$, (i = 1, 2, ..., n) of a random experiment then for any arbitrary event A of the sample space of the above experiment with P(A) > 0, we have

$$P(E_i/A) = \frac{P(E_i)P(A/E_i)}{\sum_{i=1}^{n} P(E_i)P(A/E_i)}.$$

Proof. Let S be the sample space of the random experiment.

The events $E_1, E_2, ..., E_n$ being exhaustive

$$S = E_1 \cup E_2 \cup ...... \cup E_n$$

$$\therefore \quad A = A \cap S \qquad [\because \ A \subset S]$$

$$= A \cap (E_1 \cup E_2 \cup ...... \cup E_n)$$

$$= (A \cap E_1) \cup (A \cap E_2) \cup ...... \cup (A \cap E_n) \qquad \text{[Distributive Law]}$$

$$\Rightarrow \quad P(A) = P(A \cap E_1) + P(A \cap E_2) + ... + P(A \cap E_n)$$

$$= P(E_1)P(A/E_1) + P(E_2)P(A/E_2) + ... + P(E_n)P(A/E_n)$$

$$= \sum_{i=1}^{n} P(E_i)P(A/E_i) \qquad ...(1)$$

Now $\quad P(A \cap E_i) = P(A)P(E_i/A)$

$$\Rightarrow \quad P(E_i/A) = \frac{P(A \cap E_i)}{P(A)} = \frac{P(E_i)P(A/E_i)}{\sum_{i=1}^{n} P(E_i)P(A/E_i)} \qquad \text{[Using (1)]}$$

Note. The significance of Baye's Theorem may be understood in the following manner:

$P(E_i)$ is the probability of occurrence of E_i. The experiment is performed and we are told that the event A has occurred. With this information, the probability $P(E_i)$ is changed to $P(E_i/A)$. Baye's Theorem enables us to evaluate $P(E_i/A)$ if all the $P(E_i)$ and the conditional probabilities $P(A/E_i)$ are known.

ILLUSTRATIVE EXAMPLES

Example 1. *A bag X contains 2 white and 3 red balls and a bag Y contains 4 white and 5 red balls. One ball is drawn at random from one of the bags and is found to be red. Find the probability that it was drawn from bag Y.*

Sol. Let E_1: the ball is drawn from bag X; E_2: the ball is drawn from bag Y and A: the ball is red.

We have to find $P(E_2/A)$.

By Baye's Theorem,

$$P(E_2/A) = \frac{P(E_2)P(A/E_2)}{P(E_1)P(A/E_1) + P(E_2)P(A/E_2)} \qquad ...(1)$$

Since the two bags are equally likely to be selected, $P(E_1) = P(E_2) = \frac{1}{2}$

Also $P(A/E_1) = P(\text{a red ball is drawn from bag X}) = \frac{3}{5}$

$P(A/E_2) = P(\text{a red ball is drawn from bag Y}) = \frac{5}{9}$

$\therefore$ From (1), we have $P(E_2/A) = \dfrac{\frac{1}{2}\times\frac{5}{9}}{\frac{1}{2}\times\frac{3}{5}+\frac{1}{2}\times\frac{5}{9}} = \frac{25}{52}.$

Example 2. *In a bolt factory, machines A, B and C manufacture respectively 25%, 35% and 40% of the total. Of their output 5, 4 and 2 percent are defective bolts. A bolt is drawn at random from the product and is found to be defective. What is the probability that it was manufactured by machine B?*

Sol. Let E_1, E_2 and E_3 denote the events that a bolt selected at random is manufacture by the machines A, B and C respectively and let H denote the event of its being defective. Then

$P(E_1) = 0.25$, $P(E_2) = 0.35$, $P(E_3) = 0.40$

The probability of drawing a defective bolt manufactured by machine A is $P(H/E_1) = 0.05$

Similarly, $P(H/E_2) = 0.04$ and $P(H/E_3) = 0.02$

By Baye's Theorem, we have

$$P(E_2/H) = \frac{P(E_2)P(H/E_2)}{P(E_1)P(H/E_1) + P(E_2)P(H/E_2) + P(E_3)P(H/E_3)}$$

$$= \frac{0.35 \times 0.04}{0.25 \times 0.05 + 0.35 \times 0.04 + 0.40 \times 0.02} = \frac{0.0140}{0.0345} = 0.41.$$

Example 3. *The contents of urns I, II and III are as follows:*

1 white, 2 black and 3 red balls,

2 white, 1 black and 1 red balls, and

4 white, 5 black and 3 red balls.

One urn is chosen at random and two balls drawn. They happen to be white and red. What is the probability that they come from urns I, II or III?

Sol. Let E_1: urn I is chosen; E_2: urn II is chosen; E_3: urn III is chosen

and A: the two balls are white and red.

We have to find $P(E_1/A)$, $P(E_2/A)$ and $P(E_3/A)$.

Now $P(E_1) = P(E_2) = P(E_3) = \frac{1}{3}$

$P(A/E_1) = P\text{ (a white and a red ball are drawn from urn I)} = \dfrac{{}^1C_1 \times {}^3C_1}{{}^6C_2} = \dfrac{1}{5}$

$$P(A/E_2) = \frac{{}^2C_1 \times {}^1C_1}{{}^4C_2} = \frac{1}{3};\ P(A/E_3) = \frac{{}^4C_1 \times {}^3C_1}{{}^{12}C_2} = \frac{2}{11}$$

By Baye's Theorem, we have

$$P(E_1/A) = \frac{P(E_1)P(A/E_1)}{P(E_1)P(A/E_1) + P(E_2)P(A/E_2) + P(E_3)P(A/E_3)} = \frac{\frac{1}{3} \times \frac{1}{5}}{\frac{1}{3} \times \frac{1}{5} + \frac{1}{3} \times \frac{1}{3} + \frac{1}{3} \times \frac{2}{11}} = \frac{33}{118}$$

Similarly, $P(E_2/A) = \frac{55}{118}$, $P(E_3/A) = \frac{15}{59}$.

TEST YOUR KNOWLEDGE

1. Two urns contain 4 white, 6 blue and 4 white, 5 blue balls respectively. One of the urns is selected at random and a ball is drawn from it. If the ball drawn is white, find the probability that it is drawn from the

   (*i*) first urn (*ii*) second urn.

2. Three urns contain 6 red, 4 black; 4 red, 6 black and 5 red, 5 black balls respectively. One of the urns is selected at random and a ball is drawn from it. If the ball drawn is red, find the probability that it is drawn from the first urn.

3. A factory has two machines A and B. Past record shows that machine A produced 60% of the items of output and machine B produced 40% of the items. Further, 2% of the items produced by machine A were defective and 1% produced by machine B were defective. If a defective item is drawn at random, what is the probability that it was produced by machine A?

4. An insurance company insured 2000 scooter drivers, 4000 car drivers and 6000 truck drivers. The probability of accident is 0.01, 0.03 and 0.15 respectively. One of the insured persons meets an accident. What is the probability that he is a scooter driver?

5. A company has two plants to manufacture scooters. Plant I manufactures 70% of scooters and plant II manufactures 30%. At plant I, 80% of the scooters are rated standard quality and at plant II, 90% of the scooters are rated standard quality. A scooter is chosen at random and is found to be of standard quality. What is the chance that it has come from plant II?

Answers

1. (*i*) $\frac{9}{19}$ (*ii*) $\frac{10}{19}$ 2. $\frac{2}{5}$ 3. $\frac{3}{4}$

4. $\frac{1}{52}$ 5. $\frac{27}{83}$.

5.46. RANDOM VARIABLE

If the numerical values assumed by a variable are the result of some chance factors, so that a particular value cannot be exactly predicted in advance, the variable is then called a *random variable*. A random variable is also called '*chance variable*' or '*stochastic variable*'.

Random variables are denoted by capital letters, usually, from the last part of the alphabet, for instance, X, Y, Z etc.

Continuous and Discrete Random Variables

A *continuous random variable* is one which can assume any value within an interval, *i.e.*, all values of a continuous scale. For example (*i*) the weights (in kg) of a group of individuals, (*ii*) the heights of a group of individuals.

A *discrete random variable* is one which can assume only isolated values. For example,

(*i*) the number of heads in 4 tosses of a coin is a discrete random variable as it cannot assume values other than 0, 1, 2, 3, 4.

(*ii*) the number of aces in a draw of 2 cards from a well shuffled deck is a random variable as it can take the values 0, 1, 2 only.

5.47. DISCRETE PROBABILITY DISTRIBUTION

Let a random variable X assume values $x_1, x_2, x_3, \ldots\ldots, x_n$ with probabilities $p_1, p_2, p_3, \ldots\ldots, p_n$ respectively, where $P(X = x_i) = p_i \geq 0$ for each x_i and $p_1 + p_2 + p_3 + \ldots\ldots + p_n = \sum_{i=1}^{n} p_i = 1$.

Then

X:	x_1,	x_2,	$x_3, \ldots\ldots, x_n$
P(X):	p_1,	p_2,	$p_3, \ldots\ldots, p_n$

is called the discrete probability distribution for X and it spells out how a total probability of 1 is distributed over several values of the random variable.

5.48. MEAN AND VARIANCE OF RANDOM VARIABLES

Let

X:	x_1,	x_2,	$x_3, \ldots\ldots, x_n$
P(X):	p_1,	p_2,	$p_3, \ldots\ldots, p_n$

be a discrete probability distribution.

We denote the *mean* by μ and define $\mu = \frac{\Sigma p_i x_i}{\Sigma p_i} = \Sigma p_i x_i$ $\quad (\because \ \Sigma p_i = 1)$

Other names for the mean are *average or expected value E(X)*.

We denote the *variance* by σ^2 and define $\sigma^2 = \Sigma p_i (x_i - \mu)^2$

If μ is not a whole number, then $\sigma^2 = \Sigma p_i x_i^2 - \mu^2$

Standard deviation $\sigma = + \sqrt{\text{Variance}}$.

ILLUSTRATIVE EXAMPLES

Example 1. *Five defective bulbs are accidentally mixed with twenty good ones. It is not possible to just look at a bulb and tell whether or not it is defective. Find the probability distribution of the number of defective bulbs, if four bulbs are drawn at random from this lot.*

Sol. Let X denote the number of defective bulbs in 4. Clearly X can take the values 0, 1, 2, 3 or 4.

Number of defective bulbs = 5

Number of good bulbs = 20

Total number of bulbs = 25

P(X = 0) = P (no defective) = P (all 4 good ones)

$$= \frac{^{20}C_4}{^{25}C_4} = \frac{20 \times 19 \times 18 \times 17}{25 \times 24 \times 23 \times 22} = \frac{969}{2530}$$

$$P(X = 1) = P(\text{one defective and 3 good ones}) = \frac{{}^5C_1 \times {}^{20}C_3}{{}^{25}C_4} = \frac{1140}{2530}$$

$$P(X = 2) = P\ (\text{2 defectives and 2 good ones}) = \frac{{}^5C_2 \times {}^{20}C_2}{{}^{25}C_4} = \frac{380}{2530}$$

$$P(X = 3) = P(\text{3 defectives and 1 good one}) = \frac{{}^5C_3 \times {}^{20}C_1}{{}^{25}C_4} = \frac{40}{2530}$$

$$P(X = 4) = P\ (\text{all 4 defectives}) = \frac{{}^5C_4}{{}^{25}C_4} = \frac{1}{2530}$$

$\therefore$ The probability distribution of the random variable X is

X:	0	1	2	3	4
P(X):	$\frac{969}{2530}$	$\frac{1140}{2530}$	$\frac{380}{2530}$	$\frac{40}{2530}$	$\frac{1}{2530}$.

Example 2. *A die is tossed thrice. A success is 'getting 1 or 6' on a toss. Find the mean and the variance of the number of successes.*

Sol. Let X denote the number of success. Clearly, X can take the values 0, 1, 2 or 3.

Probability of success $= \frac{2}{6} = \frac{1}{3}$; Probability of failure $= 1 - \frac{1}{3} = \frac{2}{3}$

$$P(X = 0) = P\ (\text{no success}) = P\ (\text{all 3 failures}) = \frac{2}{3} \times \frac{2}{3} \times \frac{2}{3} = \frac{8}{27}$$

$$P(X = 1) = P\ (\text{one success and 2 failures}) = {}^3C_1 \times \frac{1}{3} \times \frac{2}{3} \times \frac{2}{3} = \frac{12}{27}$$

$$P(X = 2) = P\ (\text{two successes and one failure}) = {}^3C_2 \times \frac{1}{3} \times \frac{1}{3} \times \frac{2}{3} = \frac{6}{27}$$

$$P(X = 3) = P\ (\text{all 3 successes}) = \frac{1}{3} \times \frac{1}{3} \times \frac{1}{3} = \frac{1}{27}$$

$\therefore$ The probability distribution of the random variable X is

X:	0	1	2	3
P(X):	$\frac{8}{27}$	$\frac{12}{27}$	$\frac{6}{27}$	$\frac{1}{27}$

To find the mean and variance

x_i	p_i	$p_i x_i$	$p_i x_i^2$
0	$\frac{8}{27}$	0	0
1	$\frac{12}{27}$	$\frac{12}{27}$	$\frac{12}{27}$
2	$\frac{6}{27}$	$\frac{12}{27}$	$\frac{24}{27}$
3	$\frac{1}{27}$	$\frac{3}{27}$	$\frac{9}{27}$
		1	$\frac{5}{3}$

Mean $\mu = \Sigma p_i x_i = 1$

Variance $\sigma^2 = \Sigma p_i x_i^2 - \mu^2 = \frac{5}{3} - 1 = \frac{2}{3}$.

Example 3. *A random variable X has the following probability function:*

Values of X,	*x:*	*0*	*1*	*2*	*3*	*4*	*5*	*6*	*7*
	p(x):	*0*	*k*	*2k*	*2k*	*3k*	k^2	$2k^2$	$7k^2 + k$

(i) Find k, *(ii) Evaluate* $P(X < 6)$, $P(X \geq 6)$, $P(3 < X \leq 6)$

(iii) Find the minimum value of x so that $P(X \leq x) > \frac{1}{2}$.

Sol. (*i*) Since $\sum_{x=0}^{7} p(x) = 1$, we have

$$0 + k + 2k + 2k + 3k + k^2 + 2k^2 + 7k^2 + k = 1$$

$\Rightarrow$ $10k^2 + 9k - 1 = 0$ $\Rightarrow$ $(10k - 1)(k + 1) = 0$

$\Rightarrow$ $k = \frac{1}{10}$ $[\because\ p(x) \geq 0]$

(*ii*) $P(X < 6) = P(X = 0) + P(X = 1) + \ldots\ldots + P(X = 5)$

$$= 0 + k + 2k + 2k + 3k + k^2 = 8k + k^2 = \frac{8}{10} + \frac{1}{100} = \frac{81}{100}$$

$$P(X \geq 6) = P(X = 6) + P(X = 7)$$

$$= 2k^2 + 7k^2 + k = \frac{9}{100} + \frac{1}{10} = \frac{19}{100}$$

$$P(3 < X \leq 6) = P(X = 4) + P(X = 5) + P(X = 6)$$

$$= 3k + k^2 + 2k^2 = \frac{3}{10} + \frac{3}{100} = \frac{33}{100}$$

(*iii*) $P(X \leq 1) = k = \frac{1}{10} < \frac{1}{2}$; $P(X \leq 2) = k + 2k = \frac{3}{10} < \frac{1}{2}$

$P(X \leq 3) = k + 2k + 2k = \frac{5}{10} = \frac{1}{2}$; $P(X \leq 4) = k + 2k + 2k + 3k = \frac{8}{10} > \frac{1}{2}$

$\therefore$ The maximum value of x so that $P(X \leq x) > \frac{1}{2}$ is 4.

Example 4. *The probability density function of a random variate X is given by*

$$p(x) = \begin{cases} k & \text{if } x = 0 \\ 2k & \text{if } x = 1 \\ 3k & \text{if } x = 2 \\ 0 & \text{otherwise.} \end{cases}$$

(a) Find the values of k.

(b) Evaluate $P(X < 2)$, $P(X \leq 2)$, $P(0 < X < 2)$.

(c) Find distribution function of X.

Sol. By the properties of probability density function, we have

(*i*) $p(x) \geq 0$.

(*ii*) $\Sigma p(x) = 1$, therefore, we have

$k + 2k + 3k + 0 = 1$ or $6k = 4$

$\therefore$ $k = 1/6 > 0$.

Consequently the probability density function $p(x)$ is defined as:

$$p(x) = \begin{cases} 1/6 & \text{if } x = 0 \\ 2/6 & \text{if } x = 1 \\ 3/6 & \text{if } x = 2 \\ 0 & \text{otherwise.} \end{cases}$$

So $\quad P(X < 2) = P(X = 0 \text{ or } X = 1)$

$$= P(X = 0) + P(X = 1).$$

$$= \frac{1}{6} + \frac{2}{6} = \frac{3}{6} = \frac{1}{2}$$

$$P(X \le 2) = P(X = 0 \text{ or } X = 1 \text{ or } X = 2$$

$$= P(X = 0) + P(X = 1) + P(X = 2)$$

$$= \frac{1}{6} + \frac{2}{6} + \frac{3}{6} = \frac{6}{6} = 1$$

and $\quad P(0 < X < 2) = P(X = 1) = 2/6 = 1/3.$

(*i*) Since distribution, $F(x) = \sum_{X \le x} p(x)$

$\therefore \quad F(x) = 0, \text{ if } x < 0$

$$= \sum_{X \le 0} p(x) = 0 + \frac{1}{6} = \frac{1}{6}, \text{ if } x = 0$$

$$= \sum_{X \le 1} p(x) = 0 + \frac{1}{6} + \frac{2}{6} = \frac{3}{6}\ \frac{1}{2}, \text{ if } x = 1$$

$$= \sum_{X \le 2} p(x) = 0 + \frac{1}{6} + \frac{2}{6} + \frac{3}{6} = 1, \text{ if } x = 2$$

$$= 1, \text{ if } x > 2.$$

So $$F(x) = \begin{cases} 0 & \text{if } x < 0 \\ 1/6 & \text{if } 0 \le x < 1 \text{ or } x = 0 \\ \frac{1}{2} & \text{if } 1 \le x < 2 \text{ or } x = 1 \\ 1 & \text{if } x \ge 2. \end{cases}$$

CONTINUOUS PROBABILITY DISTRIBUTIONS

5.49. INTRODUCTION—PROBABILITY DENSITY FUNCTION (PDF) OR FREQUENCY FUNCTION

The subject matter of this chapter will be to study the detailed characteristics of the various continuous probaility distributions with the help of the knowledge of the mathematical expectation and gencrating functions etc. The introduction of a continuous probability distribution has already been given. It has also been mentioned there that in dealing with the distribution of a continuous variate (*i.e.* with a variate which may assume any fractional value within its range) it will be necessary to express the probabilities in the form of intervals.

Let us describe the idea of a continuous probability distribution with the following simple example.

Suppose there is a squared target of one unit dimensions and a rifle is perfectly aimed at it which is fired several times after being clamped in that position. After some shots the target will appear as shown in Fig. 5.1. When a bullet is fired it is equally likely to strike anywhere on the squared board. If we assume the target's left side as the Y-axis and the bottom as the X-axis, obviously the variate X would definite the horizontal distance of a hit from the vertical axis (X = 0), and as such it may have any value between 0 and 1 and will be called a **continuous random variate.**

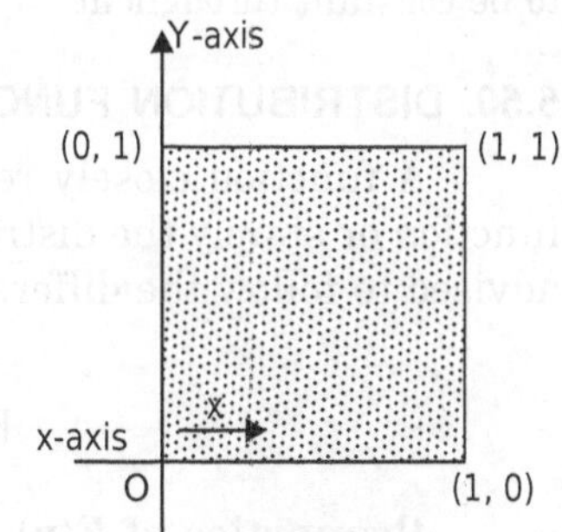

Fig. 5.1

It follows from the knowledge of the Geometric probability that the chance of a hit striking in to any interval is equal to the horizontal length of that interval divided by the whole length of the board which is unity here. For example the prob. that the hit strikes between 0.3 and 0.8 horizontal distances = o.5/1.0 = 0.5. In case of the continuous distribution it is not possible to have a finite probability associated with single point as in the discrete distribution. Here, even if we give the smallest possible values to the individual probabilities, still their sum will assume an infinite value (and not 1 as it must have been); because infinite points are possible within any smallest interval.

Instead, the probabilities in a continuous distribution are assigned to the infinitely small interval and are obtained by the integration of the density function within the desired interval limits. If dx is an infinitely small interval then the density function is defined as follows over the entire range of X.

$$\left.\begin{aligned} P\{\mu - \frac{1}{2}dx \le X \le x + \frac{1}{2}dx\} = f(x)\,dx \\ P\{x \le X \le x + dx\} = f(x)\,dx \end{aligned}\right\} \qquad \text{(Shaded area in Fig. 5.2)}$$

The function $f(x)$ is a single valued and a non-negative function within its range. *i.e.*, $f(x) \ge 0$. The range may be finite or infinite as well. For definition purposes, it is advisable to assume that the range is an infinite one (*i.e.* $-\infty \le X \le \infty$). In case it happens to be finite one then the function can be defined as zero outside that range. The curve $y = f(x)$ is called the probability curve which can be plotted with a smooth hand above the x-axis. The area of the curve will be one:

i.e. $$\int_{-\infty}^{\infty} f(x)\,dx = 1.$$

The probability that the variate falls in the interval (x_1, x_2) is given by

$$\int_{x_2}^{x_2} f(x)\,dx = P\{x_1 \le X \le x_2\} \rightarrow \text{lined area in Fig. 5.2.}$$

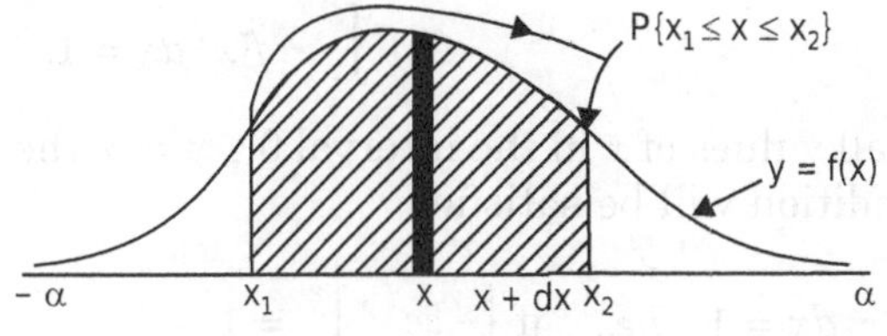

Fig. 5.2. Probability

Remark. The variate is said to have a rectangular distribution of probability when $f(x)$ happens to be constant throughout.

5.50. DISTRIBUTION FUNCTION

A function closely related to the density function $f(x)$ is the cumulative distribution function or simply the distribution function F(x) defined and explained as below. Students are advised to follow the difference between $f(x)$ and F (x) very clearly. We define

$$F(x) = P(X \leq x) = \int_{-\infty}^{x} f(x)\, dx$$

Properties of F(x)

(*i*) F $(x) \geq 0$ *i.e.* it is a non-negative and non-decreasing function with value lying between 0 and x (both inclusive). The less than type cumulative frequency curve is obtained if we plot x on x-axis and F(x) on y-axis. See Fig. 5.3 (*ii*). Here for simplicity it has been assumed that the variate lies between a and b. For a discrete distribution $F(x) = \Sigma\, p(x_1)$ is plotted like Fig. 5.3 (*i*) to obtain $x_1 \leq x$ the cumulative frequency polygon.

(*ii*) F(x) is 1 for x greater than or equal to ∞ *i.e.* $F(\infty) = 1$.

(*iii*) $P\{x_1 \leq X \leq x_2\} = P\{X \leq x_2\} - P(X \leq x_1)$

$$\int_{-\infty}^{x_2} f(x)\, dx - \int_{-\infty}^{x} f(x)\, dx = F(x_2) - F(x_1).$$

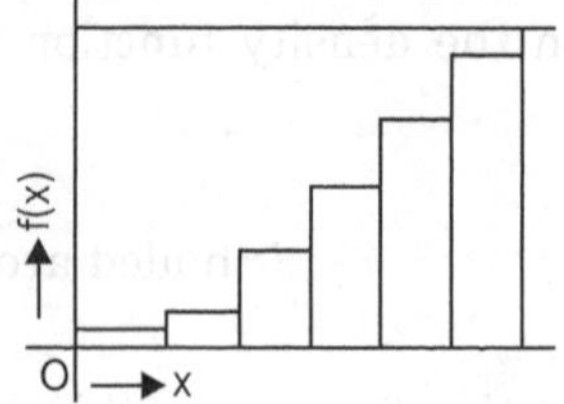

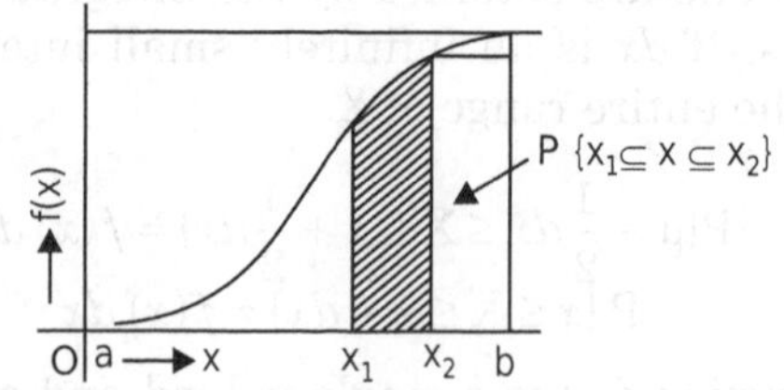

Fig. 5.3

(*i*) Discrete distribution function (*ii*) Continuous distribution function

(*iv*) F(x) is obtained by integrating the probability density function $f(x)$. Conversely $f(x)$ can be obtained from the distribution Function F(x) by differentiation *i.e.* $f(x) = \dfrac{d}{dx}$ F(x).

Example 1. *If the function f(x) is defined by*

$$f(x) = ce^{-x}, \quad 0 \leq x < \infty$$

find the value of c which changes f(x) to probability density function.

Sol. By property of probability density function, we have

(*i*) $f(x) \geq 0$ for all x. (*ii*) $\int_{-\infty}^{\infty} f(x)\, dx = 1.$

Since e^{-x} is +ve for all values of x in the interval $0 \leq x < \infty$, the condition will be satisfied if $c \geq 0$. Again the 2nd condition will be satisfied

if $$\int_0^{\infty} ce^{-x}\, dx = 1 \quad i.e., \quad \text{if } \left[-ce^{-x}\right]_0^{\infty} = 1$$

i.e., if $c = 1$.

Example 2. *If $f(x) = cx^2$, $0 < x < 1$, find the value of c and determine the probability that $\frac{1}{3} < x < \frac{1}{2}$.*

Sol. By property of probability density function, we have

$$\int_0^1 f(x)\,dx = 1$$

$$\therefore \quad \int_0^1 cx^2\,dx = 1 \quad \text{or} \quad c\left[\frac{x^3}{3}\right]_0^1 = 1$$

or $$c/3 = 1; \quad \therefore \quad c = 3.$$

Consequently $f(x) = 3x^2$, $0 < x < 1$.

Again $$P\left(\frac{1}{3} < X < \frac{1}{2}\right) = \int_{1/3}^{1/2} f(x)\,dx$$

$$= \int_{1/3}^{1/2} 3x^2\,dx = \left[3\frac{x^3}{3}\right]_{1/3}^{1/2} = \frac{1}{8} - \frac{1}{27} = \frac{19}{216}$$

Problem 1. *If a function f(x) of x is defined as follows:*

$$f(x) = 0 \text{ for } x\ x < 2$$

$$= \frac{1}{18}(3 + 2x) \text{ for } 2 \le x \le 4$$

$$= 0 \text{ for } x > 4.$$

then show that it is a density function and find the probability that a variate with this density will lie in the interval $2 \le x \le 3$.

If $f(x)$ is a density function, it must satisfy the two conditions:

(i) $f(x) \ge 0$ for every x. From definition it is clear that $f(x) \ge 0$ for every x within its given range. So the condition stands satisfied.

(ii) $\int_{-\infty}^{\infty} f(x)\,dx = 1.$ Let us examine it.

$$\because \quad \int_{-\infty}^{\infty} f(x)\,dx = \int_{-\infty}^{2} f(x)\,dx + \int_{2}^{4} f(x)\,dx + \int_{4}^{\infty} f(x)\,dx$$

$$= \int_{-\infty}^{2} 0.\,dx + \int_{0}^{4} \frac{3+2x}{18}\,dx + \int_{0}^{\infty} 0.\,dx$$

$$= 0 + \frac{1}{18}\left[3x + x^2\right]_0^4 + 0 = \frac{1}{18}[3(4-2) + (16-4) = 1$$

Thus the second condition is also satisfied and as such the function $f(x)$ is a density function.

Again, $P(2 \le x \le 3) = \int_2^3 f(x)\, dx = \int_2^3 \frac{3+2x}{18}\, dx$

$$= \frac{1}{18}\left[3x + x^2\right]_0^3 = \frac{1}{18}[3 + 5] = \frac{8}{18} = \frac{4}{9}.$$

Problem 2. *Verify that the following is a distribution function.*

$$F(x) = \begin{cases} 0 & x < -a \\ \frac{1}{2}\left(\frac{a}{x} + 1\right), & -a \le x \le a \\ 1 & x > 0. \end{cases}$$

The function F(x) as a distribution function, must satisfy the conditon:

(*i*) $F(x) \ge 0$, (*ii*) $F(-\infty) = 0$, (*iii*) $F(+\infty) = 1$.

Here the first condition is obviously satisfied *i.e.*, $F(x) \ge 0$.

From $F(x) = 0$ for $x < -a$, it is clear that $F(-\infty) = 0$

and $F(x) = 1$ for $x > a$, shows that $F(\infty) = 1$. Hence verified.

Example 3. *If the random variate X has the values 0, 1, 2..., then for what value of t, $p(x) = (1 - t)\, t^x$, $x = 0, 1, 2,...$ is a probability density function of X?*

Sol. By property of probability function, we have

(*i*) $p(x) \ge 0$, (*ii*) $\Sigma\, p(x) = 1$

where $p(x) = P(X = x)$, therefore the value of $p(x)$ should lie between 0 and 1. Clearly:

$$0 \le p\,(x) \le 1 \qquad \text{if } 0 \le t \le 1.$$

Now $\Sigma p(x) = \sum_{x=0}^{\infty} (1 - t)\, t^x = (1 - t)\,[1 + t + t^2 + ...] = (1 - t)\,(1 - t)^{-1} = 1.$

Example 4. *Function f(x) is defined as:*

$$f(x) = e^{-x}.\quad x \ge 0$$
$$= 0, \quad x < 0$$

(i) is the function f(x) a density fucntion?

(ii) If so, determine the probability that the variate having this density will fall in the interval (1, 2)?

(iii) Also find the cumulative probability function F(2).

Sol. (*i*) From the definition of $f(x)$ it is clear that $f(x) \ge 0$ for every x in (1, 2) and

$$\int_{-\infty}^{\infty} f(x)\, dx = \int_{-\infty}^{0} 0\,.\, dx + \int_0^{\infty} e^{-x}\, dx = \left[-e^{-x}\right]_0^{\infty} = 1.$$

Hence the function $f(x)$ is a density function.

(*ii*) Required probability $= P(1 \le x \le 2)$

$$= \int_1^2 e^{-x} dx = e^{-1} - e^{-2} = 0.368 - 0.135 = 0.233.$$

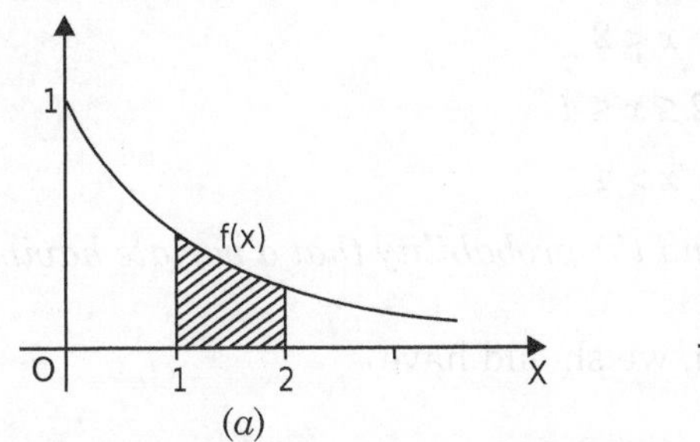

(a)

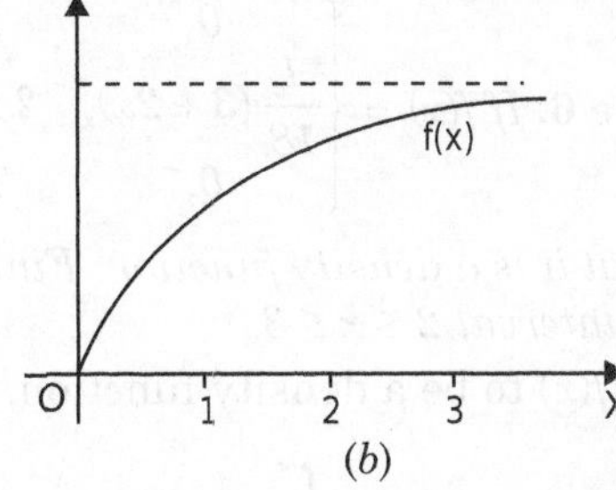

(b)

Fig. 5.4

This probability is equal to the shaded area in Fig. 5.4 (*a*).

(*iii*) Cumulative probability function F(2) is given by

$$F(2) = \int_{-\infty}^{2} f(x)\, dx = \int_{-\infty}^{0} 0 \,.\, dx + \int_{0}^{2}$$

$$= 1 - 0.135 = 0.865$$

which is shown in Fig. 5.4 (*b*).

Example 5. *If* $f(x) = \dfrac{c}{1+x^2}$, $-\infty < x < \infty$, *then find c and show that its corresponding distribution function is*

$$F(x) = (1/\pi) \tan^{-1} x + \frac{1}{2}.$$

Sol. By property of probability function, we have

$$\int_{-\infty}^{\infty} f(x)\, dx = 1 \quad \text{or} \quad c \int_{-\infty}^{\infty} \frac{dx}{1+x^2} = 1 \text{ or } 2c \int_{-\infty}^{\infty} \frac{dx}{1+x^2} = 1$$

or $$2c \left[\tan^{-1} x\right]_0^{\infty} = 1 \quad \text{or} \quad 2c\left(\frac{\pi}{2} - 0\right) = 1.$$

∴ $$c = 1/\pi.$$

So $$f(x) = \frac{1}{\pi} \cdot \frac{1}{1+x^2}, -\infty < x < \infty.$$

Now $$F(x) = P(X \le x)$$

$$\int_{-\infty}^{x} \frac{1}{\pi} \cdot \frac{1}{1+x^2}\, dx = \frac{1}{\pi} \left[\tan^{-1} x\right]_{-\infty}^{x}$$

$$= (1/\pi)\, [\tan^{-1} x - \tan^{-1}(-\infty)] = (1/\pi)\, [\tan^{-1} x + \tan^{-1}(\infty)]$$

$$= (1/\pi)\, [\tan^{-1} x + \pi/2] = (1/\pi) \tan^{-1} x + \frac{1}{2}$$

Example 6. *If* $f(x) = \begin{cases} 0, & x < 2 \\ \frac{1}{18}(3 + 2x), & 2 \le x \le 4 \\ 0, & x > 4 \end{cases}$

Prove that it is a density function. Find the probability that a variate having this density will fall in the interval $2 \le x \le 3$.

Sol. For $f(x)$ to be a density function, we should have

(i) $f(x) \ge 0$, and (ii) $\int_{-\infty}^{\infty} f(x)\, dx = 1$

By definition of $f(x)$, $f(x)$, $\ge 0\ \forall\ x$. So condition (i) is satisfied.

$$\text{Again} \quad \int_{-\infty}^{\infty} f(x)\, dx = \int_{-\infty}^{2} f(x)\, dx + \int_{2}^{4} f(x)\, dx + \int_{4}^{\infty} f(x)\, dx$$

$$= \int_{-\infty}^{2} 0\, dx + \int_{2}^{4} \frac{3+2x}{18}\, dx + \int_{4}^{\infty} 0\, dx$$

$$= 0 + \frac{1}{18}\left[3x + x^2\right]_2^4 + 0 = \frac{1}{18}(6 + 12) = 1.$$

Thus condition (ii) is also satisfied.

$$\text{Now} \quad P(2 \le x \le 3) = \int_2^3 f(x)\, dx = \int_2^3 f(x)\, \frac{3+2x}{18}\, dx$$

$$= \frac{1}{18}\left[3x + x^2\right]_2^3 = \frac{1}{18}(3 + 5) = \frac{8}{18} = \frac{4}{9}.$$

Example 7. *A continuous random variable X has the density function*

$$f(x) = 3x^2,\ 0 \le x < 1$$

find a and b, when

(i) $P(X \le a) = P(X > a)$; *(ii)* $P(X > b) = 0.05.]$

Sol. Since total probability is always one (1), so

$$P(X \le a) = \frac{1}{2} = P(X > a).$$

$$\text{Now} \quad P(X \le a) = \frac{1}{2} \Rightarrow \int_0^a f(x)\, dx = \frac{1}{2}$$

$$\Rightarrow \int_0^a 3x^2\, dx = \frac{1}{2} \Rightarrow a^3 = 1/2; \text{ or } a = (1/2)^{1/3}$$

$$(ii) \quad P(X > b) = 0.05 \Rightarrow \int_b^1 f(x)\, dx = \frac{1}{20} \Rightarrow \int_b^1 3x^2\, dx = \frac{1}{20}$$

$$\Rightarrow \left[x^3\right]_b^1 = \frac{1}{20} \Rightarrow 1 - b^3 = \frac{1}{20}$$

$$\Rightarrow b^3 = 1 - \frac{1}{20} = \frac{19}{20} \text{ or } b = \left(\frac{19}{20}\right)^{1/3}.$$

Example 8. *A random variable X has the probability density function*

$$f(x) = \begin{cases} x \ , & \text{if } 0 \le x < 1 \\ 2 - x, & \text{if } 1 \le x < 2 \\ 0 \ , & \text{if } x \ge 2. \end{cases}$$

Find cumulative distribution function of X.

Sol. By def. of cumulative distribution function, probability density function

$$F(x) = 0, \text{ for } x < 0$$

$$F(x) = \int_0^x x\, dx = \frac{x^2}{2}, \text{ for } 0 \le x < 1$$

$$F(x) = \int_1^x (2 - x)\, dx = \left[2x - \frac{x^2}{2} \right]_1^x = 2x - \frac{x^2}{2} - \frac{3}{2}, \text{ for } 1 \le x < 2$$

and $F(x) = 0$, for $x \ge 2$. But $f(\infty) = 1$.

5.51. MEAN, MEDIAN, MODE AND MOMENTS FOR A CONTINUOUS DISTRIBUTIONS DEFINITIONS

(1) (*i*) Mean $\overline{M} = X = E(X) = \int_{-\infty}^{\infty} x\, f(x)\, dx.$

Also mean $= \int_a^b x\, f(x)\, dx$, if $a \le x \le b$.

(2) Geometirc Mean. If G is geometric mean, then

$$\log G = \int_{-\infty}^{\infty} \log x\, f(x)\, dx = E(\log x).$$

(3) Harmonic Mean. Let H be the harmonic mean, then

$$\frac{1}{H} = \int_{-\infty}^{\infty} \frac{1}{x} f(x)\, dx = E\left(\frac{1}{x}\right).$$

(4) Median. The median M_d is given by

$$\int_{-\infty}^{M_d} f(x)\, dx = \int_{M_d}^{\infty} f(x)\, dx = \frac{1}{2} \int_{-\infty}^{\infty} f(x)\, dx = \frac{1}{2}.$$

(*ii*) **The lower quartile Q_1 and upper quartile Q_3 are given by**

$$\int_{-\infty}^{Q_1} f(x)\, dx = \frac{1}{4} \text{ and } \int_{Q_3}^{\infty} f(x)\, dx = \frac{1}{4}$$

(5) Mode. The mode is the value of the variate for which

$$\frac{d}{dx} f(x) = 0 \text{ and } \frac{d^2 f(x)}{dx^2} < 0.$$

[In other words, the mode is the value of the variate for which probability $f(x)$ is maximum]. The condition for which is that the values obtained from $(d/dx)\, f(x) = 0$ lies within the given range of x.

(6) Moment. The rth moment about any given arbitrary value A is given by:

$$\mu_r = \int_{-\infty}^{\infty} (x - A)^r f(x)\, dx.$$

The rth moment about the mean m is given by:

$$\mu_r' = \int_{-\infty}^{\infty} (x - m)^r f(x)\, dx.$$

(7) The mean deviation about the mean m. It is given by

$$\int_{-\infty}^{\infty} |\, x - m \,|\, f(x)\, dx.$$

Example 9. *X is a continuous random variable with probability density function given by*

$$\begin{aligned} f(x) &= kr && (0 \le x < 2) \\ &= 2k && (2 \le x < 4) \\ &= -kx + 6k && (4 \le x < 6). \end{aligned}$$

Find k and mean value of X.

Sol. For a probability density function, the total probability is unity.

$$\therefore \qquad \int_0^6 f(x)\, dx = 1$$

i.e., $$\int_0^2 kx\, dx + \int_2^4 2k\, dx + \int_4^6 (-kx + 6k)\, dx = 1$$

or $$k\left[\frac{x^2}{2}\right]_0^2 + 2k\,[x]_2^4 + \left[-k\,.\frac{x^2}{2} + 6kx\right]_4^6 = 1$$

or $$2k + 4k + (-10k + 12k) = 1$$

or $$k = \frac{1}{8}$$

$$\text{Mean of X} = \int_0^6 x\, f(x)\, dx$$

$$= \int_0^2 kx^2\, dx + \int_2^4 2kx\, dx + \int_4^6 x(-kx + 6k)\, dx$$

$$= k\left[\frac{x^3}{3}\right]_0^2 + 2k\left[\frac{x^2}{2}\right]_2^4 + \left[-k\,.\frac{x^3}{3} + 6k\,.\frac{x^2}{2}\right]_4^6$$

$$= k\,.\,\frac{8}{3} + k\,.\,12 - k\,.\,\frac{152}{3} + 3k\,.\,20 = \frac{1}{8}\,(24) = 3.$$

Example 10. *For the distribution $dF = 6(x - x^2)\, dx$, $0 \le x \le 1$. find arithmetic mean, harmonic mean, median, mode and mean deviation. Is it a symmetrical distribution?*

Sol. For the given distribution, we have

$$\int_0^1 6\,(x - x^2)\,dx = \left[6\left(\frac{1}{2}x^2 - \frac{1}{3}x^3\right)\right]_0^1 = 1.$$

$\therefore$ $f(x) = 6\,(x - x^2)$ is a probability density function

(1) Arithmetic Mean. $\quad M = \int_0^1 x\,.\,f(x)\,dx$

$$= \int_0^1 x\,.6\,(x - x^2)\,dx = 6\int_0^1 (x^2 - x^3)\,dx$$

$$= 6\left[\frac{1}{3}x^3 - \frac{1}{4}x^4\right]_0^1 = 6\left(\frac{1}{3} - \frac{1}{4}\right) = \frac{1}{2}.$$

(2) Harmonic Mean. It is given by

$$\frac{1}{H} = \int_0^1 \frac{1}{x}\,.\,6\,(x - x^2)\,dx \quad \text{or} \quad \frac{1}{H} = 6\left[x - \frac{x^2}{2}\right]_0^1 = 3$$

$\therefore$ $\quad H = \frac{1}{3}.$

(3) Median. M_d is given by:

$$\int_0^{M_d} f(x)\,dx = \frac{1}{2} \quad \text{or} \quad \int_0^{M_d} 6\,(x - x^2)\,dx = \frac{1}{2}$$

or $$6\left[\frac{1}{2}x^2 - \frac{1}{3}x^3\right]_0^{M_d} = \frac{1}{2} \quad \text{or} \quad 4M_d^3 - 6M_d^2 + 1 = 0$$

or $\quad (2M_d - 1)\,(2M_d^2 - 2M_d - 1) = 0 \quad$ or $\quad M_d = \frac{1}{2}$

or $$M_d = \frac{1}{2}\,(1 \pm \sqrt{3}).$$

Since M_d lies between 0 and 1.

$\therefore$ $$M_d = \frac{1}{2}.$$

(4) Mode. It is that value of variate x for which

$$dy/dx = 0 \quad \text{and} \quad d^2y/dx^2 < 0.$$

Here $\quad y = 6\,(x - x^2).$

$\therefore$ $\quad dy/dx = 6\,(1 - 2x) \quad$ and $\quad d^2y/dx^2 = -12.$

$\therefore$ $\quad dy/dx = 0 \;\Rightarrow\; 6(1 - 2x) = 0 \;\Rightarrow\; x = \frac{1}{2}.$

So mode to given distribution is $x = \frac{1}{2}$.

Since mean, median and mode coincide, therefore the distribution is symmetrical.

(5) Mean deviation about the mean

$$= \int_0^1 \left(x - \frac{1}{2}\right)^2 6\,(x - x^2)\,dx = 6 \int_0^1 \left(x^2 - x + \frac{1}{4}\right)(x - x^2)\,dx$$

$$= 6 \int_0^1 \left(-x^4 + 2x^3 - \frac{5}{4}x^2 + \frac{1}{4}x\right) dx$$

$$= 6 \left[-\frac{1}{5}x^5 + \frac{2}{4}x^4 - \frac{5}{15}x^3 + \frac{1}{8}x^2\right]_0^1 = 6 \times \frac{1}{120} = \frac{1}{120} = 0.05.$$

Example 11. *The frequency distribution of a measurable characteristic varying between 0 and 2 is as under*

$$f(x) = x^3, \qquad 0 \le x \le 1$$
$$= (2 - x)^3, \qquad 1 \le x \le 2.$$

Calculate the standard deviation and also the mean of deviation about the mean.

Sol. Total frequency, $N = \int_0^2 f(x)\,dx = \int_0^1 x^3\,dx + \int_1^2 (2 - x^3)\,dx = \frac{1}{4} + \frac{1}{4} = \frac{1}{2}$.

$$\therefore \quad \mu_1' \text{ (about the origin)} = \frac{1}{N}\left[\int_0^1 (x-0)\,.\,x^3\,dx + \int_1^2 (x-0)(2-x)^3\,dx\right]$$

$$= 2\left\{\left[\frac{x^5}{5}\right]_0^1 + \left[-x\,.\,\frac{(2-x)^4}{4}\right]_1^2 - \left[\frac{(2-x)^5}{20}\right]_1^2\right\}$$

$$= 2\left(\frac{1}{5} + \frac{1}{4} + \frac{1}{20}\right) = 1$$

$$\mu_2' \text{ (about the origin)} = \frac{1}{N}\left[\int_0^1 (x-0)^2\,.\,x^3\,dx + \int_1^2 (x-0)^2(2-x)^3\,dx\right]$$

$$= 2\left\{\left[\frac{x^6}{6}\right]_0^1 + \left[-x^2\frac{(2-x)^4}{4}\right]_1^2 + \frac{1}{2}\int_1^2 x\,(2-x)^4\,dx\right\}$$

$$= 2\left\{\frac{1}{6} + \frac{1}{4} + \frac{1}{2}\left[\frac{1}{5} + \frac{1}{30}\right]\right\} = \frac{16}{15}$$

Hence $\qquad \sigma^2 = \mu_2 = \mu_2' - (\mu_1')^2 = \frac{1}{15}$

$\therefore \quad$ Standard deviation, $\sigma = \frac{1}{\sqrt{15}}$.

Mean deviation about the mean

$$= \frac{1}{N}\left\{\int_0^1 |x-1|x^3\,dx + \int_1^2 |x-1|(2-x)^3\,dx\right\}$$

$$= 2\left\{\int_0^1 (1-x)x^3\,dx + \int_1^2 (x-1)(2-x)^3\,dx\right\}$$

$$= 2\left\{\left(\frac{1}{4}-\frac{1}{5}\right)+\left(0+\frac{1}{20}\right)\right\} = \frac{1}{5}.$$

Example 12 *A frequency function in the range (– 3, 3) is defined by*

$$y = (1/16)(3+x)^2, \quad -3 < x \le -1,$$
$$= (1/6)(6-2x^2), \quad -1 < x \le 1,$$
$$= (1/16)(3-x)^2, \quad 1 < x \le 3.$$

Find the mean and the standard deviation of the distribution defined by the above distribution.

Sol. The total frequency for the given distribution

$$= \int_{-3}^{-1} \frac{1}{16}(3+x)^2\,dx + \int_{-1}^{1} \frac{1}{16}(6-2x^2)\,dx + \int_{1}^{3} \frac{1}{16}(3-x)^2\,dx$$

$$= \frac{1}{16}\left[\frac{(3+x)^3}{3}\right]_{-3}^{-1} + \frac{1}{16}\left[6x-\frac{2}{3}x^3\right]_{-1}^{1} + \frac{1}{16}\left[\frac{-(3-x)^3}{3}\right]_{1}^{3}$$

$$= \frac{1}{16}\left(\frac{8}{3}+\frac{32}{3}+\frac{8}{3}\right) = 1.$$

Thus the given function is a probability density function

$$\text{Mean} = \mu_1' = \frac{1}{16}\left[\int_{-3}^{-1} x(3+x)^2\,dx + \int_{-1}^{1} x(6-2x^2)\,dx + \int_{1}^{3} x(3-x)^2\,dx\right]$$

$$= \frac{1}{16}\left\{\left[\frac{9}{2}x^2+2x^3+\frac{1}{4}x^4\right]_{-3}^{-1} + \left[\frac{6}{2}x^2-\frac{x^4}{2}\right]_{-1}^{1} + \left[\frac{9}{2}x^2-2x^3+\frac{1}{4}x^4\right]_{1}^{2}\right\}$$

$$= \frac{1}{16}\left\{\left[\left(\frac{9}{2}-2+\frac{1}{4}\right)-\left(\frac{81}{2}-54+\frac{81}{4}\right)\right] + 0 + \left[\left(\frac{81}{2}-54+\frac{81}{4}\right)-\left(\frac{9}{2}-2+\frac{1}{4}\right)\right]\right\} = 0$$

and $$\mu_2' = \frac{1}{16}\left[\int_{-3}^{-1} x^2(3+x)^2\,dx + \int_{-1}^{1} x^2(6-2x^2)\,dx + \int_{-3}^{3} x^2(3-x)^2\,dx\right]$$

= 1, (on simplifying).

$\therefore \quad \sigma^2 = \mu_2' - \mu_1'^2 = 1 - 0 = 1.$

TEST YOUR KNOWLEDGE

1. Find the probability distribution of the number of doublets in four throws of a pair of dice.
2. Two bad eggs are mixed accidently with 10 good ones. Find the probability distribution of the number of bad eggs in 3, drawn at random, without replacement, from this lot.
3. A die is tossed twice. Getting a number greater than 4 is considered a success. Find the variance of the probability distribution of the number of successes.
4. Two cards are drawn simultaneously from a well-shuffled deck of 52 cards. Compute the variance for the number of aces.
5. An urn contains 4 white and 3 red balls. Three balls are drawn, with replacement, from this urn. Find μ, σ^2 and σ for the number of red balls drawn.
6. A random variable X has the following probability distribution:

Values of X, x:	0	1	2	3	4	5	6	7	8
$p(x)$:	a	$3a$	$5a$	$7a$	$9a$	$11a$	$13a$	$15a$	$17a$

(*i*) Determine the value of a. (*ii*) Find $P(X < 3)$, $P(X \geq 3)$, $P(2 \leq X < 5)$

(*iii*) What is the smallest value of x for which $P(X \leq x) > 0.5$?

7. Find the standard deviation for the following discrete distribution:

x:	8	12	16	20	24
$p(x)$:	$\frac{1}{8}$	$\frac{1}{6}$	$\frac{3}{8}$	$\frac{1}{4}$	$\frac{1}{12}$.

Answers

1.

X:	0	1	2	3	4
P(X):	$\frac{625}{1296}$	$\frac{500}{1296}$	$\frac{150}{1296}$	$\frac{20}{1296}$	$\frac{1}{1296}$

2.

X:	0	1	2
P(X):	$\frac{12}{22}$	$\frac{9}{22}$	$\frac{1}{22}$

3. $\frac{4}{9}$ 4. $\frac{400}{2873}$ 5. $\frac{9}{7}, \frac{36}{49}, \frac{6}{7}$

6. (*i*) $a = \frac{1}{81}$ (*ii*) $\frac{1}{9}, \frac{8}{9}, \frac{7}{27}$ (*iii*) 5 7. $2\sqrt{5}$.

5.52. MATHEMATICAL EXPECTATION OR THE EXPECTED VALUE

Definition. Let x be a discrete random variable and let its frequency distribution be as follows:

variate:	x_1	x_2	x_3	...	x_n
probability:	p_1	p_2	p_3	...	p_n

where $p_1 + p_2 + p_3 + \ldots + p_n = \Sigma p = 1$, then the mathematical expectation of x (or simply expection of x) is denoted by E(x) and is defined by

$$E(x) = p_1 x_1 + p_2 x_2 + p_3 x_3 + \ldots + p_n x_n = \sum_{i=1}^{n} p_i x_i = \Sigma px.$$

If $\phi(x)$ is a **probability density function*** corresponding to the variate x, then

$$E(x) = \int_{-\infty}^{\infty} x\, \phi(x)\, dx.$$

If $\phi(x)$ is such function of x that it takes values $\phi(x_1)$, $\phi(x_2)$, ... when x takes the values of $x_1, x_2, \ldots$

Let $p_1, p_2 \ldots$ be their respective probabilities, then the mathematical expection of $\phi(x)$, denoted by E $[\phi(x)]$, is defined as

$$E\,[\phi(x)] = p_1\,\phi(x_1) + p_2\,\phi(x_2) + \ldots + p_n\,\phi(x_n) \qquad \ldots(1)$$

where $\Sigma p = 1$.

If $\phi(x) = x^r$. Then (1) gives,

$$E\,(x^r) = p_1 x_1^r + p_2 x_2^r + \ldots + p_n x_n^r \qquad \ldots(2)$$

where $\Sigma p = 1$.

(2) is defined as the rth moment of the discrete probability distribution about $x = 0$ and it is denoted by μ_r'; and μ_r' is the expected value of the rth power of the variate.

In particular,

$$\mu_1' = E\,(x) = p_1 x_1 + p_2 x_2 + \ldots + p_n x_n = \sum_{i=1}^{n} p_i x_i\,.$$

Here if p_i is replaced by $\frac{f_i}{N'}$ where $\sum_{i=1}^{n} f_i = N$, then

$$E\,(x) = \frac{\sum_{i=1}^{n} f_i x_i}{N} = \frac{\sum fx}{N} = \text{Mean}.$$

Here E(x) denotes the mean in random experiments when x take values $x_1, x_2, \ldots, x_n$ with frequencies $f_1, f_2, \ldots, f_n$.

This moment E(x) is called the **mean value** of the variate or the distribution (just as for frequency).

5.53. VARIANCE

The relation $\mu_2 = E\,[\{x - E(x)\}^2] = E(x^2) - [E(x)]^2$

denoted by Var(x), or V(x) is defined as the **variance** of the distribution of x. μ_r and μ_r' have the same relation as of frequency distribution. But here the expected value of the deviation of the variate x from its mean vanishes *i.e.,* E $[x - E(x)] = 0$.

5.54. MATHEMATICAL EXPECTATION FOR CONTINUOUS RANDOM VARIABLE

Suppose x is a continuous random variable with probability density function $f(x)$, then mathematical expectation E(x) of x with certain restrictions is given by

---

* Let the probability of the variate x falling in the infinitesimal interval $\left(x - \frac{1}{2}dx, x + \frac{1}{2}dx\right)$ be expressed in the form $\phi(x)\,dx$, where $\phi(x)$ is a continuous function of x. The function $\phi(x)$ is called the **probability density function.**

$$E(x) = \int_{-\infty}^{\infty} x\, f(x)\, dx\,.$$

Example. *Show that* $| E(x) | \le E(| x |)$.

Sol. We know that

$$E(x) = \sum_{i=1}^{n} p_i x_i = \Sigma\, px \Rightarrow \quad | E(x) | \le | \Sigma px | \le \Sigma \,| px |$$

$$\Rightarrow \qquad E(x) \,| \le \Sigma\, pE(| x |) \;\Rightarrow\; |E(x)| \le E\,| x | \qquad [\because\; p \ge 0]$$

5.53. EXPECTATION OF A SUM

Theorem. *The expectation of the sum of two variates is equal to the sum of their expectations, i.e., if x and y are two variates then* $E(x + y) = E(x) + E(y)$.

Proof. Let $x_1, x_2, ..., x_m$ be the values of the variate p and $p_1, p_2, ..., p_m$ be the corresponding probabilities. Also let $y_1, y_2, ..., y_n$ be the n values of y and $p_1', p_2', ..., p_n'$ be the corresponding probabilities then the sum $x + y$ is a random variable which can assume mn values $x_i + y_j\,(i = 1, 2, ..., m; j = 1, 2 ..., n)$. Now let p_{ij} be the probability of the variate x when x assumes the value x_i and the variate y assumes the value y_j.

Since x takes a definite value x_i, y takes one of the n values $y_1, y_2, ..., y_n$ so that the sum $\sum_{j=1}^{n} p_{ij}$ represents the probability p_i of x assuming the values x_i, that is, we shall have

$$\sum_{j=1}^{n} p_{ij} = pi. \qquad ...(1)$$

Similarly, discussing as above, $\sum_{i=1}^{m} p_{ij}$ represents the probability of y assuming the value y_j, that is, we shall have

$$\sum_{i=1}^{m} p_{ij} = p'j. \qquad ...(2)$$

Thus, by definition, we have

$$E(x + y) = \sum_{i=1}^{m} \sum_{j=1}^{n} p_{ij}(x_i + y_j) = \sum_{i=1}^{m} \sum_{j=1}^{n} p_{ij}x_i + \sum_{i=1}^{m} \sum_{j=1}^{n} p_{ij}y_j$$

$$= \sum_{i=1}^{m} x_i \left(\sum_{j=1}^{n} p_{ij} \right) + \sum_{j=1}^{n} y_j \left(\sum_{i=1}^{m} p_{ij} \right)$$

$$= \sum_{i=1}^{m} x_i p_i + \sum_{j=1}^{n} y_j p'_j \qquad \text{[use (1) and (2)]}$$

$$= E(x) + E(y).$$

Hence $E(x + y) = E(x) + E(y)$. ...(3)

The **general form** of the above theorem is $E(x + y + z + ...) = E(x) + E(y) + E(z) + ...$

Example 1. *Show that $E(ax + by) = a\,E(x) + b\,E(y)$ where a and b are constants.*

Sol. With the notations, we have

$$E(ax + by) = \sum_{i=1}^{m} \sum_{j=1}^{n} p_{ij}\ (ax_i + by_j)$$

$$= a \sum_{i=1}^{m} x_i \left(\sum_{j=1}^{n} p_{ij} \right) + b \sum_{j=1}^{n} y_j \left(\sum_{i=1}^{m} p_{ij} \right)$$

$$= a \sum_{i=1}^{m} x_i p_i + b \sum_{j=1}^{n} y_i p_j{}' = aE\,(x) + bE\,(y).$$

Example 2. *$E(ax + b) = aE(x) + b$, a and b being constants.*

Sol. $$E(ax + b) = \sum_{i=1}^{m} p_i (ax_i + b) = a \sum_{i=1}^{m} p_i x_i + b = a\,E(x) + b.$$

$i = 1$

5.56. JOINT PROBABILITY DISTRIBUTION

Let x and y be two random variables. If x assume values $x_1, x_2, ...x_m$ with corresponding probabilities $f(x_1), f(x_2), ..., f(x_n)$ and y assume values $y_1, y_2, ..., y_n$ with corresponding probabilities $g(x_1), g(x_2), ..., g(x_n)$, the system of relation $p(x) = x_i, y = y_j) = (x_i, y_j)$ is defined as the joint (or compound) probability distribution.

5.57. INDEPENDENT VARIATES

Definition. Two random variates x and y on a sample space are called independent if the probability that either of them will assume a prescribed value does not depend on the value assumed by the other.

5.58. PRODUCT OF EXPECTATIONS

Theorem. *The expectation of the product of two independent variates is equal to the product of their expectations*

i.e., $E(xy) = E(x) \,.\, E(y)$.

Proof. Let x and y be two independent random variables. Let x assume m values $x_1, x_2, ..., x_m$, with corresponding probabilities $p_1, p_2, ..., p_m$. Let y assume n values $y_1, y_2, ..., y_n$ with corresponding probabilities $p_1', p_2', ...,p_n'$. Thus p_i is the probability of x assuming the value x_i and p_j' is the probability of y assuming the value y_j. Since x and y are independent, the probability that x assumes the value x_i and at the same time y assumes the value y_j is $= p_i p_j'$ (by theorem of compound probability).

∴ By definition, we have

$$E(xy) = \sum_{i=1}^{m} \sum_{j=1}^{n} p_i p_j{}' x_i y_j$$

$$= \sum_{i=1}^{m} p_i x_i \sum_{j=1}^{n} p_j' y_j = \sum_{i=1}^{m} p_i x_i \, E(y) = E(y) \sum_{i=1}^{m} p_i x_i = E(y)\, E(x).$$

$\therefore \quad E(xy) = E(x)\, E(y).$

The **general form** of the above theorem is

$$E(xyz\ ...) = E(x)\, E(y)\, E(z)\ ...$$

where $x, y, z, ...$ are independent variates.

5.59. COVARIANCE DEFINITION

Let x and y be two random variables and $\bar{x}$ and $\bar{y}$ be their expected values (or means) repectively. The co-variance between x and y, denoted by cov (x, y), is defined as

$$\text{cov}\,(x, y) = E[(x - \bar{x})(y - \bar{y})].$$

Thus the covariance between x and y is the product of their deviations from the means.

Theorem. *The co-variance of two independent variates is zero.*

Proof: Let x and y be two independent variates, and $\bar{x}$, $\bar{y}$ be their expected values (or means) repectively, then

$$E(x - \bar{x}) = E(x) - E(\bar{x}) = \bar{x} - \bar{x} = 0$$

$$E\,(y - \bar{y}) = E(y) - E(\bar{y}) = \bar{y} - \bar{y} = 0.$$

$\therefore \quad \text{cov}\,(x, y) = E\,\{(x - \bar{x})(y - \bar{y})\} = E\,(x - \bar{x})\, E\,(y - \bar{y}) = 0\,.\,0 = 0.$

Note : $\text{Cov}\,(x, y) = E\,\{(x - \bar{x})(y - \bar{y})\} = E\,(xy - \bar{x}y - \bar{y}x + \bar{x}\bar{y})$

$$= E\,(xy) - E\,(\bar{x}y) - E\,(\bar{y}x) + E\,(\bar{x}\bar{y}) = E\,(xy) - \bar{x}\,E(y) - \bar{y}\,E\,(x) + \bar{x}\,\bar{y}$$

$$= E\,(xy) - \bar{x}\,\bar{y} - \bar{y}\,\bar{x} + \bar{x}\,\bar{y} = E(xy) - E(x)\,E(y).$$

ILLUSTRATIVE EXAMPLES

Example 1. (*a*) *What is the expected value of the number of points that will be obtained in a single throw with an ordinary die? Find variance also.*

Sol. The variate *i.e.*, numbers showing on a die assumes the values 1, 2, 3, 4, 5, 6 and probability in each case is $\frac{1}{6}$.

$\therefore$ given probability distributions is as follows:

x:	1	2	3	4	5	6
p:	1/6	1/6	1/6	1/6	1/6	1/6

$$\therefore \quad E(x) = \sum_{i=1}^{6} p_i x_i = p_1 x_1 + p_2 x_2 + p_3 x_3 + ... + p_6 x_6$$

$$= \frac{1}{6}\,.\,1 + \frac{1}{6}\,.\,2 + \frac{1}{6}\,.\,3 + \frac{1}{6}\,.\,4 + \frac{1}{6}\,.\,5 + \frac{1}{6}\,.\,6$$

$$= \frac{1}{6}[1+2+3+4+5+6] = \frac{1}{6} \times 21 = \frac{7}{2}.$$

Also $\quad \text{var}(x) = E(x^{2)} - [E(x)]^2$

$$= \frac{1}{6}(1^2 + 2^2 + \ldots + 6^2) - (7/2)^2 = 35/82.$$

Example 1. (*b*) *Two fair dice are tossed. Find the expected value of sum of the obtained number of points.*

Sol. Let x and y denote the number on first and second die respectively. *Then similar to Ex.* 1 (*a*) *we have*

$$E(x) = 7/2 \quad \text{and} \quad E(y) = 7/2.$$

$\therefore$ The required expected value $= E(x + y) = E(x) + E(y) = \frac{7}{2} + \frac{7}{2} = 7.$

Example 2. *A bag contains two one rupee coins and 3 twenty paise coins. A person is asked to draw two coins randomly. Find the value of his expectation.*

Sol. The possible ways of drawing two coins randomly are as follows:

(*i*) Both are one rupee coins, its probability $= \frac{{}^2C_2}{{}^5C_2} = \frac{1}{20} = \frac{1}{10}$

(*ii*) Both are 20 paise coins, its probability $= \frac{{}^3C_2}{{}^5C_2} = \frac{3.2}{5.4} = \frac{6}{20} = \frac{3}{10}.$

(*iii*) One is 1 rupee coin and other is 20 paise coin, its probability

$$= \frac{{}^2C_1 \,.\, {}^3C_1}{{}^5C_2} = \frac{2 \times 3 \times 2 \times 1}{5 \times 4} = \frac{3}{5}.$$

We know that two, one rupee coins = 10 twenty paise coins.

$\therefore$ 1, One rupee coin and 1, twenty paise coin.

$\therefore$ The required expectation $= \Sigma p_i x_i$

$$= \left[\left(\frac{1}{10} \times 10\right) + \left(\frac{3}{10} \times 2\right) + \left(\frac{3}{5} \times 6\right)\right] \text{ twenty paise coins}$$

$$= 1 + \frac{3}{5} + \frac{18}{5} = \frac{26}{5} \text{ twenty paise coin}$$

$$= (26/5) \times 20 \text{ Paise} = 104 \text{ Paise} = ₹\, 1.04.$$

Example 3. *Thirteen cards are drawn simultaneously from a pack of 52. If aces count 1, face cards 10 and others according to their denomination, find the expectations of the total score on 13 cards.*

Sol. Thirteen cards are drawn out of 52 cards. Let x_i be the number corresponding to the ith card. Therefore x_i assumes the values 1, 2, 3, 4, 5, 6, 7, 8, 9, 10, 10, 10, 10 (since knave, queen, king count 10), each with probability 1/13. Let $E(x_i)$ be the expectation of drawing a single card in a single draw.

Then $E(x_i) = \sum_{i=1}^{13} p_i x_i$

$$= \frac{1}{13}.1 + \frac{1}{13}.2 + \frac{1}{13}.3 + ... + \frac{1}{13}.9 + \frac{1}{13}.10 + \frac{1}{13}.10 + \frac{1}{13}.10 + \frac{1}{13}.10$$

(1/13) (1 + 2 + 3 + ... + 9 + 10 + 10 + 10 + 10) = 85/13.

Hence the required expectation of drawing 13 cards = 13 × (85/13) = 85.

Example 4. *In four tosses of a coin. Let x be the number of heads. Calculate the expected values of x.*

Sol. In four tosses of a coin, there is only one way of getting all heads, *i.e.,*

$$P\,(x = 4) = \frac{1}{2} \times \frac{1}{2} \times \frac{1}{2} \times \frac{1}{2} = \frac{1}{16}$$

$$P\,(x = 3) = \frac{^4C_3}{16} = \frac{4}{16} = \frac{1}{4}$$

$$P\,(x = 2) = \frac{^4C_2}{16} = \frac{6}{16} = \frac{3}{8}$$

$$P\,(x = 1) = \frac{^4C_1}{16} = \frac{4}{16} = \frac{1}{4}$$

$$P\,(x = 0) = \frac{^4C_0}{16} = \frac{1}{16}.$$

Therefore the probability distribution of x is as follows:

x:	0	1	2	3	4
P (x):	1/16	1/4	3/8	1/4	1/16.

∴ The required expectation

$$= E(x) = \sum_{x=0}^{4} x\,p(x) = 0\,.\,\frac{1}{16} + 1\,.\,\frac{1}{4} + 2\,.\,\frac{3}{8} + 3\,.\,\frac{1}{4} + 4\,.\,\frac{1}{16} = 2.$$

Example 5. *Find E(x), E(x^2), E {$(x - \bar{x})^2$} for the following probability distribution:*

x:	*8*	*12*	*16*	*20*	*24*
P (x):	*1/8*	*1/6*	*3/8*	*1/4*	*1/12.*

Sol. The mean of probability distributlion is

$$E(x) = \Sigma\,x\,.\,p\,(x) = 8\,.\,\frac{1}{8} + 12\,.\,\frac{1}{6} + 16\,.\,\frac{3}{8} + 20\,.\,\frac{1}{4} + 24\,.\,\frac{1}{12} = 16.$$

$$E(x^2) = \Sigma x^2\,.\,p(x) = (8)^2\,.\,\frac{1}{8} + (12)^2\,.\,\frac{1}{6} + (16)^2\,.\,\frac{3}{8} + (20)^2\,.\,\frac{1}{4} + (24)^2\,.\,\frac{1}{12}$$

= 276 = The second moment about the origin

$$E\{(x-\bar{x})^2\} = E(x-\bar{x})^2 p(x)$$

$$= (8-16)^2 \cdot \frac{1}{8} + (12-16)^2 \cdot \frac{1}{6} + (16-16)^2 \cdot \frac{3}{8} + (20-16)^2 \cdot \frac{1}{4} + (24-16)^2 \cdot \frac{1}{12}$$

= 20 = variance of the distribution.

Example 6. *A bag contains 5 black, 6 white and 7 red balls. Four balls are drawn at random from it. If x denote the numbers of white balls then find E(x).*

Sol. The total number of balls in the bag = 5 + 6 + 7 = 18.

Now x assumes the values 0, 1, 2, 3, 4.

Thus $$P(x) = \frac{{}^6C_x \times {}^{12}C_{4-x}}{{}^{18}C_4}, \text{ where } x = 0, 1, 2, 3, 4$$

Then $$P(x=0) = \frac{{}^6C_0 \times {}^{12}C_4}{{}^{18}C_4} = \frac{495}{3060}$$

$$P(x=1) = \frac{{}^6C_1 \times {}^{12}C_3}{{}^{18}C_4} = \frac{1320}{3060}$$

$$P(x=2) = \frac{{}^6C_2 \times {}^{12}C_2}{{}^{18}C_4} = \frac{990}{3060}$$

$$P(x=3) = \frac{{}^6C_3 \times {}^{12}C_1}{{}^{18}C_4} = \frac{240}{3060}$$

$$P(x=4) = \frac{{}^6C_4 \times {}^{12}C_0}{{}^{18}C_4} = \frac{15}{3060}$$

$$\therefore \quad E(x) = \sum_{x=0}^{4} x \cdot P(x) = 0 \cdot \frac{495}{3060} + 1 \cdot \frac{1320}{3060} + 2 \cdot \frac{990}{3060} + 3 \cdot \frac{240}{3060} + 4 \cdot \frac{15}{3060}$$

$$= \frac{4080}{3060} = \frac{68}{51}.$$

Example 7. *A box contains a white and b black balls; c balls are drawn. Show that the expectation of the number of white balls drawn is ca / (a + b).*

Sol. Consider a variate x_i defined as follows

$x_i = x_1, x_2, x_3, \ldots, x_c$. In this case, we have

$x_i = 1$ if the ith ball drawn is white

$= 0$ if i^{th} ball drawn is black.

If x is the number of white balls out of c drawn balls, then

$$x = x_1 + x_2 + x_3 + \ldots + x_c. \quad \ldots(1)$$

Then the probability to draw 1 white ball from the box = $P(x_i = 1) = a/(a+b)$ and the probability to draw the black ball from the box = $P(x_i = 0) = b/(a+b)$.

$\therefore \qquad E(x_i) = \{P(x_i = 1)\}\ 1 + \{P(x_i = 0)\}\ .\ 0$

$$= \frac{a}{a+b}\ .\ 1 + \frac{b}{a+b}\ .\ 0 = \frac{a}{a+b}$$

i.e., $\qquad E(x_i) = \dfrac{a}{a+b}$ and it is free from i.

Now from (1), we get

$$E(x) = E(x_1) + E(x_2) + E(x_3) + \ldots + E(x_c)$$

$$= \left(\frac{a}{a+b}\right) + \left(\frac{a}{a+b}\right) + \ldots\ c \text{ terms}$$

$$= \frac{ca}{a+b}.$$

Example 8. *If n fair dice are tossed and X denotes the sum of points on n dice, then find the expectation of X.*

Sol. Let x_i denote the number of the ith dice, then

$$i = 1, 2, 3, \ldots, n.$$

The sum of points on n dice is

$$X = x_1 + x_2 + x_3 + \ldots + x_n.$$

Therefore, $\qquad E(x) = E(x_1 + x_2 + \ldots + x_n) = E(x_1) + E(x_2) + \ldots + E(x_n).$

But for i^{th} dice, x_i can assume the values 1, 2, 3, 4, 5, 6, each with probability $\dfrac{1}{6}$, thus

$$E(x_i) = \frac{1}{6}\ .\ 1 + \frac{1}{6}\ .\ 2 + \frac{1}{6}\ .\ 3 + \frac{1}{6}\ .\ 4 + \frac{1}{6}\ .\ 5 + \frac{1}{6}\ .\ 6 = \frac{21}{6} = \frac{7}{2}$$

$\therefore \qquad E(x_1) = E(x_2) = \ldots = E(x_n) = 7/2.$

Hence $\qquad E(x) = \dfrac{7}{2} + \dfrac{7}{2} + \ldots\ n \text{ times} = \dfrac{7n}{2}.$

Example 9. *Find the expected value of the product of points on n dice.*

Sol. Let x_i denote the number of points on ith dice, then the product of points on n dice is $= x_1 . x_2 . x_3 . \ldots . x_n$.

For every single dice, we have

$$E(x_i) = \frac{7}{2},\ i = 1, 2, 3, \ldots, n$$

$\therefore \qquad E(x_1 x_2 \ldots x_n) = E(x_1)\ .\ E(x_2) \ldots E(x_n)$

$$= \frac{7}{2} \cdot \frac{7}{2} \cdot \frac{7}{2} \ldots n \text{ factors}$$

$$= (7/2)^n.$$

5.60. THEORETICAL DISTRIBUTIONS

Frequency distributions can be classified under two heads:

(*i*) Observed Frequency Distributions.

(*ii*) Theoretical or Expected Frequency Distributions.

Observed frequency distributions are based on actual observation and experimentation. If certain hypothesis is assumed, it is sometimes possible to derive mathematically what the frequency distribution of certain universe should be. Such distributions are called **Theoretical Distributions**.

There are many types of theoretical frequency distributions but we shall consider only three which are of great importance:

(*i*) Binomial Distribution (or Bernoulli's Distribution);

(*ii*) Poisson's Distribution;

(*iii*) Normal Distribution.

BINOMIAL (OR BERNOULLI'S) DISTRIBUTION

5.61. BINOMIAL PROBABILITY DISTRIBUTION

Let there be n independent trials in an experiment. Let a random variable X denote the number of successes in these n trials. Let p be the probability of a success and q that of a failure in a single trial so that $p + q = 1$. Let the trials be independent and p be constant for every trial.

Let us find the probability of r successes in n trials.

r successes can be obtained in n trials in ${}^nC_r$ ways.

$$\therefore \quad P(X = r) = {}^nC_r P(\underbrace{S\,S\,S\,......\,S}_{r \text{ times}} \quad \underbrace{F\,F\,F\,......\,F}_{(n-r) \text{ times}})$$

$$= {}^nC_r \underbrace{P(S)\,P(S)\,......\,P(S)}_{r \text{ factors}} \quad \underbrace{P(F)\,P(F)\,......\,P(F)}_{(n-r) \text{ factors}}$$

$$= {}^nC_r \underbrace{p\,p\,p\,......\,p}_{r \text{ factors}} \quad \underbrace{q\,q\,q\,......\,q}_{(n-r) \text{ factors}}$$

$$= {}^nC_r p^r q^{n-r} \qquad ...(1)$$

Hence $\mathbf{P(X = r) = {}^nC_r\, q^{n-r}\, p^r}$, **where $\mathbf{p + q = 1}$ and $\mathbf{r = 0, 1, 2, ..., n}$.**

The distribution (1) is called the *binomial probability distribution* and X is called the *binomial variate*.

Note 1. $P(X = r)$ is usually written as $P(r)$.

Note 2. The successive probabilities $P(r)$ in (1) for $r = 0, 1, 2, ..., n$ are

$${}^nC_0 q^n,\ {}^nC_1 q^{n-1} p,\ {}^nC_2 q^{n-2} p^2,\ ......,\ {}^nC_n p^n$$

which are the successive terms of the binomial expansion of $(q + p)^n$. That is why this distribution is called "binomial" distribution.

Note 3. n and p occurring in the binomial distribution are called the *parameters* of the distribution.

Note 4. In a binomial distribution:

(*i*) n, the number of trials is finite.

(*ii*) each trial has only two possible outcomes usually called success and failure.

(*iii*) all the trials are independent.

(*iv*) p (and hence q) is constant for all the trials.

5.62. RECURRENCE OR RECURSION FORMULA FOR THE BINOMIAL DISTRIBUTION

In a binomial distribution,

$$P(r) = {}^nC_r\, q^{n-r}\, p^r = \frac{n!}{(n-r)!\,r!}\, q^{n-r}\, p^r$$

$$P(r+1) = {}^nC_{r+1}\, q^{n-r-1}\, p^{r+1} = \frac{n!}{(n-r-1)!\,(r+1)!}\, q^{n-r-1}\, p^{r+1}$$

$$\therefore \quad \frac{P(r+1)}{P(r)} = \frac{(n-r)!}{(n-r-1)!} \times \frac{r!}{(r+1)!} \times \frac{p}{q}$$

$$= \frac{(n-r) \times (n-r-1)!}{(n-r-1)!} \times \frac{r!}{(r+1) \times r!} \times \frac{p}{q} \times = \left(\frac{n-r}{r+1}\right) \cdot \frac{p}{q}$$

$$\Rightarrow \quad P(r+1) = \frac{n-r}{r+1} \cdot \frac{p}{q}\, P(r)$$

which is the required recurrence formula. Applying this formula successively, we can find P(1), P(2), P(3), ..., if P(0) is known.

5.63. MEAN AND VARIANCE OF THE BINOMIAL DISTRIBUTION

For the binomial distribution, $P(r) = {}^nC_r\, q^{n-r}\, p^r$

$$\text{Mean } \mu = \sum_{r=0}^{n} rP(r) = \sum_{r=0}^{n} r \,.\, {}^nC_r q^{n-r}\, p^r$$

$$= 0 + 1 \,.\, {}^nC_1\, q^{n-1}\, p + 2 \,.\, {}^nC_2\, q^{n-2}\, p^2 + 3 \,.\, {}^nC_3\, q^{n-3}\, p^3 + \ldots\ldots + n \,.\, {}^nC_n\, p^n$$

$$= nq^{n-1}\, p + 2 \,.\, \frac{n(n-1)}{2.1}\, q^{n-2}p^2 + 3 \,.\, \frac{n(n-1)(n-2)}{3.2.1}\, q^{n-3}\, p^3 + \ldots\ldots + n.p^n$$

$$= nq^{n-1}p + n(n-1)\, q^{n-2}p^2 + \frac{n(n-1)(n-2)}{2.1}\, q^{n-3}p^3 + \ldots\ldots + np^n$$

$$= np\left[q^{n-1} + (n-1)q^{n-2}\, p + \frac{(n-1)(n-2)}{2.1}\, q^{n-3}p^2 + \ldots\ldots + p^{n-1}\right]$$

$$= np[{}^{n-1}C_0 q^{n-1} + {}^{n-1}C_1 q^{n-2}p + {}^{n-1}C_2 q^{n-3}p^2 + \ldots + {}^{n-1}C_{n-1} p^{n-1}]$$

$$= np(q+p)^{n-1} = np \qquad (\because\ p + q = 1)$$

Hence the mean of the binomial distribution is np.

$$\text{Variance } \sigma^2 = \sum_{r=0}^{n} r^2 P(r) - \mu^2 = \sum_{r=0}^{n} [r + r(r-1)]\, P(r) - \mu^2$$

$$= \sum_{r=0}^{n} rP(r) + \sum_{r=0}^{n} r(r-1)\, P(r) - \mu^2 = \mu + \sum_{r=2}^{n} r(r-1)\, {}^nC_r q^{n-r} p^r - \mu^2$$

(Since the contribution due to $r = 0$ and $r = 1$ is zero)

$= \mu + [2\,.1\,.\,{}^nC_2 q^{n-2}p^2 + 3.2\,.\,{}^nC_3 q^{n-3}p^3 + \ldots + n(n-1)\,{}^nC_n p^n] - \mu^2$

$$= \mu + \left[2\,.\,1\,.\frac{n(n-1)}{2\,.\,1}q^{n-2}p^2 + 3\,.\,2\,.\frac{n(n-1)(n-2)}{3\,.\,2\,.\,1}q^{n-3}p^3 + \ldots + n(n-1)p^n\right] - \mu^2$$

$= \mu + [n(n-1)q^{n-2}p^2 + n(n-1)(n-2)q^{n-3}p^3 + \ldots + n(n-1)p^n] - \mu^2$

$= \mu + n(n-1)\,p^2[q^{n-2} + (n-2)q^{n-3}p + \ldots + p^{n-2}] - \mu^2$

$= \mu + n(n-1)p^2\,[{}^{n-2}C_0 q^{n-2} + {}^{n-2}C_1 q^{n-3}p + \ldots + {}^{n-2}C_{n-2}p^{n-2}] - \mu^2$

$= \mu + n(n-1)p^2(q+p)^{n-2} - \mu^2 = \mu + n(n-1)p^2 - \mu^2$ $[\because\ q + p = 1]$

$= np + n(n-1)p^2 - n^2p^2$ $[\because\ \mu = np]$

$= np[1 + (n-1)p - np] = np[1 - p] = npq.$

Hence the variance of the binomial distribution is npq.

5.64. APPLICATIONS OF BINOMIAL DISTRIBUTION

The binomial distribution is applied to problems concerning:

(*i*) Number of defective items in a sample from production live,

(*ii*) Number of round fired from a gun hitting a target,

(*iii*) Estimation of reliability of systems,

(*iv*) Radar detection.

Example 10. *During war, 1 ship out of 9 was sunk on an average in making a certain voyage. What was the probability that exactly 3 out of a convoy of 6 ships would arrive safely?*

Sol. p, the probability of a ship arriving safely $= 1 - \frac{1}{9} = \frac{8}{9}$; $q = \frac{1}{9}$, $n = 6$

Binomial distribution is $\left(\frac{1}{9} + \frac{8}{9}\right)^6$

The probability that exactly 3 ships arrive safely $= {}^6C_3\left(\frac{1}{9}\right)^3\left(\frac{8}{9}\right)^3 = \frac{10240}{9^6}$.

Example 11. *Assume that on the average one telephone number out of fifteen called between 2 P.M. and 3 P.M. on week-days is busy. What is the probability that if 6 randomly selected telephone numbers are called (i) not more than three, (ii) at least three of them will be busy?*

Sol. p, the probability of a telephone number being busy between 2 P.M. and 3 P.M. on week-days $= \frac{1}{15}$

$q = 1 - \frac{1}{15} = \frac{14}{15}$, $n = 6$; Binomial distribution is $\left(\frac{14}{15} + \frac{1}{15}\right)^6$

The probability that not more than three will be busy

$= p(0) + p(1) + p(2) + p(3)$

$$= {}^6C_0\left(\frac{14}{15}\right)^6 + {}^6C_1\left(\frac{14}{15}\right)^5\left(\frac{1}{15}\right) + {}^6C_2\left(\frac{14}{15}\right)^4\left(\frac{1}{15}\right)^2 + {}^6C_3\left(\frac{14}{15}\right)^3\left(\frac{1}{15}\right)^3$$

$$= \frac{(14)^3}{(15)^6}\,[2744 + 1176 + 210 + 20] = \frac{2744 \times 4150}{(15)^6} = 0.9997$$

The probability that at least three of them will be busy

$= p(3) + p(4) + p(5) + p(6)$

$$= {}^6C_3 \left(\frac{14}{15}\right)^3 \left(\frac{1}{15}\right)^3 + {}^6C_4 \left(\frac{14}{15}\right)^2 \left(\frac{1}{15}\right)^4 + {}^6C_5 \left(\frac{14}{15}\right)\left(\frac{1}{15}\right)^5 + {}^6C_6 \left(\frac{1}{15}\right)^6 = 0.005.$$

Example 12. *Six dice are thrown 729 times. How many times do you expect at least three dice to show a five or six?*

Sol. p = the chance of getting 5 or 6 with one die $= \frac{2}{6} = \frac{1}{3}$

$$q = 1 - \frac{1}{3} = \frac{2}{3},\ n = 6,\ N = 729$$

since dice are in sets of 6 and there are 729 sets.

The binomial distribution is $N(q + p)^n = 729 \left(\frac{2}{3} + \frac{1}{3}\right)^6$

The expected number of times *at least three* dice showing five or six

$$= 729 \left[{}^6C_3 \left(\frac{2}{3}\right)^3 \left(\frac{1}{3}\right)^3 + {}^6C_4 \left(\frac{2}{3}\right)^2 \left(\frac{1}{3}\right)^4 + {}^6C_5 \left(\frac{2}{3}\right)\left(\frac{1}{3}\right)^5 + {}^6C_6 \left(\frac{1}{3}\right)^6 \right]$$

$$= \frac{729}{3^6} [160 + 60 + 12 + 1] = 233.$$

Example 13. *Out of 800 families with 4 children each, how many families would be expected to have (i) 2 boys and 2 girls (ii) at least one boy (iii) no girl (iv) at most two girls? Assume equal probabilities for boys and girls.*

Sol. Since probabilities for boys and girls are equal

$$p = \text{probability of having a boy} = \frac{1}{2};\quad q = \text{probability of having a girl} = \frac{1}{2}$$

$n = 4$, $N = 800$ $\therefore$ The binomial distribution is $800 \left(\frac{1}{2} + \frac{1}{2}\right)^4$.

(*i*) The expected number of families having 2 *boys* and 2 girls

$$= 800\ {}^4C_2 \left(\frac{1}{2}\right)^2 \left(\frac{1}{2}\right)^2 = 800 \times 6 \times \frac{1}{16} = 300.$$

(*ii*) The expected number of families having *at least one boy*

$$= 800 \left[{}^4C_1 \left(\frac{1}{2}\right)^3 \left(\frac{1}{2}\right) + {}^4C_2 \left(\frac{1}{2}\right)^2 \left(\frac{1}{2}\right)^2 + {}^4C_3 \left(\frac{1}{2}\right)\left(\frac{1}{2}\right)^3 + {}^4C_4 \left(\frac{1}{2}\right)^4 \right]$$

$$= 800 \times \frac{1}{16} [4 + 6 + 4 + 1] = 750.$$

(*iii*) The expected number of families having no girl, *i.e.*, having 4 boys

$$= 800 \,.\, {}^4C_4 \left(\frac{1}{2}\right)^4 = 50.$$

(*iv*) The expected number of families having *at most two girls i.e.*, having *at least 2 boys*

$$= 800\left[{}^4C_2\left(\frac{1}{2}\right)^2\left(\frac{1}{2}\right)^2 + {}^4C_3\left(\frac{1}{2}\right)\left(\frac{1}{2}\right)^3 + {}^4C_4\left(\frac{1}{2}\right)^4\right] = 800 \times \frac{1}{16}\,[6 + 4 + 1] = 550.$$

Example 14. *The probability that a pen manufactured by a company will be defective is* $\frac{1}{10}$. *If 12 such pens are manufactured, find the probability that*

(*a*) *exactly two will be defective,*

(*b*) *at least two will be defective,*

(*c*) *none will be defective.*

Sol. It is given that the probability of a defective pen is $\frac{1}{10} = 0.1$.

∴ The probability of a non-defective pen is = 1 – 0.1 = 0.9.

(*a*) The probability that exactly two will be defective $= {}^{12}C_2\,(0.1)^2\,(0.9)^{10} = 0.2301.$

(*b*) The probability that at least two will be defective

= 1 – (Probability that either none or one is non-defective)

$= 1 - [{}^{12}C_0\,(0.9)^{12} + {}^{12}C_1(0.1)\,(0.9)^{11}] = 0.3412.$

(*c*) The probability that none will be defective $= {}^{12}C_{12}\,(0.9)^{12} = 0.2823.$

Example 15. *Six dice are thrown 729 times. How many times do you expect at least three dice to show a five or six?*

Sol. We know that when a die is thrown, the probability to show a five or six = 2/6 = 1/3 = p (say).

$$\therefore \qquad q = 1 - p = 1 - \frac{1}{3} = \frac{2}{3}.$$

The probability to show a five or six in at least three dice

$$= \sum_{x=3}^{6} p(x) = p(3) + p(4) + p(5) + p(6),$$

where $p(x)$ is the probability to show five or six

$$= {}^6C_3q^3p^3 + {}^6C_4p^4q^2 + {}^6C_5p^5q + {}^6C_6p^6$$

$$= {}^6C_3\left(\frac{1}{3}\right)^3\left(\frac{2}{3}\right)^3 + {}^6C_4\left(\frac{1}{3}\right)^4\left(\frac{2}{3}\right)^2 + {}^6C_5\left(\frac{1}{3}\right)^5(2/3) + {}^6C_6\left(\frac{1}{3}\right)^6$$

$$= \frac{1}{(3)^6}\left[\frac{6\times5\times4}{3\times2\times1}(2)^3 + \frac{6\times5}{2\times1}(2)^2 + 6(2) + 1\right]$$

= (1/729) [160 + 60 + 12 + 1] = 233/729 = p (say).

∴ The required number = np = 729 (233/729) = 233.

Example 16. *In 256 sets of 12 tosses of a coin, in how many cases one can expect 8 heads and 4 tails.*

Sol. P (head) = $\frac{1}{2}$ and P (tail) = $\frac{1}{2}$.

By binomial distribution, probability of 8 heads and 4 tails in 12 trials is

$$P(X = 8) = {}^{12}C_8 \left(\frac{1}{2}\right)^8 \left(\frac{1}{2}\right)^4 = \frac{12!}{8!4!} \cdot \frac{1}{2^{12}} = \frac{495}{4096}.$$

∴ The expected number of such cases in 256 sets

$$= 256 \times P(X = 8) = 256 \frac{495}{4096} = 30.9 = 31 \text{ (approx.)}.$$

Example 17. *The number of males in each 106 eight pig litters was found and they are given by the following frequency distribution:*

Number of male per litter:	*0*	*1*	*2*	*3*	*4*	*5*	*6*	*7*	*8*	*Total*
Frequency:	*0*	*5*	*9*	*22*	*25*	*26*	*14*	*4*	*1*	*106*

Assuming that the probability of an animal being male of female is even i.e., $p = q = \frac{1}{2}$ *and frequency distribution follows the binomial law, calculate the expected frequencies.*

Sol. Here the frequency distribution = $106\ (\frac{1}{2} + \frac{1}{2})^8$ [Using $N(p + q)^n$]

Therefore the expected frequencies are the respective terms of this expansion which are respectively:

0.4, 3.3, 11.6, 23.2, 29, 23.2, 11.6. 3.3, 0.4.

Example 18. *In litters of 4 mice the number of litters which contained 0, 1, 2, 3, 4 females were noted. The figures are given in the table below:*

Number of female mice:	*0*	*1*	*2*	*3*	*4*	*Total*
Number of litters:	*8*	*32*	*34*	*24*	*5*	*103*

If the chance of obtaining a female in a single trial is assumed constant, estimate this constant of unknown probability. Find also expected frequencies.

Sol. Here the mean of observations

$$= \frac{8 \times 0 \times 32 \times 1 + 34 \times 2 + 24 \times 3 + 5 \times 4}{8 + 32 + 34 + 24 + 5}$$

$$= \frac{32 + 68 + 72 + 20}{103} = \frac{192}{103} 1.864.$$

∵ $n = 4$ and np = mean = 1.864.

Thus $p = \frac{0.864}{4} = 0.466.$

∴ $q = 1 - p = 1 - 0.466 = 0.534.$

Hence the expected frequencies are the respective terms of the binomial expansion of $103\ (0.534 + 0.466)^4$.

Example 19. *The following data are the number of seeds germinating out of 10 on damp filter for 80 sets of seeds. Fit a binomial distribution to these data:*

x:	0	1	2	3	4	5	6	7	8	9	10	Total
f:	6	20	28	12	8	6	0	0	0	0	0	80

Sol. Hence, we have

$$n = 10, \text{N} = \Sigma f = 80.$$

$$\therefore \quad \text{A.M.} = \frac{\Sigma fx}{\text{N}} = \frac{1\times 20 + 2\times 28 + 3\times 12 + 4\times 8 + 5\times 6}{80} = \frac{174}{80}$$

But $\quad \text{mean} = np = \dfrac{174}{80} \Rightarrow 10p = \dfrac{174}{80} \Rightarrow p = \dfrac{174}{800} = 0.2175.$

$$\therefore \quad q = 1 - p = 1 - 0.2175 = 0.7825.$$

Hence the binomial distribution to be fitted to the given data is $80\,(0.7825 + 0.2175)^{10}$ [using $\text{N}(q + p)^n$] and from this expansion the successive frequencies of 0, 1, 2, ..., 10 successes are respectively

6.9, 19.1, 24.0, 17.8, 8.6, 2.9, 0.7, 0.1, 0, 0, 0, *i.e.*, 7, 19, 24, 18, 9, 3, 1, 0, 0, 0, 0.

Example 20. *In sampling a large number of parts manufactured by a machine, the mean number of defectives in a sample of 20 is 2. Out of 1000 such samples, how many would be expected to contain at least 3 defective parts.*

Sol. Mean number of defectives, $\quad np = 20p = 2.$

$\therefore$ The probability of defective part is $p = \dfrac{2}{20} = 0.1$

and the probability of a non-defective part $= 0.9.$

$\therefore$ The probability of at least three defectives in a sample of 20

= 1 – [Probability of none is non-defective part + Probability of one is non-defective part + Probability of two are non-defective part]

$$= 1 - [\,^{20}C_0\,(0.9)^{20} + \,^{20}C_1\,(0.1)\,(0.9)^{19} + \,^{20}C_2\,(0.1)^2\,(0.9)^{18}]$$

$$= 1 - (0.9)^{18} \times 4.51 = 0.323.$$

Example 21. *Assuming that half the population are consumers of chocolate, so that the chance of an individual being a consumer is* $\dfrac{1}{2}$ *and assuming that 100 investigators each take 10 individuals to see whether they are consumers, how many investigators would you expect to report that three people or less were consumers?*

Sol. Here $\quad p$ = probability of an individual being a consumer $= \dfrac{1}{2}$ (Given).

$$\therefore \quad q = 1 - \frac{1}{2} = \frac{1}{2}.$$

Again, we have, $n = 10$, N = 100.

$\therefore$ The binomial distribution is $100\left(\dfrac{1}{2} + \dfrac{1}{2}\right)^{10}$

∴ The required number of investigators to report that 3 or less than 3 persons are consumers of chocolate

$$= \text{first four terms in the expansion } 100\left(\frac{1}{2}+\frac{1}{2}\right)^{10};$$

$$= 100\left[\left(\frac{1}{2}\right)^{10} + {}^{10}C_1\left(\frac{1}{2}\right)^{9}\left(\frac{1}{2}\right) + {}^{10}C_2\left(\frac{1}{2}\right)^{8}\left(\frac{1}{2}\right)^{2} + {}^{10}C_3\left(\frac{1}{2}\right)^{7}\left(\frac{1}{2}\right)^{3}\right]$$

$$= 100\left(\frac{1}{2}\right)^{10}\left[1+10+\frac{10\times 9}{2\times 1}+\frac{10\times 9\times 8}{3\times 2\times 1}\right]$$

$$= 100 \times \left(\frac{1}{2}\right)^{10} [1 + 10 + 45 + 120]$$

$$= \frac{100}{2^{10}} \times 176 = \frac{25}{2^8} \times 2^4 \times 11 = \frac{275}{16} = 17 \text{ nearly.}$$

Example 22. *An irregular six faced die is the own, and the expectation that in 10 throws it will give five even numbers is twice the expectations that it will give four even numbers. How many times in 10,000 sets of 10 throws would you expect it to give no even number?*

Sol. Consider that the probability of getting an even number is p.

∴ The probability of getting 5 even numbers in 10 throws $= {}^{10}C_5 p^5 q^5$,

and the probability of getting 4 even numbers in 10 throws $= {}^{10}C_4 p^4 q^6$.

∴ $${}^{10}C_5 p^5 q^5 = 2 \times {}^{10}C_4 p^4 q^6$$

⇒ $$(3/5)\, p = q \Rightarrow (3/5)\,(1 - q) = q,\ q = (3/8)$$

Thus the required number of times of not getting an even number in 10,000 sets of 10 throws = $10{,}000\,(3/8)^{10}$ = 1 nearly.

Example 23. *The probability of head in a single tossing of a biased coin is 3/5. Find the most probable number of heads and the mean of number of heads in 99 tossing of a coin.*

Sol. Let X be the number of head in 99 tossings 0, 1, 2, ..., 99.

∴ $$P(X = x) = {}^nC_x p^x q^{n-x}, \text{ where } p = 3/5,\ n = 99$$

$$= {}^{99}C_x\ (3/5)^x\ (2/5)^{99-x},\ x = 0, 1, 2, ..., 99$$

∴ mean = $np = 99 \times (3/5) = 59.4$.

Now $(n + 1)p = (99 + 1)\,(3/5) = 60$ [an integer]

Hence $x = np + p = 60$ and $x = np + p - 1 = 59$ are the most probable numbers of heads, where

$$p(X = 59) = {}^{99}C_{59}\ (3/5)^{59}\ (2/5)^{40}$$

and $$p(X = 60) = {}^{99}C_{60}\ (3/5)^{60}\ (2/5)^{39} \text{ are equal.}$$

Example 24. *In a precision bombing attack there is a 50% chance that any one bomb will strike the target. Two direct hits are required to destroy the target completely. How many bombs must be dropped to give a 99% chance or better of completely destroying the target?*

Sol. We have, $p = 50/100 = 1/2$.

The meaning of 99% chance or better is that the probability must be greater than 0.99. If n bombs are dropped, then out of them 2 succeed, 3 may succeed and so on,

$$\therefore \quad {}^nC_2\left(\frac{1}{2}\right)^n + {}^nC_3\left(\frac{1}{2}\right)^n + \ldots + {}^nC_n\left(\frac{1}{2}\right)^n \geq 0.99$$

$$i.e., \quad \left(\frac{1}{2}\right)^n [{}^nC_2 + {}^nC_3 + \ldots + {}^nC_n] \geq 0.90$$

$$\text{or} \quad \frac{2^n - n - 1}{2^n} \geq 0.99$$

$$\text{or} \quad 1 - \frac{1+n}{2^n} \geq 1 - 0.01 \quad \text{or} \quad \frac{1+n}{2^n} \leq 0.01$$

$$\text{or} \quad 100 + 100n - 2^n \leq 0$$

$$\text{or} \quad 2^n \geq 100n + 100.$$

By trial method, $n = 11$ will give this inequality

Example 25 *Prove that in the binomial distribution difference of mean and mode is not greater than unity.*

Sol. We have mean $= np$ (for binomial distribution).

From Ex. 8(*a*) and 8(*b*) above, we have three possibilities for mode:

(*i*) If np is a positive integer then mode and mean are equal.

$$\therefore \quad |\text{ mean} - \text{mode }| = 0 < 1.$$

(*ii*) If $np + p$ is a positive integer then mode is $np + p$.

$$\therefore \quad |\text{ mean} - \text{mode }| = |\, np - np - p \,| = p < 1.$$

(*iii*) If $np + p$ is not a positive integer, then

$$np + p - 1 \leq r \leq np + p$$

$$\Rightarrow \quad p - 1 \leq r - np \leq p$$

$$\Rightarrow \quad -p \leq \text{mode} - \text{mean} \leq p$$

$$\Rightarrow \quad |\text{mode} - \text{mean}| \leq 1.$$

TEST YOUR KNOWLEDGE

1. Ten coins are tossed simultaneously. Find the probability of getting at least seven heads.
2. The probability of any ship of a company being destroyed on a certain voyage is 0.02. The company owns 6 ships for the voyage. What is the probability of:

   (*i*) losing one ship (*ii*) losing at most two ships (*iii*) losing none.
3. The probability that a man aged 60 will live to be 70 is 0.65. What is the probability that out of ten men now 60, at least 7 would live to be 70?
4. The incidence of occupational disease in an industry is such that the workers have a 20% chance of suffering from it. What is the probability that out of six workers chosen at random, four or more will suffer from the disease?

5. The probability that a pen manufactured by a company will be defective is $\frac{1}{10}$. If 12 such pens are manufactured, find the probability that

(*i*) exactly two will be defective (*ii*) at least two will be defective

(*iii*) none will be defective.

6. If the chance that one of the ten telephone lines is busy at an instant is 0.2.

(*i*) What is the chance that 5 of the lines are busy?

(*ii*) What is the probability that all the lines are busy?

7. If on an average 1 vessel in every 10 is wrecked, find the probability that out of 5 vessels expected to arrive, at least 4 will arrive safely.

8. A product is 0.5% defective and is packed in cartons of 100. What percentage contains not more than 3 defectives?

9. A bag contains 5 white, 7 red and 8 black balls. If four balls are drawn, one by one, with replacement, what is the probability that

(*i*) none is white (*ii*) all are white

(*iii*) at least one is white (*iv*) only 2 are white?

10. In a hurdle race, a player has to cross 10 hurdles. The probability that he will clear each hurdle is $\frac{5}{6}$. What is the probability that he will knock down fewer than 2 hurdles?

11. Fit a binomial distribution for the following data and compare the theoretical frequencies with the actual ones:

x:	0	1	2	3	4	5
f:	2	14	20	34	22	8

12. If the sum of mean and variance of a binomial distribution is 4.8 for five trials, find the distribution.

13. If the mean of a binomial distribution is 3 and the variance is $\frac{3}{2}$, find the probability of obtaining at least 4 success.

14. In 800 families with 5 children each, how many families would be expected to have (*i*) 3 boys and 2 girls, (*ii*) 2 boys and 3 girls, (*iii*) no girl (*iv*) at the most two girls. (Assume probabilities for boys and girls to be equal.)

15. In 100 sets of ten tosses of an unbiased coin, in how many cases do you expect to get

(*i*) 7 heads and 3 tails (*ii*) at least 7 heads?

16. The following data are the number of seeds germinating out of 10 on damp filter for 80 sets of seeds. Fit a binomial distribution to this data:

x:	0	1	2	3	4	5	6	7	8	9	10	Total
f:	6	20	28	12	8	6	0	0	0	0	0	80

[**Hint.** Here n = 10, N = 80, Mean = $\frac{\Sigma fx}{\Sigma f}$ = 2.175 $\therefore$ np = 2.175 etc.]

17. A bag contains 10 balls each marked with one of the digits 0 to 9. If four falls are drawn successively with replacement from the bag, what is the probability that none is marked with the digit 0?

18. A box contains 100 tickets each bearing one of the numbers from 1 to 100. If 5 tickets are drawn successively with replacement from the box, find the probability that all the tickets bear numbers divisible by 10.

19. The probability that a bomb dropped from a plane will strike the target is $\frac{1}{5}$. If six bombs are dropped, find the probability that (*i*) exactly two will strike the target, (*ii*) at least two will strike the target.

20. In sampling a large number of parts manufactured by a machine, the mean number of defectives in a sample of 20 is 2. Out of 1000 such samples, how many would be expected to contain at least 3 defective parts.

Answers

1. $\frac{11}{64}$ 2. (*i*) 0.1085, (*ii*) 0.9997, (*iii*) 0.8858 3. 0.514

4. $\frac{53}{3125}$ 5. (*i*) 0.2301 (*ii*) 0.3412 (*iii*) 0.2833

6. (*i*) 0.02579 (*ii*) 1.024×10^{-7} 7. 0.91854 8. 99.83

9. (*i*) $\frac{81}{256}$ (*ii*) $\frac{1}{256}$ (*iii*) $\frac{175}{256}$ (*iv*) $\frac{27}{128}$ 10. $\frac{5}{2}\left(\frac{5}{6}\right)^9$

11. $100\ (0.432 + 0.568)^5$ 12. $\left(\frac{1}{5}+\frac{4}{5}\right)^5$ 13. $\frac{11}{32}$

14. (*i*) 250, (*ii*) 250, (*iii*) 25, (*iv*) 400 15. (*i*) 12 nearly (*ii*) 17 nearly

16. $80\ (0.7825 + 0.2175)^{10}$ 17. $\left(\frac{9}{10}\right)^4$ 18. 0.00001

19. (*i*) 0.246 (*ii*) 0.345 20. 323.

POISSON DISTRIBUTION

5.65. POISSON DISTRIBUTION AS A LIMITING CASE OF BINOMIAL DISTRIBUTION

If the parameters n and p of a binomial distribution are known, we can find the distribution. But in situations where n is very large and p is very small, application of binomial distribution is very labourious. However, if we assume that as $n \to \infty$ and $p \to 0$ such that np always remains finite, say λ, we get the Poisson approximation to the binomial distribution.

Now, for a binomial distribution

$$P(X = r) = {}^nC_r q^{n-r} p^r$$

$$= \frac{n(n-1)(n-2)\ldots(n-r+1)}{r!} \times (1-p)^{n-r} \times p^r$$

$$= \frac{n(n-1)(n-2)\ldots(n-r+1)}{r!} \times \left(1-\frac{\lambda}{n}\right)^{n-r} \times \left(\frac{\lambda}{n}\right)^r \qquad \text{since } np = \lambda \quad \therefore \quad p = \frac{\lambda}{n}$$

$$= \frac{\lambda^r}{r!} \times \frac{n(n-1)(n-2)\ldots(n-r+1)}{n^r} \times \frac{\left(1-\frac{\lambda}{n}\right)^n}{\left(1-\frac{\lambda}{n}\right)^r}$$

$$= \frac{\lambda^r}{r!}\left(\frac{n}{n}\right)\left(\frac{n-1}{n}\right)\left(\frac{n-2}{n}\right)\ldots\ldots\left(\frac{n-r+1}{n}\right) \times \frac{\left(1-\frac{\lambda}{n}\right)^n}{\left(1-\frac{\lambda}{n}\right)^r}$$

$$= \frac{\lambda^r}{r!}\left(1-\frac{1}{n}\right)\left(1-\frac{2}{n}\right)\ldots\ldots\left(1-\frac{r-1}{n}\right) \times \frac{\left[\left(1-\frac{\lambda}{n}\right)^{-\frac{n}{\lambda}}\right]^{-\lambda}}{\left(1-\frac{\lambda}{n}\right)^r}$$

As $n \to \infty$, each of the $(r-1)$ factors

$$\left(1-\frac{1}{n}\right), \left(1-\frac{2}{n}\right), \ldots\ldots, \left(1-\frac{r-1}{n}\right) \text{ tends to 1. Also } \left(1-\frac{\lambda}{n}\right)^r \text{ tends to 1.}$$

Since $\underset{x \to \pm\infty}{\text{Lt}} \left(1+\frac{1}{x}\right)^x = e$, the Naperian base. $\therefore \left[\left(1-\frac{\lambda}{n}\right)^{-\frac{n}{\lambda}}\right]^{-\lambda} \longrightarrow e^{-\lambda}$ as $n \to \infty$

Hence in the limiting case when $n \to \infty$, we have

$$P(X = r) = \frac{\lambda^r e^{-\lambda}}{r!} \quad (r = 0, 1, 2, 3, \ldots) \qquad \ldots(A)$$

where λ is a finite number $= np$.

(A) represents probability distribution which is called the Poisson probability distribution.

Note 1. λ is called the parameter of the distribution.

Note 2. $e^x = 1 + \frac{x}{1!} + \frac{x^2}{2!} + \ldots\ldots + \frac{x^n}{n!} + \ldots\ldots$ to ∞.

Note 3. The sum of the probabilities $P(r)$ for $r = 0, 1, 2, 3, \ldots\ldots$ is 1, since

$$P(0) + P(1) + P(2) + P(3) + \ldots\ldots = e^{-\lambda} + \frac{\lambda e^{-\lambda}}{1!} + \frac{\lambda^2 e^{-\lambda}}{2!} + \frac{\lambda^3 e^{-\lambda}}{3!} + \ldots\ldots$$

$$= e^{-\lambda}\left(1 + \frac{\lambda}{1!} + \frac{\lambda^2}{2!} + \frac{\lambda^3}{3!} + \ldots\ldots\right) = e^{-\lambda} \cdot e^{\lambda} = 1.$$

5.66. RECURRENCE FORMULA FOR THE POISSON DISTRIBUTION

For Poisson distribution, $P(r) = \frac{\lambda^r e^{-\lambda}}{r!}$ and $P(r+1) = \frac{\lambda^{r+1} e^{-\lambda}}{(r+1)!}$

$$\therefore \quad \frac{P(r+1)}{P(r)} = \frac{\lambda r!}{(r+1)!} = \frac{\lambda}{r+1} \quad \text{or} \quad P(r+1) = \frac{\lambda}{r+1} P(r), \; r = 0, 1, 2, 3, \ldots\ldots$$

This is called the recurrence formula for the Poisson distribution.

5.67. MEAN AND VARIANCE OF THE POISSON DISTRIBUTION

For the Poisson distribution, $P(r) = \frac{\lambda^r e^{-\lambda}}{r!}$

$$\text{Mean } \mu = \sum_{r=0}^{\infty} rP(r) = \sum_{r=0}^{\infty} r \cdot \frac{\lambda^r e^{-\lambda}}{r!}$$

$$= e^{-\lambda} \sum_{r=1}^{\infty} \frac{\lambda^r}{(r-1)!} = e^{-\lambda}\left(\lambda + \frac{\lambda^2}{1!} + \frac{\lambda^3}{2!} + \ldots\ldots\right)$$

$$= \lambda e^{-\lambda}\left(1 + \frac{\lambda}{1!} + \frac{\lambda^2}{2!} + \ldots\ldots\right) = \lambda e^{-\lambda} \cdot e^{\lambda} = \lambda$$

Thus, the mean of the Poisson distribution is equal to the parameter λ.

$$\text{Variance } \sigma^2 = \sum_{r=0}^{\infty} r^2 P(r) - \mu^2 = \sum_{r=0}^{\infty} r^2 \cdot \frac{\lambda^r e^{-\lambda}}{r!} - \lambda^2 = e^{-\lambda} \sum_{r=1}^{\infty} \frac{r^2 \lambda^r}{r!} - \lambda^2$$

$$= e^{-\lambda}\left[\frac{1^2 \cdot \lambda}{1!} + \frac{2^2 \cdot \lambda^2}{2!} + \frac{3^2 \lambda^3}{3!} + \frac{4^2 \lambda^4}{4!} + \ldots\ldots\right] - \lambda^2$$

$$= \lambda e^{-\lambda}\left[1 + \frac{2\lambda}{1!} + \frac{3\lambda^2}{2!} + \frac{4\lambda^3}{3!} + \ldots\ldots\right] - \lambda^2$$

$$= \lambda e^{-\lambda}\left[1 + \frac{(1+1)\lambda}{1!} + \frac{(1+2)\lambda^2}{2!} + \frac{(1+3)\lambda^3}{3!} + \ldots\ldots\right] - \lambda^2$$

$$= \lambda e^{-\lambda}\left[\left(1 + \frac{\lambda}{1!} + \frac{\lambda^2}{2!} + \frac{\lambda^3}{3!} + \ldots\ldots\right) + \left(\frac{\lambda}{1!} + \frac{2\lambda^2}{2!} + \frac{3\lambda^3}{3!} + \ldots\ldots\right)\right] - \lambda^2$$

$$= \lambda e^{-\lambda}\left[e^{\lambda} + \lambda\left(1 + \frac{\lambda}{1!} + \frac{\lambda^2}{2!} + \ldots\ldots\right)\right] - \lambda^2$$

$$= \lambda e^{-\lambda}\,[e^{\lambda} + \lambda e^{\lambda}] - \lambda^2 = \lambda e^{-\lambda} \cdot e^{\lambda}(1+\lambda) - \lambda^2 = \lambda(1+\lambda) - \lambda^2 = \lambda.$$

Hence, the variance of the Poisson distribution is also λ.

Thus, the mean and the variance of the Poisson distribution are each equal to the parameter λ.

Note. The mean and the variance of the Poisson distribution can also be derived from those of the binomial distribution in the limiting case when $n \to \infty$, $p \to 0$ and $np = \lambda$.

Mean of binomial distribution is np.

$\therefore$ Mean of Poisson distribution $= \underset{n \to \infty}{\text{Lt}}\ np = \underset{n \to \infty}{\text{Lt}}\ \lambda = \lambda$

Variance of binomial distribution is $npq = np(1-p)$

$\therefore$ Variance of Poisson distribution $= \underset{n \to \infty}{\text{Lt}}\ np(1-p) = \underset{n \to \infty}{\text{Lt}}\ \lambda\left(1 - \frac{\lambda}{n}\right) = \lambda.$

5.68. APPLICATIONS OF POISSION DISTRIBUTION

Poisson distribution is applied to problems concerning:

(*i*) Arrival pattern of 'telephone calls', 'defective blades' in a packet, 'patients in a hospital'.

(*ii*) Demand pattern for certain commodities in market.

(*iii*) Spatial distribution of bomb hits.

(*iv*) Number of fragments from a shell hitting a target.

ILLUSTRATIVE EXAMPLES

Example 1. *If the variance of the Poisson distribution is 2, find the probabilities for r = 1, 2, 3, 4 from the recurrence relation of the Poisson distribution.*

Sol. λ, the parameter of Poisson distribution = Variance = 2

Recurrence relation for the Poisson distribution is

$$P(r+1) = \frac{\lambda}{r+1} P(r) = \frac{2}{r+1} P(r) \qquad \ldots(1)$$

Now $\quad P(r) = \frac{\lambda^r e^{-\lambda}}{r!} \Rightarrow P(0) = \frac{e^{-2}}{0!} = e^{-2} = 0.1353$

Putting $r = 0, 1, 2, 3$ in (1), we get

$$P(1) = 2P(0) = 2 \times 0.1353 = 0.2706; \qquad P(2) = \frac{2}{2} P(1) = 0.2706$$

$$P(3) = \frac{2}{3} P(2) = \frac{2}{3} \times 0.2706 = 0.1804; \qquad P(4) = \frac{2}{4} P(3) = \frac{1}{2} \times 0.1804 = .0902.$$

Example 2. *Assume that the probability of an individual coalminer being killed in a mine accident during a year is* $\frac{1}{2400}$. *Use Poisson's distribution to calculate the probability that in a mine employing 200 miners there will be at least one fatal accident in a year.*

Sol. Here $\quad p = \frac{1}{2400}, n = 200; \quad \therefore \lambda = np = \frac{200}{2400} = \frac{1}{12} = 0.083$

$$\therefore \quad P(r) = \frac{\lambda^r e^{-\lambda}}{r!} = \frac{(0.083)^r e^{-.083}}{r!}$$

P(at least one fatal accident) = 1 – P (no fatal accident)

$$= 1 - P(0) = 1 - \frac{(0.083)^0 e^{-0.83}}{0!} = 1 - .92 = 0.08.$$

Example 3. *Data was collected over a period of 10 years, showing number of deaths from horse kicks in each of the 200 army corps. The distribution of deaths was as follows:*

No. of deaths:	*0*	*1*	*2*	*3*	*4*	*Total*
Frequency:	*109*	*65*	*22*	*3*	*1*	*200*

Fit a Poisson distribution to the data and calculate the theoretical frequencies:

Sol. Mean of given distribution = $\frac{\Sigma fx}{\Sigma f} = \frac{65 + 44 + 9 + 4}{200} = \frac{122}{200} = 0.61$

This is the parameter (m) of the Poisson distribution.

$\therefore$ Required Poisson distribution is $N \cdot \frac{m^r e^{-m}}{r!}$ where $N = \Sigma f = 200$

$$= 200e^{-0.61} \cdot \frac{(0.61)^r}{r!} = 200 \times 0.5435 \frac{(0.61)^r}{r!} = 108.7 \times \frac{(0.61)^2}{r!}.$$

r	$P(r)$	*Theoretical Frequency*
0	108.7	109
1	$108.7 \times 0.61 = 66.3$	66
2	$108.7 \times \frac{(0.61)^2}{2!} = 20.2$	20
3	$108.7 \times \frac{(0.61)^3}{3!} = 4.1$	4
4	$108.7 \times \frac{(0.61)^4}{4!} = 0.7$	1
		Total = 200

Example 4. *A car-hire firm has two cars, which it hires out day by day. The number of demands for a car on each day is distributed as a Poisson distribution with mean 1.5. Calculate the proportion of days on which neither car is used and the proportion of days on which some demand is refused.* ($e^{-1.5} = 0.2231$)

Sol. Since the number of demands for a car is distributed as a Poisson distribution with mean $m = 1.5$.

$\therefore$ Proportion of days on which neither car is used

= Probability of there being no demand for the car

$$= \frac{m^0 e^{-m}}{0!} = e^{-1.5} = 0.2231$$

Proportion of days on which some demand is refused

= probability for the number of demands to be more than two

$$= 1 - P(x \le 2) = 1 - \left(e^{-m} + \frac{me^{-m}}{1!} + \frac{m^2 e^{-m}}{2!} \right)$$

$$= 1 - e^{-1.5} \left(1 + 1.5 + \frac{(1.5)^2}{2} \right) = 1 - 0.2231\,(1 + 1.5 + 1.125)$$

$$= 1 - 0.2231 \times 3.625 = 1 - 0.8087375 = 0.1912625.$$

Example 5. *Six coins are tossed 6400 times. Using the Poisson distribution, determine the approximate probability of getting six heads x times.*

Sol. Probability of getting one head with one coin = $\frac{1}{2}$.

$\therefore$ The probability of getting six heads with six coins = $\left(\frac{1}{2}\right)^6 = \frac{1}{64}$

$\therefore$ Average number of six heads with six coins in 6400 throws = $np = 6400 \times \frac{1}{64} = 100$

$\therefore$ The mean of the Poisson distribution = 100.

Approximate probability of getting six heads x times when the distribution is Poisson

$$= \frac{m^x e^{-m}}{x!} = \frac{(100)^x \cdot e^{-100}}{(100)!}.$$

Example 6. *Fit Poisson's distribution to the following and calculate theoretical frequencies ($e^{-0.5}$ = 0.61):*

Deaths:	*0*	*1*	*2*	*3*	*4*
Frequency:	*122*	*60*	*15*	*2*	*1*

Sol. Here total frequency

$$(N) = \Sigma f = 122 + 60 + 15 + 2 + 1 = 200$$

$$\text{Mean } (m) = \frac{\Sigma fx}{N} = \frac{122 \times 0 + 60 \times 1 + 15 \times 2 + 2 \times 3 + 1 \times 4}{122 + 60 + 15 + 2 + 1}$$

$$= \frac{60 + 30 + 6 + 4}{200} = \frac{1}{2} = 0.5.$$

Now $\qquad e^{-m} = e^{-0.5} = 1 - (0.5) + \frac{1}{2}(0.5)^2 - \frac{1}{6}(0.5)^3 + \ldots$

$$= 1 - 0.5 + 0.125 - 0.0208 + \ldots = 0.61 \quad \text{(approx.)}$$

∴ Therefore frequencies of r deaths

$$= Ne^{-m}\frac{m^r}{r!} = 200 \times (0.61) \times \frac{(0.5)^r}{r!}$$

It gives frequencies 122, 61, 15, 2 and 4 respectively for r = 0, 1, 2, 3 and 4.

Example 7. *Find the probability that at most 5 defective fuses will be found in a box of 200 fuses, if experience shows that 2 percent of such fuses are defective.*

Sol. Here p = 2% = 2/100 = 1/50, n = 200.

∴ $\qquad np = 200/50 = 4.$

Maximum number of defective fuses = 5.

∴ $\qquad r \le 5.$

Also $\qquad e^{-4} = 1 - 4 + \frac{1}{2}.4^2 - \frac{1}{3!}.4^3 + \frac{1}{4!}.4^4 - \ldots. = 0.0183$

∴ Required probability ($r \le 5$)

$$= \sum_{r=0}^{5} \frac{4^r . e^{-4}}{r!} = e^{-4}\left[1 + \frac{4}{1!} + \frac{4^2}{2!} + \frac{4^3}{3!} + \frac{4^4}{4!} + \frac{4^5}{5!}\right]$$

$$= 0.0183\,[1 + 4 + 8 + 10.6667 + 10.6667 + 8.5333]$$

$$= 0.0183 \times 42.8667 = 0.7845.$$

Example 8. *If the probability of a bad reaction from a certain injection is 0.001, determine the chance that out of 2,000 individuals more than two will get a bad reaction.*

Sol. Since the probability of occurrence is very small, it follows a Poisson distribution.

$$\text{Mean } m = np = 2000\ (0.001) = 2.$$

∴ Probability that more than 2 will get a bad reaction

= 1 – [Probability that no one gets a bad reaction

+ Probability that one gets a bad reaction + Probability that two get bad reaction]

$$= 1 - \left[e^{-m} + \frac{m^1 e^{-m}}{1!} + \frac{m^2 e^{-m}}{2!}\right] = 1 - \left[\frac{1}{e^2} + \frac{2}{e^2} + \frac{2}{e^2}\right] \qquad [\because\ m = 2]$$

$$= 1 - \frac{5}{e^2} = 0.32 \qquad [\because\ e = 12.718]$$

Example 9. *In 1,000 extensive sets of trials for an event of small probability the frequencies f of the number x of successes are found to be*

x:	*0*	*1*	*2*	*3*	*4*	*5*	*6*	*7*
f:	*305*	*365*	*210*	*80*	*28*	*9*	*2*	*1*

Assuming it to be a Poissonian distribution calculate its mean, variance and expected frequencies for the Poissonian distribution with same mean.

Sol. Here mean $(m) = \dfrac{\Sigma fx}{\Sigma f} = \dfrac{1201}{1000} = 1.2$ (approx.)

x	f	fx	fx^2
0	305	0	0
1	365	365	365
2	210	420	840
3	80	240	720
4	28	112	448
5	9	45	225
6	2	12	72
7	1	7	49
Total	1000	1201	2719

$$\text{Variance} = \sigma^2 = \frac{2719}{1000} - (1.2)^2 = 2.719 - 1.44 = 1.279$$

$$e^{-m} = e^{-1.2} = 1 - 1.2 + \frac{(1.2)^2}{2!} - \frac{(1.2)^3}{3!} + \ldots = 0.3012.$$

The expected frequencies for $x = 0, 1, 2, 3, 4, 5, 6, 7$ are respectively calculated below:

$$Ne^{-m} = 0.3012 \times 1000 = 301.2 \qquad [\because \ N = 1{,}000]$$

$$Ne^{-m}.\ m = 301.2 \times 1.2 = 361.4$$

$$Ne^{-m}\frac{m^2}{2!} = 301.2 \times \frac{(1.2)^2}{2!} = 216.8$$

$$Ne^{-m}\frac{m^3}{3!} = 301.2 \times \frac{(1.2)^3}{3!} = 86.7$$

$$Ne^{-m}\frac{m^4}{4!} = 301.2 \times \frac{(1.2)^4}{4!} = 26$$

$$Ne^{-m}\frac{m^5}{5!} = 301.2 \times \frac{(1.2)^5}{5!} = 6.2$$

$$Ne^{-m}\frac{m^6}{6!} = 301.2 \times \frac{(1.2)^6}{6!} = 1.2$$

$$Ne^{-m}\frac{m^7}{7!} = 301.2 \times \frac{(1.2)^7}{7!} = 0.2.$$

Example 10. *In a certain factory turning razor blades, there is a small chance (1/500) for any blade to be defective. The blades are in packets of 10. Use Poisson's distribution to calculate the approximate number of packets containing no defective, one defective and two defective blades respectively in a consignment of 10,000 packets.*

Sol. Here $p = 1/500,\ n = 10,\ N = 10{,}000.$

$\therefore \quad m = np = 10\ (1/500) = 0.02.$

Now $\quad e^{-m} = e^{-0.02} = 1 - 0.02 + \frac{1}{2!}(0.02)^2 - ... = 0.9802$ (approx).

The respective frequencies (*i.e.*, number of packets) containing no defective, one defective and two defective blades are given respectively as follows:

$$Ne^{-m},\ Ne^{-m}\ .\ m,\ Ne^{-m}\ .\ \frac{1}{2}m^2$$

i.e., $\quad 10000 \times 0.9802;\ 10000 \times 0.9802 \times 0.02;\ 10000 \times 0.9802 \times \frac{1}{2}\ (0.02)^2$

i.e., $\quad 9802;\ 196;\ 2.$

Example 11. *Find the mean and standard deviation for the table of deaths of women over 80 years old recorded in a three year period. Number of deaths recorded*

In a day:	*0*	*1*	*2*	*3*	*4*	*5*	*6*	*7*
No. of days:	*364*	*376*	*218*	*89*	*33*	*13*	*2*	*1*

Find the expected number of days with one death recorded for the Poisson series fitted to the data.

Sol.

x	f	fx	fx^2	x	f	fx	fx^2
0	364	0	0	4	33	132	528
1	376	376	376	5	13	65	325
2	218	436	872	6	2	12	72
3	89	267	801	7	1	7	49
			Total		1096	1295	3023

Now N (the total frequency) = 1096.

The mean (m) of the series = $\Sigma fx/\Sigma f = 1295/1096 = 1.18$

$$\sigma^2 = \frac{3023}{1096} - (1.18)^2 = 2.758 - 1.392 = 1.366$$

$\therefore \quad \sigma = 1.17$ (approx.)

Again $\quad e^{-m} = e^{-1.18} = 1 - 1.18 + \frac{(1.18)^2}{2!} - \frac{(1.18)^3}{3!} + ...$

$= 0.316$ (on simplification).

$\therefore$ The required frequency for one death

$= Ne^{-m}\ .\ m = 1096 \times (0.316) \times (1.18) = 397.$

Example 12. *A car-hire firm has two cars, which it hires out day by day. The number of demands for a car on each day is distributed as a Poisson distribution with mean 1.5. Calculate the proportion of days on which neither car is used and the proportion of days on which some demand is refused.*

(given $e^{-1.5} = 0.2231$.)

Sol. Given $m = 1.5, e^{-m} = e^{-1.5} = 0.2231.$

The proportion of days when neither car is used $= e^{-m} \cdot \dfrac{m^0}{0!} = e^{-m} = 0.2231.$

The probability when no car, one car and two cars will be used

$$= e^{-m}\frac{m^0}{0!} + e^{-m}\cdot\frac{m}{1!} + e^{-m}\frac{m^2}{2!} = e^{-m}\left[1 + m + \frac{m^2}{2}\right]$$

$$= 0.2231\ [1 + 1.5 + 1.25] = 0.8087375.$$

∴ The proportion of days on which some demand is refused = 1 – 0.8087375

= 0.1912625.

Example 13. *A telephone switch handles 600 calls on the average during a rush hour. The board can make a maximum 20 connections per minute. Use Poisson distribution to estimate the probability that the board will be over-taxed during any given minute.*

$[e^{-10} = 0.00004539]$.

Sol. Mean (m) = number of calls per minute

= 600/60 = 10.

∴ The probability for using 0 to 20 calls per minute

$$p\ (x \le 20) = \sum_{r=0}^{20} e^{-m}\frac{m^r}{r!} = e^{-10}\sum_{r=0}^{20}\frac{10^r}{r!} = 0.00004539\sum_{r=0}^{20}\frac{10^r}{r!}.$$

Thus the probability that the board will be over-taxed during any given minute

= the probability when the calls are more than 20 connections per minute during any minute

$$= 1 - 0.00004539\sum_{r=0}^{20}\frac{10^r}{r!}.$$

Example 14. *In a book of 300 pages, a proof reader finds no error in 200 pages, in 75 pages one error on each page, in 20 pages to errors on each page and in 5 pages 3 errors on each page. Use Poisson distribution to these data and calculate theoretical frequency.*

$[e^{-0.43} = 0.6505]$.

Sol. The frequency distribution is

No. of errors x:	0	1	2	3	Total
No. of Pages f:	200	75	20	5	300

∴ mean $m = \Sigma fx/\Sigma f$

$$= \frac{0\times 200 + 1\times 75 + 2\times 20 + 3\times 5}{300} = 0.43 \quad \text{(nearly)}.$$

∴ From Poisson's distribution

$$P(x) = e^{-m}\ (m^x/x!)$$

$$= e^{-0.43}\frac{(0.43)^x}{x\,!} = 0.6505\,\frac{(0.43)^x}{x\,!}.$$

$\therefore$ The theoretical frequency for x errors

$$= \text{N}e^{-m}\,\frac{m^x}{x\,!} = 300\times 0.6505\,\frac{(0.43)^x}{x\,!}. \qquad ...(1)$$

For x = 0, 1, 2, 3, ... the theoretical frequency are [putting for x in (1)]:

195.15, 83.91, 18.04, 2.90.

TEST YOUR KNOWLEDGE

1. Fit a Poisson distribution to the following:

x:	0	1	2	3	4
f:	192	100	24	3	1

2. If the probability of a bad reaction from a certain injection is 0.001, determine the chance that out of 2000 individuals more than two will get a bad reaction.

3. If X is a Poisson variate such that P(X = 2) = 9P(X = 4) + 90P(X = 6), find the standard deviation.

4. If a random variable has a Poisson distribution such that P(1) = P(2), find

(*i*) mean of the distribution (*ii*) P(4).

5. Suppose that X has a Poisson distribution. If P(X = 2) = $\frac{2}{3}$ P(X = 1) find, (*i*) P(X = 0) (*ii*) P(X = 3).

6. A certain screw making machine produces on average 2 defective screws out of 100, and packs them in boxes of 500. Find the probability that a box contains 15 defective screws.

7. The incidence of occupational disease in an industry is such that the workmen have a 10% chance of suffering from it. What is the probability that in a group of 7, five or more will suffer from it?

8. Fit a Poisson distribution to the following and calculate theoretical frequencies:

x:	0	1	2	3	4
f:	122	60	15	2	1

9. Fit a Poisson distribution to the following data given the number of yeast cells per square for 400 squares:

No. of cells per sq:	0	1	2	3	4	5	6	7	8	9	10
No. of squares:	103	143	98	42	8	4	2	0	0	0	0

10. Show that in a Poisson distribution with unit mean, mean deviation about mean is $\left(\frac{2}{e}\right)$ times the standard deviation.

11. In a certain factory turning razor blades, there is a small chance of 0.002 for any blade to be defective. The blades are supplied in packets of 10. Use Poisson distribution to calculate the approximate number of packets containing no defective, one defective and two defective blades respectively in a consignment of 10000 packets.

12. The probability that a man aged 35 years will die before reaching the age of 40 years may be taken as 0.018. Out of a group of 400 men, now aged 35 years, what is the probability that 2 men will die within the next 5 years?

13. Suppose a book of 585 pages contains 43 typographical errors. If these errors are randomly distributed throughout the book, what is the probability that 10 pages, selected at random, will be free from errors?

Answers

1. $320 \times \dfrac{e^{0.503}(9.503)^r}{r!}$ **2.** 0.32 **3.** 1 **4.** (*i*) 2 (*ii*) $\dfrac{2}{3e^2}$

5. (*i*) e^{-4} (*ii*) $4e^{-4}$ **6.** $\dfrac{(10)^{15}e^{-10}}{(15)!} = 0.035$ **7.** 0.0008

8. $121.36 \times \dfrac{(0.5)^r}{r!}$, where $r = 0, 1, 2, 3, 4$

Theoretical frequencies are 121, 61, 15, 3, 0 respectively

9. Theoretical frequencies are 109, 142, 92, 40, 13, 3, 1, 0, 0, 0, 0

11. 9802, 196, 2 **12.** 0.01936 **13.** 0.4795.

NORMAL DISTRIBUTION

5.69. NORMAL DISTRIBUTION

The normal distribution is a continuous distribution. It can be derived from the binomial distribution in the limiting case when n, the number of trials is very large and p, the probability of a success, is close to $\frac{1}{2}$. The general equation of the normal distribution is given by

$$f(x) = \frac{1}{\sigma\sqrt{2\pi}} e^{-\frac{1}{2}\left(\frac{x-\mu}{\sigma}\right)^2}$$

where the variable x can assume all values from $-\infty$ to $+\infty$. μ and σ, called the parameters of the distribution, are respectively the mean and the standard deviation of the distribution and $-\infty < \mu < \infty$, $\sigma > 0$. x is called the normal variate and $f(x)$ is called probability density function of the normal distribution.

If a variable x has the normal distribution with mean μ and standard deviation σ, we briefly write $x : N(\mu, \sigma^2)$.

The graph of the normal distribution is called the *normal curve*. It is bell-shapped and symmetrical about the mean μ. The two tails of the curve extend to $+\infty$ and $-\infty$ towards the positive and negative directions of the x-axis respectively and gradually approach the x-axis without ever metting it. The curve is unimodal and the mode of the normal distribution coincides with its mean μ. The line $x = \mu$ divides the area under the normal curve above x-axis into two equal parts. Thus, the median of the distribution also coincides with its mean and mode. The area under the normal curve between any two given ordinates $x = x_1$ and $x = x_2$ represents the probability of values falling into the given interval. The total area under the normal curve above the x-axis is 1.

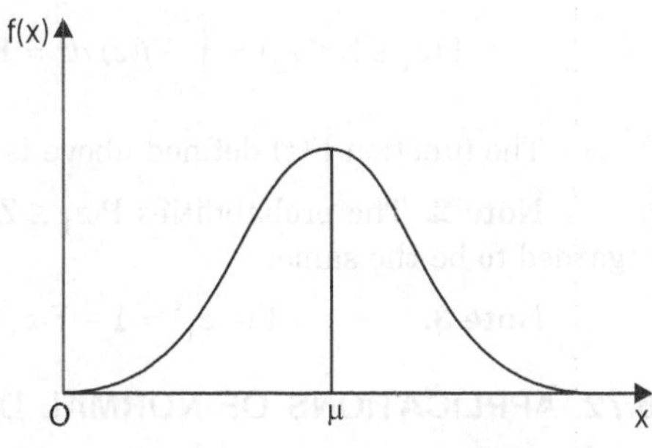

5.70. BASIC PROPERTIES OF THE NORMAL DISTRIBUTION

The probability density function of the normal distribution is given by

$$f(x) = \frac{1}{\sigma\sqrt{2\pi}} e^{-\frac{1}{2}\left(\frac{x-\mu}{\sigma}\right)^2}$$

(*i*) $f(x) \geq 0$ (*ii*) $\int_{-\infty}^{\infty} f(x)\, dx = 1$,

i.e., the total area under the normal curve above the x-axis is 1.

(*iii*) The normal distribution is symmetrical about its mean.

(*iv*) It is a unimodal distribution. The mean, mode and median of this distribution coincide.

5.71. STANDARD FORM OF THE NORMAL DISTRIBUTION

If X is a normal random variable with mean μ and standard deviation σ, then the random variable $Z = \frac{X - \mu}{\sigma}$ has the normal distribution with mean 0 and standard deviation 1. The random variable Z is called the *standardized* (or *standard*) normal random variable.

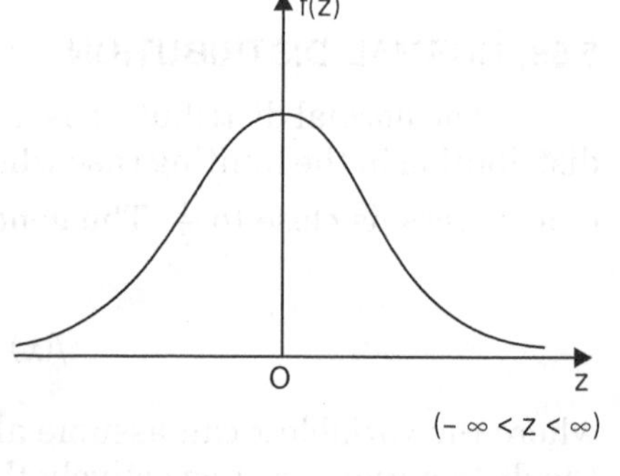

The probability density function for the normal distribution in standard form is given by

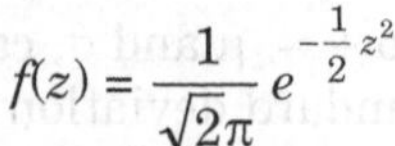

$$f(z) = \frac{1}{\sqrt{2\pi}} e^{-\frac{1}{2}z^2}$$

It is free from any parameter. This helps us to compute areas under the normal probability curve by making use of standard tables.

Note 1. If $f(z)$ is the probability density function for the normal distribution, then

$$P(z_1 \leq Z \leq z_2) = \int_{z_1}^{z_2} f(z)\, dz = F(z_2) - F(z_1), \quad \text{where} \quad F(z) = \int_{-\infty}^{z} f(z)\, dz = P(Z \leq z)$$

The function F(z) defined above is called the *distribution function* for the normal distribution.

Note 2. The probabilities $P(z_1 \leq Z \leq z_2)$, $P(z_1 < Z \leq z_2)$, $P(z_1 \leq Z < z_2)$ and $P(z_1 < Z < z_2)$ are all regarded to be the same.

Note 3. $F(-z_1) = 1 - F(z_1)$.

5.72. APPLICATIONS OF NORMAL DISTRIBUTION

Normal distribution is commonly applied to problems concerning:

(*i*) Statistical inference in almost every branch of science.

(*ii*) Calculation of errors made by chance in experimental measurements.

(*iii*) Computation of hit probability of a shot.

ILLUSTRATIVE EXAMPLES

Example 1. *A sample of 100 dry battery cells tested to find the length of life produced the following results:*

$$\bar{x} = 12 \text{ hours}, \sigma = 3 \text{ hours}.$$

Assuming the data to be normally distributed, what percentage of battery cells are expected to have life

(i) more than 15 hours *(ii) less than 6 hours*

(iii) between 10 and 14 hours?

Sol. Here x denotes the length of life of dry battery cells.

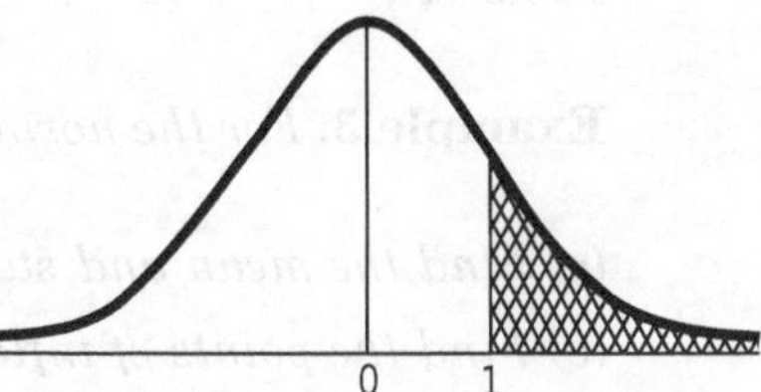

Also $$z = \frac{x - \bar{x}}{\sigma} = \frac{x - 12}{3}.$$

(*i*) When $x = 15, z = 1$

$$\therefore \quad P(x > 15) = P(z > 1)$$
$$= P(0 < z < \infty) - P(0 < z < 1)$$
$$= 0.5 - 0.3413 = 0.1587 = 15.87\%.$$

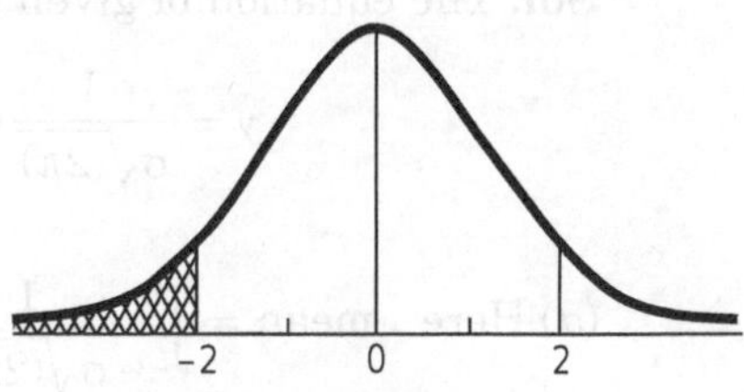

(*ii*) When $x = 6, \ z = -2$

$$\therefore \ P(x < 6) = P(z < -2)$$
$$= P(z > 2) = P(0 < z < \infty) - P(0 < z < 2)$$
$$= 0.5 - 0.4772 = 0.0228 = 2.28\%.$$

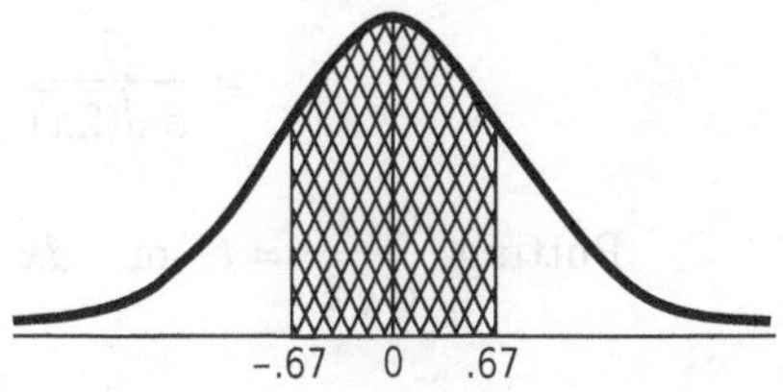

(*iii*) When $x = 10, z = -\frac{2}{3} = -0.67$

When $x = 14, z = \frac{2}{3} = 0.67$

$$P(10 < x < 14)$$
$$= P(-0.67 < z < 0.67)$$
$$= 2P(0 < z < 0.67) = 2 \times 0.2487$$
$$= 0.4974 = 49.74\%.$$

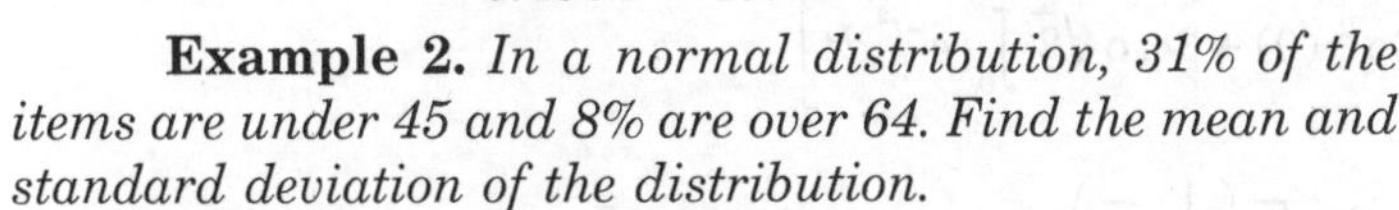

Example 2. *In a normal distribution, 31% of the items are under 45 and 8% are over 64. Find the mean and standard deviation of the distribution.*

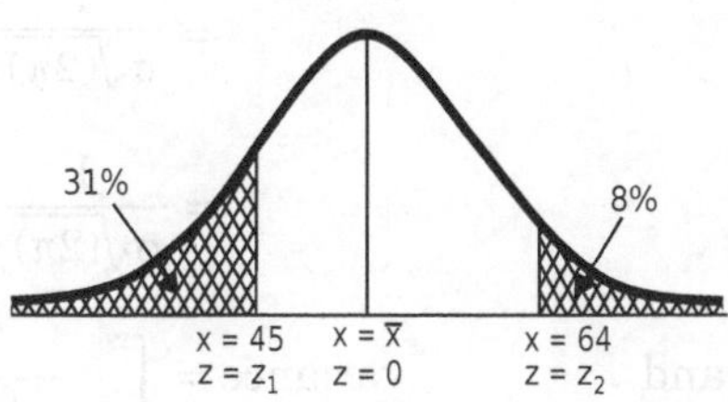

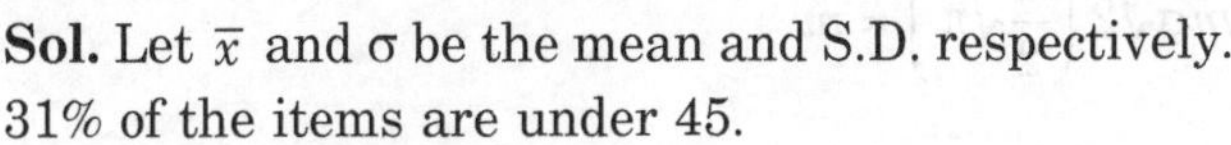

Sol. Let $\bar{x}$ and σ be the mean and S.D. respectively.

31% of the items are under 45.

$\Rightarrow$ Area to the left of the ordinate $x = 45$ is 0.31

When $x = 45$, let $z = z_1$

$$P(z_1 < z < 0) = 0.5 - 0.31 = 0.19$$

From the tables, the value of z corresponding to this area is 0.5

$$\therefore \quad z_1 = -0.5[z_1 < 0]$$

When $x = 64$, let $z = z_2$

$$P(0 < z < z_2) = 0.5 - 0.08 = 0.42$$

From the tables, the value of z corresponding to this area is 1.4.

$$z_2 = 1.4$$

Since $z = \dfrac{x - \bar{x}}{\sigma}$

$$-0.5 = \frac{45 - \bar{x}}{\sigma} \quad \text{and} \quad 1.4 = \frac{64 - \bar{x}}{\sigma}$$

$\Rightarrow$ $45 - \bar{x} = -0.5\sigma$ ...(1)

and $64 - \bar{x} = 1.4\sigma$ ...(2)

Subtracting $-19 = -1.9\sigma \quad \therefore \quad \sigma = 10$

From (1), $45 - \bar{x} = -0.5 \times 10 = -5 \quad \therefore \quad \bar{x} = 50.$

Example 3. *For the normal curve* $y = \dfrac{1}{\sigma\sqrt{(2\pi)}} e^{-(x-m)^2/2\sigma^2}$

(a) Find the mean and standard deviation.

(b) Find the points of inflexion.

Sol. The equation of given normal curve is as follows:

$$y = \frac{1}{\sigma\sqrt{(2\pi)}} e^{-(x-m)^2/2\sigma^2} \quad \text{...(1)}$$

(*a*) Here mean $= \displaystyle\int_{-\infty}^{\infty} \frac{1}{\sigma\sqrt{(2\pi)}} e^{-\frac{(x-m)^2}{2\sigma^2}} x\,dx$

$$= \frac{1}{\sigma\sqrt{(2\pi)}} - \int_{-\infty}^{\infty} e^{-t^2} (t\sigma\sqrt{2} + m)\,\sigma\sqrt{2}\,dt.$$

Putting $\dfrac{x-m}{\sigma\sqrt{2}} = t$ or $dx = \sigma\sqrt{2}\,dt$

$$= \frac{1}{\sigma\sqrt{(2\pi)}} \left[2\sigma^2 \int_{-\infty}^{\infty} t e^{-t^2} dt + m\sigma\sqrt{2} \int_{-\infty}^{\infty} e^{-t^2} dt \right]$$

$$= \frac{1}{\sigma\sqrt{(2\pi)}} \left[2\sigma^2 (0) + 3m\sigma\sqrt{2} \int_{0}^{\infty} e^{-t^2} dt \right]$$

$$= \frac{1}{\sigma\sqrt{(2\pi)}} \cdot 2m\sigma\sqrt{2} \left(\frac{1}{2}\sqrt{\pi} \right) = m$$

and variance $= \displaystyle\int_{-\infty}^{\infty} \frac{1}{\sigma\sqrt{(2\pi)}} e^{-(x-m)^2/2\sigma^2} (x-m)^2\,dx.$

Putting $\dfrac{x-m}{\sigma\sqrt{2}} = t$ or $dx = \sigma\sqrt{2}\,dt$

$$= \frac{1}{\sigma\sqrt{(2\pi)}} \int_{-\infty}^{\infty} e^{-t^2} (t\sigma\sqrt{2})^2\,\sigma\sqrt{2}\,dt$$

$$= \frac{1}{\sigma\sqrt{(2\pi)}} \sigma^3 \cdot \sqrt{2} \int_{-\infty}^{\infty} t^2 e^{-t^2} dt = \frac{2\sigma^2}{\sqrt{\pi}} \int_{-\infty}^{\infty} t \cdot (te^{-t^2})\,dt$$

$$= \frac{2\sigma^2}{\pi}\left[\left\{t.\left(-\frac{1}{2}e^{-t^2}\right)\right\}_{-\infty}^{\infty} - \int_{-\infty}^{\infty}\left(-\frac{1}{2}e^{-t^2}\right)dt\right]$$

[Integrating by parts, we have]

$$= \frac{2\sigma^2}{\sqrt{\pi}}\left[\{0\} + \frac{1}{2}\int_{-\infty}^{\infty} e^{-t^2}\, dt\right] = \frac{2\sigma^2}{\sqrt{\pi}}\left(\frac{1}{2}\sqrt{\pi}\right) = \sigma^2$$

$\therefore$ Standard deviation $= \sqrt{\sigma^2} = \sigma$.

(*b*) We know that, at the point of inflexion

$$d^2y/dx^2 = 0 \quad \text{and} \quad d^3y/dx^3 \neq 0.$$

Differentiating (1) two times and equating to zero, we have

$$d^2y/dx^2 = 0$$

$$\Rightarrow \quad \frac{1}{\sigma\sqrt{(2\pi)}}\left[\left(-\frac{x-m}{\sigma^2}\right)^2 e^{-(x-m)^2/2\sigma^2} - \frac{1}{\sigma^2}e^{-(x-m)^2/2\sigma^2}\right] = 0$$

$$\Rightarrow \quad \left(\frac{x-m}{\sigma^2}\right)^2 - 1 = 0 \;\Rightarrow\; x = m \pm \sigma.$$

Again $$\frac{d^3y}{dx^3} = \frac{1}{\sigma\sqrt{(2\pi)}}\left[\left\{-\frac{x-m}{\sigma^2}\right\}^3 e^{-(x-m)^2/2\sigma^2} + \frac{3(x-m)}{\sigma^4}e^{-(x-m)^2/2\sigma^2}\right]$$

$$= \pm\frac{2}{\sigma^4\sqrt{(2\pi)}}e^{-1/2}, \text{ on writing } m = x \pm \sigma \neq 0.$$

$\therefore$ When $x = m \pm \sigma$, $y = \dfrac{1}{\sigma\sqrt{(2\pi)}}e^{-1/2}$.

$\therefore$ The points of inflexion are $\left(m \pm \sigma, \dfrac{1}{\sigma\sqrt{(2\pi)}}e^{-1/2}\right)$, which are equidistant from the mean.

Example 4. *Assume the mean height of soldiers to be 68.22 inches with a variance of 10.8 (in)^2. How many soldiers in a regiment of 1000 would you expect to be over 6 feet tall? Given that the area under the standard normal curve between t = 0 and t = 0.35 is 0.1368, and between t = 0 and t = 1.15 is 0.3746.*

Sol. Here $m = 68.22$ inches, $\sigma^2 = 10.8$ (inch)^2 *i.e.,*

$$\sigma = \sqrt{10.8} \text{ inch} = 3.28 \text{ inches}.$$

$$\therefore \quad t = \frac{x-m}{\sigma} = \frac{72 - 68.22}{3.28} \text{ for } x = 6' = 72''$$

$$= 3.78/3.28 = 1.15 \text{ (nearly)}.$$

But it is given that the area between $t = 0$ and $t = 1.15$ is 0.3746. Thus the area between $t = 1.15$ and $t = \infty$ is $0.5 - 0.3746$ *i.e.,* 0.1254.

$\therefore$ The number of soldiers who are over 6 feet tall $= 1000 \times 0.1254 = 125.4$ nearly.

Example 5. *For a normal distribution with mean 2 and standard deviation 3, find the value of a variate such that the probability of the interval from the mean to that value is 0.4115.*

Sol. Given $\dfrac{2}{3\sqrt{(2\pi)}}\displaystyle\int_2^x e^{-\frac{1}{2}\frac{(x-2)^2}{9}}\,dx = 0.4115.$

Now substituting $t = (x - 2)/3$ or $dt = \dfrac{1}{3}dx$

$$\frac{1}{\sqrt{(2\pi)}}\int_0^t e^{-\frac{1}{2}t^2}\,dt = 0.4115.$$

But from table, $t = 1.35$, hence $x = 3t + 2 = 6.05$.

Example 6. *A random variable x is normally distributed with mean = 12 and standard deviation 2, find P(9.6 < x < 13.8) given that for $x/\sigma = 0.9$, A = 0.3159 and for $x/\sigma = 1.2$, A = 0.3849.*

Sol. We are given that

mean = m = 12 and 6 = 2.

$$\therefore \quad t = \frac{x-m}{\sigma} = \frac{9.6-12}{2} = \frac{-2.4}{2} = -1.2, \text{ when } x = 9.6$$

and $$t = \frac{13.8-12}{12} = \frac{1.8}{2} = 0.9, \text{ when } x = 13.8$$

$\therefore$ The required probability = p (9.6 < x < 13.8)

$$= \int_{9.6}^{13.8} \frac{1}{2\sqrt{(2\pi)}} e^{-\frac{1}{2}\frac{(x-12)^2}{4}}\,dx$$

$$= \frac{1}{\sqrt{(2\pi)}}\int_{-1.2}^{0.9} e^{-\frac{1}{2}t^2}\,dt \quad \text{where} \quad t = \frac{x-12}{2}$$

$$= \frac{1}{\sqrt{(2\pi)}}\int_{-1.2}^{0} e^{-\frac{1}{2}t^2}\,dt + \frac{1}{\sqrt{(2\pi)}}\int_0^{0.9} e^{-\frac{1}{2}t^2}\,dt$$

$$= \frac{1}{\sqrt{(2\pi)}}\int_0^{1.2} e^{-\frac{1}{2}t^2}\,dt + \frac{1}{\sqrt{(2\pi)}}\int_0^{0.9} e^{-\frac{1}{2}t^2}\,dt$$

$$\left[\because \text{ under the normal curve}: \int_{-1.2}^{0}\phi(t)\,dt = \int_0^{1.2}\phi(t)\,dt, \text{ where } \phi(t) = \frac{1}{\sqrt{2\pi}}e^{-\frac{1}{2}t^2}\right]$$

$$= 0.3849 + 0.3159 = 0.7008.$$

Example 7. *In a normal distribution, 31% of the items are under 45 and 8% are over 64. Find the mean and S.D. of the distribution.*

Sol. Let $\bar{x}$ and σ denote the mean and S.D. of the distribution respectively. 31% of the items are 9 under 45 means are to the left of the ordinate $x = 45$.

When $x = 45$, let $t = t_1$ so that

$$t_1 = \frac{45 - \bar{x}}{\sigma} \quad ...(1)$$

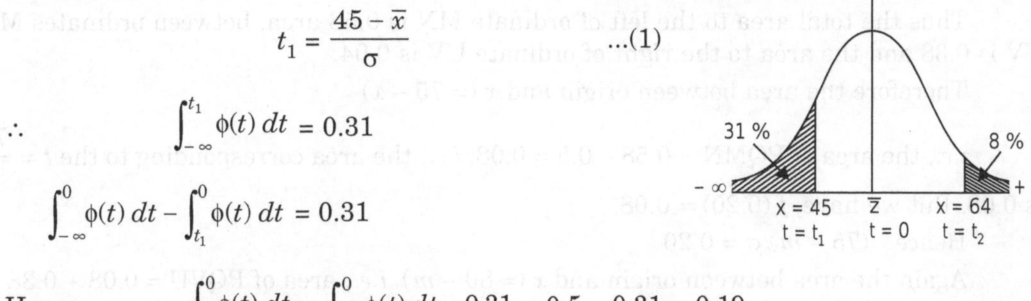

$$\therefore \quad \int_{-\infty}^{t_1} \phi(t)\, dt = 0.31$$

or $$\int_{-\infty}^{0} \phi(t)\, dt - \int_{t_1}^{0} \phi(t)\, dt = 0.31$$

Hence $$\int_{t_1}^{0} \phi(t)\, dt = \int_{-\infty}^{0} \phi(t)\, dt - 0.31 = 0.5 - 0.31 = 0.19$$

[∵ Under the normal curve, we have

$$\int_{-\infty}^{0} \phi(t)\, dt = \int_{0}^{\infty} \phi(t)\, dt = 0.5, \text{where } \phi(t) = \frac{1}{\sqrt{2\pi}} e^{-(1/2)t^2}\Bigg]$$

From the table of area under normal curve,

$$t_1 = -0.5 \quad ...(2)$$

When $x = 64$, let $t = t_2$ so that $t_2 = \dfrac{64 - \bar{x}}{\sigma}$. ...(3)

$$\therefore \quad \int_{t_2}^{\infty} \phi(t)\, dt = 0.08 \quad \text{or} \quad \int_{0}^{\infty} \phi(t)\, dt - \int_{0}^{t_2} \phi(t)\, dt = 0.08.$$

Hence $$\int_{0}^{t_2} \phi(t)\, dt = \int_{0}^{\infty} \phi(t) - 0.08 = 0.5 - 0.08 = 0.42.$$

From the table of area under normal curve,

$$t_2 = 1.4. \quad ...(4)$$

From (1) and (2), we have

$$45 - \bar{x} = -0.5\,\sigma$$

From (3) and (4), we have

$$64 - \bar{x} = 1.4\sigma.$$

Solving these equations, we get $\bar{x} = 50$ and $\sigma = 10$.

Example 8. *If skills are classified A, B, C according as the length, breadth index as under 75, between 75 and 80, or over 80, find approximately (assuming that distribution is normal) the mean and standard deviation of a series in which A are 58%, B = 38% and C are 4% being given that*

$$f(t) = \frac{1}{\sqrt{(2\pi)}} \int_0^t exp\left(-\frac{1}{2}x^2\right) dx$$

then $\quad f(0.20) = 0.80$ *and* $f(1.75) = 0.46.$

Sol. Consider that the mean and standard deviation of the distribution are m and σ respectively. Since the total frequency is taken as unity, thus frequency of skill A, whose length and breadth index is under 75 is 0.58, the frequency of skill B whose index lies between 75 and 80 is 0.38; and frequency of skill C, whose index is over 80 is 0.04.

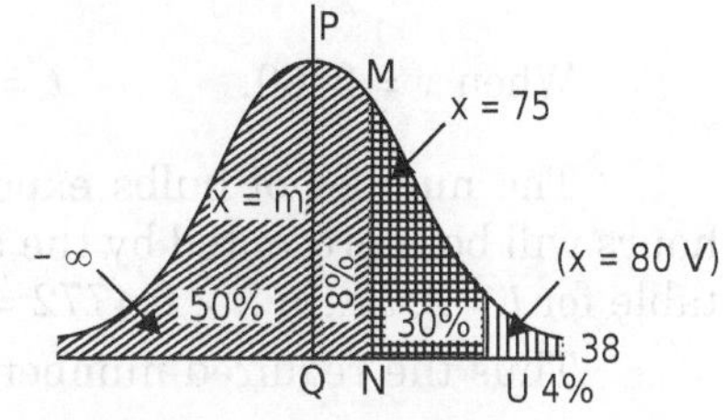

Thus the total area to the left of ordinate MN is 0.58 area, between ordinates MN and UV is 0.38 and the area to the right of ordinate UV is 0.04.

Therefore the area between origin and x (= 75 – x)

i.e., the area of PQMN = 0.58 – 0.5 = 0.08, *i.e.*, the area corresponding to the $t = \dfrac{75 - m}{\sigma}$ is 0.08. But we have, $f(0.20) = 0.08$.

Hence $\quad (75 - m)/\sigma = 0.20$ ...(1)

Again the area between origin and x (= 80 – m) *i.e.*, area of PQVU = 0.08 + 0.38 = 0.46 *i.e.*, the area corresponding to t (= 80 – m)/σ is 0.46. But $f(1.75) = 0.46$.

Hence $\quad (80 - m)/\sigma = 1.75.$ ...(2)

Dividing (1) by (2) and solving, we get

$$m = 74.3, \sigma = 3.23.$$

Example 9. *In a test on 2000 electric bulbs, it was found that the life of a particular make, was normally distributed with an average life of 2040 hours and S.D. of 60 hours. Estimate the number of bulbs likely to burn for*

(a) more than 2150 hours,

(b) less than 1950 hours and

(c) more than 1920 hours and but less than 2160 hours.

Sol. Here mean μ = 2040 hours and S.D. = σ = 60 hours

(*a*) For $x = 2150$, $\quad t = \dfrac{x - \mu}{\sigma} = \dfrac{2150 - 2040}{60} = 1.833$

∴ Area against $t = 1.83$ in the table of area under normal curve = 0.4664.

In this case, we require the area to the right of the ordinate at $z = 1.83$. So, the required area

$$= 0.5 - 0.4664 = 0.0336.$$

Thus the number of bulbs expected to burn for more than 2150 hours

$$= 0.0336 \times 2000 = 67 \text{ approx.}$$

(*b*) For $x = 1950$, $\quad t = \dfrac{x - \mu}{\sigma} = \dfrac{1950 - 2040}{60} = 1.5.$

In this case, the required area is to the left of $t = -1.33$

$$= 0.5 - 0.4082 \text{ (table value for } t = 1.33)$$
$$= 0.0918.$$

∴ The number of bulbs expected to burn for less than 1950 hours

$$= 0.0918 \times 2000 = 184 \text{ approx.}$$

(*c*) When $x = 1920$, $\quad t = \dfrac{x - \mu}{\sigma} = \dfrac{1920 - 2040}{60} = -2$

When $x = 2160$, $\quad t = \dfrac{x - \mu}{\sigma} = \dfrac{2160 - 2040}{60} = 2.$

The number of bulbs expected to burn for more than 1920 hours but less than 2160 hours will be represented by the area between $t = -2$ and $t = 3$. This is twice the area from the table for $t = 2$; *i.e.*, = 2 × 0.4772 = 0.9544.

Thus the required number of bulbs = 0.9544 × 2000 = 1909 nearly.

Example 10. *Fit a normal curve to the following data:*

length of line (in cm):	*8.60*	*8.59*	*8.58*	*8.57*	*8.56*	*8.55*	*8.54*	*8.53*	*8.52*
frequency:	*2*	*3*	*4*	*9*	*10*	*8*	*4*	*1*	*1*

Sol. Let the required normal curve be:

$$y = \frac{N}{\sigma\sqrt{(2\pi)}} e^{-\frac{(x-m)^2}{2\sigma^2}}.$$

Here $N = 42.$

x	f	$\zeta = x - A$	ζ^2	$f\zeta$	$f\zeta^2$
8.60	2	0.04	0.0016	0.08	0.0032
8.59	3	0.03	0.0009	0.09	0.0027
8.58	4	0.02	0.0004	0.08	0.0016
8.57	9	0.01	0.0001	0.09	0.0009
8.56	10	0	0	0	0
8.55	8	− 0.01	0.0001	− 0.08	0.0008
8.54	4	− 0.02	0.0004	− 0.08	0.0016
8.53	1	− 0.03	0.0009	− 0.03	0.0009
8.52	1	− 0.04	0.0016	− 0.04	0.0016
Total	42		0.0060	0.11	0.0133

$$\therefore \quad \text{mean } (m) = A + \frac{\Sigma f\zeta}{\Sigma f} = 8.56 + \frac{0.11}{42}$$

$$= 8.56 + 0.0026 = 8.5626 = 8.563 \quad \text{(nearly)}$$

and

$$\text{S. D. } (\sigma) = \sqrt{\left[\left\{\frac{\Sigma f\zeta^2}{\Sigma f} - \left(\frac{\Sigma f\zeta}{\Sigma f}\right)^2\right\}\right]} = \sqrt{\left[\left\{\frac{0.0133}{42} - \left(\frac{0.11}{42}\right)^2\right\}\right]}$$

$$= \sqrt{[\{0.000317 - 0.0000069\}]} = 0.0175 \text{ cm. (nearly)}$$

$$= 1.75 \text{ class intervals.}$$

Hence on simplification, the equation of normal curve to be fitted, is

$$y = 9.8\, e^{-0.163\,(x - 8.563)^2}.$$

Example 11. *Fit a normal curve to the following distribution:*

x:	2	4	6	8	10
F:	*1*	4	6	4	*1*

Sol. For the given distribution, we have

$$\text{Mean} = \frac{\Sigma fx}{\Sigma f} = \frac{2 + 16 + 36 + 32 + 10}{16} = \frac{96}{16} = 6$$

$$\text{S.D.} = \sqrt{\frac{\Sigma fx^2}{\Sigma f} - \left(\frac{\Sigma fx}{\Sigma f}\right)^2} = \sqrt{40 - 36} = 2.$$

Taking $\mu = 6$, $\sigma = 2$ and N = 16, the equation of normal curve is

$$y = \frac{1}{\sigma\sqrt{2\pi}} e^{-(x-\mu)^2/2\sigma^2} \quad \text{or} \quad y = \frac{1}{2\sqrt{2\pi}} e^{-(x-6)^2/8}. \quad ...(1)$$

Area under the normal curve (1) in (x_1, x_2) or (t_1, t_2)

$$= \frac{1}{\sqrt{2\pi}} \int_{t_1}^{t_2} e^{-t^2/2}\, dt \quad \text{where} \quad t = \frac{x-6}{2}$$

$$= \frac{1}{\sqrt{2\pi}} \int_{0}^{t_2} e^{-t^2/2}\, dt - \frac{1}{\sqrt{2\pi}} \int_{0}^{t_1} e^{-t^2/2}\, dt.$$

To evaluate these integrals, we have the following calculations:

Mid-value of x	(x_1, x_2)	(t_1, t_2)	*Area under curve (1) in* (t_1, t_2)	*Expected frequency*
2	(1, 3)	(– 2.5, – 1.5)	0.4938 – 0.4332	16 × 0.606 = 0.97
4	(3, 5)	(– 1.5, – 0.5)	0.4332 + 0.1915	16 × 0.2417 = 3.9
6	(5, 7)	(– 0.5, – 0.5)	0.1915 + 0.1915	16 × 0.383 = 6.1
8	(7, 9)	(0.5, – 1.5)	0.4332 – 0.1915	16 × 0.2417 = 3.9
10	(9, 11)	(1.5, – 2.5)	0.4938 – 0.4332	16 × 0.606 = 0.97

Hence the expected (or theoretical) frequencies corrected to nearest integer are 1, 4, 6, 4, 1 which agree with the observed frequencies. This shows that the normal curve (1) is a proper fit to the given distribution.

5.73. MOMENT GENERATING FUNCTION

Definition. *A moment generating function* (in brief m.g.f.) for a continuous variable x is denoted by M(t) and is defined by

$$M(t) = \int_a^b e^{tx} f(x)\, dx \quad ...(1)$$

where the integral is a function of the parameter t only. The limits a to b of integration can extend from – ∞ to ∞. The function M(t) generates moments and that is why it is called the moment generating function.

To show that M(t) generates functions.

If the integral (1) exists and the expansion is possible, then (1) may be written as

$$M(t) = \int_a^b \left(1 + tx + \frac{t^2 x^2}{2!} + ... + \frac{t^r x^r}{2!} + ...\right) f(x)\, dx$$

$$= \int_a^b f(x) dx + t \int_a^b x f(x)\, dx + \frac{t^2}{2!} \int_a^b x^2 f(x)\, dx + ... + \frac{t^r}{r!} \int_a^b x^r f(x)\, dx + ...$$

$$= \mu_0' + \mu_1' t + \mu_2' \frac{t^2}{2!} + ... + \mu_r' \frac{t^r}{r!} + ... \quad ...(2)$$

The co-efficient of $t^r/r!$ in relation (2) is μ_r' which is the rth moment about the origin. Hence the function M(t) generates moments with the condition that the integral exists and the expansion is possible.

5.74. M.G.F. FOR A DISCRETE FREQUENCY DISTRIBUTION

Definition. The m.g.f. of the discrete probability distribution of a random variable x about a point a, denoted by $M_a(t)$, is defined as

$$M_a(t) = E[e^{t(x-a)}] = \sum_i e^{t(x_i - a)} p_i \qquad ...(1)$$

where t is a parameter which takes only real values and p_i is the probability corresponding to the value x_i of x.

Now using exponential expansion (1) may be written as

$$M_a(t) = \sum_i \left[1 + \frac{t(x_i - a)}{1} + \frac{t^2 (x - a)^2}{2!} + \ldots + \frac{t^r (x_i - a)^r}{r!} + \ldots\right] p_i$$

$$= \sum_i p_i + t \sum_i p_i (x_i - a) + \frac{t^2}{2!} \sum p_i (x_i - a)^2 + \ldots + \frac{t^r}{r!} \sum_i p_i (x_i - a)^r + \ldots$$

$$= 1 + t\mu_1' + \frac{t^2}{2!} \mu_2' + \ldots + \frac{t^r}{r!} \mu_r' + \ldots \qquad ...(2)$$

where μ_r' = coeff. of $(t^r / r!)$ in the expansion of $M_a(t)$ when expanded in ascending power of

$$t = \sum_i p_i (x_i - a)^r$$

= Moment of order r about the point a.

5.75. PROPERTIES OF M. G. F.

Prop. 1. If u is a function of x and $z = cu$, then we have

$$M_z(t) = \int_a^b e^{tz} f(x)\,dx = \int_a^b e^{tcu} f(x)\,dx \qquad [\because \; z = cu]$$

$$= \int_a^b e^{(ct)u} f(x)\,dx = M_u(ct).$$

Prop. 2. If u is a function of x and $z = u + c$, then we have

$$M_z(t) = \int_a^b e^{tz} f(x)\,dx = \int_a^b e^{t(u+c)}\,dx$$

$$= e^{tc} \int_a^b e^{tu} f(x)\,dx = e^{ct} M_u(t).$$

Prop. 3. *The moment generating function of the sum of two independent variables is the product of their moment generating functions.*

Proof. Consider two independent variables x and y. The m.g.f. of the sum $x + y$ about the origin

$$M_{x+y}(t) = E[e^{t(x+y)}] = E[e^{tx} \cdot e^{ty}]$$

$$= E(e^{tx}) \cdot E(e^{ty}) \qquad \text{[as } x \text{ and } y \text{ are independent variables]}$$

The above property can be extended for n independent variables as follows:

$$E[e^{t(x_1 + x_2 + \ldots + x^n)}] = E(e^{tx_1}) \cdot E(e^{tx_2}) \ldots E(e^{tx_n}).$$

$$M_{x_1 + x_2 + \ldots + x_n}(t) = M_{x_1}(t) \cdot M_{x_2}(t) \ldots M_{x_n}(t).$$

5.76. M.G.F. FOR BINOMIAL DISTRIBUTION

We know that for a binomial distribution, the frequencies of 'x' success = ${}^nC_x p^x q^{n-x}$ consequently, the m.g.f. about the origin is given by

$$M_0(t) = \sum_{x=0}^{n} e^{txn} C_x p^x q^{n-x} = \sum_{x=0}^{n} {}^nC_x (pe^t)^x q^{n-x} = (q + pe^t)^n \quad ...(1)$$

$$= \left[q + p\left(1 + t + \frac{t^2}{2!} + \frac{t^3}{3!} + ...\right)\right]^n$$

$$M_0(t) = \left[(q+p) + pt + \frac{pt^2}{2!} + \frac{pt^3}{3!} + ...\right]^n$$

$$= \left[1 + pt + \frac{pt^2}{2!} + \frac{pt^3}{3!} + ...\right]^n \quad [\because \; q + p = 1]$$

Also, for binomial distribution we know that mean $m = np$ and $M_a(t) = e^{-at} M_0(t)$, where $M_a(t)$ is the m.g.f. about the point a.

$\therefore$ The m.g.f. $M_0(t)$ about the mean is given by

$$M_m(t) = e^{-mt} M_0(t), \text{ where } m = np$$
$$= e^{-npt} (q + pe^t)^n = [e^{-pt} (q + pe^t)]^n$$
$$= [qe^{-pt} + pe^{(1-p)t}]^n, \text{ where } p + q = 1$$
$$= [qe^{-pt} + pe^{qt}]^n.$$

Generation of moments from m.g.f.

We know that the moment of order r about the origin is given by

$$\mu'_r = \left[\frac{d^r}{dt^r}\{M(t)\}\right]_{t=0}$$

Now substituting for $M_0(t)$ from (1), we get

$$\mu'_r = \left[\frac{d^r}{dt^r}\{(q + qe^t)^n\}\right]_{t=0} \quad ...(2)$$

Putting $r = 1$ in (2), we get

$$\mu'_1 = \left[\frac{d}{dt}(q + pe^t)^n\right]_{t=0}$$
$$= [npe^t (q + pe^t)^{n-1}]_{t=0} = np\,(q + p)^{n-1} = np\,(1)^{n-1} \quad [\text{since } q + p = 1]$$
$$= np.$$

Hence the moment of distribution = $\mu'_1 = np$.

Putting $r = 2$ in (2), we get

$$\mu'_2 = \left[\frac{d^2}{dt^2}(q + pe^t)^n\right]_{t=0} = \left[\frac{d}{dt}\{npe^t (q + pe^t)^{n-1}\}\right]_{t=0}$$

$$= [npe^t (n-1) pe^t (q + pe^t)^{n-2} + npe^t (q + pe^t)^{n-1}]_{t=0}$$

$$= n(n-1) p^2 (q + p) + np (q + p), \text{ where } q + p = 1$$

$$= np [(n-1) p + 1] = np [np + (1 - p)]$$

$$= np(np + q) = n^2p^2 + npq.$$

$$\therefore \quad \mu_2 = \mu_2' - \mu_1'^2 = (n^2p^2 + npq) - (np)^2$$

$$= npq = \textbf{variance.}$$

Standard deviation (S. D.) $= \sigma = \sqrt{\mu_2} = \sqrt{npq}$.

Similarly μ_3, μ_4 etc. can be obtained.

5.77. M.G.F. FOR POISSON'S DISTRIBUTION

We know that for a poisson distribution, the frequencies for 'x' successes

$$= \frac{e^{-m} m^x}{x!}.$$

Consequently the m.g.f. about the origin is given by

$$M_0(t) = \sum_{x=0}^{\infty} e^{tx} \left(\frac{e^{-m} m^x}{x!} \right)$$

$$= e^{-m} \sum_{x=0}^{\infty} \frac{(me^t)^x}{x!} = e^{-m} e^{me^t} \qquad \left[\because \; e^{\theta} = \sum_{r=0}^{\infty} \left(\frac{\theta^r}{r!} \right) \right]$$

or $$M_0(t) = e^{m(e^t - 1)}.$$

Also, we know that the mean of this distribution is m, and

$$M_a(t) = e^{-at} M_0(t)$$

where $M_a(t)$ is the m.g.f. about the point 'a'.

$\therefore$ m.g.f. about the mean m is given by

$$M_m(t) = e^{-mt} M_0(t) = e^{-mt} e^{m(e^t - 1)} \qquad \text{[Using (1)]}$$

$$= e^{m(e^t - 1 - t)} \qquad ...(2)$$

Generation of moments from moment generation function

We know that the moment of order r about the origin is given by

$$\mu_r' = \left[\frac{d^r}{dt^r} \{M_0(t)\} \right]_{t=0} \qquad \text{[Using (1)]}$$

or $$\mu_r' = \left[\frac{d^r}{dt^r} \{e^{m(e^t - 1)}\} \right]_{t=0} \qquad ...(3)$$

Put $r = 1$, in (3), we get

$$\mu_1' = \left[\frac{d}{dt} \{e^{m(e^t - 1)}\} \right]_{t=0} = \left[me^t e^{m(e^t - 1)} \right]_{t=0} = m.$$

i.e., Poisson's mean is m.

Putting $r = 2$ in (3), we get

$$\mu_2' = \left[\frac{d^2}{dt}\{e^m(e^t-1)\}\right]_{t=0} = \left[\frac{d}{dt}\{me^t e^{m(e^t-1)}\}\right]_{t=0}$$

$$= [me^t e^{m(e^t-1)} + me^t\, me^t e^{m(e^t-1)}]_{t=0}$$

or $\qquad \mu_2' = m + m^2.$

$\therefore \qquad \mu_2 = \mu_2' - (\mu_1')^2 = (m + m^2) - (m^2) = m = \text{variance}.$

$\therefore \qquad \text{S.D.} = \sqrt{\mu_2} = \sqrt{m}.$

5.78. MOMENTS OF NORMAL DISTRIBUTION

We know that, in normal distribution the probability function (when mean is at origin) is given by

$$f(x) = \frac{1}{\sigma\sqrt{(2\pi)}} e^{-x^2/2\sigma^2}.$$

$$\therefore \quad M_0(t) = \frac{1}{\sigma\sqrt{(2\pi)}}\int_{-\infty}^{\infty} e^{tx}.e^{-x^2/2\sigma}\,dx$$

$$= \frac{1}{\sigma\sqrt{(2\pi)}}\int_{-\infty}^{\infty}\left[\exp.\left(tx - \frac{x^2}{2\sigma^2}\right)\right]dx$$

$$= \frac{1}{\sigma\sqrt{(2\pi)}}\int_{-\infty}^{\infty}\left[\exp.\left\{\frac{-(x-t\sigma^2)^2}{2\sigma^2} + \frac{1}{2}t^2\sigma^2\right\}\right]dx$$

$$= \frac{1}{\sigma\sqrt{(2\pi)}} e^{t^2\sigma^2/2}\int_{-\infty}^{\infty}\left[\exp.\left\{-\frac{(x-t\sigma^2)^2}{2\sigma^2}\right\}\right]dx$$

$$= \frac{1}{\sigma\sqrt{(2\pi)}} e^{t^2\sigma^2/2}\int_{-\infty}^{\infty} e^{-z^2}.\sqrt{2}\,dz \qquad \left[\text{Putting } \frac{x-t\sigma^2}{\sigma\sqrt{2}} = z\right]$$

$$= \frac{1}{\sqrt{\pi}} e^{t^2\sigma^2/2}\int_{-\infty}^{\infty} e^{-z^2}\,dz = e^{t^2\sigma^2/2} \qquad \left[\because \int_{-\infty}^{\infty} e^{-z^2}\,dz = \sqrt{\pi}\right]$$

or
$$M_0(t) = e^{t^2\sigma^2/2} = 1 + \left(\frac{t^2\sigma^2}{2}\right) + \frac{1}{2!}\left(\frac{t^2\sigma^2}{2}\right)^2 + \ldots + \frac{1}{n!}\left(\frac{t^2\sigma^2}{2}\right)^n \qquad \ldots(1)$$

From eqn. (1), it follows that $M_0(t)$ contains only even powers of t and so the coeff. of t^{2n+1} in $M_0(t)$ is zero.

Hence $\qquad \mu_{2n+1} = 0. \qquad \ldots(2)$

Also $\qquad \mu_{2n} = \text{coeff. of } \frac{t^{2n}}{(2n)!} \text{ in } M_0(t) \text{ in (1)}$

$$= \left(\frac{1}{2}\sigma^2\right)^n \frac{(2n)!}{n!} = \frac{1}{2^n}\sigma^{2n} \cdot \frac{(2n)(2n-1)(2n-2)\ldots 3 \,.\, 2 \,.\, 1}{n(n-1)(n-2)\ldots 2 \,.\, 1}$$

$$= \frac{1}{2^n}\sigma^{2n} \cdot \frac{[(2n-1)(2n-3) \,.\, 3 \,.\, 1]\,[(2n)(2n-2)\ldots 4 \,.\, 2]}{n(n-1)(n-2)\ldots 2 \,.\, 1}$$

$$= \frac{1}{2^n}\sigma^{2n} \cdot \frac{[(2n-1)(2n-3)\ldots 3 \,.\, 1]\,[2^n \,.\, n(n-1)\ldots 2 \,.\, 1]}{n(n-1)(n-2)\ldots 2 \,.\, 1}$$

or $\mu_{2n} = 1 \,.\, 3 \,.\, 5 \ldots (2n-1)\,\sigma^{2n}$. ...(3)

In particular,

$$\mu_2 = 1 \,.\, \sigma^2 = \sigma^2; \quad \mu_4 = 1 \,.\, 3\sigma^4 = 3\sigma^4.$$

$$\therefore \quad \beta_2 = \frac{\mu_4}{\mu_2^2} = \frac{3\sigma^4}{(\sigma^2)^2} = 3.$$

If the probability function $f(x)$ is given by $f(x) = \frac{1}{\sigma\sqrt{(2\pi)}} e^{-\frac{(x-m)^2}{2\sigma^2}}$, then

$$M_0(t) = \int_{-\infty}^{\infty} e^t \frac{1}{\sigma\sqrt{(2\pi)}} e^{-\frac{(x-m)^2}{2\sigma^2}} dx$$

$$= \frac{1}{\sigma\sqrt{(2\pi)}} \int_{-\infty}^{\infty} e^{tx} e^{-\frac{(x-m)^2}{2\sigma^2}} dx$$

$$= \frac{1}{\sigma\sqrt{(2\pi)}} \int_{-\infty}^{\infty} e^{r(m+z\sigma)} \,.\, e^{-z^2/2}\, \sigma dz$$

$$\left[\text{Putting } z = \frac{x-m}{\sigma} \text{ or } x = m + z\sigma\right]$$

$$= \frac{e^{mt}}{\sqrt{(2\pi)}} \int_{-\infty}^{\infty} e^{-(1/2)z^2 + t\sigma z} dz$$

$$= \frac{e^{mt}}{\sqrt{(2\pi)}} \int_{-\infty}^{\infty} e^{-(1/2)(z^2 - 2t\sigma z + t^2\sigma^2) + (1/2)t^2\sigma^2} dz$$

$$= \frac{e^{mt + (1/2)t^2\sigma^2}}{\sqrt{(2\pi)}} \int_{-\infty}^{\infty} e^{-(1/2)(z - t\sigma)^2} dz$$

$$= \frac{e^{mt + (1/2)t^2\sigma^2}}{\sqrt{(2\pi)}} \int_{-\infty}^{\infty} e^{-u^2} \sqrt{2}\, du \qquad [\text{Putting } z - t\sigma = u\sqrt{2}\,]$$

$$= e^{mt + (1/2)t^2\sigma^2} \qquad \left[\because \int_{-\infty}^{\infty} e^{-u^2}\, du = \sqrt{\pi}\right]$$

$$\therefore \quad M_0(t) = e^{mt + (1/2)t^2\sigma^2}$$

Moment generation function about the mean m.

We know that

$$M_m(t) = e^{-mt}\,M_0(t) = e^{-mt}\,\{e^{mt + (1/2)t^2\sigma^2}\}$$

$$= e^{(1/2)t^2\sigma^2}$$

Example 12. *Find the moment generating function of the random variable x having the probability function given by*

$$f(x) = \begin{cases} x\ , & when\ 0 \le x < 1 \\ 2 - x, & when\ 1 \le x < 2 \\ 0\ , & otherwise. \end{cases}$$

Also find the mean and variance of x.

Sol. $\because$ $$\int_{-\infty}^{\infty} f(x)\,dx = \int_{-\infty}^{0} 0.dx + \int_{0}^{1} x dx + \int_{1}^{2} (2-x)\,dx + \int_{2}^{\infty} 0.dx$$

$$= \left[\frac{1}{2}x^2\right]_0^1 + \left[2x - \frac{1}{2}x^2\right]_1^2 = \frac{1}{2} + 2 - \frac{3}{2} = 1.$$

hence it is probability density function.

Now $$M_x(t) = E(e^{tx}) = \int_0^1 xe^{tx}\,dx + \int_1^2 (2-x)e^{tx}\,dx$$

$$= \left[\frac{e^t}{t} - \left(\frac{e^t}{t^2} - \frac{1}{t^2}\right)\right] + \left[-\frac{e^t}{t} + \frac{e^{2t}}{t^2} - \frac{e^t}{t^2}\right]$$

$$= \frac{e^{2t}}{t^2} - \frac{2e^t}{t^2} + \frac{1}{t^2} = \frac{1}{t^2}(e^{2t} - 2e^t + 1) \qquad ...(1)$$

or $$M_x(t) = \frac{(e^t - 1)^2}{t^2} \qquad ...(2)$$

which is the required moment generating function

Now writing expansions in (1), we get

$$M_x(t) = \frac{1}{t^2}\left[\left(1 + 2t + \frac{(2t)^2}{2!} + \frac{(2t)^3}{3!} + \frac{(2t)^4}{4!} + ...\right) - 2\left(1 + t + \frac{t^2}{2!} + \frac{t^3}{3!} + \frac{t^4}{4!} + ...\right) + 1\right]$$

$$= \frac{1}{t^2}\left[t^2 + t^3 + \frac{7}{12}t^4 + ...\right] = 1 + t + \frac{7}{12}t^2 + ... \qquad ...(3)$$

Thus from (3), we have

μ_1' = then co-efficient of t in the expansion of $M_n(t)$ = 1 = mean

μ_2' = then co-efficient of $t^2/2\,!$ = 7/6.

$\therefore$ variance = $\mu_2' - (\mu_1')^2 = 7/6 - 1 = 1/6$.

TEST YOUR KNOWLEDGE

1. The mean height of 500 male students in a certain college is 151 cm and the standard deviation is 15 cm. Assuming the heights are normally distributed, find how many students have heights between 120 and 155 cm?
2. An aptitude test for selecting officers in a bank is conducted on 1000 candidates. The average score is 42 and the standard deviation of score is 24. Assuming normal distribution for the scores, find
   (*i*) The number of candidates whose scores exceed 60
   (*ii*) The number of candidates whose scores lie between 30 and 60.
3. In a normal distribution, 7% of the items are under 35 and 89% are under 63. What are the mean and standard deviation of the distribution?
4. Let X denote the number of scores on a test. If X is normally distributed with mean 100 and standard deviation 15, find the probability that X does not exceed 130.
5. It is known from the past experience that the number of telephone calls made daily in a certain community between 3 P.M. and 4 P.M. have a mean of 352 and a standard deviation of 31. What percentage of the time will there be more than 400 telephone calls made in this community between 3 P.M. and 4 P.M.?
6. Students of a class were given a mechanical aptitude test. Their marks were found to be normally distributed with mean 60 and standard deviation 5. What percent of students scored
   (*i*) more than 60 marks? (*ii*) less than 56 marks?
   (*iii*) between 45 and 65 marks?
7. In an examination taken by 500 candidates, the average and the standard deviation of marks obtained (normally distributed) are 40% and 10%. Find approximately:
   (*i*) How many will pass, if 50% is fixed as a minimum ?
   (*ii*) What should be the minimum if 350 candidates are to pass ?
   (*iii*) How many have scored marks above 60%?

Answers

1. 300 2. (*i*) 252 (*ii*) 533 3. $\bar{x}$ = 50.3, σ = 10.33
4. 0.9772 5. 6.06% 6. (*i*) 50% (*ii*) 21.2% (*iii*) 84%
7. (*i*) 79 (*ii*) 35% (*iii*) 11.

6

Series Solution of Differential Equations and Special Functions

6.1. SERIES SOLUTION OF DIFFERENTIAL EQUATIONS (Method of Frobenius)

We have already discussed the solution of linear differential equations *with constant co-efficients*. The solution involves elementary functions such as e^{ax}, $\sin ax$, $\cos ax$ etc. However, linear differential equations *with variable co-efficients*, which arise from physical problems, do not permit such solutions. Such equations can be solved by numerical methods. There are situations where we can find a solution in the form of an infinite convergent series.

The method of solving equations of the form

$$P_0(x)\,\frac{d^2y}{dx^2} + P_1(x)\,\frac{dy}{dx} + P_2(x)\,y = 0 \quad \text{...(1)}$$

where $p_0(x)$, $P_1(x)$, $P_2(x)$ are polynomials in x, in terms of infinite convergent series.

6.2. VALIDITY OF SERIES SOLUTION

Every differential equation of the form (1) does not have series solution. As such, we find the conditions under which the above equation admits of the series solution.

Dividing (1) by $P_0(x)$, we have $\dfrac{d^2y}{dx^2} + p(x)\,\dfrac{dy}{dx} + q(x)\,y = 0$ ...(2)

where $p(x) = \dfrac{P_1(x)}{P_0(x)}$ and $q(x) = \dfrac{P_2(x)}{P_0(x)}$.

Ordinary point. $x = 0$ is called an ordinary point of (1) if $P_0(0) \neq 0$, otherwise it is called a singular point.

When $x = 0$ is an ordinary point of (1), its every solution can be expressed as a series of the form

$$y = a_0 + a_1x + a_2x^2 + \ldots\ldots = \sum_{k=0}^{\infty} a_k\,x^k.$$

Singular point. $x = 0$ is called a singular point of (1) if $P_0(0) = 0$. If $xp(x)$ and $x^2q(x)$ possess derivatives of all orders in the neighbourhood of $x = 0$, then $x = 0$ is called a *regular singular point* of (1).

When $x = 0$ is a regular singular point of (1), at least one of its solution can be expressed as

$$y = x^m(a_0 + a_1x + a_2x^2 + \ldots\ldots) = \sum_{k=0}^{\infty} a_0 x^{m+k}$$

where m may be a positive or a negative integer or a fraction.

When $x = 0$ is an *irregular singular point* of (1), then the differential equation (1) has no series solution of the form

$$y = \sum_{k=0}^{\infty} a_0 x^{m+k}.$$

6.3. SERIES SOLUTION WHEN x = 0 IS AN ORDINARY POINT OF THE EQUATION

$$\mathbf{P_0(x)\frac{d^2y}{dx^2} + P_1(x)\frac{dy}{dx} + P_2(x)y = 0} \quad \text{...(1)}$$

Let $$y = a_0 + a_1x + a_2x^2 + \text{......} + a_nx^n + \text{......} = \sum_{k=0}^{\infty} a_k x^k \quad \text{...(2)}$$

be the solution of (1).

Then, $$\frac{dy}{dx} = \sum_{k=1}^{\infty} ka_kx^{k-1} \quad \text{and} \quad \frac{d^2y}{dx^2} = \sum_{k=2}^{\infty} k(k-1)\, a_kx^{k-2}$$

Substitute the values of y, $\frac{dy}{dx}$ and $\frac{d^2y}{dx^2}$ in (1).

Equate to zero the co-efficients of various powers of x and find $a_2, a_3, a_4,$ ...... in terms of a_0 and a_1.

Equate to zero the co-efficient of x^n. The relation so obtained is called the *recurrence relation*.

Give different values to n in the recurrence relation to determine various a_i's in terms of a_0 and a_1.

Substitute the values of $a_2, a_3, a_4,$ ...... in (2) to get the series solution of (1) having a_0 and a_1 as arbitrary constants.

ILLUSTRATIVE EXAMPLES

Example 1. *Solve in series the equation* $\frac{d^2y}{dx^2} + x^2y = 0$ ...(1)

Sol. Since $x = 0$ is an ordinary point of (1), let its series solution be

$$y = a_0 + a_1x + a_2x^2 + \text{......} + a_nx^n + \text{......} = \sum_{k=0}^{\infty} a_k x^k \quad \text{...(2)}$$

Then $$\frac{dy}{dx} = \sum_{k=1}^{\infty} ka_kx^{k-1} \quad \text{and} \quad \frac{d^2y}{dx^2} = \sum_{k=2}^{\infty} k(k-1)a_kk^{k-2}$$

Substituting the values of y and $\frac{d^2y}{dx^2}$ in the given equation (1), we get

$$\sum_{k=2}^{\infty} k(k-1)a_kx^{k-2} + x^2 \sum_{k=0}^{\infty} a_kx^k = 0$$

or $$[2.1a_2 + 3.2a_3x + ... + (n+2)(n+1)a_{n+2}\,x^n + ...] + x^2[\,a_0 + a_1x + a_2x^2 + ... + a_{n-2}\,x^{n-2} + ...] = 0$$

or $$2a_2 + 6a_3x + ... + [(n+2)(n+1)a_{n+2} + a_{n-2}]x^n + ... = 0$$

Equating to zero the co-efficients of various powers of x, $a_2 = 0$, $a_3 = 0$ and so on.

Equating to zero the co-efficients of x^n, we have

$$(n+1)(n+2)a_{n+2} + a_{n-2} = 0$$

or $$a_{n+2} = -\frac{a_{n-2}}{(n+1)(n+2)} \qquad ...(3)$$

which is the recurrence relation.

Putting $n = 2, 3, 4, 5, ...$ in (3) successively,

$$a_4 = -\frac{a_0}{3.4}; \qquad a_5 = -\frac{a_1}{4.5}; \qquad a_6 = -\frac{a_2}{5.6} = 0$$

$$a_7 = -\frac{a_3}{6.7} = 0; \quad a_8 = -\frac{a_4}{7.8} = \frac{a_0}{3.4.7.8}; \quad a_9 = -\frac{a_5}{8.9} = \frac{a_1}{4.5.8.9} = \text{and so on.}$$

Substituting the values of a_i's in (2), we get

$$y = a_0 + a_1x - \frac{a_0}{3.4}x^4 - \frac{a_1}{4.5}x^5 + \frac{a_0}{3.4.7.8}x^8 + \frac{a_1}{4.5.8.9}x^9 + ...$$

or $$y = a_0\left(1 - \frac{x^4}{3.4} + \frac{x^8}{3.4.7.8} - ...\right) + a_1\left(x - \frac{x^5}{4.5} + \frac{x^9}{4.5.8.9} - ...\right)$$

which is the required general solution of (1) containing two arbitrary constants a_0 and a_1.

Example 2. *Solve in series the equation*

$$(1+x^2)\frac{d^2y}{dx^2} + x\frac{dy}{dx} - y = 0 \qquad ...(1)$$

Sol. Since $x = 0$ is an ordinary point of (1), let its series solution be

$$y = a_0 + a_1x + a_2x^2 + ... + a_nx^n + ... = \sum_{k=0}^{\infty} a_k\,x^k \qquad ...(2)$$

Then $$\frac{dy}{dx} = \sum_{k=1}^{\infty} ka_kx^{k-1} \quad \text{and} \quad \frac{d^2y}{dx^2} = \sum_{k=2}^{\infty} k(k-1)\,a_kx^{k-2}$$

Substituting the values of y, $\frac{dy}{dx}$ and $\frac{d^2y}{dx^2}$ in the given equation (1), we get

$$(1+x^2)\sum_{k=2}^{\infty} k(k-1)\,a_kx^{k-2} + x\sum_{k=1}^{\infty} ka_kx^{k-1} - \sum_{k=0}^{\infty} a_kx^k = 0$$

$$(1+x)^2\,[2.1\,a_2 + 3.2\,a_3x + ... + n\,(n-1)\,a_n\,x^{n-2} + (n+1)\,na_{n+1}\,x^{n-1} + (n+2)\,(n+1)\,a_{n+2}\,x^n + ...]$$
$$+\,x\,[a_1 + 2a_2x + ... + na_nx^{n-1} + ...] - [a_0 + a_1x + a_2x^2 + ... + a_nx^n + ...] = 0$$

or $$(2a_2 - a_0) + 6a_3x + \ldots [(n+1)(n+2)a_{n+2} + n(n-1)a_n + na_n - a_n]x^n + \ldots = 0$$

or $$(2a_2 - a_0) + 6a_3x + \ldots + [(n+1)(n+2)a_{n+2} + (n^2-1)a_n]x^n + \ldots = 0$$

Equating to zero the co-efficients of various powers of x, $a_2 = \frac{a_0}{2}$, $a_3 = 0$

Equating to zero the co-efficient of x^n, we have

$$(n+1)(n+2)a_{n+2} + (n^2-1)a_n = 0$$

or $$a_{n+2} = -\frac{n-1}{n+2}a_n \qquad \ldots(3)$$

since $n + 1 \neq 0$

Putting $n = 2, 3, 4, 5, \ldots$ in (3) successively

$$a_4 = -\frac{1}{4}a_2 = -\frac{a_0}{8}; \quad a_5 = -\frac{2}{5}a_3 = 0; \quad a_6 = -\frac{1}{2}a_4 = \frac{a_0}{16}$$

$$a_7 = -\frac{4}{7}a_5 = 0; \quad a_8 = -\frac{5}{8}a_6 = -\frac{5a_0}{128} \text{ and so on.}$$

Substituting the values of a_i's in (2), we get

$$y = a_0 + a_1x + \frac{a_0}{2}x^2 - \frac{a_0}{8}x^4 + \frac{a_0}{16}x^6 - \frac{5a_0}{128}x^8 + \ldots\ldots$$

or $$y = a_0\left(1 + \frac{x^2}{2} - \frac{x^4}{8} + \frac{x^6}{16} - \frac{5x^8}{128} + \ldots\ldots\right) + a_1x$$

which is the required general solution of (1) containing two arbitrary constants a_0 and a_1.

Example 3. *Solve in series the equation:* $\frac{d^2y}{dx^2} - y = 0$ *or* $y'' - y = 0$.

[Raipur, B.E. (Third Sem.), July, 2006]

Sol. Let the solution of the equation be assumed as

$$y = a_0 + a_1x + a_2x^2 + a_3x^3 + \ldots$$

Then, $y' = a_1 + 2a_2x + 3a_3x^2 + \ldots$ and $y'' = 2.1a_2 + 3.2a_3x + 4.3a_4x^2 + \ldots$.

Substituting for y, y' and y'' in the given equation, we get

$$2\cdot 1a_2 + 3\cdot 2a_3x + 4\cdot 3a_4x^2 + \ldots + n(n-1)a_nx^{n-2} + \ldots$$
$$- (a_0 + a_1x + a_2x^2 + \ldots + a_nx^n + \ldots) = 0$$

$$(2\cdot 1a_2 - a_0) + (3\cdot 2a_3 - a_1)x + (4\cdot 3a_4 - a_2)x^2 + (5\cdot 4a_5 - a_3)x^3 + \ldots$$
$$+ [(n+2)(n+1)a_{n+2} - a_n]x^n + \ldots = 0.$$

Equating the coefficients of various powers of x to zero, we have

$$2\cdot 1a_2 - a_0 = 0 \quad \text{or} \quad a_2 = \frac{a_0}{2!};$$

$$3\cdot 2a_3 - a_1 = 0 \quad \text{or} \quad a_3 = \frac{a_1}{3!};$$

$$4\cdot 3a_4 - a_2 = 0 \quad \text{or} \quad a_4 = \frac{2!\,a_2}{4!} = \frac{a_0}{4!};$$

$$5\cdot 4a_5 - a_3 = 0 \quad \text{or} \quad a_5 = \frac{3!\,a_3}{5!} = \frac{a_1}{5!};$$

$$\vdots \qquad\qquad \vdots$$

$$(n+2)(n+1)\,a_{n+2} - a_n = 0 \quad \text{or} \quad a_{n+2} = \frac{n!\,a_n}{(n+2)!} \qquad ...(1)$$

Thus, on substituting $n = 4, 5, 6, ...$ in (1), we get

$$a_6 = \frac{4!\,a_4}{6!} = \frac{a_0}{6!}; \qquad a_7 = \frac{5!\,a_5}{7!} = \frac{a_1}{7!};$$

$$a_8 = \frac{6!\,a_6}{8!} = \frac{a_0}{8!}; \qquad a_9 = \frac{7!\,a_7}{9!} = \frac{a_1}{9!}; \text{ etc.}$$

Hence on substituting these values in $y = \sum_{n=0}^{\infty} a_n x^n$, we get

$$y = a_0 + a_1 x + \frac{a_0}{2!}x^2 + \frac{a_1}{3!}x^3 + \frac{a_0}{4!}x^4 + ...$$

or
$$y = a_0\left(1 + \frac{x^2}{2!} + \frac{x^4}{4!} + ...\right) + a_1\left(x + \frac{x^3}{3!} + \frac{x^5}{5!} + ...\right)$$

or
$$y = \frac{1}{2}a_0\,(e^x + e^{-x}) + \frac{1}{2}a_1\,(e^x - e^{-x})$$

or
$$y = \frac{1}{2}(a_0 + a_1)\,e^x + \frac{1}{2}(a_0 - a_1)e^{-x}$$

or
$$y = c_1 e^x + c_2 e^{-x}, \text{ where } c_1 = \frac{1}{2}(a_0 + a_1) \text{ and } c_2 = \frac{1}{2}(a_0 - a_1).$$

Hence the required general solution of the given differential equation is

$$y = c_1 e^x + c_2 e^{-x}.$$

Example 4. *Solve by power series method* $\frac{d^2y}{dx^2} + xy = 0$ *or* $y'' + xy = 0$.

Sol. The given differential equation is

$$y'' + xy = 0. \qquad ...(1)$$

Let the solution of (1) be assumed as $y = a_0 + a_1 x + a_2 x^2 + ... = \sum_{n=0}^{\infty} a_n x^n$. Then

$$y' = a_1 + 2a_2 x + 3a_3 x^2 + ... = \sum_{n=1}^{\infty} n a_n x^{n-1}$$

and
$$y'' = 2.1a_2 + 3.2a_3 x + 4.3\,a_4 x^2 + ... = \sum_{n=2}^{\infty} n(n-1)a_n x^{n-2}.$$

Substituting for y, y' and y'' in equation (1), we get

$$2.1a_2 + 3.2a_3x + 4.3a_4x^2 + \ldots + n(n-1)\,a_nx^{n-2} + \ldots + x\,(a_0 + a_1x + a_2x^2 + \ldots + a_nx^n + \ldots) = 0$$

or

$$2.1a_2 + (3.2a_3 + a_0)x + (4.3a_4 + a_1)x^2 + (5.4a_5 + a_2)\,x^3 + \ldots + [(n+2)(n+1)\,a_{n+2} + a_{n-1}]\,x^n + \ldots = 0$$

Equating the coefficients of various powers of x to zero, we have

$$a_2 = 0;\ 3.2a_3 + a_0 = 0 \quad \text{or} \quad a_3 = -\frac{a_0}{3!};\ 4.3a_4 + a_1 = 0$$

or

$$a_4 = \frac{-2a_1}{4!};\ 5.4a_5 + a_2 = 0 \quad \text{or} \quad a_5 = -\frac{a_2}{5.4} = 0 \text{ etc.}$$

Thus, $(n+2)(n+1)\,a_{n+2} + a_{n-1} = 0$ or $a_{n+2} = \dfrac{-a_{n-1}}{(n+2)(n+1)}$ ...(2)

Substituting $n = 4, 5, 6, \ldots$ in (2), we get

$$a_6 = \frac{-a_3}{6.5} = \frac{4a_0}{6!};\qquad a_7 = \frac{-a_4}{7.6} = \frac{5.2a_1}{7!};$$

$$a_8 = \frac{-a_5}{8.7} = 0;\qquad a_9 = \frac{-a_6}{9.8} = \frac{-7.4a_0}{9!} \text{ etc.}$$

Thus, substituting these values in $y = \sum_{n=0}^{\infty} a_nx^n$, we get

$$y = a_0 + a_1x + 0x^2 - \frac{a_0}{3!}x^3 - \frac{2a_1}{4!}x^4 + 0x^5 + \frac{4a_0}{6!}x^6 + \frac{5.2a_1}{7!}x^7 + 0x^8 - \frac{7.4a_0}{9!}x^9 + \ldots$$

or

$$y = a_0\left(1 - \frac{x^3}{3!} + \frac{4}{6!}x^6 - \frac{7.4}{9!}x^9 + \ldots\right) + a_1\left(x - \frac{2x^4}{4!} + \frac{5.2}{7!}x^7 - \ldots\right)$$

which is the required general solution of the given differential equation.

Example 5. *Find the power series solution of $(1 - x^2)\,y'' - 2xy' + 2y = 0$ about $x = 0$.*

Sol. Let $y = a_0 + a_1x + a_2x^2 + a_3x^3 + \ldots + a_nx^n + \ldots$ be the required solution. Since $x = 0$ is an ordinary point of the given equation, this can be written as

$$y = \sum_{k=0}^{\infty} a_kx^k$$

Then,

$$\frac{dy}{dx} = \sum_{k=1}^{\infty} a_k \,.\, kx^{k-1},\qquad \frac{d^2y}{dx^2} = \sum_{k=2}^{\infty} a_k \,.\, k(k-1)x^{k-2}$$

Substituting the values of $y \,.\, \dfrac{dy}{dx}$, and $\dfrac{d^2y}{dx^2}$ in the given equation, we get

$$(1 - x^2)\,\Sigma a_k\, k \,.\, (k-1)\,x^{k-2} - 2x\,\Sigma a_k \,.\, k\,x^{k-1} + 2\,\Sigma a_k\,x^k = 0$$

$$\Rightarrow\quad \Sigma a_k \,.\, k \,.\, (k-1)\,x^{k-2} - \Sigma a_k\, k(k-1)\,x^k - 2\,\Sigma a_k \,.\, kx^k + 2\Sigma a_kx^k = 0$$

$$\Rightarrow \quad \Sigma a_k \,.\, k \,.\, (k-1)\, x^{k-2} - \Sigma\, [k(k-1) + 2k - 2]\, a_k x^k = 0$$

$$\Rightarrow \quad \Sigma a_k \,.\, k \,.\, (k-1) x^{k-2} - \Sigma\, (k^2 + k - 2)\, a_k x^k = 0$$

where the first summation extends over all values of k from 2 to ∞ and the second from $k = 0$ to ∞.

Now equating the coefficient of x^k equal to zero, we have

$$(k+2)(k+1) a_{k+2} - (k^2 + k - 2)\, a_k = 0$$

$$\Rightarrow \quad a_{k+2} = \frac{k^2 + k - 2}{(k+2)(k+1)} a_k = \frac{(k+2)(k-1)}{(k+2)(k+1)} a_k$$

$$\Rightarrow \quad a_{k+2} = \frac{k-1}{k+1} a_k$$

For $k = 0$ $\quad a_2 = -a_0,\ a_3 = 0,\ a_4 = \dfrac{a_2}{3} = -\dfrac{a_0}{3},\ a_5 = \dfrac{2}{4} a_3 = 0$

For $k = 4$ $\quad a_6 = \dfrac{3}{5} a_4 = \dfrac{3}{5}\left(-\dfrac{a_0}{3}\right) = -\dfrac{a_0}{5},\ a_7 = \dfrac{4}{5} a_5 = 0$, etc.

$$y = a_0 + a_1 x + a_2 x^2 + a_3 x^3 + a_4 x^4 + a_5 x^5 + a_6 x^6 + a_7 x^7 + \ldots$$

$$\Rightarrow \quad y = a_0 + a_0 x - a_0 x^2 + 0 - \frac{a_0}{3} x^4 + 0 - \frac{a_0}{5} x^6 + 0 + \ldots$$

$$\Rightarrow \quad y = a_0 \left[1 - x^2 - \frac{x^4}{3} - \frac{x^6}{5} + \ldots\right] + a_1 x.$$

TEST YOUR KNOWLEDGE

Solve the following equations in series:

1. $\dfrac{d^2y}{dx^2} - y = 0.$
2. $\dfrac{d^2y}{dx^2} + xy = 0.$
3. $y'' + xy' + y = 0.$
4. $y'' - xy' + x^2 y = 0.$
5. $(1 - x^2)\dfrac{d^2y}{dx^2} - 2x\dfrac{dy}{dx} + 2y = 0.$
6. $(2 - x^2) y'' + 2xy' - 2y = 0.$

Answers

1. $y = a_0\left(1 + \dfrac{x^2}{2!} + \dfrac{x^4}{4!} + \dfrac{x^6}{6!} + \ldots\right) + a_1\left(x + \dfrac{x^3}{3!} + \dfrac{x^5}{5!} + \ldots\right) = a_0 \cosh x + a_1 \sinh x$

2. $y = a_0\left(1 - \dfrac{x^3}{3!} + \dfrac{1.4x^6}{6!} - \dfrac{1.4.7x^9}{9!} + \ldots\right) + a_1\left(x - \dfrac{2x^4}{4!} + \dfrac{2.5x^7}{7!} - \ldots\right)$

3. $y = a_0\left(1 - \dfrac{x^2}{2} + \dfrac{x^4}{2.4} - \dfrac{x^6}{2.4.6} + \ldots\right) + a_1\left(x - \dfrac{x^3}{3} + \dfrac{x^5}{3.5} - \dfrac{x^7}{3.5.7} + \ldots\right)$

4. $y = a_0\left(1 - \dfrac{x^4}{12} - \dfrac{x^6}{90} \ldots\right) + a_1\left(x + \dfrac{x^3}{6} - \dfrac{x^5}{40} - \dfrac{x^7}{144} \ldots\right)$

5. $y = a_0\left(1 - x^2 - \frac{x^4}{3} - \frac{x^6}{5}\ldots\right) + a_1 x$

6. $y = a_0\left(1 + \frac{x^4}{2!} + \frac{x^8}{4!} + \ldots\right) + a_2\left(x^2 - \frac{x^6}{3!} + \frac{x^{10}}{5!}\ldots\right).$

6.4. SERIES SOLUTION WHEN x = 0 IS A REGULAR SINGULARITY OF THE EQUATION FROBENIUS METHOD

$$\mathbf{P_0(x)\frac{d^2y}{dx^2} + P_1(x)\frac{dy}{dx} + P_2(x)y = 0} \qquad \ldots(1)$$

Let $y = x^m(a_0 + a_1x + a_2x^2 + \ldots\ldots + a_nx^n + \ldots\ldots) = \sum_{k=0}^{\infty} a_k\, x^{m+k}$ ...(2)

be the solution of (1).

Then, $\frac{dy}{dx} = \sum_{k=0}^{\infty} (m+k)\, a_k x^{m+k-1}$

and $\frac{d^2y}{dx^2} = \sum_{k=0}^{\infty} (m+k)(m+k-1)\, a_k x^{m+k-2}$

Substitute the values of y, $\frac{dy}{dx}$ and $\frac{d^2y}{dx^2}$ in (1).

Equate to zero the co-efficient of *lowest power* of x. This gives a quadratic equation in m, which is known as the **indicial equation**.

Equate to zero the co-efficients of other powers of x to find $a_1, a_2, a_3, \ldots$ in terms of a_0.

Substitute the values of $a_1, a_2, a_3, \ldots\ldots$ in (2) to get the series solution of (1) having a_0 as arbitrary constant. Obviously, this is not the complete solution of (1), since the complete solution must have two independent arbitrary constants. The method of complete solution depends on the nature of roots of the indicial equation.

Case I. *When the roots m_1, m_2 of the indicial equation are distinct and not differing by an integer.*

The complete solution is $y = c_1(y)_{m_1} + c_2(y)_{m_2}$.

Case II. *When the roots m_1, m_2 of the indicial equation are equal.*

The complete solution is $y = c_1(y)_{m_1} + c_2\left(\frac{\partial y}{\partial m}\right)_{m_1}$.

Case III. *When the roots m_1, m_2, $(m_1 < m_2)$ of the indicial equation are distinct and differ by an integer.*

Let some of the co-efficients of y series become infinite when $m = m_1$ (smaller of the two roots). Replace a_0 by $b_0(m - m_1)$ in the series for y.

The complete solution is $y = c_1(y)_{m_1} + c_2\left(\frac{\partial y}{\partial m}\right)_{m_1}$.

The solution corresponding to $m = m_2$ (greater of the two roots) is a constant multiple of the solution corresponding to $m = m_1$.

All the above cases are illustrated in the following examples.

ILLUSTRATIVE EXAMPLES

Example 1. *Solve in series the equation* $2x^2 \dfrac{d^2y}{dx^2} + (2x^2 - x)\dfrac{dy}{dx} + y = 0.$

Sol. Since $x = 0$ is a regular singular point of the given equation, let its series solution be

$$y = x^m (a_0 + a_1 x + a_2 x^2 + \ldots\ldots) = \sum_{k=0}^{\infty} a_k x^{m+k}$$

Then,
$$\frac{dy}{dx} = \sum_{k=0}^{\infty} (m+k)\, a_k x^{m+k-1}$$

and
$$\frac{d^2y}{dx^2} = \sum_{k=0}^{\infty} (m+k)(m+k-1)\, a_k x^{m+k-2}$$

Substituting the values of y, $\dfrac{dy}{dx}$ and $\dfrac{d^2y}{dx^2}$ in the given equation, we get

$$2x^2 \sum_{k=0}^{\infty} (m+k)(m+k-1)\, a_k\, x^{m+k-2} + (2x^2 - x)\sum_{k=0}^{\infty} (m+k)\, a_k\, x^{m+k-1} + \sum_{k=0}^{\infty} a_k x^{m+k} = 0.$$

or
$$\begin{aligned} 2x^2\, [m(m-1)\, a_0 x^{m-2} &+ (m+1)(m)\, a_1 x^{m-1} + (m+2)(m+1)\, a_2 x^m \\ &+ (m+3)(m+2)\, a_3 x^{m+1} + \ldots\ldots] + (2x^2 - x)[m a_0 x^{m-1} \\ &+ (m+1)\, a_1 x^m + (m+2) a_2 x^{m+1} + (m+3)\, a_3 x^{m+2} + \ldots\ldots] \\ &+ [a_0 x^m + a_1 x^{m+1} + a_2 x^{m+2} + a_3 x^{m+3} + \ldots\ldots] = 0. \end{aligned}$$

The lowest power of x is x^m. Equating to zero the co-efficient of x^m, we get

$$2m(m-1)\, a_0 - m a_0 + a_0 = 0 \quad \text{or} \quad (2m^2 - 3m + 1)\, a_0 = 0$$

or
$$(2m-1)(m-1) = 0, \text{ since } a_0 \neq 0$$

which is the indicial equation.

Its roots are $m = \dfrac{1}{2}, 1$ which are **distinct and not differing by an integer.**

Equating to zero the co-efficient of x^{m+1}, we get

$$2m(m+1)\, a_1 + 2m a_0 - (m+1) a_1 + a_1 = 0 \quad \text{or} \quad (2m^2 + m) a_1 + 2m a_0 = 0$$

or
$$(2m+1) a_1 + 2a_0 = 0, \quad \text{since } m \neq 0$$

$$\therefore \qquad a_1 = -\frac{2}{2m+1}\, a_0$$

Equating to zero the co-efficient of x^{m+2}, we get

$$2(m+2)(m+1) a_2 + 2(m+1) a_1 - (m+2) a_2 + a_2 = 0$$

or
$$(2m^2 + 5m + 3) a_2 + 2(m+1) a_1 = 0$$

or $$(2m + 3)(m + 1)a_2 + 2(m + 1)a_1 = 0$$

or $$(2m + 3)a_2 + 2a_1 = 0, \text{ since } m + 1 \neq 0$$

or $$a_2 = -\frac{2}{2m+3}a_1 = \frac{4}{(2m+1)(2m+3)}a_0$$

Equating to zero the co-efficient of x^{m+3}, we get

$$2(m + 3)(m + 2)a_3 + 2(m + 2)a_2 - (m + 3)a_3 + a_3 = 0$$

or $$(2m^2 + 9m + 10)a_3 + 2(m + 2)a_2 = 0$$

or $$(2m + 5)(m + 2)a_3 + 2(m + 2)a_2 = 0$$

or $$(2m + 5)a_3 + 2a_2 = 0$$

or $$a_3 = -\frac{2}{2m+5},\ a_2 = -\frac{8}{(2m+1)(2m+3)(2m+5)}a_0 \text{ and so on.}$$

When m = $\frac{1}{2}$, $a_1 = -a_0, a_2 = -\frac{1}{2}a_0, a_3 = -\frac{1}{6}a_0$ etc.

$\therefore$ The first solution is $y_1 = a_0x^{1/2}\left(1 - x + \frac{1}{2}x^2 - \frac{1}{6}x^3 + ...\right)$.

When m = 1, $a_1 = -\frac{2}{3}a_0, a_2 = \frac{2^2}{3.5}, a_0, a_3 = -\frac{2^3}{3.5.7}a_0$ etc.

$\therefore$ The second independent solution is

$$y_2 = a_0x\left(1 - \frac{2}{3}x + \frac{2^2}{3.5}x^2 - \frac{2^3}{3.5.7}x^3 ......\right)$$

Hence the complete solution is

$$y = c_1y_1 + c_2y_2$$

$$= \text{A}\, x^{1/2}\left(1 - x + \frac{1}{2}x^2 - \frac{1}{6}x^3 ...\right) + \text{B}x\left(1 - \frac{2}{3}x + \frac{2^2}{3.5}x^2 - \frac{2^3}{3.5.7}x^3 ......\right)$$

where, $\text{A} = c_1a_0$, $\text{B} = c_2a_0$.

Example 2. *Solve in series the equation* $9x(1 - x)\frac{d^2y}{dx^2} - 12\frac{dy}{dx} + 4y = 0.$

Sol. Substituting $y = a_0x^m + a_1x^{m+1} + a_2x^{m+2} + ...$

$$\frac{dy}{dx} = ma_0x^{m-1} + (m + 1)a_1x^m + (m + 2)a_2x^{m+1} + ...$$

and $$\frac{d^2y}{dx^2} = m(m - 1)a_0x^{m-2} + (m + 1)ma_1x^{m-1} + (m + 2)(m + 1)a_2x^m + ...$$

in the given equation, we obtain

$$9x(1 - x)[m(m - 1)a_0x^{m-2} + (m + 1)ma_1x^{m-1} + (m + 2)(m + 1)a_2x^m + ...]$$
$$- 12[ma_0x^{m-1} + (m + 1)a_1x^m + (m + 2)a_2x^{m+1} + ...] + 4[a_0x^m + a_1x^{m+1} + a_2x^{m+2} + ...] = 0.$$

The lowest power of x is x^{m-1}. Its coefficient equated to zero gives

$$a_0[9m(m - 1) - 12m] = 0, \textit{ i.e.}, m(3m - 7) = 0 \text{ as } a_0 \neq 0.$$

Thus the roots of the *indicial equation* are $m = 0, 7/3$.

The coefficient of x^m equated to zero gives

$$a_1 \{9(m+1)m - 12(m+1) + a_0\{4 - 9m(m-1)\} = 0$$

i.e., $$3a_1(3m-4)(m+1) - a_0(3m-4)(3m+1) = 0$$

i.e., $$3a_1(m+1) = a_0(3m+1).$$

Similarly, $3a_2(m+2) = a_1(3m+4)$, $3a_3(m+3) = a_2(5m+7)$ and so on.

$$\therefore \quad a_1 = \frac{3m+1}{3(m+1)} a_0, \; a_2 = \frac{(3m+4)\,a_1}{3(m+2)} = \frac{(3m+4)(3m+1)}{3^2(m+2)(m+1)} a_0,$$

$$a_3 = \frac{(3m+7)(3m+4)(3m+1)}{3^3(m+3)(m+2)(m+1)} a_0 \text{ etc.}$$

When $m = 0$, $a_1 = \frac{1}{3}a_0$, $a_2 = \frac{1.4}{3.6}a_0$, $a_3 = \frac{1.4.7}{3.6.9}a_0$ etc. giving the particular solution

$$y_1 = a_0\left[1 + \frac{1}{3}x + \frac{1.4}{3.6}x^2 + \frac{1.4.7}{3.6.9}x^3 + \ldots\right]$$

When $m = 7/3$, the particular solution is

$$y_2 = a_0 x^{7/3}\left[1 + \frac{8}{10}x + \frac{8.11}{10.13}x^2 + \frac{8.11.14}{10.13.16}x^3 + \ldots\right]$$

Thus the complete solution is $y = c_1 y_1 + c_2 y_2$

i.e., $$y = C_1\left[1 + \frac{1}{3}x + \frac{1.4}{3.6}x^2 + \frac{1.4.7}{3.6.9}x^3 + \ldots\right]$$

$$+ C_2 x^{7/3}\left[1 + \frac{8}{10}x + \frac{8.11}{10.13}x^2 + \frac{8.11.14}{10.13.16}x^3 + \ldots\right],$$

where $C_1 = c_1 a_0$, $C_2 = c_2 a_0$.

Example 3. *Find solution in generalized series form about x = 0 of the differential equation*

$$3x\frac{d^2y}{dx^2} + 2\frac{dy}{dx} + y = 0$$

Sol. $$3x\frac{d^2y}{dx^2} + 2\frac{dy}{dx} + y = 0 \qquad \ldots(1)$$

Since $x = 0$ is a regular singular point, we assume the solution in the form

$$y = \sum_{k=0}^{\infty} a_k x^{m+k}$$

Such that $$\frac{dy}{dx} = \sum_{k=0}^{\infty} a_k (m+k) x^{m+k-1}, \quad \frac{d^2y}{dx^2} = \sum_{k=0}^{\infty} a_k (m+k)(m+k-1) x^{m+k-2}$$

Substituting for y, $\frac{dy}{dx}$ and $\frac{d^2y}{dx^2}$ in the given equation (1) we get

$$3\Sigma a_k (m+k)(m+k-1)x^{m+k-1} + 2\Sigma a_k (m+k) x^{m+k-1} + \Sigma a_k x^{m+k} = 0$$

or $$\Sigma a_k\,[3(m+k)(m+k-1)+2\,(m+k)]\,x^{m+k-1}+\Sigma a_k x^{m+k}=0 \qquad ...(2)$$

The co-efficient of the lowest degree term x^{m-1} in the identity (2) is obtained by putting $k=0$ in first summation only and equating it to zero. Then the **indicial equation** is

$$a_0\,[3m\,(m-1)+2m]=0 \quad \text{or} \quad a_0\,[3m^2-m]=0 \quad \text{or} \quad a_0 m\,(3m-1)=0$$

Since $a_0 \neq 0,\ m=0$, or $1/3$

The coefficient of next lowest degree term x^m in the identity (2) is obtained by putting $k=1$ in first summation and $k=0$ in the second summation and equating it to zero.

$$a_1\,[3(m+1)\,m+2\,(m+1)]+a_0=0$$

or $$a_1\,[3m^2+5m+2]+a_0=0 \quad \text{or} \quad a_1\,(3m+2)\,(m+1)+a_0=0$$

$$a_1=-\frac{1}{(3m+2)\,(m+1)}a_0$$

Equating to zero the coefficient of x^{m+k}, the recurrence relation is given by

$$a_{k+1}\,[3\,(m+k+1)\,(m+k)+2\,(m+k+1)]+a_k=0.$$

or $$a_{k+1}\,(m+k+1)\,(3m+3k+2)+a_k=0 \quad \text{or} \quad a_{k+1}=\frac{-1}{(m+k+1)\,(3m+3k+2)}a_k$$

This gives

For $k=0$, $$a_1=\frac{-1}{(m+1)\,(3m+2)}a_0$$

For $k=1$, $$a_2=\frac{-1}{(m+2)\,(3m+5)}a_1=\frac{1}{(m+1)\,(m+2)\,(3m+2)\,(3m+5)}a_0$$

For $k=2$, $$a_3=\frac{-1}{(m+3)\,(3m+8)}a_2=\frac{-1}{(m+1)(m+2)(m+3)(3m+2)(3m+5)(3m+8)}a_0$$

For $m=0$

$$a_1=-\frac{1}{2}a_0, \qquad a_2=\frac{1}{20}a_0, \qquad a_3=-\frac{1}{480}a_0$$

Hence for $$m=0,\ y_1=a_0\left(1-\frac{1}{2}x+\frac{1}{20}x^2-\frac{1}{480}x^3+...\right)$$

For $m=\dfrac{1}{3}$

$$a_1=-\frac{1}{4}a_0, \qquad a_2=\frac{1}{56}a_0, \qquad a_3=\frac{-1}{1680}a_0$$

Hence for $m=\dfrac{1}{3}$, the second solution is

$$y_2=a_0\left(x^{\frac{1}{3}}-\frac{1}{4}x^{\frac{4}{3}}+\frac{1}{56}x^{\frac{7}{3}}-\frac{1}{1680}x^{\frac{10}{3}}+...\right)$$

Thus the complete solution is

$$y=\mathrm{A}y_1+\mathrm{B}y_2$$

$$y=a_0\left(1-\frac{x}{2}+\frac{x^2}{20}-\frac{x^3}{480}+...\right)+b_0x^{1/3}\left(1-\frac{x}{4}+\frac{x^2}{56}-\frac{x^3}{1680}+...\right)$$

Example 4. *Solve in series the equation* $x\dfrac{d^2y}{dx^2}+\dfrac{dy}{dx}-y=0$.

Sol. Since $x = 0$ is a regular singular point of the given equation, let its series solution be

$$y = x^m (a_0 + a_1x + a_2x^2 + ...) = \sum_{k=0}^{\infty} a_k x^{m+k}$$

Then, $$\frac{dy}{dx} = \sum_{k=0}^{\infty} (m+k)\, a_k x^{m+k-1}$$

and $$\frac{d^2y}{dx^2} = \sum_{k=0}^{\infty} (m+k)(m+k-1)\, a_k x^{m+k-2}$$

Substituting the values of y, $\dfrac{dy}{dx}$ and $\dfrac{d^2y}{dx^2}$ in the given equation, we get

$$x\sum_{k=0}^{\infty} (m+k)(m+k-1)\, a_k x^{m+k-2} + \sum_{k=0}^{\infty} (m+k)\, a_k x^{m+k-1} - \sum_{k=0}^{\infty} a_k x^{m+k} = 0$$

or $$x[m(m-1)\, a_0\, x^{m-2} + (m+1)(m)a_1x^{m-1} + (m+2)(m+1)a_2\, x^m + (m+3)(m+2)a_3\, x^{m+1} + ...]$$
$$+ [ma_0\, x^{m-1} + (m+1)a_1\, x^m + (m+2)a_2\, x^{m+1} + (m+3)a_3\, x^{m+2} + ...]$$
$$- [a_0x^m + a_1x^{m+1} + a_2x^{m+2} + a_3x^{m+3} + ...] = 0$$

The lowest power of x is x^{m-1}. Equating to zero the co-efficient of x^{m-1}, we get

$$m(m-1)a_0 + ma_0 = 0 \quad \text{or} \quad m^2a_0 = 0 \quad \text{or} \quad m^2 = 0, \text{ since } a_0 \neq 0$$

which is the indicial equation.

Its roots are $m = 0, 0$ which are **equal.**

Equating to zero the co-efficient of x^m, we get

$$m(m+1)a_1 + (m+1)a_1 - a_0 = 0$$

or $$(m+1)^2\, a_1 - a_0 = 0 \quad \text{or} \quad a_1 = \frac{a_0}{(m+1)^2}$$

Equating to zero the co-efficient of x^{m+1}, we get

$$(m+1)(m+2)a_2 + (m+2)a_2 - a_1 = 0 \quad \text{or} \quad (m+2)^2\, a_2 - a_1 = 0$$

or $$a_2 = \frac{a_1}{(m+2)^2} = \frac{a_0}{(m+1)^2\,(m+2)^2}$$

Equating to zero the co-efficient of x^{m+2}, we get

$$(m+2)(m+3)a_3 + (m+3)a_3 - a_2 = 0 \quad \text{or} \quad (m+3)^2\, a_3 - a_2 = 0$$

or $$a_3 = \frac{a_2}{(m+3)^2} = \frac{a_0}{(m+1)^2\,(m+2)^2\,(m+3)^2} \text{ and so on.}$$

∴ The solution is given by

$$y = a_0x^m\left[1 + \frac{x}{(m+1)^2} + \frac{x^2}{(m+1)^2\,(m+2)^2} + \frac{x^3}{(m+1)^2+(m+2)^2\,(m+3)^2} + ...\right] \quad ...(1)$$

Putting $m = 0$ in (1), the first solution is

$$y_1 = a_0\left(1 + x + \frac{x^2}{1^2 . 2^2} + \frac{x^3}{1^2 . 2^2 . 3^3} + ...\right) = a_0\left[1 + x + \frac{x^2}{(2!)^2} + \frac{x^3}{(3!)^2} + ...\right]$$

To get the second independent solution, differentiate (1) partially w.r.t. m.

$$\frac{\partial y}{\partial m} = a_0 x^m \log x\left[1 + \frac{x}{(m+1)^2} + \frac{x^2}{(m+1)^2 (m+2)^2} + \frac{x^3}{(m+1)^2 (m+2)^2 (m+3)^2} + ...\right]$$

$$+ a_0 x^m\left[-\frac{2x}{(m+1)^3} - \frac{2}{(m+1)^2 (m+2)^2}\left\{\frac{1}{m+1} + \frac{1}{m+2}\right\}x^2\right.$$

$$\left. - \frac{2}{(m+1)^2 (m+2)^2 (m+3)^2}\left\{\frac{1}{m+1} + \frac{1}{m+2} + \frac{1}{m+3}\right\}x^3 ......\right]$$

The second solution is $y_2 = \left(\frac{\partial y}{\partial m}\right)_{m=0}$

$$= a_0 \log x\left[1 + x + \frac{x^2}{(2!)^2} + \frac{x^3}{(3!)^2} + ...\right]$$

$$- 2a_0\left[x + \frac{1}{(2!)^2}\left(1 + \frac{1}{2}\right)x^2 + \frac{1}{(3!)^2}\left(1 + \frac{1}{2} + \frac{1}{3}\right)x^3 + ...\right]$$

$$= y_1 \log x - 2a_0\left[x + \frac{1}{(2!)^2}\left(1 + \frac{1}{2}\right)x^2 + \frac{1}{(3!)^2}\left(1 + \frac{1}{2} + \frac{1}{3}\right)x^3 + ...\right]$$

Hence the complete solution is

$$y = c_1 y_1 + c_2 y_2$$

$$= (c_1 a_0 + c_2 a_0 \log x)\left[1 + x + \frac{x^2}{(2!)^2} + \frac{x^3}{(3!)^2} + ...\right]$$

$$- 2c_2 a_0\left[x + \frac{1}{(2!)^2}\left(1 + \frac{1}{2}\right)x^2 + \frac{1}{(3!)^2}\left(1 + \frac{1}{2} + \frac{1}{3}\right)x^3 + ...\right]$$

$$y = (\mathrm{A} + \mathrm{B}\log x)\left[1 + x + \frac{x^2}{(2!)^2} + \frac{x^3}{(3!)^2} + ...\right]$$

$$- 2\mathrm{B}\left[x + \frac{1}{(2!)^2}\left(1 + \frac{1}{2}\right)x^2 + \frac{1}{(3!)^2}\left(1 + \frac{1}{2} + \frac{1}{3}\right)x^3 + ...\right]$$

where, $c_1 a_0 = \mathrm{A}$, $c_2 a_0 = \mathrm{B}$.

Example 5. *Solve* $x(x-1)y'' + (3x-1)y' + y = 0$

Sol. $x(x-1)y'' + (3x-1)y' + y = 0$ ...(1)

Since, $x = 0$ is a regular singular point, we assume the solution in the form

$$y = \sum_{k=0}^{\infty} a_k x^{m+k}$$

Such that $\dfrac{dy}{dx} = \sum_{k=0}^{\infty} a_k (m+k) x^{m+k-1}$, $\dfrac{d^2y}{dx^2} = \sum_{k=0}^{\infty} a_k (m+k)(m+k-1) x^{m+k-2}$

Substituting the expressions for $y, \dfrac{dy}{dx}, \dfrac{d^2y}{dx^2}$ in (1), we have

$$\Sigma x(x-1) a_k (m+k)(m+k-1) x^{m+k-2} + (3x-1) \Sigma a_k (m+k) x^{m+k-1} + \Sigma a_k x^{m+k} = 0$$

or $\Sigma a_k (m+k)(m+k-1) x^{m+k} - \Sigma a_k (m+k)(m+k-1) x^{m+k-1}$

$$+ 3 \Sigma a_k (m+k) x^{m+k} - \Sigma a_k (m+k) x^{m+k-1} + \Sigma a_k x^{m+k} = 0$$

or $\Sigma a_k [(m+k)(m+k-1) + 3(m+k) + 1] x^{m+k}$

$$- \Sigma a_k [(m+k)(m+k-1) + (m+k)] x^{m+k-1} = 0$$

or $$\Sigma a_k [(m+k)(m+k+2)+1] x^{m+k} - \Sigma a_k (m+k)^2 x^{m+k-1} = 0 \qquad ...(2)$$

The coefficient of lowest degree term x^{m-1} in (2) is obtained by putting $k = 0$ in the second summation only of (2) and equating it to zero. Then the **indicial equation** is

$$a_0 (m+0)^2 = 0 \;\Rightarrow\; m = 0, 0 \text{ as } a_0 \neq 0$$

The coefficient of the next lowest degree term x^m in (2) is obtained by putting $k = 0$ in the first summation and $k = 1$ in the second summation only of (2) and equating it to zero, we get

$$a_1 [(m+0)(m+2)+1] - a_0 (m+0)^2 = 0$$

$$a_1 - a_0 = 0 \;\Rightarrow\; a_1 = a_0 \text{ (as, } m = 0)$$

Equation the coefficient of x^{m+k} to zero, the recurrence relation is given by

$$a_k [(m+k)(m+k+2)+1] - a_{k+1} (m+k+1)^2 = 0$$

$$a_k (m+k+1)^2 - a_{k+1} (m+k+1)^2 = 0$$

Hence $a_{k+1} = a_k$

$$y = x^m [a_0 + a_1 x + a_2 x^2 + ...]$$

$$y = a_0 x^m [1 + x + x^2 + x^3 + ...]$$

$(m = 0)$

when $m = 0, 0$, this gives only one solution instead of two.

Second solution is given by

$$\left(\frac{\partial y}{\partial m}\right)_{m=0} \quad \text{and} \quad y_1 = a_0 (1 + x + x^2 + x^3)$$

$$\frac{\partial y}{\partial m} = a_0 x^m \log x \, [1 + x + x^2 + x^3 + ...]$$

$$y_2 = a_0 \log x \, [1 + x + x^2 + x^3 + ...] \quad m = 0$$

$$y_1 = a_0 [1 + x + x^2 + x^3 + ...] \quad m = 0$$

$$y = Ay_1 + By_2$$

$$y = A[1 + x + x^2 + x^3 + ...] + B \log x \, (1 + x + x^2 + x^3 + ...)$$

Example 6. *Using extended power series method find one solution of the differential equation $xy'' + y' + x^2y = 0$. Indicate the form of a second solution which is linearly independent of the first obtained above.*

Sol. $$x\frac{d^2y}{dx^2} + \frac{dy}{dx} + x^2y = 0 \quad \text{...(1)}$$

Let $$y = \Sigma a_k\, x^{m+k}$$

$$\frac{dy}{dx} = \Sigma a_k (m+k)\, x^{m+k-1}, \quad \frac{d^2y}{dx^2} = \Sigma a_k (m+k)(m+k-1)\, x^{m+k-2}$$

Substituting the values of y, $\frac{dy}{dx}$ and $\frac{d^2y}{dx^2}$ in (1), we get

$$x\, \Sigma a_k\, (m+k)(m+k-1)\, x^{m+k-2} + \Sigma a_k\, (m+k)\, x^{m+k-1} + x^2\, \Sigma a_k\, x^{m+k} = 0$$

or $$\Sigma a_k\, (m+k)(m+k-1)\, x^{m+k-1} + \Sigma a_k\, (m+k)\, x^{m+k-1} + \Sigma a_k\, x^{m+k+2} = 0$$

or $$\Sigma a_k\, [(m+k)(m+k-1) + (m+k)]\, x^{m+k-1} + \Sigma a_k\, x^{m+k+2} = 0$$

or $$\Sigma a_k (m+k)^2\, x^{m+k-1} + \Sigma a_k\, x^{m+k+2} = 0$$

The coefficient of lowest degree term x^{m-1} in (2) is obtained by putting $k = 0$ in first summation of (2) only and equating it to zero. Then the **indicial equation** is

$$a_0 m^2 = 0 \;\Rightarrow\; m^2 = 0 \;\text{ or }\; m = 0, 0$$

The coefficient of the next lowest degree term x^m in (2) is obtained by putting $k = 1$ in first summation only and equating it to zero.

$$a_1\, (m+1)^2 = 0 \;\text{ or }\; \Rightarrow a_1 = 0$$

Equating the coefficient of x^{m+1} for $k = 2$ we get

$$a_2\, (m+2)^2 = 0 \;\Rightarrow\; a_2 = 0$$

Equating the coefficient of x^{m+k+2} to zero, we have

$$a_{k+3}\, (m+k+3)^2 + a_k = 0$$

$$a_{k+3} = -\frac{a_k}{(m+k+3)^2}$$

$$k = 0, \quad a_3 = -\frac{1}{(m+3)^2}\, a_0$$

$$k = 1, \quad a_4 = -\frac{1}{(m+4)^2}\, a_1 = 0,\ a_7 = 0,\ a_{10} = 0$$

$$k = 2, \quad a_5 = -\frac{1}{(m+5)^2}\, a_2 = 0,\ a_8 = 0,\ a_{11} = 0$$

$$k = 3, \quad a_6 = -\frac{1}{(m+6)^2}\, a_3 = \frac{1}{(m+3)^2\,(m+6)^2}\, a_0$$

$$a_9 = -\frac{1}{(m+9)^2}\, a_6 = -\frac{1}{(m+3)^2\,(m+6)^2\,(m+9)^2}$$

$$y = x^m a_0 \left[1 - \frac{x^3}{(m+3)^2} + \frac{x^6}{(m+3)^2 (m+6)^2} - \frac{x^9}{(m+3)^2 (m+6)^2 (m+9)^2} + \ldots\right] \quad \ldots(3)$$

To get the first solution, let $m = 0$ in (3), then

$$y_1 = a_0 \left[1 - \frac{x^3}{3^2} + \frac{x^6}{3^2 \times 6^2} - \frac{x^9}{3^2 \times 6^2 \times 9^2} + \ldots\right] \quad \ldots(4)$$

To get the second independent solution, differentiate (3) w.r.t. m. Then

$$\frac{\partial y}{\partial m} = (x^m \log x)\, a_0 \left[1 - \frac{x^3}{(m+3)^2} + \frac{x^6}{(m+3)^2 (m+6)^2} - \frac{x^9}{(m+3)^2 (m+6)^2 (m+9)^2} + \ldots\right]$$

$$+ x^m a_0 \left[\frac{2x^3}{(m+3)^3} - \frac{2x^6}{(m+3)^3 (m+6)^2} - \frac{2x^6}{(m+3)^2 (m+6)^3}\right.$$

$$+ \frac{2x^9}{(m+3)^2 (m+6)^2 (m+9)^2} + \frac{2x^9}{(m+3)^2 (m+6)^3 (m+9)^2}$$

$$\left. + \frac{2x^9}{(m+3)^2 (m+6)^2 (m+9)^3} + \ldots\right] \quad \ldots(5)$$

Putting $m = 0$ in (5), we get

$$y_2 = (\log x) a_0 \left[1 - \frac{x^3}{3^2} + \frac{x^6}{3^2 \times 6^2} - \frac{x^9}{3^2 \times 6^2 \times 9^2} + \ldots\right]$$

$$+ a_0 \left[\frac{2x^3}{3^3} - \frac{2x^6}{3^3 \times 6^2} - \frac{2x^6}{3^2 \times 6^3} + \frac{2x^9}{3^3 \times 6^2 \times 9^2} + \frac{2x^9}{3^2 \times 6^3 \times 9^2} + \frac{2x^9}{3^2 \times 6^2 \times 9^3} + \ldots\right] \quad \ldots(6)$$

Hence the general solution is given by (4) and (6)

$$y = c_1 y_1 + c_2 y_2$$

$$y = c_1 a_0 \left[1 - \frac{x^3}{3^2} + \frac{x^6}{3^2 \times 6^2} - \frac{x^9}{3^2 \times 6^2 \times 9^2} + \ldots\right]$$

$$+ c_2 (\log x)\, a_0 \left[1 - \frac{x^3}{3^2} + \frac{x^6}{3^2 \times 6^2} - \frac{x^9}{3^2 \times 6^2 \times 9^2} + \ldots\right]$$

$$+ c_2 a_0 \left[\frac{2x^3}{3^3} - \frac{2x^6}{3^2 \times 6^6}\left(\frac{1}{3} + \frac{1}{6}\right) + \frac{2x^9}{3^2 \times 6^2 \times 9^2}\left(\frac{1}{3} + \frac{1}{6} + \frac{1}{9}\right) + \ldots\right]$$

$$= (c_1 + c_2 \log x)\, a_0 \left[1 - \frac{x^3}{3^2} + \frac{x^6}{3^2 \times 6^2} - \frac{x^9}{3^2 \times 6^2 \times 9^2} + \ldots\right]$$

$$+ 2c_2 a_0 \left[\frac{x^3}{3^3} - \frac{x^6}{3^5 \times 2^2}\left(1 + \frac{1}{2}\right) + \frac{2x^9}{3^9 \times 2^2}\left(1 + \frac{1}{2} + \frac{1}{3}\right) + \ldots\right]$$

Case III: When m_1 and m_2 are distinct and differing by an integer, then

$$y = C_1(y)_{m_1} + C_2\left(\frac{\partial y}{\partial m}\right)_{m_2} \qquad \left[\begin{array}{r}\text{if coefficient} = \infty \\ \text{when } m = m_2\end{array}\right]$$

Example 7. *Obtain the series solution of the equation*

$$x^2 \frac{d^2y}{dx^2} + x\frac{dy}{dx} + (x^2 - 4)y = 0.$$

Sol. Since $x = 0$ is a regular singular point of the given equation, let its series solution be

$$y = x^m (a_0 + a_1x + a_2x^2 + \ldots) = \sum_{k=0}^{\infty} a_k x^{m+k}$$

Then $\qquad \dfrac{dy}{dx} = \sum_{k=0}^{\infty} (m+k)a_k x^{m+k-1}$ and $\dfrac{d^2y}{dx^2} = \sum_{k=0}^{\infty} (m+k)(m+k-1)\, a_k x^{m+k-2}$

Substituting the values of y, $\dfrac{dy}{dx}$ and $\dfrac{d^2y}{dx^2}$ in the given equation, we get

$$x^2 \sum_{k=0}^{\infty} (m+k)(m+k-1)\, a_k x^{m+k-2} + x\sum_{k=0}^{\infty} (m+k)a_k x^{m+k-1} + (x^2-4)\sum_{k=0}^{\infty} a_k x^{m+k} = 0$$

or $\qquad x^2 [m(m-1)a_0x^{m-2} + (m+1)(m)a_1x^{m-1} + (m+2)(m+1)a_2x^m + (m+3)(m+2)a_3x^{m+1} + \ldots]$

$$+ x[ma_0x^{m-1} + (m+1)a_1x^m + (m+2)a_2x^{m+1} + (m+3)a_3x^{m+2} + \ldots]$$

$$+ (x^2-4)[a_0x^m + a_1x^{m+1} + a_2x^{m+2} + a_3x^{m+3} + \ldots] = 0$$

The lowest power of x is x^m. Equating to zero the co-efficient of x^m, we get

$$m(m-1)a_0 + ma_0 - 4a_0 = 0$$

or $\qquad (m^2 - 4)a_0 = 0 \quad \text{or} \quad m^2 - 4 = 0$, since $a_0 \neq 0$

which is the indicial equation.

Its roots are $m = -2, 2$ which **differ by an integer.**

Equating to zero the co-efficient of x^{m+1}, we get

$$m(m+1)a_1 + (m+1)a_1 - 4a_1 = 0$$

or $\qquad (m^2 + 2m - 3)a_1 = 0 \quad \text{or} \quad (m+3)(m-1)a_1 = 0$

or $\qquad a_1 = 0, \quad$ since $m \neq 1$ or -3

Equating to zero the co-efficient of x^{m+2}, we get

$$(m+2)(m+1)a_2 + (m+2)a_2 + a_0 - 4a_2 = 0$$

or $\qquad (m^2 + 4m)a_2 + a_0 = 0 \quad \text{or} \quad m(m+4)a_2 + a_0 = 0$

or $\qquad a_2 = -\dfrac{a_0}{m(m+4)}$

Similarly, $\quad (m+1)(m+5)a_3 + a_1 = 0$

$\qquad (m+2)(m+6)a_4 + a_2 = 0$ etc.

whence $\quad a_3 = 0.$

Similarly, $a_5 = a_7 = \ldots\ldots = 0$

and
$$a_4 = -\frac{a_2}{(m+2)(m+6)} = \frac{a_0}{m(m+2)(m+4)(m+6)}$$

$$a_6 = \frac{-a_0}{m(m+2)(m+4)^2(m+6)(m+8)} \text{ etc.}$$

$\therefore$ The solution is given by

$$y = a_0 x^m \left[1 - \frac{x^2}{m(m+4)} + \frac{x^4}{m(m+2)(m+4)(m+6)} - \frac{x^6}{m(m+2)(m+4)^2(m+6)(m+8)} + \ldots\ldots\right] \quad \ldots(1)$$

If we put $m = -2$ in (1), the co-efficients become infinite due to the presence of the factor $(m + 2)$ in the denominator. To overcome this difficulty, let $a_0 = b_0 (m + 2)$ so that

$$y = b_0 x^m \left[(m+2) - \frac{(m+2)x^2}{m(m+4)} + \frac{x^4}{m(m+4)(m+6)} - \frac{x^6}{m(m+4)^2(m+6)(m+8)} + \ldots\ldots\right] \quad \ldots(2)$$

Put $m = -2$

$$y_1 = b_0 x^{-2}\left[\frac{x^4}{(-2)(2)(4)} - \frac{x^6}{(-2)(2^2)(4)(6)} + \ldots\ldots\right]$$

Differentiating (2) partially w.r.t. m, we get

$$\frac{\partial y}{\partial m} = b_0 x^m \log x\left[(m+2) - \frac{(m+2)x^2}{m(m+4)} + \frac{x^4}{m(m+4)(m+6)} - \ldots\ldots\right]$$

$$+ b_0 x^m \left[1 - \frac{(m+2)}{m(m+4)}\left\{\frac{1}{m+2} - \frac{1}{m} - \frac{1}{m+4}\right\}x^2 + \frac{1}{m(m+4)(m+6)}\left\{-\frac{1}{m} - \frac{1}{m+4} - \frac{1}{m+6}\right\}x^4 \ldots\ldots\right]$$

The second solution is $y_2 = \left(\frac{\partial y}{\partial m}\right)_{m=-2}$

$$= b_0 x^{-2}\log x\left[\frac{x^4}{(-2)(2)(4)} - \frac{x^6}{(-2)(2)^2(4)(6)} \ldots\ldots\right]$$

$$+ b_0 x^{-2}\left[1 - \frac{x^2}{(-2)(2)} + \frac{1}{(-2)(2)(4)}\left(\frac{1}{2} - \frac{1}{2} - \frac{1}{4}\right)x^4 \ldots\ldots\right]$$

$$= b_0 x^2 \log x \left[-\frac{1}{2^2 . 4} + \frac{x^4}{2^3 . 4 . 6} ...... \right] + b_0 x^{-2} \left[1 + \frac{x^2}{2^2} + \frac{x^4}{2^2 . 4^2} + ... \right]$$

Hence the complete solution is

$$y = c_1 y_1 + c_2 y_2$$

$$= Ax^2 \left(1 - \frac{x^2}{2.6} + \frac{x^4}{2.4.6.8} - \frac{x^6}{2.4.6^2.8.10} + ... \right) + B \left[x^2 \log x \left(-\frac{1}{2^2.4} + \frac{x^4}{2^3.4.6} ...... \right) \right.$$

$$\left. + x^{-2} \left(1 + \frac{x^2}{2^2} + \frac{x^4}{2^2 . 4^2} + ... \right) \right]$$

where, $A = \dfrac{c_1 b_0}{(-2)(2)(4)}$, $B = c_2 b_0$.

Example 8. *Obtain the series solution of the equation*

$$x(1-x) \frac{d^2y}{dx^2} - (1+3x) \frac{dy}{dx} - y = 0.$$

Sol. Substituting $y = a_0 x^m + a_1 x^{m+1} + a_2 x^{m+2} + ...$

$$\frac{dy}{dx} = m a_0 x^{m-1} + (m+1) a_1 x^m + (m+2) a_2 x^{m+1} + ...$$

and $$\frac{d^2y}{dx^2} = m(m-1) a_0 x^{m-2} + (m+1) m a_1 x^{m-1} + (m+2)(m+1) a_2 x^m + ...$$

in the given equation, we obtain

$$x(1-x)[m(m-1)a_0 x^{m-2} + (m+1)m a_1 x^{m-1} + (m+2)(m+1) a_2 x^m + ...]$$
$$- (1+3x)[m a_0 x^{m-1} + (m+1) a_1 x^m + (m+2) a_2 x^{m+1} + ...]$$
$$- [a_0 x^m + a_1 x^{m+1} + a_1 x^{m+2} + ...] = 0.$$

Equating to zero the coefficients of the lowest power of x, we get $a_0[m(m-1) - m] = 0$,

i.e., $m(m-2) = 0$, *i.e.,* $m = 0, 2$ $[\because a_0 \neq 0]$

Equating to zero the coefficients of successive powers of x, we get

$(m-1)a_1 = (m+1)a_0$, $m a_2 = (m+2)a_1$, $(m+1) a_3 = (m+3) a_2$ and so on.

i.e., $$a_1 = \frac{m+1}{m-1} a_0, a_2 = \frac{(m+1)(m+2)}{(m-1)m} a_0, a_3 = \frac{(m+1)(m+2)(m+3)}{(m-1)m(m+1)} a_0 \text{ etc.}$$

These give

$$y = a_0 x^m \left[1 + \frac{m+1}{m-1} x + \frac{(m+1)(m+2)}{(m-1)m} x^2 + \frac{(m+1)(m+2)(m+3)}{(m-1)m(m+1)} x^3 + ... \right]$$

If we put $m = 0$ in this series, the coefficients become infinite. To obviate this difficulty, put $a_0 = b_0 (m - 0)$ so that

$$y_1 = b_0 x^m \left\{ m + \frac{m(m+1)}{m-1} x + \frac{(m+1)(m+2)}{m-1} x^2 + \frac{(m+1)(m+2)(m+3)}{(m-1)(m+1)} x^3 + ... \right\}$$

is a solution if $m = 0$.

This gives only one solution instead of two and the second solution is given by $\frac{\partial y_1}{\partial m}$ when, $m = 0$.

$$\therefore \quad y_2 = \left(\frac{\partial y_1}{\partial m} \right)_{m=0}$$

$$= y_1 \log x + b_0 x^m \left[1 + \frac{m^2 - 2m - 1}{(m-1)^2} x + \frac{m^2 - m - 5}{(m-1)^2} x^2 + \frac{m^2 - 2m - 11}{(m-1)^2} x^3 + ... \right]$$

when $\quad m = (y_1)_{m=0} \log x + b_0[1 - x - 5x^2 - 11x^3 + ...]$

Hence the complete solution is $y = c_1 y_1 + c_2 y_2$

i.e., $\quad y = (C_1 + C_2 \log x)(1 . 2x^2 + 2 . 3x^3 + 3 . 4x^4 + ...)$

$+ C_1(- 1 + x + 5x^2 + 11x^3 + ...)$ [where $C_1 = - b_0 c_1$, $C_2 = - b_0 c_2$]

TEST YOUR KNOWLEDGE

Solve in series the following differential equations:

1. $2x^2 \frac{d^2y}{dx^2} - x \frac{dy}{dx} + (x^2 + 1)y = 0$

2. $x(2 + x^2) \frac{d^2y}{dx^2} - \frac{dy}{dx} - 6xy = 0$

3. $9x(1 - x) \frac{d^2y}{dx^2} - 12 \frac{dy}{dx} + 4y = 0$

4. $2x(1 - x) \frac{d^2y}{dx^2} + (5 - 7x) \frac{dy}{dx} - 3y = 0$

5. $4x \frac{d^2y}{dx^2} + 2 \frac{dy}{dx} + y = 0$

6. $4x \frac{d^2y}{dx^2} + 2(1 - x) \frac{dy}{dx} - y = 0$

7. $2x^2 \frac{d^2y}{dx^2} - x \frac{dy}{dx} + (1 - x^2)y = 0$

8. $x \frac{d^2y}{dx^2} + \frac{dy}{dx} + xy = 0$

9. $x \frac{d^2y}{dx^2} + (1 + x) \frac{dy}{dx} + 2y = 0$

10. $(x - x^2) \frac{d^2y}{dx^2} + (1 - 5x) \frac{dy}{dx} - 4y = 0$

11. $x(1 - x) \frac{d^2y}{dx^2} - (1 + 3x) \frac{dy}{dx} - y = 0$

12. $x(1 - x) \frac{d^2y}{dx^2} - 3x \frac{dy}{dx} - y = 0$

13. $2x(1 - x) \frac{d^2y}{dx^2} + (1 - x) \frac{dy}{dx} + 3y = 0.$

Answers

1. $y = Ax \left(1 - \frac{x^2}{10} + \frac{x^4}{360} \ldots\ldots \right) + Bx^{1/2} \left(1 - \frac{x^2}{6} + \frac{x^4}{168} \ldots\ldots \right)$

2. $y = A\left(1 + 3x^2 + \frac{3}{5}x^4 - \frac{1}{15}x^6 \ldots\ldots\right) + Bx^{3/2}\left(1 + \frac{3}{8}x^2 - \frac{3.1}{8.16}x^4 + \frac{5.3.1}{8.16.24}x^6 \ldots\ldots\right)$

3. $y = A\left(1 + \frac{1}{3}x + \frac{1.4}{3.6}x^2 + \frac{1.4.7}{3.6.9}x^3 \ldots\ldots\right) + Bx^{7/3}\left(1 + \frac{8}{10}x + \frac{8.11}{10.13}x^2 + \frac{8.11.14}{10.13.16}x^3 + \ldots\ldots\right)$

4. $y = A\left(1 + \frac{3}{5}x + \frac{3}{7}x^2 + \frac{3}{9}x^3 \ldots\ldots\right) + Bx^{-3/2}$ **5.** $y = A\cos\sqrt{x} + B\sin\sqrt{x}$

6. $y = A\left(1 + \frac{x}{2(1)!} + \frac{x^2}{2^2\,(2)!} + \frac{x^3}{2^3\,(3)!} + \ldots\ldots\right) + Bx^{1/2}\left(1 + \frac{x}{1.3} + \frac{x^2}{1.3.5} + \frac{x^3}{1.3.5.7} + \ldots\ldots\right)$

7. $y = Ax\left(1 + \frac{x^2}{2.5} + \frac{x^4}{2.4.5.9} + \ldots\ldots\right) + Bx^{1/2}\left(1 + \frac{x^2}{2.3} + \frac{x^4}{2.4.3.7} + \ldots\ldots\right)$

8. $y = A\left[1 - \frac{x^2}{2^2} + \frac{x^4}{2^2.4^2} - \frac{x^6}{2^2.4^2.6^2}\ldots\ldots\right] + B\left[y_1\log x + a_0\left\{\frac{x^2}{2^2} - \frac{1}{2^2.4^2}\left(1+\frac{1}{2}\right)x^4 + \frac{1}{2^2.4^2.6^2}\left(1+\frac{1}{2}+\frac{1}{3}\right)x^6\ldots\ldots\right\}\right]$

9. $y = A\left(1 - 2x + \frac{3}{2}x^2 - \frac{2}{3}x^3 \ldots\ldots\right) + B\left[y_1\log x + a_0\left(3x - \frac{13}{4}x^2 + \ldots\ldots\right)\right]$

10. $y = A\,(1^2 + 2^2 x + 3^2x^2 + 4^2x^3 + \ldots\ldots) + B\,[y_1 \log x - 2a_0\,(1.2x + 2.3x^2 + 3.4x^3 + \ldots\ldots)]$

11. $y = (A + B\log x)\,(1.2x^2 + 2.3x^3 + 3.4x^4 + \ldots\ldots) + B\,(-1 + x + 5x^2 + 11x^3 + \ldots\ldots)$

12. $y = Ax\,(1 + 2x + 3x^2 + \ldots\ldots) + B[y_1 \log x + a_0\,(1 + x + x^2 + \ldots\ldots)]$

13. $y = a_0\sqrt{x}(1 - x) + a_1\left(1 - 3x + \frac{3x^2}{1.3} + \frac{3x^3}{3.5} + \frac{3x^4}{5.7} + \ldots\ldots\right)$.

6.5. BESSEL'S EQUATION

The differential equation $x^2\frac{d^2y}{dx^2} + x\frac{dy}{dx} + (x^2 - n^2)y = 0$ ...(1)

is called *Bessel's equation of order n* and its particular solutions are called *Bessel functions of order n*.

Since $x = 0$ is a regular singularity of the equation, let its solution be

$$y = x^m(a_0 + a_1x + a_2x + \ldots\ldots) = \sum_{k=0}^{\infty} a_k x^{m+k}$$

Then $$\frac{dy}{dx} = \sum_{k=0}^{\infty} (m+k)\,a_k x^{m+k-1}$$

and $$\frac{d^2y}{dx^2} = \sum_{k=0}^{\infty} (m+k)(m+k-1)\,a_k x^{m+k-2}$$

Substituting for y, $\frac{dy}{dx}$ and $\frac{d^2y}{dx^2}$ in (1), we get

$$x^2 \sum_{k=0}^{\infty} (m+k)(m+k-1)\, a_k x^{m+k-2} + x \sum_{k=0}^{\infty} (m+k)\, a_k x^{m+k-1} + (x^2 - n^2) \sum_{k=0}^{\infty} a_k x^{m+k} = 0$$

or
$$\sum_{k=0}^{\infty} [(m+k)^2 - (m+k) + (m+k) - n^2]\, a_k x^{m+k} + \sum_{k=0}^{\infty} a_k x^{m+k+2} = 0$$

or
$$\sum_{k=0}^{\infty} [(m+k)^2 - n^2]\, a_k x^{m+k} + \sum_{k=0}^{\infty} a_k x^{m+k+2} = 0$$

The lowest power of x is x^m corresponding to $k = 0$. Equating to zero the co-efficient of x^m, we get the indicial equation

$$m^2 - n^2 = 0, \text{ since } a_0 \neq 0 \quad \text{whence } m = \pm n$$

Equating to zero the co-efficient of next term *i.e.*, x^{m+1}, we get

$$[(m+1)^2 - n^2]a_1 = 0$$

$\Rightarrow$ $a_1 = 0$, since $(m+1)^2 - n^2 \neq 0$ for $m = \pm n$

Equating to zero the co-efficient of x^{m+k+2}, we get the recurrence relation

$$[(m+k+2)^2 - n^2]\, a_{k+2} + a_k = 0 \quad \text{or} \quad a_{k+2} = -\frac{a_k}{(m-n+k+2)(m+n+k+2)}$$

Putting $k = 1, 3, 5, \ldots\ldots$, we get $a_3 = a_5 = a_7 = \ldots\ldots = 0$

Putting $k = 0, 2, 4, \ldots\ldots$, we get

$$a_2 = -\frac{a_0}{(m-n+2)(m+n+2)}$$

$$a_4 = -\frac{a_2}{(m-n+4)(m+n+4)} = \frac{a_0}{(m-n+4)(m+n+4)(m-n+2)(m+n+2)}$$

and so on.

$$\therefore \quad y = a_0 x^m \left[1 - \frac{x^2}{(m+n)^2 - n^2} + \frac{x^4}{[(m+2)^2 - n^2][(m+4)^2 - n^2]} - \ldots\ldots\right] \qquad \ldots(2)$$

Depending upon the values of n, we get different types of solutions.

Case I. *When $n \neq 0$ or $n \neq$ an integer.*

In this case, we get two independent solutions for $m = n$ and $m = -n$.

For $m = n$, we get

$$y_1 = a_0 x^n \left[1 - \frac{x^2}{2(2n+2)} + \frac{x^4}{2.4(2n+2)(2n+4)} - \ldots\ldots\right]$$

$$= a_0 x^n \left[1 + (-1)^1 \frac{x^2}{2^2 (1)!\,(n+1)} + (-1)^2 \frac{x^4}{2^4 (2)!\,(n+1)(n+2)} + \ldots\ldots\right]$$

$$= a_0 x^n \sum_{k=0}^{\infty} \frac{(-1)^k}{2^{2k} (k)!\,(n+1)(n+2)\ldots\ldots(n+k)} x^{2k}$$

$$= a_0 x^n \sum_{k=0}^{\infty} \frac{(-1)^k \Gamma(n+1)}{2^{2k} \cdot k!\, \Gamma(n+k+1)} x^{2k} \qquad ...(3)$$

Replacing n by $-n$, the second independent solution corresponding to $m = -n$ is

$$y_2 = a_0 x^n \sum_{k=0}^{\infty} \frac{(-1)^k \Gamma(-n+1)}{2^{2k}\, (k)!\, \Gamma(-n+k+1)} x^{2k} \qquad ...(4)$$

$\therefore$ The complete solution of (1) is $y = c_1 y_1 + c_2 y_2$

Since a_0 is arbitrary, we can choose it in any manner.

Choose $a_0 = \dfrac{1}{2^n\, \Gamma(n+1)}$, then (3) takes the form

$$y_1 = \frac{x^n}{2^n\, \Gamma(n+1)} \sum_{k=0}^{\infty} \frac{(-1)^k \Gamma(n+1)}{2^{2k} \cdot k!\, \Gamma(n+k+1)} x^{2k} = \sum_{k=0}^{\infty} \frac{(-1)^k}{(k)!\, \Gamma(n+k+1)} \left(\frac{x}{2}\right)^{n+2k}$$

This is called **Bessel function of the first kind of order n** and is denoted by $J_n(x)$.

Thus,

$$J_n(x) = \sum_{k=0}^{\infty} \frac{(-1)^k}{(k)!\, \Gamma(n+k+1)} \left(\frac{x}{2}\right)^{n+2k}$$

The solution corresponding to $m = -n$ is

$$J_{-n}(x) = \sum_{k=0}^{\infty} \frac{(-1)^k}{(k)!\, \Gamma(-n+k+1)} \left(\frac{x}{2}\right)^{-n+2k}$$

which is called **Bessel function of the first kind of order – n.**

When n is not an integer, $J_{-n}(x)$ is distinct from $J_n(x)$. Hence the complete solution of the Bessel's equation may be expressed as

$y = AJ_n(x) + BJ_{-n}(x)$ where A and B are arbitrary constants.

Case II. *When n = 0,* the Bessel's equation (1) takes the form

$$x \frac{d^2y}{dx^2} + \frac{dy}{dx} + xy = 0.$$

This is called **Bessel's equation of order zero.**

The two roots of the indicial equation are equal, each = 0.

From equation (2), putting $n = 0$, we have (assuming $a_0 = 1$)

$$y = x^m \left[1 - \frac{x^2}{(m+2)^2} + \frac{x^4}{(m+2)^2(m+4)^2} - \frac{x^6}{(m+2)^2(m+4)^2(m+6)^2} + \ldots\ldots\right]$$

which is a solution if $m = 0$.

The first solution is given by

$$J_0(x) = \sum_{k=0}^{\infty} \frac{(-1)^k}{(k!)^2} \left(\frac{x}{2}\right)^{2k}, \quad \text{since } \Gamma(k+1) = k!$$

which is *Bessel function of the first kind of order zero.*

Now $\dfrac{\partial y}{\partial m} = x^m \log x \left[1 - \dfrac{x^2}{(m+2)^2} + \dfrac{x^4}{(m+2)^2(m+4)^2} - \ldots\ldots\right]$

$$+ x^m \left[\frac{x^2}{(m+2)^2} \cdot \frac{2}{m+2} - \frac{x^4}{(m+2)^2(m+4)^2}\left\{\frac{2}{m+2} + \frac{2}{m+4}\right\} + \ldots\ldots\right]$$

The second independent solution is given by $\left(\dfrac{\partial y}{\partial m}\right)_{m=0}$

$$= J_0(x) \log x + \left[\frac{1}{2^2} x^2 - \frac{1}{2^2 . 4^2}\left(1 + \frac{1}{2}\right) x^4 + \frac{1}{2^2 . 4^2 . 6^2}\left(1 + \frac{1}{2} + \frac{1}{3}\right) x^6 - \ldots\ldots\right]$$

$$= J_0(x) \log x + \left[\left(\frac{x}{2}\right)^2 - \frac{1}{(2\,!)^2}\left(1 + \frac{1}{2}\right)\left(\frac{x}{2}\right)^4 + \frac{1}{(3\,!)^2}\left(1 + \frac{1}{2} + \frac{1}{3}\right)\left(\frac{x}{2}\right)^6 - \ldots\ldots\right]$$

$$= J_0(x) \log x + \sum_{k=1}^{\infty} \frac{(-1)^{k+1}}{(k\,!)^2}\left(1 + \frac{1}{2} + \frac{1}{3} + \ldots\ldots + \frac{1}{k}\right)\left(\frac{x}{2}\right)^{2k}$$

It is denoted by $Y_0(x)$ and is called *Bessel function of the second kind of order zero or Neumann function.*

Thus the complete solution of the Bessel's equation of order zero is

$$y = AJ_0(x) + BY_0(x).$$

Case III. *When n is an integer,* the two functions $J_n(x)$ and $J_{-n}(x)$ are not independent but are connected by the relation.

$$\mathbf{J_{-n}(x) = (-1)^n J_n(x).}$$

$$J_{-n}(x) = \sum_{k=0}^{\infty} \frac{(-1)^k}{k\,!\,\Gamma(-n+k+1)}\left(\frac{x}{2}\right)^{-n+2k}$$

Since Γ(a negative integer or zero) $\to \infty$, each term in the summation is zero so long as

$$-n + k + 1 \le 0, \textit{ i.e.}, k \le n - 1.$$

$\Gamma(-n + k + 1)$ is finite when $k \ge n$.

Thus, $$J_{-n}(x) = \sum_{k=0}^{n-1} \frac{(-1)^k}{k\,!\,\Gamma(-n+k+1)}\left(\frac{x}{2}\right)^{-n+2k} + \sum_{k=n}^{\infty} \frac{(-1)^k}{k\,!\,\Gamma(-n+k+1)}\left(\frac{x}{2}\right)^{-n+2k}$$

$$= \sum_{k=n}^{\infty} \frac{(-1)^k}{k\,!\,\Gamma(-n+k+1)}\left(\frac{x}{2}\right)^{-n+2k}$$

Putting $k = n + r$, we observe that when $k = n$, $r = 0$ and when $k \to \infty$, $r \to \infty$.

$$\therefore \quad J_{-n}(x) = \sum_{r=0}^{\infty} \frac{(-1)^{n+r}}{(n+r)\,!\,\Gamma(r+1)}\left(\frac{x}{2}\right)^{n+2r} = (-1)^n \sum_{r=0}^{\infty} \frac{(-1)^r}{(n+r)\,!\,\Gamma(r+1)}\left(\frac{x}{2}\right)^{n+2r}$$

Since r and n are integers, $\Gamma(r + 1) = r\,!$ and $(n + r)\,! = \Gamma(n + r + 1)$

$$\therefore \qquad J_{-n}(x) = (-1)^n \sum_{r=0}^{\infty} \frac{(-1)^r}{r\,!\,\Gamma(n+r+1)} \left(\frac{x}{2}\right)^{n+2r} = (-1)^n J_n(x)$$

Now, when n is an integer, y_2 fails to give a solution for positive values of n and y_1 fails to give a solution for negative values of n. Let us find an independent solution of Bessel's equation (1), when n is an integer.

Let $y = u(x)\,J_n(x)$ be a solution of (1) when n is integral.

Then $\quad \dfrac{dy}{dx} = u' J_n + u J_n' \quad$ and $\quad \dfrac{d^2y}{dx^2} = u'' J_n + 2u' J' + u J_n''$

Substituting the values of y, $\dfrac{dy}{dx}$ and $\dfrac{d^2y}{dx^2}$ in (1), we get

$$x^2(u''J_n + 2u'J_n' + uJ_n'') + x(u'J_n + uJ_n') + (x^2 - n^2)\,uJ_n = 0$$

or $$u[x^2J_n'' + xJ_n' + (x^2 - n^2)J_n] + x^2u''J_n + 2x^2u'J_n' + xu'J_n = 0$$

or $$x^2u''J_n + 2x^2u'J_n' + xu'J_n = 0, \text{ since } J_n \text{ is a solution of (1).}$$

Dividing throughout by $x^2u'J_n$, we get

$$\frac{u''}{u'} + 2\frac{J_n'}{J_n} + \frac{1}{x} = 0 \quad \text{or} \quad \frac{d}{dx}[\log u'] + 2\frac{d}{dx}[\log J_n] + \frac{d}{dx}(\log x) = 0$$

or $$\frac{d}{dx}[\log u' + 2\log J_n + \log x] = 0 \quad \text{or} \quad \frac{d}{dx}[\log(u' J_n^{\,2} x)] = 0$$

Integrating w.r.t. x, we get

$$\log(u'J_n^{\,2}x) = \log B \qquad \text{or} \qquad u'J_n^{\,2}x = B$$

or $$u' = \frac{B}{xJ_n^{\,2}} \qquad \text{or} \qquad u = B\int \frac{dx}{xJ_n^{\,2}} + A$$

Substituting the value of u in the assumed solution $y = u(x)\,J_n(x)$, we have

$$y = \left[B\int \frac{dx}{xJ_n^{\,2}(x)} + A\right] J_n(x) \quad \text{or} \quad y = AJ_n(x) + BY_n(x)$$

where $$Y_n(x) = J_n(x) \int \frac{dx}{xJ_n^{\,2}(x)}$$

The function $Y_n(x)$ is called the *Bessel function of the second kind of order n or Neumann function*.

6.6. SERIES REPRESENTATION OF BESSEL FUNCTIONS

Since $$J_n(x) = \sum_{k=0}^{\infty} \frac{(-1)^k}{k\,!\,\Gamma(n+k+1)} \left(\frac{x}{2}\right)^{n+2k}$$

$$\therefore \qquad J_0(x) = \sum_{k=0}^{\infty} \frac{(-1)^k}{k\,!\,\Gamma(k+1)} \left(\frac{x}{2}\right)^{2k}$$

$$= \sum_{k=0}^{\infty} \frac{(-1)^k}{(k!)^2}\left(\frac{x}{2}\right)^{2k}, \qquad \because \Gamma(k+1) = k\,!$$

$$= 1 - \left(\frac{x}{2}\right)^2 + \frac{1}{(2!)^2}\left(\frac{x}{2}\right)^4 - \frac{1}{(3!)^2}\left(\frac{x}{2}\right)^6 + \ldots\ldots$$

$$= 1 - \frac{x^2}{2^2} + \frac{x^4}{2^2 . 4^2} - \frac{x^6}{2^2 . 4^2 . 6^2} + \ldots\ldots$$

$$J_1(x) = \sum_{k=0}^{\infty} \frac{(-1)^k}{k!\,\Gamma(k+2)}\left(\frac{x}{2}\right)^{1+2k} = \sum_{k=0}^{\infty} \frac{(-1)^k}{k!\,(k+1)!}\left(\frac{x}{2}\right)^{1+2k}$$

$$= \frac{x}{2} - \frac{1}{2!}\left(\frac{x}{2}\right)^3 + \frac{1}{2!\,3!}\left(\frac{x}{2}\right)^5 - \ldots\ldots = \frac{x}{2} - \frac{x^3}{2^2 . 4} + \frac{x^5}{2^2 . 4^2 . 6} - \ldots\ldots$$

In particular, $J_0(0) = 1$ and $J_1(0) = 0$.

The values of $J_0(x)$ and $J_1(x)$ are given in 'Jahnke Emde's tables' to four decimal places at intervals of 0.1.

$$J_{1/2}(x) = \sum_{k=0}^{\infty} \frac{(-1)^k}{k!\,\Gamma(k+3/2)}\left(\frac{x}{2}\right)^{1/2+2k}$$

$$= \frac{1}{\Gamma(3/2)}\left(\frac{x}{2}\right)^{1/2} - \frac{1}{\Gamma(5/2)}\left(\frac{x}{2}\right)^{5/2} + \frac{1}{2!\,\Gamma(7/2)}\left(\frac{x}{2}\right)^{9/2} - \ldots\ldots$$

$$= \left(\frac{x}{2}\right)^{1/2}\left[\frac{1}{\frac{1}{2}\Gamma\left(\frac{1}{2}\right)} - \frac{1}{\frac{3}{2}.\frac{1}{2}\Gamma\left(\frac{1}{2}\right)}\left(\frac{x}{2}\right)^2 + \frac{1}{2.\frac{5}{2}.\frac{3}{2}.\frac{1}{2}\Gamma\left(\frac{1}{2}\right)}\left(\frac{x}{2}\right)^4 \ldots\ldots\right]$$

$$= \frac{\sqrt{x}}{\sqrt{2}\,\Gamma(1/2)}\left[\frac{2}{1!} - \frac{2x^2}{3!} + \frac{2x^4}{5!} - \ldots\ldots\right] = \frac{\sqrt{x}}{\sqrt{2}\sqrt{\pi}}.\frac{2}{x}\left[\frac{x}{1!} - \frac{x^3}{3!} + \frac{x^5}{5!}\ldots\ldots\right]$$

$$= \sqrt{\frac{2}{\pi x}}\,\sin x$$

Since $$J_{-n}(x) = \sum_{k=0}^{\infty} \frac{(-1)^k}{k!\,\Gamma(-n+k+1)}\left(\frac{x}{2}\right)^{-n+2k}$$

$\therefore$ $$J_{-1/2}(x) = \sum_{k=0}^{\infty} \frac{(-1)^k}{k!\,\Gamma(k+\frac{1}{2})}\left(\frac{x}{2}\right)^{-\frac{1}{2}+2k}$$

$$= \frac{1}{\Gamma\left(\frac{1}{2}\right)}\left(\frac{x}{2}\right)^{-1/2} - \frac{1}{\Gamma\left(\frac{3}{2}\right)}\left(\frac{x}{2}\right)^{3/2} + \frac{1}{2!\,\Gamma\left(\frac{5}{2}\right)}\left(\frac{x}{2}\right)^{7/2} - \ldots\ldots$$

$$= \left(\frac{x}{2}\right)^{-1/2}\left[\frac{1}{\Gamma\left(\frac{1}{2}\right)} - \frac{1}{\frac{1}{2}\Gamma\left(\frac{1}{2}\right)}\left(\frac{x}{2}\right)^2 + \frac{1}{2.\frac{3}{2}.\frac{1}{2}\Gamma\left(\frac{1}{2}\right)}\left(\frac{x}{2}\right)^4 - \ldots\ldots\right]$$

$$= \frac{\sqrt{2}}{\sqrt{x}\,\Gamma\left(\frac{1}{2}\right)}\left[1 - \frac{x^2}{2!} + \frac{x^4}{4!} - \ldots\ldots\right] = \frac{\sqrt{2}}{\sqrt{x}\,\sqrt{\pi}}\left[1 - \frac{x^2}{2!} + \frac{x^4}{4!} - \ldots\ldots\right]$$

$$= \sqrt{\frac{2}{\pi x}}\cos x.$$

6.7. RECURRENCE RELATIONS OF BESSEL FUNCTIONS

The following recurrence relations connect Bessel functions of different orders and are very helpful in the solution of problems involving Bessel functions.

(I) $\frac{d}{dx}[x^n J_n(x)] = x^n J_{n-1}(x)$

Since $$J_n(x) = \sum_{k=0}^{\infty} \frac{(-1)^k}{k!\,\Gamma(n+k+1)}\left(\frac{x}{2}\right)^{n+2k}$$

$$\therefore \quad x^n J_n(x) = \sum_{k=0}^{\infty} \frac{(-1)^k x^{2n+2k}}{2^{n+2k}\,k!\,\Gamma(n+k+1)}$$

$$\frac{d}{dx}[x^n J_n(x)] = \sum_{k=0}^{\infty} \frac{(-1)^k.(2n+2k)\,x^{2n+2k-1}}{2^{n+2k}\,k!\,\Gamma(n+k+1)} = x^n \sum_{k=0}^{\infty} \frac{(-1)^k (n+k)\,x^{n+2k-1}}{2^{n+2k-1}\,k!\,(n+k)\,\Gamma(n+k)}$$

$$= x^n \sum_{k=0}^{\infty} \frac{(-1)^k}{k!\,\Gamma(n+k)}\left(\frac{x}{2}\right)^{n+2k-1} = x^n \sum_{k=0}^{\infty} \frac{(-1)^k}{k!\,\Gamma(n-1+k+1)}\left(\frac{x}{2}\right)^{n-1+2k}$$

$$= x^n J_{n-1}(x).$$

(II) $\frac{d}{dx}[x^{-n} J_n(x)] = -x^{-n} J_{n+1}(x)$

Since $$J_n(x) = \sum_{k=0}^{\infty} \frac{(-1)^k}{k!\,\Gamma(n+k+1)}\left(\frac{x}{2}\right)^{n+2k}$$

$$\therefore \quad x^{-n} J_n(x) = \sum_{k=0}^{\infty} \frac{(-1)^k x^{2k}}{2^{n+2k}.k!\,\Gamma(n+k+1)}$$

$$\frac{d}{dx}[x^{-n} J_n(x)] = \sum_{k=1}^{\infty} \frac{(-1)^k . 2kx^{2k-1}}{2^{n+2k} . k . (k-1)! \Gamma(n+k+1)}$$

$$= -x^{-n} \sum_{k=1}^{\infty} \frac{(-1)^{k-1} x^{n+2k-1}}{2^{n+3k-1} (k-1)! \Gamma(n+k+1)}$$

$$= -x^{-n} \sum_{k=0}^{\infty} \frac{(-1)^r x^{n+2k+1}}{2^{n+2r+1} . r! \Gamma(n+r+2)} \quad \text{where } r = k-1$$

$$= -x^{-n} \sum_{r=0}^{\infty} \frac{(-1)^r}{r! \Gamma(n+1+r+1)} \left(\frac{x}{2}\right)^{n+1+2r} = -x^{-n} J_{n+1}(x)$$

In particular, when $n = 0$, we have

$$\frac{d}{dx}[J_0(x)] = -J_1(x) \quad \text{or} \quad \mathbf{J_0' = -J_1}.$$

(III) $\mathbf{J_n'(x) + \frac{n}{x} J_n(x) = J_{n-1}(x)}$

(IV) $\mathbf{J_n'(x) - \frac{n}{x} J_n(x) = -J_{n+1}(x)}$

(V) $\mathbf{2J_n'(x) = J_{n-1}(x) - J_{n+1}(x)}$

(VI) $\mathbf{\frac{2n}{x} J_n(x) = J_{n-1}(x) + J_{n+1}(x).}$

Since $$\frac{d}{dx}[x^n J_n(x)] = x^n J_{n-1}(x)$$

$\therefore$ $$x^n J_n'(x) + nx^{n-1} J_n(x) = x^n J_{n-1}(x)$$

Dividing by x^n, we have $$J_n'(x) + \frac{n}{x} J_n(x) = J_{n-1}(x) \quad \text{...(III)}$$

Also $$\frac{d}{dx}[x^{-n} J_n(x)] = -x^{-n} J_{n+1}(x)$$

$\therefore$ $$x^{-n} J_n'(x) - nx^{-n-1} J_n(x) = -x^{-n} J_{n+1}(x)$$

Multiplying by x^n, we have $$J_n'(x) - \frac{n}{x} J_n(x) = -J_{n+1}(x) \quad \text{...(IV)}$$

Adding (III) and (IV), we get $$2J_n'(x) = J_{n-1}(x) - J_{n+1}(x) \quad \text{...(V)}$$

Subtracting (IV) from (III), we get $$\frac{2n}{x} J_n(x) = J_{n-1}(x) + J_{n+1}(x) \quad \text{...(VI)}$$

Another Form of (VI)

$$J_n(x) = \frac{x}{2n}[J_{n-1}(x) + J_{n+1}(x)].$$

ILLUSTRATIVE EXAMPLES

Example 1. *Show that:*

(a) $J_{5/2}(x) = \sqrt{\frac{2}{\pi x}}\left(\frac{3-x^2}{x^2}\sin x - \frac{3}{x}\cos x\right)$

(b) $J_4(x) = \left(\frac{48}{x^3} - \frac{8}{x}\right)J_1(x) + \left(1 - \frac{24}{x^2}\right)J_0(x).$

Sol. (*a*) We know that $\quad J_{n+1}(x) = \frac{2n}{x}J_n(x) - J_{n-1}(x) \qquad ...(1)$

Putting $n = \frac{3}{2}$ in (1), we have $\quad J_{5/2}(x) = \frac{3}{x}J_{3/2}(x) - J_{1/2}(x)$

and putting $n = \frac{1}{2}$, we have $\quad J_{3/2}(x) = \frac{1}{x}J_{1/2}(x) - J_{-1/2}(x)$

$$\therefore \quad J_{5/2}(x) = \frac{3}{x}\left[\frac{1}{x}J_{1/2}(x) - J_{-1/2}(x)\right] - J_{1/2}(x)$$

$$= \left(\frac{3}{x^2} - 1\right)J_{1/2}(x) - \frac{3}{x}J_{-1/2}(x) = \frac{3-x^2}{x^2}\cdot\sqrt{\frac{2}{\pi x}}\sin x - \frac{3}{x}\cdot\sqrt{\frac{2}{\pi x}}\cos x$$

$$= \sqrt{\frac{2}{\pi x}}\left(\frac{3-x^2}{x^2}\sin x - \frac{3}{x}\cos x\right).$$

(*b*) Putting $n = 1$ in (1), we get $\quad J_2(x) = \frac{2}{x}J_1(x) - J_0(x) \qquad ...(2)$

Putting $n = 2$ in (1), we get $\quad J_3(x) = \frac{4}{x}J_2(x) - J_1(x) \qquad ...(3)$

$\therefore$ Putting $n = 3$ in (1), $\quad J_4(x) = \frac{6}{x}J_3(x) - J_2(x)$

$$= \frac{6}{x}\left[\frac{4}{x}J_2(x) - J_1(x)\right] - J_2(x) \qquad \text{[Using (3)]}$$

$$= \left(\frac{24}{x^2} - 1\right)J_2(x) - \frac{6}{x}J_1(x) = \left(\frac{24}{x^2} - 1\right)\left(\frac{2}{x}J_1(x) - J_0(x)\right) - \frac{6}{x}J_1(x) \qquad \text{[Using (2)]}$$

$$= \left(\frac{48}{x^3} - \frac{2}{x} - \frac{6}{x}\right)J_1(x) + \left(1 - \frac{24}{x^2}\right)J_0(x) = \left(\frac{48}{x^3} - \frac{8}{x}\right)J_1(x) + \left(1 - \frac{24}{x^2}\right)J_0(x).$$

Example 2. *Prove that:* $4J_n''(x) = J_{n-2}(x) - 2J_n(x) + J_{n+2}(x)$.

Sol. We know that $\quad 2J_n'(x) = J_{n-1}(x) - J_{n+1}(x) \qquad ...(1)$

Differentiating (1), we have $\quad 2J_n''(x) = J'_{n-1}(x) - J'_{n+1}(x) \qquad ...(2)$

Changing n to $(n - 1)$ and $(n + 1)$ in (1), we get

$$2J'_{n-1}(x) = J_{n-2}(x) - J_n(x)$$

$$2J'_{n+1}(x) = J_n(x) - J_{n+2}(x)$$

Subtracting $\quad 2[J'_{n-1}(x) - J'_{n+1}(x)] = J_{n-2}(x) - 2J_n(x) + J_{n+2}(x)$

$\therefore$ From (2), we have $\quad 4J_n''(x) = J_{n-2}(x) - 2J_n(x) + J_{n+2}(x).$

Example 3. *Show that:* $4J_0'''(x) + 3J_0'(x) + J_3(x) = 0.$

Sol. We know that $J_0'(x) = - J_1(x)$

Differentiating w.r.t. x, we get

$$J_0''(x) = - J_1'(x) = - \frac{1}{2} [J_0(x) - J_2(x)]$$

[From $2J_n'(x) = J_{n-1}(x) - J_{n+1}(x)$, with $n = 1$]

Differentiating again w.r.t. x, we get

$$J_0''(x) = - \frac{1}{2} J_0'(x) + \frac{1}{2} J_2'(x)$$

$$= - \frac{1}{2} J_0'(x) + \frac{1}{2} . \frac{1}{2} [J_1(x) - J_3(x)]$$

[From $2J_n'(x) = J_{n-1}(x) - J_{n+1}(x)$, with $n = 2$]

$$= - \frac{1}{2} J_0'(x) - \frac{1}{4} J_0'(x) - \frac{1}{4} J_3(x) \qquad [\because \; J_1(x) = - J_0'(x)]$$

$\Rightarrow \quad 4J_0'''(x) + 3J_0'(x) + J_3(x) = 0.$

Example 4. *Prove that:*

(*i*) $\dfrac{d}{dx} [xJ_n(x) J_{n+1}(x)] = x [J_n^2(x) - J_{n+1}^2(x)]$

(*ii*) $\dfrac{d}{dx} [J_n^2(x) + J_{n+1}^2(x)] = 2\left[\dfrac{n}{x} J_n^2(x) - \dfrac{n+1}{x} J^2{}_{n+1}(x)\right].$

Sol. (*i*) $\dfrac{d}{dx} [x\, J_n(x)\, J_{n+1}(x)] = \dfrac{d}{dx} [x^{-n}J_n(x) \,.\, x^{n+1} J_{n+1}(x)]$

$$= x^{-n} J_n(x) \,.\, \frac{d}{dx} [x^{n+1} J_{n+1}(x)] + x^{n+1} J_{n+1}(x) \,.\, \frac{d}{dx} [x^{-n} J_n(x)] \quad ...(1)$$

Now, $\quad \dfrac{d}{dx} [x^n J_n(x)] = x^n J_{n-1}(x)$

Changing n to $(n + 1)$, we have

$$\frac{d}{dx} [x^{n+1}J_{n+1}(x)] = x^{n+1}J_n(x)$$

Also, $\quad \dfrac{d}{dx} [x^{-n}J_n(x)] = - x^{-n}J_{n+1}(x)$

$\therefore$ From (1), we have

$$\frac{d}{dx}[xJ_n(x)\,J_{n+1}(x)] = x^{-n}J_n(x)\,.\,x^{n+1}J_n(x) + x^{n+1}J_{n+1}(x)\,.\,[-x^{-n}J_{n+1}(x)]$$

$$= x\,[J_n^2(x) - J_{n+1}^2(x)].$$

(*ii*) $$\frac{d}{dx}[J_n^2(x) + J_{n+1}^2(x)] = 2J_n(x)\,J_n'(x) + 2J_{n+1}(x)J'_{n+1}(x) \quad ...(1)$$

Now, $$J_n'(x) = \frac{n}{x}J_n(x) - J_{n+1}(x) \quad ...(2)$$

Also, $$J_n'(x) = -\frac{n}{x}J_n(x) + J_{n-1}(x)$$

Replacing n by $(n + 1)$, we have

$$J'_{n+1}(x) = -\frac{n+1}{x}J_{n+1}(x) + J_n(x) \quad ...(3)$$

Substituting the values of $J_n'(x)$ and $J'_{n+1}(x)$ from (2) and (3) in (1), we get

$$\frac{d}{dx}[J_n^2(x) + J_{n+1}^2(x)] = 2J_n(x)\left[\frac{n}{x}J_n(x) - J_{n+1}(x)\right] + 2J_{n+1}(x)\left[-\frac{n+1}{x}J_{n+1}(x) + J_n(x)\right]$$

$$= 2\left[\frac{n}{x}J_n^2(x) - \frac{n+1}{x}J_{n+1}^2(x)\right].$$

Example 5. *Prove that:*

(*i*) $$\int J_3(x)\,dx = -J_2(x) - \frac{2}{x}J_2(x)$$

(*ii*) $$\int xJ_0^2\,dx = \frac{1}{2}x^2[J_0^2(x) + J_1^2(x)] + c,$$ *where c is a constant.*

Sol. (*i*) We know that $\frac{d}{dx}[x^{-n}J_n(x)] = -x^{-n}J_{n+1}(x)$

$$\therefore \quad \int x^{-n}J_{n+1}(x)\,dx = -x^{-n}J_n(x) \quad ...(1)$$

Now, $$\int J_3(x)\,dx = \int x^2\,[x^{-2}J_3(x)]\,dx$$

Integrating by parts $$= x^2\,[-x^{-2}J_2(x)] - \int 2x\,[-x^{-2}J_2(x)]\,dx \quad \text{[By putting } n = 2 \text{ in (1)]}$$

$$= -J_2(x) + 2\int x^{-1}J_2(x)\,dx$$

$$= -J_2(x) + 2\,[-x^{-1}J_1(x)] \quad \text{[By putting } n = 1 \text{ in (1)]}$$

$$= -J_2(x) - \frac{2}{x}J_1(x).$$

(ii) $$\int x J_0^2(x)\,dx = \int J_0^2(x)\,.\,x dx$$

Integrating by parts $$= J_0^2(x)\,.\,\frac{x^2}{2} - \int 2 J_0(x) J_0^1(x)\,.\,\frac{x^2}{2} dx + c$$

$$= \frac{1}{2}x^2 J_0^2(x) - \int x^2 J_0(x)[-J_1(x)]\,dx + c \qquad [\because\ J_0'(x) = -J_1(x)]$$

$$= \frac{1}{2}x^2 J_0^2(x) + \int xJ_1(x)\,.\,xJ_0(x)\,dx + c$$

$$= \frac{1}{2}x^2 J_0^2(x) + \int xJ_1(x)\,\frac{d}{dx}[xJ_1(x)]\,dx + c$$

$$= \frac{1}{2}x^2 J_0^2(x) + \frac{[x J_1(x)^2]}{2} + c = \frac{1}{2}x^2 [J_0^2(x) + J_1^2(x)] + c.$$

6.8. GENERATING FUNCTION FOR $J_n(x)$

We shall now show that Bessel functions of various orders can be derived as co-efficients of various powers of t in the expansion of the function $e^{\frac{x}{2}\left(t-\frac{1}{t}\right)}$.

i.e., $$\mathbf{e^{\frac{x}{2}\left(t-\frac{1}{t}\right)} = \sum_{n=-\infty}^{\infty} t^n J_n(x).}$$

$$\frac{x}{2}\left(t-\frac{1}{t}\right) = e^{\frac{xt}{2}}\,.\,e^{-\frac{x}{2t}}$$

$$= \left[1+\left(\frac{xt}{2}\right)+\frac{1}{2!}\left(\frac{xt}{2}\right)^2+\frac{1}{3!}\left(\frac{xt}{2}\right)^3+\ldots\ldots\right]$$

$$\times\left[1+\left(\frac{x}{2t}\right)+\frac{1}{2!}\left(\frac{x}{2t}\right)^2-\frac{1}{3!}\left(\frac{x}{2t}\right)^3+\ldots\ldots\right]$$

The co-efficient of t^n in the product

$$= \frac{1}{n!}\left(\frac{x}{2}\right)^n - \frac{1}{(n+1)!}\left(\frac{x}{2}\right)^{n+2} + \frac{1}{2!(n+2)}!\left(\frac{x}{2}\right)^{n+4} - \ldots\ldots$$

$$= \sum_{k=0}^{\infty} \frac{(-1)^k}{k!(n+k)!}\left(\frac{x}{2}\right)^{n+2k} = J_n(x)$$

Similarly, the co-efficient of $t^{-\pi}$ in the product

$$= \frac{(-1)^n}{n!}\left(\frac{x}{2}\right)^n + \frac{(-1)^{n+1}}{(n+1)!}\left(\frac{x}{2}\right)^{n+2} + \frac{(-1)^{n+2}}{2!(n+2)!}\left(\frac{x}{2}\right)^{n+4} + \ldots\ldots$$

$$= (-1)^n \left[\frac{1}{n!}\left(\frac{x}{2}\right)^n - \frac{1}{(n+1)!}\left(\frac{x}{2}\right)^{n+2} + \frac{1}{2!(n+2)!}\left(\frac{x}{2}\right)^{n+4} \ldots\ldots\right]$$

$$= (-1)^n \sum_{k=0}^{\infty} \frac{(-1)^k}{k!(n+k)!}\left(\frac{x}{2}\right)^{n+2k} = (-1)^n J_n(x) = J_{-n}(x)$$

$$\therefore \quad e^{\frac{x}{2}\left(t-\frac{1}{t}\right)} = J_0(x) + tJ_1(x) + t^2J_2(x) + \ldots\ldots + t^nJ_n(x) + \ldots\ldots$$
$$+ t^{-1}J_{-1}(x) + t^{-2}J_{-2}(x) + \ldots\ldots t^{-n}J_{-n}(x) + \ldots\ldots$$
$$= \sum_{n=-\infty}^{\infty} t^n J_n(x)$$

The function $e^{\frac{x}{2}\left(t-\frac{1}{t}\right)}$ is called the **generating function**. It must be noted that the above result is true if n is an integer.

6.9. INTEGRAL FORM OF BESSEL FUNCTION

We know that

$$e^{\frac{x}{2}\left(t-\frac{1}{t}\right)} = \sum_{n=-\infty}^{\infty} t^n J_n(x)$$
$$= J_0(x) + tJ_1(x) + t^2J_2(x) + t^3J_3(x) + \ldots\ldots + t^{-1}J_{-1}(x) + t^{-2}J_{-2}(x) + t^{-3}J_{-3}(x) + \ldots\ldots$$
$$= J_0(x) + tJ_1(x) + t^2J_2(x) + t^3J_3(x) + \ldots\ldots - t^{-1}J_1(x) + t^{-2}J_2(x) - t^{-3}J_3(x) + \ldots\ldots$$
$$[\because \; J_{-n}(x) = (-1)^n J_n(x)]$$

$$= J_0(x) + \left(t - \frac{1}{t}\right)J_1(x) + \left(t^2 + \frac{1}{t^2}\right)J_2(x) + \left(t^3 - \frac{1}{t^3}\right)J_3(x) + \ldots\ldots \qquad \ldots(1)$$

Put $\quad t = \cos\theta + i\sin\theta$

By De Moivre's theorem,

$$t^n = \cos n\theta + i\sin n\theta \quad \text{and} \quad \frac{1}{t^n} = \cos n\theta - i\sin n\theta$$

so that $t^n + \dfrac{1}{t^n} = 2\cos n\theta$ and $t^n - \dfrac{1}{t^n} = 2i\sin n\theta$

Substituting these values in (1), we have

$$e^{ix\sin\theta} = J_0(x) + 2i\sin\theta\, J_1(x) + 2\cos 2\theta\, J_2(x) + 3i\sin 3\theta\, J_3(x) + \ldots\ldots \qquad \ldots(2)$$

Since $\quad e^{ix\sin\theta} = \cos(x\sin\theta) + i\sin(x\sin\theta)$

$\therefore$ Equating the real and imaginary parts in (2), we get

$$\mathbf{\cos(x\sin\theta) = J_0(x) + 2[J_2(x)\cos 2\theta + J_4(x)\cos 4\theta + \ldots\ldots]} \qquad \ldots(3)$$

$$\mathbf{\sin(x\sin\theta) = 2(J_1(x)\sin\theta + J_3(x)\sin 3\theta + \ldots\ldots]} \qquad \ldots(4)$$

These are known as **Jacobi series.**

Multiplying both sides of (3) by cos $n\theta$ and integrating w.r.t. θ between the limits 0 and π (when n is odd, all terms on the R.H.S. vanish ; when n is even, all terms on the R.H.S. except the one containing cos $n\theta$ vanish), we get

$$\int_0^{\pi} \cos(x \sin\theta)\cos n\theta \, d\theta = \begin{cases} 0 & \text{, when } n \text{ is odd} \\ \pi J_n(x), & \text{when } n \text{ is even} \end{cases}$$

Similarly, multiplying (4) by sin $n\theta$ and integrating w.r.t. θ between the limits 0 and π, we get

$$\int_0^{\pi} \sin(x \sin\theta)\sin n\theta \, d\theta = \begin{cases} \pi J_n(x), & \text{when } n \text{ is odd} \\ 0, & \text{when } n \text{ is even} \end{cases}$$

Adding, we get $\int_0^{\pi} [\cos(x\sin\theta)\cos n\theta + \sin(x\sin\theta)\sin n\theta]\, d\theta = \pi J_n(x)$

$$\Rightarrow \qquad J_n(x) = \frac{1}{\pi}\int_0^{\pi} \cos(n\theta - x\sin\theta)\, d\theta$$

for all integral values of n.

Example 6. *Use Jacobi series to prove that:*

$$[J_0(x)]^2 + 2[J_1(x)]^2 + 2I(J_2(x)]^2 + 2[J_3(x)]^2 + ... = 1.$$

Sol. The Jacobi series are

$$J_0(x) + 2J_2(x)\cos 2\theta + 2J_4(x)\cos 4\theta + ... = \cos(x\sin\theta) \qquad ...(1)$$

and
$$2J_1(x)\sin\theta + 2J_3(x)\sin 3\theta + ... = \sin(x\sin\theta) \qquad ...(2)$$

Squaring (1) and (2) and integrating w.r.t. θ between the limits 0 and π, and remembering that if m, n are integers then

$$\int_0^{\pi} \cos^2 n\theta \, d\theta = \int_0^{\pi} \sin^2 n\theta \, d\theta = \frac{\pi}{2}$$

and
$$\int_0^{\pi} \cos m\theta \cos n\theta \, d\theta = \int_0^{\pi} \sin m\theta \sin n\theta \, d\theta = 0, \; m \neq n, \text{ we get}$$

$$[J_0(x)]^2 \pi + 2\,[J_2(x)]^2 \pi + 2[J_4(x)]^2 \pi + ... = \int_0^{\pi} \cos^2(x\sin\theta)\, d\theta$$

$$2\,[J_1(x)]^2 \pi + 2\,[J_3(x)]^2 \pi + ... = \int_0^{\pi} \sin^2(x\sin\theta)\, d\theta$$

Adding, we have $\pi\,\{[J_0(x)]^2 + 2[J_1(x)]^2 + 2[J_2(x)]^2 + ...\}$

$$= \int_0^{\pi} [\cos^2(x\sin\theta) + \sin^2(x\sin\theta)]\, d\theta = \int_0^{\pi} d\theta = \pi$$

$$\therefore \quad [J_0(x)]^2 + 2[J_1(x)]^2 + 2[J_2(x)]^2 + ... = 1.$$

TEST YOUR KNOWLEDGE

1. Show that:

(i) $J_{3/2}(x) = \sqrt{\frac{2}{\pi x}}\left(\frac{\sin x}{x} - \cos x\right)$ (ii) $J_{1/2}(x) = J_{-1/2}(x) \tan x$

(iii) $J_{-3/2}(x) = -\sqrt{\frac{2}{\pi x}}\left(\sin x - \frac{\cos x}{x}\right)$ (iv) $J_{-5/2}(x) = \sqrt{\frac{2}{\pi x}}\left(\frac{3}{x}\sin x + \frac{3-x^2}{x^2}\cos x\right)$

(v) $[J_{1/2}(x)]^2 + [J_{-1/2}(x)]^2 = \frac{2}{\pi x}$.

2. Show that:

(i) $J_0'' = \frac{1}{2}(J_2 - J_0)$ (ii) $J_2 = J_0'' - x^{-1}J_0'$ (iii) $\frac{J_2}{J_1} = \frac{1}{x} - \frac{J_0''}{J'_0}$.

3. Prove that:

(i) $\frac{d}{dx}[xJ_1(x)] = xJ_0(x)$ (ii) $\frac{d}{dx}[x^nJ_n(ax)] = ax^n J_{n-1}(ax)$

(iii) $\frac{d}{dx}[J_n^2(x)] = \frac{x}{2n}[J_{n-1}^2(x) - J_{n+1}^2(x)]$ (iv) $x^2J_n''(x) = (n^2 - n - x^2)J_n(x) + xJ_{n+1}(x)$.

4. Prove that:

(i) $\int J_0(x)J_1(x)\,dx = -\frac{1}{2}J_0^2(x)$ (ii) $\int x^2 J_0(x)J_1(x)\,dx = \frac{1}{2}x^2J_1^2(x)$

(iii) $\int \frac{J_4(x)}{x}dx = -\frac{1}{x}J_3(x) - \frac{2}{x^2}J_2(x)$ (iv) $\int J_5(x)\,dx = -J_4(x) - \frac{4}{x}J_3(x) - \frac{8}{x^2}J_2(x)$

(v) $\int x^3J_0(x)\,dx = x^3J_1(x) - 2x^2J_2(x)$ (vi) $\int_0^{\alpha} x J_0(\lambda x)\,dx = \frac{\alpha}{\lambda}J_1(\alpha\lambda)$.

5. Prove that:

(i) $J_n(x) = \frac{1}{2\pi}\int_0^{2\pi}\cos(x\sin\theta - n\theta)\,d\theta$

(ii) $\cos x = J_0 - 2J_2 + 2J_4 - \ldots\ldots$; $\sin x = 2J_1 - 2J_3 + 2J_5 - \ldots\ldots$

[**Hint.** Put $\theta = \frac{\pi}{2}$ in Jacobi series.]

(iii) $\cos(x\cos\theta) = J_0 - 2J_2\cos 2\theta + 2J_4\cos 4\theta - \ldots\ldots$
$\sin(x\cos\theta) = 2J_1\cos\theta - 2J_3\cos 3\theta + 2J_5\cos 5\theta - \ldots\ldots$

[**Hint.** Replace θ by $\frac{\pi}{2} - \theta$ in Jacobi series.]

(iv) $J_0(x) = \frac{1}{\pi}\int_0^{\pi}\cos(x\sin\theta)\,d\theta = \frac{2}{\pi}\int_0^{\pi/2}\cos(x\sin\theta)\,d\theta = \frac{2}{\pi}\int_0^{\pi/2}\cos(x\cos\theta)\,d\theta$

(v) $J_0 + 2J_2 + 2J_4 + 2J_6 + \ldots\ldots = 1$. (vi) $\int_0^{\pi/2}\sqrt{\pi x}\,J_{1/2}(2x)\,dx = 1$.

6.10. EQUATIONS REDUCIBLE TO BESSEL'S EQUATION

A number of second order differential equations with variable co-efficients can be reduced to Bessel's equation by a suitable transformation and, hence, can be solved in terms of Bessel functions.

Consider the differential equation

$$x^2 \frac{d^2y}{dx^2} + (1 - 2\alpha)x \frac{dy}{dx} + [\beta^2\gamma^2x^{2r} + (\alpha^2 - n^2\gamma^2)]y = 0 \qquad ...(1)$$

where, α, β, γ and n are constants.

Putting $X = \beta x^r$ and $Y = x^{-\alpha} y$, equation (1) reduces to

$$X^2 \frac{d^2Y}{dX^2} + X\frac{dY}{dX} + (X^2 - n^2)Y = 0 \qquad ...(2)$$

which is Bessel's equation.

When n is not an integer, the solution of (2) is

$$Y = c_1J_n(X) + c_2J_{-n}(X)$$

and, hence, the solution of (1) is

$$x^{-\alpha}y = c_1J_n(\beta x^\gamma) + c_2J_{-n}(\beta x^\gamma)$$

or $$y = x^\alpha [c_1J_n(\beta x^\gamma) + c_2J_{-n}(\beta x^\gamma)].$$

When n is an integer, the solution of (2) is

$$Y = c_1J_n(X) + c_2Y_n(X)$$

and, hence, the solution of (1) is $y = x^\alpha [c_1J_n(\beta x^\gamma) + c_2Y_n(\beta x^\gamma)]$.

Equation (1) is a general form of Bessel's equation with α, β, γ and n as parameters. Comparing the given equation with (1), we get specific values of the parameters and hence the solution.

Example. *Solve the following differential equations in terms of Bessel functions*:

(*i*) $y'' - \frac{2}{x}y' + 4\left(x^2 - \frac{1}{x^2}\right)y = 0$ (*ii*) $xy'' - 3y' + xy = 0$.

Sol. (*i*) The given equation is $x^2y'' - 2xy' + (4x^4 - 4)y = 0$

Comparing with the general form, we get

$$1 - 2\alpha = -2, \quad \beta^2\gamma^2 = 4, \quad 2y = 4, \quad \alpha^2 - n^2\gamma^2 = -4.$$

i.e., $$\alpha = \frac{3}{2}, \quad \beta = 1, \quad \gamma = 2, \quad n = \frac{5}{4}.$$

Here n is not integral and the solution is

$$y = x^{3/2} [c_1J_{5/4}(x^2) + c_2J_{-5/4}(x^2)]$$

(*ii*) Multiplying by x, the given equation becomes

$$x^2y'' - 3xy' + x^2y = 0$$

Comparing with the general form, we get

$$1 - 2\alpha = -3, \quad \beta^2\gamma^2 = 1, 2y = 2, \alpha^2 - n^2\gamma^2 = 0$$

i.e., $$\alpha = 2, \quad \beta = \gamma = 1, n = 2$$

Here n is integral and the solution is $y = x^2[c_1J_2(x) + c_2Y_2(x)]$.

6.11. MODIFIED BESSEL'S EQUATION

The differential equation $$x^2 \frac{d^2y}{dx^2} + x\frac{dy}{dx} - (x^2 + n^2)\, y = 0 \qquad ...(1)$$

is called modified Bessel's equation of order n.

Equation (1) can be re-written as $$x^2 \frac{d^2y}{dx^2} + x\frac{dy}{dx} + (i^2x^2 - n^2)\, y = 0$$

When n is not an integer, its solution is given by $y = c_1 J_n(ix) + c_2 J_{-n}(ix)$

Now, $$J_n(ix) = \sum_{k=0}^{\infty} \frac{(-1)^k}{k!\,\Gamma(n+k+1)} \left(\frac{ix}{2}\right)^{n+2k}$$

$$= i^n \sum_{r=0}^{\infty} \frac{1}{k!\,\Gamma(n+k+1)} \left(\frac{x}{2}\right)^{n+2k} \qquad [\because\ (-1)^k (i)^{2k} = i^{2k}\,.\,i^{2k} = i^{4k} = 1]$$

The series $$\sum_{k=0}^{\infty} \frac{1}{k!\,\Gamma(n+k+1)} \left(\frac{x}{2}\right)^{n+2k}$$

is a real function with all terms positive. It is denoted by $I_n(x)$ and is called the *modified Bessel function of the first kind of order n.* Thus

$$I_n(x) = i^{-n} J_n(ix)$$

Since i^{-n} is a constant, $I_n(x)$ is also a solution of (1).

If n is not an integer, a second independent solution of (1) is $I_{-n}(x)$,

where $$I_{-n}(x) = \sum_{r=0}^{\infty} \frac{1}{k!\,\Gamma(-n+k+1)} \left(\frac{x}{2}\right)^{-n+2k}$$

Thus, if n is not an integer, the complete solution of (1) is given by

$$y = c_1 I_n(x) + c_2 I_{-n}(x)$$

If n is a non-zero integer, a second independent solution of (1) is given by

$$K_n(x) = \frac{\pi/2}{\sin n\pi}\, [I_{-n}(x) - J_n(x)]$$

and is called modified Bessel function of second kind of order n.

In this case, the complete solution of (1) is given by

$$y = c_1 I_n(x) + c_2 K_n(x).$$

6.12. BER AND BEI FUNCTIONS

Consider the differential equation $$x\frac{d^2y}{dx^2} + \frac{dy}{dx} - ixy = 0 \qquad ...(1)$$

Comparing it with equation (1) of Art. 6.10, we have

$$\alpha = 0, \quad n = 0, \quad \gamma = 1, \quad \text{and} \quad \beta^2 = -i \quad \text{or} \quad \beta^2 = i^3$$

so that $$\beta = i^{3/2}.$$

Hence a solution of (1) is given by $J_0(i^{3/2}x)$

Replacing x by $i^{3/2}\,x$ in the series for $J_0(x)$, we have

$$J_0(i^{3/2}x) = 1 - \frac{i^3x^2}{2^2} + \frac{i^6x^4}{(2\,!)^2\,.\,2^4} - \frac{i^9x^6}{(3\,!)^2\,.\,2^6} + \frac{i^{12}x^8}{(4\,!)^2\,.\,2^8} - \ldots\ldots$$

$$= \left[1 - \frac{x^4}{2^2\,.\,4^2} + \frac{x^8}{2^2\,.\,4^4\,.6^2\,.\,8^2} - \ldots\ldots\right]$$

$$+ i\left[\frac{x^2}{2^2} - \frac{x^6}{2^2\,.\,4^2\,.\,6^2} + \frac{x^{10}}{2^2\,.\,4^2\,.\,6^2\,.\,8^2\,.\,10^2} - \ldots\ldots\right]$$

Thus $J_0\,(i^{2/3}x)$ is a complex function for real values of x. The real and the imaginary parts are denoted by ber (x) (Bessel-real) and bei (x) (Bessel-imaginary) respectively. Thus

$$ber(x) = 1 + \sum_{k=1}^{\infty} \frac{(-1)^k\,x^{4k}}{2^2\,.\,4^2\,.\,6^2\,\ldots\ldots\,(4k)^2}$$

and

$$bei(x) = 1 - \sum_{k=1}^{\infty} \frac{(-1)^k\,x^{4k-2}}{2^2\,.\,4^2\,.\,6^2\,\ldots\ldots\,(4k-2)^2}$$

Hence a solution of (1) is

$$y = J_0(i^{3/2}x) = ber\,(x) + i\;bei\,(x).$$

Example. *Show that:*

(*a*) $\dfrac{d}{dx}\,[x\;ber'(x)] = -\,x\;bei\,(x)$ (*b*) $\dfrac{d}{dx}\,[x\;bei'(x)] = x\;ber\,(x).$

Sol. We know that $ber(x) = 1 - \dfrac{x^4}{2^2\,.\,4^2} + \dfrac{x^8}{2^2\,.\,4^2\,.\,6^2\,.\,8^2} - \ldots\ldots$

$$bei(x) = \frac{x^2}{2^2} - \frac{x^6}{2^2\,.\,4^2\,.\,6^2} + \frac{x^{10}}{2^2\,.\,4^2\,.\,6^2\,.\,8^2\,.\,10^2} - \ldots\ldots$$

(*a*) Now, $ber'(x) = -\dfrac{x^3}{2^2\,.\,4} + \dfrac{x^7}{2^2\,.\,4^2\,.\,6^2\,.\,8} - \ldots\ldots$

$\therefore$ $x\;ber'(x) = -\dfrac{x^4}{2^2\,.\,4} + \dfrac{x^8}{2^2\,.\,4^2\,.\,6^2\,.\,8} - \ldots\ldots$

$$\frac{d}{dx}\,[x\;ber'(x)] = -\frac{x^3}{2^2} + \frac{x^7}{2^2\,.\,4^2\,.\,6^2} - \ldots\ldots = -\,x\left(\frac{x^2}{2^2} - \frac{x^6}{2^2\,.\,4^2\,.\,6^2} + \ldots\ldots\right)$$

$$= -\,x\;bei\,(x).$$

(*b*) Also, $bei'(x) = \dfrac{x}{2} - \dfrac{x^5}{2^2\,.\,4^2\,.\,6} + \dfrac{x^9}{2^2\,.\,4^2\,.\,6^2\,.\,8^2\,.\,10} - \ldots\ldots$

$$\therefore \qquad x\, bei'(x) = \frac{x^2}{2} - \frac{x^6}{2^2 . 4^2 . 6} + \frac{x^{10}}{2^2 . 4^2 . 6^2 . 8^2 . 10} - \ldots\ldots$$

$$\frac{d}{dx}[x\, bei'(x)] = x - \frac{x^5}{2^2 . 4^2} + \frac{x^9}{2^2 . 4^2 . 6^2 . 8^2} - \ldots\ldots$$

$$= x\left(1 - \frac{x^4}{2^2 . 4^2} + \frac{x^8}{2^2 . 4^2 . 6^2 . 8^2} - \ldots\ldots\right) = x\, ber(x).$$

6.13. ORTHOGONALITY OF BESSEL FUNCTIONS

If α and β are the roots of $J_n(x) = 0$, then

$$\int_0^1 x\, J_n(\alpha x) . J_n(\beta x)\, dx = \begin{Bmatrix} 0, & \text{when } \alpha \neq \beta \\ \frac{1}{2} J_{n+1}^2(\alpha), & \text{when } \alpha = \beta \end{Bmatrix}$$

Consider the Bessel's equations

$$x^2 u'' + xu' + (\alpha^2 x^2 - n^2) u = 0 \qquad \ldots(1)$$

and

$$x^2 v'' + xv' + (\beta^2 x^2 - n^2)\, v = 0 \qquad \ldots(2)$$

Their solutions are $u = J_n(\alpha x)$ and $v = J_n(\beta x)$ respectively.

Multiplying (1) by $\frac{v}{x}$ and (2) by $\frac{u}{x}$ and subtracting, we get

$$x(u''v - uv'') + (u'v - uv') + (\alpha^2 - \beta^2)\, xuv = 0$$

or

$$\frac{d}{dx}[x(u'v - uv')] = (\beta^2 - \alpha^2)\, xuv$$

Integrating both sides w.r.t. x between the limits 0 and 1, we get

$$(\beta^2 - \alpha^2) \int_0^1 xuv\, dx = \Big[x(u'v - uv') \Big]_0^1 = \Big[u'v - uv' \Big]_{x=1} \qquad \ldots(3)$$

Since $\qquad u = J_n(\alpha x)$

$$\therefore \qquad u' = \frac{d}{dx}[J_n(\alpha x)] = \frac{d}{d(\alpha x)}[J_n(\alpha x)] . \frac{d(\alpha x)}{dx} = \alpha J_n'(\alpha x)$$

Similarly, $\qquad v = J_n(\beta x) \Rightarrow v' = \beta J_n'(\beta x)$

Substituting for u, v, u' and v' in (3), we get

$$\int_0^1 x J_n(\alpha x)\, J_n(\beta x)\, dx = \frac{\alpha J_n'(\alpha) J_n(\beta) - \beta J_n(\alpha) J_n'(\beta)}{\beta^2 - \alpha^2} \qquad \ldots(4)$$

If α and β are *distinct* roots of $J_n(x) = 0$, then $J_n(\alpha) = 0$ and $J_n(\beta) = 0$.

Hence, from (4), we have

$$\int_0^1 xJ_n(\alpha x)\, J_n(\beta x)\, dx = 0$$

However, if $\alpha = \beta$, the value of the integral is $\frac{0}{0}$, which is indeterminate.

To evaluate the integral, we assume that α is a root of $J_n(x) = 0$ so that $J_n(\alpha) = 0$ and β is a variable approaching α. Thus, from (4), we have

$$\underset{\beta\to\alpha}{\text{Lt}} \int_0^1 x\, J_n(\alpha x)\, J_n(\beta x)\, dx = \underset{\beta\to\alpha}{\text{Lt}} \frac{\alpha\, J'_n(\alpha)\, J_n(\beta)}{\beta^2 - \alpha^2}$$

or

$$\int_0^1 x\, J_n^2(\alpha x)\, dx = \underset{\beta\to\alpha}{\text{Lt}} \frac{\alpha\, J'_n(\alpha)\, J_n(\beta)}{2\beta} \qquad \text{[By L' Hospital's rule]}$$

$$= \frac{1}{2}\, [J_n'(\alpha)]^2 \qquad ...(5)$$

But $\qquad xJ_n'(x) = nJ_n(x) - xJ_{n+1}(x) \qquad ...(6)$

$\therefore$ Putting $x = \alpha$ in (6), $xJ_n'(\alpha) = nJ_n(\alpha) - \alpha J_{n+1}(\alpha) = -\alpha J_{n+1}(\alpha)$, since $J_n(\alpha) = 0$

$\Rightarrow \qquad J_n'(\alpha) = -J_{n+1}(\alpha)$

Hence, from (5), we get $\displaystyle\int_0^1 x\, J_n^2(\alpha x)\, dx = \frac{1}{2} J_{n+1}^2(\alpha)$.

Note. If the interval is from 0 to a, it can be shown that

$$\int_0^a xJ_n^2(\alpha x)\, dx = \frac{a^2}{2} J_{n+1}^2(\alpha a), \text{ where } \alpha \text{ is a root of } J_n(\alpha a) = 0.$$

6.14. LEGENDRE'S EQUATION

The differential equation $\quad (1 - x^2)\dfrac{d^2y}{dx^2} - 2x\dfrac{dy}{dx} + n(n+1)\, y = 0 \qquad ...(1)$

where n is real number, is called Legendre's differential equation. This equation is of considerable importance in applied mathematics, particularly in boundary value problems involving spherical configurations.

Though n is a real number, in most physical applications only integral values of n are required. Also, equation (1) can be solved in series of ascending or descending powers of x. The solution in descending powers of x is more important than the one in ascending powers.

Let, $\qquad y = \displaystyle\sum_{k=0}^{\infty} a_k x^{m-k}$

Then $\qquad \dfrac{dy}{dx} = \displaystyle\sum_{k=0}^{\infty} (m-k)\, a_k x^{m-k-1}$ and $\dfrac{d^2y}{dx^2} = \displaystyle\sum_{k=0}^{\infty} (m-k)(m-k-1)\, a_k x^{m-k-2}$

Substituting for y, $\frac{dy}{dx}$ and $\frac{d^2y}{dx^2}$ in (1), we get

$$(1-x^2)\sum_{k=0}^{\infty}(m-k)(m-k-1)\,a_k x^{m-k-2} - 2x\sum_{k=0}^{\infty}(m-k)\,a_k x^{m-k-1} + n(n+1)\sum_{k=0}^{\infty}a_k x^{m-k} = 0$$

or
$$\sum_{k=0}^{\infty}(m-k)(m-k-1)\,a_k x^{m-k-2} - \sum_{k=0}^{\infty}[(m-k)(m-k-1)+2(m-k)-n(n+1)]a_k x^{m-k} = 0$$

or
$$\sum_{k=0}^{\infty}(m-k)(m-k-1)\,a_k x^{m-k-2} - \sum_{k=0}^{\infty}(m-k)^2-n^2+(m-k)-n]\,a_k x^{m-k} = 0$$

or
$$\sum_{k=0}^{\infty}(m-k)(m-k-1)\,a_k x^{m-k-2} - \sum_{k=0}^{\infty}[(m-k-n)(m-k+n+1)\,a_k x^{m-k} = 0.$$

Equating to zero the co-efficient of highest power of x, *i.e.*, x^m, we get the indicial equation

$$(m-n)(m+n+1)a_0 = 0$$

whence $m = n$ or $m = -(n+1)$ since $a_0 \neq 0$

Equating to zero the co-efficient of the next lower power of x, *i.e.*, x^{m-1}, we get

$$(m+n)(m-n-1)\,a_1 = 0 \quad \text{or} \quad a_1 = 0,$$

since $(m+n)$ and $(m-n-1)$ are not zero for $m = n$ or $-(n+1)$.

Equating to zero the co-efficient of x^{m-k}, we get the recurrence relation

$$[m-(k-2)]\,[m-(k-2)-1]\,a_{k-2} - (m-k-n)(m-k+n+1)a_k = 0$$

or
$$a_k = -\frac{(m-k+2)(m-k+1)}{(n-m+k)(n+m-k+1)}\,a_{k-2} \qquad \text{...(2)}$$

Since $a_1 = 0$, therefore, from (2), we get $a_3 = a_5 = a_7 = \ldots\ldots = 0$.

When m = n, the recurrence relation (2) reduces to $a_k = -\frac{(n-k+2)(n-k+1)}{k(2n-k+1)}a_{k-2}$

Putting $k = 2, 4, 6, \ldots\ldots$, we get $a_2 = -\frac{n(n-1)}{2(2n-1)}\,a_0$,

$$a_4 = -\frac{(n-2)(n-3)}{4(2n-3)}\,a_2 = \frac{n(n-1)(n-2)(n-3)}{2.4.(2n-1)(2n-3)}\,a_0, \text{ etc.}$$

Therefore, one solution of Legendre's equation is given by

$$y_1 = a_0\left[x^n - \frac{n(n-1)}{2(2n-1)}x^{n-2} + \frac{n(n-1)(n-2)(n-3)}{2.4.(2n-1)(2n-3)}x^{n-4} - \ldots\ldots\right]$$

When m = - (n + 1), the recurrence relation (2) reduces to $a_k = \frac{(n+k-1)(n+k)}{k(2n+k+1)} a_{k-2}$

Putting $k = 2, 4, 6, \ldots\ldots.$, we get

$$a_2 = \frac{(n+1)(n+2)}{2(2n+3)} a_0,$$

$$a_4 = \frac{(n+3)(n+4)}{4(2n+5)} a_2 = \frac{(n+1)(n+2)(n+3)(n+4)}{2.4.(2n+3)(2n+5)} a_0, \text{ etc.}$$

Therefore, the second solution of Legendre's equation is given by

$$y_2 = a_0 \left[x^{-n-1} + \frac{(n+1)(n+2)}{2(2n+3)} x^{-n-3} + \frac{(n+1)(n+2)(n+3)(n+4)}{2.4.(2n+3)(2n+5)} x^{-n-5} + \ldots\ldots \right]$$

When n is *a positive integer* and $a_0 = \frac{1.3.5 \ldots\ldots (2n-1)}{n!}$,

the first solution is denoted by $P_n(x)$ and is called the **Legendre's function of the first kind.**

Thus,

$$P_n(x) = \frac{1.3.5 \ldots\ldots (2n-1)}{n!} \left[x^n - \frac{n(n-1)}{2(2n-1)} x^{n-2} + \frac{n(n-1)(n-2)(n-3)}{2.4.(2n-1)(2n-3)} x^{n-4} - \ldots\ldots \right]$$

$P_n(x)$ is a terminating series and gives what are called **Legendre's Polynomials** for different values of n such that $P_n(1) = 1$.

We can write $$P_n(x) = \sum_{k=0}^{\infty} \frac{(-1)^k (2n-2k)!}{2^n . k!(n-k)!(n-2k)!} x^{n-2k}$$

where, $$N = \begin{cases} \frac{n}{2}, & \text{if } n \text{ is even} \\ \frac{n-1}{2}, & \text{if } n \text{ is odd} \end{cases}$$

Again, when n is a positive integer and $a_0 = \frac{n!}{1.3.5\ldots\ldots(2n+1)}$,

the second solution is denoted by $Q_n(x)$ and is called the **Legendre's function of the second kind.** Thus

$$Q_n(x) = \frac{n!}{1.3.5\ldots\ldots(2n+1)}$$

$$\left[x^{-n-1} + \frac{(n+1)(n+2)}{2(2n+3)} x^{-n-3} + \frac{(n+1)(n+2)(n+3)(n+4)}{2.4.(2n+3)(2n+5)} x^{-n-5} + \ldots\ldots \right]$$

which is an infinite series.

Hence the solution of Legendre's equation is given by $y = c_1 P_n(x) + c_2 Q_n(x)$ where c_1 and c_2 are arbitrary constants.

6.15. RODRIGUE'S FORMULA

The relation $P_n(x) = \dfrac{1}{2^n \, . \, n!} \dfrac{d^n}{dx^n} (x^2 - 1)^n$ is known as Rodrigue's formula.

To prove it, let $v = (x^2 - 1)^n$ then $v_1 = \dfrac{dv}{dx} = n(x^2 - 1)^{n-1} . \, 2x$

Multiplying both sides by $(x^2 - 1)$, we get

$$(x^2 - 1)v_1 = 2nx(x^2 - 1)^n = 2nxv \quad \text{or} \quad (1 - x^2)v_1 + 2nxv = 0$$

Differentiating $(n + 1)$ times by Leibnitz's theorem, we have

$$\left[(1-x^2)v_{n+2} + (n+1)(-2x)v_{n+1} + \frac{(n+1)n}{2!}(-2)v_n\right] + 2n\,[xv_{n+1} + (n+1)v_n] = 0$$

or $$(1 - x^2)v_{n+2} - 2xv_{n+1} + n(n + 1)v_n = 0$$

or $$(1 - x^2)\frac{d^2(v_n)}{dx^2} - 2x\frac{d(v_n)}{dx} + n(n + 1)v_n = 0$$

which is Legendre's equation and v_n is its solution. But the solutions of Legendre's equation are $P_n(x)$ and $Q_n(x)$.

Since $v_n = \dfrac{d^n}{dx^n}(x^2 - 1)^n$ contains only positive powers of x, it must be a constant multiple of $P_n(x)$.

i.e., $$v_n = cP_n(x)$$

or $$cP_n(x) = \frac{d^n}{dx^n}(x^2 - 1)^n \qquad \text{...(1)}$$

$$= \frac{d^n}{dx^n}[(x - 1)^n(x + 1)^n]$$

$$= (x - 1)^n \frac{d^n}{dx^n}(x + 1)^n + {}^nC_1 \, . \, n(x - 1)^{n-1}\frac{d^{n-1}}{dx^{n-1}}(x + 1)^n + \ldots\ldots$$

$$+ (x + 1)^n \frac{d^n}{dx^n}(x - 1)^n$$

$$= (x - 1)^n \, . \, n! + {}^nC_1 \, . \, n(x - 1)^{n-1} . \frac{d^{n-1}}{dx^{n-1}}(x + 1)^n + \ldots\ldots + (x + 1)^n \, n!$$

$$= n!\,(x + 1)^n + \text{terms containing powers of } (x - 1)$$

Putting $x = 1$ on both sides, we get

$$cP_n(1) = n! \, . \, 2^n \text{ or } c = 2^n \, n!, \text{ since } P_n(1) = 1$$

Substituting in (1), we get $P_n(x) = \dfrac{1}{2^n \, n!} \dfrac{d^n}{dx^n}(x^2 - 1)^n$

Putting $n = 0, 1, 2, 3, ...$ in Rodrigue's formula, we get Legendre's polynomials. Thus

$$P_0(x) = 1$$

$$P_1(x) = \frac{1}{2}\frac{d}{dx}(x^2 - 1) = x$$

$$P_2(x) = \frac{1}{2^2\,(2)!}\frac{d^2}{dx^2}(x^2 - 1)^2 = \frac{1}{8}\frac{d^2}{dx^2}(x^4 - 2x^2 + 1) = \frac{1}{2}(3x^2 - 1)$$

$$P_3(x) = \frac{1}{2^3\,(3)!}\frac{d^3}{dx^3}(x^2 - 1)^3 = \frac{1}{48}\frac{d^3}{dx^3}(x^6 - 3x^4 + 3x^2 - 1) = \frac{1}{2}(5x^3 - 3x)$$

Similarly, $P_4(x) = \frac{1}{8}(35x^4 - 30x^2 + 3)$

$$P_5(x) = \frac{1}{8}(63x^5 - 70x^3 + 15x)$$

$$P_6(x) = \frac{1}{6}(231x^6 - 351x^4 + 105x^2 - 5) \text{ etc.}$$

ILLUSTRATIVE EXAMPLES

Example 1. *Show that* $x^4 = \frac{1}{35}[8P_4(x) + 20P_2(x) + 7P_0(x)]$.

Sol. We know that

$$P_4(x) = \frac{1}{8}(35x^4 - 30x^2 + 3)$$

$$P_2(x) = \frac{1}{2}(3x^2 - 1),\ P_0(x) = 1$$

$$\therefore \quad \frac{1}{35}[8P_4(x) + 20P_2(x) + 7P_0(x)] = \frac{1}{35}[35x^4 - 30x^2 + 3) + 10(3x^2 - 1) + 7] = x^4.$$

Example 2. *Express* $f(x) = x^3 - 5x^2 + x + 2$ *in terms of Legendre's polynomials.*

Sol. We know that

$$P_3(x) = \frac{1}{2}(5x^3 - 3x)$$

$$\therefore \quad x^3 = \frac{2}{5}P_3(x) + \frac{3}{5}x$$

$$\therefore \quad f(x) = \left[\frac{2}{3}P_3(x) + \frac{3}{5}(x)\right] - 5x^2 + x + 2$$

$$= \frac{2}{5} P_3(x) - 5x^2 + \frac{8}{5} x + 2 = \frac{2}{5} P_3(x) - 5\left[\frac{2}{3} P_2(x) + \frac{1}{3}\right] + \frac{8}{5} x + 2$$

$$\left[\because \; P_2(x) = \tfrac{1}{2}(3x^2 - 1) \;\; \therefore \;\; x^2 = \tfrac{2}{3} P_2(x) + \tfrac{1}{3}\right]$$

$$= \frac{2}{5} P_3(x) - \frac{10}{3} P_2(x) + \frac{8}{5} x + \frac{1}{3}$$

$$= \frac{2}{5} P_3(x) - \frac{10}{3} P_2(x) + \frac{8}{5} P_1(x) + \frac{1}{3} P_0(x) \quad [\because \; x = P_1(x) \text{ and } 1 = P_0(x)]$$

Example 3. *Express $f(x) = x^4 + 3x^3 - x^2 + 5x - 2$ in terms of Legendre polynomials.*

Sol. Since $P_4(x) = \frac{35}{8} x^4 - \frac{15}{4} x^2 + \frac{3}{8}$. $\therefore$ $x^4 = \frac{8}{35} P_4(x) + \frac{6}{7} x^2 - \frac{3}{35}$

$$\therefore \quad f(x) = \left[\frac{8}{35} P_4(x) + \frac{6}{7} x^2 - \frac{3}{35}\right] + 3x^3 - x^2 + 5x - 2$$

$$= \frac{8}{35} P_4(x) + 3x^3 - \frac{1}{7} x^2 + 5x - \frac{73}{35}$$

$$\left[\because \; x^3 = \frac{2}{5} P_3(x) + \frac{3}{5} x \,;\, x^2 = \frac{2}{3} P_2(x) + \frac{1}{3}\right]$$

$$= \frac{8}{35} P_4(x) + 3\left[\frac{2}{5} P_3(x) + \frac{3}{5} x\right] - \frac{1}{7}\left[\frac{2}{3} P_2(x) + \frac{1}{3}\right] + 5x - \frac{73}{35}$$

$$= \frac{8}{35} P_4(x) + \frac{6}{5} P_3(x) - \frac{2}{21} P_2(x) + \frac{34}{5} x - \frac{224}{105} \qquad [\because \; x = P_1(x),\; 1 = P_0(x)]$$

$$= \frac{8}{35} P_4(x) + \frac{6}{5} P_3(x) - \frac{2}{21} P_2(x) + \frac{34}{5} P_1(x) - \frac{224}{105} P_0(x).$$

Example 4. *Express $f(x) = 4x^3 + 6x^2 + 7x + 2$ in terms of Legendre Polynomials:*

Sol. Let

$$4x^3 + 6x^2 + 7x + 2 \equiv a\, P_3(x) + b\, P_2(x) + cP_1(x) + d\, P_0(x) \qquad \ldots(1)$$

$$\equiv a\left(\frac{5x^3}{2} - \frac{3x}{2}\right) + b\left(\frac{3x^2}{2} - \frac{1}{2}\right) + c(x) + d(1)$$

$$\equiv \frac{5ax^3}{2} - \frac{3ax}{2} + \frac{3bx^2}{2} - \frac{b}{2} + cx + d$$

$$\equiv \frac{5ax^3}{2} - \frac{3bx^2}{2} + \left(\frac{-3a}{2} + c\right) x - \frac{b}{2} + d.$$

Equating the coefficients of like powers of x, we have

$$4 = \frac{5a}{2}, \quad \text{or} \quad a = \frac{8}{5}$$

$$6 = \frac{3b}{2} \quad \text{or} \quad b = 4$$

$$7 = \frac{-3a}{2} + c \quad \text{or} \quad 7 = \frac{-3}{2}\left(\frac{8}{5}\right) + c \quad \text{or} \quad c = \frac{47}{5}$$

$$2 = \frac{-b}{2} + d \quad \text{or} \quad 2 = \frac{-4}{2} + d \quad \text{or} \quad d = 4$$

Putting the values of a, b, c, d, in (1), we get

$$4x^3 + 6x^2 + 7x + 2 = \frac{8}{5} P_3(x) + 4 P_2(x) + \frac{47}{5} P_1(x) + 4P_0(x)$$

6.16. GENERATING FUNCTION FOR $P_n(x)$

We shall show that $P_n(x)$ is the co-efficient of z^n in the expansion of $(1 - 2xz + z^2)^{-1/2}$ in ascending powers of z.

i.e.,
$$(1 - 2xz + z^2)^{-1/2} = \sum_{n=0}^{\infty} P_n(x) \,.\, z^n$$

Using binomial theorem,

$$(1 - t)^{-1/2} = 1 + \frac{1}{2} t + \frac{\frac{1}{2}\cdot\frac{3}{2}}{2!} t^2 + \frac{\frac{1}{2}\cdot\frac{3}{2}\cdot\frac{5}{2}}{3!} t^3 + \ldots\ldots$$

$$= 1 + \frac{1}{2} t + \frac{1.3}{2.4} t^2 + \frac{1.3.5}{2.4.6} t^3 + \ldots\ldots + \frac{1.3.5 \ldots\ldots (2n-1)}{2.4.6 \ldots\ldots 2n} t^n + \ldots\ldots$$

$$\therefore \quad (1 - 2xz + z^2)^{-1/2} = [1 - z(2x - z)]^{-1/2}$$

$$= 1 + \frac{1}{2} z(2x - z) + \frac{1.3}{2.4} z^2 (2x - z)^2 + \ldots\ldots$$

$$+ \frac{1.3.5 \ldots\ldots (2n-5)}{2.4.6 \ldots\ldots (2n-4)} z^{n-2} (2x - z)^{n-2}$$

$$+ \frac{1.3.5 \ldots\ldots (2n-3)}{2.4.6 \ldots\ldots (2n-2)} z^{n-1} (2x - z)^{n-1}$$

$$+ \frac{1.3.5 \ldots\ldots (2n-1)}{2.4.6 \ldots\ldots (2n)} z^n (2x - z)^n + \ldots\ldots$$

Now, the co-efficient of z^n in $\frac{1.3.5 \ldots\ldots (2n-3)}{2.4.6 \ldots\ldots (2n)} z^n (2x - z)^n$ is

$$\frac{1.3.5 \ldots\ldots (2n-3)}{2.4.6 \ldots\ldots (2n)} (2x)^n = \frac{1.3.5 \ldots\ldots (2n-1)}{2^n (n)!} (2x)^n = \frac{1.3.5 \ldots\ldots (2n-1)}{n!} x^n$$

The co-efficient of z^n in $\frac{1.3.5 \ldots\ldots (2n-3)}{2.4.6 \ldots\ldots (2n-2)} z^{n-1} (2x - z)^{n-1}$ is

$$\frac{1.3.5 \ldots\ldots (2n-3)}{2.4.6 \ldots\ldots (2n-2)} (-{}^{n-1}C_1 (2x)^{n-2}] = -\frac{1.3.5 \ldots\ldots (2n-3)}{2.4.6 \ldots\ldots (2n-2)} (n-1)\, 2^{n-2} \,.\, x^{n-2}$$

$$= -\frac{1.3.5 \ldots\ldots (2n-3)}{2^{n-1} (n-1)!} (n-1)\, 2^{n-2} \,.\, x^{n-2}$$

$$= -\frac{1.3.5 \ldots\ldots (2n-3)(2n-1)}{2(n)!} \cdot \frac{n(n-1)}{2n-1} x^{n-2}$$

$$= -\frac{1.3.5 \ldots\ldots (2n-1)}{n!} \cdot \frac{n(n-1)}{2(2n-1)} x^{n-2}$$

Similarly, the co-efficient of z^n in $\frac{1.3.5 \ldots\ldots (2n-5)}{2.4.6 \ldots\ldots (2n-4)} z^{n-2} (2x-z)^{n-2}$ is

$$\frac{1.3.5 \ldots\ldots (2n-1)}{n!} \cdot \frac{n(n-1)(n-2)(n-3)}{2.4.(2n-1)(2n-3)} x^{n-4} \text{ and so on.}$$

$\therefore$ The co-efficient of z^n in $(1 - 2xz + z^2)^{-1/2}$ is given by

$$\frac{1.3.5 \ldots\ldots (2n-1)}{n!} \left[x^n - \frac{n(n-1)}{2(2n-1)} x^{n-2} + \frac{n(n-1)(n-2)(n-3)}{2.4.(2n-1)(2n-3)} x^{n-4} - \ldots\ldots \right]$$

$$= P_n(x)$$

Thus, in the expansion of $(1 - 2xz + z^2)^{-1/2}$, $P_1(x), P_2(x), P_3(x), \ldots\ldots, P_n(x), \ldots\ldots$ are the co-efficients of $z^1, z^2, z^3, \ldots\ldots, z^n, \ldots\ldots$ respectively.

$$\therefore \quad (1 - 2xz + z^2)^{-1/2} = 1 + P_1(x) \,.\, z + P_2(x) \,.\, z^2 + \ldots\ldots + P_n(x) \,.\, z^n + \ldots\ldots = \sum_{n=0}^{\infty} P_n(x) \,.\, z^n$$

The function $(1 - 2xz + z^2)^{-1/2}$ is called the **generating function for $P_n(x)$.**

Example 5. *Show that:*

(i) $P_n(1) = 1$ *(ii)* $P_n(-x) = (-1)^n P_n(x)$.

Sol. We know that $\sum_{n=0}^{\infty} z^n P_n(x) = (1 - 2xz + z^2)^{-1/2}$ ...(1)

(*i*) Putting $x = 1$ in (1), we get $\sum_{n=0}^{\infty} z^n P_n(1) = (1 - 2z + z^2)^{-1/2}$

or $$= (1 - z)^{-1} = 1 + z + z^2 + \ldots\ldots + z^n + \ldots.. = \sum_{n=0}^{\infty} z^n$$

Equating the co-efficients of z^n, we have $P_n(1) = 1$.

(*ii*) Replacing x by $(-x)$ in (1), we have

$$\sum_{n=0}^{\infty} z^n P_n(-x) = (1 + 2xz + z^2)^{-1/2} \quad \ldots(2)$$

Again, replacing z by $(-z)$ in (1), we have

$$\sum_{n=0}^{\infty} (-z)^n P_n(x) = (1 + 2xz + z^2)^{-1/2}$$

or $$\sum_{n=0}^{\infty} (-1)^n z^n P_n(x) = (1 + 2xz + z^2)^{-1/2} \quad \ldots(3)$$

From (2) and (3), $\sum_{n=0}^{\infty} z^n P_n(-x) = \sum_{n=0}^{\infty} (-1)^n z^n P_n(x)$

Equating the co-efficients of z^n, we have $P_n(-x) = (-1)^n P_n(x)$.

Example 6. *Show that:*

(*i*) $P_{2n}(0) = (-1)^n \dfrac{2n!}{2^{2n}(n!)^2}$ (*ii*) $P_{2n+1}(0) = 0$.

Sol. We know that $\sum_{n=0}^{\infty} z^n P_n(x) = (1 - 2xz + z^2)^{-1/2}$

Putting $x = 0$, we get $\sum_{n=0}^{\infty} z^n P_n(0) = (1 + z^2)^{-1/2}$

$$= 1 - \frac{1}{2} z^2 + \frac{1.3}{2.4} z^4 - \ldots\ldots + (-1)^r . \frac{1.3.5 \ldots\ldots (2r-1)}{2.4.6 \ldots\ldots (2r)} z^{2r} + \ldots\ldots$$

(*i*) Equating the co-efficients of z^{2n} on both sides, we get

$$P_{2n}(0) = (-1)^n \frac{1.3.5 \ldots\ldots (2n-1)}{2.4.6 \ldots\ldots (2n)} = (-1)^n \frac{1.2.3.4 \ldots\ldots (2n-1)(2n)}{[2.4.6 \ldots\ldots (2n)]^2}$$

$$= (-1)^n \frac{2!}{[2^n \, 1.2.3 \ldots\ldots n]^2} = (-1)^n \frac{(2n)!}{2^{2n}(n!)^2}.$$

(*ii*) Equating the co-efficients of z^{2n+1} on both sides, we get $P_{2n+1}(0) = 0$, since the right-hand side contains only even powers of z.

6.17. RECURRENCE RELATIONS

Now we establish the following recurrence relations from the generating function for $P_n(x)$.

(*i*) $\mathbf{(2n + 1)\, x\, P_n(x) = (n + 1)\, P_{n+1}(x) + nP_{n-1}(x)}$.

(*ii*) $\mathbf{nP_n(x) = xP_n'(x) - P'_{n-1}(x)}$.

(*iii*) $\mathbf{(2n + 1)\, P_n(x) = P'_{n+1}(x) - P'_{n-1}(x)}$.

(*iv*) $\mathbf{P_n'(x) = x\, P'_{n-1}(x) + nP_{n-1}(x)}$.

(*v*) $\mathbf{(1 - x^2)\, P_n'(x) = n\, [P_{n-1}(x) - xP_n(x)]}$.

(*vi*) $\mathbf{(1 - x^2)\, P_n'(x) = (n + 1)\, [xP_n(x) - P_{n+1}(x)]}$.

We know that $(1 - 2xz + z^2)^{-1/2} = \sum_{n=0}^{\infty} z^n P_n(x)$ ...(1)

(*i*) Differentiating both sides of (1) partially w.r.t. z, we get

$$-\frac{1}{2}(1-2xz+z^2)^{-3/2}(-2x+2z)=\sum_{n=0}^{\infty} nz^{n-1}P_n(x)$$

or
$$(x-z)(1-2xz+z^2)^{-1/2}=(1-2xz+z^2)\sum_{n=0}^{\infty} nz^{n-1}P_n(x)$$

or
$$(x-z)\sum_{n=0}^{\infty} z^nP_n(x)=(1-2xz+z^2)\sum_{n=0}^{\infty} nz^{n-1}P_n(x)$$

Equating the co-efficients of z^n on both sides, we have

$$xP_n(x)-P_{n-1}(x)=(n+1)P_{n+1}(x)-2nxP_n(x)+(n-1)P_{n-1}(x)$$

or
$$(2n+1)xP_n(x)=(n+1)P_{n+1}(x)+nP_{n-1}(x)$$

(*ii*) Differentiating (1) partially w.r.t. z, we get

$$(x-z)(1-2xz+z^2)^{-3/2}=\sum_{n=0}^{\infty} nz^{n-1}P_n(x) \quad ...(2)$$

Differentiating (1) partially w.r.t. x, we get

$$-\frac{1}{2}(1-2xz+z^2)^{-3/2}(-2z)=\sum_{n=0}^{\infty} z^nP_n'(x)$$

or
$$z(x-z)(1-2xz+z^2)^{-3/2}=(x-z)\sum_{n=0}^{\infty} z^nP_n'(x)$$

or
$$z\sum_{n=0}^{\infty} nz^{n-1}P_n(x)=(x-z)\sum_{n=0}^{\infty} z^nP_n'(x) \quad \text{[Using (2)]}$$

Equating the co-efficients of z^n on both sides, we have

$$nP_n(x)=xP_n'(x)-P'_{n-1}(x).$$

(*iii*) From recurrence relation (*i*), we have

$$(2n+1)xP_n(x)=(n+1)P_{n+1}(x)+nP_{n-1}(x)$$

Differentiating w.r.t. x, we have

$$(2n+1)xP_n'(x)+(2n+1)P_n(x)=(n+1)P'_{n+1}(x)+nP'_{n-1}(x) \quad ...(3)$$

From recurrence relation (*ii*), we have

$$xP_n'(x)=nP_n(x)+P'_{n-1}(x)$$

Substituting the value of $xP_n'(x)$ in (3), we get

$$(2n+1)[nP_n(x)+P'_{n-1}(x)]+(2n+1)P_n(x)=(n+1)P'_{n+1}(x)+nP'_{n-1}(x)$$

or
$$(2n+1)(n+1)P_n(x)=(n+1)P'_{n+1}(x)-(n+1)P'_{n-1}(x)$$

or
$$(2n+1)P_n(x)=P'_{n+1}(x)-P'_{n-1}(x).$$

(*iv*) From recurrence relation (*iii*), we have

$$(2n+1)P_n(x)=P'_{n+1}(x)-P'_{n-1}(x)$$

From recurrence relation (*ii*), we have

$$nP_n(x) = xP_n'(x) - P'_{n-1}(x)$$

Subtracting $P'_{n-1}(x)$ is eliminated and, we get

$$(n + 1)\, P_n(x) = P'_{n+1}(x) - xP_n'(x)$$

Replacing n by (-1) $$nP_{n-1}(x) = P_n'(x) - xP'_{n-1}(x)$$

or $$P_n'(x) = xP'_{n-1}(x) + nP'_{n-1}(x)$$

(*v*) From recurrence relation (*iv*), we have

$$P_n'(x) - xP'_{n-1}(x) = nP'_{n-1}(x) \qquad ...(4)$$

From recurrence relation (*ii*), we have

$$xP_n'(x) - P'_{n-1}(x) = nP_n(x) \qquad ...(5)$$

Multiplying (5) by x and subtracting from (4), we get

$$(1 - x^2)\, P_n'(x) = n[P_{n-1}(x) - xP_n(x)].$$

(*vi*) Recurrence relation (*i*) may be written as

$$(n + 1)\, xP_n(x) + nxP_n(x) = (n + 1)\, P_{n+1}(x) + nP_{n-1}(x)$$

or $$(n + 1)\, [xP_n(x) - P_{n+1}(x)] = n[P_{n-1}(x) - xP_n(x)]$$
$$= (1 - x^2)\, P_n'(x)$$

[Using recurrence relation (*v*)]

$$\therefore \quad (1 - x^2)\, P_n'(x) = (n + 1)\, [xP_n(x) - P_{n+1}(x)].$$

6.18. ORTHOGONALITY OF LEGENDRE POLYNOMIALS

We shall show that

$$\int_{-1}^{1} P_m(x)\, P_n(x)\, dx = \begin{cases} 0, & \text{if } m \neq n \\ \dfrac{2}{2n+1}, & \text{if } m = n \end{cases}$$

Case I. When m ≠ n

We know that $P_m(x)$ and $P_n(x)$ are the solutions of the equations

$$(1 - x^2)u'' - 2xu' + m(m + 1)u = 0 \qquad ...(1)$$

and $$(1 - x^2)v'' - 2xv' + n(n + 1)v = 0 \qquad ...(2)$$

Multiplying (1) by v and (2) by u and subtracting, we get

$$(1 - x^2)\,(u''v - v''u) - 2x(u'v - v'u) + [m(m + 1) - n(n + 1)]uv = 0$$

or $$\frac{d}{dx}\, [(1 - x^2)(u'v - v'u)] + (m - n)(m + n + 1)uv = 0$$

or $$(n - m)(n + m + 1)\, uv = \frac{d}{dx}\, [(1 - x^2)(u'v - v'u)]$$

Integrating w.r.t. x from -1 to 1, we get

$$(n - m)(n + m + 1) \int_{-1}^{1} uv\, dx = \left[(1 - x^2)(u'v - v'u)\right]_{-1}^{1} = 0$$

Hence $\int_{-1}^{1} P_m(x)\, P_n(x)\, dx = 0$, since $m \neq n$.

Case II. When m = n

We know that $(1 - 2xz + z^2)^{-1/2} = \sum_{n=0}^{\infty} z^n P_n(x)$

Squaring both sides, we get

$$(1 - 2xz + z^2)^{-1} = \sum_{n=0}^{\infty} [z^n P_n(x)]^2 = \sum_{n=0}^{\infty} z^{2n} [P_n(x)]^2 + 2 \sum_{\substack{m=0\\n=0\\m\neq 0}}^{\infty} z^{m+n} P_m(x)P_n(x)$$

Integrating w.r.t. x between the limits -1 to 1, we have

$$\sum_{n=0}^{\infty} \int_{-1}^{1} z^{2n} [P_n(x)]^2 \, dx + 2 \sum_{\substack{m=0\\n=0\\m\neq 0}}^{\infty} \int_{-1}^{1} z^{m+n} P_m(x)P_n(x) \, dx = \int_{-1}^{1} \frac{dx}{1 - 2xz + z^2}$$

or
$$\sum_{n=0}^{\infty} \int_{-1}^{1} z^{2n} [P_n(x)]^2 \, dx = \int_{-1}^{1} \frac{dx}{1 - 2xz + z^2}$$

Since other integrals on the L.H.S. vanish by Case I as $m \neq n$

$$= -\frac{1}{2z}\left[\log(1 - 2xz + z^2)\right]_{-1}^{1} = -\frac{1}{2z}[\log(1-z)^2 - \log(1+z)^2]$$

$$= \frac{1}{z}[\log(1+z) - \log(1-z)]$$

$$= \frac{1}{z}\left[\left(z - \frac{z^2}{2} + \frac{z^3}{3} - \frac{z^4}{4} + \ldots\ldots\right) + \left(z + \frac{z^2}{2} + \frac{z^3}{3} + \frac{z^4}{4} + \ldots\ldots\right)\right]$$

$$= \frac{2}{z}\left[z + \frac{z^3}{3} + \frac{z^5}{5} + \ldots\ldots\right]$$

or
$$\sum_{n=0}^{\infty} z^{2n} \int_{-1}^{1} [P_n(x)]^2 \, dx = 2\left(1 + \frac{z^2}{3} + \frac{z^4}{5} + \ldots\ldots + \frac{z^{2n}}{2n+1} + \ldots\ldots\right)$$

Equating the co-efficients of z^{2n} on the two sides, we get

$$\int_{-1}^{1} [P_n(x)]^2 \, dx = \frac{2}{2n+1}.$$

6.19. EXPANSION OF A FUNCTION IN A SERIES OF LEGENDRE POLYNOMIALS

The orthogonality property of Legendre polynomials enables us to expand a function $f(x)$, defined from $x = -1$ to $x = 1$, in a series of Legendre polynomials.

Let $$f(x) = \sum_{n=0}^{\infty} a_n P_n(x) \qquad \ldots(1)$$

To determine a_n, multiplying both sides of (1) by $P_n(x)$ and integrating w.r.t. x from -1 to 1, we have

$$\int_{-1}^{1} f(x)\, P_n(x)\, dx = a_n \int_{-1}^{1} [P_n(x)]^2\, dx = a_n \left(\frac{2}{2n+1}\right)$$

$$\Rightarrow \qquad a_n = \left(n + \frac{1}{2}\right)\int_{-1}^{1} f(x)\, P_n(x)\, dx$$

The expansion of $f(x)$ given by (1) is known as **Fourier-Legendre series.**

Example 7. *Prove that:* $\int_{-1}^{1} xP_n(x)\, P_{n-1}(x)\, dx = \dfrac{2n}{(2n-1)(2n+1)} = \dfrac{2n}{4n^2-1}$.

Sol. From recurrence relation I, we have

$$(2n+1)\, xP_n(x) = (n+1)P_{n+1}(x) + nP_{n-1}(x)$$

i.e.,

$$xP_n(x) = \frac{n+1}{2n+1} P_{n+1}(x) + \frac{n}{2n+1} P_{n-1}(x).$$

Multiplying both sides by $P_{n-1}(x)$ and integrating w.r.t. x from limits – 1 to 1, we get

$$\int_{-1}^{1} xP_n(x)\, P_{n-1}(x)\, dx = \frac{n+1}{2n+1}\int_{-1}^{1} P_{n+1}(x)\, P_{n-1}(x)\, dx + \frac{n}{2n+1}\int_{-1}^{1}\{P_{n-1}(x)\}^2\, dx \qquad ...(1)$$

But by orthogonality property of $P_n(x)$, we know that

$$\int_{-1}^{1} P_m(x)\, P_n(x)\, dx = \begin{cases} 0, & \text{when } m \neq n \\ \dfrac{2}{2n+1}, & \text{when } m = n. \end{cases} \qquad ...(2)$$

Using equation (2) in equation (1), we get

$$\int_{-1}^{1} xP_n(x)\, P_{(n-1)}(x) dx \frac{n+1}{2n+1} \times 0 + \frac{n}{2n+1} \cdot \frac{2}{2(n-1)+1} = \frac{2n}{(2n-1)(2n+1)} = \frac{2n}{4n^2-1}.$$

Example 8. *Prove that:* $\int_{-1}^{1} (x^2-1)\, P_{n+1}P_n'\, dx = \dfrac{2n(n+1)}{(2n+1)(2n+3)}$, *where* $P_n' = \dfrac{d}{dx} P_n(x)$.

Sol. From recurrence relation V, we have

$$(x^2-1)\, P_n' = n(xP_n - P_{n-1}).$$

Multiplying both sides by P_{n+1} and integrating w.r.t. x from limits – 1 to 1, we get

$$\int_{-1}^{1} (x^2-1)\, P_{n+1}P_n'\, dx = n\int_{-1}^{1} (xP_n - P_{n-1})\, P_{n+1} dx$$

$$= n\left[\int_{-1}^{1} xP_nP_{n+1}\, dx - \int_{-1}^{1} P_{n-1}P_{n+1}\, dx\right]$$

$$= n\left[\int_{-1}^{1} xP_nP_{n+1} dx - 0\right] \qquad \text{[By orthogonality property of } P_n(x)]$$

$$= n\int_{-1}^{1} \frac{(n+1)\, P_{n+1} + nP_{n-1}}{2n+1} P_{n+1}\, dx, \qquad \text{[By recurrence relation I]}$$

$$= \frac{n(n+1)}{2n+1}\int_{-1}^{1}(P_{n+1})^2\,dx + \frac{n^2}{2n+1}\int_{-1}^{1}P_{n-1}P_{n+1}\,dx$$

$$= \frac{n(n+1)}{2n+1}\cdot\frac{2}{2(n+1)+1} + 0 \qquad \text{[By orthogonality property of } P_n(x)]$$

$$= \frac{2n(n+1)}{(2n+1)(2n+3)}$$

Example 9. *Prove that:* $\int_{-1}^{1} x^2 P_{n-1}(x)\, P_{n+1}(x)\, dx = \dfrac{2n(n+1)}{(2n-1)(2n+1)(2n+3)}$.

Sol. From recurrence relation I, we have

$$(2n+1)\, xP_n(x) = (n+1)\, P_{n+1}(x) + nP_{n-1}(x).$$

Putting $(n-1)$ and $(n+1)$ in place of n in the above relation, we get

$$(2n-1)\, xP_{n-1}(x) = nP_n(x) + (n-1)P_{n-2}(x) \qquad ...(1)$$

and

$$(2n+3)\, xP_{n+1}(x) = (n+2)P_{n+2}(x) + (n+1)\, P_n(x). \qquad ...(2)$$

Multiplying equations (1) and (2), we get

$$(2n-1)(2n+3)\, x^2\, P_{n-1}(x)\, P_{n+1}(x) = n(n+1)(P_n(x))^2 + n(n+2)\, P_n(x)\, P_{n+2}(x)$$
$$+ (n-1)(n+2)\, P_{n-2}(x)\, P_{n+2}(x) + (n^2-1)\, P_{n-2}(x)\, P_n(x) \qquad ...(3)$$

Integrating both sides w.r.t x from limits -1 to 1, we get

$$(2n-1)(2n+3)\int_{-1}^{1} x^2\, P_{n-1}(x)\, P_{n+1}(x)\, dx$$

$$= n(n+1)\int_{-1}^{1} P_n(x)^2\, dx + n(n+2)\int_{-1}^{1} P_n(x)\, P_{n+2}(x)\, dx$$

$$+ (n-1)(n+2)\int_{-1}^{1} P_{n-2}P_{n+2}\, dx + (n^2-1)\int_{-1}^{1} P_{n-2}P_n\, dx$$

$$= n(n+1)\cdot\frac{2}{2n+1} + 0 + 0 + 0, \qquad \text{[By orthogonality property of } P_n(x)$$

$$= \frac{2n(n+1)}{2n+1}$$

Therefore,

$$\int_{-1}^{1} x^2 P_{n-1}(x)\, P_{n+1}(x)\, dx = \frac{2n(n+1)}{(2n-1)(2n+1)(2n+3)}$$

Example 10. *Prove that:*

(i) $\int_{-1}^{1} (1-x^2)\, P_m' P_n'\, dx = 0$, *where m and n are distinct integers;*

(ii) $\int_{-1}^{1} (1-x^2)(P_n')^2\, dx = \dfrac{2n(n+1)}{2n+1}$.

Sol. (*i*) Integrating by parts, we have

$$\int_{-1}^{1}(1-x^2)\underset{I}{P_m'}\,\underset{II}{P_n'}\,dx = [(1-x^2)P_m'P_n]_{-1}^{1} - \int_{-1}^{1}\frac{d}{dx}\{(1-x^2)P_m'\}P_n\,dx$$

$$= 0 - \int_{-1}^{1}\frac{d}{dx}\{(1-x^2)P_m'\}P_n\,dx. \qquad ...(1)$$

But we know that P_m is a solution of the Legender's equation

$$\frac{d}{dx}\left\{(1-x^2)\frac{dy}{dx}\right\} + m(m+1)y = 0$$

Thus $\quad \dfrac{d}{dx}\{(1-x^2)P_m'\} = -m(m+1)P_m. \qquad ...(2)$

Now using equation (2) in the right hand side of equation (1), we get

$$\int_{-1}^{1}(1-x^2)P_m'P_n'\,dx = m(m+1)\int_{-1}^{1}P_mP_n\,dx \qquad ...(3)$$

$$= m(m+1) \times 0 \text{ [By orthogonality property of } P_n(x)$$

$$= 0.$$

(*ii*) Again, where $m = n$, then from equation (3), we have

$$\int_{-1}^{1}(1-x^2)(P_n')^2\,dx = n(n+1)\int_{-1}^{1}(P_n)^2\,dx$$

$$= n(n+1) \times \frac{2}{2n+1} \qquad \text{[By orthogonality property of } P_n(x)]$$

$$= \frac{2n(n+1)}{2n+1}$$

By unifying the above two results, we can also write as

$$\int_{-1}^{1}(1-x^2)P_m'P_n'\,dx = \begin{cases} 0, & \text{if } m \neq n \\ \dfrac{2n(n+1)}{2n+1}, & \text{if } m = n \end{cases}.$$

Example 11. *Prove that:* $\displaystyle\int_{-1}^{1} x^2P_n^2\,dx = \frac{1}{8(2n-1)} + \frac{3}{4(2n+1)} + \frac{1}{8(2n+3)}.$

Sol. From recurrence relation I, we have

$$(2n+1)xP_n = nP_{n-1} + (n+1)P_{n+1}.$$

Squaring both sides, we get

$$(2n+1)^2x^2P_n^2 = n^2P_{n-1}^2 + 2n(n+1)P_{n-1}P_{n+1} + (n+1)^2P_{n+1}^2$$

or

$$x^2P_n^2 = \frac{n^2P_{n-1}^2 + 2n(n+1)P_{n-1}P_{n+1} + (n+1)^2P_{n+1}^2}{(2n+1)^2}$$

Integrating both sides w.r.t. x from limits – 1 to 1, we get

$$\int_{-1}^{1} x^2 P_n^2 \, dx = \frac{n^2}{(2n+1)^2} \int_{-1}^{1} P_{n-1}^2 \, dx + \frac{2n(n+1)}{(2n+1)^2} \int_{-1}^{1} P_{n-1} P_{n+1} \, dx + \frac{(n+1)^2}{(2n+1)^2} \int_{-1}^{1} P_{n+1}^2 \, dx$$

$$= \frac{n^2}{(2n+1)^2} \cdot \frac{2}{2(n-1)+1} + 0 + \frac{(n+1)^2}{(2n+1)^2} \cdot \frac{2}{2(n+1)+1}$$

[By orthogonality property of $P_n(x)$]

$$= \frac{2}{(2n+1)^2} \left[\frac{n^2}{2n-1} + \frac{(n+1)^2}{2n+3} \right]$$

$$= \frac{2}{(2n+1)^2} \cdot \frac{(\overline{2n+1}+2) n^2 + (n+1)^2 (\overline{2n+1}-2)}{(2n-1)(2n+3)}$$

$$= \frac{2}{(2n+1)^2} \cdot \frac{(2n+1)(2n^2+2n-1)}{(2n-1)(2n+3)} = \frac{2(2n^2+2n-1)}{(2n+1)(2n-1)(2n+3)}$$

$$= \frac{1}{8(2n-1)} + \frac{3}{4(2n+1)} + \frac{1}{8(2n+3)},$$ on factorizing into partial fraction.

Example 12. *Show that:* $\int_{-1}^{1} x^2 P_{n-1} P_{n+1} \, dx = \dfrac{2n(n+1)}{(2n-1)(2n+1)(2n+3)}$.

Sol. We have from the recurrence relation (1),

$$(2n+1) x P_n = (n+1) P_{n+1} + n P_{n-1}$$

$$\therefore \quad x P_{n-1} = \frac{1}{2n-1} \{ n P_n + (n-1) P_{n-2} \}$$

and

$$x P_{n+1} = \frac{1}{2n+3} \{ (n+2) P_{n+2} + (n+1) P_n \} P$$

$$\therefore \quad x^2 P_{n-1} P_{n+1} = \frac{1}{(2n-1)(2n+3)} \{ n(n+2) P_n P_{n+2} + n(n+1) P_n^2 + (n-1)(n+2) P_{n-2} P_{n+2} + (n^2-1) P_n P_{n-2} \}$$

Integrating both sides from – 1 to 1 and using orthogonality of Legendre polynomials, we get

$$\int_{-1}^{1} x^2 P_{n-1} P_{n+1} \, dx = \frac{n(n+1)}{(2n-1)(2n+3)} \int_{-1}^{1} P_n^2 dx$$

$$= \frac{2n(n+1)}{(2n-1)(2n+1)(2n+3)}.$$

Example 13. *If* $f(x) = 0, \quad -1 < x \le 0$
$\quad = x, \quad 0 < x < 1,$

Show that $f(x) = \frac{1}{4} P_0(x) + \frac{1}{2} P_1(x) + \frac{5}{16} P_2(x) - \frac{3}{32} P_4(x) + \dots$

Sol. Let $\quad f(x) = \sum_{n=0}^{\infty} c_n P_n(x)$

Then c_n is given by $c_n = \left(n + \frac{1}{2}\right) \int_{-1}^{1} f(x) P_n(x)\, dx$

$$= \left(n + \frac{1}{2}\right)\left[\int_{-1}^{0} 0 . P_n(x)\, dx + \int_0^1 x\, P_n(x)\, dx\right] = \left(n + \frac{1}{2}\right) \int_0^1 x\, P_n(x)\, dx$$

$$\therefore \quad c_0 = \frac{1}{2}\int_0^1 x\, P_0(x)\, dx = \frac{1}{2}\int_0^1 x\, dx = \frac{1}{4}$$

$$c_1 = \frac{3}{2}\int_0^1 x\, P_1(x)\, dx = \frac{3}{2}\int_0^1 x^2\, dx = \frac{1}{2}$$

$$c_2 = \frac{5}{2}\int_0^1 x\, P_2(x)\, dx = \frac{5}{2}\int_0^1 x . \frac{3x^2 - 1}{2}\, dx = \frac{5}{4}\left|\frac{3x^4}{4} - \frac{x^2}{2}\right|_0^1 = \frac{5}{16}$$

$$c_3 = \frac{7}{2}\int_0^1 x\, P_3(x)\, dx = \frac{7}{2}\int_0^1 x . \frac{5x^3 - 3x}{2}\, dx = \frac{7}{4}\left|5\frac{x^5}{5} - 3\frac{x^3}{3}\right|_0^1 = 0$$

$$c_4 = \frac{9}{2}\int_0^1 x\, P_4(x)\, dx = \frac{9}{2}\int_0^1 x . \frac{35x^4 - 30x^2 + 3}{8}\, dx$$

$$= \frac{9}{16}\left|35\frac{x^6}{6} - 35\frac{x^4}{4} + 3\frac{x^2}{2}\right|_0^1 = -\frac{3}{32}$$

and so on. Hence $f(x) = \frac{1}{4} P_0(x) + \frac{1}{2} P_1(x) + \frac{5}{16} P_2(x) - \frac{3}{32} P_4(x) + \dots$

Example 14. *Prove that:*

$$\int_{-1}^{+1} x^2 P_{n+1}(x) . P_{n-1}(x)\, dx = \frac{2n(n+1)}{(2n-1)(2n+1)(2n+3)}$$

Sol. The recurrence formula *I* is

$$(2n + 1)x\, P_n = (n + 1)\, P_{n+1} + nP_{n-1}$$

Replacing n by $(n + 1)$ and $(n - 1)$, we have

$$(2n + 3)\, xP_{n+1} = (n + 2)\, P_{n+2} + (n + 1)\, P_n \qquad \dots(1)$$

$$(2n - 1)xP_{n-1} = nP_n + (n - 1)\, P_{n-2} \qquad \dots(2)$$

Multiplying (1) and (2) and integrating in the limits – 1 to + 1, we have

$$(2n + 3)(2n - 1)\int_{-1}^{+1} x^2\, P_{n+1} . P_{n-1}\, dx = n(n+1)\int_{-1}^{1} P_n^2\, dx + n(n + 2)\int_{-1}^{+1} P_n . P_{n+2}\, dx$$

$$+ (n^2 - 1)\int_{-1}^{+1} P_n\, P_{n-2}\, dx + (n - 1)(n + 2)\int_{-1}^{+1} P_{n+2} . P_{n-2}\, dx$$

$$= n(n+1)\int_{-1}^{1} P_n^2\, dx + 0 + 0 + 0 \qquad \text{(Orthogonality Property)}$$

$$= n(n+1)\cdot\frac{2}{(2n+1)} \quad \text{or} \quad \int_{-1}^{+1} x^2 \cdot P_{n+1}\cdot P_{n-1}\, dx = \frac{2n(n+1)}{(2n-1)(2n+1)(2n+3)}$$

TEST YOUR KNOWLEDGE

1. Show that:

   (*i*) $\frac{2}{5}P_3(x) + \frac{3}{5}P_1(x) = x^3$ (*ii*) $x^5 = \frac{8}{63}\left[P_5(x) + \frac{7}{2}P_3(x) + \frac{27}{8}P_1(x)\right]$.

2. Prove that:

   (*i*) $P_n(-1) = (-1)^n$ (*ii*) $P_n'(1) = \frac{1}{2}n(n+1)$

   (*iii*) $P_n'(-1) = \frac{1}{2}(-1)^{n-1}\cdot n(n+1)$.

3. Express the following in terms of Legendre polynomials:

   (*i*) $1 + x - x^2$ (*ii*) $5x^3 + x$

   (*iii*) $x^3 - 5x^2 + 6x + 1$ (*iv*) $x^4 + 3x^3 - x^2 + 5x - 2$.

4. Using Rodrigue's formula, show that $P_n(x)$ satisfies the differential equation

$$\frac{d}{dx}\left[(1-x^2)\frac{d}{dx}\{P_n(x)\}\right] + n(n+1)P_n(x) = 0.$$

5. Show that $P_5'(x) = 9P_4(x) + 5P_2(x) + P_0(x)$.

6. Prove that $P'_{n+1} + P_n' = P_0 + 3P_1 + 5P_2 + \ldots\ldots + (2n+1)P_n$.

   [**Hint.** In recurrence relation (*iii*), put $n = 1, 2, 3, \ldots\ldots$ and add.]

7. Show that: $\frac{1-z^2}{(1-2xz+z^2)^{3/2}} = \sum_{n=0}^{\infty}(2n+1)z^n P_n(x)$.

8. Prove that:

   (*i*) $\int_{-1}^{1} P_n(x)\, dx = 0$ if $n \neq 0$ (*ii*) $\int_{-1}^{1} P_0(x)\, dx = 2$

   (*iii*) $\int_{-1}^{1} x^m P_n(x)\, dx = 0$, m, n are positive integers and $m < n$.

   (*iv*) $\int_{0}^{1} P_{2n}(x)\, P_{2n+1}(x)\, dx = \int_{0}^{1} P_{2n}(x)\, P_{2n-1}(x)\, dx$

   (*v*) $\int_{-1}^{1} x^3 P_4(x)\, dx = 0$ (*vi*) $\int_{-2}^{1} x^6 P_4(x)\, dx = \frac{16}{231}$

   (*vii*) $\int_{0}^{1} P_n(x)\, dx = \frac{1}{n+1}P_{n-1}(0)$ (*viii*) $\int_{-1}^{1} xP_n(x)\, P_{n-1}(x)\, dx = \frac{2n}{4n^2-1}$

(ix) $\int_{-1}^{1} x P_n(x) P_n'(x)\, dx = \dfrac{2n}{2n+1}$

(x) $\int_{-1}^{1} x^2 P_{n+1}(x) P_{n-1}(x)\, dx = \dfrac{2n(n+1)}{(2n-1)(2n+1)(2n+3)}$

(xi) $\int_{-1}^{1} P_n(x)(1 - 2xz + z^2)^{-1/2}\, dx = \dfrac{2z^n}{2n+1}$

(xii) $\int_{-1}^{1} P'_{n-1}(x) P'_{n+1}(x)\, dx = n(n-1)$

(xiii) $\int_{0}^{1} P_{2n}(x)\, dx = 0$

(xiv) $\int_{0}^{1} P_{2n+1}(x)\, dx = (-1)^n$

(xv) $\int_{-1}^{1} (1 - x^2) P_n'(x) P_m'(x)\, dx = 0, m \neq n.$

9. Obtain the Legendre expansion of $f(x)$ defined as $f(x) = \begin{cases} 0, & -1 < x < 0 \\ 1, & 0 < x < 1 \end{cases}$

10. If $f(x) = \begin{cases} 0, & -1 < x \leq 0 \\ x, & 0 < x < 1 \end{cases}$. Show that $f(x) = \dfrac{1}{4} P_0(x) + \dfrac{1}{2} P_1(x) + \dfrac{5}{16} P_2(x) - \dfrac{3}{32} P_4(x)$ ......

Answers

3. (i) $\dfrac{2}{3} P_0(x) + P_1(x) - \dfrac{2}{3} P_2(x)$ (ii) $2P_3(x) + 4P_1(x)$

(iii) $\dfrac{2}{5} P_3(x) - \dfrac{10}{3} P_2(x) + \dfrac{33}{5} P_1(x) - \dfrac{2}{3} P_0(x)$

(iv) $\dfrac{8}{35} P_4(x) + \dfrac{6}{5} P_3(x) - \dfrac{2}{21} P_2(x) + \dfrac{34}{5} P_1(x) - \dfrac{224}{105} P_0(x)$

9. $f(x) = \dfrac{1}{2} P_0(x) + \dfrac{3}{4} P_1(x) - \dfrac{7}{16} P_3(x)$ ....

APPENDIX—I

Table 1 : NORMAL TABLE

AREAS UNDER THE STANDARD NORMAL

$$\text{CURVE} = \frac{1}{\sqrt{2\pi}} \int_0^z e^{-\frac{z^2}{2}} dz$$

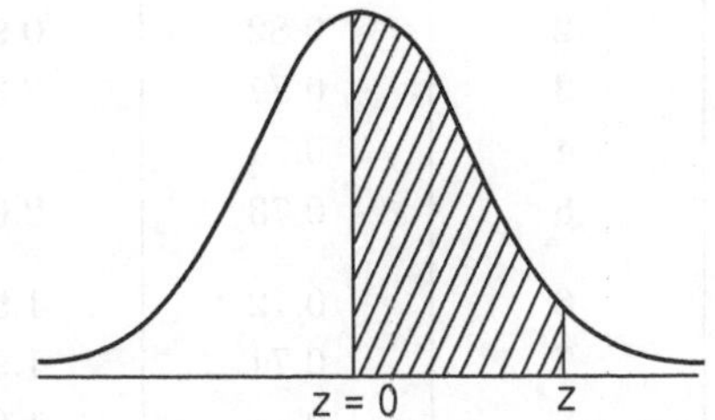

z	*.00*	*.01*	*.02*	*.03*	*.04*	*.05*	*.06*	*.07*	*.08*	*.09*
0.0	.0000	.0040	.0080	.0120	.0160	.0199	.0239	.0279	.0319	.0359
0.1	.0398	.0438	.0478	.0517	.0557	.0596	.0636	.0675	.0714	.0754
0.2	.0793	.0832	.0871	.0910	.0948	.0987	.1026	.1064	.1103	.1141
0.3	.1179	.1217	.1255	.1293	.1331	.1368	.1406	.1443	.1480	.1517
0.4	.1554	.1591	.1628	.1664	.1700	.1736	.1772	.1808	.1844	.1879
0.5	.1915	.1950	.1985	.2019	.2054	.2088	.2123	.2157	.2190	.2224
0.6	.2257	.2291	.2324	.2357	.2389	.2422	.2454	.2485	.2517	.2549
0.7	.2580	.2611	.2642	.2673	.2704	.2734	.2764	.2794	.2823	.2852
0.8	.2881	.2910	.2939	.2967	.2995	.3023	.3051	.3078	.3106	.3133
0.9	.3159	.3186	.3212	.3238	.3264	.3289	.3315	.3340	.3365	.3389
1.0	.3413	.3438	.3461	.3485	.3508	.3531	.3554	.3577	.3599	.3621
1.1	.3643	.3665	.3686	.3708	.8729	.3749	.3770	.3790	.3810	.3830
1.2	.3849	.3869	.3888	.3907	.3925	.3944	.3962	.3980	.3997	.4015
1.3	.4032	.4049	.4066	.4082	.4099	.4115	.4131	.4147	.4162	.4177
1.4	.4192	.4207	.4222	.4236	.4251	.4255	.4279	.4292	.4306	.4319
1.5	.4332	.4345	.4357	.4370	.4382	.4394	.4406	.4418	.4429	.4441
1.6	.4452	.4463	.4474	.4484	.4495	.4505	.4515	.4525	.4535	.4545
1.7	.4554	.4564	.4573	.4582	.4591	.4599	.4608	.4616	.4625	.4633
1.8	.4641	.4649	.4656	.4664	.4671	.4678	.4686	.4693	.4699	.4706
1.9	.4713	.4719	.4726	.4732	.4738	.4744	.4750	.4756	.4761	.4767
2.0	.4772	.4778	.4783	.4788	.4793	.4798	.4803	.4808	.4812	.4817
2.1	.4821	.4826	.4830	.4834	.4838	.4842	.4846	.4850	.4854	.4857
2.2	.4861	.4864	.4868	.4871	.4875	.4878	.4881	.4884	.4887	.4890
2.3	.4893	.4896	.4898	.4901	.4904	.4906	.4909	.4911	.4913	.4916
2.4	.4918	.4920	.4922	.4925	.4927	.4929	.4931	.4932	.4934	.4936
2.5	.4938	.4940	.4941	.4943	.4945	.4946	.4948	.4949	.4951	.4952
2.6	.4953	.4955	.4956	.4957	.4959	.4930	.4961	.4962	.4963	.4964
2.7	.4965	.4966	.4967	.4968	.4969	.4970	.4971	.4972	.4973	.4974
2.8	.4974	.4975	.4976	.4977	.4977	.4978	.4979	.4979	.4980	.4981
2.9	.4981	.4982	.4982	.4983	.4984	.4984	.4985	.4985	.4986	.4986
3.0	.4987	.4987	.4987	.4988	.4988	.4989	.4999	.4989	.4990	.4990
3.1	.4990	.4991	.4991	.4991	.4992	.4992	.4992	.4992	.4993	.4993

Table 2 : SIGNIFICANT VALUES t_v (α) OF t-DISTRIBUTION (TWO TAIL AREAS) $[\,|\,t\,| > t_v(\alpha)] = \alpha$

d.f.	*Probability (Level of Significance)*					
(ν)	*0.50*	*0.10*	*0.05*	*0.02*	*0.01*	*0.001*
1	1.00	6.31	12.71	31.82	63.66	636.62
2	0.82	0.92	4.30	6.97	6.93	31.60
3	0.77	2.32	3.18	4.54	5.84	12.94
4	0.74	2.13	2.78	3.75	4.60	8.61
5	0.73	2.02	2.57	3.37	4.03	6.86
6	0.72	1.94	2.45	3.14	3.71	5.96
7	0.71	1.90	2.37	3.00	3.50	5.41
8	0.71	1.80	2.31	2.90	3.36	5.04
9	0.70	1.83	2.26	2.82	3.25	4.78
10	0.70	1.81	2.23	2.76	3.17	4.59
11	0.70	1.80	2.20	2.72	3.11	4.44
12	0.70	1.78	2.18	2.68	3.06	4.32
13	0.69	1.77	2.16	2.05	3.01	4.22
14	0.69	1.76	2.15	2.62	2.98	4.14
15	0.69	1.75	2.13	2.60	2.95	4.07
16	0.69	1.75	2.12	2.58	2.92	4.02
17	0.69	1.74	2.11	2.57	2.90	3.97
18	0.69	1.73	2.10	2.55	2.88	3.92
19	0.69	1.73	2.09	2.54	2.86	3.88
20	0.69	1.73	2.09	2.53	2.85	3.85
21	0.69	1.72	2.08	2.52	2.83	3.83
22	0.69	1.72	2.07	2.51	2.42	3.79
23	0.69	1.71	2.07	2.50	2.81	3.77
24	0.69	1.71	2.06	2.49	2.80	3.75
25	0.68	1.71	2.06	2.49	2.79	3.73
26	0.68	1.71	2.06	2.48	2.78	3.71
27	0.68	1.70	2.05	2.47	2.77	3.69
28	0.68	1.70	2.05	2.47	2.76	3.67
29	0.68	1.70	2.05	2.46	2.76	3.66
30	0.68	1.70	2.04	2.46	2.75	3.65
∞	0.67	1.65	1.96	2.33	2.58	3.29

Table 3 : F-Distribution

Values of F for F-Distributions with 0.05 of the Area in the Right Tail

	Degrees of freedom for numerator																		
	1	*2*	*3*	*4*	*5*	*6*	*7*	*8*	*9*	*10*	*12*	*15*	*20*	*24*	*30*	*40*	*60*	*120*	∞
1	161	200	216	225	230	234	237	239	241	242	244	246	248	249	250	251	252	253	254
2	18.5	19.0	19.2	19.2	19.3	19.3	19.4	19.4	19.4	19.4	19.4	19.4	19.4	19.5	19.5	19.5	19.5	19.5	19.5
3	10.1	9.55	9.28	9.12	9.01	9.94	8.89	8.85	8.81	8.79	8.74	8.70	8.66	8.64	8.62	8.59	8.57	8.55	8.53
4	7.71	6.94	6.59	6.39	6.26	6.16	6.09	6.04	6.00	5.96	5.91	5.86	5.80	5.77	5.75	5.72	5.69	5.66	5.63
5	6.61	5.79	5.41	5.19	5.05	4.95	4.88	4.82	4.77	4.74	4.68	4.62	4.56	4.53	4.50	4.46	4.43	4.40	4.37
6	5.99	5.14	4.76	4.53	4.39	4.28	4.21	4.15	4.10	4.06	4.00	3.94	3.87	3.84	3.81	3.77	3.74	3.70	3.67
7	5.59	4.74	4.35	4.12	3.97	3.87	3.79	3.73	3.68	3.64	3.57	3.51	3.44	3.41	3.38	3.34	3.30	3.27	3.23
8	5.32	4.46	4.07	3.84	3.69	3.58	3.50	3.44	3.39	3.35	3.28	3.22	3.15	3.12	3.08	3.04	3.01	2.97	2.93
9	5.12	4.26	3.86	3.63	3.48	3.37	3.29	3.23	3.18	3.14	3.07	3.01	2.94	2.90	2.86	2.83	2.79	2.75	2.71
10	4.96	4.10	3.71	3.48	3.33	3.22	3.14	3.07	3.02	2.98	2.91	2.85	2.77	2.74	2.70	2.66	2.62	2.58	2.54
11	4.84	3.98	3.59	3.36	3.20	3.09	3.01	2.95	2.90	2.85	2.79	2.72	2.65	2.61	2.57	2.53	2.49	2.45	2.40
12	4.75	3.89	3.49	3.26	3.11	3.00	2.91	2.85	2.80	2.75	2.69	2.62	2.54	2.51	2.47	2.43	2.38	2.34	2.30
13	4.67	3.81	3.41	3.18	3.03	2.92	2.83	2.77	2.71	2.67	2.60	2.53	2.46	2.42	2.38	2.34	2.30	2.25	2.21
14	4.60	3.74	3.34	3.11	3.96	2.85	2.76	2.70	2.65	2.60	2.53	2.46	2.39	2.35	2.31	2.27	2.22	2.18	2.13
15	4.54	3.68	3.29	3.06	3.90	2.79	2.71	2.64	2.59	2.54	2.48	2.40	2.33	2.29	2.25	2.20	2.16	2.11	2.07
16	4.49	3.63	3.24	3.01	2.85	2.74	2.66	2.59	2.54	2.49	2.42	2.35	2.28	2.24	2.19	2.15	2.11	2.06	2.01
17	4.45	3.59	3.20	2.96	2.81	2.70	2.61	2.55	2.49	2.45	2.38	2.31	2.23	2.19	2.15	2.10	2.06	2.01	1.96
18	4.41	3.55	3.16	2.93	2.77	2.66	2.58	2.51	2.46	2.41	2.34	2.27	2.19	2.15	2.11	2.06	2.02	1.97	1.92
19	4.38	3.52	3.13	2.90	2.74	2.63	2.54	2.48	2.42	2.38	2.31	2.23	2.16	2.11	2.07	2.03	1.98	1.93	1.88
20	4.35	3.49	3.10	2.87	2.17	2.60	2.51	2.45	2.39	2.35	2.28	2.20	2.12	2.08	2.04	1.99	1.95	1.90	1.84
21	4.32	3.47	3.07	2.84	2.68	2.57	2.49	2.42	2.37	2.32	2.25	2.18	2.10	2.05	2.01	1.96	1.92	1.87	1.81
22	4.30	3.44	3.05	2.82	2.66	2.55	2.46	2.40	2.34	2.30	2.23	2.15	2.07	2.03	1.98	1.94	1.89	1.84	1.78
23	4.28	3.42	3.03	2.80	2.64	2.53	2.44	2.37	2.32	2.27	2.20	2.13	2.05	2.01	1.96	1.91	1.86	1.81	1.76
24	4.26	3.40	3.01	2.78	2.62	2.51	2.42	2.36	2.30	2.25	2.18	2.11	2.03	1.98	1.94	1.98	1.84	1.79	1.73
25	4.24	3.39	2.99	2.76	2.60	2.94	2.40	2.34	2.28	2.24	2.16	2.29	2.01	1.96	1.92	1.87	1.82	1.77	1.71
30	4.17	3.32	2.92	2.69	2.53	2.42	2.33	2.27	2.21	2.16	2.09	2.01	1.93	1.89	1.84	1.79	1.74	1.64	1.62
40	4.08	3.23	2.84	2.61	2.45	2.34	2.25	2.18	2.12	2.08	2.00	1.92	1.84	1.79	1.74	1.69	1.64	1.58	1.51
60	4.00	3.15	2.76	2.53	2.37	2.25	2.17	2.10	2.04	1.99	1.92	1.84	1.75	1.70	1.65	1.59	1.53	1.47	1.39
120	3.92	3.07	2.68	2.45	2.29	2.18	2.09	2.02	1.96	1.91	1.83	1.75	1.66	1.61	1.55	1.50	1.43	1.35	1.25
∞	3.84	3.00	2.60	2.37	2.21	2.10	2.01	1.94	1.88	1.83	1.75	1.67	1.57	1.52	1.46	1.39	1.32	1.22	1.00

Table 4 : CHI-SQUARE (χ^2)
Significant Values χ^2 (α) of χ^2 Distribution Right Tail Areas
for Given Probability α,
$P = P_r(\chi^2 > \chi^2(\alpha)) = \alpha$
And is Degrees of Freedom (d.f.)

Degree of freedom (ν)	*Probability (Level of Significance)*						
	0 = .99	*0.95*	*0.50*	*0.10*	*0.05*	*0.02*	*0.01*
1	.000157	.00393	.455	2.706	3.841	5.214	6.635
2	.0201	.103	1.386	4.605	5.991	7.824	9.210
3	.115	.352	2.366	6.251	7.815	9.837	11.341
4	.297	.711	3.357	7.779	9.488	11.668	13.277
5	.554	1.145	4.351	9.236	11.070	13.388	15.086
6	.872	2.635	5.348	10.645	12.592	15.033	16.812
7	1.239	2.167	6.346	12.017	14.067	16.622	18.475
8	1.646	2.733	7.344	13.362	15.507	18.168	20.090
9	2.088	3.325	8.343	14.684	16.919	19.679	21.669
10	2.558	3.940	9.340	15.987	18.307	21.161	23.209
11	3.053	4.575	10.341	17.275	19.675	22.618	24.725
12	3.571	5.226	11.340	18.549	21.026	24.054	26.217
13	4.107	5.892	12.340	19.812	22.362	25.472	27.688
14	4.660	6.571	13.339	21.064	23.685	26.873	29.141
15	4.229	7.261	14.339	22.307	24.996	28.259	30.578
16	5.812	7.962	15.338	23.542	26.296	29.633	32.000
17	6.408	8.672	15.338	24.769	27.587	30.995	33.409
18	7.015	9.390	17.338	25.989	28.869	32.346	34.805
19	7.633	10.117	18.338	27.204	30.144	33.687	36.191
20	8.260	10.851	19.337	28.412	31.410	35.020	37.566
21	8.897	11.591	20.337	29.615	32.671	36.343	38.932
22	9.542	12.338	21.337	30.813	33.924	37.659	40.289
23	10.196	13.091	22.337	32.007	35.172	38.968	41.638
24	10.856	13.848	23.337	32.196	36.415	40.270	42.980
25	11.524	14.611	24.337	34.382	37.65	41.566	44.314
26	12.198	15.379	25.336	35.363	38.885	41.856	45.642
27	12.879	16.151	26.336	36.741	40.113	41.140	46.963
28	13.565	16.928	27.336	37.916	41.337	45.419	48.278
29	14.256	17.708	28.336	39.087	42.557	46.693	49.588
30	14.933	18.493	29.336	40.256	43.773	47.962	50.892

Note. For degrees of freedom (ν) greater than 30, the quantity $\sqrt{2\chi^2} - \sqrt{2\nu - 1}$ may be used as a normal variate with unit variance.

APPENDIX—II

SOME IMPORTANT CURVES

Cissoid (Imp.)

$y^2(2a - x) = x^3$

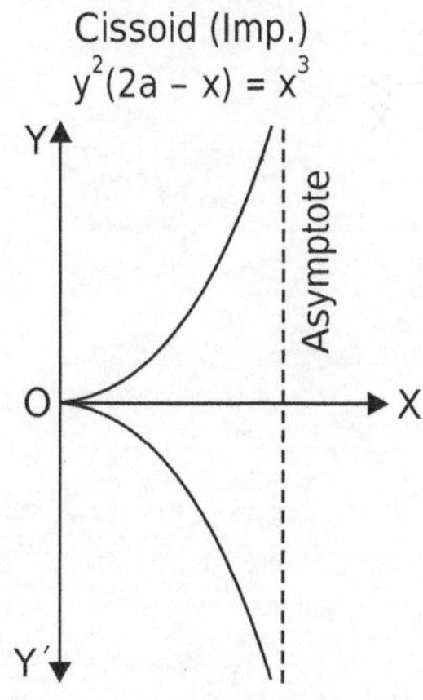

Catenary

$y = a \cosh \frac{x}{a}$

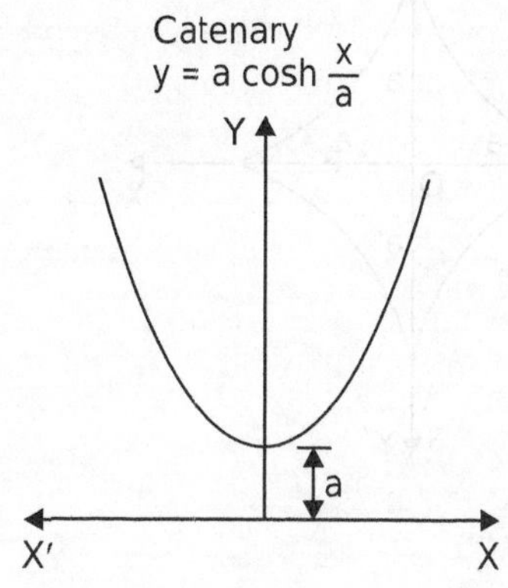

Cubical Parabola

$a^2y = x^3$

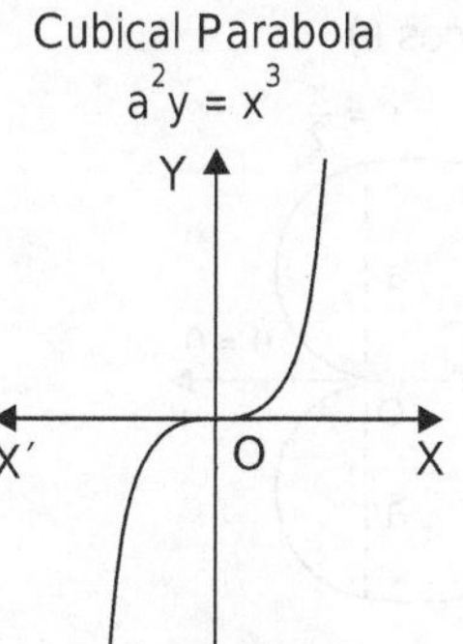

Semicubical Parabola

$ay^2 = x^3$

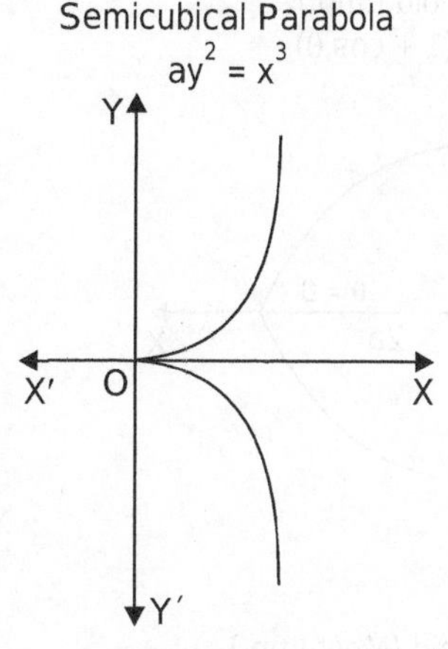

Strophoid

$y^2(a - x) = x^2(a + x)$

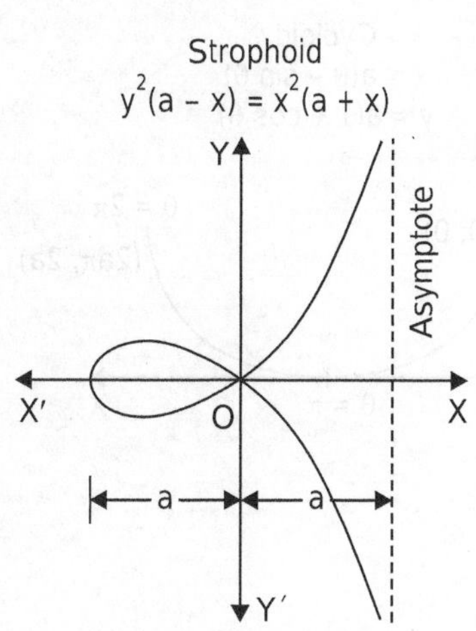

Folium of Descartes

$x^3 + y^3 = 3axy$

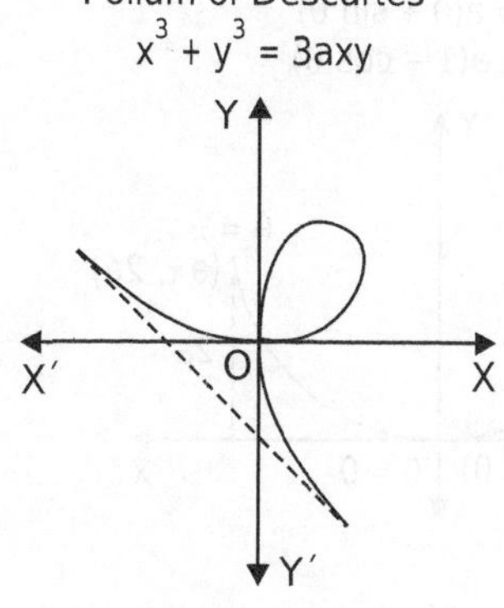

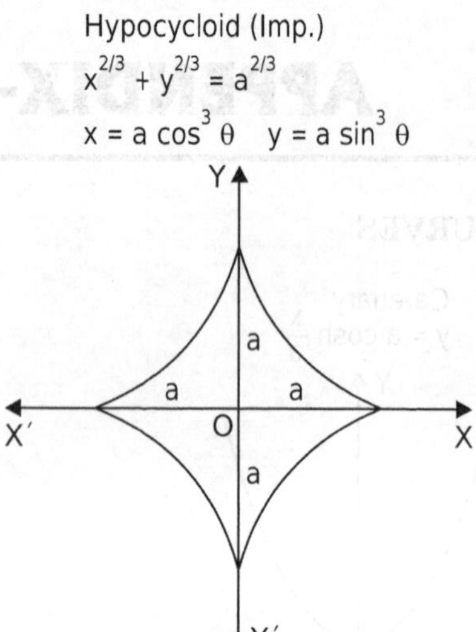
Hypocycloid (Imp.)
$x^{2/3} + y^{2/3} = a^{2/3}$
$x = a\cos^3\theta \quad y = a\sin^3\theta$
Y
a
a
a
X′
O
X
a
Y′

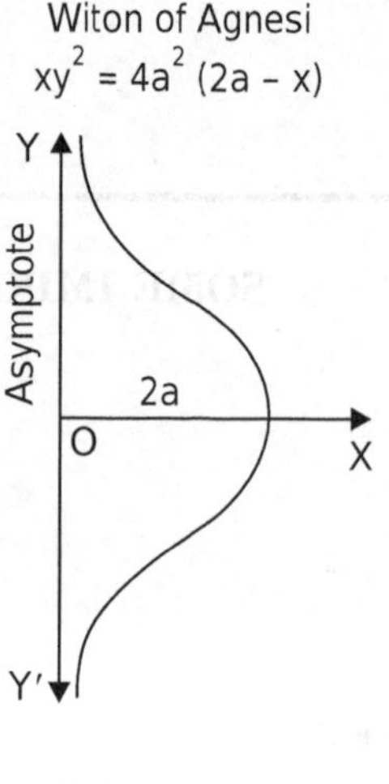
Witon of Agnesi
$xy^2 = 4a^2(2a - x)$
Y
Asymptote
2a
O
X
Y′

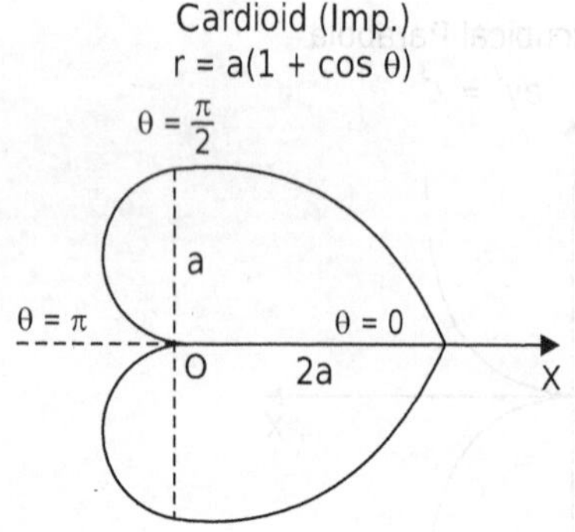
Cardioid (Imp.)
$r = a(1 + \cos\theta)$
$\theta = \frac{\pi}{2}$
a
$\theta = \pi$
$\theta = 0$
O
2a
X

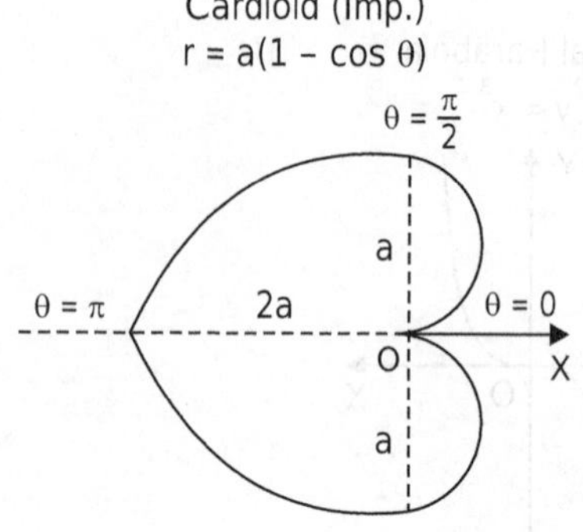
Cardioid (Imp.)
$r = a(1 - \cos\theta)$
$\theta = \frac{\pi}{2}$
a
$\theta = \pi$
2a
$\theta = 0$
O
X
a

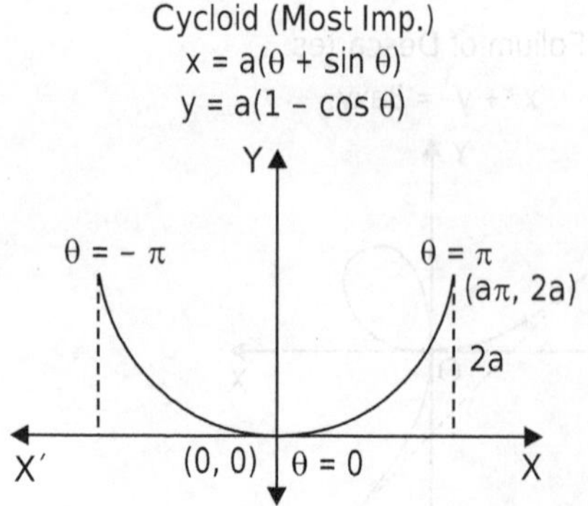
Cycloid (Most Imp.)
$x = a(\theta + \sin\theta)$
$y = a(1 - \cos\theta)$
Y
$\theta = -\pi$
$\theta = \pi$
$(a\pi, 2a)$
2a
X′
(0, 0)
$\theta = 0$
X

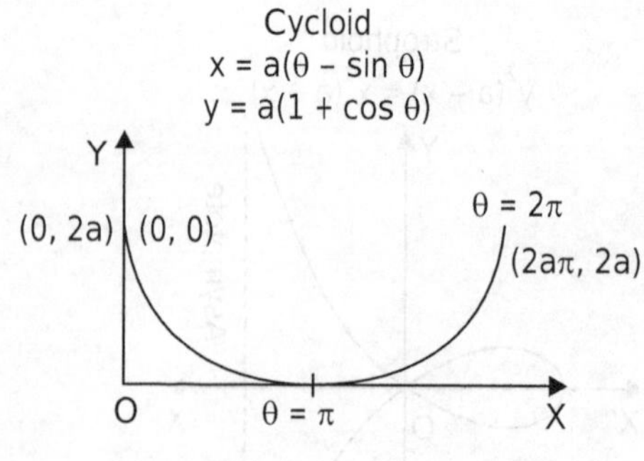
Cycloid
$x = a(\theta - \sin\theta)$
$y = a(1 + \cos\theta)$
Y
$\theta = 2\pi$
(0, 2a)
(0, 0)
$(2a\pi, 2a)$
O
$\theta = \pi$
X

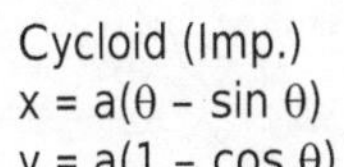

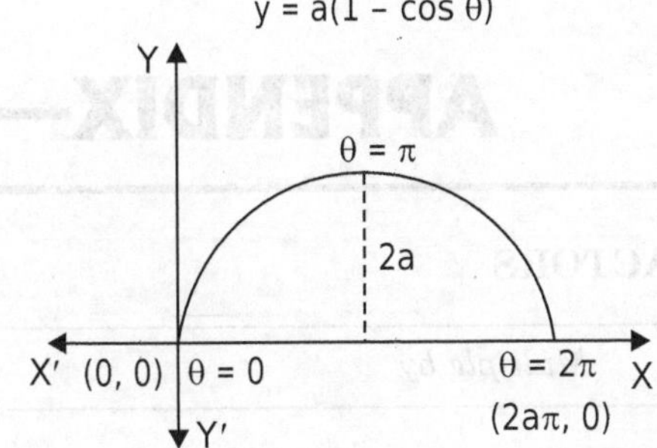

Cycloid

$x = a(\theta + \sin\theta)$

$y = a(1 + \cos\theta)$

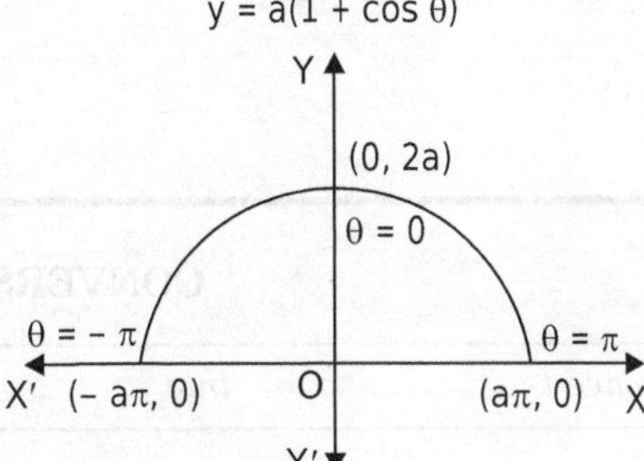

Lemniscate

$r^2 = a^2 \cos 2\theta$

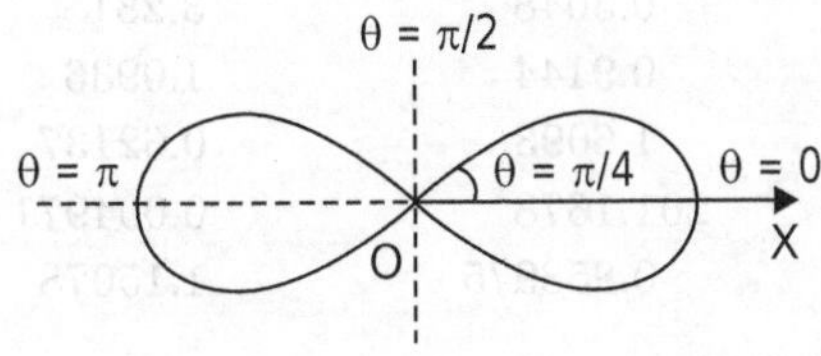

The three leaved rose

$r = a \sin 3\theta$

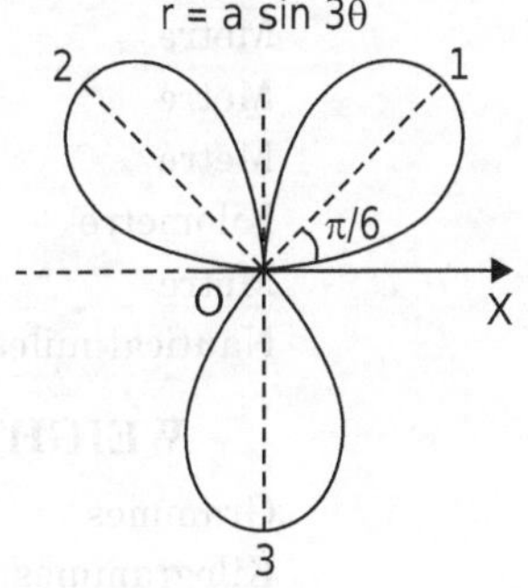

APPENDIX—III

CONVERSION FACTORS

To convert	*into*	*Multiple by*	
	LINEAR MEASURE		
Inch	Millimetre	25.40	0.03937
Inch	Centimetre	2.54	0.3937
Inch	Metre	0.0254	39.37
Feet	Metre	0.3048	3.281
Yard	Metre	0.9144	1.0936
Mile	Kilometre	1.6093	0.62137
Furlong	Metre	201.1678	0.004971
Miles	Nautical miles (int.)	0.858975	1.15078
	WEIGHT MEASURE		
Ounces	Grammes	28.3495	0.035274
Pounds	Kilogrammes	0.4536	2.20462
Cwt.	Quintal (100 kg.)	0.50802	1.9684
Ton	Kilogrammes	1016.05	0.0009842
U.S. Ton	Kilogrammes	907.18	0.0011
Ton (2240 lb)	Metric ton (Tonne)	1.01605	0.9842
Maunds	Ton	0.036735	27.2222
Tola	Ounce	0.411429	2.43056
Tola	Gramme	11.6628	0.087535
	SQUARE MEASURE		
Square inches	Square millimetres	645.16	0.00155
Square inches	Square centimetres	6.4516	0.155
Square feet	Square metre	0.0929	10.764
Square Yard	Square metre	0.836	1.196
Acre	Square metre	4046.9	0.00025
Square mile	Square kilometre	2.59	0.3861
Acre (4840 sq.yd.)	Hectare (10,000 sq. m)	0.40468	2.4711
	CUBIC MEASURE		
Cubic inch	Cubic centimetre	16.387	0.061
Cubic feet	Cubic metre	0.02832	35.315
Cubic Yard	Cubic metre	0.76455	1.308
Cubic foot	Gallon (Imp.)	6.2288	0.16054

Cubic foot	Litres (1000 cc)	28.3161	0.0353156
Gollon (Imp.)	Litres	4.54596	0.219976
Pint (0.124 Imp. Gln)	Litres	0.568	1.76
U.S. Gallon	Litres	3.78533	0.264178
	PRESSURES AND DENSITIES		
Pounds per sq in.	Kg. per sq. cm.	0.070307	14.2233
Pounds per sq. ft.	Kg. per sq. metre	4.88243	0.204816
Tonnes per sq. ft.	Tonnes per sq. metre	10.937	0.091436
Atmospheres	Pounds per sq. in.	14.6959	0.06804
Inch of Hg	Pounds per sq. in.	0.491154	2.036
Inch of Hg	Pounds per sq. in.	0.0345316	28.959
Pounds per sq. ft.	mm. of Hg.	0.035913	2.78450
Pounds per sq. in.	Feet of water	2.30672	0.433516
Feet of water	Kg. per sq. cm	0.0304792	32.81
Pounds per cu. m.	Grammes per cc	27.6799	0.0361273
Pounds per cft.	Kg. per cu. m.	16.02	0.052422
	MISCELLANEOUS		
Pounds per foot	Kg. per metre	1.48816	0.67197
Miles per hour	Feet per metre	1.46667	0.68182
Feet per Second	Km per hour	1.09728	0.91134
H.P. (550 ft. lb/sec.)	Metric H.P.	1.01387	0.98632
To obtain	From		Multiple by